T0251545

Decision Diagram Techniques for Micro- and Nanoelectronic Design

HANDBOOK

Decision Diagram Techniques for Micro- and Nanoelectronic Design

HANDBOOK

Svetlana N. Yanushkevich
D. Michael Miller
Vlad P. Shmerko
Radomir S. Stanković

A CRC title, part of the Taylor & Francis imprint, a member of the Taylor & Francis Group, the academic division of T&F Informa plc.

Published in 2006 by
CRC Press
Taylor & Francis Group
6000 Broken Sound Parkway NW, Suite 300
Boca Raton, FL 33487-2742

Printed in the United States of America on acid-free paper
10 9 8 7 6 5 4 3 2 1

International Standard Book Number-10: 0-8493-3424-1 (Hardcover)
International Standard Book Number-13: 978-0-8493-3424-5 (Hardcover)
Library of Congress Card Number 2005051439

Library of Congress Cataloging-in-Publication Data

Decision diagram techniques for micro- and nanoelectric design handbook / Svetlana N. Yanushkevich ... [et al.].
p. cm.
Includes bibliographical references and index.
ISBN 0-8493-3424-1 (alk. paper)
1. Decision making--Mathematical models. 2. Decision logic tables. 3. Microelectronics. 4. Nanotechnology. I. Yanushkevich, Svetlana N.

T57.95.I549 2004
621.381--dc22 2005051439

Taylor & Francis Group
is the Academic Division of T&F Informa plc.

Visit the Taylor & Francis Web site at
http://www.taylorandfrancis.com

and the CRC Press Web site at
http://www.crcpress.com

Brief Contents

Contents

Preface

This handbook presents the fundamental theory and practice of decision diagrams as applied to the representation and manipulation of logic functions. It is primarily intended for computer science, and electrical and computer engineering students, researchers and practitioners. However, individuals in other fields with interest in these and related areas will find it a comprehensive source of relevant information.

The handbook includes decision diagram techniques that have been collected by the authors during preparation of courses and lectures on advanced logic circuit design, research projects, and various student projects. The focus is on the use of decision diagram techniques for analysis and design applicable to micro and nanoelectronic integrated circuits (nanoICs).

Rationale and audience

Algorithms used in CAD systems for micro and nano-devices typically require the manipulation of large amounts of data. Graphical representations such as networks, flow-graphs, hypercubes, trees, and decision diagrams are foundational data structures used in logic design. Among these, decision diagrams are of a specific interest, since they support:

Representation of logic functions. Diverse representations such as sum-of-products expressions, Reed-Muller expansions, arithmetic expansions, and word-level expressions can be represented using decision diagrams compactly in terms of both space and time.

Optimization of logic functions. Both exact and heuristic minimization of logic functions can be performed by the manipulation of decision diagrams.

Estimates of switching circuit performance. Measures such as switching activity, power dissipation, and effects of noise in circuits can be estimated using decision diagrams.

Verification of circuits can be accomplished by comparing decision diagrams.

Decision diagrams have two key properties that make their use both attractive and efficient. These include:

(*a*) *Canonical form.* For a given variable ordering, reduced ordered decision diagrams represent each logic function uniquely.

(*b*) *Compactness.* Decision diagrams can in many instances compactly represent large functions and large sets of combinatorial data.

Decision diagram techniques have been used at the following levels of abstraction in various CAD systems for micro and nano-devices:

- *Logic synthesis* (decomposition, testing, minimization, timing optimization),
- *Behavioral synthesis* (pipelining, partitioning, etc.),
- *Sequential synthesis* (state assignments and minimization, testability, state machine verification),
- *Technology mapping* (mapping to a library of logic gates, synthesis with macro-cells, different programmable structures, etc.), and
- *Physical design mapping* (cell placements, layout, routing, fabrication).

Specific applications of decision diagrams in digital design include:

Circuit optimization. Reduction of the size and complexity of decision diagrams can be directly relevant to minimizing the area of logic networks. Choosing a variable ordering that significantly reduces the decision diagram size can be done by dynamic programming, sifting, and information-theoretic criterion.

Pass-transistor logic circuit optimization. The advantage of this method is that pass-transistor logic circuits derived from decision diagrams are path-free, i.e., there is no assignment to the inputs that produce a conducting path from power supply to the ground.

Testing. Decision diagrams can be used to calculate changes that are described by differences which are applicable for testing.

Fault simulation. Simulation of multiple stuck-at faults can be accomplished by representing sets of fault combinations with decision diagrams.

Time-driven analysis. The testability of circuits synthesized by mapping a decision diagram to a multiplexor circuit using the path delay fault model allows detection of static and dynamic faults.

Estimation of power dissipation. Power dissipation reductions can be considered with respect to the minimum size decision diagram. The power dissipation is estimated as a function of switching activity and heuristic optimization targets the cost expressed in terms of the probability of signal as derived from the decision diagram.

Decomposition. The logic decomposition methods based in the past on algebraic techniques have been replaced with new decision diagram based decomposition techniques.

The verification problem is formulated as follows. Given two specifications, e.g two circuits, with the same number of inputs and outputs, verify that both produce the same output for each input assignment. Complete ordered decision trees can be generated from each specification and compared for equivalency. Alternatively, reduced ordered decision diagrams can be generated and compared. This approach is possible since both the complete ordered decision tree and the reduced ordered decision diagram are canonical forms.

Nanodevices design. Decision diagrams are a viable candidate of topological models for nanoelectronic circuits since nanoelectronics requires switch-type, not gate-level, logic modelling (similar to pass-transistor logic). In particular, a type of quantum-dot network, decision diagram nanowire wrap-gate networks, has been developed. Decision diagrams and trees can be embedded into various topologies, for example, into hexagonal, lattice, or hypercube-like topologies, suitable for nanoelectronic design.

Our aim has been to make this handbook both a comprehensive review and a suitable starting point for developing applications of decision diagram techniques in logic circuit design and related areas. To achieve this goal, the presentation is organized in four parts consisting of self-contained chapters written in a tutorial style. Selected methods of contemporary logic design are presented using decision diagrams, as specific methods for the design of nanoICs.

When applied to nano-device design, decision diagram techniques offer two attractive features:

(*a*) Nodes of decision diagrams can be implemented by multiplexers; the latter, viewed as one-input two-output switches, are good candidates for various nanotechnologies, in particular, wrap-gate nanowire networks.

(*b*) Decision diagrams can be embedded into a number of other topological structures.

The following key features distinguish this handbook:

- ► This handbook presents a comprehensive catalog of decision diagram techniques. Standard textbooks on logic circuit design generally do not discuss decision diagram techniques in entirely or in a tutorial manner.
- ► The central role of topological models, for example, embedding decision trees and diagrams in various spatial structures and assembling the topology, is discussed.
- ► The key recent trends in the theory and practice of decision diagram techniques are highlighted.

► Novel techniques of advanced logic design (algebraic, matrix, and graph-based) are presented with respect to new possibilities of processing in spatial dimensions.

► Solutions inspired by recent developments in nano-technologies are given such as the information theoretic approach and stochastic computing for the modelling and simulation of logic circuits.

This volume should be especially valuable to those engineers, scientists, and students who are interested in the design of discrete devices in the coming era of nano-technologies. It is our hope that this handbook will inspire a new generation of innovators to continue the development of decision diagram techniques for the design and analysis of micro and nano-devices.

Svetlana N. Yanushkevich[†*]

D. Michael Miller[†]

Vlad P. Shmerko[†]

Radomir S. Stanković[‡]

Calgary (Canada)

Victoria (Canada)

Niš (Serbia)

[*]The work by S. N. Yanushkevich was partially supported by Natural Sciences and Engineering Research Council of Canada (NSERC), grant 239025-02.

[†]The joint work by S. N. Yanushkevich, D. M. Miller, and V. P. Shmerko was partially supported by NATO Collaborative Linkage Grant PST-CLG.979071.

[‡]A part of the work by R. S. Stankovic has been done during his stay at Tampere International Center for Signal Processing, Tampere University of Technology, Tampere, Finland, and was supported by the Academy of Finland, Finish Center of Excellence Programme, Grant No. 44876.

Part I

FUNDAMENTALS OF DECISION DIAGRAM TECHNIQUES

- Digital Circuit Technologies and Decision Diagrams Technique
- Data Structures for Logic Functions
- Graph Theory and Decision Trees
- Reed-Muller Representation
- Boolean and Arithmetic Differences
- Arithmetic Representations
- Word-Level Representations
- Spectral Technique
- Walsh and Haar Arithmetic Representations
- Information-Theoretical Measures

1

Introduction

This part includes eight chapters covering the basic theory of decision diagram techniques:

- Data structures for representations of discrete and switching functions, including algebraic and graphical data structures,
- Spectral techniques for analysis and synthesis of discrete systems, and
- The basics of stochastic (probabilistic) techniques and information theoretical measures for logic functions, and
- Event-driven analysis.

The basics of decision diagram techniques include methods from various areas (Figure 1.1) such as: graph theory, discrete mathematics, group theory, as well as Boolean and multivalued algebras, information theory, etc. The choice of the representation form and corresponding manipulation technique depends heavily on the application. Most applications of decision diagrams considered in this book lie in the areas of digital system design, verification, and testing. Prior to their application to switching theory, decision trees were applied in the programming of decision tables (branching programs), databases, pattern recognition, taxonomy and identification of objects (system faults, biological specimens, etc.), machine diagnosis, testing, and analysis of algorithms, complexity measures in particular. Many methods developed for the above applications have been adopted for solving the problems of logic design. This application eventually evolved towards VLSI circuit design, and further towards nanoelectronic design.

1.1 Data structures for the representation of discrete functions

In this section, various representation forms of discrete functions are considered with respect to their applications and their relationship to decision trees and diagrams.

Discrete functions can be represented:

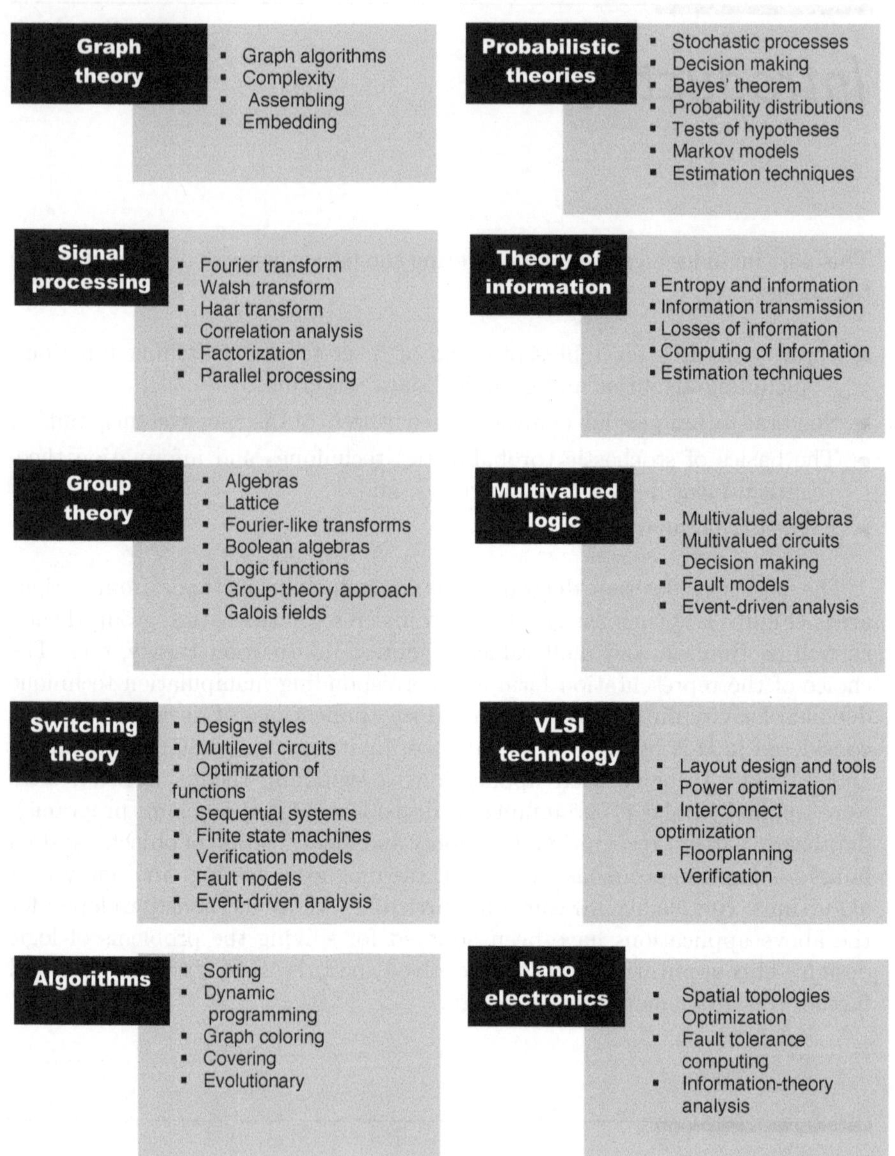

FIGURE 1.1
The basics of decision diagram techniques include selected topics from several theoretical areas.

- Algebraically, as an expression in terms of an algebra (Boolean algebra for switching functions, in particular),
- In a tabular form, as a decision table,
- In graphic form, as a graph or hypercube.

A discrete function is a function of discrete variables, $f(x_1, ..., x_n)$, where each variable, x_i, $i = 1, ..., n$, takes exactly m values, which are denoted $0, ..., m-1$. A discrete function is *completely specified* if it is defined everywhere, and *incompletely specified* otherwise. When a variable is evaluated, given $x_i = c_i$, the function is denoted as

$$f(x_1, ..., x_{i-1}, c_i, x_{i+1}, ..., x_n),$$

or $f|_{x_i=c_i}$. A variable, x_i is *redundant* if, and only if,

$$f|_{x_i=0} = ... = f|_{x_i=m-1}.$$

Table 1.1 shows different classes of discrete functions where a finite discrete function of n variables is defined as a mapping

$$f : \times_{i=1}^{n} \mathcal{D}_i \rightarrow \mathcal{R}$$

where $\times$ is the Cartesian product, $\mathcal{D}_i$ and $\mathcal{R}$ are finite sets, and variable $x_i \in \mathcal{D}_i$.

1.1.1 Switching theory

A *Boolean algebra* $\mathcal{B} = \{\mathcal{A};\ +, \times;\ ^{-};\ 0, 1\}$ is defined as a set $\mathcal{A}$, two elements 0 and 1, two binary operations $+, \times$, and a unary operation $^{-}$. *Boolean functions* are described in terms of expressions over a Boolean algebra.

Switching functions. The special case of Boolean algebra denoted by $\mathbb{B} = \{(0, 1);\ +, \times;\ ^{-};\ 0, 1\}$ with two elements is called *switching algebra.* Switching algebra is the theoretical foundation of circuit design and decision diagram techniques. In switching algebra, an n-variable function $f : \mathbb{B}^n \rightarrow \mathbb{B}$ is called a *switching function.* A switching function of n variables is a discrete function,

$$f{:}\{0, 1\}^n \rightarrow \{0, 1\},$$

where $\{0, 1\}^n$ denotes the n-fold Cartesian product of $\{0, 1\}$, that is, the set of all binary n-tuples. Each n-tuple, $(x_1, ..., x_n)$, mapped to 1 by the function, is a minterm of the function. A switching function of n variables can be specified by its $2^n \times 1$ truth table or by a *Boolean formula.* A switching function of n variables can be expressed in terms of Shannon's expansion theorem:

$$f = \overline{x}_i f|_{x_i=0} \vee x_i f|_{x_i=1}.$$

TABLE 1.1
Discrete functions.

Name	Formal representation
Integer	$f : \overset{n}{\underset{i=1}{\times}} \{0, 1, \ldots, m_i - 1\} \rightarrow \{0, 1, \ldots, \tau - 1\}$
Multivalued	$f : \{0, 1, \ldots, \tau - 1\}^n \rightarrow \{0, 1, \ldots, \tau - 1\}$
Switching (Boolean)	$f : \{0, 1\}^n \rightarrow \{0, 1\}$
Pseudo-logic	$f : \overset{n}{\underset{i=1}{\times}} \{0, 1, \ldots, m_i - 1\} \rightarrow \{0, 1\}$
	$f : \{0, 1\}^n \rightarrow \{0, 1, \ldots, \tau - 1\}$
Pseudo-Boolean	$f : \{0, 1\}^n \rightarrow \mathcal{R}$
Galois	$f : \{GF(p)\}^n \rightarrow \{GF(p)\}$
Fuzzy	$f : I^n \rightarrow I, \ I = [0, 1]$

Example 1.1 *A function f of two variables x_1 and x_2 can describe various data structures:*

- *Table 1.2 specifies a two-output switching function,*
- *Table 1.3 describes an incompletely specified ternary logic function.*

There are several useful properties of a switching function:

- There are exactly four switching functions of a single variable x: two of them are the constant functions $f_0(x) = 0$, and $f_1(x) = 1$, $f_2(x) = x$, and $f_3(x) = \overline{x}$. Note, a *literal* denotes a variable x or its complement $\overline{x}$.
- There are 16 switching functions of two variables. Among these are six trivial functions (constants and literals). The remaining functions are used in elementary logic gate descriptions (AND, OR, EXOR, etc).
- The number of n-variable switching functions with one output and m outputs are equal to 2^{2^n} and 2^{m2^n} correspondingly.
- A circuit with n inputs and one output can be described in terms of an n-variable switching function $f = (x_1, x_2, \ldots x_n)$.
- A circuit with n inputs and m outputs can be described in terms of m-tuple $f = (f_1, f_2, \ldots f_m)$.

TABLE 1.2
Binary input two-output switching function (Example 1.1).

	x_1	x_2	f
0	0	0	$f_0(0)f_1(0)$
1	0	1	$f_0(1)f_1(1)$
2	1	0	$f_0(2)f_1(2)$
3	1	1	$f_0(3)f_1(3)$

TABLE 1.3
Incompletely specified ternary logic function (Example 1.1).

	x_1	x_2	f
0	0	0	$f(0)$
1	0	1	$f(1)$
2	0	2	$f(2)$
3	1	0	$f(3)$
4	1	1	$f(4)$
5	1	2	$f(5)$

Switching theory offers efficient algorithms for minimization, analysis, testing, and verification of switching circuits. The choice of data structure for representation and manipulation of logic functions is important for implementation or mapping these algorithms into hardware.

Logic function minimization. Specialized methods of switching theory for the reduction of switching functions to canonical or minimal forms and decomposition methods are well developed.

In terms of graphical data structures, logic function minimization means the construction of decision trees and diagrams with minimal implementation cost.

Verification.

Formal verification in contemporary digital design flow is accomplished through advanced techniques such as Boolean satisfiability (SAT) and binary decision diagrams.

The aim of formal verification is a formal mathematical proof that the functional behavior of the *specification* and the *implementation* coincide. The representation of the original circuit is interpreted as the specification. The representation of the optimized circuit is considered an implementation. It has to be proven formally that specification and implementation are functionally equivalent, i.e., that both compute exactly the same switching function(s). Functional representations for both networks are derived and tested for equivalence. By this method, two sequential systems can also be compared. The two systems must have identical encoding, i.e., output and next-state functions.

Event-driven analysis stands for sensitivity and observability analysis, testability and related applications for logic design. The theoretical basis of this field is *Boolean differential calculus.*

Boolean differential operators analyze the behavior of computing structures in terms of change. Boolean differences can be computed on decision trees and diagrams, as considered in Chapter 9.

The Boolean difference is a certain analog of the Taylor cofactor of an algebraic function. The analog of the Taylor expansion in switching theory is the Reed–Muller expansion, as well as arithmetic and Walsh forms. Thus, Boolean difference can be utilized to calculate Reed–Muller, arithmetical and Walsh coefficients, and, therefore, is relevant to spectral techniques.

One application of this analysis is in automatic test generation. The Boolean difference describes the set of all tests for each single fault. Another application is in the area of combinational logic optimization. The "don't care set" for a signal line of the circuit, i.e., those cases where the circuit outputs are independent of the signal value on this line, can be determined. This information can be used by the circuit optimizer, which can apply transformations such as eliminating a signal line or moving a line to a different gate output. To avoid recomputing the Boolean difference every time the optimizer modifies the circuit, a set of "compatible don't care" functions can be found, which remain valid even as the circuit structure is changed.

Analysis of sequential circuits. A finite-state system over a sequence of state transitions is traditionally used to model a sequential digital system. For very small systems, an explicit representation using state graph methods and analysis of the graph is used. For larger systems, symbolic state graph methods with the state transition represented as a switching function are more appropriate.

1.1.2 Multivalued functions

Besides the well-known two-valued Boolean algebra or switching algebra, there are other Boolean algebras.

Any m-valued function can be represented by a multivalued decision tree and decision diagram, which are canonical graph-based data structures.

A multivalued tree and diagram is specified in the same way as a binary decision tree and BDD, except the nodes become more complex due to the use of Shannon and Davio expansion for multivalued functions. Multivalued decision diagrams are considered in Part III as an extension of decision diagram technique to multivalued functions.

1.2 Decision tables

A *decision table* is an organizational or programming tool for the representation of discrete functions. It can be viewed as a matrix where the columns specify sets of conditions and the rows sets of actions to be taken when the corresponding conditions are satisfied. The columns specify the rules of the type "if conditions, then actions."

If some columns specify identical actions, they are redundant. Some rules can overlap, i.e., a combination of condition values can be found that satisfies the condition sets of both rules. Two overlapping rules that specify different actions make the table ambiguous.

Example 1.2 *An unambiguous decision table is a special case of a multivalued discrete function, where the conditions correspond to the variables and the action sets to the function values. In particular, a complete decision table corresponds to a completely specified discrete function.*

A decision table can be evaluated sequentially, that is, it can be implemented, or programmed by "if-then-else" constructs. This is also known as an ID3 algorithm. Evaluation leads to classification of event objects represented by the decision table into one of a finite number of known categories. Such a classification is based upon the outcome of a number of tests (assignment of variables).

Example 1.3 *In decision making and pattern recognition, tests are known as features and objects as classes.*

Decision tables can be easily represented by a tree, and a probability distribution can be evaluated using the information specified on the set of objects. A priori probability that an unknown object will be identified as each object in the set can be given.

1.3 Graphical data structures

This section investigates the genesis of decision trees and diagrams and the ontology with respect to data structures for discrete function representation.

> Decision trees *and* decision diagrams *are graphical data structures for logic functions. Manipulation of decision tables or algebraic expressions is replaced by manipulation of graphical components.*

Other graphical data structures, such as *directed acyclic graphs* of Boolean networks, are used for logic functions and circuit representation.

Decision trees and diagrams. Decision trees and diagrams are models of the evaluation of discrete functions, wherein the value of a variable is determined and the next action (to choose another variable to evaluate or to output the value of the function) is chosen accordingly. Since switching functions are a sub-class of discrete functions, decision trees and diagrams are applicable to logic design tasks formulated in terms of such evaluation (see Chapter 2).

While a decision tree is derived from the simple mapping of variable assignment to function values, it implicitly reflects the relationship between variables and function values. A simpler way of mapping assignments of variables to the function values is a singular hypercube. Many algorithms for solving various computational problems of logic analysis and synthesis employ decision trees and diagrams.

Analysis of algorithms. Decision diagrams are well suited to composition and recursion, while these operations are difficult to carry out using formulas. Many graph algorithms have been developed for computations based upon a matrix representation. Generally, any algorithm that can be formulated or refined to the evaluation of a variable on a function f of k-ary variables can be resolved using a decision tree. Evaluation of a k-ary variable provides at most $\log_2 k$ bits of information, and, thus, the minimum height of any tree for the function f must obey the relation

$$h \geq H(f)/log_2 k,$$

where $H(f)$ is the functional entropy. This relation has been used to provide lower bounds on the complexity of several combinational problems, such as sorting and various set operations.

Decision tree optimization. The problem of constructing optimal decision trees and diagrams has been addressed by many researchers using *branch-and-bound* techniques and *dynamic programming.*

- ▶ The dynamic programming approach is embodied in an algorithm designed to convert limited-entry decision tables to decision trees with minimal expected testing cost.
- ▶ The branch-and-bound method is aimed at optimization of diagram storage cost. A branch-and-bound algorithm proceeds by always developing the partial solution that is potentially less expensive than any other, often switching from one partial solution to another when lower bounds change, until one solution has been completely developed. A non-redundant variable must appear at least once in any diagram. The principal disadvantage of the branch-and-bound method is that it may result in a near-exhaustive search of the possible trees and diagrams.

1.4 Spectral techniques

Spectral techniques is a discipline of applied mathematics, representing a subarea of abstract harmonic analysis. Spectral representations of logic functions are canonical. Spectral techniques express the principle of linearity and superposition.

Switching functions are considered as elements of the Hilbert space consisting of complex-valued functions defined on finite dyadic groups. Functional expressions for switching and multivalued functions include spectral representations such as sum-of-products, Reed–Muller, word-level arithmetic, and Walsh expressions.

Spectral representations are based on matrix procedures, which, on the other hand, are easily represented by decision trees and diagrams. That is why the spectral transforms of switching and multivalued functions considered in this book have their representation as spectral decision diagrams *related to matrix procedures of spectral transforms.*

1.4.1 Group theory

An algebraic structure is defined by the tuple

$$\langle \mathcal{A}, o_1, \ldots, o_k; R_1, \ldots, R_m; c_1, \ldots, c_k \rangle,$$

where $\mathcal{A}$ is a non-empty set, o_i is a function $\mathcal{A}^{p_i} \rightarrow \mathcal{A}$, R_j is a relation in $\mathcal{A}$, p_i is a positive integer, and c_i is an element of $\mathcal{A}$.

Various algebraic systems are used in decision diagram techniques (Figure 1.2).

Example 1.4 *The algebraic system* $\langle Z, +\rangle$, *where* Z *is a set of integers and* $+$ *is addition is used in edge-valued binary decision diagrams (EVBDDs).*

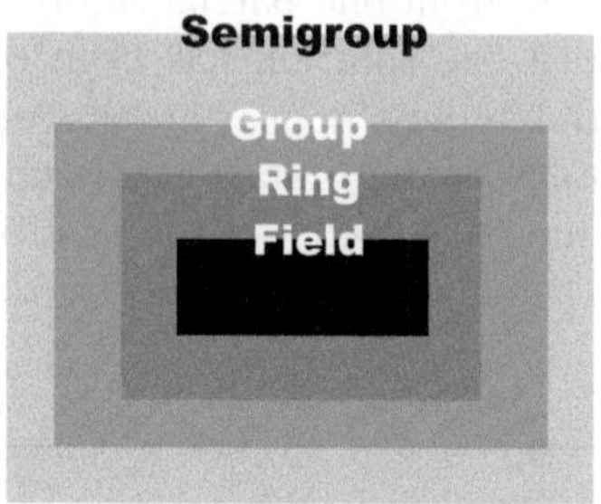

FIGURE 1.2
Relation of algebraic systems.

A lattice is an algebraic system with two binary operations. In a lattice, any two elements have both meet and joint. All finite lattices have a greatest element, denoted **1** and a least element (**0**). Various algebraic systems satisfy the axioms of a lattice.

Example 1.5 *Let* $B = \{0, 1\}$. *Then* $\{B, \oplus, \cdot, 1\}$ *is a field where symbols* $\oplus$ *and* $\cdot$ *are modulo 2 addition and multiplication. This is the field* $GF(2)$, *also known as Reed–Muller algebra. A field must satisfy associative and commutative laws for addition and multiplication, distributive law, and the following: for an arbitrary non-zero element* z, *there exists* y *such that* $x \cdot y = y \cdot x = 1$.

A function f is a mapping $f{:}G \rightarrow P$, $x \in G$ and $f \in P$, where G is an Abelian group of order $|G| = N$, and P is a field. Switching functions are a particular example, when $G = C_2 \ldots C_2$, $C_2 \in (0, 1, \oplus)$, and the group operation is the addition modulo 2, and P is the finite Galois field $GF(2)$, Fourier and Fourier-like transforms on a finite group G of order $|G| = N$ are defined using the group characters of G as the set of basic functions.

1.4.2 Abstract Fourier analysis

Functional expressions for switching and multivalued functions include spectral representations such as sum-of-products, Reed–Muller, word-level arithmetic, and Walsh expressions. The technique for generation, mutual transform and manipulation of these expression is based on *abstract Fourier analysis*, which covers various spectral transforms and correlation analysis.

Fourier transform. Classical Fourier analysis is applicable for functions defined on a particular Abelian group, i.e., for functions of real-valued variables that are to be transformed into the complex field C. Extensions to other classes of functions, i.e., to functions defined on other groups in other fields, are possible. This is easily done through the group-theoretical approach and group representation theory. The real group R is replaced by an arbitrary, not necessarily Abelian group. Fourier representations for functions in these groups are defined in terms of group representations. This leads to a mathematical discipline known as *abstract harmonic analysis.*

Fourier and Fourier-like transforms on dyadic groups. The Fourier series represents the signal as a sum of cosines with given frequencies. The Fourier transform of a signal is a function of frequency. The signal representations depend on frequency. This concept is known as the *spectral content* of a signal. This concept plays an important role in signal analysis. In particular, the frequencies that play a dominant role in the composition of the signal give us clues to the properties of the signal.

Functional analysis is a mathematical discipline dealing with functions, function spaces, different operators in these spaces, and related subjects. The main statement of functional analysis is that each function can be represented (expanded) in terms of a basis in vector space. The basis is a set of linearly independent functions, which means that any function from a set of linearly independent functions cannot be reduced to some other function from this set by some simple linear operations over this function and other functions.

The Reed–Muller transform can be expressed as:

$$f\ :\ \{GF(2)\}^{n} \rightarrow GF(2).$$

The Reed–Muller transforms a switching function from its original domain to the Reed–Muller domain, an algebra defined on logic product and exclusive OR of Boolean variables and constant 1. The Kronecker transforms considered below are a subclass of the Reed–Muller transform.

Arithmetic transform. Assume that the values 0 and 1 of a logic function are interpreted as integers 0 and 1. To have a vector space as the range, we need to consider these values 0 and 1 as elements of a field containing integers. The smallest possible would be the field of rational numbers Q, but it will be convenient to go further and use the field of complex numbers. Thus, we replace $GF(2)$ by C and consider functions

$$f\ :\ C_2^n \rightarrow C.$$

Arithmetic polynomials are used for efficient representation and calculation of multi-output functions, and can be represented using word-level decision diagrams.

Walsh transform. In the Reed–Muller transform, we viewed the values 0,1 of a logic function as elements of $GF(2)$. By viewing these values as integers 0,1 as elements of C, we obtain an arithmetic (functions) transform. Instead of the values 0 and 1, we could use some other complex numbers α, β and consider functions $f : C_2^n \rightarrow \{\alpha, \beta\}$. By using $\alpha = 1$ and $\beta = -1$ we obtain the Walsh functions and the Walsh transform. This transform has many useful properties. In particular, under multiplication of complex numbers, $\{0, -1\}$ forms a group with exactly the same structure as the group $\{0, 1\}$ under modulo 2 addition.

It can be shown that the Walsh transform is the Fourier transform defined on the dyadic groups C_n^2. It expresses properties corresponding to those of the classical Fourier transform on R. Therefore, it is often called the *Walsh–Fourier* transform.

Reed–Muller and arithmetic transforms do not have the properties of the Walsh transform, and cannot be derived from the group representation theory as a generalization of the classical Fourier transform. However, they are closely related and share some of the useful properties of Fourier transforms defined on groups. Therefore, they are referred to as *Fourier-like* transforms.

Kronecker transforms. The transforms we discussed can be generalized into Kronecker product representable transforms. Denote by $P(C_2^n)$ the set of all switching functions on C_n^2 into P, where P is a field that may be a finite (Galois) field or the complex field C.

Partial Kronecker transforms. The fast Fourier transform (FFT) is an efficient algorithm for calculation of the discrete Fourier transform. The algorithm is based on the property that a function on a group C_2^n of order 2^n can be considered as an n-variable function on C_n^2 .

A discrete Fourier transform of order 2^n can be performed as n discrete Fourier transforms of orders 2. Each of these transforms is performed with respect to a variable in f. The method defined for the discrete Fourier transform can be extended to any Kronecker product representable transform due to the properties of the Kronecker product.

Fixed polarity transforms. In a *fixed polarity* Reed–Muller transform, the Davio expansion rule of a given polarity is applied to each variable in f. The fixed polarity Reed–Muller spectrum with the minimum number of non-zero coefficients defines the coefficients in the minimum Reed–Muller expansion for f. Fixed polarity arithmetic and Walsh transforms can be defined by analogy.

Walsh polynomials expressed in terms of switching variables reduce, after simplification, to arithmetic polynomials. Walsh polynomials can be used to represent multi-output functions in the same way as is done with arithmetic polynomials.

Haar transform. The discrete Haar transform is an important example of non-Kronecker transforms in $C(C_2^n)$. It is defined in terms of the discrete Haar functions, which can be considered as a discrete counterpart of Haar functions on the interval [0; 1] derived by sampling these functions at 2^n points, or as a set of functions in $C(C_2^n)$ defined in such a way as to share or resemble properties of the Haar functions.

1.4.3 Correlation analysis

The degree of similarity between two signals can be described using correlation coefficients, which should logically take the values +1 for two identical signals, zero for two signals that have no relationship to each other, and -1 for signals in opposition to each other.

Autocorrelation is a useful concept in spectral methods for analysis and synthesis of networks realizing logic functions. For a given n-variable switching function f, the autocorrelation function B_f is defined as

$$B_f(\tau) = \sum_{x=0}^{2^n-1} f(x)f(x \oplus \tau),$$

where $\tau \in \{0, \ldots, 2^n - 1\}$ is the Hamming distance. The Winer–Khintchine theorem states a relationship between the autocorrelation function and Walsh (Fourier) coefficients. The autocorrelation function is invariant to the shift operator $\oplus$ in terms of which B_f is defined.

There are also various results that are useful in spectral techniques, in particular, composition properties of Walsh–Hadamard spectra of switching functions derived as convolutions of the corresponding spectra, properties of correlation matrices and their applications in permutations of switching functions, and cross-correlation and autocorrelation including the Winer–Khintchine theorem for switching functions.

Example 1.6 *One of the effective methods in decision diagram technique, linear transformed decision diagrams, can be formulated in terms of the Wiener–Khintchine theorem* [8] *(Chapter 14).*

1.5 Stochastic models and information theoretic measures

Stochastic, or probabilistic, techniques constitute the theoretical basis for fault-tolerant computing.

This is important for large scale integrated circuits, and especially for computing in nanodimensions. Probabilistic models of switching networks can be related to decision diagrams, as shown in Chapter 20.

1.5.1 Stochastic models and probability

A stochastic process $X = \{X(t),\ t \in T\}$ is a collection of random variables. That is, for each t in the set T, $X(t)$ is a random variable. The parameter t is often interpreted as time. Any realization of X is called a sample path.

Statistical inference concerns generalizations based on sample data. It applies to such problems in decision diagram technique as estimation of variable activity (see "Further Reading" section in Chapter 20).

Markov models. Stochastic models for noise-making signals include Markov chain models and stochastic pulse stream models. Advanced models such as the Markov random field can be deployed toward logic design.

Example 1.7 *Given a circuit, a Markov random field is a graph indicating the neighborhood relations of the circuit nodes. This graph is used in probability maximization processes, aimed at characterization of circuit configurations for the best thermal noise reliability (expressed in terms of logic signal errors).*

Example 1.8 *Let the inputs of the OR gate* $x_1 \in \{0,1\}$ *and* $x_2 \in \{0,1\}$ *be mutually independent with probabilities* $p_1 = p(x_1)$ *and* $p_2 = p(x_2)$ *correspondingly (Figure b). The output probability can be evaluated as the probability of at least one event* x_1 *and* x_2*, i.e.,*

$$p = 1 - (1 - p_1)(1 - p_2) = p_1 + p_2 - p_1 p_2.$$

Since the input signals are independent, then $p_1 p_2 = 0$*, and, thus, the OR gate with deterministic input and output signals can be considered as an adder of probabilities* p_1 *and* p_2*:* $p = p_1 + p_2$.

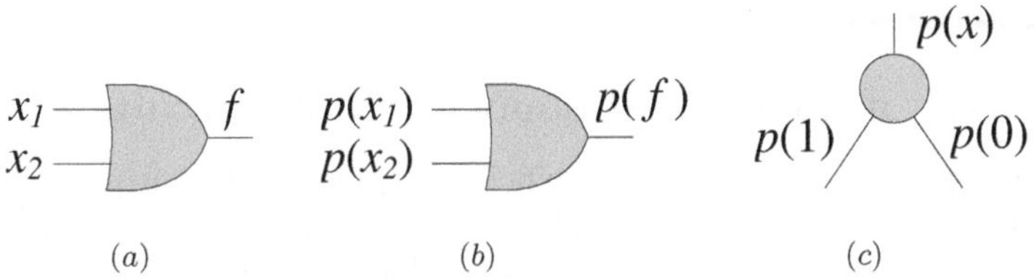

(a) (b) (c)

FIGURE 1.3
Deterministic input signals (a), random input signals (b), and probabilistic interpretation of the node of a decision diagram (Example 1.8).

These methods are used for probabilistic computing and probabilistic decision diagram design (see Chapter 20). Probabilities of variables calculated on decision trees and diagrams are used to estimate power dissipation (Chapter 21).

1.5.2 Information theoretical measures of logic functions and decision trees

A circuit can be seen as a process of communication between circuit components (gates). The source of information is defined in terms of the probability distribution for signals from this source.

> *Entropy is a measure of the information content of logic functions that can be computed using decision diagrams.*

In this strategy, commonly used (near) minimization of the expected testing cost, the problem is viewed as one of refining an initial uncertainty about the function's value in a certitude. At each step, the test of a variable diminishes the universe of possibilities, thereby removing a certain amount of ambiguity. In terms of information theory, the initial ambiguity of a function is expressed by the entropy of the function. The ambiguity remaining after testing a variable can be computed as the average ambiguity among the restrictions.

Details of entropy computations using decision trees are considered in Chapter 24.

1.6 CAD of micro- and nanoelectronic circuits

VLSI design. CAD tools provide efficient synthesis and optimization techniques at every step of microelectronic circuit design. Synthesis is aimed at the generation of models of a circuit. These models are created for the following levels of abstraction:

- Architectural, or high level,
- Logic level, and
- Physical design level.

Some of the CAD tools for microelectronics based on decision diagram are referred to in Part 2. As for nanoelectronic design, the CAD tools, especially those based on decision diagram techniques, are still in their infancy. The reason for this is a current gap between the possibilities that nanoelectronic and molecular technology offer, and the design methodology developed in this area. Some technological aspects and problems are mentioned below.

Nanoelectronics and nanotechnologies include, in particular, the following directions:

- Solid-state nanoelectronics
- Molecular electronics, and
- Bioelectronics.

Logic design of nanodevices is being developed in two directions:

- Using advanced logic design techniques and methods from other disciplines, in particular, fault-tolerant computing.
- Development of a new theory and technique for logic design in spatial dimensions.

The first direction adapts architectural approaches such as array-based logic, parallel and distributed architectures, methods from fault-tolerant computing and adaptive structures such as neural networks. The design of nanostructure topologies based on decision diagram techniques is considered in Chapter 34. The second direction can be justified, in particular, by nanotechnologies that implement devices on the reversibility principle. This is quite relevant to the design of adiabatic circuits, which is not associated only with nanotechnology. This is where the usefulness of multivalued logic models such as decision diagrams has been shown too (see Chapter 35).

1.7 Further reading

Discrete functions. For an overview of the theory of discrete and switching functions, see the book by Davio et al. [4]. References and overview of event-driven analysis are presented in Chapter 9. Decision tables are used to formalize many approaches and to represent data [1, 13, 16]. Multivalued counterpart of decision tables are used to represent multivalued relations (see, for example, [12]).

Decision trees and diagrams. Decision trees were introduced first to represent discrete sets in 1935 by Kurepa [9]. The applications of decision trees include decision table programming [13, 16], databases, pattern recognition [1], taxonomy and identification [6], and analysis of algorithms [17]. A particular case, binary decision trees are used in many areas of data mining and processing. Binary decision diagrams were first considered with respect to switching functions by Lee [10]. An excellent survey of decision tree and diagram techniques and applications was written by Moret [11].

Hypercube-based structures are at the forefront of massive parallel computation because of the unique characteristics of hypercubes (fault tolerance, ability to efficiently permit the embedding of various topologies, such as lattices and trees) [2, 3]. A number of topologies can be considered relevant to the problems of spatial logic design, for example, hypercube topology [15], cube-connected cycles known as CCC-topology [14], and pyramid topology [13].

Probabilistic techniques and information-theoretical measures Models based on reliable gates with stochastic input streams were first considered by Gaines [5] and Yakovlev [18]. The bibliography for information-theoretical measures is given in Chapter 8. Spectral techniques and related references are considered in Chapter 7.

References

[1] Bell DA. Decision trees, tables, and lattices. In Batchelor BG, Ed., *Pattern Recognition: Ideas in Practice*, Chapter 5, Plenum Press, New York, 1978.

[2] Chen HL and Tzeng NF. A Boolean expression-based approach for maximum incomplete subcubes identification in faulty hypercubes. *IEEE Transactions on Parallel and Distributed Systems*, 8(11):1171–1183, 1997.

[3] Chen HL and Tzeng NF. Subcube determination in faulty hypercubes. *IEEE Transactions on Computers*, 46(8):871–879, 1997.

[4] Davio M, Deschamps JP, and Thayse A. Discrete and Switching Functions. *McGraw Hill*, New York, 1978.

[5] Gaines BR. Stochastic computing systems. In Tou JT, Ed., *Advances in Information Systems Science*, Plenum, New York, vol. 2, chapter 2, pp. 37–172, 1969.

[6] Garey MR. Optimal binary identification procedures. *SIAM Joirnal of Applied Mathematics*, 23(2), 173–186, 1972.

[7] Greenberg RI. The fat-pyramid and universal parallel computation independent of wire delay. *IEEE Transactions on Computers*, 43(12):1358–1364, 1994.

[8] Karpovsky MG, Stanković RS, and Astola JT. Reduction of sizes of decision diagrams by autocorrelation functions. *IEEE Transactions on Computers*, 52(5):592–606, 2003.

[9] Kurepa DJR. *Ensembles ordonné et ramifiès (Thèse)*. PhD thesis, Paris, 1935.

[10] Lee CY. Binary decision diagrams. *Bell System Technical Journal,* 38(4):985–999, 1959.

[11] Moret BME. Decision trees and diagrams. *Computing Surveys*, 14(4):593–623, 1982.

[12] Pleban M, Yanushkevich S, Shmerko V, and Luba T. Argument reduction algorithms for multi-valued relations, In *Proceedings of the IEEE International Conference on Artificial Intelligence and Soft Computing*, Banff, Canada, pp. 699–614, 2002.

[13] Pooch UW. Translation of decision tables. *ACM Computing Surveys*, 6(2):125–151, 1974.

[14] Preparata FR and Vuillemin J. The cube-connected cycles: a versatile network for parallel computation. *Communications ACM*, 24(5):300–309, 1981.

[15] Roth JP. Algebraic topological methods for the synthesis of switching systems. *Transactions of the American Mathematical Society*, 88(2):301–326, July 1958.

[16] Silberg B, Ed., Decision tables. *SIGPLAN Notes (ACM)*, 6(8), 1971.

[17] Weide B. A survey of analysis techniques for discrete algorithms. *ACM Computing Surveys*, 9(4):291–313, Dec. 1977.

[18] Yakovlev VV and Fedorov RF. *Stochastic Computing.* "Mashinostroenie" Publishers, Moscow, 1974 (In Russian).

2

Data Structures

The focus of this chapter is various algebraic expressions for switching function representation that can be mapped into graph data structures. The algebraic expressions include sum-of-products, Reed–Muller, Walsh, and other forms.

2.1 Introduction

Switching functions have a corresponding implementation in terms of interconnected gates. This is a *circuit*, or *schematic*. The composition of primitive components is accomplished by physically wiring the gates together. A collection of wires is called a *switching net*. The tabulation of gate inputs and outputs and the nets to which they are connected is called a *netlist*. The *fan-in* of a gate is its number of inputs. The *fan-out* of a gate is the number of inputs to which the gate's output is connected.

A *data structure* representing switching functions is a collection of Boolean variables connected in various ways. It is a mathematical model of a switching function. *Data type* is defined by classification of properties of the mathematical model. A data structure for a switching function f of n variables $x_1, x_2, \ldots x_n$ can be represented in different mathematical *forms* or *descriptions*:

- ▶ General (algebraic) expansion,
- ▶ Matrix form,
- ▶ Cube representation,
- ▶ Data flowgraphs,
- ▶ Decision trees, and
- ▶ Decision diagrams.

The algebraic data structures for switching functions considered in this book are the following:

- ▶ Sum-of-products expressions,
- ▶ Reed–Muller expressions,

- ▶ Arithmetic expressions including Walsh expansions, and
- ▶ Word-level expressions.

Choosing the appropriate data structure for a switching function is a crucial part in designing implementable circuits.

2.2 Truth tables

Switching functions are conveniently defined by tables showing elements of the domain in the left part, and the corresponding function values in the right part. These tables are called *truth-tables*, and function values are represented by *truth-vectors.* In tabular representation, all the function values are explicitly shown, without taking into account their possible relationships. A decision table with binary values ("yes" and "no" answers to conditions) corresponds to a function of binary variables. The tests are binary variables and the objects are values of a partial bijective function from the variables' space to the set of objects.

The size of truth tables is exponentially dependent on the number of variables. This representation of switching functions is unsuitable for functions of a large number of variables.

In representing switching functions by decision diagrams, the reduction is achieved due to the particular properties of functions and the corresponding relationships among function values presented in the truth table.

Example 2.1 *A two-output switching function is represented in Figure 2.1 by the circuit, algebraic description (see Chapter 5 on arithmetical forms), and the truth table.*

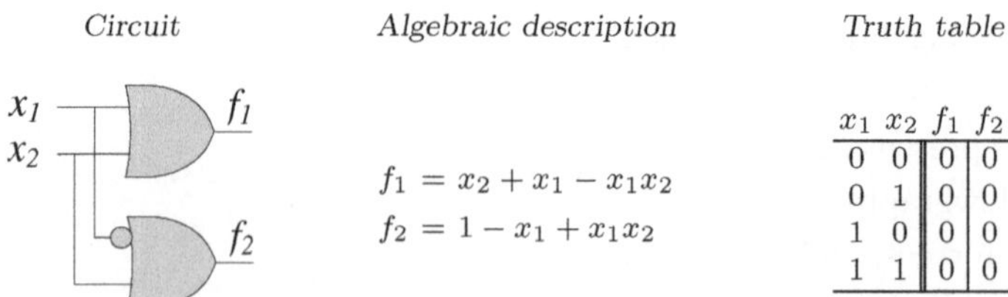

$$f_1 = x_2 + x_1 - x_1 x_2$$
$$f_2 = 1 - x_1 + x_1 x_2$$

x_1	x_2	f_1	f_2
0	0	0	0
0	1	0	0
1	0	0	0
1	1	0	0

FIGURE 2.1
A two-output switching function: circuit implementation, arithmetic expressions of the outputs, and truth tables (Example 2.1).

2.3 Algebraic representations

An algebraic representation of a switching function is a formal description of a data structure carrying information through the algebraic relations between the variables. The chosen algebraic operation determines the type of expression (logic or arithmetic).

2.3.1 Algebraic structure

Each variable in the product term is called a *space coordinate*, and the number of variables specifies the number of space dimensions.

> **Example 2.2** *A one-variable function corresponds to a one-dimensional space, two-variables means a two-dimensional space, etc.*

The product $(x_1^{i_1} \cdots x_n^{i_n})$ generates two kinds of expressions for $i = 0, 1, 2, \ldots, 2^n - 1$. If $x_j^{i_j} = \overline{x}_j$ for $i_j = 0$, and $x_j^{i_j} = x_j$ for $i_j = 1$, then

$$\underbrace{(x_1^{i_1} \cdots x_n^{i_n})}_{i-th\ product} = \overline{x}_1 \ldots \overline{x}_n \odot \overline{x}_1 \overline{x}_2 \ldots \overline{x}_{n-1} x_n \odot \ldots \odot x_1 x_2 \ldots x_n.$$

Let $x_j^{i_j} = 1$ for $i_j = 0$, and $x_j^{i_j} = x_j$ for $i_j = 1$, then

$$\underbrace{(x_1^{i_1} \cdots x_n^{i_n})}_{i-th\ product} = 1 \odot x_n \odot x_{n-1} x_n \odot \ldots \odot x_1 x_2 \ldots x_{n-1} \odot x_1 x_2 \ldots x_n.$$

The general equation for representing a switching function in various forms is the following:

$$f = \bigodot_{i=0}^{2^n-1} \Omega_i \underbrace{(x_1^{i_1} \cdots x_n^{i_n})}_{i-th\ product\ term}, \tag{2.1}$$

where

$$\odot = \begin{cases} \vee, & \text{for sum-of-products expression;} \\ \oplus, & \text{for Reed–Muller expression;} \\ +, & \text{for arithmetic and word-level arithmetic expression.} \end{cases}$$

The coefficients Ω_i are 0 or 1 for sum-of-product and Reed–Muller expressions, and integer numbers for arithmetic and word-level representations.

2.3.2 Matrix representation

A data structure can be described in matrix form. The matrix form (Figure 2.2) is based on:

- The truth-vector **F** of a given switching function f,
- A vector of coefficients $\mathbf{\Omega}$, and
- A transform matrix $\mathbf{\Omega}_{2^n}$.

Given the truth-vector **F**, the result of the direct transformation is the vector of coefficients $\mathbf{\Omega}$. The inverse transform is used to restore the truth-vector **F** given a vector of coefficients $\mathbf{\Omega}$. This matrix form can be mapped into a data flowgraph that carries useful algorithmic properties (parallel computing and complexity).

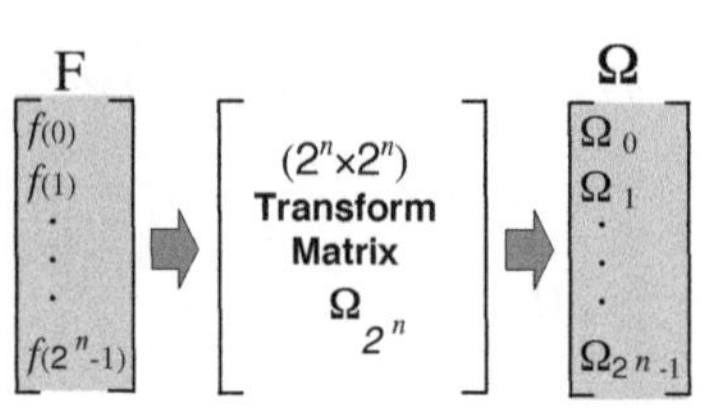

- *Truth-vector* **F** *of a switching function* f
- *Transform matrix* $\mathbf{\Omega}_{2^n}$
- *Vector of coefficients* $\mathbf{\Omega}$
- *The result of the direct transformation is the vector of coefficients* $\mathbf{\Omega}$
- *Inverse transform is used to restore the truth-vector* **F** *given a vector of coefficients* $\mathbf{\Omega}$

FIGURE 2.2
The matrix representation of a switching functions.

The *polarity* of an expression is one of the possible ways to represent a function. The polarity of a variable indicates:

- A complemented variable ($\overline{x}$ or polarity 1) or
- An uncomplemented variable (x).

All the above forms can be interpreted and modified with respect to 2^n polarities. Moreover, in this modification the *fixed* and *mixed* polarity are distinguished.

2.4 AND-OR expressions

AND-OR, called *sum-of-products* (SOPs) expressions, are derived as canonical polynomials of variables over OR, AND, NOT operations. Sum-of-products expressions correspond to two-level AND-OR logic networks.

2.4.1 Sum-of-products expressions

Given a switching function f of n variables, the sum-of-products expression is specified by

$$f = \bigvee_{i=0}^{2^n-1} s_i \cdot \underbrace{(x_1^{i_1} \cdots x_n^{i_n})}_{i-th\ product}, \tag{2.2}$$

where $s_i \in \{0,1\}$ is a coefficient, i_j is the j-th bit, $j = 1, 2, \ldots, n$, in the binary representation of the index $i = i_1 i_2 \ldots i_n$, and $x_j^{i_j}$ is defined as

$$x_j^{i_j} = \begin{cases} \overline{x}_j, \ i_j = 0; \\ x_j, \ i_j = 1. \end{cases} \tag{2.3}$$

A product in Equation 2.2 is referred to as a *minterm*. Note that a literal is a variable or its complement. A minterm of n variables is a product of n literals in which each variable appears exactly once in either true or complemented form (but not both).

Example 2.3 *An arbitrary switching function of two variables ($n = 2$) is represented in sum-of-products form (Equations 2.2 and 2.3) as follows:*

$$f = s_0(\overline{x}_1\overline{x}_2) \vee s_1(\overline{x}_1 x_2) \vee s_2(\overline{x}_1 x_2) \vee s_3(x_1 x_2).$$

2.4.2 Computing the SOP coefficients

Given the truth vector $\mathbf{F} = [f(0)\ f(1) \ldots f(2^n-1)]^T$, the vector of coefficients $\mathbf{S} = [s_0\ s_1 \ldots s_{2^n-1}]^T$ is derived by the matrix equation specified on AND and OR operations

$$\mathbf{S} = \mathbf{S}_{2^n} \cdot \mathbf{F}, \tag{2.4}$$

where the $2^n \times 2^n$ matrix $\mathbf{S}_{2^n}$ is formed by the Kronecker (tensor) product $\mathbf{S}_{2^1}$:

$$\mathbf{S}_{2^n} = \bigotimes_{i=1}^{n} \mathbf{S}_{2^1}, \qquad \mathbf{S}_{2^1} = \begin{bmatrix} 1 & 0 \\ 0 & 1 \end{bmatrix}. \tag{2.5}$$

Note that the Kronecker product of 2×2 matrixes $\mathbf{A}_2 = \begin{bmatrix} a_{11} & a_{12} \\ a_{21} & a_{22} \end{bmatrix}$ and $\mathbf{B}_2 = \begin{bmatrix} b_{11} & b_{12} \\ b_{21} & b_{22} \end{bmatrix}$ results in 4×4 matrix

$$\mathbf{C}_{2^2} = \mathbf{A}_2 \otimes \mathbf{B}_2 = \begin{bmatrix} a_{11}\mathbf{B}_2 & a_{12}\mathbf{B}_2 \\ a_{21}\mathbf{B}_2 & a_{22}\mathbf{B}_2, \end{bmatrix} = \begin{bmatrix} a_{11}b_{11} & a_{11}b_{12} & a_{12}b_{13} & a_{12}b_{14} \\ a_{11}b_{21} & a_{11}b_{22} & a_{12}b_{23} & a_{12}b_{24} \\ a_{21}b_{31} & a_{21}b_{32} & a_{22}b_{33} & a_{22}b_{34} \\ a_{21}b_{41} & a_{21}b_{42} & a_{22}b_{43} & a_{22}b_{44} \end{bmatrix}.$$

Matrix $\mathbf{A}_2$ controls the block structure of the Kronecker product matrix $\mathbf{C}_{2^2}$.

Since, $\mathbf{S}_{2^n}$ is the $2^n \times 2^n$ identity matrix, $\mathbf{F} = \mathbf{S}$. This sum-of-products is the particular case of other forms. In Figure 2.3a, the general scheme of computing is illustrated.

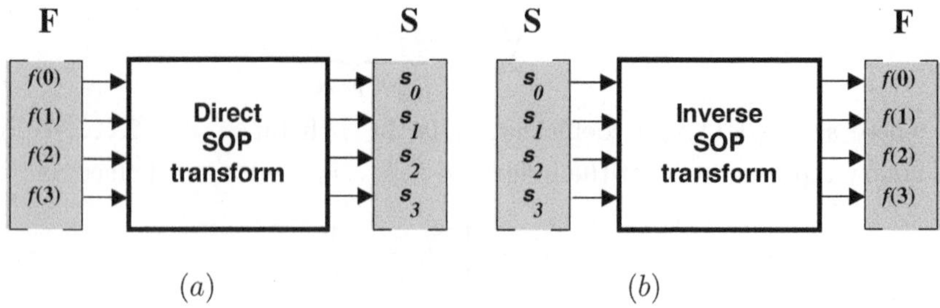

FIGURE 2.3
Direct (a) and inverse (b) sum-of-products (SOP) transforms for a switching function of two variables.

Example 2.4 *Computing the coefficients of the sum-of-products by Equations 2.4 and 2.5 for the elementary function* $f = x_1 \oplus x_2$ *given by the truth vector* $\mathbf{F} = [0\ 1\ 1\ 0]^T$ *is illustrated in Figure 2.4.*

2.4.3 Restoration

The following matrix equation using AND and OR operations restores the truth-vector $\mathbf{F}$ from the vector of coefficients $\mathbf{S}$ (Figure 2.3b):

$$\mathbf{F} = \mathbf{S}_{2^n}^{-1} \cdot \mathbf{S}, \tag{2.6}$$

where $\mathbf{S}_{2^1}^{-1} = \mathbf{S}_{2^1}$. Notice that the matrix $\mathbf{S}_{2^1}$ is self-inverse.

Example 2.5 *A restoration the truth-vector* $\mathbf{F}$ *of the sum-of-products of a switching function* f *given by the vector of coefficients* $\mathbf{S} = [0\ 1\ 1\ 0]^T$ *is given in Figure 2.4.*

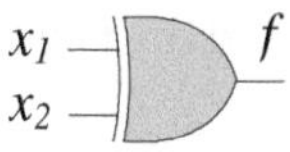

x_1, x_2, f

$f = x_1 \oplus x_2$

Computing the vector of coefficients

$$\mathbf{S}{=}\mathbf{S}_{2^3}\cdot\mathbf{F} = \left[\begin{array}{cc|cc} 1 & & & \\ & 1 & & \\ \hline & & 1 & \\ & & & 1 \end{array}\right]\begin{bmatrix}0\\1\\1\\0\end{bmatrix} = \begin{bmatrix}0 = s_0\\1 = s_1\\1 = s_2\\0 = s_3\end{bmatrix}$$

Restoring the truth-vector **F**

$$\mathbf{F}{=}\mathbf{S}_{2^3}^{-1}\cdot\mathbf{S} = \left[\begin{array}{cc|cc} 1 & & & \\ & 1 & & \\ \hline & & 1 & \\ & & & 1 \end{array}\right]\begin{bmatrix}0\\1\\1\\0\end{bmatrix} = \begin{bmatrix}0\\1\\1\\0\end{bmatrix}.$$

Sum-of-products expression

$f = s_0(\overline{x}_1\overline{x}_2) \vee s_1(\overline{x}_1 x_2) \vee s_2(x_1\overline{x}_2) \vee s_3(x_1x_2) = \overline{x}_1x_2 \vee x_1\overline{x}_2$

FIGURE 2.4
Computation of the sum-of-products expression for an EXOR gate (Examples 2.4 and 2.5).

Because the vector of coefficients **S** and the truth-vector **F** are equal, the sum-of-products expression can be derived directly from the truth vector **F**.

Example 2.6 *Given the truth vector* $\mathbf{F} = [0\ 1\ 0\ 0]^T$*, the vector of coefficients is* $\mathbf{S} = [0\ 1\ 0\ 0]^T$*, which corresponds to the algebraic form of the switching function* $f = \overline{x}_1 x_2$.

2.5 Relevance to other data structures

The other data structures used in digital circuits design include:

- Data flowgraphs,
- Direct acyclic graphs (DAGs),
- Hypercubes, and
- Decision trees and decision diagrams.

These can be embedded into n-dimensional space. Embedding these data structures into spatial dimensions and representing 3D computing architectures, with a particular focus on embedding decision trees and diagrams into so-called $\mathcal{N}$-hypercubes, is considered in Chapter 34.

2.5.1 Dataflow graphs

A data flowgraph represents the data flow in a transform algorithm for a switching function. The graph edges correspond to parallel streams of computing. The data flowgraph of a signal transform can be derived from a *factorization* of the transform matrix $\mathbf{\Omega}_{2^n}$. The nodes of the data flowgraph implement the operation Ω.

Example 2.7 *In the graph of a Reed–Muller transform, the nodes implement modulo-2 operations. Correspondingly, an arithmetic sum operation is implemented in the nodes of data flowgraphs for arithmetic and word-level arithmetic transforms. For Reed–Muller and arithmetic transforms, the basic configuration of the data flowgraph is a "butterfly," well-known from the fast Fourier transform used in digital signal processing.*

2.5.2 Hypercube

The hypercube is a topological representation of a switching function by an n-dimensional graph, where n is a number of variables. This representation is aimed at:

- *Interpretation* (representation) of the function in a form useful for manipulation, and
- *Mapping* of the function to 3D space.

In switching theory, a hypercube is defined as a collection of 2^m minterms $(m \leq n)^*$, therefore, the vertices of the hypercube are assigned with the minterms,

$$x_1^{i_1} x_2^{i_2} \cdots x_{n-1}^{i_{n-1}} x_n^{i_n}, \tag{2.7}$$

where $x_j^{i_j} = 1$ for $i_j = 0$, and $x_j^{i_j} = x_j$ for $i_j = 1$,

Example 2.8 *In Figure 2.5, edges of the hypercubes for $n = 1, \ldots, 5$, carry information about the product terms in sum-of-products form.*

The information in a hypercube is encoded as shown in Figure 2.6. The hypercube encodes a switching function by the rules given in Figure 2.6, assigning the codes over 2^n vertices and $n2^{n-1}$ edges. The operations between two hypercubes produce a new hypercube (product) that is useful in optimization problems.

---

*For a switching function of n variables, a product term in which each of the n variables appears once is called a *minterm*. The uncomplemented and complemented variables can appear in a minterm.

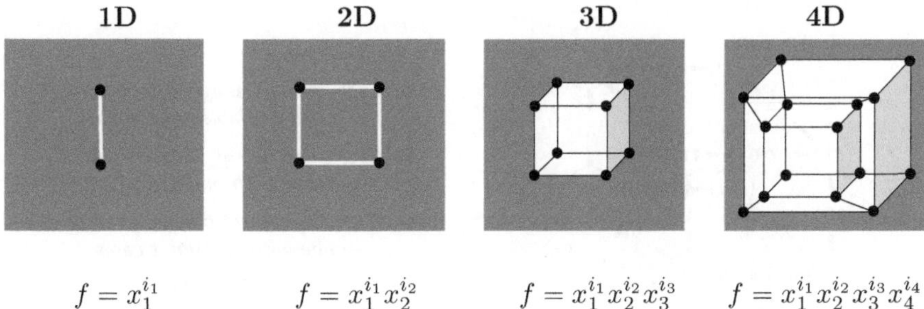

1D
2D
3D
4D
$f = x_1^{i_1}$
$f = x_1^{i_1} x_2^{i_2}$
$f = x_1^{i_1} x_2^{i_2} x_3^{i_3}$
$f = x_1^{i_1} x_2^{i_2} x_3^{i_3} x_4^{i_4}$

FIGURE 2.5
A product term of n variables and its spatial interpretation by an n-dimensional hypercube, $n = 1, 2, 3, 4$ (Example 2.8).

The problem of implementation is defined as embedding a 2D graphical structure (decision tree or decision diagram of a switching function, see details in Chapter 3) into a hypercube that carries *functional* and *topological* information about the switching function:

- ▶ The vertices of the hypercube denote the minterms, thus, the hypercube is a collection of minterms;
- ▶ The number of minterms is a power of two, 2^m, for some $m \leq n$;
- ▶ The number of edges in a hypercube is $3 \cdot s^{n-1}$.

Information in hypercubes are carried by labels of vertices and labels of links. Manipulation of hypercubes consists of several operations, for example, *merging the vertex.*

In Figure 18.3, the hypercubes for elementary switching functions of three variables are represented.

Example 2.9 *Examples of the relationship between algebraic representation, truth-table and hypercubes are as follow:*

(*a*) *A switching function $f = x_1 \vee x_2$, given by a truth table is represented by the hypercube shown in Figure 2.7).*

(*b*) *Let $f = \sum m(0, 2, 4, 5, 6)$. The truth table and the 3D hypercube are given in Figure 2.9.*

A 4D hypercube consists of two 3D hypercubes with their corners connected. The simplest way to visualize a 4D hypercube is to have one hypercube placed inside the other hypercube.

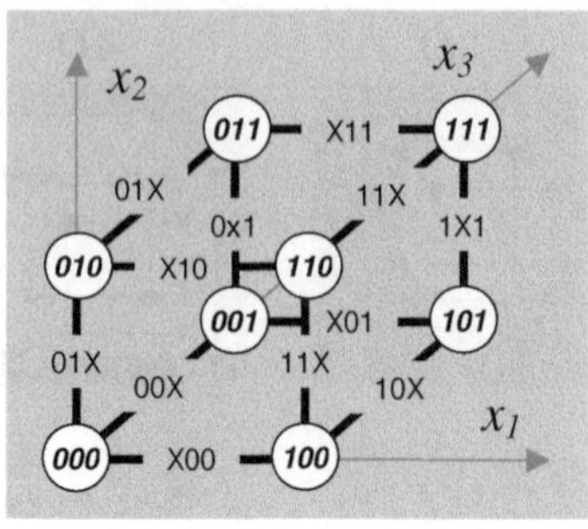

3D hypercube:

- *Each node is assigned to one of 8 codes (variable assignment)*
- *Each edge out of 12 is assigned to a cube with one don't care (x)*
- *Each face out of 6 is assigned to a cube with two don't cares*

FIGURE 2.6
A hypercube data structure for representing and manipulating switching functions.

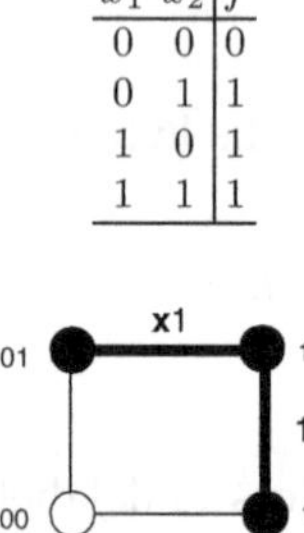

x_1	x_2	f
0	0	0
0	1	1
1	0	1
1	1	1

The four corners are called vertices. They correspond to the four rows of a truth table. Each vertex is identified by two coordinates. The horizontal coordinate is assumed to correspond to variable x_1, and the vertical coordinate to x_2.

- *The function f is equal to 1 for vertices 01, 10, and 11. The function f can be expressed as a set of vertices, $f = \{01, 10, 11\}$.*
- *The edge joins two vertices for which the labels differ in the value of only one variable. For example, $f = 1$ for vertices 10 and 11. They are joined by the edge that is labeled* **1x**. *The letter* **x** *is used to denote the fact that the corresponding variable can be either 0 or 1. Hence* **1x** *means that $x_1 = 1$, while x_2 can be either 0 or 1. Similarly, vertices 01 and 11 are joined by the edge labeled* **x1**. *The edge* **1x** *means a merger of vertices 10 and 11.*
- *The term x_1 is the sum of minterms $x_1\overline{x}_2$ and x_1x_2. It follows that $x_1\overline{x}_2 \vee x_1x_2 = x_1$. The edges* **1x** *and* **x1** *define in a unique way the function f, hence we can write $f = \{\mathbf{1x}, \mathbf{x1}\}$. This corresponds to the function $f = x_1 \vee x_2$.*

FIGURE 2.7
Representation of the switching function $f = \overline{x}_3 \vee x_1\overline{x}_2$ by the hypercube: the truth table and 2D hypercube representation (Example 2.9a).

2.5.3 Decision trees and diagrams

Decision trees can be derived from a functional equation or truth table. This provides a canonical representation of functions in graphical form, so that for a fixed variable order there is a bijection between switching functions and decision diagrams. The definitions for decision trees and diagrams are given in Chapter 2.

AND	*NAND*	*OR*	*EXOR*
$f = x_1x_2x_3$	$f = \overline{x_1x_2x_3}$	$f = x_1 \vee x_2 \vee x_3$	$f = x_1 \oplus x_2 \oplus x_3$

$f = [1\ 1\ 1]$ $\quad f = \begin{bmatrix} 0\,\text{x}\,\text{x} \\ \text{x}\,0\,\text{x} \\ \text{x}\,\text{x}\,0 \end{bmatrix}$ $\quad f = \begin{bmatrix} \text{x}\,\text{x}\,1 \\ \text{x}\,1\,\text{x} \\ 1\,\text{x}\,\text{x} \end{bmatrix}$ $\quad f = \begin{bmatrix} 1\,0\,0 \\ 0\,1\,0 \\ 0\,0\,1 \\ 1\,1\,1 \end{bmatrix}$

FIGURE 2.8
A 3D hypercube and cube-based representation of AND, NAND, OR and EXOR 3-input gates.

x_1	x_2	x_3	f
0	0	0	1
0	0	1	0
0	1	0	1
0	1	1	0
1	0	0	1
1	0	1	1
1	1	0	1
1	1	1	0

- *There are five vertices that correspond to $f = 1$: 000,010,100,101, and 110*
- *Merging the vertex assignments yield* x00, 0x0, x10, 1x0 *and* 10x.
- *Four of these vertices can be merged to face, or term* xx0. *This term means that $f = 1$ if $x_3 = 0$, regardless of the values of x_1 and x_2.*
- *The function f can be represented in several ways. Some of the possibilities are*

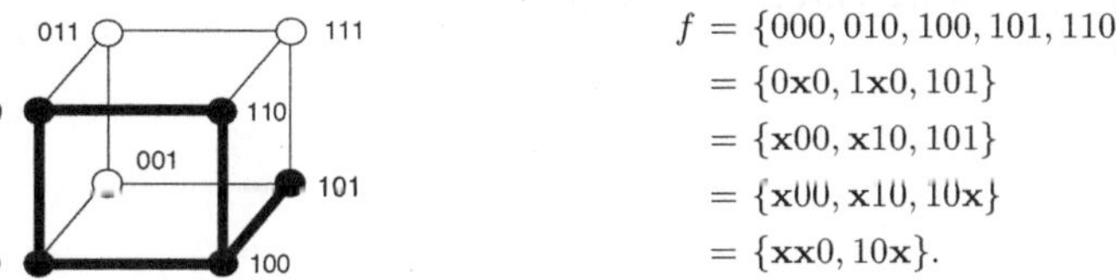

$$\begin{aligned} f &= \{000, 010, 100, 101, 110\} \\ &= \{0\text{x}0, 1\text{x}0, 101\} \\ &= \{\text{x}00, \text{x}10, 101\} \\ &= \{\text{x}00, \text{x}10, 10\text{x}\} \\ &= \{\text{xx}0, 10\text{x}\}. \end{aligned}$$

FIGURE 2.9
Representation of the switching function $f = \overline{x}_3 \vee x_1\overline{x}_2$ by the hypercube: truth table and 3D hypercube representation for the function (Example 2.9b).

2.5.4 Direct acyclic graph

A logic network is modelled by a direct acyclic graph (DAG). A DAG $G(V, E)$ where V is a set of nodes and $E(\subseteq V \times V)$ is a set of edges is defined as follows:

- For an edge $(n_i, n_j) \in E$, n_i is a *fanin* of n_j and n_j is a *fanout* of n_i. The set of all fanins of n_i is denoted as FI_i, and the set of all fanouts of n_i

is denoted as FO_i.

- A Boolean network represents a multi-level circuit. Nodes with no fanin are *primary inputs*, and nodes with no fanout are *primary outputs*. We denote the set of all primary inputs of a network η as PI, and that of all primary outputs of η as PO $(PI \subset V, PO \subset V)$.
- If there is a directed path from a node n_i to a node $n_j (n_i \rightarrow n_j)$ in η, n_i is a *transitive fanin* of n_j and n_j is a *transitive fanout* of n_i.

Example 2.10 *A DAG shown in Figure 2.10(b) has been derived from the circuit given in Figure 2.10(a).*

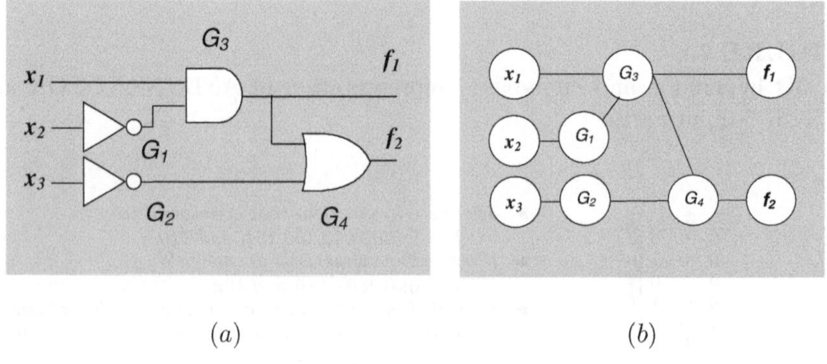

FIGURE 2.10
Logic circuit (a) and its DAG (b) (Example 2.10).

2.6 Relationship of data structures

Tabular (truth tables) and graphical (hypercubes, decision trees and diagrams) representations of switching functions have a one-to-one relationship.

Example 2.11 *Relationship of schematic, algebraic, truth table, hypercube, decision tree, decision diagram, and spatial ($\mathcal{N}$-hypercube) representations for the switching function $f = x_1\overline{x}_2 \vee \overline{x}_3$ are given in Figure 18.5.*

Chapter 3 focuses on decision trees and diagrams, and Chapter 34 considers embedding DAGs, decision trees and diagrams into $\mathcal{N}$-hypercubes.

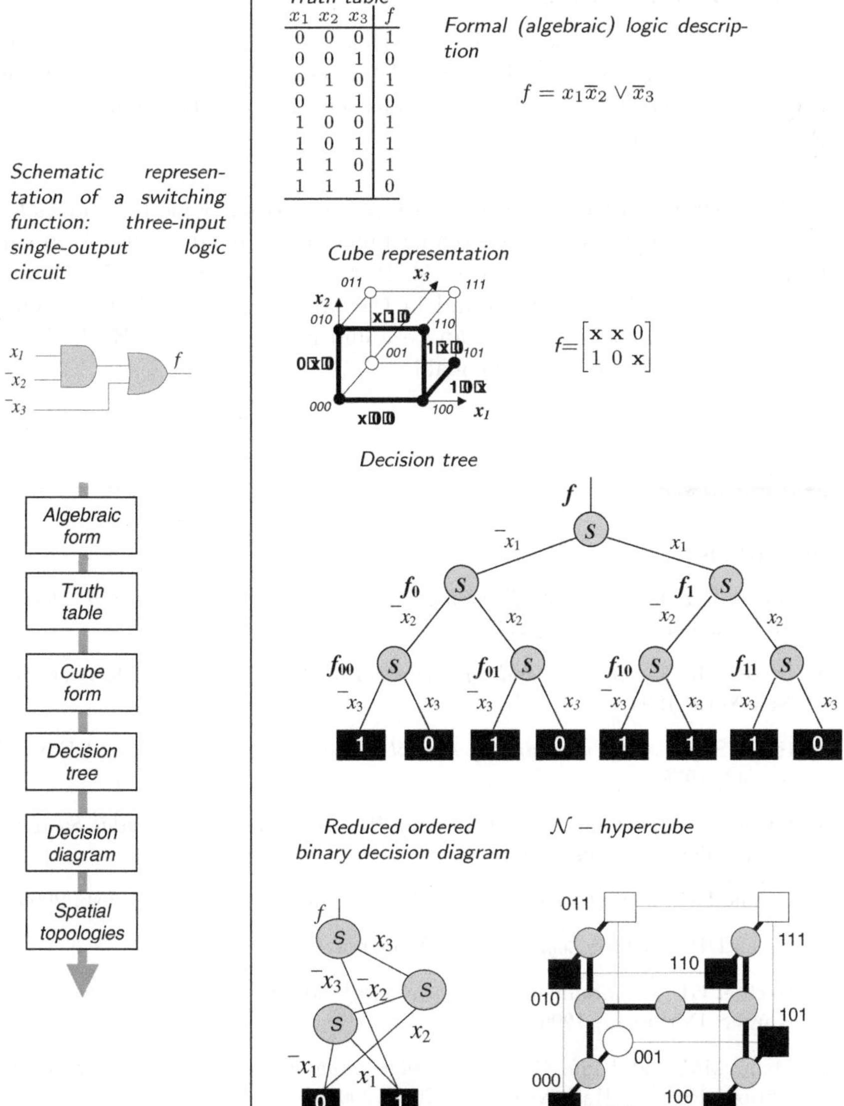
Truth table

x_1	x_2	x_3	f
0	0	0	1
0	0	1	0
0	1	0	1
0	1	1	0
1	0	0	1
1	0	1	1
1	1	0	1
1	1	1	0

FIGURE 2.11
The relationship between various data structures in circuit description (Example 18.16).

2.7 Further reading

Textbooks. There are many excellent textbooks on the basics of logic design, in particular, the textbook by Brown and Vranesic [1], by DeMicheli [2], by Ercegovac et al. [4], by Givone [5], by Marcovitz [9], by Mano and Kime [8] by Roth Jr. [11], and Sasao [27].

Techniques for manipulating switching functions. Some techniques for representing and manipulating switching functions in Reed–Muller form can be found in the textbooks by Green [6], Hurst [13], and Sasao [27]. Details on cube representation can be found in the book by Roth JP [13]. Many useful algorithms for logic design can be found in textbooks by Hachtel and Sommenzi [7] and Meinel and Theobald [10].

References

[1] Brown S and Vranesic Z. *Fundamentals of Digital Logic with VHDL Design.* McGraw-Hill, New York, 2000.

[2] DeMicheli G. *Synthesis and Optimization of Digital Circuits.* McGraw Hill, New York, 1994.

[3] Hurst SL. *Logic Processing of Digital Signals.* Grawe Russak and Edward Arnold, 1978.

[4] Ercegovac MD, Lang T, and Moreno JH. *Introduction to Digital Systems.* John Wiley and Sons, 1999.

[5] Givone DD. *Digital Principles and Design.* McGraw-Hill, New York, 2003.

[6] Green DH. *Modern Logic Design.* Academic Press, 1986.

[7] Hachtel GD and Sommenzi F. *Logic Synthesis and Verification Algorithms.* Kluwer, Dordrecht, 1996.

[8] Mano MM and Kime C. *Logic and Computer Design Fundamentals.* 3rd edition. Prentice Hall, New York, 2004. 2005.

[9] Marcovitz AB. *Introduction to Logic Design.* 2nd edition. McGraw-Hill, New York, 2005.

[10] Meinel C and Theobald T. *Algorithms and Data Structures in VLSI Design.* Springer, Heidelberg, 1998.

[11] Roth Jr.CH. *Fundamentals of Logic Design.* 5th Edition. Thomson Brooks/Cole, 2004.

[12] Roth JP. *Mathematical Design. Building Reliable Complex Computer Systems.* IEEE Press, 1999.

[13] Sasao T. *Switching Theory for Logic Synthesis.* Kluwer, Dordrecht, 1999.

3

Graphical Data Structures

Decision trees and diagrams are graph data structures that are suitable for representing logic functions*. These graphs show the relationship between variables of the function. The manipulation of literals and terms in an algebraic representation of a logic function is replaced in decision trees and diagrams by the manipulation of graphical components (nodes, links, and paths), i.e.,

$$< \texttt{Algebraic data structure} > \Leftrightarrow < \texttt{Graph data structure} >.$$

In this chapter the elements of graph theory are introduced. The focus is the design of tree structures, as well as their structural and topological properties.

We start below with notation and definitions that introduce decision trees and diagrams and the various measures associated with them. The main fields of application and related results are surveyed.

3.1 Introduction

Information about logic functions represented as graphical data structures are carried in the following components:

- ▶ Nodes that operate on data,
- ▶ Edges between nodes that are associated with data flow, and
- ▶ Topology (local and global), that specifies the geometrical properties and spatial relations of these nodes and edges, and that is unaffected by changes in the shape or size of the graph.

There are various algebraic forms of switching functions. These forms correspond to certain types of decision trees and diagrams. The manipulation of switching functions in algebraic forms is based on mathematical relations, theorems, etc. These rules have graphical equivalents in the form of topological objects such as sets of nodes, levels, subtrees, shapes, etc., and also

*The term "logic functions" refers to switching and multivalued logic functions.

functional characteristics as distribution of nodes, weights, balance, homogeneity, etc. Many problems of logic analysis and synthesis can be efficiently solved by manipulating these graphical objects.

The manipulation of graphical structures is especially beneficial for large size problems. Moreover, during the logic synthesis of circuits manufactured using certain technologies (for example, pass-transistor logic and wrap-gate quantum networks) it is not necessary to convert the graph data structure into a circuit. This is because the topology of the obtained graphical structure is itself a circuit solution.

In this chapter, basic techniques for using graphical data structures to representing switching functions are given.

3.2 Graphs

A *graph* is defined as $G = (V, E)$, where V is the vertex (node) set and E is the edge set. We say $(i, j) \in E$, where $i, j \in V$. Functional elements (gates, circuits) correspond to nodes and communication links correspond to edges in the graph. In constructing decision trees and diagrams[†] for the representation of a logic function, the following characteristics of graphs are used:

- ▶ The number of nodes in G is $n = |V|$.
- ▶ The *degree* of a vertex v, written $deg(v)$, is the number of edges incident with v.
- ▶ An edge in E between nodes v and u is specified as an unordered pair (v, u), and v and u are said to be *adjacent* to each other or are called *neighbors*.
- ▶ The *distance* between two nodes i and j in a graph G is the number of edges in G on the shortest path connecting i and j.
- ▶ The *diameter* of a graph G is the maximum distance between two nodes in G.
- ▶ A graph G is *connected* if a path always exists between any pair of nodes i and j in G.

The terms *edge*, *link*, *connection*, and *interconnection* are used here interchangeably, and the terms *graph* and *logic network* are considered synonyms.

Directed graphs. A *directed* graph G consists of a finite vertex set V (nodes) and a set E of directed edges between the vertices, and characterized by:

[†]A decision diagram is a reduced decision tree.

- The *indegree* of a node (vertex) i, the number of edges in G leading to i. The *outdegree* of a node (vertex) i is the number of edges in G starting in i.
- A node is called a *sink* if it has an outdegree of 0. If the outdegree of v is bigger than 0, v is called an *internal* node. A node is called a *root* if it has an indegree of 0.
- The adjacency matrix $A = (a_{ij})$ is defined as $|G| \times |G|$ size matrix such that $a_{ij} = 1$ if $(i, j) \in E$, and $a_{ij} = 0$ otherwise.

The adjacency matrix contains only two elements, 0 and 1. The graph and its adjacency matrix contain the same information, they are simply two alternative representations or data structures. Permutation of any two rows or columns in an adjacency matrix corresponds to relabeling the vertices and edges of the same graph.

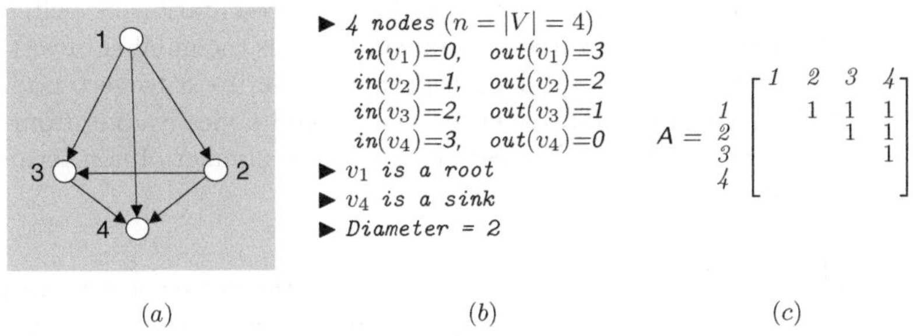

FIGURE 3.1
The directed graph (a), its properties (b), and an adjacency matrix (c) (Example 3.1).

> **Example 3.1** *The properties of the directed graph with four nodes and its adjacency matrix A are illustrated in Figure 3.1, where* $in(v_i)$ *and* $out(v_i)$ *are indegrees and outdegrees of a node* v_i.

Undirected graphs. In the case of undirected graphs the edges are considered unordered pairs and therefore have no distinguished direction. The *degree* of a node i in a graph G is the number of edges in G that are incident with i, i.e., where the outdegree and the indegree coincide.

> **Example 3.2** *Figure 3.2 illustrates the properties of the undirected graph. Its adjacency matrix A is equal to the transposed matrix A,* $A = A^T$.

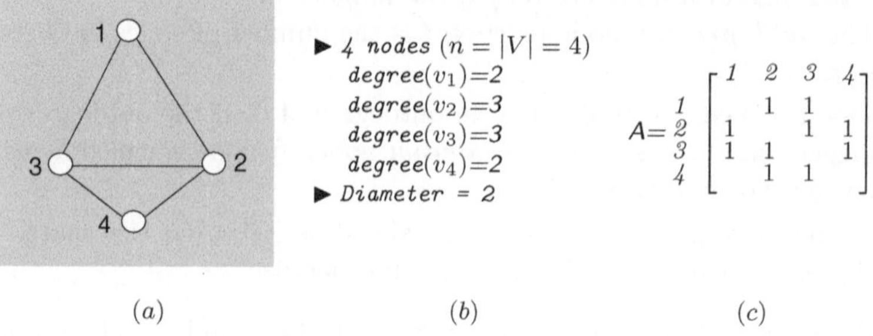

FIGURE 3.2
The undirected graph (a), its properties (b), and an adjacency matrix (c) (Example 3.2).

A path carries local information for algebraic representation of switching functions. A path can be measured by a *length* t that is the number of vertices in the path. There is always a 0-length path from vertex u to vertex u'. A *path* in a decision diagram is a sequence of edges and nodes leading from the root node to a terminal node. The *path length* is the number of non-terminal nodes in the path.

A reachability. The vertex u' is *reachable* from the vertex u if there is a path from u to u'. A *reachability* in graph G can be described by a *reachaility matrix* (see details for decision diagrams in section 3.4).

Isomorphism. Two graphs $G(V, E)$ and $G^{'}(V^{'}, E^{'})$ are isomorphic if there exists a bijection $f : V \rightarrow V^{'}$ such that $(u, v) \in E$ if and only if $(f(u), f(v)) \in E^{'}$.

> **Example 3.3** *Figure 3.3a,b shows two isomorphic graphs, where* $f(1) = D, f(2) = B, f(3) = C, f(4) = A, f(5) = E$. *Graph* $G^{'}$ *is obtained by relabeling the vertices of* G*, maintaining the corresponding edges in* G *and* $G^{'}$.

If G and $G^{'}$ are isomorphic graphs, then they have the same number of vertices, the same number of vertices with a given degree, and the same number of edges. Two graphs G and $G^{'}$ are isomorphic if their vertices can be labeled in such way that the corresponding adjacency matrices are equal.

Bipartite graphs. A graph G is called *bipartite* if its vertex set V can be decomposed into two disjoint subsets V_1 and V_2 such that every edge in G joints a vertex in V_1 with a vertex in V_2. Every tree is a bipartite graph.

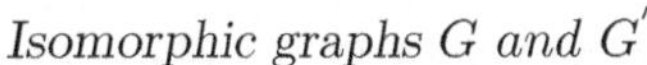

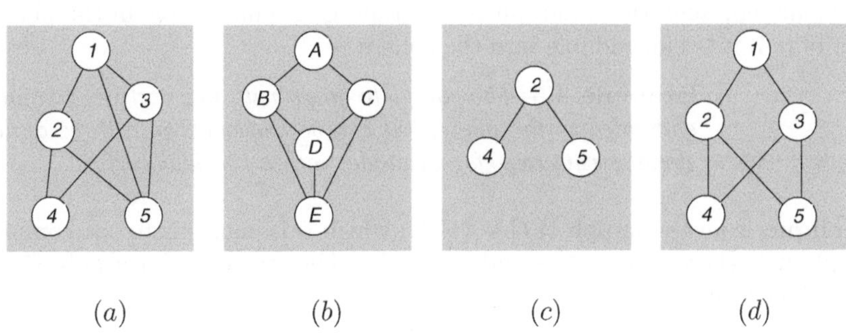

FIGURE 3.3
Examples of isomorphic graphs (a,b) and a subgraph (c) of a graph (d) (Examples 3.3 and 3.4).

Subgraphs. A graph $G^{'}(V^{'}, E^{'})$ is a *subgraph* of $G(V, E)$ if $V^{'} \subseteq V$ and $E^{'} \subseteq E$.

> **Example 3.4** *Figure 3.3 shows a graph (c) and a subgraph (d).*

Cartesian product graphs provide a framework in which it is convenient to analyze as well as to construct new graphs. Let $G_1 = (V_1, E_1)$ and $G_2 = (V_2, E_2)$ be two graphs. The product of G_1 and G_2, denoted $G_1 \times G_2 = (V_1 \times V_2, E)$, is a graph where the set of nodes is the product set

$$V_1 \times V_2 = \{x_1 x_2 | x_1 \in V_1,\ x_2 \in V_2\}$$
$$E = \{\langle x_1 x_2, y_1, y_2 \rangle | (x_1 = y_1,\ \langle x_2, y_2 \rangle \in E_2) \ or \ (x_2 = y_2,\ \langle x_1, y_1 \rangle \in E_1)\}.$$

It can be shown that the hypercube can be defined as the product of n copies of the complete graph on two vertices, K_2. That is

$$H_n = H_{n-1} \times K_2.$$

Planarity. A graph G is *planar* if it is isomorphic to a graph G' such that:

(*a*) The vertices and edges of G' are contained in the same plane and
(*b*) At most one vertex occupies or at most one edge passes through any point of the plane.

In other words, a graph is planar if it can be drawn on a plane with no two edges intersecting. Graph G' is said to be *embedded* in the plane and to be a planar representation of graph G. A graph G is embeddable in the plane if and only if it is embeddable into the sphere.

Example 3.5 *The complete graph with five vertices cannot be embedded in the plane, but can be embedded into three dimensional structures, for example, into a toroidal surface.*

A clique is any subgraph of $G = (V, E)$ which is isomorphic to the complete graph K_i, where $1 \leq i \leq n$ and $n = |V|$. The vertices of a graph can be partitioned into cliques.

A matching of a graph $G = (V, E)$ is any subset of edges $M \subseteq E$ such that no two elements of M are adjacent. A *covering* C is any set of edges such that any vertex of the graph is an end-point of at least one edge of C. A *minimum-cardinality covering* is a covering with the smallest number of edges.

A coloring of a graph is an assignment of colors to vertices so that adjacent vertices have different colors.

Coverings. In a graph G, a set g of edges is said to *cover* G if every vertex in G is incident on at least one edge in g.

Example 3.6 *A graph G is trivially its own covering.*

Operations on graphs. The *union* of two graphs $G_1 = (V_1, E_1)$ and $G_2 = (V_2, E_2)$ is another graph $G_3 = G_1 \cup G_2$ whose vertex set $V_3 = V_1 \cup V_2$ and the edge set $E_3 = E_1 \cap E_2$. The *intersection* of two graphs $G_1 = (V_1, E_1)$ and $G_2 = (V_2, E_2)$ is another graph $G_4 = G_1 \cap G_2$ whose vertex set $V_4 = V_1 \cap V_2$ and the edge set $E_4 = E_1 \cap E_2$, i.e., consisting only of those vertices and edges that are in both G_1 and G_2. The *ring sum* of graphs G_1 and G_2, $G_1 \oplus G_2$, is a graph consisting of the vertex set $V_1 \cup V_2$ and of edges that are either in G_1 or G_2, but not in both. These operations can be extended to include any finite number of graphs.

A pair of vertices x, y in a graph G are said to be *merged* if the two vertices are replaced by a single new vertex z such that every edge that was incident on either x or y or on both is incident on the vertex z. This operation does not alter the number of edges, but it reduces the number of vertices by one.

If graphs G_1 and G_2 are edge disjoint, then $G_1 \cap G_2$ is a null graph, and $G_1 \oplus G_2 = G_1 \cup G_2$. If G_1 and G_2 are vertex disjoint, then $G_1 \cap G_2$ is empty. For any graph G $G \cup G = G \cap G = G$ and $G \oplus G$ is a null graph.

If g is a subgraph of G, then $G \oplus g$ is that subgraph of G which remains after all the edges in g have been removed from G, i.e., $G \oplus g = G - g$.

Decomposition. A graph G is to have been *decomposed* into two subgraphs g_1 and g_2 if $g_1 \cup g_2 = G$ and $g_1 \cap g_2$ is a null graph. A graph G can be decomposed into more than two subgraphs.

Vector space associated with graphs. There is a vector space associated with every graph G. This vector space consists of:

- ▶ A Galois field $GF(2)$ that is, a set $\{0, 1\}$ with operation addition modulo 2 (EXOR) and multiplication (AND),
- ▶ 2^e vectors (e-tuples), where e is the number of edges in G,
- ▶ An addition operation between two vectors $\mathbf{X}$ and $\mathbf{Y}$ in this space, defined as the vector sum $\mathbf{X} \oplus \mathbf{Y} = [(x_1 \oplus y_1), (x_2 \oplus y_1 2), \ldots, (x_e \oplus y_e)]^T$, and
- ▶ A scalar multiplication between a scalar c in Z_2 and a vector $\mathbf{X}$ defined as $c \cdot \mathbf{X} = [c \cdot x_1, c \cdot x_2, \ldots, c \cdot x_e]^T$.

3.3 Decision trees

A class of graphical data structures, decision trees, are the predecessors to decision diagrams.

3.3.1 Tree-like graphs

A graph is called *rooted* if there exists exactly one node with an indegree of 0, the root. A *tree* is a rooted acyclic graph in which every node but the root has an indegree of 1. A special class of rooted trees, called *binary* trees. *The path length* of a tree can be defined as the sum of the path lengths from root to all terminal vertices. There are several useful characteristics and properties of a tree, in particular:

- ▶ For every vertex v there exists a unique path from the root to v. The length of this path is called the *depth* or *level* of v.
- ▶ The *height* of a tree is equal to the greatest depth of any node in the tree.
- ▶ A node with no children is a *terminal* (*external*) node or *leaf.* A nonleaf node is called an *internal* node.
- ▶ A *complete* $n-$level p-tree, is a tree with p^k nodes on level k for $k = 0, \ldots, n-1$. A p^n-leaf complete tree has a level hierarchy (levels $0, 1, \ldots, n$), the root is associated with level zero and its p children are on level one. This edge type describes the direction of data transmission between the child and the parent, so that data is sent in only one direction at a time, up or down.

A balanced tree is a tree where no leaf is much far than from the root than any other leaf.

Example 3.7 *A complete binary tree is a balanced tree.*

The weight of a subtree is defined as the number of nodes in this subtree. The weight-balanced tree is a tree whose subtrees are the same weights.

Example 3.8 *Figure 3.4 shows examples of balanced and non-balanced trees, and their weights. Tree (a) is balanced because all leaves are on the length tree from the root. Tree (b) is not, because two leaves are on the length two, and the others are on the length three from the root. In both trees, subtrees "c" and "d" are weight-balanced.*

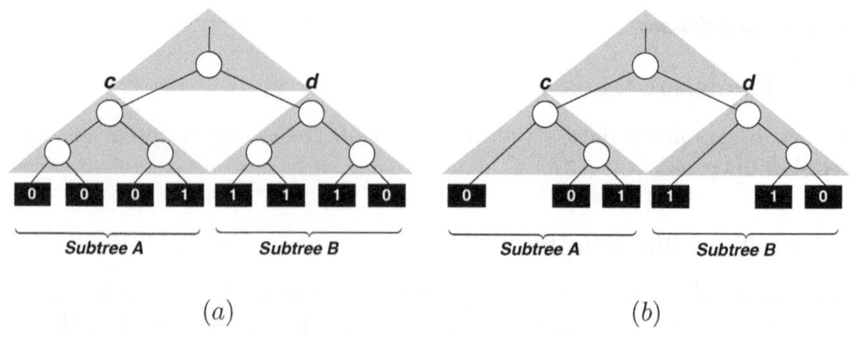

FIGURE 3.4
Balanced (a) and non-balanced trees (b) (Example 3.8).

Example 3.9 *In machine learning, the paradigm of learning classification rules from a set of training examples is used. Each training example is described by giving values for a fixed number of attributes. The rules that algorithms learn are represented as decision trees. A node on the tree can represents a test on an attribute and each outgoing branch corresponds to a possible result of this test. Each leaf node represents a classification to be assigned to an example. To classify a new example, a path from the root of the decision tree to a leaf node is traced. The class at the leaf node represents the class prediction for that example.*

The distance $d(v_i, v_j)$ between two vertices v_i and v_j is the length of the shortest path between them, i.e., the number of edges in the shortest path. The *diameter* of a tree is defined as the length of the longest path in this tree.

3.3.2 H-trees

An H-tree is a recursive topological construction of H_1-trees, where H_1 is defined in Figure 3.5a. An H_{k+1}-tree can be constructed by replacing the leaves of H_k with H_1 trees. The number of terminal nodes in an H_k-tree is equal to 4^k. An H-tree makes optimal use of area and wire length if the n nodes are allowed to be arranged as an $\sqrt{n} \times \sqrt{n}$ array.

Example 3.10 *Balanced trees and H-trees are the most common styles of physical clocking networks. A regular structure of the H-tree allows predictable delay in a network.*

Example 3.11 *In Figure 3.5, the H_{k+1} tree is constructed for $k = 1$.*

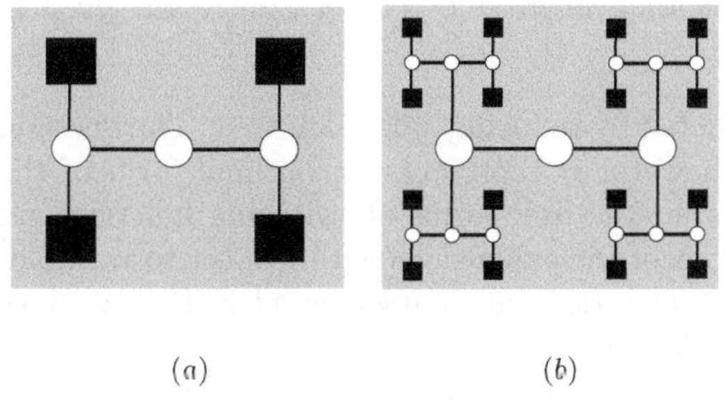

The H_1 tree includes:

- *4 terminal nodes,*
- *3 nonterminal nodes*

The H_2 tree is constructed by replacing terminal nodes with H_1 trees; the H_2 tree includes:

- $4 \times 4 = 16$ *terminal nodes,*
- $4 \times 3 + 3 = 15$ *nonterminal nodes*

FIGURE 3.5
The H_1 (a) and H_2 (b) trees (Example 3.11).

3.3.3 Definition of a decision tree

A decision tree is defined as a rooted, ordered, vertex-labeled tree, each node having either m outgoing edges or none, and being a leaf.

A decision tree can be regarded as a deterministic algorithm for deciding which variable to test next, based on the previously tested variables and the results of their evaluation, until the function's value can be determined. In this respect, recursive definition is more appropriate.

Definition 3.1 *Let f be a function of n discrete variables $x_1, \ldots, x_n$. For each variable x_i, $i = 1, ..., n$, the decision tree of f is composed of a root labeled x_i, and m subtrees, which are decision trees corresponding to the sub-functions $f|_{x_i=0}, \ldots, f|_{x_i=m-1}$.*

The evaluation of a discrete function represented as a decision tree starts by assigning the value of the variable associated with the root of the tree. This process is repeated on each subtree, until a constant node, a leaf, is reached; this constant value of the leaf gives the value of the function.

Since switching functions are a sub-class of discrete functions, decision trees and diagrams are applicable to logic design tasks formulated in terms of such evaluation. The task of minimization of a switching function is formulated in terms of the optimization of some measurements done on the trees.

A node of the decision tree implements decomposition with respect to a variable. There are 2^n terminal nodes (exactly the same as the number of values in the truth vector) in the complete binary decision tree. It is a canonical data structure.

Example 3.12 *Shannon, Davio, and arithmetic Davio expansion for the switching function $f = \overline{x}_1 x_3 \vee x_2 \overline{x}_3$ are given in Figure 3.6.*

A decision diagram is constructed by reducing a decision tree. The decision diagram is a canonical form, however it represents an optimal form of the given switching function. In Table 3.1 three forms of switching function and the corresponding decompositions of the expansions with respect to variables are given. These decomposition equations are implemented in the nodes of the decision trees and decision diagrams.

There are three types of expansions with respect to a variable:

- Shannon expansion,
- Davio expansion, and
- Arithmetic Davio expansion.

These expansions are the special forms of a Boolean formula f of n variables x_i, $i = 1, 2, \ldots n$. They are said to be expansions with respect to a variable x_i.

The binary decision tree implements Shannon expansion and consists of S-nodes. According to Table 3.1,a binary decision tree corresponds to an SOP expression. A product of SOPs corresponds to a *path* in the tree. A path is described by the equation

$$x_{k_1}^{i_1} x_{k_2}^{i_2} \cdots x_{k_n}^{i_n}.$$

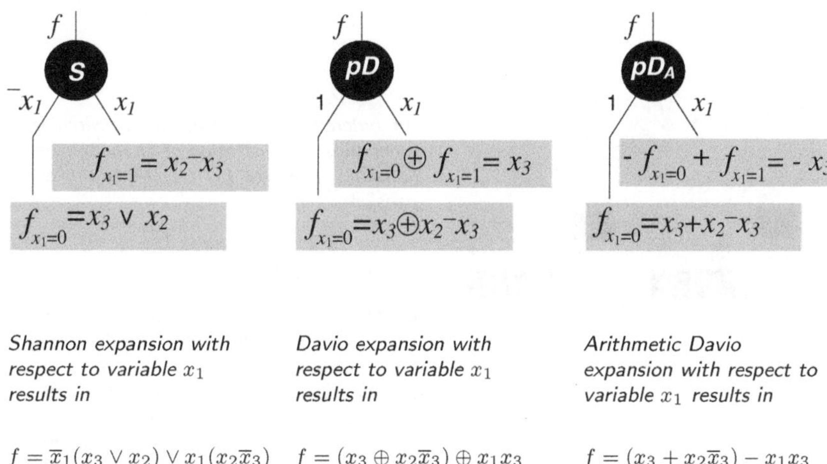

FIGURE 3.6
Shannon, Davio, and arithmetic Davio expansion of the switching function $\overline{x}_1 x_3 \vee x_2 \overline{x}_3$ with respect to variable x_1 (Example 3.12).

TABLE 3.1
Expansion of a switching function and corresponding decomposition rules for decision trees and diagrams.

Expansion	Formal description	$x_j^{i_j}$	Decomposition
Sum-of-products	$\bigvee_{i=0}^{2^n-1} s_i \cdot (x_1^{i_1} \cdots x_n^{i_n})$	$\begin{cases} \overline{x}_j, & i_j = 0; \\ x_j, & i_j = 1. \end{cases}$	*Shannon*
Reed-Muller	$\bigoplus_{i=0}^{2^n-1} r_i \cdot (x_1^{i_1} \cdots x_n^{i_n})$	$\begin{cases} 1, & i_j = 0, \\ x_j, & i_j = 1. \end{cases}$	*Davio*
Arithmetic	$\sum_{i=0}^{2^n-1} p_i \cdot (x_1^{i_1} \cdots x_n^{i_n})$	$\begin{cases} 1, & i_j = 0; \\ x_j, & i_j = 1. \end{cases}$	*Arithmetic Davio*

Example 3.13 *A switching function of three variables is given in the SOP form*

$$f = \overline{x}_1 \overline{x}_2 \vee x_2 \overline{x}_3.$$

A possible ordered decision tree for that function is shown in Figure 3.7a, a reduced ordered tree is shown in Figure 3.7b, and a free (non-ordered) tree is shown in Figure 3.7c.

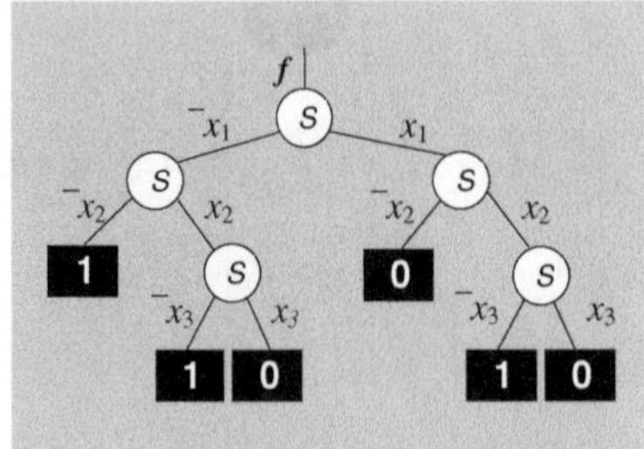

(a)

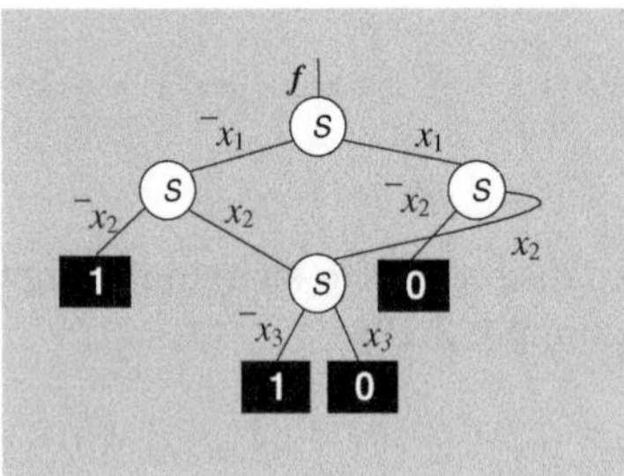

(b)

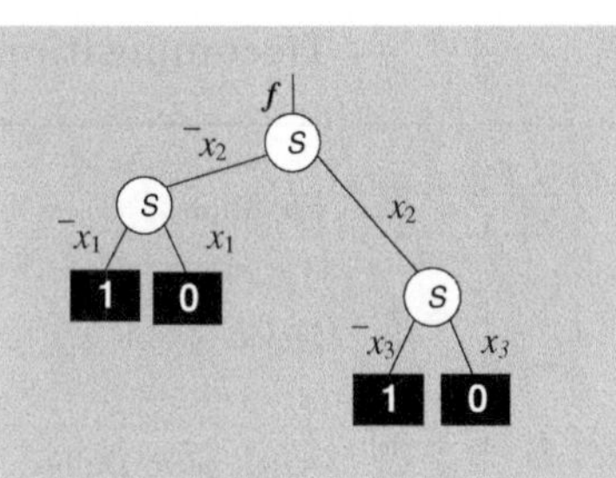

(c)

An ordered decision tree for a switching function is an illustration of Shannon's expansion. The tree (a) represents the expansion

$$f = \overline{x}_1(\overline{x}_2 \cdot 1 \vee x_2(\overline{x}_3 \cdot 1 \vee x_3 \cdot 0)) \\ \vee\ x_1(\overline{x}_2 \cdot 0 \vee x_2 \cdot (\overline{x}_3 \cdot 1 \vee x_3 \cdot 0))$$

- ▶ *The left subtree of each node corresponds to the node's variable being evaluated at 0,*
- ▶ *The right subtree corresponds to the variable evaluated at 1.*
- ▶ *The left subtree of the root corresponds to*

$$f|_{x_1=0} = 1,$$

and the right subtree corresponds to

$$f|_{x_1=1} = x_2\overline{x}_3$$

For example, an evaluation on the tree given $x_1, x_2, x_3 = (1, 1, 0)$:

- ▶ *On finding* $x_1 = 1$, *it proceeds to the right subtree,*
- ▶ *On finding* $x_2 = 1$, *it proceeds to the right subtree,*
- ▶ *After evaluation of* $x_3 = 0$ *we find* $f(1, 1, 0) = 1$.

The diagram (b) is an ordered reduced diagram derived from the tree (a) by:

- ▶ *Merging the identical subtrees rooted in nodes labeled* S *for* x_2 *and the identical leaves.*

The ordered tree is optimal in the number of nodes (except for leaves). The tree (c) has a free order of variables, but it provides a minimal number of nodes

FIGURE 3.7
Designing a decision tree given the switching function $f = \overline{x}_1\overline{x} \vee x_2\overline{x}_3$: an ordered decision tree (a), an optimal ordered decision tree (b), and an optimal free decision diagram (c) (Example 3.13).

To every decision tree for a completely specified function there corresponds a unique "minimal" diagram, that is, one in which every possible reduction has been accomplished.

Note that in terms of circuit implementation, any function represented by a decision tree can be implemented on a multiplexer-based network. The evaluation of a function then proceeds from the "leaves" (the constant values) to the "root" multiplexer; the function variables, used as control variables, select a unique path from the root to one leaf, and the value assigned to that leaf propagates along the path to the output of the "root" multiplexer.

A spanning tree of a connected undirected graph G is a subgraph which is both a tree and which contains all the vertices of G. A spanning tree can be obtained by removing edges only and such that any pair of vertices remain connected.

Example 3.14 *A spanning tree uses wire segments to connect the gate inputs and outputs. If two paths intersect at an intermediate point, that point is called a* Steiner point *and the tree is referred to as a* Steiner tree. *A Steiner tree adds nodes to the graph so that wires can join at a Steiner point.*

3.3.4 Definition of a binary decision diagram

In definition of binary decision diagram, we follow the results and terminology of a fundamental work by Bryant [5]. Decision diagrams are a graphical data structure for representing discrete functions. Decision diagrams for switching functions are called *ordered binary decision diagrams* (OBDDs).

Decision diagrams are derived by reduction of decision trees, by exploiting the relationship between function values. Given a particular decision tree, for example, Shannon or Davio, a corresponding decision diagram can be derived.

Given a switching function f of n variables $x_1, x_2, \ldots, x_n$. Let π be the order on the set of these variables.

An OBDD is a direct acyclic graph which specified as follows:

- ▶ *It contains one root, non-terminal, and terminal nodes.*
- ▶ *Each non-terminal node is labeled by a variable x_i. In this book, for the purpose of unification of various decision diagrams, a non-terminal node of a OBDD is labeled by S, Shannon decomposition, and corresponds to a variable x_i. Each non-terminal node of a OBDD has two outgoing edges, which are labeled by $\overline{x}_i$ (left) and x_i (right).*
- ▶ *The order in which the variables appear on a path of a OBDD is consistent with the variable order π.*

An OBDD with the above properties represents a given switching function f. *Reduced* OBDDs (ROBDDs) provide a canonical representation of switching functions. Reducing OBDDs is based on *elimination* and *merging* rules.

3.3.5 Measures on decision trees and diagrams

To evaluate a decision tree, some criteria have been developed which are based on measurements of the important properties of the tree. Such measurements are implementation cost and testing cost.

The implementation cost is the tree storage cost (the sum of the storage costs of the internal nodes of the tree). In the case of decision diagram, it is the diagram storage cost (the internal nodes of the minimal diagram). In addition, a probability distribution can be specified on the variables' space (which is assumed uniform if not otherwise known) and evaluated on the tree. Other measures are information-theoretic, which also uses probability distribution (see Chapters 8 and 24). These measures are used for manipulation, optimization, and transforms of the decision trees and diagrams.

The properties of an algebraic representation of a switching function are related to the properties of its graphical model. Figure 3.8 illustrates, with some simplification, the relationship between algebraic and topological parameters. Such a graph, in particular, a decision tree, is characterized by *topological* and *functional* characteristics. The testing cost is measured for each variable and incurs the expense each time that variable is evaluated.

Topological characteristics of decision trees and diagrams include the following of parameters (Figure 3.8):

- ▶ The *size* is the number of nodes,
- ▶ The *depth* is the number of levels,
- ▶ The *width* is the maximum number of nodes at any level, and
- ▶ The *area* is `Depth` × `Width`.

► The *path length* is computed as a number of edges on the path leading from the root node to a given terminal node. A useful characteristic is the average path length. In [6, 56], the average path length is the sum of path lengths for all assignments of values to the variables divided by the number of all assignments equal to 2^n.

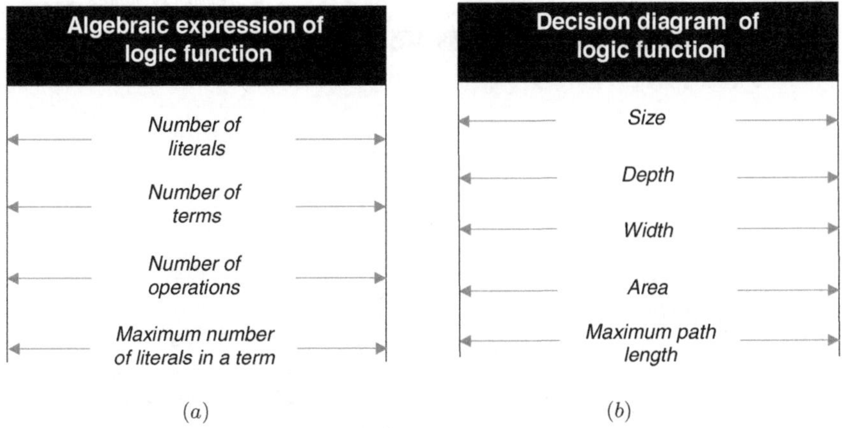

FIGURE 3.8
Measures of algebraic expression (a) and decision diagram for logic function (b).

Example 3.15 *Figure 3.9 shows the complete ternary ($p = 3$) three-level ($n = 3$) tree. The root corresponds to the level (depth) 0. Its three children are associated with level 1 ($3^1 = 3$). Level 2 includes $3^2 = 9$ nodes. Finally, there are $3^3 = 27$ terminal nodes.*

Functional characteristics of decision trees that represent switching functions include:

► The *weight* of a tree (subtree) is the number of ones in the terminal nodes.

► The *balance* of subtrees A and B is defined as the difference of their weights.

► *Homogeneity* of a tree (subtree) is defined as the absolute value of difference between number of 1s, n_1, and the number of 0s, n_0, in terminal nodes. By analogy, the homogeneity of a level of a tree can be defined.

Example 3.16 *In Figure 14.7, several functional characteristics of a decision tree are given.*

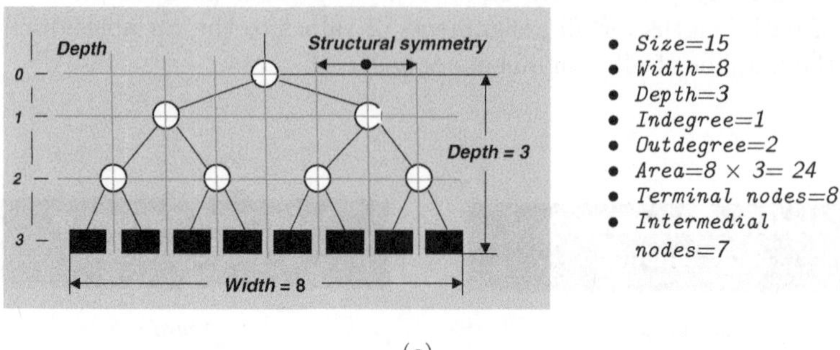

(*a*)

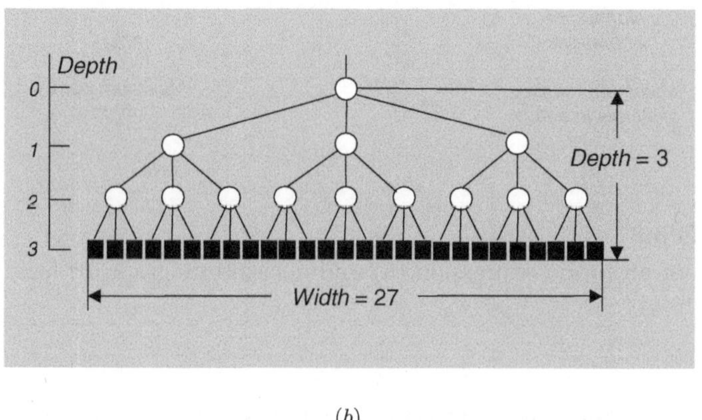

(*b*)

FIGURE 3.9
The complete binary (a) and ternary (b) tree (Example 3.15).

3.4 Shape of decision diagrams

Stanković [32, 33, 29] showed that the information content of a decision diagram is encoded in its shape and developed methods for evaluating the shape of decision diagrams.

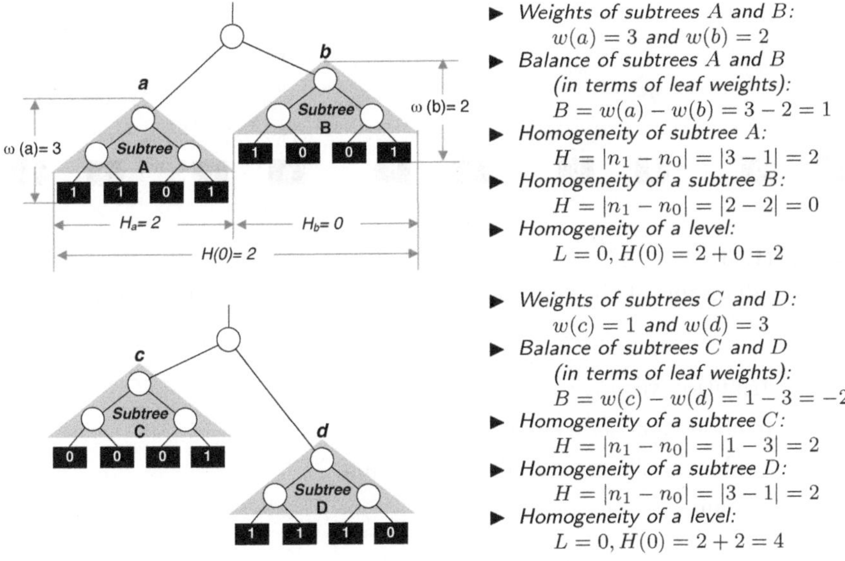

a
b
Subtree A
Subtree B
ω (a)= 3
ω (b)= 2
1 1 0 1
1 0 0 1
H_a= 2
H_b= 0
H(0)= 2
Weights of subtrees A and B:
w(a) = 3 and w(b) = 2
Balance of subtrees A and B
(in terms of leaf weights):
B = w(a) − w(b) = 3 − 2 = 1
Homogeneity of subtree A:
H = |n1 − n0| = |3 − 1| = 2
Homogeneity of a subtree B:
H = |n1 − n0| = |2 − 2| = 0
Homogeneity of a level:
L = 0, H(0) = 2 + 0 = 2
c
d
Subtree C
Subtree D
0 0 0 1
1 1 1 0
Weights of subtrees C and D:
w(c) = 1 and w(d) = 3
Balance of subtrees C and D
(in terms of leaf weights):
B = w(c) − w(d) = 1 − 3 = −2
Homogeneity of a subtree C:
H = |n1 − n0| = |1 − 3| = 2
Homogeneity of a subtree D:
H = |n1 − n0| = |3 − 1| = 2
Homogeneity of a level:
L = 0, H(0) = 2 + 2 = 4

FIGURE 3.10
Functional characteristics of binary decision trees (Example 14.12).

> *The shape is an invariant characteristic of decision diagrams, since by choosing different interpretations of nodes and labels at the edges, we can read different information from the decision diagram. There are decision diagrams that have the same depth, size and width, and even identical distributions of nodes at each level, but with different interconnections. Thus, these BDDs are different, but not distinguished by the mentioned characteristics. Therefore, these characteristics are not sufficient to precisely characterize decision diagrams.*

The type of a decision diagram for a switching function is specified by the type of expansion used to represent this function. The most common are Shannon and Davio expansions. The diagrams can have the same shape, but represent different functions because the nodes are assigned with different expansions.

Example 3.17 *Figure 3.11 shows that the information content of a decision diagram is coded in its shape.*

The following formal notations are used in description of a shape of decision diagram:

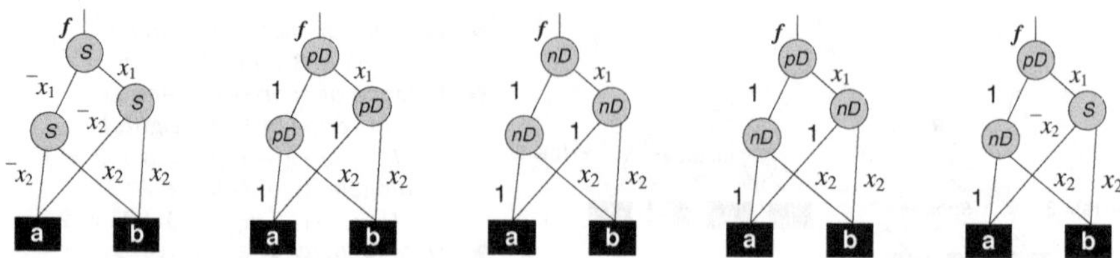

FIGURE 3.11
Decision diagrams of the same shape represent different switching functions.

- ▶ A distribution vector,
- ▶ The symbolic reachaility matrix,
- ▶ The total symbolic reachability matrix, and
- ▶ *S*-polynomial.

Distribution vector of a decision diagram describes a total topology!indexTopolog of decision diagrams excepting connections between nodes.

Definition 3.2 *Distribution of nodes per level is described by a distribution vector*

$$\mathbf{D}_n = [n_1, n_2, \ldots n_{n+1}] \tag{3.1}$$

whose elements are numbers of nodes per level starting from the root node at the first level up to the terminal nodes at the $n+1$*-st level in the decision diagram.*

There are several useful properties of the distribution vector $\mathbf{D}_n$ (Equation 3.1):

- ▶ For bit-level decision diagrams, $n_{n+1} = 2$.
- ▶ The number of elements in distribution vector $\mathbf{D}_n$ minus 1 shows the *depth* of a decision diagram, i.e., `Depth`$= n$.
- ▶ The maximum value of an element in $\mathbf{D}_n$ is the *width* of a decision diagram.

Example 3.18 *Distribution vector* $\mathbf{D}_2 = [1, 2, 4]$ *describes the topology of a decision diagram without connections with one root node, two non-terminal nodes, and four terminal nodes (Figure 3.12).*

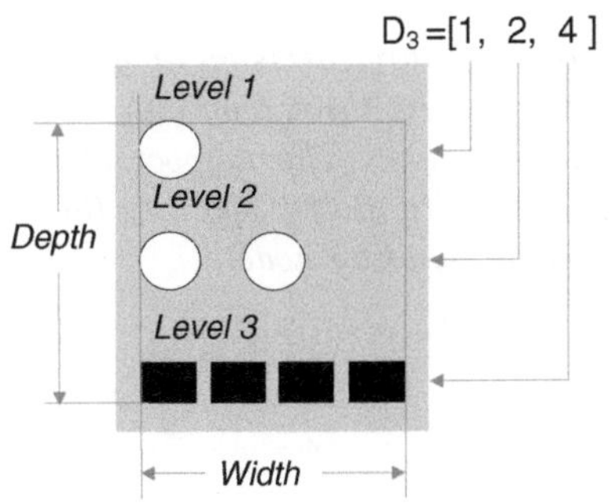

FIGURE 3.12
Description of the topology of a decision diagram without connections using distribution vector (Example 3.18).

The symbolic reachability matrix describes a local topology of a decision diagram. Denote by q_i and q_{i+1} the total number of nodes and cross points at the i-th and $i+1$-th level in a decision diagram, respectively.

Definition 3.3 *The symbolic reachaility matrix (SR-matrix)* $\mathbf{R}_{i,i+1}$ *describing the connection of the nodes between i-th and* $i+1$*-th levels in decision diagram is defined as a* $q_i \times q_{i+1}$ *matrix*

$$\mathbf{R}_{i,i+1} = [r_{jk}^{i,i+1}]. \tag{3.2}$$

The element $r_{jk}^{i,i+1}$ *is defined as follows:*

(a) If j is a non-terminal node, and k and j are connected, then

$$r_{jk}^{i,i+1} = \begin{cases} x, & \textit{if } k \textit{ is a non-terminal node or cross point;} \\ c_y x, & \textit{if } k \textit{ is a terminal node with the value } c_y; \\ 0, & \textit{otherwise.} \end{cases} \tag{3.3}$$

(b) If j is a cross point, and k and j are connected, then

$$r_{jk}^{i,i+1} = \begin{cases} 1, & \textit{if } k \textit{ is a non-terminal node or cross point;} \\ c_y 1, & \textit{if } k \textit{ is a terminal node with the value } c_y; \\ 0, & \textit{otherwise.} \end{cases} \tag{3.4}$$

It follows from Equations 3.3 and 3.4 that in description of a local topology of decision diagram the following features must be encoded

- ► Non-terminal nodes and their connections with terminal and nonterminal nodes, denoted by x and $c_y x$, and

▶ The cross points and their connections with terminal and non-terminal nodes, denoted by 1 and c_y1.

Example 3.19 *In Figure 3.13, three decision diagrams of different topologies are given. In the third diagram, nodes have different number of outgoing edges. The number of non-zero entries in a row of the reachability matrix specifies the number of outgoing edges of the corresponding node.*

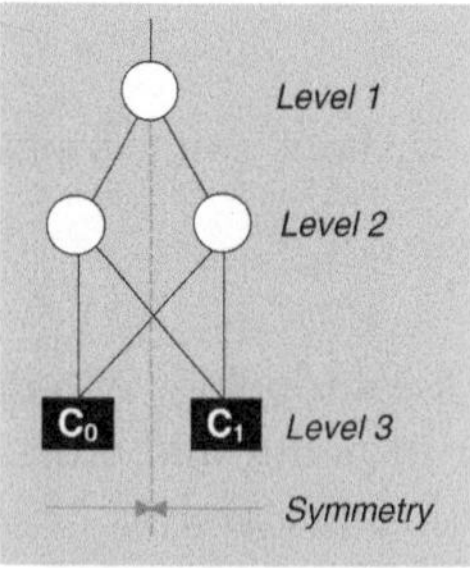

$$\mathbf{R}_{12} = [\, x\ x \,],$$
$$\mathbf{R}_{23} = \begin{bmatrix} x \cdot c_0 & x \cdot c_1 \\ x \cdot c_0 & x \cdot c_1 \end{bmatrix}$$

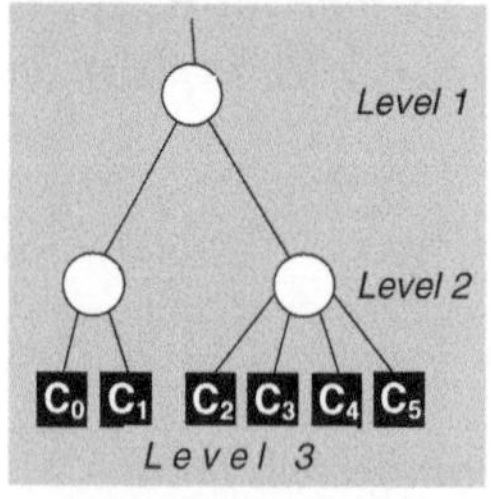

$$\mathbf{R}_{12} = \begin{bmatrix} x & x \end{bmatrix},$$
$$\mathbf{R}_{23} = \begin{bmatrix} x \cdot c_0 & x \cdot c_1 & 0 & 0 & 0 & 0 \\ 0 & 0 & x \cdot c_2 & x \cdot c_3 & x \cdot c_4 & x \cdot c_5 \end{bmatrix}$$

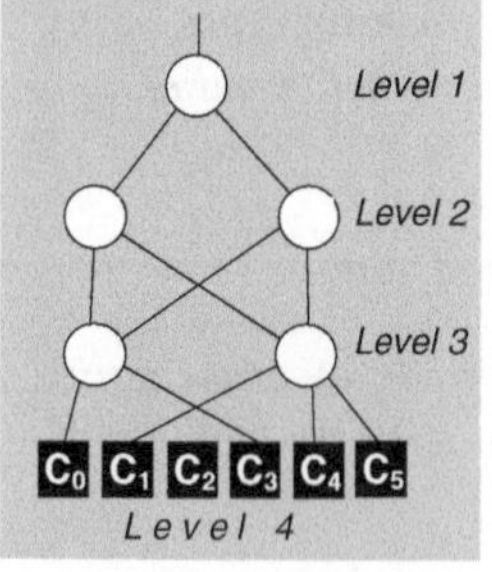

$$\mathbf{R}_{12} = \begin{bmatrix} x & x \end{bmatrix},$$
$$\mathbf{R}_{23} = \begin{bmatrix} x & x \\ x & x \end{bmatrix},$$
$$\mathbf{R}_{34} = \begin{bmatrix} x \cdot c_0 & 0 & 0 & x \cdot c_3 & 0 & 0 \\ 0 & x \cdot c_1 & 0 & 0 & x \cdot c_4 & x \cdot c_5 \end{bmatrix}$$

FIGURE 3.13
The symbolic reachability matrices for various decision diagrams (Example 3.19).

The total symbolic reachability matrix represent the total topology of a decision diagram. This matrix combine local descriptions using the multiplicative rule for symbolic reachability matrices.

Definition 3.4 *The total symbolic reachability matrix (total SR-matrix) for a decision diagram is defined by*

$$\mathbf{R}_{1,n+1} = \mathbf{R}_{1,2} \cdot \mathbf{R}_{2,3} \cdot \mathbf{R}_{n,n+1} \tag{3.5}$$

Example 3.20 *Given a decision diagram (Figure 3.14), the total symbolic reachability matrix is calculated by multiplication of reachability matrices derived for each level:*

$$\mathbf{R}_{1,4} = \mathbf{R}_{1,2} \cdot \mathbf{R}_{2,3} \cdot \mathbf{R}_{3,4}$$

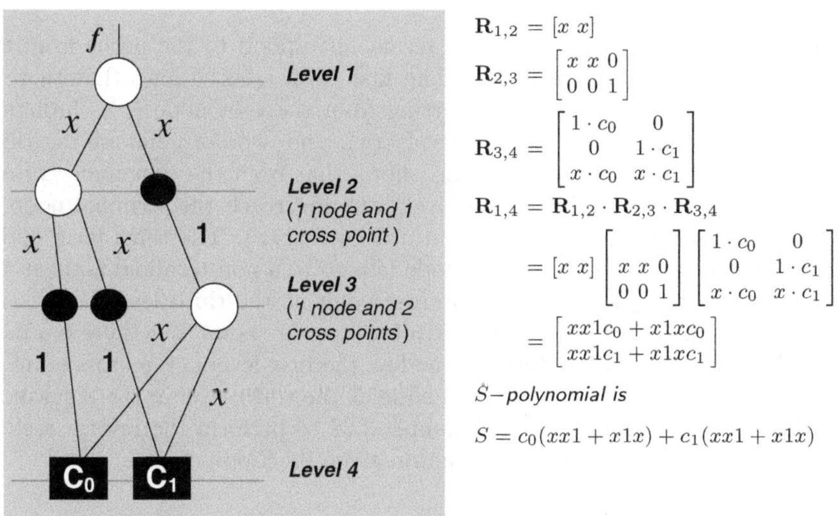

FIGURE 3.14
Computing S-polynomial for description of decision diagram topology (Examples 3.20 and 3.21).

S-polynomial of a decision diagram. An algebraic equation which describes the shape of decision diagram is called *S-polynomial.* An S-polynomial is derived from the information obtained from a decision diagram:

- Enumerating the paths from the root node to the terminal nodes, and
- Values of the terminal nodes.

Definition 3.5 *The S-polynomial for a decision diagram is defined as the sum of elements of* $\mathbf{R}_{1,n+1}$ *(Equation 3.5):*

$$S = \sum_{i=1}^{q_{n+1}} r_i^{1,n+1}. \tag{3.6}$$

An S-polynomial (Equation 3.6) completely describes the shape of a decision diagram.

Example 3.21 *Given a decision diagram (Figure 3.14). The* S*–polynomial for description this diagram topology is computed using Equation 3.6*

$$S = c_0(xx1 + x1x) + c_1(xx1 + x1x).$$

In an S-polynomial, the product terms correspond to the paths from the root node to the terminal nodes. The first term c_0xx1 shows that there is a path to the constant node c_0 starting from the root node (x), through a non-terminal node at the second level (xx), and a cross point at the third level ($xx1$). The second term c_0x1x, shows that from the root node (x), we go to a cross point at the second level ($x1$) and reach the terminal node c_0 through a non-terminal node at the third level ($x1x$). The third term c_1xx1 shows that we start from the root node (x), go to a non-terminal node at the second level (xx), and then over a cross point at the third level ($xx1$), and reach the terminal node c_1. The fourth term c_1x1x shows that there is a path to c_1 that goes over a non-terminal node a the first level (x), a cross point at the second level and a non-terminal node at the third level in the diagram.

This analysis will be used in Example 3.22 to perform the inverse task of determination of the shape of a diagram given its S-polynomial.

The inverse problem is formulated as follows: given S-polynomial, draw the shape of the decision diagram and calculate the topological characteristics.

Example 3.22 *Given an* S*-polynomial*

$$S = c_0(xxx) + c_1(xxx) + c_2(xx1 + x1x) + c_3(x1x).$$

The shape of the corresponding decision diagram is determined in Figure 3.15.

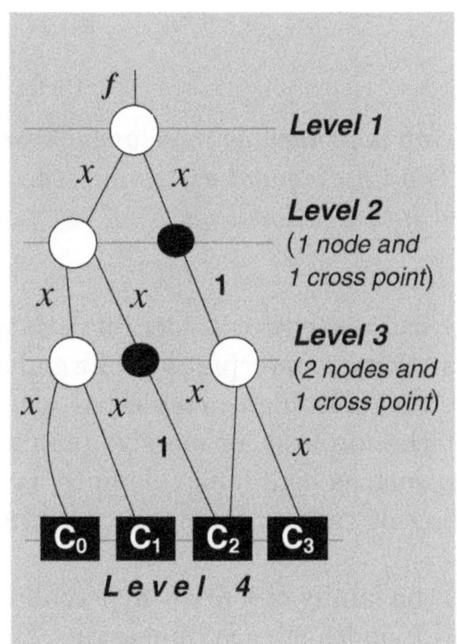

► *Given an S-polynomial*

$$S = c_0(xxx) + c_1(xxx) + c_2(xx1 + x1x) + c_3(x1x),$$

Determine the shape of the decision diagram.
There are four coefficients c_0, c_1, c_2, and c_3, which means that these would be four terminal nodes in the diagram.

► *The first term*

$$c_0(xxx)$$

means that

(a) *We start from the root node (x),*
(b) *Reach the terminal node c_0 through a non-terminal node at the second level (xx), and*
(c) *Reach a non-terminal node at the third level in the diagram.*

It is similar for the term c_1xxx.

► *The term*

$$c_2(xx1 + x1x)$$

shows that the terminal node c_2 can be reach through two paths:

(a) *The first path goes through a non-terminal node at the second level (xx) and a cross point at the third level $(xx1)$.*
(b) *The second path goes through a cross point at the second level $(x1)$ and a non-terminal node at the third level $(x1x)$.*
It is similar for the term c_3x1x which determines the path that reach the node c_3.

FIGURE 3.15
Recovering the decision diagram from the S-polynomial (Example 3.22).

3.5 Embedding decision trees and diagrams into various topological structures

An embedding of a *guest* graph G into a *host* graph H is a one-to-one mapping φ: $V(G) \to V(H)$, along with a mapping α that maps an edge $(u, v) \in E(G)$ to a path between $\varphi(u)$ and $\varphi(v)$ in H. In this section, an introduction to graph embedding techniques is given. The embedding of decision trees into the following topological structures are considered:

- Hypercubes,
- Cycle connected cubes (CCC-hypercubes),
- Pyramids,
- $\mathcal{N}$-hypercubes, and
- Hexagonal configuration.

Example 3.23 *An Akers array is defined as a rectangular array of identical cells. All cells on a diagonal are connected. A decision tree can be embedded into this topology.*

The hypercube topology (Figure 34.1a) has received considerable attention in classical logic design due mainly to its ability to interpret logic formulas and its good implementibility (small diameter, regular, highly connected, symmetric). Hypercube-based structures are at the forefront of massive parallel computation because of their characteristics, such as high fault tolerance, i.e., an ability to efficiently permit the embedding of various topologies, such as lattices and trees.

The binary n-hypercube is a special case of the family of k-ary n-hypercubes, which are hypercubes with n dimensions and k nodes in each dimension. The total number of nodes in such a hypercube is $N = k^n$. Parallel computers with direct-connect topologies are often based on k-ary n-cubes or isomorphic structures such as rings, meshes, tori, and direct binary n-cubes.

The ***CCC*****-hypercube** is created from a hypercube by replacing each node with a cycle of s nodes (Figure 34.1b). This hence increases the total number of nodes from 2^n to $s \cdot 2^n$ and preserves all features of the hypercube. The CCC-hypercube is closely related to the butterfly network. As has been shown in the previous sections, "butterfly" flowgraphs are the nature of most transforms of switching functions in matrix form.

Pyramid topology (Figure 34.1c) is suitable for many computations based on the principle of hierarchical control, for example, decision trees and decision diagrams. An arbitrarily large pyramid can be embedded into the hypercube

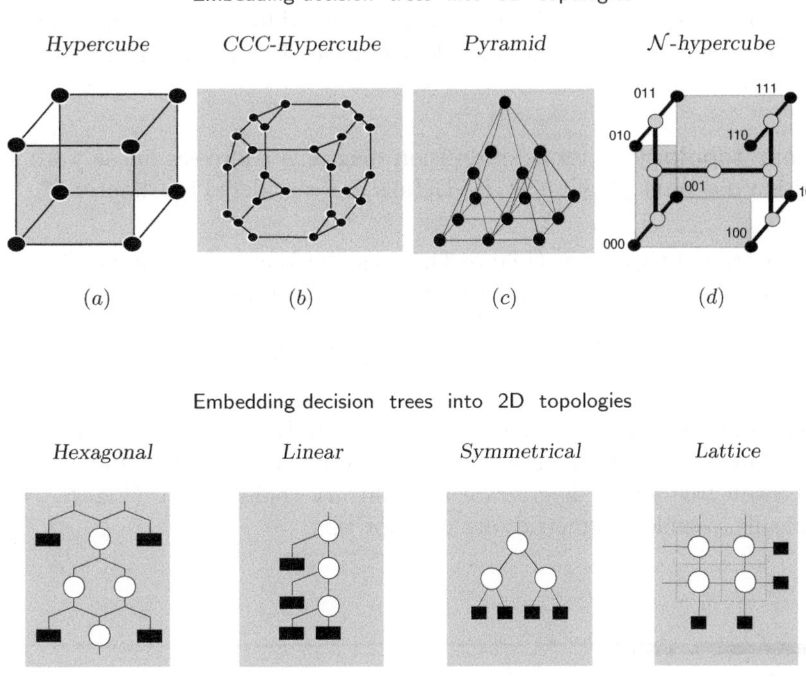

FIGURE 3.16
Embedding decision trees into 3D and 2D spatial configurations.

with a minimal load factor. Pyramid P_n has nodes on levels $0, 1, \ldots, n$. The number of nodes on level i is 4^i. So, the number of nodes of P_n is $(4^{n+1}/3)$. The unique node on level 0 is called the *root* of the pyramid. The subgraph of P_n induced by the nodes on level i is isomorphic to mesh $2^i \times 2^i$. The subgraph of P_n induced by the edges connecting different levels is isomorphic to a 4^n-leaf quad-tree. This structure is very flexible for extension. Pyramid topology is relevant also to fractal-based computation, which is effective for symmetric functions and is used in digital signal processing, image processing, and pattern recognition.

$\mathcal{N}$-hypercube topology is a hypercube-like topology that is more suitable for hosting decision trees and diagrams of switching functions (Figure 34.1d). Details are given in Chapter 34.

Hexagonal planar topology is often relevant to some technologies (Figure 34.1e). One reason is that this topology is suitable for a close-packed layout, including a hexagonal lattice of nanowires. In addition, three-fold symmetry for the node can be utilized in implementation. Details are given in Chapter 34.

Linear topology in terms of decision diagrams addresses linear word-level decision diagrams (Figure 34.1f). Details are considered in Chapter 37.

Symmetrical topology (Figure 34.1g) can be found in various spatial structures that represent objects with symmetrical properties. Particular techniques have been developed for embedding an arbitrary decision tree or diagram using so-called *pseudo-symmetry* (see details in Chapter 34).

Lattice structures. Decision diagrams of totally symmetric switching functions can be embedded into a lattice directly. Lattice structures can host a decision tree and a diagram of an arbitrary configuration (Figure 34.1h). Typically, pseudo-symmetries are used for this.

3.6 Examples with hierarchical FPGA configuration

A hierarchical, or binary-tree-like FPGA is based on single-input two-output switches. A 2×2 cluster of processing elements can be connected using switches, and four copies of the cluster organized into a "macro" cluster. The structure of this FPGA is represented by a binary decision tree of depth 4, in which the root and levels correspond to switching blocks, and 16 terminal nodes correspond to processing elements.

Another architecture of a hierarchical FPGA is based on a cluster of logic blocks connected by single-input 4-output switch blocks. Possible configurations can be described by a complete binary tree or a multirooted k-ary tree. This tree can be embedded into the hypercube-like structures.

Example 3.24 *In Figure 3.17 two topologies of FPGA are illustrated, where ■ denotes a processing element and ○ denotes a switch block. Four copies of the cluster are organized into a "macro cluster" of different configurations. The first topology (Figure 3.17a) is based on H-tree construction (see Section 3.3.2) and described by a complete binary tree. The spatial structure includes two 3D hypercubes. The second topology (Figure 3.17b) is described by a complete quaternary tree and also by a hypercube-like structure in three dimensions.*

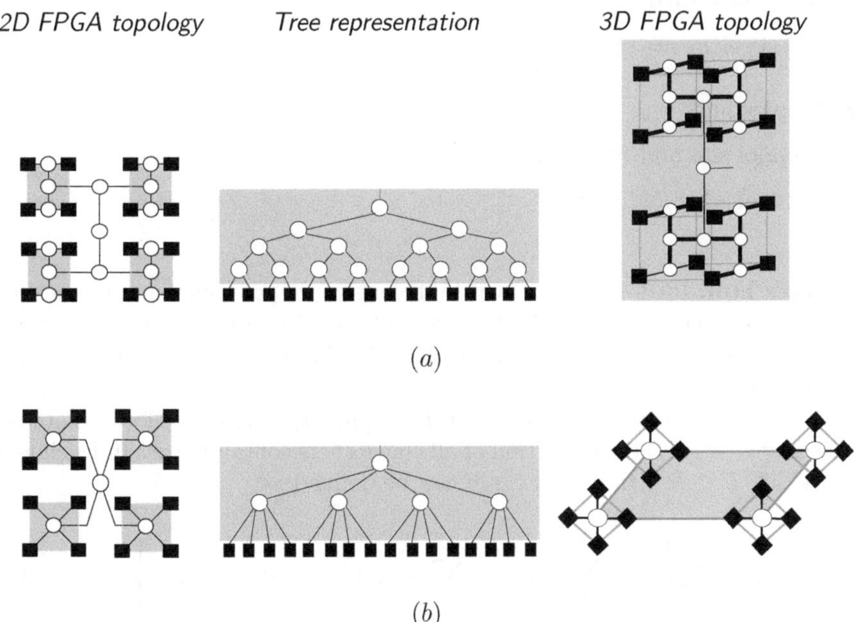

FIGURE 3.17
Topological representations of hierarchical FPGAs by 2D structure, tree, and hypercube-like structure (Example 3.24).

3.7 Voronoi diagrams and Delaunay triangulations

This section[‡] is concerned with graphical data structures such as the Voronoi diagram and the Delaunay tessellation well-known in computer graphics. These structures represent topological spaces, and, therefore, are considered in this book as potential tools for analysis of topologies in nanodimensions.

Voronoi diagrams and Delaunay tessellations are known to be useful for:

- Generation of fractal patterns,
- Shape analysis and topology-based pattern recognition, and
- Location optimization.

[‡]This section was co-authored with O. Bulanov, University of Calgary, Canada

3.7.1 Voronoi diagram

The basic elements of a Voronoi diagram are

- Convex hull,
- Voronoi site and cell, and
- Voronoi edge and vertex.

Convex hull. Intuitively, a convex hull of a given set of points in n-dimensional Euclidean space $\mathbb{R}^n$ is a smallest convex set containing all given points. A set $S \subset \mathbb{R}^n$ is a convex set if the line segment joining any pair of points in S lies entirely in S.

Formally, the convex hull of a set $P = \{p_1, p_2, \ldots, p_N\}$ of N points denoted by $C(P)$ is the intersection of all convex sets containing P. In Euclidean space $\mathbb{R}^n$, $P \subset \mathbb{R}^n$, the convex hull can be defined as

$$C(P) \stackrel{\text{def}}{=} \left\{ x = \sum_{i=1}^{N} \lambda_i p_i \;\middle|\; \sum_{i=1}^{N} \lambda_i = 1,\ \lambda_i \geq 0 \ \text{ for all } \ i \right\}. \tag{3.7}$$

Voronoi cell. The basic structure in the Voronoi diagram technique is a Voronoi cell. Let $P = \{p_1, p_2, \ldots, p_N\}$ be a set of N distinct points in $\mathbb{R}^n$. The Voronoi cell of a single point $p_\alpha \in P$ denoted by $V(p_\alpha)$ is defined as

$$V(p_\alpha) \stackrel{\text{def}}{=} \{x \in \mathbb{R}^n \ : \ d(x, p_\alpha) \leq d(x, p) \ \text{ for all } \ p \in P\}. \tag{3.8}$$

The point p_α generating the Voronoi cell $V(p_\alpha)$ is called the *Voronoi site*. The Voronoi cell $V(p_\alpha)$ is a closed set and therefore contains its boundary.

Voronoi face. In 3D space, $\mathbb{R}^3$, a Voronoi cell is a 3D polyhedron. The boundary of the *Voronoi polyhedron* consists of 2D facets which are called *Voronoi faces.*

Voronoi edge and Voronoi vertex. For a 3D Voronoi diagram, the boundary of a Voronoi face consists of line segments, half lines or infinite line, which are called *Voronoi edges.*

In 2D space $\mathbb{R}^2$, a Voronoi cell is a 2D polygon. The boundary of a *Voronoi polygon* consists of Voronoi edges. The end points of a Voronoi edge are called *Voronoi vertices.*

For a 1D Voronoi diagram, the boundary of a Voronoi cell consists of Voronoi vertices.

Example 3.25 (Convex hull and Voronoi cell in $\mathbb{R}^2$)
Consider a set of points representing some 2D hypercube structure:

$$P = \{\, p_i \in \mathbb{R}^2 \mid i = 1, 2, \ldots, 9 \,\}.$$

Let the points $p_6 \in P$ and $p_8 \in P$ be two distinct points such that the point p_6 lies exactly on the border of the convex hull, $C(P)$, of the set P while p_8 is located inside $C(P)$ as shown in Figure 3.18(a). The Voronoi site p_6 located on the boundary of the convex hull generates an unbounded Voronoi cell $V(p_6)$, while the site p_8 located inside the convex hull generates a bounded Voronoi cell $V(p_8)$ as demonstrated in Figure 3.18(b).

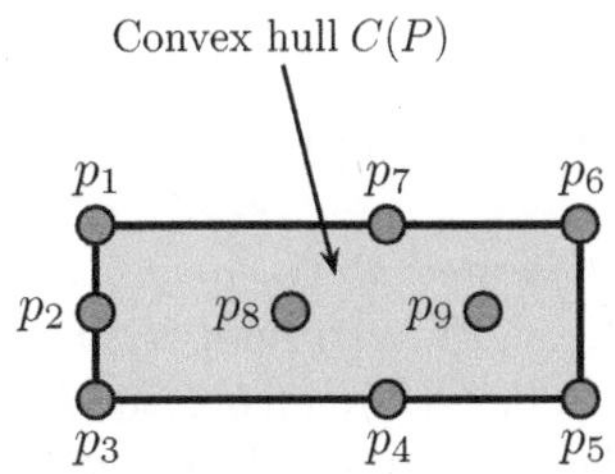

CONVEX HULL IN $\mathbb{R}^2$: $C(P)$

The dots $p_1, p_2, \ldots, p_9$ represent a finite point set P in the plane (i.e., in space $\mathbb{R}^2$). In $\mathbb{R}^2$, the convex hull $C(P)$ of the set P is a convex polygon specified by points $p_1, p_2, p_3, p_4, p_5, p_6$, and p_7. The convex hull is represented by a gray area in the picture; the border is shown as a solid line.
The point p_6 is located on the border of the convex hull polygon $C(P)$.
The point p_8 is located in the inside area of the convex hull polygon $C(P)$.

VORONOI CELLS IN $\mathbb{R}^2$: $V(p_6)$ and $V(p_8)$

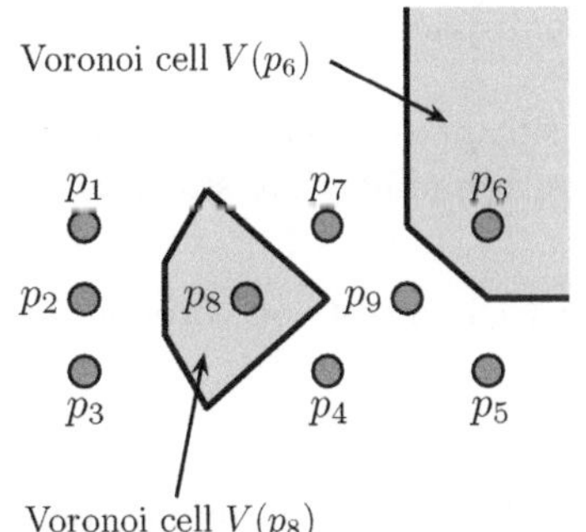

Each Voronoi cell, $V(p_6)$ and $V(p_8)$, of the sites p_6 and p_8, respectively, is a polygon surrounding the corresponding site.
The Voronoi site p_6 located on the boundary of the convex hull generates an unbounded Voronoi cell $V(p_6)$.
The Voronoi site p_8 located inside the convex hull generates a bounded Voronoi cell $V(p_8)$.

FIGURE 3.18
Convex hull $C(P)$ of the set $P = \{p_1, p_2, \ldots, p_9\}$ in $\mathbb{R}^2$ and Voronoi cells $V(p_6)$ and $V(p_8)$ generated by the sites $p_6 \in P$ and $p_8 \in P$ (Example 3.25).

Example 3.25 illustrates, that in a plane

- The Voronoi cell $V(p_8)$ generated by the corresponding Voronoi site p_8 is the polygon surrounding that site (Figure 3.18(b)). The inside region of the polygon consists of the closest points with respect to other members of the set P.
- The Voronoi site p_6 generates, like any Voronoi site located exactly on the boundary of the convex hull, an unbound Voronoi cell $V(p_6)$ (see Figure 3.18(b)).
- The boundary of the bounded Voronoi polygon $V(p_8)$ consists of Voronoi edges, which are finite line segments.
- The boundary of the unbounded Voronoi cell $V(p_6)$ consists of Voronoi edges, which are a finite line segment and half-lines.

Example 3.26 (Convex hull and Voronoi cell in $\mathbb{R}^3$)
Consider a set of points representing some 3D hypercube structure:

$$P = \{\, p_i \in \mathbb{R}^3 \mid i = 1,\, 2,\, \ldots,\, 15 \,\}.$$

Let the point $p_1 \in P$ be the point located at the origin while the other points, $p_i \in P$, $i = 2, 3, \ldots, 15$, surround it so that the point p_1 is inside the convex hull, $C(P)$, of the set P as shown in Figure 3.19(a). Then the Voronoi cell $V(p_1)$ generated by the site p_1 is a bounded polyhedron as demonstrated in Figure 3.19(b).

Example 3.26 illustrates that in 3-dimensional space $\mathbb{R}^3$:

- The Voronoi cell $V(p_1)$ generated by the site $p_1 \in P$ located inside the convex hull of P is a 3D polyhedron surrounding that site (see Figure 3.19(b)).
- The boundary of the Voronoi polyhedron $V(p_1)$ consists of Voronoi faces, which are 2D polygons.

Voronoi diagram. Let $P = \{\, p_1,\, p_2,\, \ldots,\, p_N \,\}$ be a set of N distinct points in $\mathbb{R}^n$. The Voronoi diagram of the set P denoted by $V(P)$ is a partitioning of the space $\mathbb{R}^n$ into N Voronoi cells, that is

$$V(P) \stackrel{\text{def}}{=} \{\, V(p_1),\, V(p_2),\, \ldots,\, V(p_N) \,\}, \tag{3.9}$$

where $p_i \in P$ for all i. Hence, each Voronoi cell $V(p_i)$ contains exactly one generating Voronoi site, $p_i \in P$, and the interior of the cell consists of points which are closer to p_i than to any other point in P.

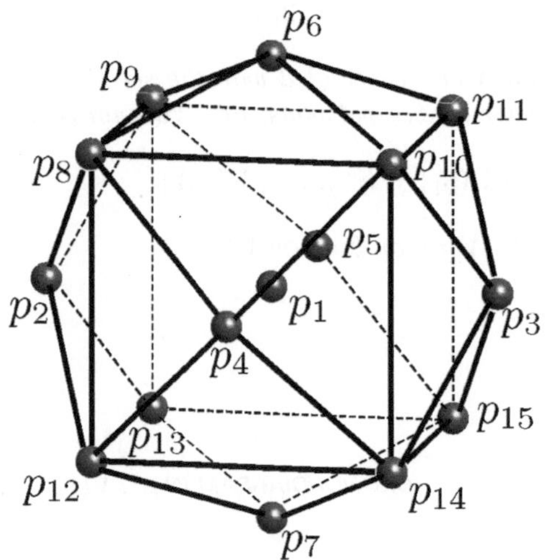

CONVEX HULL IN $\mathbb{R}^3$: $C(P)$

The dots

$$p_1, p_2, \ldots, p_{15}$$

represent a finite point set P in $\mathbb{R}^3$. The convex hull $C(P)$ of the set P is a convex polyhedron defined by the points

$$p_2, p_3, \ldots, p_{15}.$$

The point p_1 is located inside the convex hull. The edges of the convex hull facets are shown as solid and dashed lines.

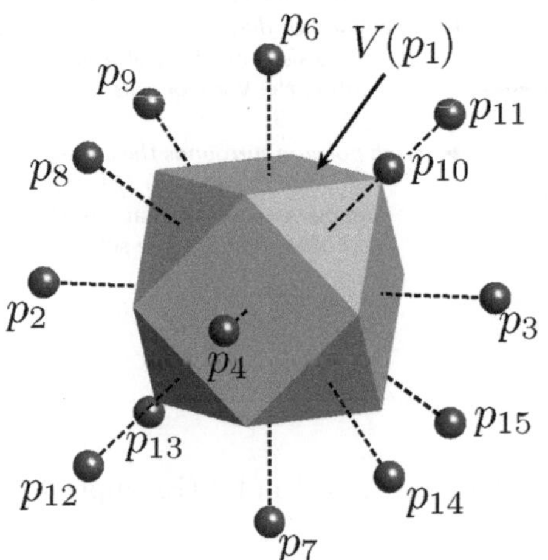

VORONOI CELL IN $\mathbb{R}^3$: $V(p_1)$

The Voronoi cell $V(p_1)$ of the site p_1 is a polyhedron surrounding the site p_1. The Voronoi site p_1 is located inside the convex hull and therefore generates a bounded Voronoi cell.

FIGURE 3.19
Convex hull $C(P)$ of the set $P = \{p_1, p_2, \ldots, p_{15}\}$ in $\mathbb{R}^3$ and Voronoi cell $V(p_1)$ generated by the site $p_1 \in P$ (Example 3.26).

Example 3.27 (Voronoi diagram in $\mathbb{R}^2$) *Consider a set of distinct points in $\mathbb{R}^2$, specifically the same set P as in Example 3.25:*

$$P = \{\, p_i \in \mathbb{R}^2 \mid i = 1, 2, \ldots, 9 \,\}.$$

The Voronoi diagram $V(P)$ associated with the set P consists of the nearest neighbor convex polygons, i.e., Voronoi cells:

$$V(P) = \{\, V(p_1),\, V(p_2),\, \ldots,\, V(p_9) \,\}.$$

Figure 3.20 shows the Voronoi diagram $V(P)$.

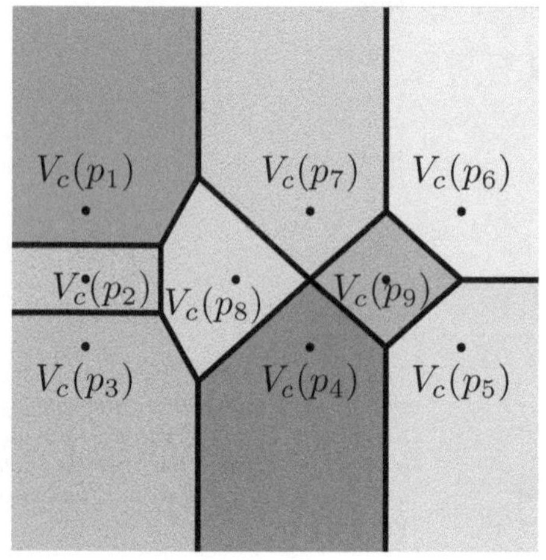

VORONOI DIAGRAM IN $\mathbb{R}^2$: $V(P)$

- *The dots $p_1, p_2, \ldots, p_9$ represent a finite point set P in the plane.*
- *The Voronoi diagram $V(P)$ of the set P consists of the polygons representing the Voronoi cells $V(p_i)$, $i = 1, 2, \ldots, 9$.*
- *Each polygon surrounds the corresponding Voronoi site with the region of the plane closer to that site than to any other point in the set P.*

FIGURE 3.20
Voronoi diagram $V(P)$ of the set $P = \{\, p_1, p_2, \ldots, p_9 \,\}$ in $\mathbb{R}^2$ (Example 3.27).

Example 3.27 shows, that:

- Each Voronoi polygon surrounds the corresponding Voronoi site with the region of the plane closer to that site than to any other site from the point set P.
- The Voronoi edges, which form the boundaries of the Voronoi polygons, are the points of the plane equidistant simultaneously to two or more Voronoi sites.

> *The 2D Voronoi diagram is a graphical data structure consisting of nearest-neighbor polygons. It is created given a set of Voronoi sites, by deriving the* Voronoi polygons *around each site so that this region on the plane is closer to the corresponding site than to any other site in the given set.*
> *The 3D Voronoi diagram is a structure consisting of nearest-neighbor polyhedra.*

Given a set of Voronoi sites in $\mathbb{R}^3$, the Voronoi diagram is created by deriving the *Voronoi polyhedra* around each site so that every point inside the polyhedron is closer to the corresponding site then to any other site.

The associated structure is Delaunay tessellation.

3.7.2 Delaunay tessellation

Delaunay tessellation in $\mathbb{R}^2$. In 2D space, $\mathbb{R}^2$, a Delaunay tessellation is called a *Delaunay triangulation.* Let $P = \{p_1, p_2, \ldots, p_N\}$ be a set of N distinct points in $\mathbb{R}^2$. The Delaunay triangulation for the set P denoted by $D(P)$ is the triangulation of the convex hull of P such that no point in P is located inside the circumcircle of any triangle in $D(P)$.

The Delaunay triangulation $D(P)$ is related to the Voronoi diagram $V(P)$ of the same set P and can be constructed from $V(P)$ by setting up line segments connecting those pairs of Voronoi sites with which Voronoi cells share a common Voronoi edge.

For non-degenerated Voronoi diagrams such procedures generate precisely a Delaunay triangulation. For degenerated Voronoi diagram this gives a tessellation, also called *Delaunay pretriangulation*, containing polygons of four or more vertices. The final Delaunay triangulation is obtained by partitioning all such polygons into triangles.

Example 3.28 (Delaunay triangulation in $\mathbb{R}^2$) *Consider the set P of distinct points and the Voronoi diagram $V(P)$, associated with P, the same as in Examples 3.25 and 3.27. The first step is to set up line segments connecting each Voronoi site $p_i \in P$ to its natural neighbors, that is, to those points in P in which Voronoi cells have common Voronoi edges with that site; the resultant tessellation is shown in Figure 3.21(a). However, this tessellation contains a square $(p_4\, p_7\, p_8\, p_9)$ which is to be triangulated by connecting, for example, the points $p_8 \in P$ and $p_9 \in P$; the final Delaunay triangulation is represented in Figure 3.21(b).*

Example 3.28 demonstrates that 2D Delaunay triangulation, $D(P)$, for a set P of points in the plane can be constructed by setting line segments with respect to Voronoi edges of the corresponding Voronoi diagram, $V(P)$.

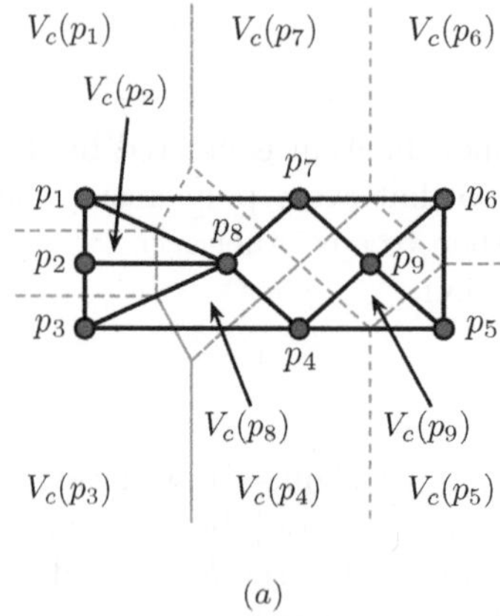

(a)

DELAUNAY PRETRIANGULATION IN $\mathbb{R}^2$

The dots $p_1, p_2, \ldots, p_9$ represent a point set P. The pretriangulation is set up by connecting each point in P to its natural neighbors, points in P which Voronoi cells have common Voronoi edges. For instance, the Voronoi cell $V(p_8)$ of the point $p_8 \in P$ has:

- *Common edges with $V(p_1)$, $V(p_2)$, $V(p_3)$, $V(p_4)$, and $V(p_7)$ Voronoi cells, and*
- *Connections (p_8, p_1), (p_8, p_2), (p_8, p_3), (p_8, p_4), and (p_8, p_7).*

This tessellation contains a square $(p_4\, p_8\, p_7\, p_9)$ which is still to be triangulated

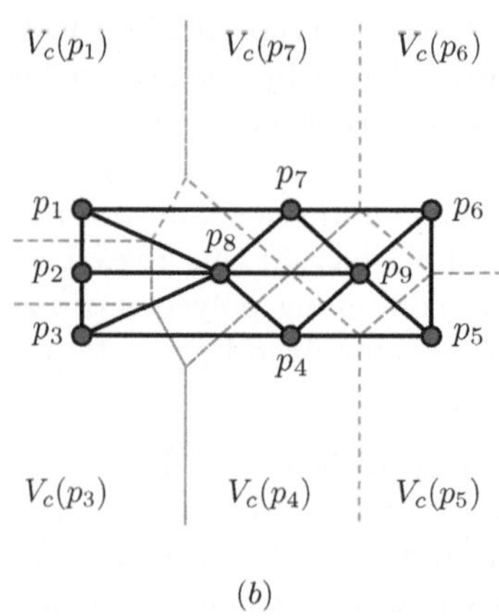

(b)

DELAUNAY TRIANGULATION IN $\mathbb{R}^2$: $D(P)$

The final Delaunay triangulation $D(P)$ is obtained by triangulating the square

$$(p_4\, p_8\, p_7\, p_9)$$

by connecting, for example, the points $p_8 \in P$ and $p_9 \in P$

FIGURE 3.21
Delaunay triangulation $D(P)$ of the set $P = \{p_1, \ldots, p_9\}$ in $\mathbb{R}^2$ (Example 3.28).

Delaunay tessellation in $\mathbb{R}^3$. In 3-dimensional space, a Delaunay tessellation is called a *Delaunay tetrahedrization.* Let $P = \{p_1, p_2, \ldots, p_N\}$ be a set of N distinct points in $\mathbb{R}^3$. The Delaunay tetrahedrization $D(P)$ of the set P is the tessellation by tetrahedra of the convex hull of P such that no point in P is inside the circumsphere of any tetrahedron in $D(P)$.

Similarly to the space $\mathbb{R}^2$, the 3D Delaunay tetrahedrization, $D(P)$, of the set P can be constructed from the corresponding Voronoi diagram, $V(P)$, by setting up line segments connecting those pairs of Voronoi sites which Voronoi cells have a common Voronoi face.

Example 3.29 (Delaunay tetrahedrization in $\mathbb{R}^3$) *Let P be a set of five distinct points in $\mathbb{R}^3$, namely*

$$P = \{(0,0,0),\ (1,1,1),\ (-1,-1,1),\ (-1,1,-1),\ (1,-1,-1)\}.$$

Setting up the connections for all points in P we obtain the Delaunay tetrahedrization $D(P)$ of the point set P which is represented in Figure 3.22.

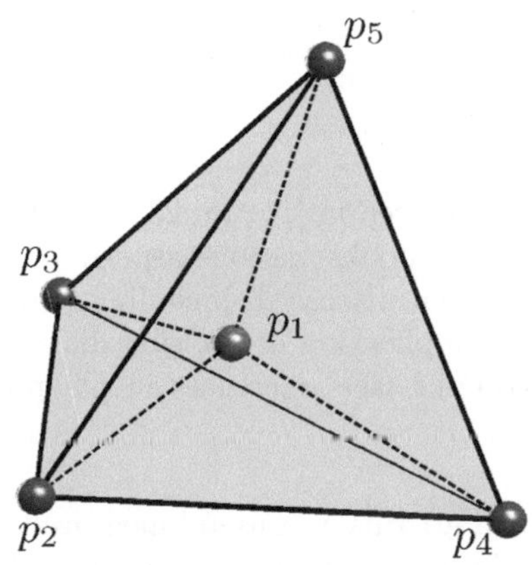

DELAUNAY TETRAHEDRIZATION IN $\mathbb{R}^3$: $D(P)$

The dots $p_1, p_2, \ldots, p_5$ represent a point set P. The Delaunay tetrahedrization $D(P)$ for the set P is obtained by setting up the connections for all points in P.

FIGURE 3.22
Delaunay tetrahedrization $D(P)$ in $\mathbb{R}^3$ (Example 3.29).

3.7.3 Decision diagrams and Delaunay tessellation

A logical formula can be represented as a tree structure and then the tree can be implemented in some geometrical structure. A given set of points in Euclidean space can be assigned many different spanning trees connecting them together. An application of the Voronoi diagram and Delaunay tessellation techniques is in analysis and synthesis of optimal connections of a set of nodes. For instance, the Euclidean minimum spanning tree, which connects the points such that the total length of all segments is minimized, is a subset of the Delaunay triangulation of the same points.

Example 3.30 *Given a decision diagram, we start the analysis from the root node of the tree and then pass consecutively all levels toward the termination nodes as shown in Figures 3.23(b), (f), (j), and (n). The corresponding Voronoi diagrams are shown in Figures 3.23(c), (g), (k), and (o). The Delaunay triangulations are presented in Figures 3.23(d), (h), (l), and (p).*

Some observations for the hypercube structure from Example 3.30:

- Each segment in the tree is represented by the unique distinct connection in the Delaunay structures.
- The Delaunay connections corresponding to the tree segments are kept when passing from one level to another.

3.8 Further reading

Decision tree design techniques are introduced in [1, 2, 8, 23] and in the textbooks [12, 14, 15, 21, 27]. In a number of books, various aspects of the decision trees and decision diagrams have been discussed, including manipulation and reduction. Techniques for the application of decision diagrams are introduced in many papers and textbooks (see references in Chapters 12,15,17,18).

The average path length in decision diagrams is a useful measure (the average number of edges traversed from the root node to a terminal node). Butler and Sasao [6] studied the average path length in decision diagrams for representing various multivalued functions. Nagayama and Sasao [56] proposed exact and heuristic algorithms for minimizing the average path length of so-called *heterogeneous* multivalued decision diagrams.

Copy properties of decision diagrams. Janković et al. [40] studied the reduction of size of multi-valued decision diagrams by copy properties (symmetry and shift).

Graph embedding techniques. Rings and linear arrays can be mapped into hypercubes (see, in particular, paper by Saad and Schultz [29]). For example,

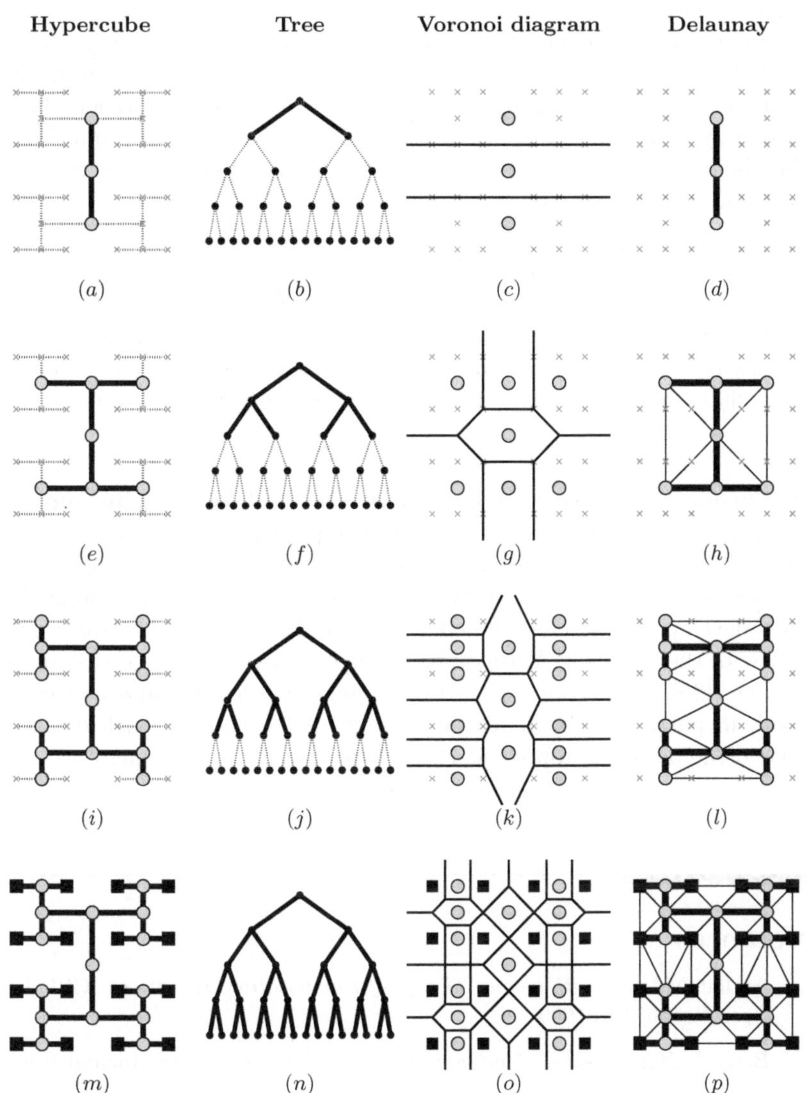

FIGURE 3.23
2D hypercube, corresponding tree, Voronoi diagram, and Delaunay triangulation (Example 3.30).

a linear array of a length k can be mapped into a hypercube of dimension $n = \log_2(l+1)$. Linear arrays and linear word-level decision diagrams have the same configuration. Meshes that represent cell-based computing use lattice decision diagrams as considered by Perkowski et al. [29]. Earlier, Sasao and Butler [31] developed a method for embedding decision diagrams into an FPGA topology. For this, so-called *pseudo Kronecker* decision diagrams were used. Various aspects of embedding decision trees into hypercubes are discussed by Cull and Larson [9], Shen et al. [35], Wagner [37], and Wu [38]. Embedding decision trees and DAGs into hypercube topologies was considered by Yanushkevich et al. [41].

FPGA topology was studied, in particular, by Lai and Wang [24] with respect to hierarchy of the interconnections. The problem of optimal topology in VLSI design was studied as mapping of particular algorithms to the computational structures (see, for example, results of Leiserson [25]) and embedding of tree-related network to hypercube structures (see paper by Öhring and Das [28]).

The Voronoi diagram and Delaunay triangulation technique. Theoretical bases of the Voronoi diagram in computational geometry have been developed by Dirichlet (1850) (as revised by Boywer [4]) and Voronoi (1908). The concept of Delauney triangulation has its origins with Voronoi, was further developed by Delone (spelled later as Delauney) [10]. An excellent overview of the theory of Voronoi diagrams is given by Fortune [13] and Okabe et al. [26]. Deployment of Voronoi diagrams for drawings of trees was studied, in particular, by Liotta and Meijer [20]. Yanushkevich et al. [42] discussed the design of topological structures using Voronoi diagrams.

References

[1] Atallah MJ. *Algorithms and Theory of Computation.* Handbook, CRC Press, Boca Raton, FL, 1999.

[2] Berg C. *Hypergraphs. Combinatorics of Finite Sets.* North-Holland, 1989.

[3] Bhuyan N and Agrawal DP. Generalized hypercube and hyperbus structures for a computer network. *IEEE Transactions on Computers*, 33(1):323–333, 1984.

[4] Bowyer A. Computing Dirichlet tessellations. *Computer Journal*, 24:162–166, 1981.

[5] Bryant RE. Graph-based algorithms for Boolean function manipulation. *IEEE Transactions on Computers*, 35(6):677–691, 1986.

[6] Butler JT, Dueck G, Shmerko VP, and Yanushkevich SN. Comments on SYMPATHY: fast exact minimization of fixed polarity Reed-Muller expansion for symmetric functions. *IEEE Transactions on Computer-Aided Design of Integrated Circuits and Systems*, 19(11):1386–1388, 2000.

[7] Butler JT and Sasao T. On the average path length in decision diagrams of multiple-valued functions. In *Proceedings of the IEEE 33rd International Symposium on Multiple-Valued Logic*, pp. 383–390, 2003.

[8] Cormen TH, Leiserson CE, Rivest RL, and Stein C. *Introduction to Algorithms.* MIT Press, 2001.

[9] Cull P and Larson S. The mobius cubes. In *Proceedings of the 6th Distributed Memory Computing Conference*, pp. 699–702, 1991.

[10] Delaunay BN. Sur la sphere vide. *Proceedings of the USSR Academy of Sciences, Natural Sciences*, 7:793–800, 1934.

[11] Dueck GW, Maslov D, Butler JT, Shmerko VP, and Yanuskevich SN. A method to find the best mixed polarity Reed-Muller expression using transeunt triangle. In *Proceedings of the 5th International Workshop on Applications of the Reed-Muller Expansion in Circuit Design.* Mississippi State University, pp. 82–92, 2001.

[12] Even S. *Graph Algorithms.* Computer Science Press, Rockville, Maryland, 1979.

[13] Fortune S. Voronoi diagrams and Delaunay triangulations. In Du DZ and Hwang F, Eds., *Computing in Euclidean Geometry*, pp. 193–233, World Scientific, 1992.

[14] Gibbons A. *Algorithmic Graph Theory.* Cambridge University Press, 1987.

[15] Goodaire EG and Parmenter MM. *Discrete Mathematics with Graph Theory.* Prentice-Hall, New York, 1998.

[16] Janković D, Stanković RS, and Drechsler R. Reduction of sizes of multi-valued decision diagrams by copy properties. In *Proceedings of the IEEE 34th International Symposium on Multiple-Valued Logic*, pp. 223–228, 2004.

[17] Kumar VKP and Tsai Y-C. Designing linear systolic arrays. *Journal on Parallel and Distributed Computing*, 7:441–463, 1989.

[18] Lai YT and Wang PT. Hierarchical interconnection structures for field programmable gate arrays. *IEEE Transactions on VLSI Systems*, 5(2):186–196, 1997.

[19] Leiserson CH. Fat-trees: universal networks for hardware-efficient supercomputing. *IEEE Transactions on Computers*, 34(10):892–901, 1985.

[20] Liotta G and Meijer H. Voronoi drawings of trees. *Computational Geometry*, 24(3):147–178, 2003.

[21] Meinel C and Theobald T. *Algorithms and Data Structures in VLSI Design.* Springer, Heidelberg, 1998.

[22] Mitchell T. Decision tree learning. In *Machine Learning*, pp. 52–78, McGraw-Hill, New York, 1997.

[23] Moret BME. Decision trees and diagrams. *Computing Surveys*, 14(4):593–623, 1982.

[24] Mukhopadhyay A and Schmitz G. Minimization of exclusive-OR and logical-equivalence switching circuits, *IEEE Transactions on Computers*, 19(2):132–140, 1970.

[25] Nagayama S and Sasao T. On the minimization of average path length for heterogeneous MDDs. In *Proceedings of the IEEE 34th International Symposium on Multiple-Valued Logic*, pp. 216–222, 2004.

[26] Okabe A, Boots B, and Sugihara K. *Spatial Tessellations. Concept and Applications of Voronoi Diagrams.* Wiley, New York, NY, 1992.

[27] Perkowski MA, Chrzanowska-Jeske M, and Xu Y. Lattice diagrams using Reed-Muller logic. In *Proceedings of the IFIP WG 10.5 International Workshop on Applications of the Reed-Muller Expansion in Circuit Design.* Japan, pp. 85–102, 1997.

[28] Öhring S and Das SK. Incomplete hypercubes: embeddings of tree-related networks. *Journal of Parallel and Distributed Computing*, 26:36–47, 1995.

[29] Saad Y and Schultz MH. Topological properties of hypercubes. *IEEE Transactions on Computers*, 37(7):867–872, 1988.

[30] Sasao T. *Switching Theory for Logic Synthesis*, Kluwer, Dordrecht, 1999.

[31] Sasao T and Butler J. A design method for look-up table type FPGA by pseudo-Kronecker expansion. In *Proceedings of the 23rd International Symposium on Multiple-Valued Logic.* pp. 97–106, 1994.

[32] Stanković RS. Simple theory of decision diagrams for representation of discrete functions. In *Proceedings of the 4th International Workshop on Applications of the Reed-Muller Expansion in Circuit Design.* University of Victoria, BC, Canada, pp. 161–178, 1999.

[33] Stanković RS. Some remarks on basic characteristics of decision diagrams. In *Proceedings of the 4th International Workshop on Applications of the Reed-Muller Expansion in Circuit Design.* University of Victoria, BC, Canada, pp. 139–148, 1999.

[34] Stanković RS. Unified view of decision diagrams for representation of discrete functions. *International Journal on Multi-Valued Logic and Soft Computing,* 8(2):237–283, 2002.

[35] Shen X, Hu Q, and Liang W. Embedding k-ary complete trees into hypercubes. *Journal of Parallel and Distributed Computing*, 24:100–106, 1995.

[36] Varadarajan R. Embedding shuffle networks into hypercubes. *Journal of Parallel and Distributed Computing*, 11:252–256, 1990.

[37] Wagner AS. Embedding the complete tree in hypercube. *Journal of Parallel and Distributed Computing*, 26:241–247, 1994.

[38] Wu AY. Embedding a tree network into hypercubes. *Journal of Parallel and Distributed Computing*, 2:238–249, 1985.

[39] Wu H, Perkowski MA, Zeng X, and Zhuang N. Generalized partially-mixed-polarity Reed-Muller expansion and its fast computation. *IEEE Transactions on Computers*, 45:1084–1088, Sep., 1996.

[40] Yanushkevich SN, Butler JT, Dueck GW, and Shmerko VP. *Experiments on FPRM expressions for partially symmetric logic functions.* In *Proceedings of the IEEE 30th International Symposium on Multiple-Valued Logic*, pp. 141–146, 2000.

[41] Yanushkevich SN, Shmerko VP. and Lyshevski SE. *Logic Design of NanoICs*, CRC Press, Boca Raton, FL, 2005.

[42] Yanushkevich SN, Stoica A, Shmerko VP, and Popel DV. *Biometric Inverse Problems*, Chapter "Biometric Data Structure Representation by Voronoi Diagrams", CRC Press/Taylor & Francis Group, Boca Raton, FL, 2005.

4

AND-EXOR Expressions, Trees, and Diagrams

Among the various forms of algebraic representations, AND-EXOR expressions play a special role in logic design. In this chapter, algebraic and graphical data structures using AND-EXOR representations are introduced. Classification of algebraic expansions, decision trees and diagrams are provided based on common criteria. In addition, an alternative graph data structure, transeunt triangles, is considered as a special structure for representation of a particular case of functions, symmetric functions.

4.1 Introduction

The term "AND-EXOR" specifies the algebraic structure in which AND products of variables are related by the EXOR operation. The AND-EXOR representation:

- ▶ Is the polynomial expansion of switching functions on a Boolean ring, or equivalently Boolean algebra, since the transition from one structure to another and vice versa is always possible. This structure meets the requirements of a Galois field of order 2, $GF(2)$.
- ▶ Meets the requirements of a Galois field of order 2, $GF(2)$.
- ▶ Is a universal basis that includes: the constant 1 and operations EXOR and AND over Boolean variables. An arbitrary AND product of literals combined by EXORs forms an exclusive-OR sum-of-product expression (ESOP).

The simplest class of EXORs is the Reed–Muller expressions with uncomplemented variables. Polarized Reed–Muller expressions (with positive or negative Davio expansion of variables) can include the NOT operation.

The AND-EXOR expression of switching functions can be interpreted as a counterpart of polynomial representations such as sum-of-products. On the other hand, they can be considered Fourier series-like expansions. The coefficients in AND-EXOR expressions for a given switching function are either:

- *Different order Boolean differences of the function, so that a Reed–Muller expression is an analog of Taylor series.*
- *Spectral coefficients resulting from a Fourier-like transform.*

These two interpretations lead to two different ways of manipulating Reed–Muller expressions and Fourier series-like expansions (see detail in Chapters 7 and 39).

The following theorems apply to EXOR: $x \oplus 0 = x, x \oplus 1 = \overline{x}$, $x \oplus x = 0$, $x \oplus \overline{x} = 1$, and also commutative, associative, and distributive laws. The *equivalence* operation (also called an exclusive-NOR operation) is defined as $x \equiv y = \overline{x \oplus y}$.

There are several features that distinguish AND-EXOR based techniques from AND-OR, or sum-of-products based techniques. In particular, for AND-EXOR expressions, Davio expansions (positive and negative) are used. These two types of expansions provide flexibility in decision tree and diagram design. For example, nodes can implement positive or negative Davio. There are various classes of ESOP expressions (fixed and mixed polarity expressions, pseudo Reed–Muller expressions, etc.).

4.2 Terminology and classification of AND-EXOR expressions

The polynomial expressions of switching functions using AND and EXOR operations are often called *Reed–Muller* expressions. Classification of AND-EXOR expressions is based on various criteria, in particular:

- Variables in the expression can be uncomplemented only,
- Each variable can appear in the expression in complemented or uncomplemented form, i.e., the polarity of each variable is *fixed* through the expression,
- The polarities of a variable can be different in various products of the expression, i.e., the polarity of each variable is *mixed* through the expression.

Based on these and other criteria, several classes of AND-EXOR expressions are defined. These classes of AND-EXOR expressions are characterized by particular properties. Designers of discrete devices can benefit from this classification in several ways, for example, by choosing AND-EXOR expressions with the minimal number of terms for a given switching function. Mapping of AND-EXOR expressions into graph data structure produces different types of decision diagrams. The name decision diagram corresponds to the name of AND-EXOR expression from a given class.

Various forms of AND-EXOR representations of switching functions are divided into the following classes (Figure 4.1):

- *Reed–Muller,**
- *Pseudo Reed–Muller* (PSRM),
- *Kronecker*
- *Pseudo Kronecker* (PSK),
- *Generalized Reed–Muller* (GRM), and
- *Exclusive OR sum-of-products* (ESOP).

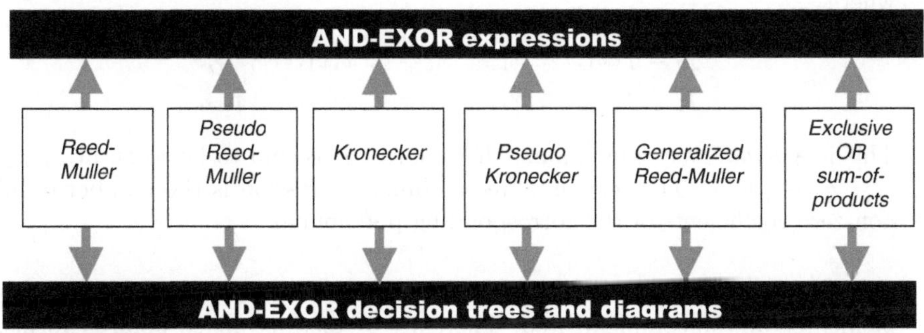

FIGURE 4.1
Classification of AND-EXOR expressions and corresponding decision trees and diagrams.

The rules for construction of various AND-EXOR expressions and corresponding decision trees are given in Section 4.4.

*The subclass of fixed polarity Reed–Muller expressions, positive polarity Reed–Muller expressions, is also called *Zhegalkin polynomials*.

Example 4.1 *Given the switching function* $f = \overline{x} \vee \overline{y}$, *the following four fixed polarity Reed–Muller forms (FPRM) can be generated:*

- $1 \oplus xy$,
- $1 \oplus x \oplus x\overline{y}$,
- $1 \oplus y \oplus \overline{x}y$, *and*
- $\overline{x} \oplus \overline{y} \oplus \overline{x}\ \overline{y}$,

Some examples of other AND-EXOR forms are given below:
(a) Pseudo Reed–Muller: $1 \oplus x \oplus x\overline{y}$,
(b) Kronecker: $\overline{x} \oplus x\overline{y}$,
(c) Pseudo Kronecker: $1 \oplus x \oplus \overline{y} \oplus \overline{x}y$,
(d) Generalized Reed–Muller: $x \oplus y \oplus \overline{x}\ \overline{y}$,
(e) ESOP: $\overline{x}\ \overline{y} \oplus \overline{x}y \oplus x\overline{y}$.

Let $P = (\rho_1, \ldots, \rho_n)$, where $\rho_1, \ldots, \rho_n \in \{0, 1, 2\}$ is the polarity of a mixed polarity AND-EXOR expression for the function f, such that x_i appears complemented if $\rho_i = 0$, x_i appears uncomplemented if $\rho_i = 1$ and x_i appears in both forms if $\rho_i = 2$. If $\rho_i = 2$, then f can be decomposed as

$$f = \overline{x}_i f_0 \oplus x_i f_1,$$

where

$$f_0 = f(x_1, \ldots, x_{i-1},\ x_i = 0,\ x_{i+1}, \ldots, x_n),$$
$$f_1 = f(x_1, \ldots, x_{i-1},\ x_i = 1,\ x_{i+1}, \ldots, x_n).$$

This is a *Kronecker* expression. The polarity P is said to be *fixed*, iff $\rho_1, \rho_2, \ldots, \rho_n \in \{0, 1\}$. The cost of a Reed–Muller expression is the number of all non-zero coefficients of the corresponding polynomial.

Mixed polarity. In AND-EXOR form, some variables can appear both complemented and uncomplemented. There are 3^n distinct mixed polarity forms. While the incomplemented variable is coded by 0 and the complemented variable is coded by 1, the mixed polarity of a variable x_i is coded by 2. So, such a polarity is coded by a polarity vector P consisting of 0,1 and 2s.

The *mixed polarity p vector* is a ternary n-tuple,

$$P = [p_1 p_2 \ldots p_n],$$

where

- $p_i = 0$, $i \in (1, n)$ means that variable x_i in the p-polarity AND-EXOR form appears uncomplemented,
- $p_i = 1$ means that variable x_i appears complemented, and
- $p_i = 2$ means that variable x_i appears both complemented and uncomplemented.

The 0-factor, p_0, of the $n \times 1$ mixed polarity vector P, whose elements take values 0, 1, 2, is an $n \times 1$ vector formed from P so that 2's are replaced by 0's, and 1-factor, p_1 is an $n \times 1$ vector formed from P so that 2's are replaced by 1's.

Example 4.2 *The mixed polarity vector* $P = [0122]$ *generates its 0-factor as* $p_0 = [0100]$ *and its 1-factor as* $p_1 = [0111]$.

Totally symmetric functions. A switching function f is *totally symmetric* if and only if it is unchanged by any permutation of variables.

Example 4.3 *The switching function*

$$f(x_1, x_2, x_3) = \overline{x}_1\overline{x}_2\overline{x}_3 \vee x_1x_2x_3$$

is symmetric, since permuting any variable labels leaves the function unchanged.

The symmetry of variable labels extends to symmetry of the assignments of values to variables. For all assignments of values to variables that have the same number of 1's, there is *exactly* one value of a symmetric function. Therefore, a symmetric function is completely specified by a *carrier vector* of logic values $A = [a_0, a_1, \ldots, a_n]$, such that $f(x_1, x_2, \ldots, x_n)$ is a_i for all assignments of values to $x_1, x_2, \ldots, x_n$ that have i 1's, where $0 \leq i \leq n$.

Example 4.4 *The symmetric function*

$$f = \overline{x}_1\overline{x}_2\overline{x}_3 + x_1x_2x_3$$

is specified by the carrier vector $[1, 0, 0, 1]$, *because the function is 1 if and only if zero or three of the variables are 1.*

Partially symmetric switching functions. A function f is *partially symmetric* in k variables $x_1, x_2, ..., x_k$, $k < n$, if and only if it is unchanged by any permutation of these k variables only.

Example 4.5 *The switching function*

$$f = x_2\overline{x}_3 + \overline{x}_2x_3 + x_1x_2x_3$$

is symmetric in variables x_2 *and* x_3.

4.3 Algebraic representation

In this section, the basics of manipulation techniques of AND-EXOR expansions are introduced. Some techniques are demonstrated for the simplest

AND-EXOR expressions with uncomplemented variables (positive polarity of variables), and also for fixed polarity of variables. These basic techniques of manipulation of AND-EXOR expressions can be applied to other forms.

Example 4.6 *Algebraic manipulation is based on the manipulation of the structural components of these expressions. Figure 4.2 illustrates the structure and the components of fixed and mixed polarity Reed–Muller expressions.*

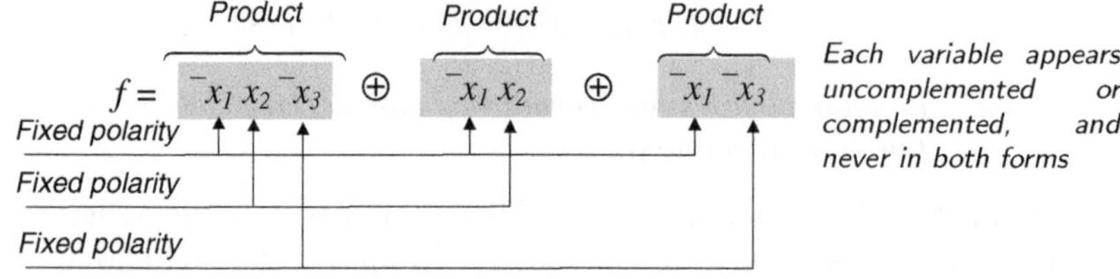

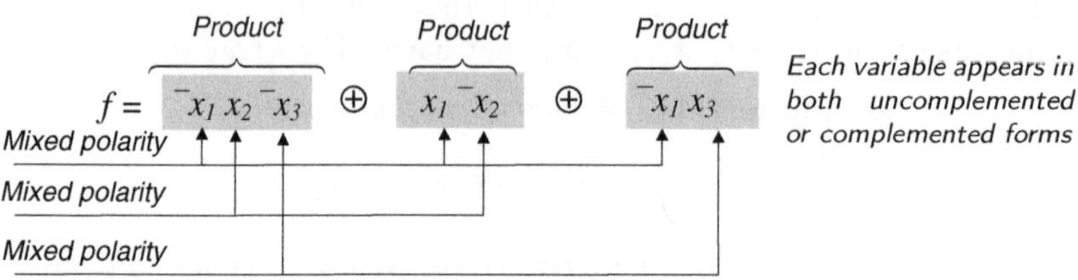

FIGURE 4.2
Structures of fixed and mixed Reed–Muller expression (Example 4.6).

4.3.1 Positive polarity Reed–Muller expressions

Positive polarity Reed–Muller expressions are a sub-class of the fixed polarity class. Given a switching function f of n variables, the Reed–Muller expression is specified by

$$f = \bigoplus_{i=0}^{2^n-1} r_i \cdot \underbrace{(x_1^{i_1} \cdots\ x_n^{i_n})}_{i-th\ product}, \tag{4.1}$$

where $r_i \in \{0, 1\}$ is a coefficient, i_j is the j-th bit, $j = 1, 2, \ldots, n$, in the binary representation of the index $i = i_1 i_2 \ldots i_n$, and $x_j^{i_j}$ is defined as

$$x_j^{i_j} = \begin{cases} 1, & \text{if } i_j = 0; \\ x_j, & \text{if } i_j = 1. \end{cases} \tag{4.2}$$

In other words, each variable appears uncomplemented or with 0 polarity.

Example 4.7 *An arbitrary switching function of two variables can be represented by the positive polarity Reed–Muller expression (Equation 4.1 and Equation 4.2):*

$$\begin{aligned} f &= r_0(x_1^0 x_2^0) \oplus r_1(x_1^0 x_2^1) \oplus r_2(x_1^1 x_2^0) \oplus r_3(x_1^1 x_2^1) \\ &= r_0 \oplus r_1 x_2 \oplus r_2 x_1 \oplus r_3 x_1 x_2. \end{aligned}$$

Figure 4.3 illustrates how to assemble the expression given the coefficients r_j, where the shadowed nodes implement the AND operation.

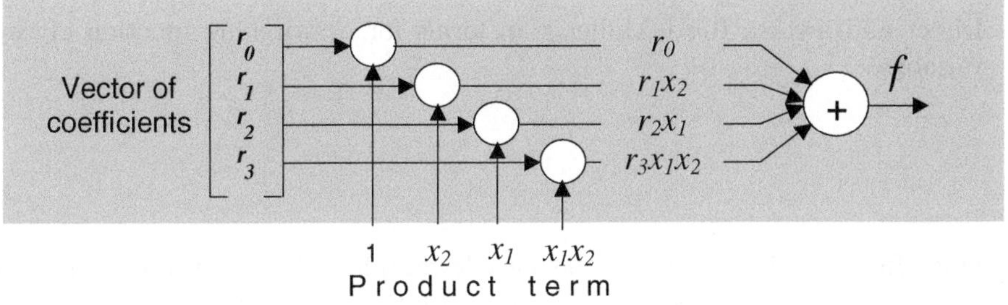

FIGURE 4.3
Deriving the Reed–Muller expression for a function of two variables (Example 4.7).

Computing the coefficients. Given a truth vector $\mathbf{F} = [f(0)\ f(1) \ldots f(2^n - 1)]^T$ of a switching function f, the vector of positive polarity Reed–Muller coefficients $\mathbf{R} = [r_0\ r_1 \ldots r_{2^n-1}]^T$ is derived by the matrix equation

$$\mathbf{R} = \mathbf{R}_{2^n} \cdot \mathbf{F} \quad (mod\ 2), \tag{4.3}$$

the $2^n \times 2^n$ matrix $\mathbf{R}_{2^n}$ is formed by the Kronecker product

$$\mathbf{R}_{2^n} = \bigotimes_{j=1}^{n} \mathbf{R}_{2^1}, \qquad \mathbf{R}_{2^1} = \begin{bmatrix} 1 & 0 \\ 1 & 1 \end{bmatrix}, \tag{4.4}$$

and *mod* 2 means that matrix-vector product is performed over Galois field $GF(2)$, i.e., using AND and EXOR operations. In Figure 4.4, a general computing scheme is shown for $n = 2$.

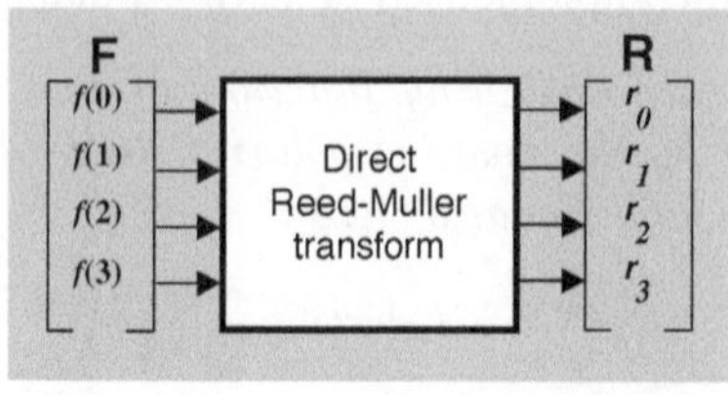

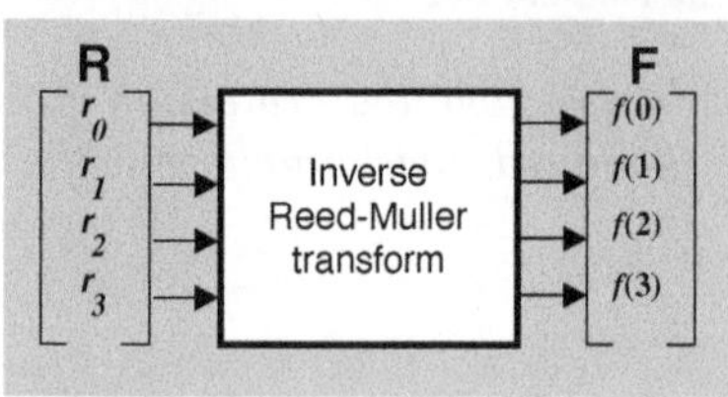

The pair of direct and inverse Reed–Muller transforms

$$\mathbf{R} = \mathbf{R}_{2^2} \cdot \mathbf{F} \quad (mod\ 2),$$
$$\mathbf{F} = \mathbf{R}_{2^2}^{-1} \cdot \mathbf{R} \quad (mod\ 2),$$

where

$$\mathbf{R}_{2^2} = \mathbf{R}_{2^2}^{-1},$$
$$\mathbf{R}_{2^2} = \bigotimes_{j=1}^{2} \mathbf{R}_{2^1},$$
$$\mathbf{R}_{2^1} = \begin{bmatrix} 1 & 0 \\ 1 & 1 \end{bmatrix}$$

FIGURE 4.4
Direct and inverse Reed–Muller transforms for a switching function of two variables.

Data flowgraphs. To design the data flowgraph of the algorithm, the matrix $\mathbf{R}_{2^n}$ must be represented in the factorized form

$$\mathbf{R}_{2^n} = \mathbf{R}_{2^n}^{(n)} \mathbf{R}_{2^n}^{(n-1)} \cdots \mathbf{R}_{2^n}^{(1)}, \tag{4.5}$$

where $\mathbf{R}_{2^n}^{(i)}$, $i = 1, 2, \ldots, n$, is formed by the Kronecker product. Substitution of Equation 4.5 into Equation 4.3 yields

$$\mathbf{R}_{2^n}^{(i)} = \mathbf{I}_{2^{n-i}} \otimes \mathbf{R}_{2^1} \otimes \mathbf{I}_{2^{i-1}} \tag{4.6}$$

and the fast Reed–Muller transform

$$\mathbf{R} = \mathbf{R}_{2^n}^{(n)} \mathbf{R}_{2^n}^{(n-1)} \cdots \mathbf{R}_{2^n}^{(1)} \mathbf{F},$$

in which the Reed–Muller coefficients can be computed in n iterations.

Example 4.8 *Computing the Reed–Muller coefficients by Equation 4.3 and Equation 4.7 for the function $f = x_1 \vee \overline{x}_2$ is illustrated in Figure 4.5. The data flowgraph includes two iterations accordingly to factorization relations (Equation 4.5 and Equation 4.6):*

$$\mathbf{R}_{2^2} = \mathbf{R}_{2^2}^{(2)}\mathbf{R}_{2^2}^{(1)} = (\mathbf{I}_{2^{2-2}} \otimes \mathbf{R}_{2^1} \otimes \mathbf{I}_{2^{2-1}})(\mathbf{I}_{2^{2-1}} \otimes \mathbf{R}_{2^1} \otimes \mathbf{I}_{2^{1-1}})$$

$$= \underbrace{(1 \otimes \mathbf{R}_{2^1} \otimes \mathbf{I}_{2^1})}_{1st\ iteration}\underbrace{(\mathbf{I}_{2^1} \otimes \mathbf{R}_{2^1} \otimes 1)}_{2nd\ iteration}$$

$$= \left[\begin{array}{c|c} \mathbf{I}_{2^1} & \\ \hline \mathbf{I}_{2^1} & \mathbf{I}_{2^1} \\ & \end{array}\right] \left[\begin{array}{c|c} \mathbf{R}_{2^1} & \\ \hline & \mathbf{R}_{2^1} \end{array}\right] = \left[\begin{array}{cc|cc} 1 & & & \\ & 1 & & \\ \hline 1 & & 1 & \\ & 1 & & 1 \end{array}\right] \left[\begin{array}{cc|cc} 1 & & & \\ 1 & 1 & & \\ \hline & & 1 & \\ & & 1 & 1 \end{array}\right].$$

This graph is known as a "butterfly-like" data flowgraph.

Restoration. The following matrix equation with AND and EXOR operations restores the truth-vector $\mathbf{F}$ from the vector of coefficients $\mathbf{R}$ (Figure 4.4):

$$\mathbf{F} = \mathbf{R}_{2^n}^{-1} \cdot \mathbf{R} \ \ (mod\ 2), \tag{4.7}$$

where $\mathbf{R}_{2^1}^{-1} = \mathbf{R}_{2^1}$. Notice that the matrix $\mathbf{R}_{2^1}$ is a self-inverse matrix over a Galois field $GF(2)$, i.e., in terms of logic operations AND and EXOR. The elements of $\mathbf{R}_{2^1}^{-1}$ are formed as follows:

$$r_{i,j} = i^j \quad (mod\ 2),$$

that is,

$$\mathbf{R}_{2^1}^{-1} = \begin{bmatrix} 1^0 & 0^1 \\ 1^0 & 1^1 \end{bmatrix} = \begin{bmatrix} 1 & 0 \\ 1 & 1 \end{bmatrix} \quad (mod\ 2).$$

Example 4.9 *Restore the truth-vector $\mathbf{F}$ of a switching function f given by the vector of Reed–Muller coefficients $\mathbf{R} = [1\ 1\ 0\ 1]^T$:*

$$\mathbf{F} = \mathbf{R}_{2^3}^{-1} \cdot \mathbf{R} = \left[\begin{array}{cc|cc} 1 & 0 & 0 & 0 \\ 1 & 1 & 0 & 0 \\ \hline 1 & 0 & 1 & 0 \\ 1 & 1 & 1 & 1 \end{array}\right] \begin{bmatrix} 1 \\ 1 \\ 0 \\ 1 \end{bmatrix} = \begin{bmatrix} 1 \\ 0 \\ 1 \\ 1 \end{bmatrix} \quad (mod\ 2).$$

Useful rules. The following rules are applicable to manipulating the positive polarity Reed–Muller expression:

x_1 x_2 f

$f = x_1 \vee \overline{x}_2$

Vector of coefficients

$$\mathbf{R}{=}\mathbf{R}_{2^2} \cdot \mathbf{F} = \left[\begin{array}{cc|cc} 1 & 0 & 0 & 0 \\ 1 & 1 & 0 & 0 \\ \hline 1 & 0 & 1 & 0 \\ 1 & 1 & 1 & 1 \end{array}\right] \begin{bmatrix} 1 \\ 0 \\ 1 \\ 1 \end{bmatrix} = \begin{bmatrix} 1 \\ 1 \\ 0 \\ 1 \end{bmatrix} \; (mod\ 2)$$

Reed–Muller spectrum computing using factorized matrices (Equation 4.5)

$$\mathbf{R} = \mathbf{R}_{2^2}^{(2)}\ \mathbf{R}_{2^2}^{(1)}\ \mathbf{F}$$

$$= \left[\begin{array}{cc|cc} 1 & 0 & 0 & 0 \\ 1 & 1 & 0 & 0 \\ \hline 0 & 0 & 1 & 0 \\ 0 & 0 & 1 & 1 \end{array}\right] \left[\begin{array}{cc|cc} 1 & 0 & 0 & 0 \\ 0 & 1 & 0 & 0 \\ \hline 1 & 0 & 1 & 0 \\ 0 & 1 & 0 & 1 \end{array}\right] \begin{bmatrix} 1 \\ 0 \\ 1 \\ 1 \end{bmatrix} = \begin{bmatrix} 1 \\ 1 \\ 0 \\ 1 \end{bmatrix}$$

Reed–Muller expression

$f = 1 \oplus x_2 \oplus x_1 x_2$

Data flowgraphs of the algorithm

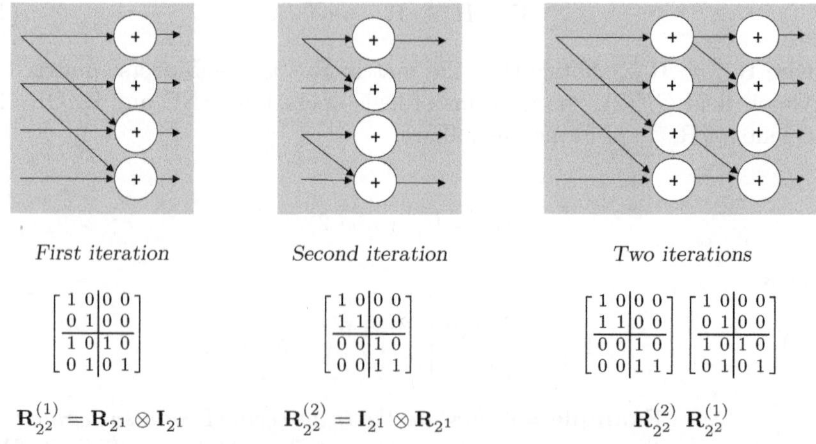

FIGURE 4.5
Computing the Reed–Muller expression for the two-input OR (Example 4.8).

Rule 1: To derive the positive polarity Reed–Muller expression from the canonical sum-of-products, the OR operation must be replaced with the EXOR operation. For example, $x_1\overline{x}_2 \vee \overline{x}_1 x_2 = x_1\overline{x}_2 \oplus \overline{x}_1 x_2$.

Rule 2: To derive the positive polarity Reed–Muller expression given a NAND form, the complement variable $\overline{x}_i$ must be replaced with $x_1 \oplus 1$, and the obtained expression must be simplified. For example,

$$
\begin{aligned}
x_1 \vee x_2 &= \overline{\overline{x_1 \vee x_2}} = \overline{\overline{x}_1\overline{x}_2} \\
&= (1 \oplus x_1)(1 \oplus x_2) \\
&= x_1 \oplus x_2 \oplus x_1 x_2.
\end{aligned}
$$

4.3.2 Fixed polarity

The product term in AND-EXOR expressions consists of complemented and uncomplemented literals (input variables). This is represented as follows: the polarity of a variable x_j can be either:

$c_j = 1$, corresponding to the uncomplemented variable x_j, or

$c_j = 0$, corresponding to the complemented variable $\overline{x}_j$.

Let the polarity of variables in an AND-EXOR expression be $c = c_1 c_2 \ldots c_n$, $c \in \{0, 1, 2, \ldots, 2^n - 1\}$, where c_j is the j-th bit in binary representation of c. For a switching function f of n variables, the Reed–Muller expression in a given polarity $c = c_1 c_2 \ldots c_n$ of variables $x_1 x_2 \ldots x_n$ is the following

$$
f = \bigoplus_{i=0}^{2^n-1} r_i \cdot \underbrace{(x_1 \oplus c_1)^{i_1} \cdots \ (x_n \oplus c_n)^{i_n}}_{i-th\ product}, \tag{4.8}
$$

where r_i is the i-th coefficient, and $(x_j \oplus c_j)^{i_j}$ is defined as

$$
a_j^{i_j} = \begin{cases} 1, \text{ if } i_j = 0; \\ a, \text{ if } i_j = 1. \end{cases} \quad x_j \oplus c_j = \begin{cases} x_j, \text{ if } c_j = 0; \\ \overline{x}_j, \text{ if } c_j = 1. \end{cases} \tag{4.9}
$$

In a fixed polarity Reed–Muller (FPRM) expansion of a given switching function f, every variable appears either complemented or uncomplemented; never in both forms. If all variables are uncomplemented (or complemented), the FPRM expansion is called a *positive (negative) polarity* Reed–Muller form (Figure 4.6). FPRM expansions are *unique.* Thus, only one representation exists for a given polarity c, that is polarities $c_1, c_2, \ldots, c_n$ of variables of f.

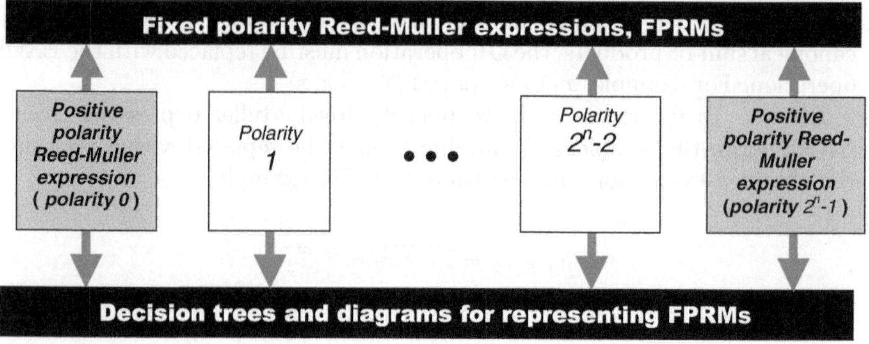

FIGURE 4.6
Positive and negative polarity Reed–Muller expressions are boundary cases of a class of fixed polarity Reed–Muller expressions.

Example 4.10 *In Example 4.7, a switching function of two variables has been represented by the Reed–Muller expression with only uncomplemented variables, that is, for the polarity* $c = 0$ $(c_1c_2 = 00)$. *This function can be re-written in the polarity* $c = 2$, $c_1c_2 = 10$ *using Equation 4.8 and Equation 4.9:*

$$\begin{aligned} f &= r_0(x_1 \oplus 1)^0(x_2 \oplus 0)^0 \oplus r_1(x_1 \oplus 1)^0(x_2 \oplus 0)^1 \\ &\oplus r_2(x_1 \oplus 1)^1(x_2 \oplus 0)^0 \oplus r_3(x_1 \oplus 1)^1(x_2 \oplus 0)^1 \\ &= r_0 \oplus r_1 x_2 \oplus r_2 \overline{x}_1 \oplus r_3 \overline{x}_1 x_2. \end{aligned}$$

Let $f = x \vee y$, *then four FPRM expressions can be derived as shown in Figure 4.7.*

FPRM expansions can be used to change a given switching function into an equivalence class of switching functions, where two switching functions are equivalent if one is transformed into the other by permuting variables, complementing variables, or complementing the switching function. This is useful for determining library cells in computer aided design (CAD) tools. This is called *Boolean matching,* and is important in the determination of library cells for use by computer-aided design tools.

Given the truth table $\mathbf{F} = [f(0)\ f(1) \ldots f(2^n-1)]^T$, the vector of Reed–Muller coefficients in the polarity c, $\mathbf{R^{(c)}} = [r_0^{(c)}\ r_1^{(c)} \ldots r_{2^n-1}^{(c)}]^T$ is derived by the matrix eqaution

$$\mathbf{R^{(c)}} = \mathbf{R}_{2^n}^{(c)} \cdot \mathbf{F} \quad (mod\ 2), \tag{4.10}$$

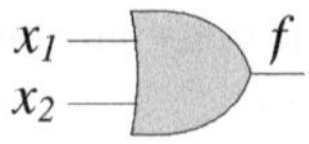

x_1, x_2 — f

$f = x_1 \vee x_2$

$0-polarity: \; f = x_1 \oplus x_2 \oplus x_1x_2$
$1-polarity: \; f = 1 \oplus \overline{x}_2 \oplus x_1\overline{x}_2$
$2-polarity: \; f = 1 \oplus \overline{x}_1 \oplus \overline{x}_1x_2$
$3-polarity: \; f = 1 \oplus \overline{x}_1\overline{x}_2$

FIGURE 4.7
Representation of a two-input OR gate by the Reed–Muller forms of $2^2 = 4$ polarities (Example 4.10).

where the $2^n \times 2^n$-matrix $\mathbf{R}_{2^n}^{(c)}$ is generated by the Kronecker product

$$\mathbf{R}_{2^n}^{(c)} = \bigotimes_{j=1}^{n} \mathbf{R}_{2^1}^{(c_j)}, \qquad \mathbf{R}_{2^1}^{(c)} = \begin{cases} \begin{bmatrix} 1 & 0 \\ 1 & 1 \end{bmatrix}, & c_j = 0; \\ \begin{bmatrix} 0 & 1 \\ 1 & 1 \end{bmatrix}, & c_j = 1. \end{cases} \tag{4.11}$$

Example 4.11 *In the matrix form, the solution to Example 4.8 can be derived by Equation 4.10 as follows:*

$$\mathbf{R}^{(2)} = \mathbf{R}_{2^2}^{(2)} \cdot \mathbf{F} = \begin{bmatrix} 0 & 0 & 0 & 1 \\ 0 & 0 & 1 & 1 \\ 0 & 1 & 0 & 1 \\ 1 & 1 & 1 & 1 \end{bmatrix} \begin{bmatrix} 1 \\ 0 \\ 1 \\ 1 \end{bmatrix} = \begin{bmatrix} 1 \\ 0 \\ 1 \\ 1 \end{bmatrix} \quad (mod\ 2)$$

where the matrix $\mathbf{R}_{2^2}^{(2)}$ given $c = 2$ is generated by Equation 4.11 as

$$\mathbf{R}_{2^2}^{(2)} = \mathbf{R}_{2^1}^{(1)} \otimes \mathbf{R}_{2^1}^{(0)} = \begin{bmatrix} 0 & 1 \\ 1 & 1 \end{bmatrix} \otimes \begin{bmatrix} 1 & 0 \\ 1 & 1 \end{bmatrix}.$$

The vector of coefficients $\mathbf{R}^{(2)} = [1\ 0\ 1\ 1]^T$ corresponds to the expression $f = 1 \oplus \overline{x}_1 \oplus \overline{x}_1x_2$.

Note that the inverse transform matrix given $n = 1$ is formed as follows:

$$r'_{i,j} = \overline{i}^{\,j} \quad (mod\ 2),$$

that is,

$$(\mathbf{R}_{2^1}^{(1)})^{-1} = \begin{bmatrix} \overline{1}^{\,0} & \overline{1}^{\,1} \\ \overline{1}^{\,0} & \overline{0}^{\,1} \end{bmatrix} = \begin{bmatrix} 1 & 1 \\ 1 & 0 \end{bmatrix} \quad (mod\ 2).$$

4.3.3 Linear Reed–Muller expressions

Linear Reed–Muller expressions can be represented in algebraic and matrix form.

Algebraic form. A *linear* switching function f of n variables can be represented by a linear positive polarity Reed–Muller expression

$$\begin{aligned} f &= r_0 \oplus \bigoplus_{i=1}^{n} r_i x_i \\ &= r_0 \oplus r_1 x_1 \oplus \cdots \oplus r_n x_n, \end{aligned} \tag{4.12}$$

where $r_i \in \{0, 1\}$ is i-th coefficient, $i = 1, 2, \ldots, n$.

Example 4.12 *The switching function:*

(a) $f = x \vee y$ *is non-linear because it is represented by the non-linear Reed–Muller expression* $f = x \oplus y \oplus xy$.
(b) $f = xy \vee \overline{x}\,\overline{y}$ *is linear because it is represented by the linear Reed–Muller expression* $f = 1 \oplus x \oplus y$.
(c) *There are eight linear functions* $f(x, y)$ *of two variables:* $f_0 = 0$, $f_1 = 1$, $f_2 = x$, $f_3 = 1 \oplus x$, $f_4 = y$, $f_5 = 1 \oplus y$, $f_6 = x \oplus y$, *and* $f_7 = 1 \oplus x \oplus y$.

The matrix form of Equation 4.12 can be derived as follows. Given a truth vector $\mathbf{F} = [f(0)\ f(1) \ldots f(2^n - 1)]^T$, of a linear switching function f, the $(n+1) \times 1$ truncated vector of Reed–Muller coefficients $\mathbf{R} = [r_0\ r_1 \ldots r_n]^T$ is derived by the matrix equation

$$\mathbf{R} = \mathbf{R}_L \mathbf{F} \quad (mod\ 2), \tag{4.13}$$

the $(n+1) \times 2^n$ matrix R_L is formed from $\mathbf{R}_{2^n}$ by taking its $2^0, 2^1, 2^2, \ldots, 2^{n-1}$ rows only. The truncated vector $\mathbf{R}$ contains only $n+1$ elements because only $n+1$ elements in the complete Reed–Muller coefficient vector are non-zero.

Given the truncated vector $\mathbf{R}$, the truth vector $\mathbf{F}$ can be be restored by

$$\mathbf{F} = \mathbf{R}_L^{-1} \mathbf{R} \quad (mod\ 2), \tag{4.14}$$

where $2^n \times (n+1)$ matrix $\mathbf{R}_L^{-1}$ is formed from $\mathbf{R}_{2^n}^{-1}$ by taking its $2^0, 2^1, 2^2, \ldots,$ 2^{n-1} columns only.

Example 4.13 *Figure 4.8 shows the matrices* $\mathbf{R}_L$ *and* $\mathbf{R}_L^{-1}$ *of direct and inverse Reed–Muller transform given a linear switching function of three variables.*

(a) *The function is given by the truth vector* $\mathbf{F} = [01101001]^T$ *and the vector of Reed–Muller coefficients* $\mathbf{R}$ *is calculated by direct Reed–Muller transform.*

(b) *Given the truncated vector of Reed–Muller coefficients* $\mathbf{R} = [0111]^T$, *that is* $x_1 \oplus x_2 \oplus x_3$, *the truth vector* $\mathbf{F}$ *is calculated by inverse transform.*

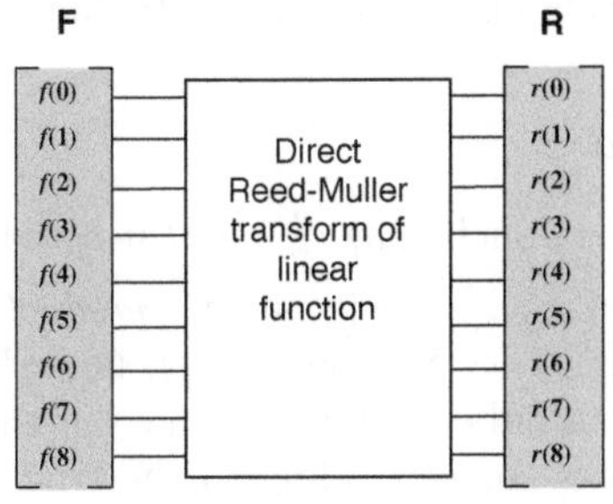

Direct transform of the linear switching function given by the truth vector

$$\mathbf{F} = [01101001]^T :$$

$$\mathbf{R} = \mathbf{R}_L\mathbf{F} = \left[\begin{array}{cccc|cccc} 1&0&0&0&0&0&0&0\\ 1&1&0&0&0&0&0&0\\ \hline 1&0&1&0&0&0&0&0\\ 1&0&0&0&1&0&0&0 \end{array}\right] \begin{bmatrix} 0\\1\\1\\0\\1\\0\\0\\1 \end{bmatrix} = \begin{bmatrix} 0\\1\\1\\1 \end{bmatrix}$$

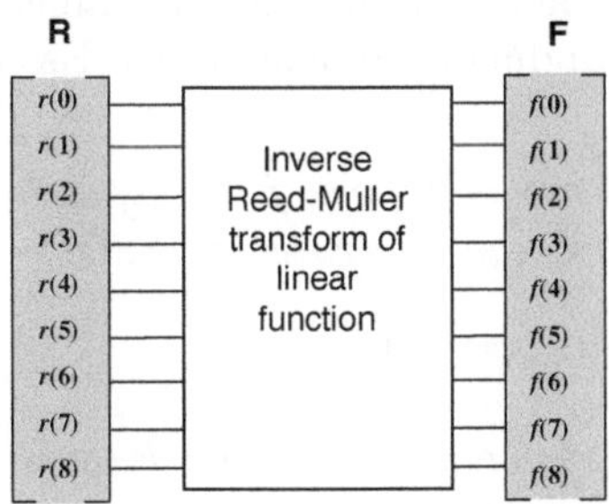

Inverse transform of the linear switching function given by the vector

$$\mathbf{F} = [0111]^T :$$

$$\mathbf{F} = \mathbf{R}_L^{-1}\mathbf{R} = \begin{bmatrix} 1&0&0&0\\ 1&1&0&0\\ 1&0&1&0\\ 1&1&1&0\\ 1&0&0&1\\ 1&1&0&1\\ 1&0&1&1\\ 1&1&1&1 \end{bmatrix} \begin{bmatrix} 0\\1\\1\\1 \end{bmatrix} = \begin{bmatrix} 0\\1\\1\\0\\1\\0\\0\\1 \end{bmatrix}$$

FIGURE 4.8
Direct and inverse transforms for a linear switching function of three variables (Example 4.13).

Linear Reed–Muller expressions have several useful properties, in particular:

- ► Expression 4.13 is a *parity* function if $r_1 = r_2 = \cdots = r_n = 1$.
- ► There are 2^{n+1} linear Reed–Muller expressions of n variables.
- ► A linear Reed–Muller expression is either a self-dual or a self-antidual switching function:

$$f = \begin{cases} f(\overline{x}_1, \overline{x}_2, \ldots, \overline{x}_n) = \overline{f}(x_1, x_2, \ldots, x_n), & \text{if } \sum_{i=0}^{n+1} r_i = 1; \\ f(\overline{x}_1, \overline{x}_2, \ldots, \overline{x}_n) = f(x_1, x_2, \ldots, x_n), & \text{if } \sum_{i=0}^{n+1} r_i = 0. \end{cases}$$

- ► The switching function obtained by the linear composition of linear Reed–Muller expressions is also a linear Reed–Muller expression.

4.4 Graphical representation

The polynomial expressions for a switching function, such as sum-of-products and AND-EXOR forms, can be expressed in terms of singular hypercubes.

Cubes, or n-tuples, correspond to the terms in polynomial expressions.

4.4.1 Hypercube representation

Let $x_j^{i_j}$ be a literal of a Boolean variable x_j such that $x_j^{i_j} = \overline{x}_j$ if $i_j = 0$, and $x_j^{i_j} = x_j$ if $i_j = 1$. A product of literals $x_1^{i_1} x_2^{i_2} \ldots x_n^{i_n}$ is called a *product term.* If the variable x_j is not present in a cube, $i_j = \mathbf{x}$ (don't care), i.e., $x_j^{\mathbf{x}} = 1$. In cube notation, a term is described by a cube that is a ternary n-tuple of components $i_j \in \{0,\ 1,\ \mathbf{x}\}$. A set of cubes corresponding to the true values of a switching function f represents the sum-of-products for this function.

A Reed–Muller expression consists of products combined by an EXOR operation.

Example 4.14 *A sum-of-products form* $f = \overline{x}_3 \vee x_1\overline{x}_2$ *given by the cubes* $[\mathbf{x}\ \mathbf{x}\ 0] \vee [1\ 0\ \mathbf{x}]$ *can be written as ESOP*

$$[\mathbf{x}\ \mathbf{x}\ 0] \oplus [1\ 0\ 1],$$

that is $\overline{x}_3 \oplus x_1\overline{x}_2x_3$. *The different cubes arise because of the different operation between the products in the expressions.*

Thus, the manipulation of the cubes involves OR, AND, and EXOR operations, applied to the appropriate literals, following the rules given in Table 39.2.

Example 4.15 *Given the cubes* $[1\ 1\ \mathbf{x}]$ *and* $[1\ 0\ \mathbf{x}]$, *the AND, OR, and EXOR operations on these cubes are shown in Figure 18.6.*

TABLE 4.1
AND, OR and EXOR operation on literals of cubes.

AND				*OR*				*EXOR*			
C_i / C_j	0	1	x	C_i / C_j	0	1	x	C_i / C_j	0	1	x
0	0	∅	0	0	0	x	x	0	0	x	1
1	∅	1	1	1	x	1	1	1	x	1	0
x	0	1	x	x	x	1	x	x	1	0	x

Suppose a sum-of-products expression for a switching function f is given by its cubes. To represent this function in Reed–Muller form, we have to generate cubes based on the equation $x \vee y = x \oplus y \oplus xy$ that can be written in cube notation as

$$[C_1] \vee [C_2] = [C_1] \oplus [C_2] \oplus [C_1][C_2]. \tag{4.15}$$

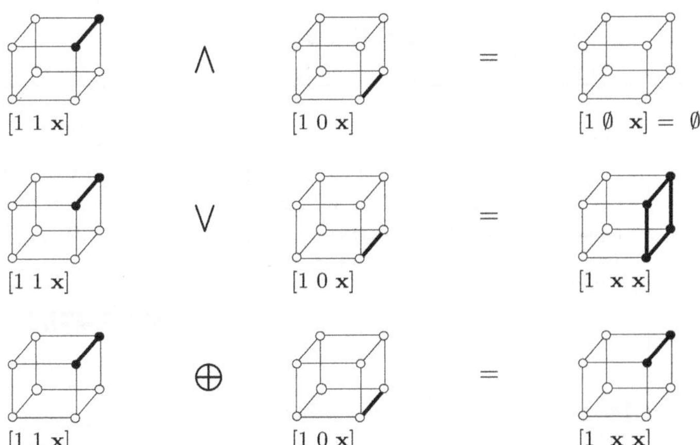

FIGURE 4.9
AND, OR, and EXOR operations on the cubes (Example 18.18).

Example 4.16 *Consider a switching function given in sum-of-products form by four cubes,*

$$f = [\mathbf{x}\ 1\ 0\ 1] \vee [1\ 0\ 0\ \mathbf{x}] \vee [0\ \mathbf{x}\ \mathbf{x}\ 0] \vee [\mathbf{x}\ \mathbf{x}\ 1\ 0].$$

To find its ESOP expression, OR must be replaced by EXOR operation and AND must be computed for each cube distinguished in only one literal (the rules for AND are given in Table 39.2). We then obtain the cube representation

$$f = [\mathbf{x}\ 1\ 0\ 1] \oplus [1\ 0\ 0\ \mathbf{x}] \oplus [0\ \mathbf{x}\ \mathbf{x}\ 0] \oplus [\mathbf{x}\ \mathbf{x}\ 1\ 0] \oplus [0\ \mathbf{x}\ 1\ 0]$$

that is

$$f = x_2\overline{x}_3x_4 \oplus x_1\overline{x}_2\overline{x}_3 \oplus \overline{x}_1\overline{x}_4 \oplus x_3\overline{x}_4 \oplus \overline{x}_1x_3\overline{x}_4.$$

Note that ESOP is a *mixed* polarity form where a variable appears in both complemented and uncomplemented form.

4.4.2 Decision trees using AND-EXOR operations

Types of decision trees using AND-EXOR operations are given in Figure 4.1 and Table 4.2. Recall that a Shannon decision tree is constructed by applying the Shannon expansion recursively to a switching function. This tree corresponds to the sum-of-products expression.

Each node in a Davio tree of a switching function f corresponds to the Davio decomposition of the function with respect to each variable x_i. There

TABLE 4.2
Classification of decision trees using the EXOR operation.

Class	Construction	Example
Shannon trees	The Shannon expansion is recursively applied to a switching function. This tree represents sum-of-products expression.	$\overline{x} \vee \overline{y} = \overline{x}\,\overline{y} \vee \overline{x}y \vee x\overline{y}$ 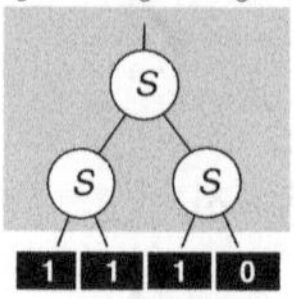 S; S, S; 1 1 1 0
Positive (negative) Davio trees	The positive or negative Davio expansion is recursively applied to a switching function. This tree represents the Reed–Muller expression of a positive polarity.	$1 \oplus xy$ 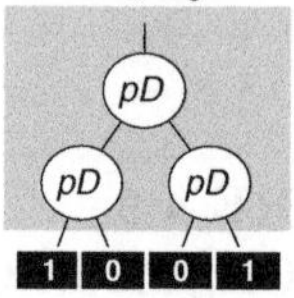 pD; pD, pD; 1 0 0 1
Reed–Muller trees (RM)	The positive and negative Davio expansions are recursively applied to a switching function. This tree represents the FPRM. Each level contains only positive Davio pD or negative Davio nD nodes.	$1 \oplus y \oplus \overline{x}y$ 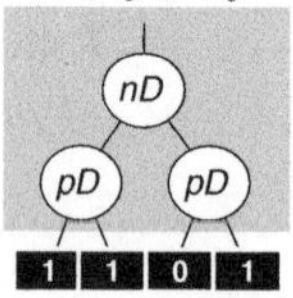 nD; pD, pD; 1 1 0 1
Pseudo Reed–Muller trees (PSRM)	The positive and negative Davio expansion is recursively applied to a switching function. Each level can contain both positive Davio pD and negative Davio nD nodes.	$1 \oplus x \oplus x\overline{y}$ 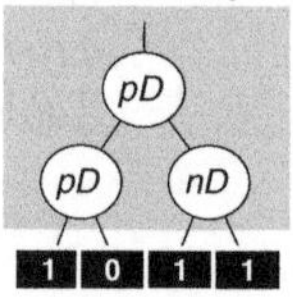 pD; pD, nD; 1 0 1 1
Kronecker trees (KRO)	Positive Davio, negative Davio, or the Shannon expansion is recursively applied to a switching function. This tree represents a Kronecker expression. Each level contains only one type of nodes.	$\overline{x} \oplus x\overline{y}$ 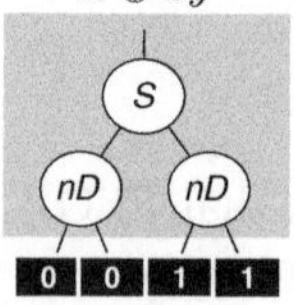 S; nD, nD; 0 0 1 1
Pseudo Kronecker trees (PSKRO)	The Shannon, and the positive and negative Davio expansions can be applied for each variable. Each level can contain any type of node.	$1 \oplus x \oplus \overline{y} \oplus \overline{x}y$ 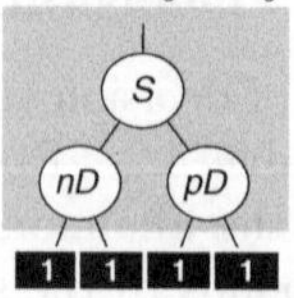 S; nD, pD; 1 1 1 1

exists: the *positive Davio* expansion

$$f = f_0 \oplus x_i f_2, \tag{4.16}$$

and the *negative Davio* expansion

$$f = \overline{x}_i f_2 \oplus f_1, \tag{4.17}$$

where $f_0 = f|_{x_i=0}$, $f_1 = f|_{x_i=1}$, and $f_2 = f|_{x_i=1} \oplus f|_{x_i=0}$.

The positive and negative Davio decompositions are labeled as pD and nD respectively (Figure 18.8). A binary decision tree that corresponds to the fixed positive polarity Reed–Muller canonical representation of a switching function is called a *positive Davio tree.* A Davio diagram can be derived from a Davio tree.

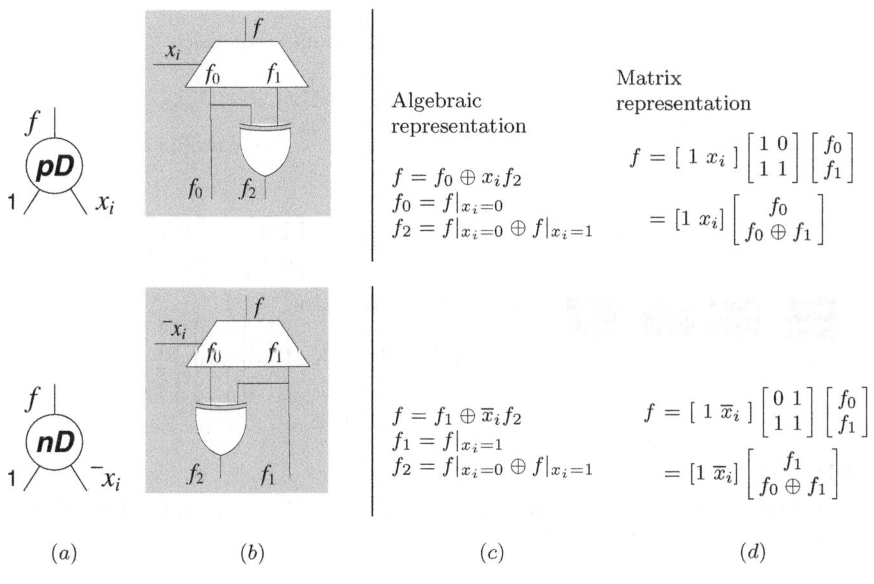

FIGURE 4.10
The node of a Davio tree (a), MUX-based implementation (b), algebraic (c) and matrix (d) descriptions.

In matrix notation, the switching function f of the Davio node is a function of a single variable x_i given by the truth vector $\mathbf{F} = [\ f(0)\ f(1)\ ]^T$ and is defined as

$$\begin{aligned} f &= [\,\overline{x}_i \; x_i\,] \begin{bmatrix} 1 & 0 \\ 1 & 1 \end{bmatrix} \begin{bmatrix} f_0 \\ f_1 \end{bmatrix} = [\,\overline{x}_i \; x_i\,] \begin{bmatrix} f_0 \\ f_1 \end{bmatrix} \\ &= \overline{x}_i f_0 \oplus x_i f_1 = (1 \oplus x_i) f_0 \oplus x_i f_1 = f_0 \oplus x_i f_2 \\ &= [1 \; x_i] \begin{bmatrix} f_0 \\ f_1 \end{bmatrix}. \end{aligned}$$

Recursive application of the positive Davio expansion to the function f given by the truth-vector $\mathbf{F} = [f(0)\; f(1) \ldots f(2^n - 1)]^T$ can be expressed in matrix notation as follows:

$$f = \widehat{\mathbf{X}}\, \mathbf{R}_{2^n}\, \mathbf{F}, \tag{4.18}$$

where

$$\widehat{\mathbf{X}} = \bigotimes_{i=1}^{n} [\,1 \; x_i\,], \quad \mathbf{R}_{2^n} = \bigotimes_{i=1}^{n} \mathbf{R}_2, \quad \mathbf{R}_2 = \begin{bmatrix} 1 & 0 \\ 1 & 1 \end{bmatrix},$$

and $\otimes$ denotes the Kronecker product.

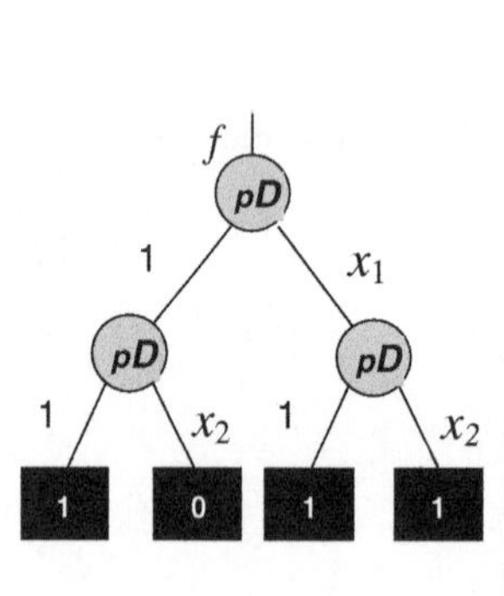

Structure of Reed–Muller expression

$$\begin{aligned} \widehat{\mathbf{X}} &= [\,1 \; x_1\,] \otimes [\,1 \; x_2\,] \\ &= [\,1,\; x_2,\; x_1,\; x_1 x_2\,] \end{aligned}$$

Transform matrix

$$\mathbf{R}_{2^2} = \mathbf{R}_2 \otimes \mathbf{R}_2 = \begin{bmatrix} 1 & 0 \\ 0 & 1 \end{bmatrix} \otimes \begin{bmatrix} 1 & 0 \\ 0 & 1 \end{bmatrix} = \left[\begin{array}{cc|cc} 1 & & & \\ 1 & 1 & & \\ \hline 1 & & 1 & \\ 1 & 1 & 1 & 1 \end{array}\right]$$

Reed–Muller expression

$$\begin{aligned} f &= \widehat{\mathbf{X}}\, \mathbf{R}_{2^2}\, \mathbf{F} = \widehat{\mathbf{X}} \left[\begin{array}{cc|cc} 1 & & & \\ 1 & 1 & & \\ \hline 1 & & 1 & \\ 1 & 1 & 1 & 1 \end{array}\right] \begin{bmatrix} 1 \\ 1 \\ 0 \\ 1 \end{bmatrix} = \widehat{\mathbf{X}} \begin{bmatrix} 1 \\ 0 \\ 1 \\ 1 \end{bmatrix} \\ &= 1 \oplus x_1 \oplus x_1 x_2 \end{aligned}$$

FIGURE 4.11
The Davio tree for the switching function $f = \overline{x}_1 \vee x_2$ and calculations in matrix form (Example 4.17).

Example 4.17 *Let us derive the positive Davio tree given the switching function $f = \overline{x}_1 \vee x_2$. Its truth-vector is $\mathbf{F} = [1\;1\;0\;1]^T$. Equation 18.4 is applied to get the solution shown in Figure 5.7. The coefficient vector of the positive polarity Reed–Muller expression is directly mapped into terminal nodes of the complete positive Davio tree.*

Structural properties. The most important structural properties of the positive Davio tree are described below:

- The nodes of the decision tree are associated with Davio expansion.
- An n-variable switching function is represented by an $(n+1)$-level Davio tree.
- The i-th level, $i = 1, \ldots, n$, includes 2^{i-1} nodes.
- Nodes at the lowest, $n+1$ level are connected to 2^n terminal nodes, which take values 0 or 1.
- The nodes corresponding to the i-th variable form the i-th level in the Davio tree; the root corresponds to the first chosen variable.
- In every path from the root node to a terminal node, the variables appear in a fixed order; it is said that this tree is ordered.
- The values of constant nodes are the values of the positive polarity Reed–Muller expression for the represented function. Thus, they are elements of the Reed–Muller coefficient vector $\mathbf{R} = [f_{000}\ f_{002}\ f_{020}\ f_{022}\ f_{200}\ f_{202}\ f_{220}\ f_{222}]$, where "0" corresponds to the value $f_0 = f|_{x_i=0}$, and "2" corresponds to the value $f_2 = f|_{x_i=1} \oplus f|_{x_i=0}$.

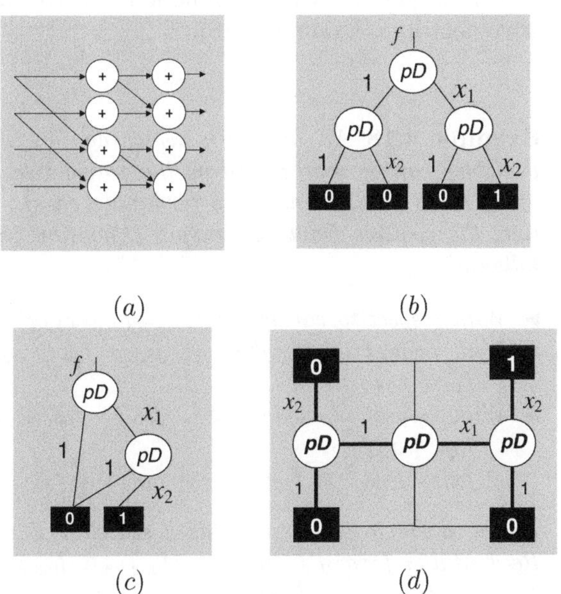

FIGURE 4.12
Graphical representation of the AND function of two variables: data flowgraph (a), decision tree (b), decision diagram (c), and decision tree embedded in the hypercube (d).

Figure 4.12 illustrates the relationship between the sum-of-products and graphical representations by the example of an AND switching function. The complete decision tree is reduced to a decision diagram. Finally, the decision tree or diagram is embedded in a $\mathcal{N}$-hypercube (see details in Chapter 34).

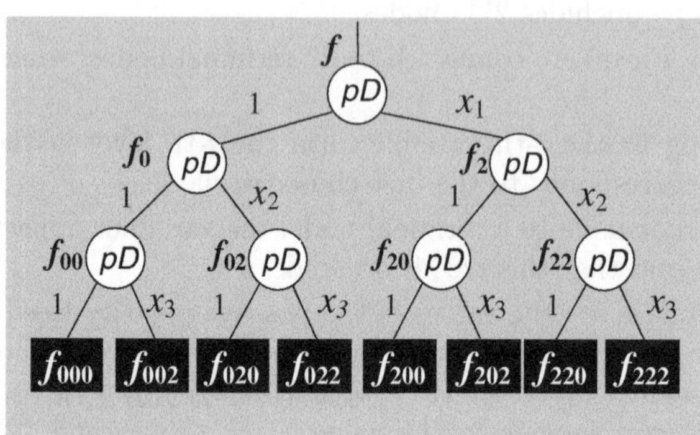

There are 8 paths from f to the terminal nodes:

Path 1: $t_1 = 1$
Path 2: $t_2 = x_3$
Path 3: $t_3 = x_2$
Path 4: $t_4 = x_2x_3$
Path 5: $t_5 = x_1$
Path 6: $t_6 = x_1x_3$
Path 7: $t_7 = x_1x_2$
Path 8: $t_8 = x_1x_2x_3$

Reed–Muller expression
$f = f_{000}t_1 \oplus f_{002}t_2 \oplus \ldots \oplus f_{222}t_8$

FIGURE 4.13
Reed–Muller representation of a switching function of three variables by the Davio tree (positive polarity) (Example 4.18).

Example 4.18 *An arbitrary switching function f of three variables can be represented by the Davio tree shown in Figure 4.13 (3 levels, 7 nodes, 8 terminal nodes). To design this tree, the positive Davio expansion (Equation 18.2) is used as follows:*

- ► *With respect to variable x_1: $f = f_0 \oplus x_1 f_2$;*
- ► *With respect to variable x_2: $f_0 = f_{00} \oplus x_2 f_{02}$ and $f_1 = f_{10} \oplus x_2 f_{22}$; and*
- ► *With respect to variable x_3: $f_{00} = f_{000} \oplus x_3 f_{002}$, $f_{02} = f_{020} \oplus x_3 f_{022}$,*
  $f_{20} = f_{200} \oplus x_3 f_{202}$, $f_{22} = f_{220} \oplus x_3 f_{222}$.

Hence, a Davio tree represents a switching function f in the Reed–Muller form $f = f_{000} = \oplus f_{002}x_3 \oplus f_{020}x_2 \oplus f_{022}x_2x_3 \oplus f_{200}x_1 \oplus f_{202}x_1x_3 \oplus f_{220}x_1x_2 \oplus f_{222}x_1x_2x_3$.

4.4.3 Decision diagrams

In decision trees and diagrams representing AND-EXOR expressions, Davio decomposition in nodes is used, thus, they are called *Davio* trees and diagrams.

> *The Davio diagram is derived from the Davio tree by deleting redundant nodes, and by sharing equivalent subgraphs.*

The rules below produce the reduced Davio diagram (Figure 4.14):

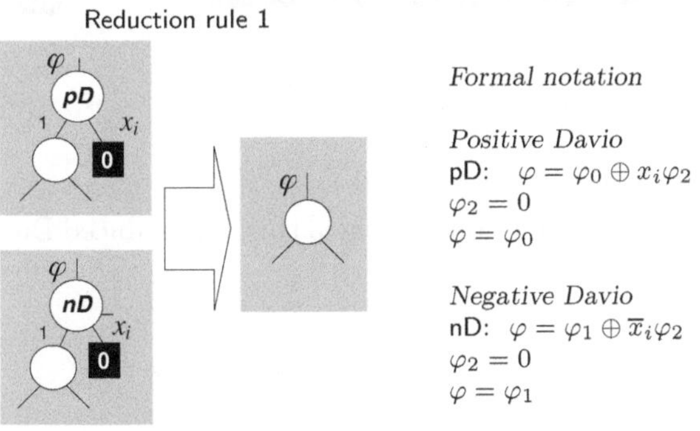

Reduction rule 2

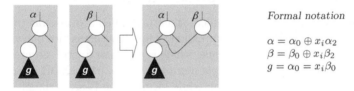

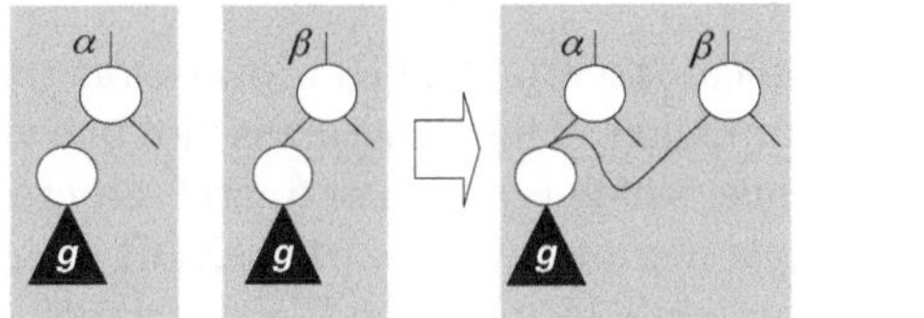

Formal notation

$\alpha = \alpha_0 \oplus x_i \alpha_2$
$\beta = \beta_0 \oplus x_i \beta_2$
$g = \alpha_0 = x_i \beta_0$

FIGURE 4.14
Reduction of a Davio tree.

Elimination rule: if the outgoing edge of a node labeled with x_i and $\overline{x}_i$ points to the constant zero, then delete the node and connect the edge to the other subgraph directly.

Merging rule: share equivalent subgraphs.

In a tree, edges longer than one, i.e., connecting nodes at nonsuccessive levels, can appear. For example, the length of an edge connecting a node at $(i-1)$-th level with a node at $(i+1)$-th level is two.

Example 4.19 *Application of the reduction rules to the three-variable NAND function is demonstrated in Figure 18.9.*

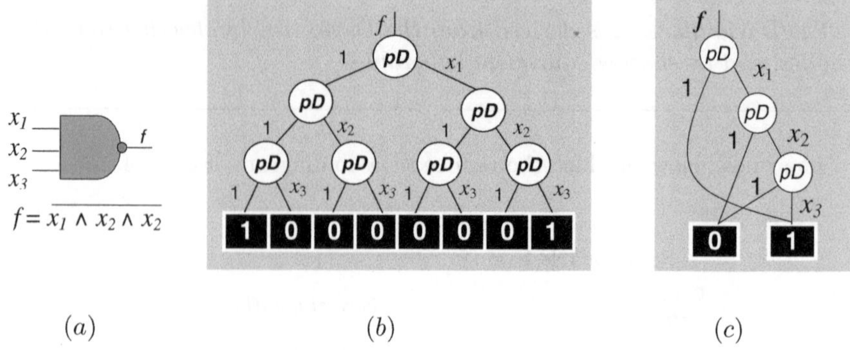

(a) (b) (c)

FIGURE 4.15
The three-variable NAND function, its Davio tree, and reduced Davio decision diagram (Example 18.21).

4.5 Transeunt triangle representation

In this section, we introduce a graphical structure called a *transeunt triangle* for representation of AND-EXOR expression. This structure is also known as a *Sierpinski gasket*, or *Sierpinski triangle*, which is a kind of fractal structure. A Sierpinski gasket is the *Pascal triangle* modulo two, i.e., the EXOR operation is used instead of arithmetic addition while forming the Pascal triangle.

> *The transeunt triangle for a switching function $f(x_1, x_2, \ldots, x_n)$ is a triangle of 0's and 1's where the bottom row is the truth vector of f and the other elements form various Kronecker, or Reed–Muller forms.*

4.6 Triangle topology

Transeunt triangle topology!indexTopology is useful in AND-EXOR expression synthesis and manipulation. A transeunt triangle is a fractal structure, also called a *Sierpinski gasket.* A Sierpinski gasket is a Pascal triangle modulo two, i.e., the EXOR operation is used instead of arithmetic addition.

The transeunt triangle for a switching function $f(x_1, x_2, ..., x_n)$ is a triangle formed by modulo 2 addition of 0's and 1's starting from the bottom row, which is the truth vector of f. A transeunt triangle for an n-variable function has a width of 2^n and a height of 2^n.

There are $(4^n + 2^n)/2$ elements in total. Besides the defining relation, there are other relations among elements in the transeunt triangle. Since the truth vector for functions with one or more variables has an even number of entries, it can be divided evenly into two parts. Each part produces, on its own, two sub-triangles.

Example 4.20 *Let T be a transeunt triangle for a switching function $f(x_1, x_2, ..., x_n)$. Representation of T by 3 triangles T_0, T_1 and T_2 is shown in Figure 4.21.*

Non-bottom elements of T_2 by the construction of T are corresponding sums. So, T_2 is the transeunt triangle that represents the switching function $f(0, x_2, ..., x_n) \oplus f(1, x_2, ..., x_n)$.

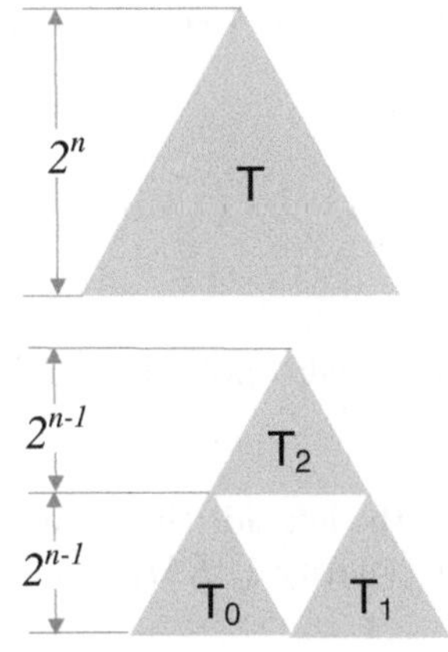

- *T_0 is the transeunt triangle for the function*

  $$f(0, x_2, ..., x_n)$$

- *T_1 is the transeunt triangle for the function*

  $$f(1, x_2, ..., x_n)$$

- *T_2 is the transeunt triangle for the function*

  $$f(0, x_2, ..., x_n) \oplus f(1, x_2, ..., x_n)$$

The bottom of T_2 is the "exclusive or" of the elements of the bottom row of transeunt triangles for functions

$$f(0, x_2, ..., x_n) \text{ and } f(1, x_2, ..., x_n),$$

FIGURE 4.16
Representation of transeunt triangle T by combination of three sub-triangles T_0, T_1, and T_2 (Example 4.27).

Example 4.21 *Switching function f of three variables x_1, x_2, x_3 given by a truth vector $\mathbf{F} = [00101011]^T$. This function can be represented by decision tree as shown in Figure 4.17 (details of the representation of switching functions by Reed–Muller expressions and corresponding decision trees are discussed in Chapter 4). A transeunt triangle can be constructed directly from a truth vector or by embedding a decision tree into this triangle.*

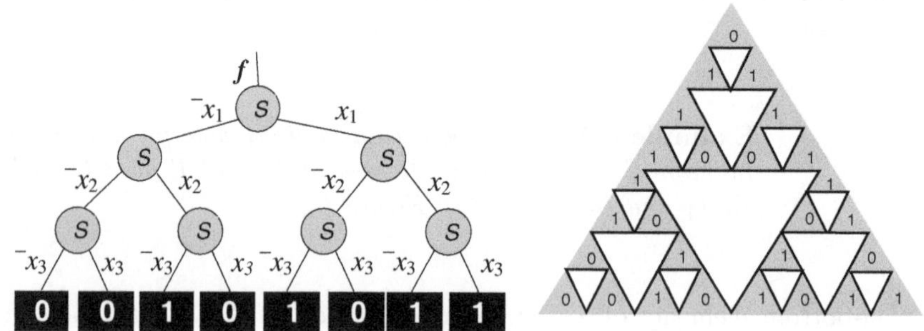

FIGURE 4.17
Relationship between a switching function $f(x_1, x_2, x_3)$ given a truth vector $\mathbf{F} = [00101011]^T$, structure of transeunt triangle, and decision tree (Example 4.21).

In general,

- The bits along triangle's left side are coefficients of zero polarity Reed–Muller form RM_0,
- The bits along the right side of each subtriangle on the left side the triangle are the coefficients of Reed–Muller form of the polarity 1, RM_1,
- The bits along left or right side of each following row of subtriangles such that the bottom element is jth element of the truth vector $\mathbf{F}$ are the coefficients of Reed–Muller forms of polarity j, RM_j, $j = 1, 2, \ldots, 2^n - 2$.
- The bits along the left side of the triangle are the coefficients of Reed–Muller form of polarity $2^n - 1$, RM_{2^n-1}.

Example 4.22 *Given the switching function* $f = x_1\overline{x}_2$ *and its truth vector* $[0010]^T$*, its complete decision tree and Sierpinski, or transeunt triangle are shown in Figure 4.18a and b, respectively. The location of coefficients of the fixed Reed–Muller expressions of polarities 0,1,2, and 3,*

$$RM(0) = [0011]^T, \quad RM(1) = [0001]^T,$$
$$RM(2) = [1111]^T, \quad RM(3) = [0101]^T,$$

is pictured in Figure 4.18c,d,e and f, respectively.

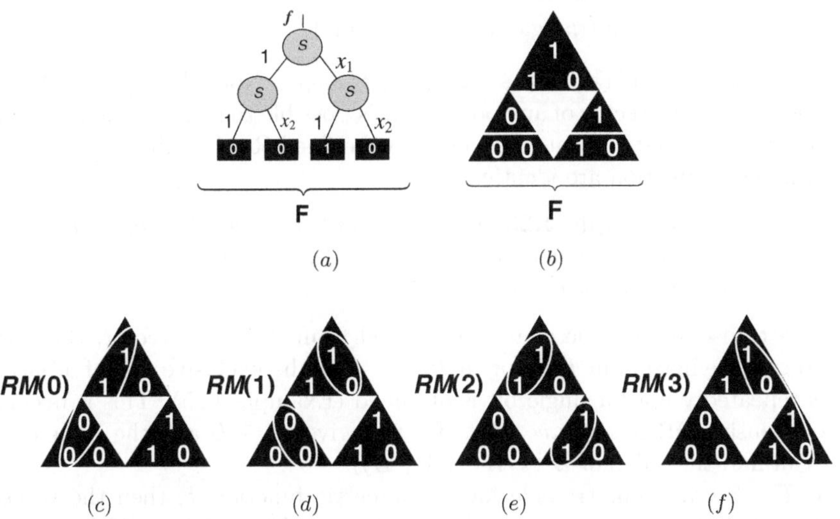

FIGURE 4.18
Representation of the switching function $f = x_1\overline{x}_2$ of two variables by the binary decision tree (a) and transeunt triangle (b-f) (Example 4.22).

4.6.1 Representation of a symmetric function

A switching function f of n variables $x_1, x_2, \ldots, x_n$ is *totally symmetric* if and only if it is unchanged by any permutation of variables.

Example 4.23 *A switching function* $f = \overline{x}_1\overline{x}_2\overline{x}_3 \vee x_1x_2x_3$ *is totally symmetric.*

A totally symmetric function is completely specified by a *carry* vector $\mathbf{A} = [a_0a_1, \ldots, a_n]$, such that switching function f is a_i for all assignments to

$x_1, x_2, \ldots, x_n$ that have i 1's, where $0 \leq i \leq n$.

Example 4.24 *A carry vector of a totally symmetric switching function* $f = \overline{x}_1\overline{x}_2\overline{x}_3 \vee x_1x_2x_3$ *is* [1 0 0 1].

Representation of a symmetric switching function consisting of 1's and 0's arranged in a triangle is based on the function's $(n+1)$-bit *carry* vector. Thus, the triangle is formed as follows:

- ▶ The carrier vector is located at the base of triangle.
- ▶ A vector of n 1's and 0's is formed by the exclusive OR of adjacent bits in the carry vector.
- ▶ A vector of $(n-1)$ 1's and 0's is formed by the exclusive OR of adjacent bits in the previous vector, etc.
- ▶ At the apex of the triangle, is a single 0 or 1.

The resulting triangle is the *transeunt triangle.* Note that given a Reed–Muller coefficient vector of any polarity, it is possible to recreate the transeunt triangle. Note that certain coefficients in the fixed Reed–Muller expansion of a symmetric function are identical.

Example 4.25 *The transeunt triangle for the symmetric switching function* $f = \overline{x}_1\overline{x}_2\overline{x}_3 \oplus x_1x_2x_3$ $(n = 3)$*, is constructed as shown in Figure 4.19.*

The transeunt triangle can be generated from the carry vector, the carry vector of coefficients in RM_0 or in RM_{2^n-1} by the exclusive OR of adjacent bits repeatedly until a single bit is obtained (Example 4.25). This is because the exclusive OR is *self-invertible.* That is, given $A \oplus B$ and the value of A, we can find the value of B $(= A \oplus (A \oplus B))$.

If T is a transeunt triangle for a symmetric function f, then the transeunt triangle $T_{\overline{f}}$ for (symmetric function) $\overline{f}$ is obtained from T by replacing the carry vector at the base of T by its complement. This is because $A \oplus B = \overline{A} \oplus \overline{B}$. In a similar manner, other rows in a transeunt triangle can be complemented.

Theorem 4.1 *Let T be the transeunt triangle of a symmetric switching function f. Then, embedded in T are all Reed–Muller coefficient matrices.*

Proof is given in [2].

Among the Reed–Muller expansions, one with a minimal number of nonzero coefficients can be found. An algorithm for computing the minimal fixed polarity Reed–Muller expansion using the transeunt triangle is as follows:

(*a*) Generate the transeunt triangle,
(*b*) Compute number of terms of each RM_i, and
(*c*) Choose an RM_i with the fewest product terms.

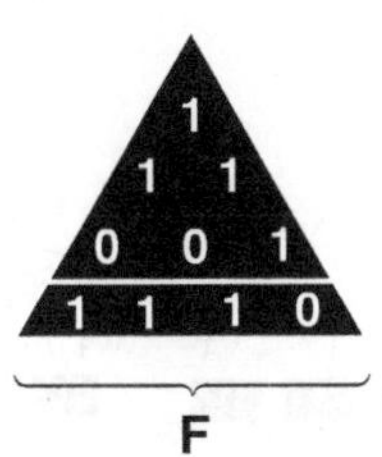

Given:
the switching function

$$f = \overline{x}_1\overline{x}_2\overline{x}_3 \oplus x_1x_2x_3.$$

- *Start from the truth vector* $[1\ 1\ 1\ 0]$ *(the base of the triangle), the* $RM(0) =$ $[1\ 0\ 1\ 1]$ *is formed as the left side of the triangle, read from the bottom to the top.*
- *Alternatively, starting from the vector of coefficients*

$$RM(0) = [1001]^T$$

placed at the base of the triangle, the triangle can be restored, so that the truth vector $F = [1\ 1\ 1\ 1\ 0]$ *will be formed as the right side of the triangle (read from the bottom to the top).*

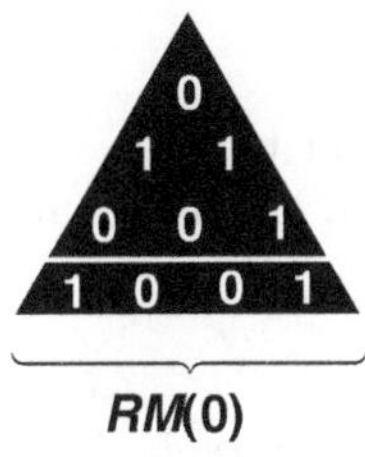

FIGURE 4.19
Construction of a transeunt triangle for the switching function $f = \overline{x}_1\overline{x}_2\overline{x}_3 \oplus x_1x_2x_3$ (Example 4.25).

Example 4.26 *The Sierpinski's and transeunt triangles for the symmetric function* $f = \overline{x_1\overline{x}_2}$ *are shown in Figure 4.20. Carry vector* $\mathbf{A} = [1\ 1\ 0]$ *is formed from the initial truth vector* $\mathbf{F} = [1\ 1\ 1\ 0]$. *The Davio tree and diagram that represent Reed–Muller expansion of zero polarity are shown as well.*

4.6.2 Measures of transeunt triangle structures

A transeunt triangle for an n-variable function (non-symmetric, in general) has a width of 2^n and a height of 2^n. There are $(4^n + 2^n)/2$ elements in total. Besides the defining relation, there are other relations among elements in the transeunt triangle. Since the truth vector for functions with one or more variables has an even number of entries, it can be divided evenly into two parts. Each part produces, on its own, two sub-triangles.

Let T be the transeunt triangle for a switching function $f(x_1, x_2, ..., x_n)$. Representation of T by 3 triangles T_0, T_1 and T_2 is shown in Figure 4.21. Here, elements of T_2 are constructed by element-wise modulo-2 sum of T_0 and T_1, i.e., T_2 is the transeunt triangle for the function

$$f(0, x_2, ..., x_n) \oplus f(1, x_2, ..., x_n).$$

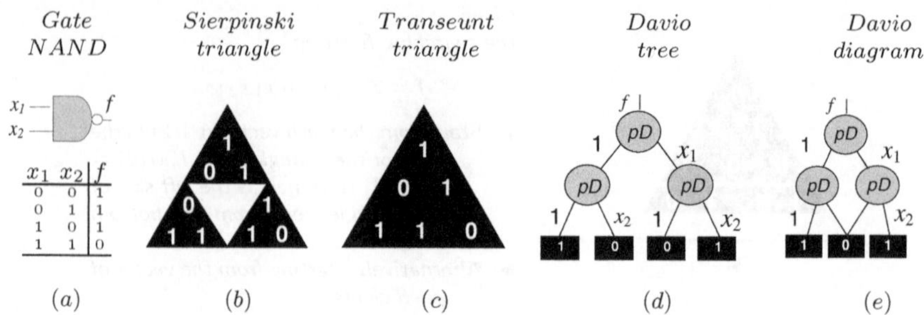

FIGURE 4.20
Graphical representation of AND gate (a): Sierpinski triangle (b), transeunt triangle formed from the carry vector (c), decision tree (d), and decision diagram (e) (Example 4.26).

Example 4.27 *Let T_0 be a transeunt triangle for a switching function f of a single variable, i.e., it is 3-element triangle $T_0 = (0, x_1)$. To form T for a function of two variables, another two 3-element triangles must be added. Here $T_0 = (0, x_2)$,*

$$T_1 = (1, x_2),$$

and

$$T_2 = T_1 \oplus T_2,$$

and T_0 turns to $T_0 = (0, x_2)$. The elements of T_2 are $f(0, x_2) \oplus f(1, x_2)$. This recursive procedure can be applied to a function of any number of variables.

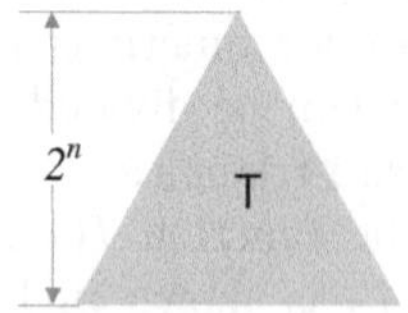

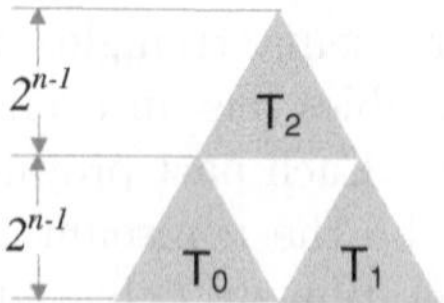

FIGURE 4.21
Representation of a transeunt triangle T by three triangles T_0, T_1 and T_2 (Example 4.27).

4.6.3 Transeunt triangles and decision trees

> *A transeunt triangle represents all positive Reed–Muller expressions of both fixed and mixed polarity. The corresponding elements of the triangle are components of the vector of Reed–Muller coefficients, and they are formed from the truth vector by modulo-2 addition of its elements.*

This process, on the other hand, can be represented by a Davio tree, in which the input (to the root node) is supplied with a sequence of the elements of the truth vector and this sequence is used to form the components of a vector of any polarity Reed–Muller expression.

Example 4.28 *Figure 4.22 illustrates the relationship between a triangle and a positive Davio tree for a switching function of three variables given its truth vector* $\mathbf{F} = [00101011]^T$.

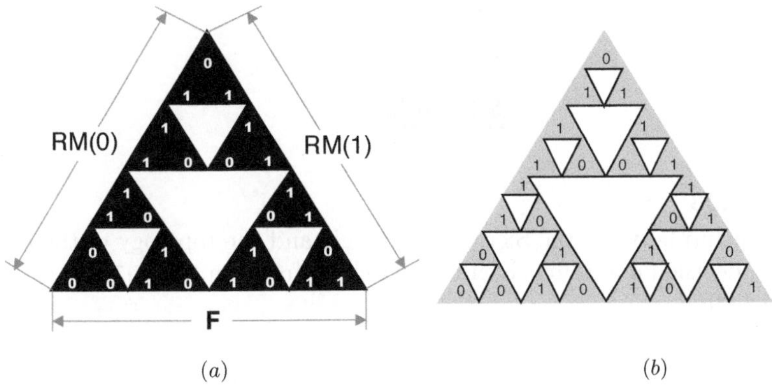

FIGURE 4.22
Representation of a switching function $f(x_1, x_2, x_3)$ of three variables given a truth vector $\mathbf{F} = [0\ 0\ 1\ 0\ 1\ 0\ 1\ 1]^T$ by Sierpinski triangle (a) and Davio diagram (b) (Example 4.28).

Example 4.29 *Figure 4.23 shows the relationship of the topology of the* 8×8×8 *Sierpinski triangle and* 4×4×4 *transeunt triangle and decision diagrams for two polarities of the family of fixed polarity Reed–Muller expressions, correspondingly, given a switching function of three variables.*

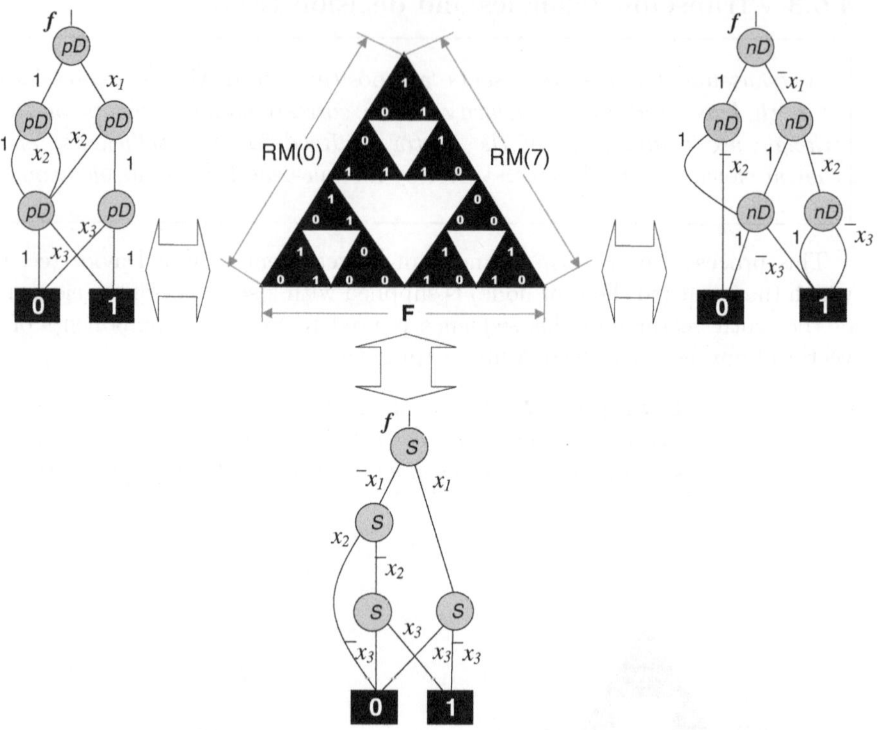

FIGURE 4.23
The relationship between the Sierpinski triangle and the topology of the corresponding decision diagrams given the switching function $f = \overline{x}_1\overline{x}_2x_3 \vee x_1\overline{x}_3$ (Example 4.29).

Example 4.30 *Figure 4.24 shows the transeunt triangle given the symmetric switching function*

$$f = \overline{x}_1\overline{x}_2x_3 \vee x_1x_2x_3 \vee x_1\overline{x}_2\overline{x}_3 \vee \overline{x}_1x_2\overline{x}_3.$$

In this triangle, only elements that correspond to a truncated truth vector of the symmetric functions, are given. The corresponding Davio trees for

$$RM(0) = x_1 \oplus x_2 \oplus x_3 \text{ and}$$
$$RM(7) = 1 \oplus \overline{x}_1 \oplus \overline{x}_2 \oplus \overline{x}_3$$

and the Shannon tree are given in Figure 4.24 as well.

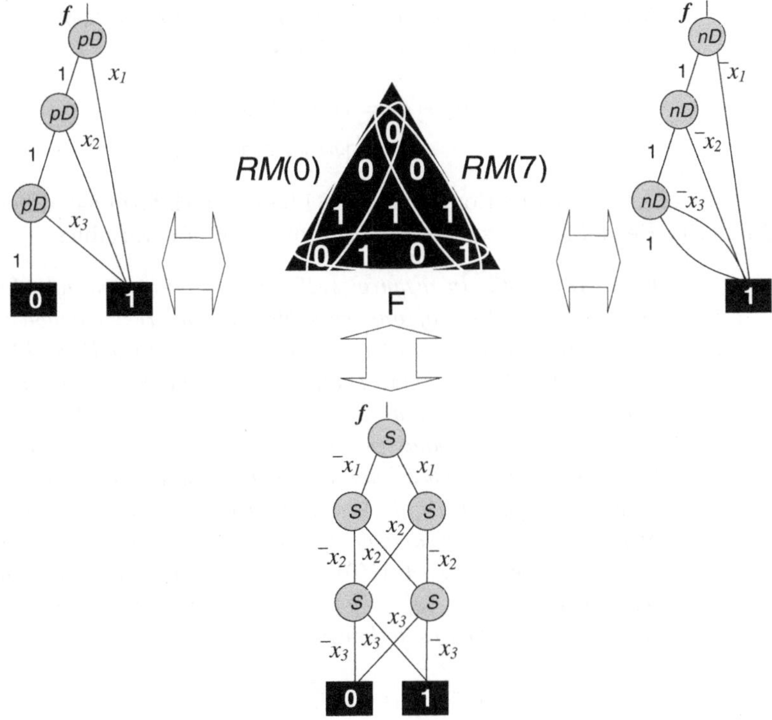

FIGURE 4.24
The relationship between the transeunt triangle and the topology of the corresponding decision diagrams for the totally symmetric switching function $f = x_1 \oplus x_2 \oplus x_3$ (Example 4.29).

4.6.4 Using transeunt triangles for representation of multivalued functions

A mixed polarity of Reed–Muller expressions is a family of 3^n Reed–Muller representations of switching functions of n variables. In this family, each binary variable x_i can be represented in one of the following *basic polarities*:

$$\mathbf{c}_0 = [1\ x],$$
$$\mathbf{c}_1 = [1\ \overline{x}],$$
$$\mathbf{c}_2 = [\overline{x}\ x].$$

Given $n > 1$, various combinations of terms of variables can be generated by the Kronecker product of basic polarities.

Example 4.31 *Product terms in the* $(c_0, c_2) = (0, 2)$*-polarity Reed–Muller expression of a switching function of two variables* x_i *and* x_j *are generated as follows*

$$\begin{aligned}\mathbf{c}_0 \otimes \mathbf{c}_2 &= [1\ x_i] \otimes [\overline{x}_j\ x_j]\\ &= [\overline{x}_j\ x_j\ x_i\overline{x}_j\ x_ix_j].\end{aligned}$$

In the case of a ternary function, there are 84 bases which form 84^n different mixed polarity Reed–Muller expressions for a function of n variables.

Example 4.32 *In Figure 4.25, a transeunt triangle for a switching function of one variable (a) and two variables (b) is pictured. The Davio tree (c) corresponds to a Reed–Muller expression* $RM(0)$ *of zero polarity and elements* $RM(0)$ *are 0's and 1's on the right side of the triangle. In case of a ternary function, possible polarities of variables are represented by a pyramid (d) in which all possible paths between nodes represent 84 different polarities of Reed–Muller expressions of a function of a single ternary variable (see details in Chapter 30). One of these expansions,* $RM(0)$*, is represented by the complete ternary Davio tree (e). The product terms of the expression* $RM(0)$ *are formed as follows:*

$$\begin{aligned}&[1\ x_1\ x_1^2] \otimes [1\ x_2\ x_2^2]\\ &= [1\ x_2\ x_2^2\ x_1\ x_1x_2\ x_1^2x_2\ x_1^2\ x_1x_2^2\ x_1^2x_2^2].\end{aligned}$$

Example 4.33 *Table 4.3 contains various graphical data structures for representation of five elementary switching functions. These functions are symmetric, therefore both Sierpinski triangle and transeunt triangle for symmetric functions are shown.*

4.7 Further reading

Fundamentals of the techniques for manipulating EXOR expressions are discussed by Davio et al. in [5], and in Sasao's textbook [27]. Remarks on the history of development of Reed–Muller techniques can be found in the book by Stanković et al. [30].

The classification of EXOR expressions and relations between them can be found in Sasao's works [25, 27]. The usefulness of manipulation with different polarities has been shown in the papers by Tsai et al. [22, 23] and Sasao [24,

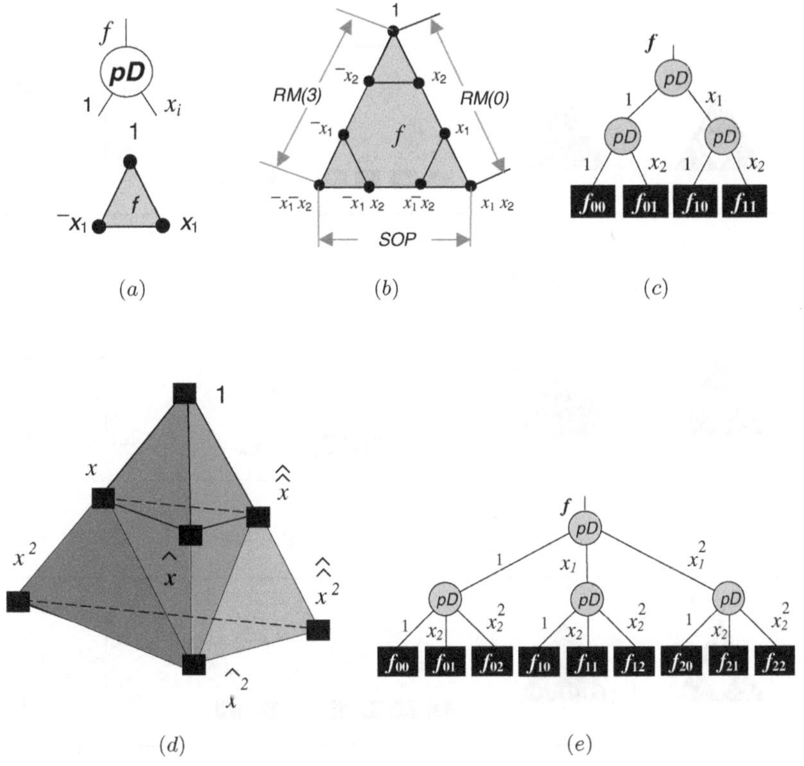

FIGURE 4.25
Transeunt triangles and Davio trees for representation of a switching function (a,b,c) of one and two variables, and a pyramid and ternary Davio tree for a ternary function (d,e) of one and two variables (Example 4.32).

TABLE 4.3

Graphical data structures: Sierpenski gasket, transeunt triangle, decision tree, decision diagram, and $\mathcal{N}$-hypercube for representation of elementary switching functions of two variables.

Sierpinski gasket	Transeunt transeunt	Decision tree	Decision diagram	$\mathcal{N}$-hypercube
	AND-function		$\mathbf{F} = [0001]^T$	
	OR-function		$\mathbf{F} = [0001]^T$	
	EXOR-function		$\mathbf{F} = [0001]^T$	
	NAND-function		$\mathbf{F} = [0001]^T$	
	NOR-function		$\mathbf{F} = [0001]^T$	
	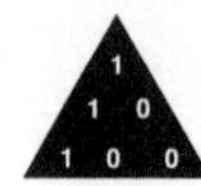	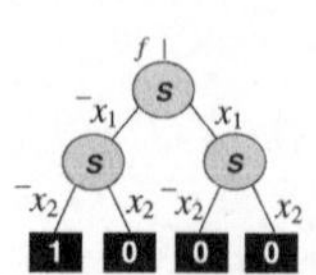	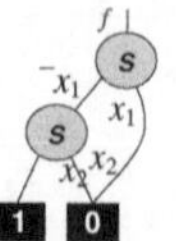	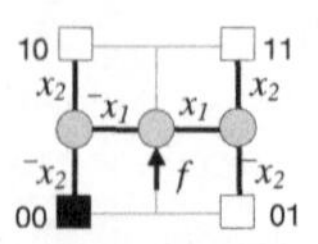

25]. Matrix (spectral) computing of AND-EXOR expressions are discussed by Stanković in [70]. The reader can find a very detailed study on fixed and mixed polarity AND-EXOR expressions and systematic classification in papers by Green, in particular, [13, 14].

Bernasconi and Codenotti [1] argued that Walsh analysis of switching functions can be viewed as a Cayley graph eigenvalue problem. In this approach, a switching function is classified as equivalent under a set of affine transformations (see papers by Edwards [8, 9] and Lechner [16] for details of Walsh spectrum of switching functions).

Optimal polarity of AND-EXOR expressions has been a subject of many papers, in particular, by Butler et al. [2], Debnath and Sasao [6], Dueck et al. [7], and Sasao [26]. Finding optimal polarity for symmetric and partially symmetric functions was studied by Butler et al. [15], Davio et al. [5], and Yanushkevich et al. [28].

Cube representation of AND-EXOR expressions. Luccio and Pagli [17] proposed to describe EXOR expressions using so-called *pseudocubes.* This work was continued by Ciriani [7] using affine spaces. In notation of [7], a pseudocube of degree m of a switching function is an affine space over a vector subspace. Earlier Perkowski [13] developed the technique for manipulation of unspecified Reed–Muller expressions. In [18], Mishchenko and Perkowski reported an improved version of the heuristic minimization of exclusive sum-of-products expressions [13, 28]. Note that decision diagrams are used in various approaches to minimization of exclusive sum-of-products expressions, in particular, in iterative cube transformations, term rewriting technique, and stair-case technique.

Spectral interpretation of AND-EXOR expressions. The coefficients of AND-EXOR expansions of switching functions can be conveniently regarded as spectral coefficients of Fourier-like transforms. In this setting, product terms appearing in Reed–Muller expansions and defined under the logical AND multiplication, are considered as a base in the space of switching functions (the base of a Reed–Muller transform). Optimization of decision diagrams using these manipulations, that as a general rule are performed on non-Abelian groups, was considered by Stanković [49].

Sierpinski triangles and transeunt triangles for switching function representation. Interest in the Pascal triangle concerns its connection with the expansion of $(x+y)^n$, for the numbers in row n of the triangle given the requisite coefficients. The Fibonacci sequence can be obtained from the Pascal triangle. Transeunt triangles have been studied, in particular, by Falkoner [10], Gardner [12], and Hilton [15]. Applications of transeunt triangles for switching function manipulation can be found in works by Butler et al. [2, 15], and Yanushkevich et al. [28]. In particular, Butler et al. [2] considered transeunt triangles for calculation of optimal polarity of Reed–Muller expressions. Dueck et al. [7] developed an algorithm to find the minimum mixed polarity Reed–Muller expression of a given switching function. This algorithm runs in $O(n^3)$ time and uses $O(n^3)$ storage space. Popel and Dani [60] used the Sierpinski gaskets to represent 4-valued logic functions. They also interpreted using Shannon and Davio expansion in this topological structure.

References

[1] Bernasconi A and Codenotti B. Spectral analysis of Boolean functions as a graph eigenvalue problem. *IEEE Transactions on Computers*, 48(3):345–351, 1999.

[2] Butler JT, Dueck GW, Shmerko VP, and Yanushkevich SN. Comments on SYMPATHY: fast exact minimization of fixed polarity Reed-Muller expansion for symmetric functions. *IEEE Transactions on Computer-Aided Design of Integrated Circuits and Systems*, 19(11):1386–1388, 2000.

[3] Butler JT, Dueck GW, Yanushkevich SN, and Shmerko VP. On the number of generators of transeunt triangles. *Discrete Applied Mathematics*, 108:309–316, 2001.

[4] Ciriani V. Synthesis of SPP three-level logic networks using affine spaces. *IEEE Transactions on Computer-Aided Design of Integrated Circuits And Systems*, 22(10):1310–1323, 2003.

[5] Davio MJ, Deschamps P, and Thayse A. Discrete and Switching Functions, *McGraw-Hill Int. Book Co*, 1978.

[6] Debnath D and Sasao T. Minimization of AND-OR-EXOR three-level networks with AND gate sharing. *IEICE Transactions on Information Systems*, E80-D(10):1001-1008, 1997.

[7] Dueck GW, Maslov D, Butler JT, Shmerko VP, and Yanushkevich SN A method to find the best mixed polarity Reed–Muller expressions using transeunt triangle. In *Proceedings of the 5th International Workshop on Applications of Reed–Muller Expansion in Circuit Design*, Mississipi State University, pp. 82–92, 2002.

[8] Edwards CR. The generalized dyadic differentiator and its application to 2-valued functions defined on an n-space. *Proceedings IEE*, Comput. and Digit. Techn., 1(4):137–142, 1978.

[9] Edwards CR. The Gibbs dyadic differentiator and its relationship to the Boolean difference. *Comput. and Elect. Eng.*, 5:335–344, 1978.

[10] Falconer K. *Fractal Geometry*. Wiley, New York, 1990.

[11] Falkowski BJ and Kannurao S. Algorithm to identify skew symmetries through Walsh transform. *Electronic Letters*, 36(5):401–402, 2000.

[12] Gardner M. Mathematical games. *Scientific American*, 215:128–132, 1966.

[13] Green DH. Families of Reed–Muller canonical forms, *International Journal of Electronics*, 2:259–280, 1991.

[14] Green DH. *Modern Logic Design,* Addison-Wesley Publishing Company, 1986.

[15] Hilton P and Pedersen J. Extending the binomial coefficients to preserve symmetry and pattern. *Computers Math. Appl.*, 17:89–102, 1989.

[16] Lechner RJ. Harmonic analysis of switching functions. In *Recent Development in Switching Theory,* pp. 122–229, Academic Press, 1971.

[17] Luccio F and Pagli L. On a new Boolean function with applications. *IEEE Transactions on Computers,* 48:296–310, Mar., 1999.

[18] Mishchenko A and Perkowski M. Fast heuristic minimization of exclusive-sums-of-products. In *Proceedings of the 5th International Workshop on Applications of the Reed–Muller Expansion in Circuit Design.* Mississippi State University, pp. 242–249, 2001.

[19] Perkowski MA. A fundamental theorem for EXOR circuits. In *Proceedings of the IFIP WG 10.5 Workshop on Applications of the Reed–Muller Expression in Circuit Design*, Hamburg, Germany, pp. 52–60, 1993.

[20] Perkowski M. A new representation of strongly unspecified switching functions and its application to multi-level AND/OR/EXOR synthesis. In *Proceedings of the IFIP WG 10.5 Workshop on Applications of the Reed–Muller Expression in Circuit Design*, Japan, pp. 143–151, 1995.

[21] Popel DV and Dani A. Sierpinski gaskets for logic function representation. In *Proceedings of the IEEE 32th International Symposium on Multiple-Valued Logic*, pp. 39–45, 2002.

[22] Tsai CC and Marek-Sadowska M. Boolean functions classification via fixed polarity Reed–Muller forms. *IEEE Transactions on Computers*, 46(2):173–186, 1997.

[23] Tsai CC and Marek-Sadowska M. Generalized Reed-Muller Forms as a tool to detect symmetries. *IEEE Transactions on Computers*, 45(1):33–40, 1996.

[24] Sasao T. EXMIN: A simplified algorithm for exclusive-OR-sum-of-products expressions for multiple-valued input two-valued output functions. *IEEE Transactions on Computer Aided Design of Integrated Circuits and Systems,* 12(5):621–632, 1993.

[25] Sasao T. Representation of logic functions using EXOR operators. In Sasao T and Fujita M, Eds., *Representations of Discrete Functions*, Kluwer, Dordrecht, pp. 29–54, 1996.

[26] Sasao T. AND-EXOR expressions and their optimization. In Sasao T, Ed., *Logic Synthesis and Optimization,* Kluwer, Dordrecht, pp. 287–312, 1993.

[27] Sasao T. *Switching Theory for Logic Synthesis*, Kluwer, Dordrecht, 1999.

[28] Song N and Perkowski M. Minimization of exclusive sum-of-products expressions for multi-output multiple-valued input incompletely specified functions. *IEEE Transactions on Computer Aided Design of Integrated Circuits and Systems*, 15(4):385–395, 1996.

[29] Stanković RS and Astola JT. *Spectral Interpretation of Decision Diagrams.* Springer, Heidelberg, 2003.

[30] Stanković RS, Moraga C, and Astola JT. Reed–Muller expressions in the previous decade. In *Proceedings of the 5th International Workshop on Applications of Reed–Muller Expansion in Circuit Design*, Mississipi State University, pp. 7–26, 2002.

[31] Stanković RS. Non-Abelian groups in optimization of decision diagrams representations of discrete functions. *Formal Methods in System Design*, 18:209–231, 2001.

[32] Yanushkevich SN, Butler JT, Dueck GW, and Shmerko VP. *Experiments on FPRM expressions for partially symmetric logic functions.* In *Proceedings of the 30th International Symposium on Multiple-Valued Logic*, pp. 141–146, 2000.

5

Arithmetic Representations

Arithmetic representations of switching functions are known as word-level forms, and are a way to describe the parallel calculation of several switching functions at once. These forms are different from logical expressions such as sum-of-products and AND-EXOR expressions that are not word-level. Another useful property of these arithmetic representations is linearization. A multi-output switching function can be represented by a linear word-level arithmetic polynomial and a linear word-level decision diagram (see Chapter 37).

5.1 Introduction

For Boolean variables x, x_1, and x_2, taking on the values of 0 or 1, the following is true, as introduced by the founder of Boolean algebra George Boolé:

$$\begin{aligned} \overline{x} &= 1 - x, \\ x_1 \vee x_2 &= x_1 + x_2 - x_1 x_2, \\ x_1 \wedge x_2 &= x_1 x_2. \end{aligned}$$

Other logic operations can be represented by arithmetic operations as well, for example, $x_1 \oplus x_2 = x_1 + x_2 - 2x_1 x_2$. The right part of the equation is called an *arithmetic expression.*

A switching function of n variables is the mapping

$$\{0,1\}^n \to \{0,1\},$$

while an integer-valued function in arithmetical logic denotes the mapping

$$\{0,1\}^n \to \{0,1,\ldots,p-1\},$$

where $p > 2$. Representing switching functions by arithmetic expressions is useful for the analysis of circuits, verification, testability, and for reliability analysis (see "Further reading" section). They are also useful for representing and manipulating multivalued functions for which logical expressions can be very complicated.

5.2 Algebraic description

There are a number of similarities between arithmetic and Reed–Muller expression of a single switching function.

The main difference between the arithmetic and Reed–Muller expression is that, in an arithmetic expression, arithmetic addition and multiplication is used, whereas with Reed–Muller polynomials a modulo-two sum and product is used.

5.2.1 General algebraic form

For a switching function f of n variables, the arithmetic expression is given by

$$f = \sum_{i=0}^{2^n-1} p_i \cdot \underbrace{(x_1^{i_1} \cdots x_n^{i_n})}_{i-th\ product} \tag{5.1}$$

where p_i is an integer coefficient, i_j is the j-th bit $1, 2, \ldots, n$, in the binary representation of the index $i = i_1 i_2 \ldots i_n$, and $x_j^{i_j}$ is defined as

$$x_j^{i_j} = \begin{cases} 1, & i_j = 0; \\ x_j, & i_j = 1. \end{cases} \tag{5.2}$$

Note that $\sum$ is the arithmetic addition.

Example 5.1 *An arbitrary switching function of three variables is represented in the arithmetic expression by Equation 5.1 and Equation 5.2:*

$$\begin{aligned} f &= a_0 + a_1 x_3 + a_2 x_2 + a_3(x_2 x_3) + a_4 x_1 + a_5(x_1 x_3) \\ &+ a_6(x_1 x_2) + a_7(x_1 x_2 x_3). \end{aligned}$$

Figure 5.1 illustrates the structure of an arithmetic expression where the shaded nodes implement AND operations and the sum is an arithmetic operation.

5.2.2 Computing the coefficients

Given the truth vector $\mathbf{F} = [f(0)\ f(1) \ldots f(2^n - 1)]^T$, the vector of arithmetic coefficients $\mathbf{P} = [p_0\ p_1 \ldots p_{2^n-1}]^T$ is derived by the matrix equation resulting in

$$\mathbf{P} = \mathbf{P}_{2^n} \cdot \mathbf{X} \tag{5.3}$$

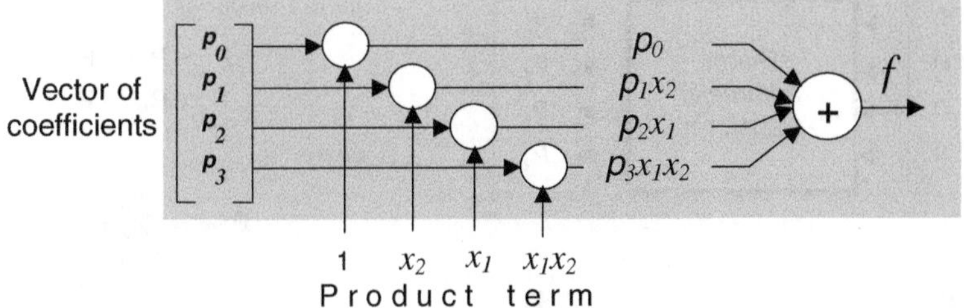

FIGURE 5.1
Deriving the arithmetic expression for a function of two variables (Example 5.1).

where the $2^n \times 2^n$-matrix $\mathbf{P}_{2^n}$ is formed by the Kronecker product

$$\mathbf{P}_{2^n} = \bigotimes_{j=1}^{n} \mathbf{P}_{2^j}, \qquad \mathbf{P}_{2^j} = \begin{bmatrix} 1 & 0 \\ -1 & 1 \end{bmatrix}. \tag{5.4}$$

Note that in the Reed–Muller matrix R_{2^n} the elements are logical values 0 and 1, and the calculation with R_{2^n} is performed in Galois field of order 2, $GF(2)$. In the arithmetic transform matrix P_{2^n}, the elements are the integers 0 and 1, and the calculation with P_{2^n} is performed on integers.

The general scheme for computing the coefficients is shown in Figure 5.2.

5.2.3 Data flowgraphs

To design the data flowgraph of the algorithm, the matrix $\mathbf{P}_{2^n}$ is represented in the factorized form

$$\mathbf{P}_{2^n} = \mathbf{P}_{2^n}^{(n)} \mathbf{P}_{2^n}^{(n-1)} \cdots \mathbf{P}_{2^n}^{(1)}, \tag{5.5}$$

where $\mathbf{P}_{2^n}^{(i)}$, $i = 1, 2, \ldots, n$, is formed by the Kronecker product

$$\mathbf{P}_{2^n}^{(i)} = \mathbf{I}_{2^{n-i}} \otimes \mathbf{P}_{2^1} \otimes \mathbf{I}_{2^{i-1}}. \tag{5.6}$$

Hence, the arithmetic coefficients are computed in n iterations.

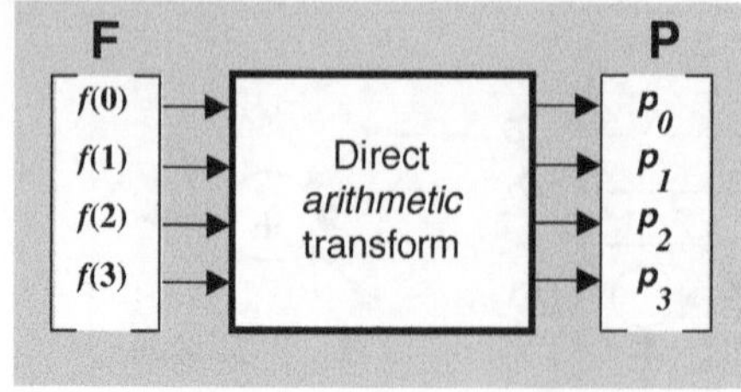

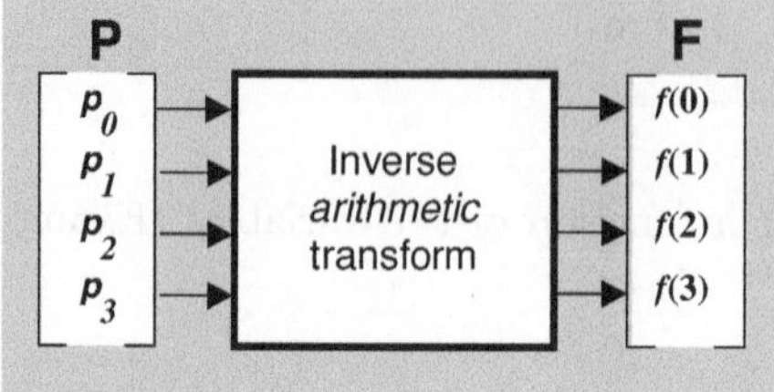

The pair of direct and inverse arithmetic transforms

$$\mathbf{P} = \mathbf{P}_{2^n} \cdot \mathbf{F},$$
$$\mathbf{F} = \mathbf{P}_{2^n}^{-1} \cdot \mathbf{P},$$

where

$$\mathbf{P}_{2^n} = \bigotimes_{j=1}^{n} \mathbf{P}_{2^j},$$

$$\mathbf{P}_{2^1} = \begin{bmatrix} 1 & 0 \\ -1 & 1 \end{bmatrix}$$

$$\mathbf{P}_{2^n}^{-1} = \bigotimes_{j=1}^{n} \mathbf{P}_{2^j}^{-1},$$

$$\mathbf{P}_{2^1}^{-1} = \begin{bmatrix} 1 & 0 \\ 1 & 1 \end{bmatrix}$$

FIGURE 5.2
Direct and inverse arithmetic transform for a switching function of two variables.

Example 5.2 *Computing the arithmetic coefficients p_i by Equation 5.3 for the elementary function $f = x_1 \vee x_2$ given its truth-vector $\mathbf{F} = [0\ 1\ 1\ 1]^T$ is illustrated in Figure 5.3. The data flowgraph includes two iterations, according to factorization rules (Equation 5.5 and Equation 5.6). The matrix $\mathbf{P}_{2^2}$ is formed as follows:*

$$\mathbf{P}_{2^2} = \mathbf{P}_{2^2}^{(2)} \mathbf{P}_{2^2}^{(1)} = (\mathbf{I}_{2^{2-2}} \otimes \mathbf{P}_{2^1} \otimes \mathbf{I}_{2^{2-1}})(\mathbf{I}_{2^{2-1}} \otimes \mathbf{P}_{2^1} \otimes \mathbf{I}_{2^{1-1}})$$

$$= \underbrace{(1 \otimes \mathbf{P}_{2^1} \otimes \mathbf{I}_{2^1})}_{1st\ iteration} \underbrace{(\mathbf{I}_{2^1} \otimes \mathbf{P}_{2^1} \otimes 1)}_{2nd\ iteration}$$

$$= \left[\begin{array}{c|c} \mathbf{I}_{2^1} & \\ \hline -\mathbf{I}_{2^1} & \mathbf{I}_{2^1} \end{array}\right] \left[\begin{array}{c|c} \mathbf{P}_{2^1} & \\ \hline & \mathbf{P}_{2^1} \end{array}\right] = \left[\begin{array}{cc|cc} 1 & & & \\ & 1 & & \\ \hline -1 & & 1 & \\ & -1 & & 1 \end{array}\right] \left[\begin{array}{cc|cc} 1 & & & \\ -1 & 1 & & \\ \hline & & 1 & \\ & & -1 & 1 \end{array}\right].$$

5.2.4 Restoration

The following matrix equation restores the truth-vector $\mathbf{F}$ from the vector of coefficients $\mathbf{P}$ (Figure 5.2)

$$\mathbf{F} = \mathbf{P}_{2^n}^{-1} \cdot \mathbf{P}, \tag{5.7}$$

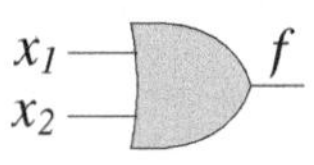

$f = x_1 \vee x_2$

Vector of coefficients

$$\mathbf{P} = \mathbf{P}_{2^2} \cdot \mathbf{F} = \left[\begin{array}{rr|rr} 1 & 0 & 0 & 0 \\ -1 & 1 & 0 & 0 \\ \hline -1 & 0 & 1 & 0 \\ 1 & -1 & -1 & 1 \end{array}\right] \begin{bmatrix} 0 \\ 1 \\ 1 \\ 1 \end{bmatrix} = \begin{bmatrix} 0 \\ 1 \\ 1 \\ -1 \end{bmatrix}$$

Arithmetic spectrum computing using factorized matrices (Equation 5.5)

$$\mathbf{P} = \mathbf{P}_{2^2}^{(2)} \, \mathbf{P}_{2^2}^{(1)} \, \mathbf{F}$$

$$= 2^{-2} \left[\begin{array}{rr|rr} 1 & 0 & 0 & 0 \\ -1 & 1 & 0 & 0 \\ \hline 0 & 0 & 1 & 0 \\ 0 & 0 & -1 & 1 \end{array}\right] \left[\begin{array}{rr|rr} 1 & 0 & 0 & 0 \\ 0 & 1 & 0 & 0 \\ \hline -1 & 0 & 1 & 0 \\ 0 & -1 & 0 & 1 \end{array}\right] \begin{bmatrix} 0 \\ 1 \\ 1 \\ 1 \end{bmatrix} = \begin{bmatrix} 0 \\ 1 \\ 1 \\ -1 \end{bmatrix}$$

Arithmetic expression

$$f = x_2 + x_1 - x_1 x_2$$

Data flowgraphs of the algorithm

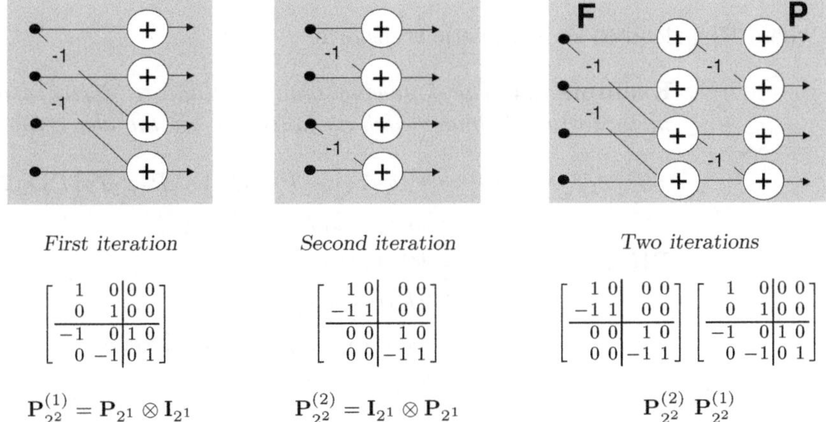

First iteration

$$\left[\begin{array}{rr|rr} 1 & 0 & 0 & 0 \\ 0 & 1 & 0 & 0 \\ \hline -1 & 0 & 1 & 0 \\ 0 & -1 & 0 & 1 \end{array}\right]$$

$$\mathbf{P}_{2^2}^{(1)} = \mathbf{P}_{2^1} \otimes \mathbf{I}_{2^1}$$

Second iteration

$$\left[\begin{array}{rr|rr} 1 & 0 & 0 & 0 \\ -1 & 1 & 0 & 0 \\ \hline 0 & 0 & 1 & 0 \\ 0 & 0 & -1 & 1 \end{array}\right]$$

$$\mathbf{P}_{2^2}^{(2)} = \mathbf{I}_{2^1} \otimes \mathbf{P}_{2^1}$$

Two iterations

$$\left[\begin{array}{rr|rr} 1 & 0 & 0 & 0 \\ -1 & 1 & 0 & 0 \\ \hline 0 & 0 & 1 & 0 \\ 0 & 0 & -1 & 1 \end{array}\right] \left[\begin{array}{rr|rr} 1 & 0 & 0 & 0 \\ 0 & 1 & 0 & 0 \\ \hline -1 & 0 & 1 & 0 \\ 0 & -1 & 0 & 1 \end{array}\right]$$

$$\mathbf{P}_{2^2}^{(2)} \, \mathbf{P}_{2^2}^{(1)}$$

FIGURE 5.3
Calculation of arithmetic expansion for OR gate (Example 5.2).

where a $2^n \times 2^n$ matrix $\mathbf{P}_{2^n}^{-1}$ is formed by the Kronecker product

$$\mathbf{P}_{2^n}^{-1} = \bigotimes_{i=0}^{n} \mathbf{P}_{2^1}^{-1}, \quad \mathbf{P}_{2^1}^{-1} = \begin{bmatrix} 1 & 0 \\ 1 & 1 \end{bmatrix} \tag{5.8}$$

Example 5.3 *To restore the truth-vector* $\mathbf{F}$ *given by its vector of arithmetical coefficients* $\mathbf{P} = [0\ 1\ 1\ -1]^T$, *Equation 5.7 can be applied:*

$$\mathbf{F} = \mathbf{P}_{2^3}^{-1} \cdot \mathbf{P} = \left[\begin{array}{cc|cc} 1 & 0 & 0 & 0 \\ 1 & 1 & 0 & 0 \\ \hline 1 & 0 & 1 & 0 \\ 1 & 1 & 1 & 1 \end{array}\right] \begin{bmatrix} 0 \\ 1 \\ 1 \\ -1 \end{bmatrix} = \begin{bmatrix} 0 \\ 1 \\ 1 \\ 1 \end{bmatrix}$$

5.2.5 Useful rules

The simple arithmetic expressions $\overline{x} = 1 - x$, $x_1 \vee x_2 = x_1 + x_2 - x_1x_2$, $x_1x_2 = x_1x_2$, $x_1 \oplus x_2 = x_1 + x_2 - 2x_1x_2$ can be generalized if x_1 and x_2 are replaced with the switching functions f_1 and f_2:

$$\begin{array}{|l|} \hline \mathcal{P}\{\overline{f}\} = 1 - \mathcal{P}\{f\} \\ \mathcal{P}\{f_1 \vee f_2\} = \mathcal{P}\{f_1\} + \mathcal{P}\{f_2\} - \mathcal{P}\{f_1\}\mathcal{P}\{f_2\} \\ \mathcal{P}\{f_1 f_2\} = \mathcal{P}\{f_1\}\mathcal{P}\{f_2\} \\ \mathcal{P}\{f_1 \oplus f_2\} = \mathcal{P}\{f_1\} + \mathcal{P}\{f_2\} - 2\mathcal{P}\{f_1\}\mathcal{P}\{f_2\} \\ \hline \end{array}$$

where $\mathcal{P}\{\cdot\}$ denotes an arithmetic transform.

Example 5.4 *The following manipulations to derive arithmetic forms of some switching functions use the above rules:*

$$\begin{aligned} \mathcal{P}\{x_1\overline{x}_2 \vee x_1\overline{x}_3\} &= \mathcal{P}\{x_1\overline{x}_2\} + \mathcal{P}\{x_1\overline{x}_3\} - \mathcal{P}\{x_1\overline{x}_2\}\mathcal{P}\{x_1\overline{x}_3\} \\ &= x_1\overline{x}_2 + x_1\overline{x}_3 - x_1\overline{x}_2\overline{x}_3 \\ \mathcal{P}\{(x_1 \vee x_2) \oplus x_3\} &= \mathcal{P}\{(x_1 \vee x_2) + \mathcal{P}\{x_3\} \\ &- 2\mathcal{P}\{(x_1 \vee x_2)\mathcal{P}\{x_3\} \\ &= x_1 + x_2 + x_3 - x_1x_2 - 2x_1x_3 \\ &- 2x_2x_3 + 2x_1x_2x_3 \end{aligned}$$

5.2.6 Polarity

The polarity of a variable x_j can take the values:

(*i*) $c_j = 1$, corresponding to the uncomplemented variable x_j, or
(*ii*) $c_j = 0$, corresponding to the complemented variable $\overline{x}_j$.

Let the polarity $c = c_1, c_2, \ldots, c_n$, $c \in \{0, 1, 2, \ldots, 2^n - 1\}$, where c_j is the j-th bit of binary representation of c. For a switching function f of n

variables, the arithmetic expression, given the polarity $c = c_1, c_2, \ldots, c_n$ of variables $x_1, x_2, \ldots, x_n$ is as follows

$$f = \sum_{i=0}^{2^n-1} p_i \cdot \underbrace{(x_1 \oplus c_1)^{i_1} \cdots\ (x_n \oplus c_n)^{i_n}}_{i-th\ product}, \tag{5.9}$$

where p_i is the coefficient, and $(x_j \oplus c_j)^{i_j}$ is defined as

$$a_j^{i_j} = \begin{cases} 1, \text{ if } i_j = 0; \\ a, \text{ if } i_j = 1. \end{cases} \quad x_j \oplus c_j = \begin{cases} x_j, \text{ if } c_j = 0; \\ \overline{x}_j, \text{ if } c_j = 1. \end{cases} \tag{5.10}$$

Example 5.5 *Let $c = 2,\ c_1, c_2 = 1, 0$. The representation of a switching function of two variables by arithmetic expression of the polarity $c = 2$ can be derived by Equation 5.9 and Equation 5.10:*

$$\begin{aligned} f &= p_0(x_1 \oplus 1)^0(x_2 \oplus 0)^0 + p_1(x_1 \oplus 1)^0(x_2 \oplus 0)^1 \\ &+ p_2(x_1 \oplus 1)^1(x_2 \oplus 0)^0 + p_3(x_1 \oplus 1)^1(x_2 \oplus 0)^1 \\ &= p_0 + p_1 x_2 + p_2 \overline{x}_1 + p_3 \overline{x}_1 x_2. \end{aligned}$$

The coefficients p_i can be derived from the function's truth values by the following technique. Given the truth vector $\mathbf{F} = [f(0)\ f(1) \ldots f(2^n - 1)]^T$, the vector of arithmetic word-level coefficients of polarity c, $\mathbf{P}^{(c)} = [p_0^{(c)}\ p_1^{(c)} \ldots p_{2^n-1}^{(c)}]^T$ is derived by the matrix equation

$$\mathbf{P}^{(c)} = \mathbf{P}_{2^n}^{(c)} \cdot \mathbf{F}, \tag{5.11}$$

where the $2^n \times 2^n$-matrix $\mathbf{P}_{2^n}^{(c)}$ is generated by the Kronecker product

$$\mathbf{P}_{2^n}^{(c)} = \bigotimes_{j=1}^{n} \mathbf{P}_{2^1}^{(c_j)}, \qquad \mathbf{P}_{2^1}^{(c)} = \begin{cases} \begin{bmatrix} 1 & 0 \\ -1 & 1 \end{bmatrix}, & c_j = 0; \\ \\ \begin{bmatrix} 0 & 1 \\ 1 & -1 \end{bmatrix}, & c_j = 1. \end{cases} \tag{5.12}$$

Example 5.6 *Figure 5.4 demonstrates the derivation of coefficients p_i to Example 5.5 by matrix Equation 5.11. Here, the matrix $\mathbf{P}_{2^2}^{(2)}$ for the polarity $c = 2$ is generated by Equation 5.12 as follows:*

$$\mathbf{P}_{2^2}^{(2)} = \mathbf{P}_{2^1}^{(1)} \otimes \mathbf{P}_{2^1}^{(0)} = \begin{bmatrix} 0 & 1 \\ 1 & -1 \end{bmatrix} \otimes \begin{bmatrix} 1 & 0 \\ -1 & 1 \end{bmatrix}.$$

Note that in this example the two-output switching function is presented by the truth vector $\mathbf{F}$, which is a word-level interpretation of two functions, f_1 and f_2. The matrix manipulation of word-level vectors is the same as that of binary ones.

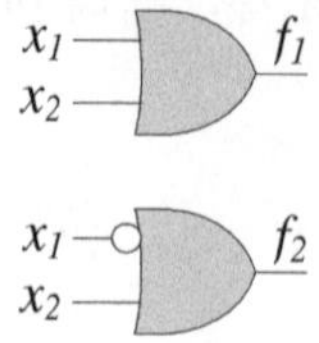

$f_1 = x_1 \vee x_2$
$f_2 = \overline{x}_1 \vee x_2$

$$\mathbf{F} = \left[\mathbf{F}_1 \middle| \mathbf{F}_2\right] = \begin{bmatrix} 1 & 0 \\ 1 & 1 \\ 0 & 1 \\ 1 & 1 \end{bmatrix} = \begin{bmatrix} 2 \\ 3 \\ 1 \\ 3 \end{bmatrix}$$

$$\mathbf{P}^{(2)} = \mathbf{P}^{(2)}_{2^2} \cdot \mathbf{F} = \left[\begin{array}{cc|cc} 0 & 0 & 1 & 0 \\ 0 & 0 & -1 & 1 \\ \hline 1 & 0 & -1 & 0 \\ -1 & 1 & 1 & -1 \end{array}\right] \begin{bmatrix} 2 \\ 3 \\ 1 \\ 3 \end{bmatrix} = \begin{bmatrix} 1 \\ 2 \\ 1 \\ -1 \end{bmatrix},$$

$$f = 1 + 2x_2 + \overline{x}_1 - \overline{x}_1 x_2$$

Alternatively,

$$\mathbf{P}^{(2)}_{\mathbf{1}} = \begin{bmatrix} 1 \\ 0 \\ -1 \\ 1 \end{bmatrix}, \quad \mathbf{P}^{(2)}_{\mathbf{2}} = \begin{bmatrix} 0 \\ 1 \\ 1 \\ -1 \end{bmatrix}$$

$$\mathbf{P}^{(2)} = 2^1 \mathbf{P}^{(2)}_{\mathbf{2}} + 2^0 \mathbf{P}^{(2)}_{\mathbf{1}} = \begin{bmatrix} 2 \\ 3 \\ 1 \\ 3 \end{bmatrix}$$

FIGURE 5.4
Computing the arithmetic expression of polarity $c = 2$ for two gates (Example 5.6).

5.3 Graphical representations

The following graphical data structures for arithmetical expressions of switching functions are considered in this section:

- Hypercube representations,
- Decision trees and diagrams,
- $\mathcal{N}$-hypercubes representations (details are given in Chapter 34).

5.3.1 Hypercube representation

Let a function be given by its cubes. To derive an arithmetic expression of the function, we can employ an algorithm similar to the one used to derive its ESOP form. However, it must be taken into account that operations over cubes in an arithmetic form are specific. The generation of a new cube is based on the equation $x \vee y = x + y - xy$. Given the cubes $[C_1]$ and $[C_2]$ in the sum-of-products expression, the cubes to be included in the arithmetical expression are derived as follows:

$$[C_1] \vee [C_2] = [C_1] + [C_2] - [C_1][C_2]. \tag{5.13}$$

Example 5.7 *Let* $f = \overline{x}_2 x_3 \vee x_1 x_3$, *i.e.*, $[C_1] = [\mathbf{x}\ 0\ 1]$ *and* $[C_2] = [1\ \mathbf{x}\ 1]$ *(Figure 5.5). To derive the arithmetic form, Equation 39.4 is applied and three cubes are produced:* $[C_1]$, $[C_2]$, *and the new cube* $-[C_1][C_2] = -[1\ 0\ 1]$. *Thus,*

$$
\begin{aligned}
f &= \overline{x}_2 x_3 + x_1 x_3 - x_1 \overline{x}_2 x_3 \\
&= (1 - x_1)\overline{x}_2 x_3 + x_1 x_3 \\
&= \overline{x}_1 \overline{x}_2 x_3 + x_1 x_3,
\end{aligned}
$$

that is, two cubes $[0\ 0\ 1]$ *and* $[0\ \mathbf{x}\ 1]$ *are needed to represent the resulting arithmetic form.*

A cube that corresponds to a product in the arithmetic expression of a switching function is composed of the components $\{0,\ 1,\ \mathbf{x},\ a, b\}$, where

$$
\begin{aligned}
a &= -\overline{x}_i + x_i = (-1)^{\overline{x}_i} \\
b &= -x_i + \overline{x}_i = (-1)^{x_i}.
\end{aligned}
$$

Also, $1 - x = \overline{x}$ and $1 - \overline{x} = x$.

Example 5.8 *Given the arithmetic expression:*
(a) $f = -\overline{x}_1 x_2 \overline{x}_3 + x_1 x_2 \overline{x}_3$, *its cube form is derived as follows:* $f = -\overline{x}_1 x_2 \overline{x}_3 + x_1 x_2 \overline{x}_3 = x_2 \overline{x}_3(-\overline{x}_1 + x_1) = (-1)^{\overline{x}_1} x_2 \overline{x}_3$, *which corresponds to* $f = [a\ 1\ 0]$.
(b) $f = \overline{x}_2 x_3 - x_1 \overline{x}_2 x_3$, *we get* $f = (1 - x_1)\overline{x}_2 x_3 = \overline{x}_1 \overline{x}_2 x_3$, *i.e.*, $f = [\mathbf{x}\ 0\ 1] - [1\ 0\ 1] = [0\ 0\ 1]$.

5.3.2 Decision trees and diagrams

A binary decision tree that corresponds to the arithmetic canonical representation of a switching function is called an *arithmetic* decision tree.

> *Algebraic decision trees are associated with algebraic decision diagrams, also called* arithmetic *or* functional *diagrams.*

A node in an algebraic decision tree of a switching function f corresponds to the arithmetic analog of the Davio decomposition of f with respect to a variable x_i. There exists:

- ► *The arithmetic analog of the positive Davio* expansion

$$f = f_0 + x_i f_2, \tag{5.14}$$

where $f_0 = f|_{x_i=0}$ and $f_2 = -f|_{x_i=1} + f|_{x_i=0}$, and

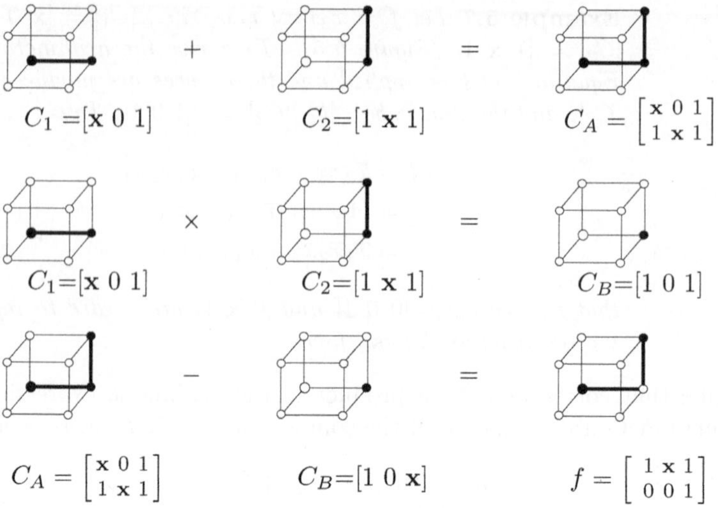

C_1 =[x 0 1] + C_2=[1 x 1] = $C_A = \begin{bmatrix} \text{x}\ 0\ 1 \\ 1\ \text{x}\ 1 \end{bmatrix}$

C_1=[x 0 1] × C_2=[1 x 1] = C_B=[1 0 1]

$C_A = \begin{bmatrix} \text{x}\ 0\ 1 \\ 1\ \text{x}\ 1 \end{bmatrix}$ − C_B=[1 0 x] = $f = \begin{bmatrix} 1\ \text{x}\ 1 \\ 0\ 0\ 1 \end{bmatrix}$

FIGURE 5.5
Computing the cubes of a switching function of three variables by the rule $f = [C_1] + [C_2] - [C_1][C_2]$ (Example 5.7).

► *The arithmetic analog of the negative Davio* expansion

$$f = f_1 + \overline{x}_i f_2, \tag{5.15}$$

where $f_1 = f|_{x_i=1}$.

The arithmetic analogs of positive and negative Davio expansion will be labeled as pD_A and nD_A correspondingly (Figure 5.6).

In matrix notation, the expansion of a switching function f given by the truth-vector $\mathbf{F} = [\ f(0)\ f(1)\ ]^T$, implemented in the non-terminal nodes, is defined as

$$f = [1\ x_i\ ] \begin{bmatrix} 1 & 0 \\ -1 & 1 \end{bmatrix} \begin{bmatrix} f_0 \\ f_1 \end{bmatrix} = [1\ x_i\ ] \begin{bmatrix} f_0 \\ -f_0 + f_1 \end{bmatrix}$$
$$= f_0 + x_i(-f_0 + f_1) = f_0 + x_i f_2,$$

where $f_0 = f|_{x_i=0}$, $f_2 = -f_0 + f_1$. Recursive application of the arithmetic analog of positive Davio expansion to a function f given by its truth-vector $\mathbf{F} = [f(0)\ f(1) \ldots f(2^n - 1)]^T$ is expressed in matrix notation as

$$f = \widehat{\mathbf{X}}\, \mathbf{P}_{2^n}\, \mathbf{F}, \tag{5.16}$$

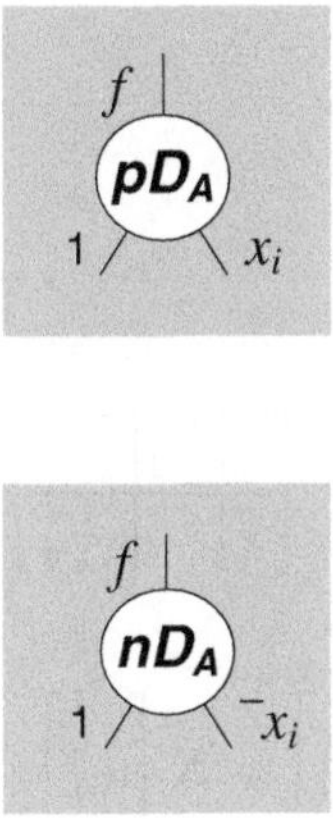

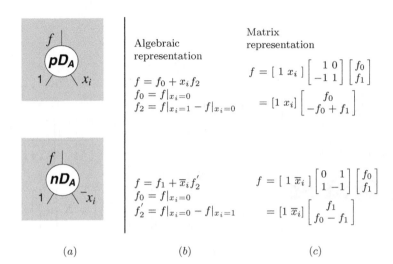

Algebraic representation

$$f = f_0 + x_i f_2$$
$$f_0 = f|_{x_i=0}$$
$$f_2 = f|_{x_i=1} - f|_{x_i=0}$$

$$f = f_1 + \overline{x}_i f_2'$$
$$f_0 = f|_{x_i=0}$$
$$f_2' = f|_{x_i=0} - f|_{x_i=1}$$

Matrix representation

$$f = [\ 1\ x_i\ ]\begin{bmatrix} 1 & 0 \\ -1 & 1 \end{bmatrix}\begin{bmatrix} f_0 \\ f_1 \end{bmatrix} = [1\ x_i]\begin{bmatrix} f_0 \\ -f_0 + f_1 \end{bmatrix}$$

$$f = [\ 1\ \overline{x}_i\ ]\begin{bmatrix} 0 & 1 \\ 1 & -1 \end{bmatrix}\begin{bmatrix} f_0 \\ f_1 \end{bmatrix} = [1\ \overline{x}_i]\begin{bmatrix} f_1 \\ f_0 - f_1 \end{bmatrix}$$

(*a*) (*b*) (*c*)

FIGURE 5.6
The node of a positive and negative Davio tree (a), algebraic (b) and matrix (c) descriptions.

where

$$\widehat{\mathbf{X}} = \bigotimes_{i=1}^{n} [\ 1\ x_i\ ],$$
$$\mathbf{P}_{2^n} = \bigotimes_{i=1}^{n} \mathbf{P}_2,$$
$$\mathbf{P}_2 = \begin{bmatrix} 1 & 0 \\ 1 & 1 \end{bmatrix},$$

and $\otimes$ denotes the Kronecker product.

> **Example 5.9** *Let us derive the algebraic decision tree of the switching function* $f = \overline{x}_1 \vee x_2$ *given by the truth-vector* $\mathbf{F} = [1\ 1\ 0\ 1]^T$. *The solution to Equation 5.16 is shown in Figure 5.7. The product terms are generated by the Kronecker product* $\widehat{\mathbf{X}}$. *The* 4×4 *transform matrix* $\mathbf{P}$ *is generated by the Kronecker product over the basic matrix* $\mathbf{P}_{2^1}$. *The final result, the arithmetic coefficients, is directly mapped into the complete algebraic decision tree.*

5.3.3 Structural properties

The structural properties of the algebraic decision tree are similar to the Reed–Muller decision tree, except:

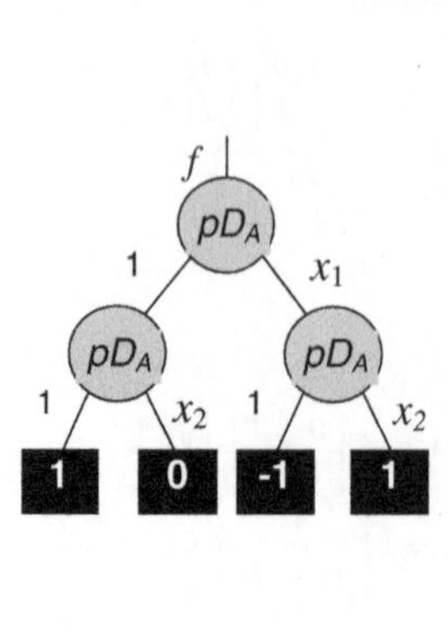

$$\widehat{\mathbf{X}} = [\,1\ x_1\,] \otimes [\,1\ x_2\,] = [\,1,\ x_2,\ x_1,\ x_1x_2\,]$$

$$\mathbf{P}_{2^2} = \mathbf{P}_2 \otimes \mathbf{P}_2 = \begin{bmatrix} 1 & 0 \\ -1 & 1 \end{bmatrix} \otimes \begin{bmatrix} 1 & 0 \\ -1 & 1 \end{bmatrix} = \left[\begin{array}{cc|cc} 1 & & & \\ -1 & 1 & & \\ \hline -1 & & 1 & \\ 1 & -1 & -1 & 1 \end{array}\right]$$

$$f = \widehat{\mathbf{X}}\,\mathbf{P}_{2^2}\,\mathbf{F} = \widehat{\mathbf{X}} \left[\begin{array}{cc|cc} 1 & & & \\ -1 & 1 & & \\ \hline -1 & & 1 & \\ 1 & -1 & -1 & 1 \end{array}\right] \begin{bmatrix} 1 \\ 1 \\ 0 \\ 1 \end{bmatrix} = \widehat{\mathbf{X}} \begin{bmatrix} 1 \\ 0 \\ -1 \\ 1 \end{bmatrix} = 1 - x_1 + x_1x_2$$

FIGURE 5.7
Derivation of the algebraic decision tree for the switching function $f = \overline{x}_1 \vee x_2$ (Example 5.9).

- The values of terminal nodes are integer numbers and correspond to the coefficients of the arithmetic expression.
- Each path from the root to a terminal node corresponds to a product in the arithmetic expression.
- The values of constant nodes are the values of the arithmetic spectrum in the positive polarity for the represented functions. Thus, they are elements of the vector of arithmetic coefficients

$$\mathbf{P}_f = [f_{000}\ f_{002}\ f_{020}\ f_{022}\ f_{200}\ f_{202}\ f_{220}\ f_{222}],$$

where "0" corresponds to the value of $f_0 = f|_{x_i=0}$, and "2" corresponds to the value of $-f|_{x_i=0} + f|_{x_i=1}$, i.e.,

$$\text{"0"} \rightarrow f_0 = f|_{x_i=0}$$
$$\text{"2"} \rightarrow -f|_{x_i=0} + f|_{x_i=1}$$

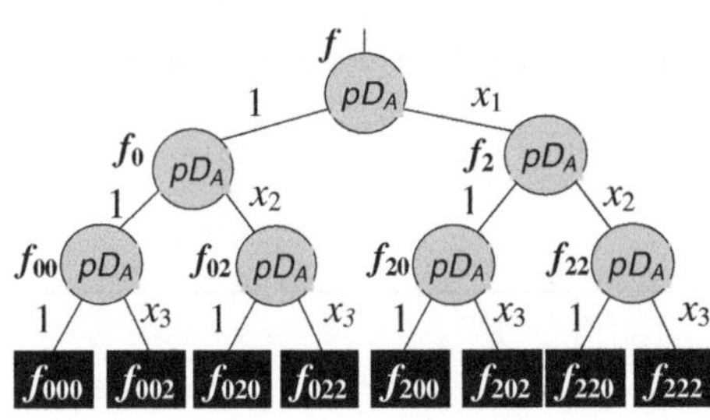

There are 8 paths from f to the terminal nodes:

Path 1: $t_1 = 1$
Path 2: $t_2 = x_3$
Path 3: $t_3 = x_2$
Path 4: $t_4 = x_2x_3$
Path 5: $t_5 = x_1$
Path 6: $t_6 = x_1x_3$
Path 7: $t_7 = x_1x_2$
Path 8: $t_8 = x_1x_2x_3$

Arithmetic expression
$f = f_{000}t_1 + f_{002}t_2 + \dots$
$+ f_{222}t_8$

FIGURE 5.8
Arithmetic representation of a switching function of three variables by an algebraic tree ($c = 000$) (Example 5.10).

Example 5.10 *An arbitrary switching function f of three variables can be represented by the algebraic decision tree shown in Figure 5.8 (3 levels, 7 nodes, 8 terminal nodes). To design this tree, the positive arithmetic expansion (Equation 5.14) is used as follows:*

(a) *with respect to the variable x_1:*

$$f = f_0 + x_1 f_2$$

(b) *with respect to the variable x_2:*

$$f_0 = f_{00} + x_2 f_{02}$$
$$f_1 = f_{10} + x_2 f_{22}$$

(c) *with respect to the variable x_3:*

$$f_{00} = f_{000} + x_3 f_{002}, \qquad f_{02} = f_{020} + x_3 f_{022},$$
$$f_{20} = f_{200} + x_3 f_{202}, \qquad f_{22} = f_{220} + x_3 f_{222}.$$

Hence, the algebraic decision tree represents the following expansion of a switching function f in the form of the logic expression

$$f = f_{000} + f_{002}x_3 + f_{020}x_2 + f_{022}x_2x_3 + f_{200}x_1 + f_{202}x_1x_3 + f_{220}x_1x_2 + f_{222}x_1x_2x_3.$$

5.3.4 Decision diagram reduction

The algebraic decision diagram also called arithmetic transform decision diagram (ACDD), is an arithmetic analog of Davio diagram. It is derived from the algebraic decision tree by deleting redundant nodes, and by sharing equivalent subgraphs. The rules are similar to the ones used for deriving Davio diagrams (AND-EXOR representation).

5.4 Further reading

Historical remarks. Apparently, the first attempts to present logic operations by arithmetical ones were taken by the founder of Boolean algebra George Boolé (1854). He did not use the Boolean operators that are well known today. Rather, he used arithmetic expressions. It is interesting to note that Aiken first found that arithmetic expressions can be useful to design circuits and used them in the Harvard MARK 3 and MARK 4 computers [1].

In today's computers, register-transfer-level descriptions specify the architecture at the bit level of the registers, even for word-level arithmetical components. However, word-level descriptions can be simulated significantly faster than bit-level descriptions. Word-level representations are also useful for reducing the complexity of component matching since the number of words is significantly smaller than the number of bits.

Relationship to AND-EXOR expressions. Arithmetic expressions are closely related to AND-EXOR expressions. However, with variables and function values interpreted as integers 0 and 1 instead of logic values, arithmetic expressions are considered as integer counterparts of AND-EXOR expressions. The computing techniques developed for AND-EXOR forms work for arithmetical forms as well.

Relationship to spectral techniques. The spectral computation of word-level expressions is discussed in [70, 17, 28]. Heidtmann [6] used the arithmetic spectra to testing of circuits. Details are discussed in Chapters 13 and 16.

Other arithmetic transforms. Discrete Walsh functions are a discrete version of the functions introduced by Walsh in 1923 for solving some problems in the approximation of square-integrable functions on the interval $0, 1$. The basic Walsh matrix is defined as $W_2 = \begin{bmatrix} 1 & 1 \\ 1 & -1 \end{bmatrix}$. The Walsh transform matrix is constructed using the Kronecker product as $W_{2^i} = \begin{bmatrix} W_{2^{i-1}} & W_{2^{i-1}} \\ W_{2^{i-1}} & -W_{2^{i-1}} \end{bmatrix}$. These are the so-called *Hadamard-ordered* Walsh functions. The Walsh functions possess symmetry properties, and due to this the Walsh matrix is orthogonal, symmetric, and self-inverse (with a normalization constant of 2^{-n}). The Walsh functions take two values, +1 and -1, and in that respect are compatible with switching functions, which are also two-valued. The discrete Walsh transform is defined as $\mathbf{W} = W_{2^n}\mathbf{F}$. The algebraic expressions for arithmetic, Walsh, and Haar spectra were obtained for switching functions by Shmerko [14]. These results were generalized for various discrete basis for multivalued functions by Kukharev et al. [8, 9]. Details are discussed in Chapters 7 and 30.

More details on techniques of manipulation of arithmetic expressions using decision diagrams can be found in [7], [3], [11], [12], [70], [17], [20].

Word-level arithmetic representations and decision diagrams are discussed in Chapters 6 and 17.

Linear word-level arithmetic expressions and linear decision diagrams are discussed in Chapter 37.

Probabilistic computing. Arithmetic logic has many applications in contemporary logic design, for example, in the computation of signal probabilities for test generators, and switching activities for power and noise analysis. Jain introduces the probabilistic computing of arithmetic transforms [7].

References

[1] Aiken IIII. Synthesis of electronic computing and control circuits. *Ann. Computation Laboraratory of Harvard University*, XXVII, Harvard University, Cambridge, MA, 1951.

[2] Davio MJ, Deschamps P, and Thayse A. *Discrete and Switching Functions.* McGraw-Hill, New York, 1978.

[3] Falkowski BJ, Shmerko VP, and Yanushkevich SN. Arithmetical logic – its status and achievements. In *Proceedings of the International Conference on Applications of Computer Systems*, Technical University of Szczecin, Poland, pp. 208–223, 1997.

[4] Falkowski BJ. A note on the polynomial form of Boolean functions and related topics. *IEEE Transactions on Computers*, 48(8):860–863, 1999.

[5] Falkowski BJ and Stanković RS. Spectral interpretation and applications of decision diagrams. *VLSI Design International Journal of Custom Chip Design, Simulation and Testing*, 11(2):85–105, 2000.

[6] Heidtmann KD. Arithmetic spectrum applied to fault detection for combinational networks, *IEEE Trans.actions on Computers,* 40(3):320–324, 1991.

[7] Jain J. Arithmetic transform of Boolean functions. In Sasao T and Fujita M, Eds., *Representations of Discrete Functions*, pp. 55–92, Kluwer, Dordrecht, 1996.

[8] Kukharev GA, Shmerko VP, and Zaitseva EN. *Algorithms and Systolic Arrays for Multivalued Data Processing,* Science and Technics Publishers, Minsk, Belarus, 1990 (In Russian).

[9] Kukharev GA, Shmerko VP, and Yanushkevich SN *Parallel Binary Processing Techniqies* Higher School Publishers, Minsk, Belarus, 1991 (In Russian).

[10] Malyugin VD. Representation of Boolean functions by arithmetical polynomials. *Automation and Remote Control*, Kluwer/Plenum Publishers, 43(4):496–504, 1982.

[11] Minato S. *Binary Decision Diagrams and Applications for VLSI Design.* Kluwer, Dordrecht, 1996.

[12] Papaioannou SG. Optimal test generation in combinational networks by pseudo-Boolean programming. *IEEE Transactions on Computers*, 26:553–560, 1977.

[13] Roth JP. *Mathematical Design. Building Reliable Complex Computer Systems.* IEEE Press, New York, 1999.

[14] Shmerko VP. Synthesis of arithmetic forms of Boolean functions using the Fourier transform. *Automation and Remote Control*, Kluwer/Plenum Publishers, 50(5):684–691, Pt2, 1989.

[15] Stanković RS and Astola JT. *Spectral Interpretation of Decision Diagrams.* Springer, Heidelberg, 2003

[16] Stanković RS, Moraga C, and Astola JT. Reed–Muller expressions in the previous decade. In *Proceedings of the 5th International Workshop on Applications of the Reed–Muller Expansion in Circuit Design*, pp. 7–26, Mississippi State University, MS, 2001.

[17] Thornton M and Nair V. Efficient calculation of spectral coefficients and their applications. *IEEE Transactions on Computer-Aided Design of Integrated Circuits and Systems*, 14(1):1328–13411, 1995.

[18] Yanushkevich SN. Computer arithmetic, In Dorf R, Ed., *The Electrical Engineering Handbook*, 3rd Edition, CRC Press, Boca Raton, FL, 2005.

[19] Yanushkevich SN. Arithmetical canonical expansions of Boolean and MVL functions as generalized Reed–Muller series. In *Proceedings of the IFIP WG 10.5 Workshop on Applications of the Reed–Muller Expansions in Circuit Design*, pp. 300–307, Japan, 1995.

[20] Yanushkevich SN. Multiplicative properties of spectral Walsh coefficients of the Boolean function. *Automation and Remote Control*, Kluwer/Plenum Publishers, 64(12):1938–1947, 2003.

6

Word - Level Representations

The word-level representations of switching functions introduce parallelism to their computation by means of the parallel processing of several functions (words) at once. Such processing (and also simulation) of the word-level description is much faster than bit-level. Given an r-output switching function f (outputs $f_1, f_2, ..., f_r$), these outputs can be grouped in a bit-string (word) by several methods. In this chapter, two approaches to word-level construction are introduced. The first approach uses the linearity property of arithmetic transforms to produce a word-level arithmetic expansion. The second approach, grouping switching functions in a word, is based on the algebra of corteges. Switching functions can be restored after processing the word-level expression used to represent them.

6.1 Introduction

Methods for grouping switching functions into a word-based representation must satisfy several requirements; in particular, efficiency of parallel processing and simulation, verification, etc. In this chapter, two methods for designing a word-based representation are introduced:

- A method based on the linear property of arithmetic transform, and
- A method based on the algebra of corteges.

In grouping, the positional number system principle is used: each symbol (from a finite set of symbols) in the word is weighted; the weight determines the location of a symbol within the word.

Example 6.1 *The decimal and binary number systems are examples of a positional system (10 symbols are used in the decimal and two symbols in the binary system). The Roman number system is an example of a nonpositional number system (different symbols are used, each having a fixed definite quantity associated with it).*

The problem is formulated as follows: given a multi-output switching function, represent these outputs by a word-level arithmetic expression using a positional principle of grouping. There are two steps in the solution to this problem:

(a) Represent each output by an arithmetic expression, and

(b) Derive a compact word-level arithmetic expression using the linearity property of the arithmetic transform.

Given an r-output switching function f of n variables, the arithmetic word-level expression is the weighted sum of the arithmetic expressions for the outputs f_j, $t = 1, \ldots, r$,

$$f = 2^{r-1} f_r + \cdots + 2^1 f_2 + 2^0 f_1. \tag{6.1}$$

Equation 6.1 can be rewritten in the form

$$f = \sum_{i=0}^{2^n-1} d_i \cdot \underbrace{(x_1^{i_1} \cdots\ x_n^{i_n})}_{i-th\ product} \tag{6.2}$$

where the coefficient d_i is an integer number, i_j, $j = 1, 2, \ldots, n$, is the j-th bit in the binary representation of the index $i = i_1 i_2 \ldots i_n$, and $x_j^{i_j}$ is defined as

$$x_j^{i_j} = \begin{cases} 1, & i_j = 0; \\ x_j, & i_j = 1. \end{cases}$$

Expression 6.2 is an *arithmetic word-level* representation of a multi-output switching function.

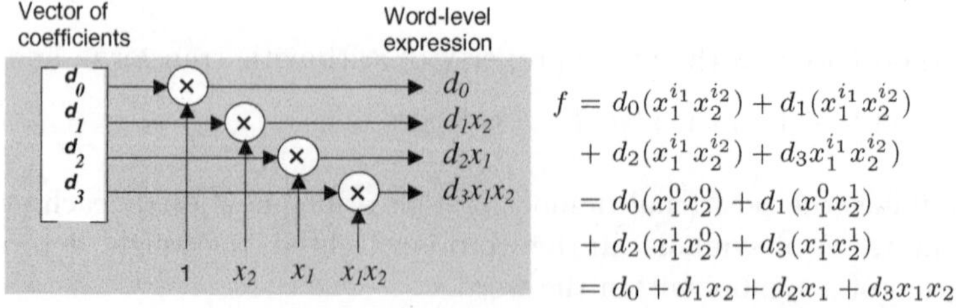

FIGURE 6.1
Determination of an arithmetic word-level expression for an r-output switching function of two variables (Example 6.2).

Example 6.2 *An arbitrary r-output switching function of two variables ($n = 2$) is represented by the arithmetic word-level expression using Equation 6.2, $i = 0, 1, 2, 3$. Figure 6.1 illustrates the structure of this expression.*

There are two ways of interpreting bits within a word:

- ▶ Direct positioning, and
- ▶ Cortege based grouping.

Direct positioning means a bit or digit is interpreted as being referred to by a fixed position number system. Regarding arithmetical forms, the position of the i-th bit f_i in an n-bit word $f_{n-1} \dots f_0$ corresponds to 2^i, since the decimal number is formed as $d = f_{n-1}2^{n-1} + \dots + f_1 2^1 + f_0 2^0$.

The inverse to the grouping procedure, *masking,* is used to extract information from positions of bits in the word-level expression. The *masking operator* $\Xi^r\{f\}$ is used to recover the single function f_r from the word-level representation of the r-output switching function f

$$f_r = \Xi^r\{f\}.$$

The masking operator is applied to extract functions from positions or values of functions if the assignments of variables are given.

Example 6.3 *Given a word-level arithmetic expression f that describes a 3-output switching function. Figure 6.2 illustrates the masking operator for this function.*

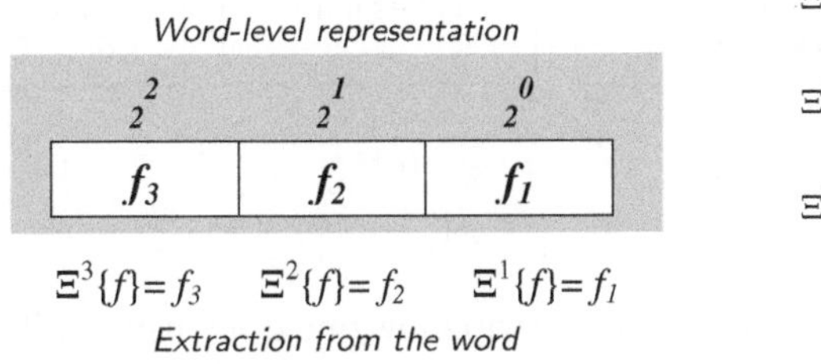

$$\begin{aligned}\Xi^1\{f\} &= x_1 \oplus x_2\\ &= f_1\\ \Xi^2\{f\} &= x_1 x_2\\ &= f_2\\ \Xi^3\{f\} &= x_1 \vee x_2\\ &= f_3\end{aligned}$$

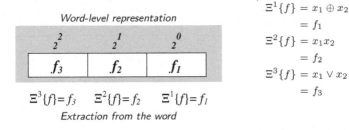

FIGURE 6.2
The masking operator extracts the function f_i from a word-level representation of the function f (Example 6.3).

Cortege-based grouping. In *cortege-based* grouping, non-direct positional grouping is used*. Each position in a word is associated with a cortege or its elements. In contrast to a direct positional interpretation of arithmetic (word-level) expressions, a cortege-based word contains Boolean expressions that describe a multi-output switching function. Computing a word-level expression for given assignments of variables requires:

- Computing an element (positive integer number) of a truth vector, and
- Recovering the output values of the switching function.

The following approaches to computing a word-level expression of a switching function are given in Figure 6.3:

- *Algebraic* (Figure 6.3a),
- *Matrix* (Figure 6.3b),
- *Arithmetic Taylor expansion* (Figure 6.3c), and
- *Cortege-based grouping* (Figure 6.3d).

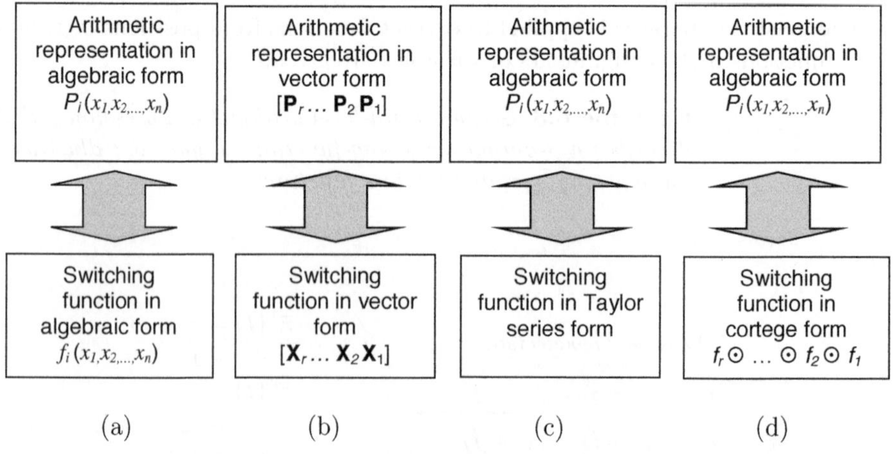

FIGURE 6.3
Methods of word-level representation of switching functions using arithmetic description in various forms.

*Examples in positional number systems are conversions between binary and octal numbers, binary and hexadecimal, etc.

6.2 Arithmetic word-level computation in matrix form

Algebraic representation is based on the linearity property of the arithmetic transform (Figure 6.3a). Let $\mathcal{A}\{f\}$ be an arithmetic transform of switching function f represented in the form

$$f = f_1 + f_2 \cdots + f_r$$

The arithmetic representation of this function $\mathcal{A}\{f\}$ can be calculated as follows

$$\mathcal{A}\{f\} = \mathcal{A}\{f_1 + f_2 \cdots + f_r\} = 1 - \prod_{i=1}^{r} \mathcal{A}\{\overline{f}_i\},$$

where $\mathcal{A}\{\overline{f}_i\} = 1 - \mathcal{A}\{f_i\}$. For two switching functions f_1 and f_2, Expression 6.2 can be rewritten in the form

$$\mathcal{A}\{f\} = \mathcal{A}\{f_1 + f_2\} = \mathcal{A}\{f_1\} + \mathcal{A}\{f_2\} - \mathcal{A}\{f_1\}\mathcal{A}\{f_1\}.$$

Example 6.4 *In Figure 6.4, arithmetic representations of elementary gates and 3-input 3-output circuits are given. The linearity property of arithmetic transform is used.*

Example 6.5 *Using Equations 6.1 and 6.2, the result of Example 6.4 can be represented in a word-level form.*

6.2.1 Application of direct and inverse arithmetic transforms to word-level expressions

The spectral technique, a direct arithmetic transform, is applied to compute coefficients of an arithmetic word-level expression. The inverse arithmetic transform is used to recover the switching function from a word-level arithmetic expression.

Direct and inverse arithmetic transforms. Let **F** and **D** be the truth vector of a multi-output switching function f and the vector of coefficients of its arithmetic representation respectively (Figure 6.3b). The relationship of **F** and **D** is defined by the direct and inverse arithmetic transforms

$$\mathbf{D} = \mathbf{A}_{2^n} \cdot \mathbf{F}, \qquad (6.3)$$

$$\mathbf{F} = \mathbf{A}_{2^n}^{-1} \cdot \mathbf{D}, \qquad (6.4)$$

where $\mathbf{A}_{2^n}$ and $\mathbf{A}_{2^n}^{-1}$ are matrix of direct and inverse arithmetic transform, respectively.

$\overline{x} = 1 - x$ | $x_1 \vee x_2 = x_1 + x_2 - x_1x_2$ | $x_1x_2 = x_1x_2$ | $x_1 \oplus x_2 = x_1 + x_2 - 2x_1x_2$

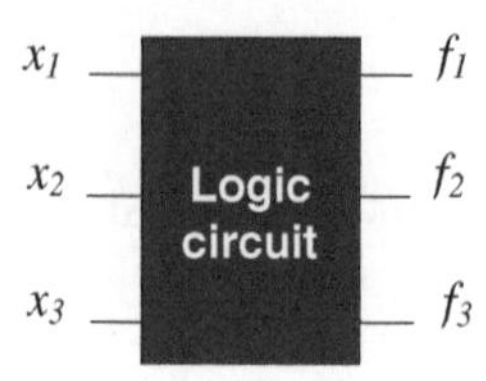

Arithmetic expressions of the outputs:

$$\mathcal{A}\{f_1\} = x_1\overline{x}_2 + x_1\overline{x}_3 - x_1\overline{x}_2\overline{x}_3$$

$$\begin{aligned}\mathcal{A}\{f_2\} &= x_1 + (x_1 \oplus x_3) - x_1(x_2 \oplus x_3)\\ &= x_1 + (x_1 + x_3 - 2x_2x_3) - x_1(x_1 + x_3 - 2x_2x_3)\\ &= x_1 + x_2 + x_3 - x_1x_2 - x_1x_3 - 2x_2x_3 + 2x_1x_2x_3\end{aligned}$$

$$\mathcal{A}\{f_3\} = x_1 + x_2 - 2x_1x_2$$

Word-level expression:

$$\begin{aligned}f &= 2^0 f_1 + 2^1 f_2 + 2^2 f_3\\ &= 7x_1 + 6x_2 + 2x_3 - 10x_1x_2 - 4x_1x_3 - 4x_2x_3 + 3x_1x_2x_3\end{aligned}$$

Outputs:

$$f_1 = x_1\overline{x}_2 \vee x_1\overline{x}_3$$

$$f_2 = x_1 \vee (x_2 \oplus x_3)$$

$$f_3 = x_1 \oplus x_2$$

FIGURE 6.4
Word-level arithmetic representation in algebraic form of elementary gates and 3-input 3-output logic circuits (Examples 6.4 and 6.5).

Example 6.6 *Figure 6.5 illustrates the relationship between the truth vector* **F** *and the vector of a coefficients* **D** *of an arithmetic word-level expression.*

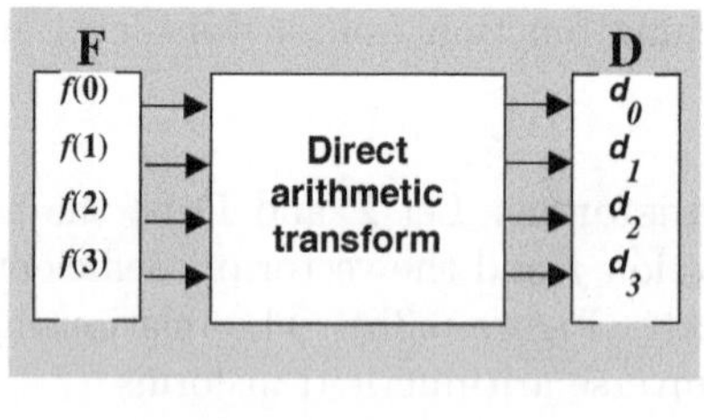

(a)

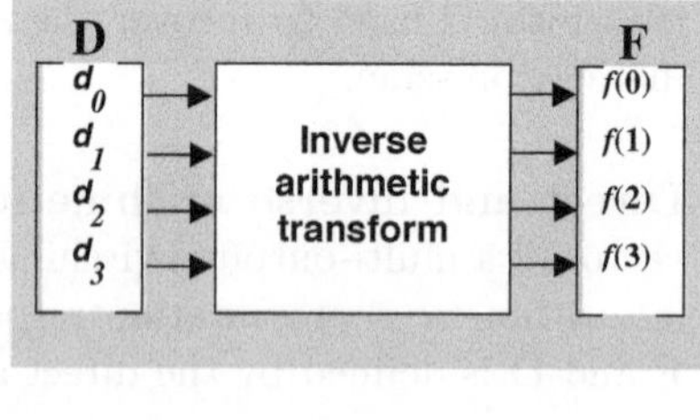

(b)

FIGURE 6.5
Direct (a) and inverse (b) arithmetic word-level transforms for an r-output switching function of two variables (Example 6.6).

Computing coefficients. A direct arithmetic transform (Equation 6.3) is used for representing a switching function by a vector of coefficients **D**.[†] The matrix of the direct arithmetic transform A_{2^n} is formed by the equation

$$\mathbf{A}_{2^n} = \bigotimes_{i=1}^{n} \mathbf{A}_2, \quad \mathbf{A}_2 = \begin{bmatrix} 1 & 0 \\ -1 & 1 \end{bmatrix},$$

where A_2 is 2×2 basis matrix and "$\otimes$" denotes the Kronecker product.

Example 6.7 *Computing the coefficients of a word-level arithmetic expression by Equations 6.3 and 6.4 for the two-output switching function*

$$f_1 = x_1 \vee x_2,$$
$$f_2 = \overline{x}_1 \vee x_2$$

given the truth-vectors $\mathbf{F_1}$ *and* $\mathbf{F_2}$ *is explained in Figure 6.6. Notice that the same result can be obtained by algebraic manipulations.*

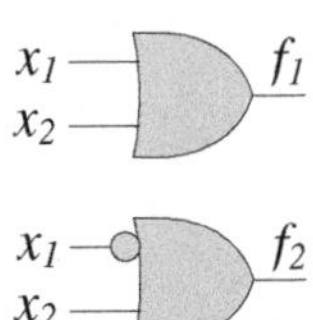

Outputs:

$$f_1 = x_1 \vee x_2 = x_2 + x_1 - x_1x_2$$
$$f_2 = \overline{x}_1 \vee x_2 = 1 - x_1 + x_1x_2$$

Computing in matrix form

The truth vector $\mathbf{F}$ *of a word-level arithmetic expression:*

$$\mathbf{F} = [\mathbf{F_2}|\mathbf{F_1}] = \begin{bmatrix} 1 & 0 \\ 1 & 1 \\ 0 & 1 \\ 1 & 1 \end{bmatrix} = \begin{bmatrix} 2 \\ 3 \\ 1 \\ 3 \end{bmatrix}.$$

$$\mathbf{D} = \mathbf{P}_{2^2} \cdot \mathbf{F} = \left[\begin{array}{rr|rr} 1 & 0 & 0 & 0 \\ -1 & 1 & 0 & 0 \\ \hline -1 & 0 & 1 & 0 \\ 1 & -1 & -1 & 1 \end{array}\right] \begin{bmatrix} 2 \\ 3 \\ 1 \\ 3 \end{bmatrix} = \begin{bmatrix} 2 \\ 1 \\ -1 \\ 1 \end{bmatrix}$$

that is

$$f = 2 + x_2 - x_1 + x_1x_2$$

Computing in algebraic form:

$$f = 2^1 f_2 + 2^0 f_1 = 2^1(1 - x_1 + x_1x_2) + 2^0(x_2 + x_1 - x_1x_2) = 2 + x_2 - x_1 + x_1x_2$$

FIGURE 6.6
Computing the arithmetic word-level expression for some gates (Example 6.7).

[†]This vector is also called an *arithmetic spectrum* of a multi-output switching function f.

Example 6.8 *Figure 6.7 shows a three-output switching function given in Example 6.5, and its truth vector* $\mathbf{F}$. *Application of the direct arithmetic transform (Equation 6.3) yields the vector of coefficient* $\mathbf{D}$ *of the arithmetic expression.*

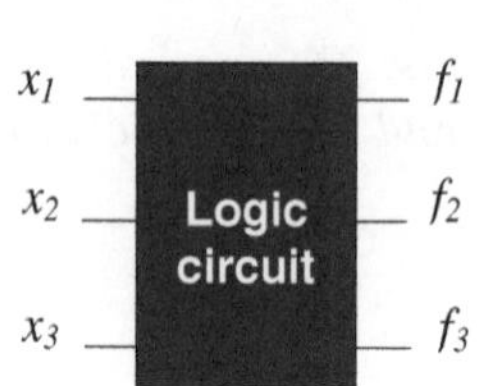

Outputs:

$$f_1 = x_1\overline{x}_2 \vee x_1\overline{x}_3 \equiv \mathbf{F}_1 = [00001110]^T$$

$$f_2 = x_1 \vee (x_2 \oplus x_3) \equiv \mathbf{F}_2 = [01101001]^T$$

$$f_3 = x_1 \oplus x_2 \equiv \mathbf{F}_2 = [00111100]^T$$

Truth vector of the 3-output switching function:

$$\mathbf{X} = \underbrace{[\mathbf{F}_3|\mathbf{F}_2|\mathbf{F}_1]}_{\text{Truth vector}} = \begin{bmatrix} 0\,0\,0 \\ 0\,1\,0 \\ 1\,1\,0 \\ 1\,0\,0 \\ 1\,1\,1 \\ 1\,0\,1 \\ 0\,0\,1 \\ 0\,1\,0 \end{bmatrix} = \begin{bmatrix} 0 \\ 2 \\ 6 \\ 4 \\ 7 \\ 5 \\ 1 \\ 2 \end{bmatrix}$$

Vector of coefficients $\mathbf{D}$ *of the word-level arithmetic expression:*

$$\mathbf{D} = \mathbf{A}_{2^3} \cdot \mathbf{X}$$

$$= \begin{bmatrix} 1 & 0 & 0 & 0 & 0 & 0 & 0 & 0 \\ -1 & 1 & 0 & 0 & 0 & 0 & 0 & 0 \\ -1 & 0 & 1 & 0 & 0 & 0 & 0 & 0 \\ 1 & -1 & -1 & 1 & 0 & 0 & 0 & 0 \\ -1 & 0 & 0 & 0 & 1 & 0 & 0 & 0 \\ 1 & -1 & 0 & 0 & -1 & 1 & 0 & 0 \\ 1 & 0 & -1 & 0 & -1 & 0 & 1 & 0 \\ -1 & 1 & 1 & -1 & 1 & -1 & -1 & 1 \end{bmatrix} \begin{bmatrix} 0 \\ 2 \\ 6 \\ 4 \\ 7 \\ 5 \\ 1 \\ 2 \end{bmatrix} = \begin{bmatrix} 0 \\ 2 \\ 6 \\ 4 \\ 7 \\ 5 \\ 1 \\ 2 \end{bmatrix} \begin{matrix} \\ x_3 \\ x_2 \\ x_2x_3 \\ x_1 \\ x_1x_3 \\ x_1x_2 \\ x_1x_2x_3 \end{matrix}$$

Word-level arithmetic expression:

$$f = 7x_1 + 6x_2 + 2x_3 - 10x_1x_2 - 4x_1x_3 - 4x_2x_3 + 3x_1x_2x_3$$

FIGURE 6.7
Word-level arithmetic representation in matrix form of a three-input and three output logic circuit (Example 6.8).

Recovering switching function. The inverse arithmetic transform (Equation 6.3) recovers the truth vector $\mathbf{F}$ from the vector of coefficients $\mathbf{D}$ of a word-level arithmetic expression. The inverse transform matrix $\mathbf{A}^{-1}$ is formed by the equation

$$\mathbf{A}_{2^n}^{-1} = \bigotimes_{i=1}^{n} \mathbf{A}_2^{-1},$$

where $\mathbf{A}_2^{-1} = \begin{bmatrix} 1 & 0 \\ 1 & 1 \end{bmatrix}$ is $2^1 \times 2^1$ basis matrix.

Example 6.9 *Given the word-level arithmetic expression*

$$f = 2 + 3x_1 + 5x_2$$

of a switching function f, recover this function. The inverse arithmetic transform (Equation 6.4) must be used to compute the truth vector from the vector of coefficients $\mathbf{D} = [2\ 5\ 3\ 0]^T$:

$$\mathbf{F} = \mathbf{A}_{2^n}^{-1} \cdot \mathbf{D} = \begin{bmatrix} 1&0&0&0 \\ 1&1&0&0 \\ 1&0&1&0 \\ 1&1&1&1 \end{bmatrix} \begin{bmatrix} 2 \\ 5 \\ 3 \\ 0 \end{bmatrix} = \begin{bmatrix} 2 \\ 7 \\ 5 \\ 10 \end{bmatrix} = \begin{bmatrix} 0&0&1&0 \\ 0&1&1&1 \\ 0&1&0&1 \\ 1&0&1&0 \end{bmatrix}$$
$$= [\mathbf{F}_4|\mathbf{F}_3|\mathbf{F}_2|\mathbf{F}_1].$$

Hence, the word-level arithmetic expression f represents a four-output switching function $f_4 = x_1x_2$, $f_3 = x_1 \oplus x_2$, $f_2 = \overline{x}_1 \vee x_2$, $f_1 = x_1 \oplus x_2$.

6.2.2 Arithmetic word-level form in a given polarity

The polarity of a variable x_j can be either:

$c_j = 1$ corresponding to the uncompliemented variable x_j, or
$c_j = 0$ corresponding to the complemented variable $\overline{x}_j$.

Let the polarity of an arithmetical expression be

$$c = c_1c_2 \dots c_n, \ c \in \{0, 1, 2, \dots, 2^n - 1\},$$

where c_j is the j-th bit of binary representation of c. For an m-output switching function f of n variables, the arithmetic word-level form of a given polarity $c = c_1, c_2, \dots, c_n$ of variables $x_1, x_2, \dots, x_n$ is written as

$$f = \sum_{i=0}^{2^n-1} d_i \cdot \underbrace{(x_1 \oplus c_1)^{i_1} \cdots (x_n \oplus c_n)^{i_n}}_{i-th\ product}, \tag{6.5}$$

where d_i is the coefficient, and $(x_j \oplus c_j)^{i_j}$ is defined as

$$a_j^{i_j} = \begin{cases} 1, \text{ if } i_j = 0; \\ a, \text{ if } i_j = 1. \end{cases} \quad x_j \oplus c_j = \begin{cases} x_j, \text{ if } c_j = 0; \\ \overline{x}_j, \text{ if } c_j = 1. \end{cases} \tag{6.6}$$

Example 6.10 *Represent the given two-output switching function of two variables by the arithmetic word-level expression of the polarity $c = 2$, $c_1c_2 = 10$. Equation 6.5 and Equation 6.6 are applied to derive a word-level expression (Figure 6.8).*

Notice that the matrix $\mathbf{P}_{2^2}^{(2)}$ for the polarity $c = 2$ in Example 6.10 is generated by Equation 6.6 as

$$\mathbf{P}_{2^2}^{(2)} = \mathbf{P}_{2^1}^{(1)} \otimes \mathbf{P}_{2^1}^{(0)} = \begin{bmatrix} 0 & 1 \\ 1 & -1 \end{bmatrix} \otimes \begin{bmatrix} 1 & 0 \\ -1 & 1 \end{bmatrix}.$$

Given a truth vector $\mathbf{F} = [f(0)\ f(1) \ldots f(2^n - 1)]^T$, the vector of arithmetic word-level coefficients in the polarity c, $\mathbf{D^{(c)}} = [d_0^{(c)}\ d_1^{(c)} \ldots d_{2^n-1}^{(c)}]^T$ is derived by the matrix equation

$$\mathbf{D^{(c)}} = \mathbf{P}_{2^n}^{(c)} \cdot \mathbf{F}, \tag{6.7}$$

where the $(2^n \times 2^n)$-matrix $\mathbf{P}_{2^n}^{(c)}$ is generated by the Kronecker product

$$\mathbf{P}_{2^n}^{(c)} = \bigotimes_{j=1}^{n} \mathbf{P}_{2^1}^{(c_j)}, \qquad \mathbf{P}_{2^1}^{(c)} = \begin{cases} \begin{bmatrix} 1 & 0 \\ -1 & 1 \end{bmatrix}, & c_j = 0; \\ \begin{bmatrix} 0 & 1 \\ 1 & -1 \end{bmatrix}, & c_j = 1. \end{cases} \tag{6.8}$$

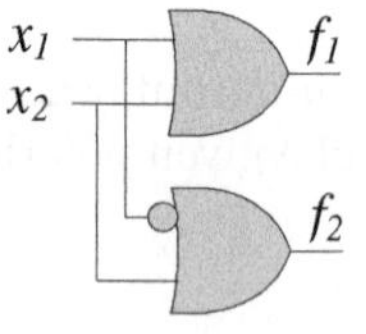

x_1 f_1
x_2
f_2

$f_1 = x_2 + x_1 - x_1x_2$
$f_2 = 1 - x_1 + x_1x_2$

$$\begin{aligned} f &= d_0(x_1 \oplus 1)^0(x_2 \oplus 0)^0 + d_1(x_1 \oplus 1)^0(x_2 \oplus 0)^1 \\ &+ d_2(x_1 \oplus 1)^1(x_2 \oplus 0)^0 + d_3(x_1 \oplus 1)^1(x_2 \oplus 0)^1 \\ &= d_0 + d_1x_2 + d_2\overline{x}_1 + d_3\overline{x}_1x_2 \end{aligned}$$

$$\mathbf{D^{(2)}} = \mathbf{P}_{2^2}^{(2)} \cdot \mathbf{F} = \left[\begin{array}{cc|cc} 0 & 0 & 1 & 0 \\ 0 & 0 & -1 & 1 \\ \hline 1 & 0 & -1 & 0 \\ -1 & 1 & 1 & -1 \end{array}\right] \begin{bmatrix} 2 \\ 3 \\ 1 \\ 3 \end{bmatrix} = \begin{bmatrix} 1 \\ 2 \\ 1 \\ -1 \end{bmatrix}$$

$$f = 1 + 2x_2 + \overline{x}_1 - \overline{x}_1x_2$$

FIGURE 6.8
Computing the arithmetic word-level expression in the polarity $c = 2$ for the two gate circuit (Example 6.10).

6.2.3 Taylor representation

To derive a word-level arithmetic representation of a multi-output function, the Taylor series can be used for deriving an arithmetic expression for each output (Figure 6.3c). Details are given in Chapter 39.

6.3 Computing word-level expressions in the form of corteges

A *cortege* is defined as an ordered concatenation of m elements $U_i, i = 1, 2, \ldots m$, and is denoted as Cor_m:

$$U_m \odot U_{m-1} \odot \cdots \odot U_1 = Cor_m = \bigodot_{i=1}^{m} U_i, \tag{6.9}$$

where $< \odot >$ is a separator. A cortege of m elements with the name A is denoted by ${}^{A}Cor_m$. In an arithmetic word-level representation of a switching function, the elements of a cortege can be the constants 0 and 1, Boolean variables, switching functions (Boolean expressions) and arithmetic expressions.

Example 6.11 *Examples of corteges are as follows:*

(a) Cortege of constants

$$Cor = 0 \odot 1 \odot 1,$$

(b) Cortege of constants and variables

$$Cor = 0 \odot x_3 \odot 1,$$

(c) Cortege of switching functions

$$Cor = f_3 \odot f_2 \odot f_1,$$

(d) Cortege of arithmetic expressions

$$Cor = AR_3 \odot AR_2 \odot AR_1,$$

(e) Cortege of switching functions and a constant

$$Cor = \underbrace{(x_2 \vee x_3)}_{f_2} \odot \underbrace{(x_1 \oplus x_2)}_{f_1} \odot 0.$$

6.3.1 Algebra of corteges

An algebra of corteges is given in Table 6.1. Let ${}^{A}Cor_m, {}^{B}Cor_s$ and ${}^{C}Cor_k$ be corteges A, B, and C. For an arbitrary cortege Cor_k, there exists a cortege $\overline{Cor_k}$ such that

$$Cor_k + \overline{Cor_k} = 2^m$$

where $2^m = 0 \pmod{2^m}$.

TABLE 6.1
Axioms of algebra of corteges.

Axiom	Addition	Multiplication
Identity	$Cor_k + 0 = Cor_k$	$Cor_k \cdot 1 = Cor_k$
Inverse	$Cor_k + \overline{Cor_k} = 2^m$	$Cor_k \cdot \overline{Cor_k} = 0$
Commutativity	$Cor_m + Cor_s = Cor_s + Cor_m$	$Cor_m \cdot Cor_s = Cor_s \cdot Cor_m$
Associativity	$(Cor_m + Cor_s) + C_k$ $= Cor_m + (Cor_s + C_k)$	$(Cor_m \cdot Cor_s) \cdot C_k$ $= Cor_m \cdot (Cor_s \cdot C_k)$
Distributivity	$(Cor_m + Cor_s) \cdot C_k$ $= Cor_m \cdot C_k + Cor_s \cdot C_k$	$C_k \cdot (Cor_m + Cor_s)$ $= C_k \cdot Cor_m + C_k \cdot Cor_s$

6.3.2 Implementation of algebra of corteges

A binary half adder performs the addition of two bits. Let A be a cortege of $a_i \in \{0, 1, x\}$ elements and B be a cortege of $b_i \in \{0, 1, y\}$ elements. The half adder of these corteges A and B can be defined by analogy with a binary half adder by replacing the binary input signals with the cortege elements $a_i \in \{0, 1, x\}$ and $b_i \in \{0, 1, y\}$. In Figure 6.9, a comparison of a binary half adder and half adder of corteges is shown. In more detail:

- Constants of the corteges in the half adder of corteges process as in binary half adder: $0+0 = 0 \odot 0$, $0+1 = 0 \odot 1$, $1+0 = 0 \odot 1$, $1+1 = 1 \odot 0$.
- Constants and variables process by the following rules:

  (*a*) $0 + x = 0 \odot x$,
  (*b*) $1 + x = x \odot \overline{x}$,
  (*c*) $y + 0 = 0 \odot y$, and
  (*d*) $y + 1 = 1 \odot \overline{y}$.

- The sum of two variables is formed similar to a binary half adder: $x + y = (x \oplus y) \odot xy$, where the sum and the carry are denoted as $SUM = x \oplus y$ and $CARRY = xy$.

Full adder of corteges. A binary full adder computes the arithmetic sum of three bits: two bits to be added and carried from the previous lower significant position. The results are two outputs: sum and carry. By analogy, a full adder of cortege processes three-input cortege A, B, and $CARRY_{i-1}$ (carry from previous cortege), and produces two output corteges of SUM and $CARRY$. In contrast to a binary full adder, which operates with binary constants, a full adder of the cortege operates on the elements $\{0, 1, x, y, z\}$.

It follows from the above that the full adder of corteges and the binary full adder have the same formal description:

$$SUM = a_i \oplus b_i \oplus CARRY_{i-1},$$
$$CARRY_i = a_i b_i \vee CARRY_{i-1}(a_i \oplus b_i),$$

where a_i, b_i, SUM, $CARRY_i$ and $CARRY_{i-1}$ can be constants (for binary adder only), variables, and functions.

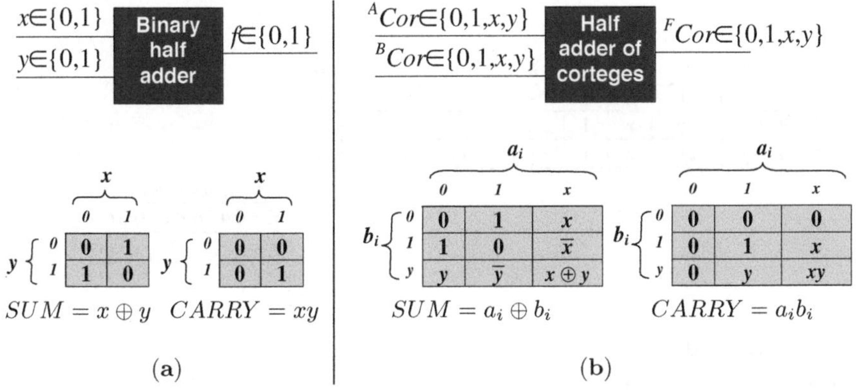

FIGURE 6.9
Truth tables of binary half adder (a) and half adder of corteges (b).

Example 6.12 *Consider an elementary adder of corteges. Constants are processed similarly to binary adder. The above four rules of processing constants 0,1 and variables x, y are used in the adder. Other useful rules are given in Figure 6.10.*

Multiplication of corteges is performed in the same way as binary numbers. The first cortege (the multiplicand) is multiplied by each element of the second cortege (the multiplier) starting from the least significant element. Each such multiplication forms a partial product. Successive partial products are shifted one position to the left. The final product is obtained from the sum of the partial products.

Truth table for SUM

$a_i\,Carry_{i-1}$

b_i	00	01	0x	10	11	1x	x0	x1	xx
0	0	1	x	1	0	$\overline{x}$	x	$\overline{x}$	0
1	1	0	$\overline{x}$	0	1	x	$\overline{x}$	x	1
y	y	$\overline{y}$	$x \oplus y$	$\overline{y}$	y	$\overline{x} \oplus y$	$x \oplus y$	$\overline{x} \oplus y$	y

Truth table for CARRY

$a_i\,Carry_{i-1}$

b_i	00	01	0x	10	11	1x	x0	x1	xx
0	0	0	0	0	1	x	0	x	x
1	0	1	xy	y	1	$x \vee y$	xy	$x \vee y$	x
y	0	y	xy	y	1	$x \vee y$	xy	$x \vee y$	x

The truth table for SUM-output

$$SUM = a_i \oplus b_i \oplus z_i$$

$1 + x + 1 = 1 \odot x$
$1 + y + 1 = 1 \odot y$
$x + 1 + y = 0 \odot (\overline{x} \oplus y)$
because

$$\begin{aligned} x + 1 + y &= (x + 1) + y \\ &= (x \odot \overline{x}) + y \\ &= (x \oplus \overline{x}y) \odot (\overline{x} \oplus y) \\ &= (x \vee y) \odot (\overline{x} \oplus y) \end{aligned}$$

Also
$x + x + y$ and $x + y + y$:

$$\begin{aligned} x + x + y &= (x + x) + y \\ &= (x \odot 0) + y \\ &= x \odot y. \end{aligned}$$

The truth table for CARRY-output

$$CARRY = a_i b_i \vee a_i z_i \vee b_i z_i$$

FIGURE 6.10
Truth tables of a two-input adder of corteges (Example 6.12).

Example 6.13 *Given the cortege* ${}^{A}Cor_2 = x \odot 1$ *and cortege* ${}^{B}Cor_2 = 0 \odot y$*, find their product.*

$$\begin{array}{rccc} \times & & x \odot & 1 \\ & & 0 \odot & y \\ \hline & & xy & 1y \\ & 0 & 0 & \\ \hline & 0 \odot & xy \odot & y \end{array}$$

The result of multiplication is the sum of the partial products ${}^{A}Cor_2 \times {}^{B}Cor_2 = 0 \odot xy \odot y$

Complement of cortege. Let Cor_k be a cortege, then an inverse cortege is defined as cortege Cor_k with inverse order of elements and is denoted as cortege $\overline{Cor}$. It can be shown that:

$$\overline{Cor_k} = 2^k - 1 - Cor_k,$$

because $Cor_k + \overline{Cor_k} = 2^k - 1$.

Example 6.14 *Given the cortege* $Cor_2 = x_1 \odot x_1x_2$*, the inverse cortege is* $\overline{x}_1 \odot \overline{x_1x_2}$ *(Figure 6.11).*

Complement of the binary variable x

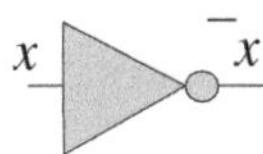

Complement of the cortege

$$x_1 \odot x_1x_2$$

$x_1 \odot x_1 x_2$ Cor $\overline{x_1} \odot \overline{x_1 x_2}$

Given the cortege $Cor_2 = x_1 \odot x_1x_2$, find the inverse cortege.
The inverse cortege in algebraic form:

$$\begin{aligned}\overline{Cor_2} &= 2^2 - 1 - Cor_2\\ &= 3 - x_1 \odot x_1x_2\\ &= (1 \odot 1) - (x_1 \odot x_1x_2)\\ &= 1 - x_1 \odot 1 - x_1x_2\\ &= \overline{x}_1 \odot \overline{x_1x_2},\end{aligned}$$

where $3 = 1 \odot 1$
The inverse cortege in vector form:

$$Cor_2 = x_1 \odot x_1x_2 = \begin{bmatrix}0\\0\\1\\1\end{bmatrix} \odot \begin{bmatrix}0\\0\\0\\1\end{bmatrix}$$

$$\overline{Cor_2} = \overline{x}_1 \odot \overline{x_1x_2} = \begin{bmatrix}1\\1\\0\\0\end{bmatrix} \odot \begin{bmatrix}1\\1\\1\\0\end{bmatrix}$$

FIGURE 6.11
Complement of both a binary variable and a cortege (Example 6.14).

Addition and multiplication of a function and constant. Let f be a switching function and a be a positive integer number. Consider two particularly useful cases: addition and multiplication of a switching function f and a constant a. Since $x_1 + x_2 = x_1x_2 \odot (x_1 \oplus x_2)$, then

$$a + f = af \odot (a \oplus f). \tag{6.10}$$

Given $a = 0$ and $a = 1$, Equation 6.10 yields $0 + f = 0f \odot (0 \oplus f) = f$ and $1 + f = 1f \odot (1 \oplus f) = f \odot \overline{f}$ respectively. If $a > 1$, the constant may be represented in the form of a cortege.

As for addition corteges, consider the cases $a = 0, 1$ and $a > 1$. Figure 6.12 illustrates addition and multiplication of a constant and a cortege. In matrix form addition and multiplication, for example, for $a = 5$ and $\mathbf{F} = [0\ 1\ 1\ 0]^T$ are calculated as follows:

$$5 + f = 5 + \begin{bmatrix}0\\1\\1\\0\end{bmatrix} = \begin{bmatrix}1\\1\\1\\1\end{bmatrix} \odot \begin{bmatrix}0\\1\\1\\0\end{bmatrix} \odot \begin{bmatrix}1\\0\\0\\1\end{bmatrix},$$

$$5f = 5\begin{bmatrix}0\\1\\1\\0\end{bmatrix} = \begin{bmatrix}0\\1\\1\\0\end{bmatrix} \odot \begin{bmatrix}0\\0\\0\\0\end{bmatrix} \odot \begin{bmatrix}0\\1\\1\\0\end{bmatrix}.$$

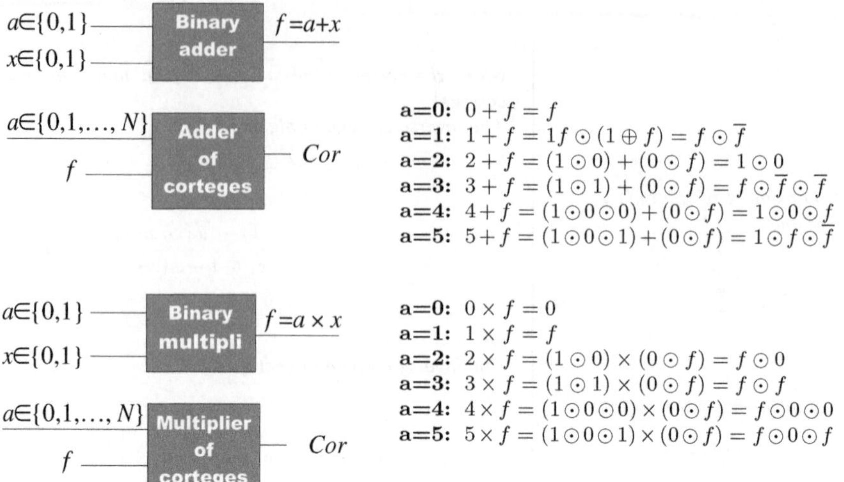

FIGURE 6.12
Addition and multiplication of corteges with a constant.

Manipulation of corteges. Given a cortege, a word-level expression that represents this cortege can be constructed.

Example 6.15 *Given the switching function*

$$f = x_1 \oplus x_2,$$

find $5+f$ *and* $5f$ *in algebraic and matrix form. Since* $x_1 \oplus x_2 = x_1 + x_2 - 2x_1x_2$*, a word-level representation in algebraic form is*

$$5 + f = 5 + x_1 + x_2 - 2x_1x_2,$$
$$5f = 5(x_1 + x_2 - 2x_1x_2).$$

Example 6.16 *Given a cortege*

$$Cor = (x_1 \vee x_2) \odot x_2$$

and constant $a = 1$*, generate a word-level expression*

$$Cor = \{(x_1 \vee x_2) \odot x_2\} + 1.$$

The solution is presented in Figure 6.13.

In this example, the following functions in positions are computed:

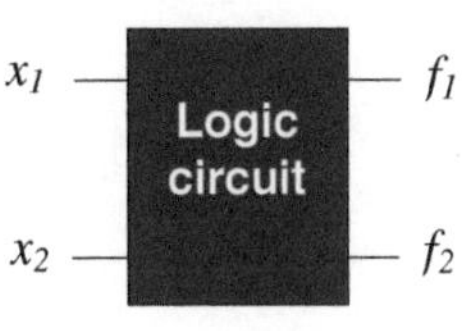

$f_1 = x_2$
$f_2 = x_1 \vee x_2$

Given the cortege

$$Cor = \underbrace{(x_1 \vee x_2)}_{f_2} \odot \underbrace{x_2}_{f_1}$$

and the constant $a = 1$, generate the word level expression in algebraic form

$$Cor = \{(x_1 \vee x_2) \odot x_2\} + 1$$

$$\begin{array}{rcl} + \quad Cor & = & x_1 \vee x_2 \odot x_2 \\ 1 & = & 0 \quad \odot 1 \\ \hline Cor + 1 & = & x_2 \odot \; x_1\overline{x}_2 \; \odot \overline{x}_2 \end{array}$$

Matrix form:

$$1 + Cor = 1 + \begin{bmatrix} 0 \\ 1 \\ 1 \\ 1 \end{bmatrix} \odot \begin{bmatrix} 0 \\ 1 \\ 0 \\ 1 \end{bmatrix} = \begin{bmatrix} 0 \\ 1 \\ 0 \\ 1 \end{bmatrix} \odot \begin{bmatrix} 0 \\ 0 \\ 1 \\ 0 \end{bmatrix} \odot \begin{bmatrix} 1 \\ 0 \\ 1 \\ 0 \end{bmatrix}$$
$$= x_2 \odot x_1\overline{x}_2 \odot \overline{x}_2$$

FIGURE 6.13
Generating a word-level expression using corteges (Example 6.16).

- The least significant position: $x_2 + 1 = x_2 \odot \overline{x}_2$.
- The intermediate position: $(x_1 \vee x_2) + 0 + x_2 = (x_1 \vee x_2) + x_2$.
- The most significant position: using equation $f_1 + f_2 = f_1 f_2 \odot (f_1 \oplus f_2)$, compute $(x_1 \vee x_2)x_2 \odot (x_1 \vee x_2) \oplus x_2 = x_2 \odot x_1\overline{x}_2$.

Interpretation results in arithmetic form. Given a cortege $f_{n-1} \odot \ldots \odot f_1 \odot f_0$, its arithmetic form is derived as

$$f = 2^{n-1} f_{n-1} + \ldots + 2^1 f_1 + 2^0 f_0.$$

Example 6.17 *Consider the cortege*

$$x_2 \odot x_1\overline{x}_2 \odot \overline{x}_2$$

of the function $f+1$ from Example 6.15. This cortege is transformed to the arithmetic expression

$$\begin{aligned} f + 1 &= 2^2 x_2 + 2^1 x_1\overline{x}_2 + 2^0\overline{x}_2 \\ &= 4x_2 + 2x_1\overline{x}_2 + \overline{x}_2. \end{aligned}$$

Inverse arithmetic transform in algebraic form. Given a multi-output switching function, each output is represented in arithmetic form. A direct arithmetic transform results in the word-level representation of a switching function in the form of cortege. The inverse arithmetic transform recovers the switching function from the cortege. The direct arithmetic transform is represented in algebraic and matrix form.

Given a word-level arithmetic expression, a cortege of switching functions can be recovered by the equation:

$$\mathcal{A}^{-1}(f) = Cor_m = \sum_{i=0}^{2^n-1} \bigodot_{j=1}^{m} \tau_{ij}, \tag{6.11}$$

where τ_{ij} is the j-th element of the i-th cortege.

Example 6.18 *(Continuation Example 6.9). Recover switching function given the word-level representation $f = 2 + 3x_1 + 5x_2$. By Equation 6.11, each product term must be represented as a cortege*

$$\begin{aligned} 2 &= 1 \odot 0, \\ 3x_1 &= x_1 \odot x_1, \\ 5x_2 &= x_2 \odot 0 \odot x_2. \end{aligned}$$

Solution is given in Figure 6.14. In matrix notation:

$$\mathcal{A}^{-1}(f) = \sum_{i=0}^{2^2-1} \bigodot_{j=1}^{3} \tau_{ij}$$

$$= 2 + \underbrace{3\begin{bmatrix}0\\0\\1\\1\end{bmatrix}}_{A} + \underbrace{5\begin{bmatrix}0\\1\\0\\1\end{bmatrix}}_{B} = \underbrace{\begin{bmatrix}0&0&1&0\\0&0&1&0\\0&0&1&0\\0&0&1&0\end{bmatrix}}_{2} + \underbrace{\begin{bmatrix}0&0&0&0\\0&0&0&0\\0&0&1&1\\0&0&1&1\end{bmatrix}}_{A} + \underbrace{\begin{bmatrix}0&0&0&0\\0&1&0&1\\0&0&0&0\\0&1&0&1\end{bmatrix}}_{B}$$

$$= \begin{bmatrix} f_4 & f_3 & f_2 & f_1 \\ 0&0&1&0\\0&1&1&1\\0&1&0&1\\1&0&1&0\end{bmatrix} \qquad \begin{aligned} f_1 &= x_1 \oplus x_2 \\ f_2 &= \overline{x}_1 \vee x_2 \\ f_3 &= x_1 \oplus x_2 \\ f_4 &= x_1 x_2 \end{aligned}$$

6.3.3 Orthogonal corteges

Two corteges ${}^{A}Cor$ and ${}^{B}Cor$ are *orthogonal* if their logical product equals the zero-cortege

$${}^{A}Cor \wedge {}^{B} Cor = 0 \odot \cdots \odot 0 \tag{6.12}$$

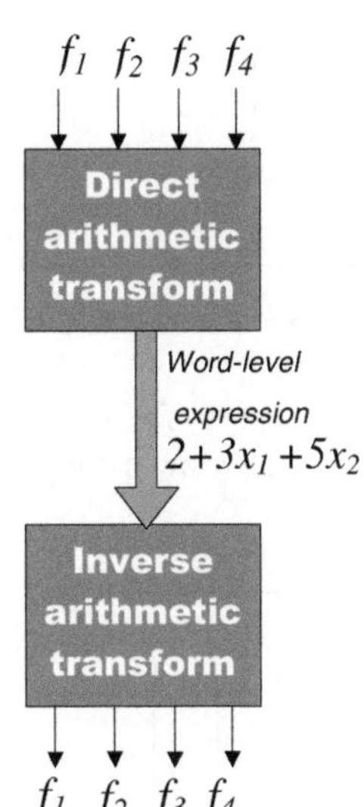

Given : *Word level representation of a switching function in arithmetic form* $f = 2 + 3x_1 + 5x_2$,
Find *Cortege representation* Cor.
Step 1. *Represent each product term by a cortege.*

$$f = \overbrace{(1 \odot 0)}^{2} + \overbrace{(x_1 \odot x_1)}^{3x_1} + \overbrace{(x_2 \odot 0 \odot x_2)}^{5x_2}$$

Step 2. *Compute the sum of corteges*

$$\begin{array}{lllllll}
+ & 2 = & & & 1 & \odot & 0 \\
 & 3x_1 = & & & x_1 & \odot & x_1 \\
\hline
+ & & x_1 & \odot & \overline{x}_1 & \odot & x_1 \\
 & 5x_2 = & x_2 & \odot & 0 & \odot & x_2 \\
\hline
\end{array}$$

$$x_1x_2 \odot (x_1 \oplus x_2) \odot (\overline{x}_1 \vee x_2 \odot (x_1 \oplus x_2)$$

The result is the cortege of four functions

$$f = \overbrace{(x_1x_2)}^{f_4} \odot \overbrace{(x_1 \oplus x_2)}^{f_3} \odot \overbrace{(\overline{x}_1 \vee x_2)}^{f_2} \odot \overbrace{(x_1 \oplus x_2)}^{f_1}$$

FIGURE 6.14
Direct and inverse transforms of corteges in algebraic and matrix forms (Example 6.18).

Example 6.19 (*a*) *The corteges* $(1 \odot 0 \odot 1)$ *and* $(0 \odot x_1 \odot 0)$ *are orthogonal because the result of bit-wise AND operation is zero:* $(1 \odot 0 \odot 1) \wedge (0 \odot x_1 \odot 0) = 0$.

(*b*) *The product terms of arithmetic word-level expressions* $5x_1$ *and* $2x_2$ *are orthogonal because the corresponding corteges* $(x_1 \odot 0 \odot x_1)$ *and* $(0 \odot x_2 \odot 0)$ *are orthogonal* $(x_1 \odot 0 \odot x_1) \wedge (0 \odot x_2 \odot 0) = (0 \odot 0 \odot 0)$.

Example 6.20 (*a*) *Given the arithmetic expression* $5x_1 + 8x_2$, *the product terms* $= 5x_1$ *and* $= 8x_2$ *are orthogonal.*

(*b*) *The word-level expression* $8 + 5x_1 + 6x_2 + 3x_3$ *can be decomposed into two orthogonal word-level expressions* $8 + 5x_1 + 2x_2$ *and* $4x_2 + 3x_3$ *(Figure 6.15)*

It follows from the above examples that the word-level expression of a switching function f of n variables can be calculated without an arithmetic operation on the assignments of variables $x_1, x_2, \ldots x_n$. In general, it is possible to calculate word-level expressions in less than $n + 1$ arithmetic operations.

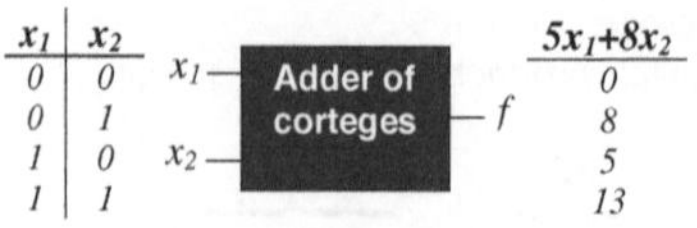

x_1	x_2	$5x_1+8x_2$
0	0	0
0	1	8
1	0	5
1	1	13

x_1, x_2 → Adder of corteges → f

Direct computing of the word-level expression, given the assignment x_1x_2:

$$x_1x_2 = 00 : 5 \times 0 + 8 \times 0 = 0$$
$$x_1x_2 = 01 : 5 \times 0 + 8 \times 1 = 8$$
$$x_1x_2 = 10 : 5 \times 1 + 8 \times 0 = 5$$
$$x_1x_2 = 11 : 5 \times 1 + 8 \times 1 = 13$$

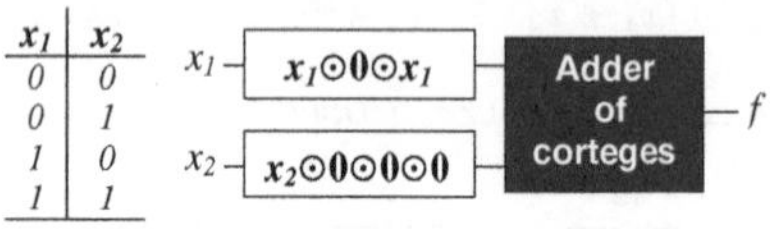

x_1	x_2
0	0
0	1
1	0
1	1

x_1 → $x_1 \odot 0 \odot x_1$; x_2 → $x_2 \odot 0 \odot 0 \odot 0$ → Adder of corteges → f

Computing word-level expression using orthogonal corteges:

$${}^{A}Cor = 5x_1 = 101x_1 = x_1 \odot 0 \odot x_1$$
$${}^{B}Cor = 8x_2 = 1000x_2 = x_2 \odot 0 \odot 0 \odot 0$$

$$\begin{array}{r} + \quad 0 \odot x_1 \odot 0 \odot x_1 \\ x_2 \odot 0 \odot 0 \odot 0 \\ \hline x_2 \odot x_1 \odot 0 \odot x_1 \\ \Downarrow \quad \Downarrow \quad \Downarrow \quad \Downarrow \end{array}$$

$$\begin{bmatrix} 0 & 0 & 0 & 0 \\ 1 & 0 & 0 & 0 \\ 0 & 1 & 0 & 1 \\ 1 & 1 & 0 & 1 \end{bmatrix} = \begin{bmatrix} 0 \\ 8 \\ 5 \\ 13 \end{bmatrix}$$

FIGURE 6.15
Computing the word-level expression $5x_1 + 8x_2$ using orthogonality (Example 6.20).

6.4 Further reading

Fundamentals of arithmetic representations. References to fundamental aspects of arithmetic representations of logic functions are given in Chapters 5, 7, 30, 37, and 39.

Cortege-based grouping and an algebra of corteges for the arithmetic representation of multi-output switching functions were introduced for the first time by Malyugin [6] and adapted to modern techniques by Shmerko [8, 6].

Discrete wavelet transforms can be used for a word-level representation of logic functions (see Chapter 7).

Application aspects of arithmetic forms were considered in [1, 39, 3, 5]. Particular applications of arithmetic forms are the probabilistic verification of switching functions [3], integer linear programming, functional decomposition [5], and verification of arithmetic functions [1]. Decision diagram based representation and manipulation of arithmetic forms were introduced in [1] and various techniques of manipulation were developed in [7, 10, 11, 12]. An alternative type of diagram called EVBDD (edge valued BDD) was introduced in [5]. The calculation and manipulation of arithmetic expressions in forms of arithmetic analogs of logic Taylor expansion was considered in [14].

The orthogonality property plays a significant role in manipulating logic functions in CAD of ICs [39]. Hence, the idea is to divide the problem into several simpler independent problems via the recognition of orthogonal properties in the circuit.

References

[1] Bryant R and Chen Y. Verification of arithmetic functions using binary moment diagrams. In *Proceedings of the Design Automation Conference*, pp. 535–541, 1995.

[2] Davio MJ, Deschamps P, and Thayse A. *Discrete and Switching Functions.* McGraw-Hill, New York, 1978.

[3] Jain J. Arithmetic transform of Boolean functions. In Sasao T and Fujita M, Eds., *Representations of Discrete Functions*, pp. 55–92, Kluwer, Dordrecht, 1996.

[4] Jain J, Bitner J, Fussell DS, and Abraham JA. Probabilistic verification of Boolean functions. *Formal Methods in System Design*, Kluwer, Dordrecht, 1:61–115, 1992.

[5] Lai Y, Pedram M, and Vrudhula S. EVBDD-based algorithms for integer linear programming, spectral transformation, and function decomposition. *IEEE Transactions on Computer-Aided Design of Integrated Circuits and Systems,* 13(8):959–975, 1994.

[6] Malyugin V. Representation of Boolean functions by arithmetical polynomials. *Automation and Remote Control*, Pt2, 43(4):496–504, 1982.

[7] Minato S. *Binary Decision Diagrams and Applications for VLSI CAD*, Kluwer, Dordrecht, 1996.

[8] Shmerko VP. Synthesis of arithmetic forms of Boolean functions using the Fourier transform. *Automation and Remote Control*, Kluwer/Plenum Publishers, 50(5):684–691, Pt2, 1989.

[9] Shmerko VP. Malyugin's theorems: a new concept in logical control, VLSI design, and data structures for new technologies. *Automation and Remote Control,* Plenum/Kluwer Publishers, Special Issue on Arithmetical Logic in Control Systems, Pt2, 65(6):893–912, 2004.

[10] Stanković RS and Astola JT. *Spectral Interpretation of Decision Diagrams*, Springer, Heidelberg, 2003.

[11] Stanković RS, Sasao T, and Moraga C. Spectral transform decision diagrams. In Sasao T and Fujita M, Eds., *Representations of Discrete Functions*, pp. 55–92, Kluwer, Dordrecht, 1996.

[12] Thornton M and Nair V. Efficient calculation of spectral coefficients and their applications, *IEEE Transactions on Computer-Aided Design of Integrated Circuits and Systems*, 14(11):1328–1341, 1995.

[13] Yanushkevich SN, Shmerko VP, and Lyshevski SE. *Logic Design of NanoICs*, CRC Press, Boca Raton, FL, 2005.

[14] Yanushkevich S. Arithmetical canonical expansions of Boolean and MVL functions as generalized Reed–Muller series. In *Proceedings of the IFIP WG 10.5 Workshop on Applications of the Reed–Muller Expansions in Circuit Design*, pp. 300–307, Japan, 1995.

7

Spectral Techniques

Spectral techniques are a discipline of applied mathematics, representing a subarea of abstract harmonic analysis devoted primarily to applications in electrical and computer engineering. Spectral techniques offer alternative methods for solving complex tasks efficiently (in terms of space and time) by converting them into manipulation and calculations with several suitably selected spectral coefficients.

7.1 Introduction

Transferring a problem from the original domain into the spectral domain may provide several advantages. We will mention two examples demonstrating this statement.

- The convolution product is often used in description and mathematical modelling of linear shift-invariant systems. The calculation of results of this operation for given functions is a rather complex task when performed directly. However, it reduces to ordinary multiplication in the transform domain, as stated by the well-known convolution theorem in classical Fourier analysis.
- Some properties of a signal or a system, which are difficult to observe in the original domain, become easily observable in the spectral domain. Examples are the determination of cut-off frequencies and sampling rates in signal processing and the detection of decomposability and symmetry properties in logic design.

There are several reasons to apply a spectral representation in decision diagram techniques, in particular:

Canonicity. Transferring to the spectral domain performs redistribution of the information content of a signal, but does not reduce it. Spectral representations are canonical, i.e., a unique spectrum corresponds to a given function, and vice versa – the function can be reconstructed from the spectrum by the inverse transform.

Compactness. In many cases, due to particular properties a signal may express, the main portion of the information content of the signal is encoded in a (relatively) small number of spectral coefficients. Worded differently, the information about a signal and its properties, sufficient for many applications, may be contained in a few spectral coefficients. This results in compact representation of signals in the spectral domain, and the complexity of the representation depends on the properties of the signals represented. A well known example is lossy compression methods, the kernel of which is the discrete cosine transform (DCT).

The fast calculation algorithms for spectral coefficients further improve performances of algorithms exploiting spectral representation of logic functions. Decision diagrams extend the applicability of these algorithms to functions defined in a large number of points. In the spectral techniques approach to switching theory, switching functions are considered as elements of the Hilbert space consisting of complex-valued functions on finite dyadic groups.

In digital device design, the cost of design grows very fast when the number of components increases. The cost of testing also grows with the increased density of components and limited number of input and output pins. Restricted controllability and observability make the problem of testing even more complex. For many digital devices the cost of testing is higher than the cost of design and manufacturing. Traditional methods of design and testing often require brute force computer searches for solving optimization problems. Unlike traditional methods, spectral methods may provide for simple analytic solutions.

7.2 Spectral transforms

In this book, a discrete function f is considered as a mapping

$$f(x):\ G \rightarrow P,\ x \in G,\ f \in P,$$

where G is an Abelian group of order $|G| = N$, and P is a field that can be the field of complex numbers or a finite (Galois) field. We denote by $P(G)$ the space of all functions thus defined.

Example 7.1 *Switching functions are a particular example, when*

- ► $G = C_2^n$, *where* C_2 *is the basic cyclic group of order 2, i.e.,* $C_2 = (\{0, 1\}, \oplus)$, *and the group operation is the addition modulo 2, usually denoted as logic EXOR.*
- ► P *is the finite Galois field* $GF(2)$.

Thus, switching functions are elements of $GF_2(C_2^n)$. Alternatively, n-variable switching functions can be considered as a subset of functions in $C(C_2^n)$, where C is the field of complex numbers, if the logic values 0 and 1 for $f(x)$ are interpreted as integers 0, and 1, respectively. In this case, it is convenient to view binary valued variables as binary coordinates of an integer $x = (x_1, \ldots, x_n)$. Similarly, a system of k switching functions $f = (f_0, \ldots, f_{k-1})$ can be represented by an *integer equivalent* function

$$f(z) = \sum_{i=0}^{k-1} 2^{k-1-i} f_i$$

A spectral transform is defined as a mapping

$$T: \ f(x) \to \hat{f}(w),$$

where in the case of Abelian groups $x, w \in G$ and $f(x), \hat{f}(w) \in P$. The mapping T is defined in terms of a set of basic functions. Different spectral transforms are defined by choosing different sets of basic functions. In this way, by an appropriate choice, the set of basis functions, i.e., the spectral transform, can be adapted to suit the features of the problem to be solved, to provide computation advantages, or to fulfil some other specific criteria important for the task considered.

7.2.1 Fourier transforms on finite groups

A Fourier transform on a finite Abelian group G of order $g = |G|$ is defined by using the group characters of G as the set of basic functions. For a given group G, the group characters χ_w over the complex field C are defined as a homomorphism $\chi_w : G \to C$ with $|\chi_w| = 1$. Thus,

$$\chi_w(x \oplus y) = \chi_w(x)\chi_w(y).$$

The multiplicative group $\{\chi_w\}$ is isomorphic to the original (additive) Abelian group G. The set of group characters forms a complete orthogonal system, thus, a basis in $C(G)$. Therefore,

$$\frac{1}{g} < \chi_w, \overline{\chi}_\nu >= \delta_{w,\nu},$$

where $< \cdot >$ denotes the inner product in $C(G)$ and $\overline{\chi}$ is the complex-conjugate of χ. Then, the Fourier transform in $C(G)$ is defined as

$$\hat{f}(w) = \frac{1}{g}\sum_{x} f(x)\overline{\chi}_w(x) = \frac{1}{g} < f, \overline{\chi}_w > .$$

The inverse transform is defined by

$$f(x) = \sum_{w} \hat{f}(w)\chi_w(x).$$

In matrix notation, for a function $f \in P(G)$, given by the vector of function values $\mathbf{F} = [f(0), \ldots, f(g-1)]^T$, the discrete Fourier transform pair is defined by the equations (Figure 7.1)

$$\mathbf{S}_f = [\chi]^{-1}\,\mathbf{F},$$
$$\mathbf{F} = [\chi]\,\mathbf{S}_f,$$

where $[\chi]$ is the matrix whose columns are group characters of G and

$$\mathbf{S}_f = [\mathbf{S}_f(0), \mathbf{S}_f(1) \ldots \mathbf{S}_f(2^n - 1)]^T$$

is the vector of spectral coefficients.

7.2.2 Discrete Walsh transform

Let $G = C_2^n$. Then, the group characters are the discrete Walsh functions

$$\chi_w(x) = W_w(x) = (-1)^{<w,x>}.$$

In matrix notation, Walsh transform is defined by the Walsh matrix

$$\mathbf{W}_{2^n} = \bigotimes_{i=1}^{n} \mathbf{W}_{2^1},$$

where $\otimes$ denotes the Kronecker product and the basic Walsh matrix is defined as

$$\mathbf{W}_{2^1} = \begin{bmatrix} 1 & 1 \\ 1 & -1 \end{bmatrix}.$$

Notice that the Walsh matrix is symmetric

$$\mathbf{W}_{2^n} = \mathbf{W}_{2^n}^T,$$

and orthogonal, thus,

$$\mathbf{W}_{2^n}\mathbf{W}_{2^n}^{-1} = \mathbf{I}_{2^n}.$$

Due to that, $\mathbf{W}_{2^n}^{-1} = 2^{-n}\mathbf{W}_{2^n}$.

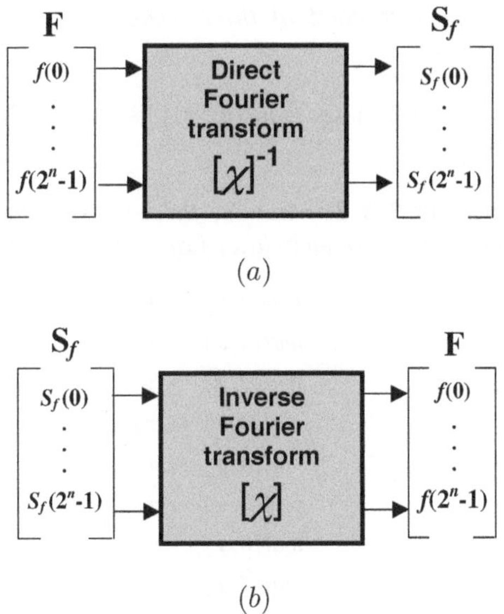

FIGURE 7.1
Direct (a) and inverse (b) discrete Fourier transforms.

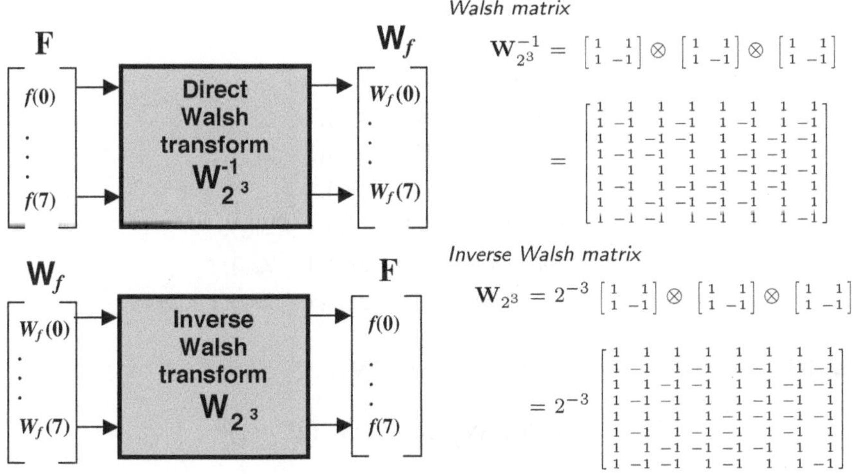

$$\mathbf{W}_{2^3}^{-1} = \begin{bmatrix} 1 & 1 \\ 1 & -1 \end{bmatrix} \otimes \begin{bmatrix} 1 & 1 \\ 1 & -1 \end{bmatrix} \otimes \begin{bmatrix} 1 & 1 \\ 1 & -1 \end{bmatrix}$$

$$= \begin{bmatrix} 1 & 1 & 1 & 1 & 1 & 1 & 1 & 1 \\ 1 & -1 & 1 & -1 & 1 & -1 & 1 & -1 \\ 1 & 1 & -1 & -1 & 1 & 1 & -1 & -1 \\ 1 & -1 & -1 & 1 & 1 & -1 & -1 & 1 \\ 1 & 1 & 1 & 1 & -1 & -1 & -1 & -1 \\ 1 & -1 & 1 & -1 & -1 & 1 & -1 & 1 \\ 1 & 1 & -1 & -1 & -1 & -1 & 1 & 1 \\ 1 & -1 & -1 & 1 & -1 & 1 & 1 & -1 \end{bmatrix}$$

$$\mathbf{W}_{2^3} = 2^{-3} \begin{bmatrix} 1 & 1 \\ 1 & -1 \end{bmatrix} \otimes \begin{bmatrix} 1 & 1 \\ 1 & -1 \end{bmatrix} \otimes \begin{bmatrix} 1 & 1 \\ 1 & -1 \end{bmatrix}$$

$$= 2^{-3} \begin{bmatrix} 1 & 1 & 1 & 1 & 1 & 1 & 1 & 1 \\ 1 & -1 & 1 & -1 & 1 & -1 & 1 & -1 \\ 1 & 1 & -1 & -1 & 1 & 1 & -1 & -1 \\ 1 & -1 & -1 & 1 & 1 & -1 & -1 & 1 \\ 1 & 1 & 1 & 1 & -1 & -1 & -1 & -1 \\ 1 & -1 & 1 & -1 & -1 & 1 & -1 & 1 \\ 1 & 1 & -1 & -1 & -1 & -1 & 1 & 1 \\ 1 & -1 & -1 & 1 & -1 & 1 & 1 & -1 \end{bmatrix}$$

FIGURE 7.2
Walsh matrix for $n = 3$ (Example 7.2).

Example 7.2 *For $n = 3$, the Walsh matrix is defined as the Kronecker product of three basic matrices $\mathbf{W}_{2^1}$ as shown in Figure 7.2.*

The group of Walsh functions $\{wal(w,x)\}$ is isomorphic to the group of linear switching functions)

Example 7.3 *For $n = 3$, the correspondence of Walsh functions and linear switching functions is as follows*

$$
\begin{aligned}
wal(0,x) &\to 1,\\
wal(1,x) &\to x_3,\\
wal(2,x) &\to x_2,\\
wal(3,x) &\to x_2 \oplus x_3,\\
wal(4,x) &\to x_1,\\
wal(5,x) &\to x_1 \oplus x_3,\\
wal(6,x) &\to x_1 \oplus x_2,\\
wal(7,x) &\to x_1 \oplus x_2 \oplus x_3.
\end{aligned}
$$

The Walsh functions can be expressed in terms of switching variables, assuming that the logic values 0 and 1 are interpreted as integers 0 and 1.

Example 7.4 *For $n = 3$, the Walsh functions can be written in terms of switching variables as follows*

$$
\begin{aligned}
wal(0,x) &= 1,\\
wal(1,x) &= (1-2x_3),\\
wal(2,x) &= (1-2x_2),\\
wal(3,x) &= (1-2x_2)(1-2x_3),\\
wal(4,x) &= (1-2x_1),\\
wal(5,x) &= (1-2x_1)(1-2x_3),\\
wal(6,x) &= (1-2x_1)(1-2x_2),\\
wal(7,x) &= (1-2x_1)(1-2x_2)(1-2x_3).
\end{aligned}
$$

Walsh transform. The *Walsh spectrum* $\mathbf{W}_f$,

$$\mathbf{W}_f = [\mathbf{W}_f(0), \mathbf{W}_f(1) \ldots, \mathbf{W}_f(2^n-1)]^T,$$

is defined by the equation

$$\mathbf{W}_f = 2^{-n}\mathbf{W}_{2^n}^{-1}\,\mathbf{F},$$

where $\mathbf{F} = [f(0), \ldots, f(2^n-1)]^T$ is the vector of function values. Since the entries of $\mathbf{W}_{2^n}$ are in $\{1, -1\}$, the Walsh spectrum can be calculated in terms of additions and subtractions.

Fast algorithms. A factorization of $\mathbf{W}_{2^n}^{-1}$ and $\mathbf{W}_{2^n}$ in terms of the Kronecker product permits derivation of fast algorithms for calculation of the Walsh transform. The complexity of fast algorithms (number of additions) for computing direct and inverse Walsh transform is $O(n2^n)$.

Example 7.5 *For the switching function given by the truth-vector*

$$\mathbf{F} = [0, 1, 1, 1]^T,$$

the Walsh spectrum is calculated as follows

$$\begin{aligned}\mathbf{W}_f &= 2^{-2}\mathbf{W}_{2^2}^{-1}\,\mathbf{F} \\ &= 2^{-2}\begin{bmatrix}1 & 1 & 1 & 1\\ 1 & -1 & 1 & -1\\ 1 & 1 & -1 & 1\\ 1 & -1 & 1 & -1\end{bmatrix}\begin{bmatrix}0\\1\\1\\1\end{bmatrix} = 2^{-2}\begin{bmatrix}3\\-1\\-1\\-1\end{bmatrix}\end{aligned}$$

Inverse Walsh transform. The following matrix equation describes the inverse transform to determine the vector of values $\mathbf{F}$ of a function f from its Walsh spectrum, given in matrix notation as a vector of spectral coefficients $\mathbf{W}_f$:

$$\mathbf{F} = \mathbf{W}_{2^n}\,\mathbf{W}_f. \tag{7.1}$$

Example 7.6 *We reconstruct the vector of function values* $\mathbf{F}$ *from the vector of Walsh coefficients*

$$\mathbf{W}_f = [3, -1, -1, -1]^T$$

as follows

$$\begin{aligned}\mathbf{F} &= \mathbf{W}_{2^2}\,\mathbf{W}_f \\ &= \begin{bmatrix}1 & 1 & 1 & 1\\ 1 & -1 & 1 & -1\\ 1 & 1 & -1 & -1\\ 1 & -1 & -1 & 1\end{bmatrix}\begin{bmatrix}3\\-1\\-1\\-1\end{bmatrix} = \begin{bmatrix}0\\1\\1\\1\end{bmatrix}\end{aligned}$$

The Fourier transform on other finite Abelian groups. The same considerations can be extended in a straightforward way to define the Fourier transform on other finite Abelian groups, and also hold when group characters are defined over some field P, different from C, provided that some relationships between $|G|$ and the characteristic* of P are satisfied. Notice that generalizations to finite non-Abelian groups are possible in terms of unitary irreducible representations instead of group characters.

*If e is the identity in P, then the smallest number p for which $p \cdot e = 0$ is called the characteristic of P. If $n \cdot e \neq 0$ for each $n \in N$, then $charP = 0$.

7.3 Haar and related transforms

In some applications, basic functions consisting of group characters are employed. They differ from the Fourier basis. In switching theory and logic design, the most often used are:

- ► The Reed-Muller transform when switching functions are viewed as elements of $GF_2(C_2^n)$ (see Chapter 4), and
- ► The arithmetic and Haar transforms; in this case switching functions $f \in C(C_2^n)$ (see Chapter 5).

Example 7.7 *For many functions $f \in C(C_2^n)$, the basis consisting of the discrete Haar functions produces compact representations, where compactness is viewed as the number of non-zero coefficients. Non-normalized discrete Haar functions can be defined as rows of the Haar matrix*

$$\mathbf{H}_{2^n} = \begin{bmatrix} \mathbf{H}_{2^{(n-1)}} \otimes \begin{bmatrix} 1 & 1 \end{bmatrix} \\ \\ \mathbf{I}_{2^{(n-1)}} \otimes \begin{bmatrix} 1 & -1 \end{bmatrix} \end{bmatrix},$$

where $H_{2^0} = [1]$.

Example 7.8 *For $n = 3$, and $G = C_2^3$, non-normalized discrete Haar functions are defined as rows of the Haar matrix (Figure 7.3).*

It should be noted that:

- ► There is a slight difference in definition of the direct Haar and the inverse Haar transforms compared to the transforms where the basis functions are written as columns of a matrix.
- ► Similarly to Walsh functions, Haar functions can be expressed in terms of switching variables assuming that logic values 0 and 1 are interpreted as integers 0 and 1.

Example 7.9 *For $n = 3$, the Haar functions can be written in terms of switching variables as follows:*

$$\begin{aligned}
har(0, x) &= 1, \\
har(1, x) &= (1 - 2x_1), \\
har(2, x) &= (1 - 2x_2)\overline{x}_1, \\
har(3, x) &= (1 - 2x_2)x_1, \\
har(4, x) &= (1 - 2x_3)\overline{x}_1\overline{x}_2, \\
har(5, x) &= (1 - 2x_3)\overline{x}_1 x_2, \\
har(6, x) &= (1 - 2x_3)x_1\overline{x}_2, \\
har(7, x) &= (1 - 2x_3)x_1 x_2.
\end{aligned}$$

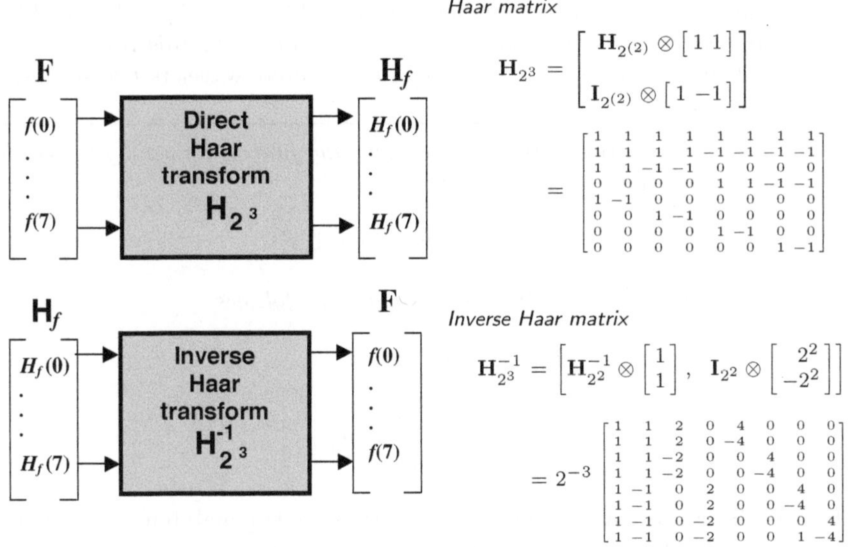

FIGURE 7.3
The non-normalized discrete Haar functions for $n = 3$ (Example 7.8).

Haar transform. By analogy to the classical Fourier analysis, using the definition of Fourier transform on finite groups, it is convenient to write the set of basis functions as columns of a matrix and use its inverse to calculate the spectral coefficients. This also applies to the definition of the Walsh (Fourier) transform. However:

- ▶ The Walsh matrix is symmetric and
- ▶ The Walsh functions may be equally viewed as either columns or rows of the Walsh matrix.

By analogy, the Haar functions are written as rows of the Haar matrix, which is usually used to determine the Haar spectral coefficients.

The *Haar spectrum* $\mathbf{H}_f$ represented as a vector

$$\mathbf{H}_f = [\hat{f}(0), \ldots, \hat{f}(2^n - 1)]^T,$$

is defined as

$$\mathbf{H}_f = 2^{-n}\mathbf{H}_{2^n}\ \mathbf{F}.$$

Due to many zero entries in the Haar transform matrix, we do not need all the function values for the calculation of each of the Haar spectral coefficients.

The Haar spectral coefficients reflect the local behavior of f. Due to this, the Haar transform is the so-called local *transform. In this respect, the discrete Haar transform is an example of the* discrete wavelets *transforms.*

Example 7.10 *For the switching function given by the truth-vector*

$$\mathbf{F} = [0, 1, 1, 1]^T$$

the Haar spectrum is calculated as follows

$$\mathbf{H}_f = 2^{-1}\mathbf{H}_{2^2}^{-1}\,\mathbf{F} = 2^{-2}\begin{bmatrix} 1 & 1 & 1 & 1 \\ 1 & -1 & 1 & -1 \\ 1 & -1 & 0 & 0 \\ 0 & 0 & 1 & -1 \end{bmatrix}\begin{bmatrix} 0 \\ 1 \\ 1 \\ 1 \end{bmatrix} = 2^{-2}\begin{bmatrix} 3 \\ -1 \\ -1 \\ 0 \end{bmatrix}$$

Inverse Haar transform. The Haar matrix is orthogonal, but not symmetric, and due to this, the inverse Haar matrix used to reconstruct a function from its spectrum is obtained by the transposition of the Haar matrix. Further, since we are considering the non-normalized Haar matrix, the columns are multiplied by the weight coefficients whose values can be related to the number of non-zero entries in the column.

The inverse Haar matrix $\mathbf{H}_{2^n}^{-1}$ is determined by

$$\mathbf{H}_{2^n}^{-1} = \left[\mathbf{H}_{2^{n-1}}^{-1} \otimes \begin{bmatrix} 1 \\ 1 \end{bmatrix},\ \mathbf{I}_{2^{n-1}} \otimes \begin{bmatrix} 2^{n-1} \\ -2^{n-1} \end{bmatrix}\right] \tag{7.2}$$

with $\mathbf{H}_{2^0}^{-1} = [1]$.

Example 7.11 *For $n = 3$ and $G = C_2^3$, the inverse Haar matrix $\mathbf{H}_{2^3}^{-1}$ is defined as shown in Figure 7.3.*

The following matrix equation defines the inverse Haar transform to determine the vector of function values $\mathbf{F}$ from the Haar spectrum $\mathbf{H}_f$

$$\mathbf{F} = \mathbf{H}_{2^n}^{-1}\,\mathbf{H}_f. \tag{7.3}$$

Example 7.12 *We reconstruct the vector of function values* $\mathbf{F}$ *of a function f given by the vector of Haar coefficients*

$$\mathbf{H}_f = [3, -1, -1, 0]^T$$

as follows

$$\mathbf{F} = \mathbf{H}_{2^2}^{-1}\,\mathbf{H}_f = 2^{-2}\begin{bmatrix} 1 & 1 & 2 & 0 \\ 1 & 1 & -2 & 0 \\ 1 & -1 & 0 & 2 \\ 1 & -1 & 0 & -2 \end{bmatrix}\begin{bmatrix} 3 \\ -1 \\ -1 \\ 0 \end{bmatrix} = \begin{bmatrix} 0 \\ 1 \\ 1 \\ 1 \end{bmatrix}$$

7.4 Computing spectral coefficients

In this section, we introduce the notion of fast algorithms for computation of spectral coefficients of discrete transforms on finite Abelian groups by the examples of the corresponding algorithms for the Walsh and Haar transforms.

7.4.1 Computing the Walsh coefficients

A direct computing of the Walsh spectrum requires multiplication of the vector of function values of the length 2^n by the $(2^n \times 2^n)$ Walsh matrix, which implies 2^{2n} arithmetic operations. However, due to the Kronecker product structures, the Walsh matrix can be factorized into a product of n sparse matrices, each of which describes a step in the fast algorithm for computing the Walsh spectrum. This algorithm corresponds to the Fast Fourier transform (FFT) and is often called the fast Walsh transform (FWT).

Data flowgraphs. To design the data flowgraph of the algorithm, the matrix $\mathbf{W}_{2^n}^{-1}$ is factorized as

$$\mathbf{W}_{2^n}^{-1} = \mathbf{W}_{2^n}^{(n)}\, \mathbf{W}_{2^n}^{(n-1)} \cdots\, \mathbf{W}_{2^n}^{(1)}, \tag{7.4}$$

where

$$\mathbf{W}_{2^n}^{(i)} = \bigotimes_{i=1}^{n} \mathbf{X}_j, \quad \mathbf{X}_j = \begin{cases} \mathbf{W}_{2^1}, & \text{if } j - i, \\ \mathbf{I}_{2^1}, & \text{otherwise.} \end{cases}$$

Hence, the Walsh spectrum is computed in n steps.

Example 7.13 *For $n = 3$, the Walsh matrix is factorized as follows*

$$\mathbf{W}_{2^3}^{-1} = \mathbf{W}_{2^3}^{(3)}\, \mathbf{W}_{2^3}^{(2)}\, \mathbf{W}_{2^3}^{(1)},$$

where matrices $\mathbf{W}_{2^3}^{(1)}$, $\mathbf{W}_{2^3}^{(2)}$, and $\mathbf{W}_{2^3}^{(3)}$ are given in Figure 7.4.

$$
\underbrace{\overset{\mathbf{W}_{2^3}^{(1)}}{\begin{bmatrix}
1&0&0&0&1&0&0&0\\
0&1&0&0&0&1&0&0\\
0&0&1&0&0&0&1&0\\
0&0&0&1&0&0&0&1\\
1&0&0&0&-1&0&0&0\\
0&1&0&0&0&-1&0&0\\
0&0&1&0&0&0&-1&0\\
0&0&0&1&0&0&0&-1
\end{bmatrix}}}_{\textit{First iteration}}
\underbrace{\overset{\mathbf{W}_{2^3}^{(2)}}{\begin{bmatrix}
1&0&1&0&0&0&0&0\\
0&1&0&1&0&0&0&0\\
1&0&-1&0&0&0&0&0\\
0&1&0&-1&0&0&0&0\\
0&0&0&0&1&0&1&0\\
0&0&0&0&0&1&0&1\\
0&0&0&0&1&0&-1&0\\
0&0&0&0&0&1&0&-1
\end{bmatrix}}}_{\textit{Second iteration}}
\underbrace{\overset{\mathbf{W}_{2^3}^{(3)}}{\begin{bmatrix}
1&1&0&0&0&0&0&0\\
1&-1&0&0&0&0&0&0\\
0&0&1&1&0&0&0&0\\
0&0&1&-1&0&0&0&0\\
0&0&0&0&1&1&0&0\\
0&0&0&0&1&-1&0&0\\
0&0&0&0&0&0&1&1\\
0&0&0&0&0&0&1&-1
\end{bmatrix}}}_{\textit{Third iteration}}
$$

FIGURE 7.4
Factorized matrices of Walsh transform expansion for $n = 3$ (Example 7.13).

Example 7.14 *Computing the Walsh coefficients for the switching function $f = x_1 \vee x_2$ given its truth vector*

$$\mathbf{F} = [0, 1, 1, 1]^T$$

is illustrated in Figure 7.5. The matrix $\mathbf{W}_{2^2}$ is factorized as follows:

$$
\begin{aligned}
\mathbf{W}_{2^2}^{-1} &= \mathbf{W}_{2^2}^{(2)}\ \mathbf{W}_{2^2}^{(1)} \\
&= \underbrace{(\mathbf{I}_{2^1} \otimes \mathbf{W}_{2^1}^{-1})}_{\textit{2nd iteration}}\ \underbrace{(\mathbf{W}_{2^1}^{-1} \otimes \mathbf{I}_{2^1})}_{\textit{1st iteration}} \\
&= \left[\begin{array}{c|c} \mathbf{W}_{2^1}^{-1} & \mathbf{0}_{2^1} \\ \hline \mathbf{0}_{2^1} & \mathbf{W}_{2^1}^{-1} \end{array}\right] \left[\begin{array}{c|c} \mathbf{I}_{2^1} & \mathbf{I}_{2^1} \\ \hline \mathbf{I}_{2^1} & -\mathbf{I}_{2^1} \end{array}\right].
\end{aligned}
$$

Then, the Walsh spectrum is computed, $\mathbf{W}_f = \mathbf{W}_{2^2}^{(2)}\mathbf{W}_{2^1}^{(1)}\ \mathbf{F}$, where the first iteration gives the vector $\mathbf{W}_f^{'} = \mathbf{W}_{2^1}^{(1)}\ \mathbf{F}$, and the second iteration results in $\mathbf{W}_f = \mathbf{W}_{2^2}^{(2)}\ \mathbf{W}_f^{'}$.

7.4.2 Computing the Haar coefficients

The Haar matrix, unlike the Walsh matrix, cannot be represented as the Kronecker product of a single basic matrix. However, it is a layered Kronecker product representable matrix as follows from the definition of it. Due to that, it is possible to factorize the Haar matrix and derive a fast FFT-like algorithm for computation of the Haar spectrum.

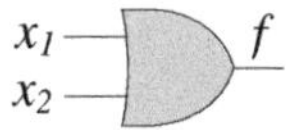

$f = x_1 \vee \overline{x}_2$

Walsh spectrum computing

$$\mathbf{W}_f = \mathbf{W}_{2^2}^{-1}\,\mathbf{F} = \left[\begin{array}{cc|cc} 1 & 1 & 1 & 1 \\ 1 & -1 & 1 & -1 \\ \hline 1 & 1 & -1 & 1 \\ 1 & -1 & 1 & -1 \end{array}\right] \begin{bmatrix} 0 \\ 1 \\ 1 \\ 1 \end{bmatrix} = \begin{bmatrix} 3 \\ -1 \\ -1 \\ -1 \end{bmatrix}$$

Walsh spectrum computing using factorized matrices (Equation 7.4)

$$\mathbf{W}_f = \mathbf{W}_{2^2}^{(2)}\,\mathbf{W}_{2^2}^{(1)}\,\mathbf{F}$$

$$= 2^{-2} \left[\begin{array}{cc|cc} 1 & 1 & 0 & 0 \\ 1 & -1 & 0 & 0 \\ \hline 0 & 0 & 1 & 1 \\ 0 & 0 & 1 & -1 \end{array}\right] \left[\begin{array}{cc|cc} 1 & 0 & 1 & 0 \\ 0 & 1 & 0 & 1 \\ \hline 1 & 0 & -1 & 0 \\ 0 & 1 & 0 & -1 \end{array}\right] \begin{bmatrix} 0 \\ 1 \\ 1 \\ 1 \end{bmatrix} = \begin{bmatrix} 3 \\ -1 \\ -1 \\ -1 \end{bmatrix}$$

Walsh expression

$f = 2^{-2}(3-(1-2x_2)-(1-2x_1)-(1-2x_1)(1-2x_2))$

Data flowgraphS of the algorithm

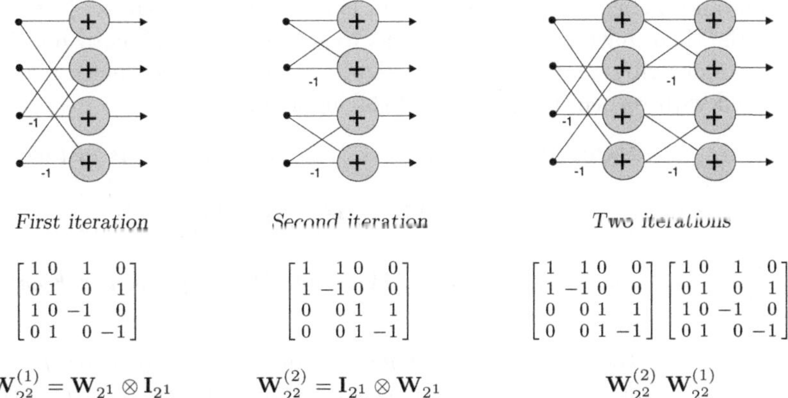

FIGURE 7.5
Calculation of Walsh expansion for the two-input OR gate (Example 7.14).

Data flowgraph. The data flowgraph of the algorithm is derived from the matrix factorization of $\mathbf{H}_{2^n}$ as

$$\mathbf{H}_{2^n} = \mathbf{H}_{2^n}^{(n)}\, \mathbf{H}_{2^n}^{(n-1)} \cdots \mathbf{H}_{2^n}^{(1)}, \qquad (7.5)$$

where the matrices $\mathbf{H}_{2^n}^{(i)}$, $i = 1, 2, \ldots, n$, describing the steps in the algorithm are defined in terms of an auxiliary matrix

$$\mathbf{R}_i = \begin{bmatrix} \mathbf{I}_{2^{n-i}} \oplus \begin{bmatrix} 1 & 1 \end{bmatrix} \\ \\ \mathbf{I}_{2^{n-i}} \oplus \begin{bmatrix} 1 & -1 \end{bmatrix} \end{bmatrix},$$

as follows

$$\begin{aligned} \mathbf{H}_{2^n}^{(1)} &= \mathbf{R}_1, \\ \mathbf{H}_{2^n}^{(2)} &= diag(\mathbf{R}_2,\ \mathbf{I}_{2^2}), \\ \mathbf{H}_{2^n}^{(3)} &= diag(\mathbf{R}_3,\ \mathbf{I}_{2^1}, \mathbf{I}_{2^2}), \\ &\ \vdots \\ \mathbf{H}_{2^n}^{(n)} &= diag(\mathbf{R}_n,\ \mathbf{I}_{2^1},\ \mathbf{I}_{2^2},\ \ldots,\ \mathbf{I}_{2^{n-1}}), \end{aligned}$$

and hence $\mathbf{H}_f = \mathbf{H}_f^{(n)}\, \mathbf{H}_f^{(n-1)} \cdots \mathbf{H}_f^{(1)}$.

Each step in the data flowgraph is determined by a matrix in this factorization and, hence, the Haar spectrum is computed in n steps. Notice that, since the Haar matrix is not symmetric, to get the Haar matrix by multiplying matrices describing steps of the fast algorithm, the order of factor matrices is reversed with respect to the steps of the algorithm.

Example 7.15 *For $n = 3$, the Haar matrix is factorized as*

$$\mathbf{H}_{2^3} = \mathbf{H}_{2^3}^{(3)}\, \mathbf{H}_{2^3}^{(2)}\, \mathbf{H}_{2^3}^{(1)},$$

where matrixes $\mathbf{H}_{2^3}^{(1)}$, $\mathbf{H}_{2^2}^{(1)}$, and $\mathbf{H}_{2^3}^{(3)}$ are given in Figure 7.6.

Example 7.16 *Computing the Haar coefficients p_i by the Equation 7.5 for the switching function $f = x_1 \oplus x_2$ given the truth-vector $\mathbf{F} = [0, 1, 1, 0]^T$ is illustrated in Figure 7.7. The flow-graph consists of two steps as specified by the corresponding factorization of the Haar matrix.*

Example 7.17 *Table 7.1 shows the Walsh and Haar spectra of some elementary switching functions. It illustrates that by selecting different spectral transforms, the same function can be represented by a different number of spectral coefficients.*

TABLE 7.1
Walsh and Haar spectra for elementary switching functions in vector and algebraic forms.

Function	Walsh spectrum	Haar spectrum
$f = x_1 x_2$	$\mathbf{W}_f = 2^{-2}[1, -1, -1, 1]$ $f = \frac{1}{4}(1 - (1 - 2x_2)$ $- (1 - 2x_1)$ $+ (1 - 2x_1)(1 - 2x_2))$	$\mathbf{H}_f = 2^{-2}[1, -1, 0, -1]$ $f = \frac{1}{4}(1 - (1 - 2x_1)$ $- 2x_1(1 - 2x_2))$
$f = x_1 \vee x_2$	$\mathbf{W}_f = 2^{-2}[3, -1, -1, -1]$ $f = \frac{1}{4}(3 - (1 - 2x_2)$ $- (1 - 2x_1)$ $- (1 - 2x_1)(1 - 2x_2))$	$\mathbf{H}_f = 2^{-2}[3, -1, -1, 0]$ $f = \frac{1}{4}(3 - (1 - 2x_1)$ $- 2\overline{x}_1(1 - 2x_2))$
$f = x_1 \oplus x_2$	$\mathbf{W}_f = 2^{-2}[2, 0, 0, -2]$ $f = \frac{1}{4}(2 - 2(1 - 2x_1)$ $\times (1 - 2x_2))$	$\mathbf{H}_f = 2^{-2}[2, 0, -1, 1]$ $f = \frac{1}{4}(2 - 2\overline{x}_1(1 - 2x_2)$ $+ 2x_1(1 - 2x_2))$
$f = \overline{x_1 x_2}$	$\mathbf{W}_f = 2^{-2}[3, 1, 1, -1]$ $f = \frac{1}{4}(3 + (1 - 2x_2)$ $+ (1 - 2x_1)$ $- (1 - 2x_1)(1 - 2x_2))$	$\mathbf{H}_f = 2^{-2}[3, 1, 0, 1]$ $f = \frac{1}{4}(3 + (1 - 2x_1)$ $+ 2x_1(1 - 2x_2))$
$f = \overline{x}_1 \vee x_2$	$\mathbf{W}_f = 2^{-2}[1, 1, 1, 1]$ $f = \frac{1}{4}(1 + (1 - 2x_1)$ $+ (1 - 2x_2)$ $+ (1 - 2x_1)(1 - 2x_2))$	$\mathbf{H}_f = 2^{-2}[1, 1, 1, 0]$ $f = \frac{1}{4}(1 + (1 - 2x_1)$ $+ 2\overline{x}_1(1 - 2x_2))$

$$\underbrace{\overset{\mathbf{H}_{2^3}^{(1)}}{\begin{bmatrix} 1 & 1 & 0 & 0 & 0 & 0 & 0 & 0 \\ 0 & 0 & 1 & 1 & 0 & 0 & 0 & 0 \\ 0 & 0 & 0 & 0 & 1 & 1 & 0 & 0 \\ 0 & 0 & 0 & 0 & 0 & 0 & 1 & 1 \\ 1 & -1 & 0 & 0 & 0 & 0 & 0 & 0 \\ 0 & 0 & 1 & -1 & 0 & 0 & 0 & 0 \\ 0 & 0 & 0 & 0 & 1 & -1 & 0 & 0 \\ 0 & 0 & 0 & 0 & 0 & 0 & 1 & -1 \end{bmatrix}}}_{\textit{First iteration}} \quad \underbrace{\overset{\mathbf{H}_{2^3}^{(2)}}{\begin{bmatrix} 1 & 1 & 0 & 0 & 0 & 0 & 0 & 0 \\ 0 & 0 & 1 & 1 & 0 & 0 & 0 & 0 \\ 1 & -1 & 0 & 0 & 0 & 0 & 0 & 0 \\ 0 & 0 & 1 & -1 & 0 & 0 & 0 & 0 \\ 0 & 0 & 0 & 0 & 1 & 0 & 0 & 0 \\ 0 & 0 & 0 & 0 & 0 & 1 & 0 & 0 \\ 0 & 0 & 0 & 0 & 0 & 0 & 1 & 0 \\ 0 & 0 & 0 & 0 & 0 & 0 & 0 & 1 \end{bmatrix}}}_{\textit{Second iteration}} \quad \underbrace{\overset{\mathbf{H}_{2^3}^{(3)}}{\begin{bmatrix} 1 & 1 & 0 & 0 & 0 & 0 & 0 & 0 \\ 1 & -1 & 0 & 0 & 0 & 0 & 0 & 0 \\ 0 & 0 & 1 & 0 & 0 & 0 & 0 & 0 \\ 0 & 0 & 0 & 1 & 0 & 0 & 0 & 0 \\ 0 & 0 & 0 & 0 & 1 & 0 & 0 & 0 \\ 0 & 0 & 0 & 0 & 0 & 1 & 0 & 0 \\ 0 & 0 & 0 & 0 & 0 & 0 & 1 & 0 \\ 0 & 0 & 0 & 0 & 0 & 0 & 0 & 1 \end{bmatrix}}}_{\textit{Third iteration}}$$

FIGURE 7.6
Factorized matrixes of Haar transform expansion for $n = 3$ (Example 7.15).

7.5 Discrete wavelet transforms

Unlike the discrete Fourier transform, which can be completely defined by two equations, the direct and inverse transform, the discrete wavelet transform refers to a class of transformations that differ not only in the transformation kernels employed, but also in the fundamental nature of those functions and in the way they are applied. The discrete wavelet transform encompasses a variety of unique but related transformations, i.e., we cannot write a single equation that completely describe them all. The discrete Haar transform is an example of the discrete wavelets transform. The properties of the discrete wavelets transform can be useful, for example, in a word-level representation of switching functions, interpretation, and construction of decision diagrams.

7.6 Further reading

The fundamentals of signal processing techniques can be found in many books, for example, by Ahmed and Rao [1], Beauchamp [2], and McClellan and Rader [18].

Applications in switching theory. The first applications of spectral techniques in switching theory and logic design are related to group theoretic approaches to the optimization problems in these areas and date back to the early 1950s (see, for example, results by Komamiya [14] and Lechner [17]). Gibbs and Millard [7] showed that the Walsh functions can be considered as

x_1, x_2, f

$f = x_1 \vee x_2$

Haar spectrum computing using Equation 7.3

$$\mathbf{H}_f = \mathbf{H}_{2^2}\ \mathbf{F}$$
$$= \begin{bmatrix} 1 & 1 & 1 & 1 \\ 1 & 1 & -1 & -1 \\ 1 & -1 & 0 & 0 \\ 0 & 0 & 1 & -1 \end{bmatrix} \begin{bmatrix} 0 \\ 1 \\ 1 \\ 1 \end{bmatrix} = \begin{bmatrix} 3 \\ -1 \\ -1 \\ 0 \end{bmatrix}$$

Haar spectrum computing using factorized matrices (Equation 7.5)

$$\mathbf{H}_f = \mathbf{H}_{2^2}^{(2)}\ \mathbf{H}_{2^2}^{(1)}\ \mathbf{F}$$
$$= 2^{-2} \left[\begin{array}{cc|cc} 1 & 1 & 0 & 0 \\ 0 & 0 & 1 & 1 \\ \hline 1 & -1 & 0 & 0 \\ 0 & 0 & 1 & -1 \end{array}\right] \left[\begin{array}{cc|cc} 1 & 1 & 1 & 0 \\ 1 & -1 & 0 & 1 \\ \hline 0 & 0 & 1 & 0 \\ 0 & 0 & 0 & 1 \end{array}\right] \begin{bmatrix} 0 \\ 1 \\ 1 \\ 1 \end{bmatrix} = \begin{bmatrix} 3 \\ -1 \\ -1 \\ 0 \end{bmatrix}$$

Haar expression

$$f = 2^{-2}(3 - (1 - 2x_1) - (1 - 2x_2)\overline{x}_1)$$

Data flowgraphs of the algorithm

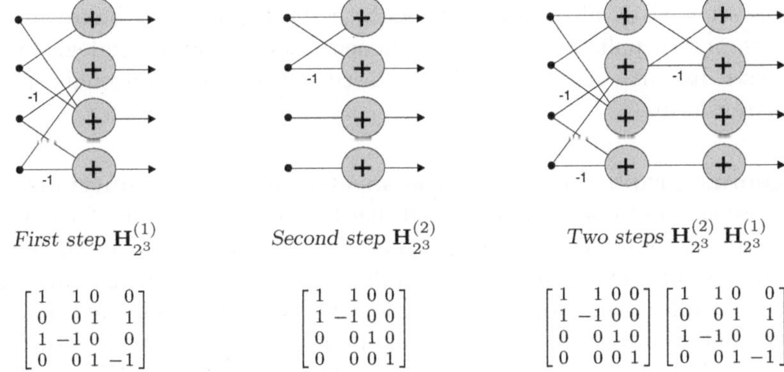

FIGURE 7.7
Calculation of Haar expansion for the EXOR gate (Example 7.16).

solutions of a logical differential equation.

The highest interest in this subject was recorded in the early 1970s, related to the interest of spectral methods in digital signal processing in general, which resulted in publication of a few monographs, in particular, by Hurst [9] and by Karpovsky [11]. In [23], a list of over 60 monographs is entirely or partially devoted to the subject. The research work has been reviewed periodically: by Falkowski [6], by Karpovsky et al. [12, 13], by Moraga et al. [19], and by Stanković et al. [9]. The present interest is due to the report by Yang and De Micheli [33] related to problems in technology mapping of large switching functions. Extension of the Haar transform to functions in a finite field yields the Haar–Galois transforms one can find in [11]. The algebraic expressions for spectral representations for various discrete basis of switching and multivalued functions and techniques for their manipulation were developed by Shmerko [21] and Kukharev et al. [15, 16].

Spectral techniques. Various spectral techniques are discussed by Hurst et al, [15], Sadykhov et al. [25], Thornton et al. [25], and Yanushkevich et al. [31]. The recent advances in this area are discussed in special issues of some journals devoted to applications of spectral techniques in binary and multivalued logic in general [24], or some particular spectral representations [30].

The Karhunen–Loève transform, also known as the Hotelling transform is a unitary transform with basic functions that are orthogonal eigenvectors of the covariance matrix of a data set. The Karhunen–Loève spectrum is defined as the set of eigenvalues associated with the basis functions. There are many papers on the applications of the Karhunen–Loève transform to logic function analysis. In particular, Thornton [26] discussed the computational aspects of this transform.

Algebraic graph theory and the spectra of discrete functions. Bernasconi and Codenotti [3] showed that the Walsh spectrum of a logic function can be computed as the spectrum of a Cayley graph over the additive Abelian group. This result was generalized by Thornton [27, 26] and Thornton and Miller [28]. Notice that Taylor-like expansions provide a technique to compute the specta of switching and multivalued functions in terms of differences that are related to graph-based interpretation (see Chapter 39 and [31, 32]).

Discrete wavelet transforms. The fundamentals of wavelet analysis can be found, for example, in the book by Chui [4] and Wickerhuser [29]. Application of the *discrete wavelet transforms* in switching theory is not properly studied yet (the details are given in the paper by Egiazarian et al. [5] and Hansen and Sekine [8].

References

[1] Ahmed N and Rao KR. *Orthogonal Transforms for Digital Signal Processing.* Springer, Heidelberg, 1975.

[2] Beauchamp KG. *Applications of Walsh and Related Functions with an Introduction to Sequency Theory.* Academic Press, Bristol, 1984.

[3] Bernasconi A and Codenotti B. Spectral analysis of Boolean functions as a graph eigenvalue problem. *IEEE Transactions on Computers*, 48(3):345–351, 1999.

[4] Chui CK. *An Introduction to Wavelets.* Academic Press, San Diego, CA, 1992.

[5] Egiazarian K, Astola JT, Stanković RS, and Stanković M. Construction of compact word-level representations of multiple-output switching functions by wavelet packages, In *Proceedings of the 6th International Symposium on Representations and Methodology for Future Computing Technologies*, Trier, Germany, March 10-11, pp. 77–84, 2003.

[6] Falkowski BJ. A note on the polynomial form of Boolean functions and related topics. *IEEE Transactions on Computers*, 48(8):860–863, 1999.

[7] Gibbs JE and Millard MS. Walsh functions as solutions of a logical differential equation. *DES Report, No.1 National Physical Laboratory* Middlesex, England, 1969.

[8] Hansen JP and Sekine M. Decision diagrams based techniques for the Haar wavelet transform. In *Proceedings of the IEEE International Conference on Information, Communications and Signal Processing*, Singapore, Vol. 1, pp. 59–63, 1997.

[9] Hurst SL. *Logical Processing of Digital Signals.* Crane Russak and Edward Arnold, London and Basel, 1978.

[10] Hurst SL, Miller DM, and Muzio JC. *Spectral Techniques in Digital Logic*, Academic Press, Bristol, 1985.

[11] Karpovsky MG. *Finite Orthogonal Series in the Design of Digital Devices.* Wiley and JUP, New York and Jerusalem, 1976.

[12] Karpovsky MG. Recent developments in applications of spectral techniques in logic design and testing of computer hardware. In *Proceedings of the International Symposium on Multiple-Valued Logic*, 1981.

[13] Karpovsky MG, Stanković RS, and Moraga C. Spectral techniques in binary and multiple-valued switching theory, a review of results in the previous decade. *International Journal Multiple-Valued Logic and Soft Computing*, 10(3):261–286, 2004.

[14] Komamiya Y. Application of logical mathematics to information theory (Application of theory of group to logical mathematics). *The Bulletin of the Electrotechnical Laboratory in Japanese Government*, 1953.

[15] Kukharev GA, Shmerko VP, and Zaitseva EN. *Algorithms and Systolic Arrays for Multivalued Data Processing.* Science and Technics Publishers, Minsk, Belarus, 1990 (In Russian).

[16] Kukharev GA, Shmerko VP, and Yanushkevich SN *Parallel Binary Processing Techniqies* Higher School Publishers, Minsk, Belarus, 1991 (In Russian).

[17] Lechner R. A transform theory for functions of binary variables. In *Theory of Switching,* Harvard Computation Lab., Cambridge, Mass., Progress Rept. BL-30, Sec-X, pp. 1–37, 1961.

[18] McClellan JH and Rader CM *Number Theory in Digital Signal Processing.* Prentice-Hall, New York, 1979.

[19] Moraga C. A decade of spectral techniques. In *Proceedings of the 21st International Symposium on Multiple-Valued Logic,* pp. 182–188, 1991.

[20] Sadykhov RCh, Chegolin PM, and Shmerko VP. *Signal Processing in Discrete Bases.* Science and Technics Publishers, Minsk, Belarus, 1986 (In Russian).

[21] Shmerko VP. Synthesis of arithmetic forms of Boolean functions using the Fourier transform. *Automation and Remote Control,* Kluwer/Plenum Publishers, 50(5):684–691, Pt2, 1989.

[22] Stanković RS, Moraga C, and Astola JT. Reed–Muller expressions in the previous decade. *International Journal Multiple Valued Logic and Soft Computing,* 10(1):5–28, 2004.

[23] Stanković RS, Stanković MS, Egiazarian K, and Yaroslavsky L. Remarks on history of FFT and related algorithms. *Proceedings of the International TICSP Workshop on Spectral Methods and Multirate Signal Processing,* Vienna, Austria, pp. 281–288, 2004.

[24] Stanković RS, Ed. Spectral techniques. *Multiple-Valied Logic and Soft Computing,* Special issue, vol. 10 no. 1 and 2, 2004.

[25] Thornton M, Drechsler R, and Miller DM. *Spectral Techniques in VLSI CAD.* Kluwer, Dordrecht, 2001.

[26] Thornton MA. The Karhunen–Loève transform of discrete MVL functions. In *Proceedings of the 35th IEEE International Symposium on Multiple-Valued Logic,* pp. 194–199, 2005.

[27] Thornton MA. Spectral tramsforms of mixed-radix MVL functions. In *Proceedings of the 33th IEEE International Symposium on Multiple-Valued Logic,* pp. 329–333, 2003.

[28] Thornton MA and Miller DM. Computation of discrete function Chrestenson spectrum using Cayley color graphs. *International Journal of Multiple-Valued Logic and Soft Computing,* 10(2):1028–1030, 2004.

[29] Wickerhuser MV. *Adapted Wavelet Analysis from Theory to Software.* IEEE Press, New York, 1994.

[30] Yanushkevich SN and Stanković RS, Eds. Arithmetic logic in contol systems. *Automation and Remote Control,* Special issue, Plenum/Kluwer Publishers, vol. 65, no 6, 2004.

[31] Yanushkevich SN, Shmerko VP, and Lyshevski SE. *Logic Design of NanoICs.* CRC Press, Boca Raton, FL, 2005.

[32] Yanushkevich SN. Multiplicative properties of spectral Walsh coefficients of Boolean functions. *Automation and Remote Control* Kluwer/Plenum Publishers, 64(12):1933–1947, 2003.

[33] Yang J and De Micheli G. Spectral transforms for technology mapping. *Technical Report CSL-TR-91-498*, Stanford University, 1991.

8

Information - Theoretical Measures

This chapter focuses on information measures. Applying the notation to a physical system (hardware), information, in a certain sense, is a measurable quantity which is independent of the physical medium by which it is conveyed. The most appropriate measure of information is mathematically similar to the measure of entropy, that is, it is measured in bits of digits.

The technique of information theory is applied to the problems of the extraction of information from systems containing an element of randomness.

The information-theoretical concept is relevant to probability estimation on decision trees and diagrams (Chapter 20) to measures of decision trees and diagrams in terms of information and entropy (Chapter 24).

8.1 Introduction

A computing system can be seen as a process of communication between computer components. The classical concept of information advocated by Shannon is the basis for this. However, some adjustment must be done to it in order to capture a number of features of the design and processing of a computing system.

The information-theoretical standpoint on computing is based on the following notations:

- *Source of information* is a stochastic process where an event occurs at time point i with probability p_i. That is, the source of information is defined in terms of the probability distribution for signals from this source. The problem is usually formulated in terms of sender and receiver of information and used by analogy with communication problems.
- *Information engine* is the machine that deals with information.

- *Quantity of information* is a value of a function that occurs with the probability p; this quantity equal to $(-\log_2 p)$.
- *Entropy*, $H(f)$, is the measure of the information content of X. The greater the uncertainty in the source output, the higher is its information content. A source with zero uncertainty would have zero information content and, therefore, its entropy would be equal to zero.

The information and entropy, in their turn, can be calculated with respect to the given sources:

Information carried by the value of a variable or function,
Conditional entropy of function f values given function g,
Relative information of the value of a function given the value of a variable,
Mutual information between the variable and function,
Joint entropy over a distribution of jointly specified functions f and g.

Many useful characteristics are derived from the entropy, namely, the conditional entropy, mutual information, joint information, and relative information. Figure 8.1 illustrates the basic principles of input and output information measures in a logic circuit, where the shared arrows mean that the value of $x_i(f)$ carries the information, the shared arrow therefore indicating the direction of the information stream. Obviously, we can compare the results of the input and output measures and calculate the loss of information. Information and entropy can be measured on decision trees and diagrams. This is implemented in the same way it is done in the entropy-based machine learning approach to optimization of decision trees, called $ID3$ algorithm (see "Further reading" section for references). This approach is different from measuring in the circuit, and is discussed in Chapter 24.

8.2 Information-theoretical measures

Information-theoretical measures include information, or information quantity, and entropy.

8.2.1 Quantity of information

Let us assume that all combinations of values of variables occur with equal probability. A value of a function that occurs with the probability p carries a quantity of information equal to

$$< \texttt{Quantity of information} > \; = \; -\log_2 p \;\; bit,$$

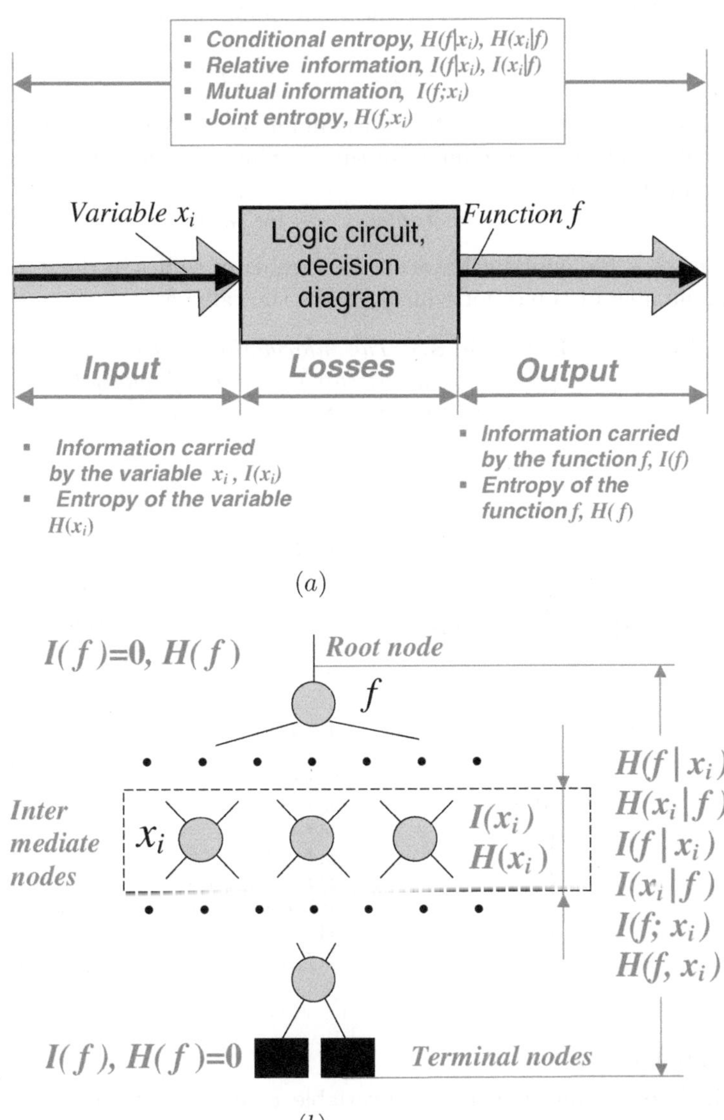

FIGURE 8.1
Information measures and computing input/output relationships of information in circuits and decision diagrams.

where p is the probability of that value occurring. Note that information is measured in bits.

The information carried by the value of a of x_i is equal to

$$I(x_i)_{|x_i=a} = -\log_2 p \ \ bit,$$

where p is the quotient between the number of tuples whose i-th components equal a and the total number of tuples. Similarly, the information carried by a value b of f is

$$I(f)_{|f=b} = -\log_2 q \ \ bit,$$

where q is the quotient between the number of tuples in the domain of f and the number of tuples for which f takes the value b.

Example 8.1 *The information carried by the values of variable x_i for a switching function f given by a truth table is calculated in Figure 8.2.*

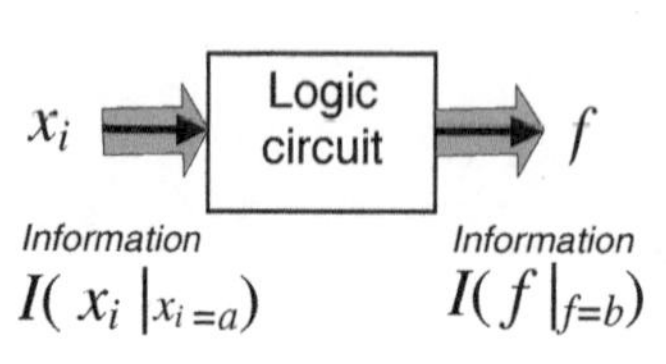

Input x_i	Output f
0	0
1	0
0	0
1	1
0	0

Probabilities of the values of variable:
$p(x_i = 0) = {}^3/_5$
$p(x_i = 1) = {}^2/_5$
The information carried by the variable:
$I(x_i)_{|x_i=0} = -\log_2 {}^3/_5 = 0.737\ bit$
$I(x_i)_{|x_i=1} = -\log_2 {}^2/_5 = 1.322\ bit$

Probabilities of the values of function:
$p(f = 0) = {}^4/_5,\ p(f = 1) = {}^1/_5$
while
$p(f = 0)|_{x_i=0} = {}^3/_5$
$p(f = 0)|_{x_i=1} = {}^1/_5$
$p(f = 1)|_{x_i=0} = 0$
$p(f = 1)|_{x_i=1} = {}^1/_5$
Information carried by the function:
$I(f)_{|f=0} = -\log_2 {}^4/_5 = 0.322\ bit$
$I(f)_{|f=1} = -\log_2 {}^1/_5 = 2.322\ bit$

FIGURE 8.2
The information carried by the variable x_i and the switching function f (Example 8.1).

8.2.2 Conditional entropy and relative information

Conditional entropy is a measure of a random variable f given a random variable x. To compute the conditional entropy, the conditional probability of f must be calculated. The conditional probability of value b of the function f, the input value a of x_i being known, is

$$p(f = b|x_i = a) = \frac{p_{\big| \substack{f=b \\ x_i=a}}}{p_{|x_i=a}},$$

where $p_{\big| \substack{f=b \\ x_i=a}}$ is the probability of f being equal to b and x_i being equal to a. Similarly, the conditional probability of value a of x_i given value b of the function f is

$$p(x_i = a|f = b) = \frac{p_{\big| \substack{f=b \\ x_i=a}}}{p_{|f=b}}$$

Conditional entropy $H(f|g)$ of the function f given the function g is

$$H(f|g) = H(f,g) - H(g), \tag{8.1}$$

Joint entropy. In Equation 8.1, $H(f,g)$ is *joint entropy*:

$$H(f,g) = -\sum_{a=0}^{m-1} \sum_{b=0}^{m-1} p_{\big| \substack{f=a \\ g=b}} \cdot \log p_{\big| \substack{f=a \\ g=b}}, \tag{8.2}$$

where $p_{\big| \substack{f=a \\ g=b}}$ denotes the probability that f takes value a and g takes value b, simultaneously*.

Chain rule. In circuit analysis and decision tree design, the so-called *chain rule* is useful (Figure 8.3)

$$H(f_1, \ldots, f_n|g) = \sum_{i=1}^{n} H(f_i|f_1, \ldots, f_{i-1}, g). \tag{8.3}$$

The relative information of value b of function f given value a_i of the input variable x_i is

$$I(f = b|x_i = a) = -\log_2 p\ (f = b|x_i = a).$$

*It follows from Equation 8.1, that the joint entropy is the sum of the source entropy, $H(g)$, and conditional entropy given g, i.e., $H(f,g) = H(g) + H(f|g)$. The conditional entropy $H(f|g)$ represents the information loss in the circuit in going from input to output. It is how much must be added to the source entropy to get the joint entropy. Since f and g play symmetric roles, Equation 8.1 can be rewritten in the form $H(f,g) = H(f) + H(g|f)$.

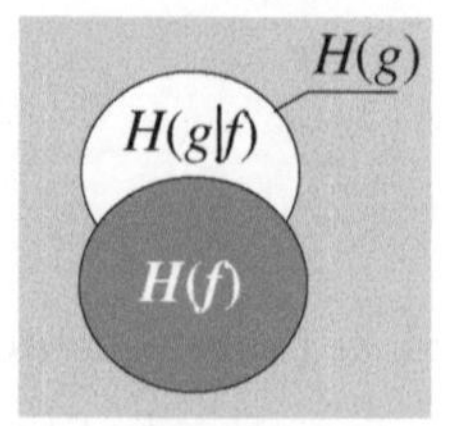

Additivity of entropy, or the chain rule for entropies, is defined as

$$H(f, g) = H(f) + H(g|f)$$

The conditional entropy $H(f|g)$:

- *Is non-negative,*
- *Is equal to zero if and only if a function u such that*

$$f = u(g)$$

exists with probability one.

FIGURE 8.3
The additivity of entropy.

The relative information of value a_i of the input variable x_i given value b of the function f is

$$I(x_i = a|f = b) = -\log_2 p\ (x_i = a|f = b).$$

Once the probability is equal to 0, we suppose that the relative information is equal to 0.

Example 8.2 *Figure 8.4 illustrates the calculation of the conditional and relative information given the truth table of a switching function.*

8.2.3 Entropy of a variable and a function

Let the input variable x_i be the outcome of a probabilistic experiment, and the random function f represent the output of some step of computation. Each experimental outcome results in different conditional probability distributions on the random f. Shannon's entropy of the variable x_i is defined as

$$H(x_i) = \sum_{l=0}^{m-1} p_{|x_i=a_l} \log_2 p_{|x_i=a_l}, \tag{8.4}$$

where m is the number of distinct values assumed by x_i. Shannon's entropy of the function f is

$$H(f) = \sum_{k=0}^{n-1} p_{|f=b_k} \log_2 p_{|f=b_k}, \tag{8.5}$$

where n is the number of distinct values assumed by f.

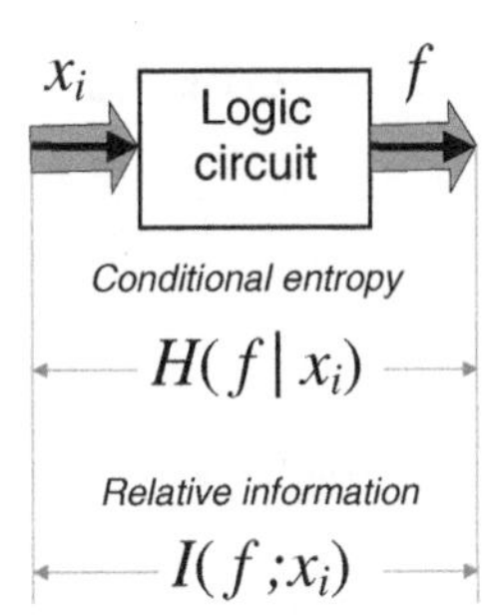

Input x_i	*Output* f
0	0
1	0
0	0
1	1
0	0

Conditional probabilities:

$p_{\substack{|f=0\\|x_i=0}} = {}^3/_5$

$p_{\substack{|f=0\\|x_i=1}} = {}^1/_5$

$p_{\substack{|f=1\\|x_i=0}} = 0$

$p_{\substack{|f=1\\|x_i=1}} = {}^1/_5$

Then

$p(f=0|x_i=0) = p_{\substack{|f=0\\|x_i=0}} : p_{|x_i=0} = {}^3/_5 : {}^3/_5 = 1$

$p(f=0|x_i=1) = p_{\substack{|f=0\\|x_i=1}} : p_{|x_i=1} = {}^1/_5 : {}^2/_5 = {}^1/_2$

$p(f=1|x_i=0) = p_{\substack{|f=1\\|x_i=0}} : p_{|x_i=0} = 0$

$p(f=1|x_i=1) = p_{\substack{|f=1\\|x_i=1}} : p_{|x_i=1} = {}^1/_5 : {}^2/_5 = {}^1/_2$

Conditional entropy:

$$\begin{aligned} H(f|x_i) &= -p(f=0|x_i=0)\log p(f=0|x_i=0) \\ &\quad - p(f=0|x_i=1)\log p(f=0|x_i=1) \\ &\quad - p(f=1|x_i=0)\log p(f=1|x_i=0) \\ &\quad - p(f=1|x_i=1)\log p(f=1|x_i=1) \\ &= -1\log 1 - {}^1/_2 \log {}^1/_2 - 0\log 0 - {}^1/_2 \log {}^1/_2 \\ &= 1 \end{aligned}$$

Relative information:

$I(f=0|x_i=0) = -\log_2 1 = 0$

$I(f=0|x_i=1) = -\log_2 {}^1/_2 = 1$

$I(f=1|x_i=0) = 0$

$I(f=1|x_i=1) = -\log_2 {}^1/_2 = 1$

FIGURE 8.4
Computing conditional entropy and relative information (Example 8.2).

This definition of the measure of information implies that the greater the uncertainty in the source output, the smaller is its information content. In a similar fashion, a source with zero uncertainty would have zero information content and, therefore, its entropy would likewise be equal to zero.

Example 8.3 *Figure 8.5 illustrates the calculation of entropy of the variable and function. The entropy of the variable x_i and function f is 0.971 bits and 0.722 bits.*

In general,

- For any variable x_i it holds that $0 \leq H(f) \leq 1$; similarly, for any function f, $0 \leq H(x_i) \leq 1$. Note that for patterns of variables it may not hold.
- The entropy of any variable in a completely specified function is 1.
- The entropy of a constant is 0.

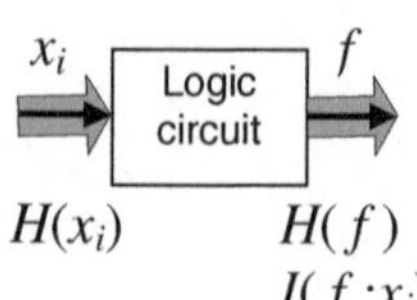

Input x_i	Output f
0	0
1	0
0	0
1	1
0	0

Shannon's entropy:

$$\begin{aligned} H(x_i) &= -{}^3/_5 \cdot \log_2 {}^3/_5 - {}^2/_5 \cdot \log_2 {}^2/_5 \\ &= 0.971\ bit \\ H(f) &= -{}^4/_5 \cdot \log_2 {}^1/_5 - {}^1/_5 \cdot \log_2 {}^4/_5 \\ &= 0.722\ bit \end{aligned}$$

The mutual information:

$$\begin{aligned} I(f;x_i) &= \sum_{k=1}^{5}\sum_{l=1}^{5} p_{\left|\substack{f=b_k \\ x_i=a_l}\right.} \times I_{\left|\substack{f=b_k \\ x_i=a_l}\right.} \\ &= {}^3/_5 \cdot \log_2 {}^3/_5 + {}^1/_5 \cdot \log_2 {}^1/_5 + {}^1/_5 \cdot \log_2 {}^1/_5 \\ &= 1.371\ bit \end{aligned}$$

FIGURE 8.5
Shannon's entropy and mutual information (Examples 8.3 and 8.4).

8.2.4 Mutual information

The mutual information is used to measure the dependence of the function f on the values of the variable x_i and vice-versa, i.e., how statically distinguishable distributions of f and x_i are. If the distributions are different, then the amount of information f carries about x_i is large. If f is independent of x_i, then f carries zero information about x_i. Figure 8.6 illustrates mutual information between two variables f and g.

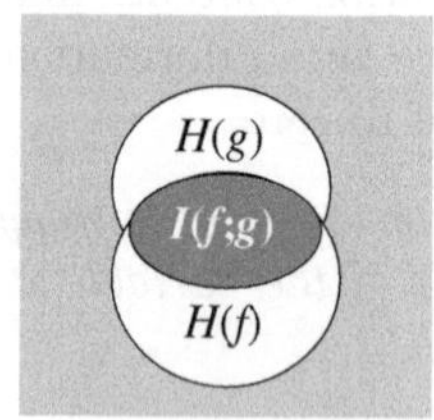

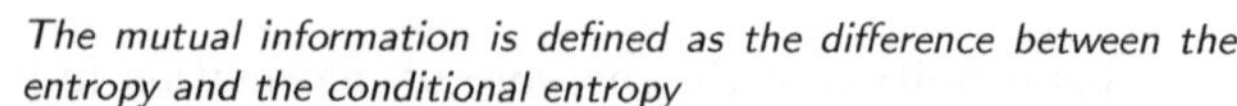

The mutual information is defined as the difference between the entropy and the conditional entropy

$$\begin{aligned} I(f;g) &= H(f) - H(f|g) \\ &= H(f) + H(g) - H(f,g), \end{aligned}$$

i.e., the difference of the uncertainty of f and the remaining uncertainty of f after knowing g. This quantity is the information of f obtained by knowing g.
The mutual information is a degree of the dependency between f and g and always takes positive values.
The additivity for the mutual information of three random variables:

$$I(f;g,z) = I(f;g) + I(f;z|g)$$

FIGURE 8.6
The mutual information.

The mutual information between the value b of the function and the value a of the input variable x_i is:

$$
\begin{aligned}
I(f;x_i) &= I(f;x_i)_{|f=b} - I(f=b|x_i=a) \\
&= -\log_2 p_{|f=b} + \log_2 \frac{p_{|\substack{f=b \\ x_i=a}}}{p_{|x_i=a}}.
\end{aligned}
$$

By analogy, the mutual information between the input variable x_i and the function f is

$$
\begin{aligned}
I(f;x_i) &= \sum_k \sum_l p_{|\substack{f=b_k \\ x_i=a_l}} \times I(f;x_i)_{|\substack{f=b_k \\ x_i=a_l}} \\
&= \sum_k \sum_l p_{|\substack{f=b_k \\ x_i=a_l}} \times \log_2 \frac{p_{|\substack{f=b_k \\ x_i=a_l}}}{p_{|x_i=a_l}}.
\end{aligned}
$$

Useful relationships are

$$
\begin{aligned}
I(g;f) = I(f;g) &= H(f) - H(f|g) \\
&= H(g) - H(g|f) \\
&= H(f) + H(g) - H(f,g); \\
I(g;f_1,\ldots,f_n|z) &= \sum_{i=1}^{n} I(g;f_i|f_1,\ldots,f_{i-1},z).
\end{aligned}
$$

Interpretation of mutual information. The mutual information can also be interpreted as follows. A circuit can be represented as a transmission system. The input and output signals are x_i and y_j, $i,j \in 1,2,\ldots,q$, respectively. Prior to reception the probability of the signal x_i was $p(x_i)$. This is the a *priori* probability of x_i. After transmission of x_i and reception of y_j, the probability that the input signal was x_i becomes $p(x_i|y_j)$. This is the a *posteriori* probability of x_i. The change in the probabilities $p(x_i)$ and $p(x_i|y_j)$ can be understood as how much the receiver (output) learned from the reception of the y_j. For example, if the a posteriori probability $p(x_i|y_j) = 1$, the output signal is exactly what was the input signal. The difference between the a priori probabilities and the a posteriori probabilities is the difference of information uncertainty before and after the reception of a signal y_j. This difference measures the gain in information and can be described using notation of a mutual information

$$
I(x_i;y_i) = \log_2 \frac{1}{p(x_i)} - \log_2 \frac{1}{p(x_i|y_j)} = \log_2 \frac{p(x_i|y_j)}{p(x_i)}.
$$

If $p(x_i) = p(x_i|y_j)$, then the mutual information is zero and no information has been transmitted. If some additional knowledge is obtained about x_i from the output signal y_j, the mutual information is a positive value.

Conditional mutual information. In Equation 8.6, $I(g; f|z)$ is the *conditional mutual* information between g and f given z. If g and f are independent, then $I(g; f) \geq 0$. The mutual information is a measure of the correlation between g and f. For example, if g and f are equal with high probability, then $I(g; f)$ is large. If f_1 and f_2 carry information about g and are independent given g then $I(z(f_1, f_2); g) \leq I(f_1; g) + I(f_2; g)$ for any switching function z.

Example 8.4 *Figure 8.5 illustrates the calculation of the mutual information. The variable x_i carries 0.322 bits of information about the function f.*

8.3 Information measures of elementary switching function of two variables

There are two approaches to information measures of elementary functions of two variables:

- The values of input variables are considered as random patterns; for a two-input elementary function there are 4 random patterns $x_1x_2 \in \{00.01, 10, 11\}$.
- The values of input variables are considered as non-correlated random signals; for a two-input elementary function there are random signals $x_1 \in \{0, 1\}$ and $x_2 \in \{0, 1\}$.

Information measures based on pattern. Consider a two-input AND function with four random combinations of input signals: 00 with probability p_{00}, 01 with probability p_{01}, 10 with probability p_{10}, and 11 with probability p_{11} (Figure 8.7a).

Using Shannon's formula (8.4), we can calculate the entropy of the input signals, denoted by H_{in} as follows

$$\begin{aligned} H_{in} = & - p_{00} \times \log_2 p_{00} - p_{01} \times \log_2 p_{01} \\ & - p_{10} \times \log_2 p_{10} - p_3 \times \log_2 p_{11} \ \textit{bit/pattern}. \end{aligned}$$

The maximum entropy of the input signals can be calculated by inserting into the above equation $p_i = 0.25$, $i = 0, 1, 2, 3$ (Figure 8.8).

The output of the AND function is equal to 0 with probability 0.25, and equal to 1 with probability 0.75. The entropy of the output signal, H_{out}, is calculated by (8.5)

$$H_{out} = -0.25 \times \log_2 0.25 - 0.75 \times \log_2 0.75 = 0.81 \ \textit{bit/pattern}.$$

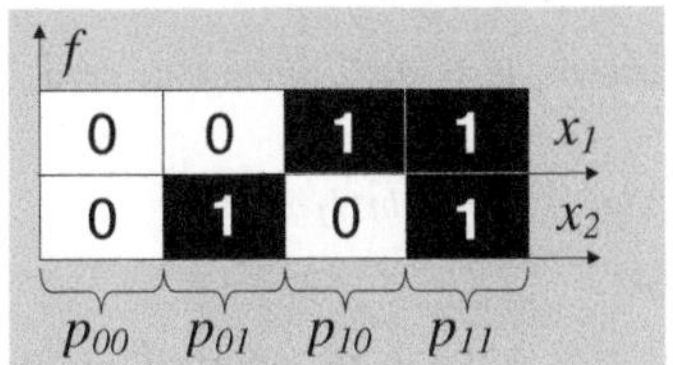

Patterns

Four random combinations of the signals:

$00 \rightarrow p_{00}$
$01 \rightarrow p_{01}$
$10 \rightarrow p_{10}$
$11 \rightarrow p_{11}$

(*a*)

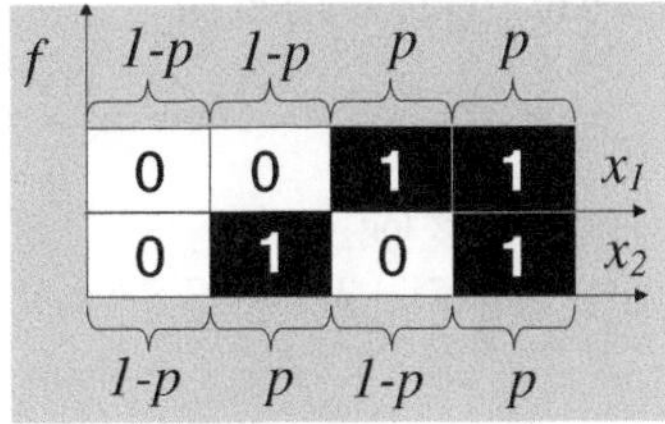

Non-correlated signals

$1 \rightarrow p$
$0 \rightarrow 1 - p$

(*b*)

FIGURE 8.7
Measurement of probabilities: random patterns (a) and non-correlated signals (b).

The example below demonstrates a technique of computing information measures for input signals that are not correlated. Information measures of the two-variable functions AND, OR, EXOR and NOT are given in Table 24.2 for $p(x_1) = p(x_2) = 0.5$.

Information measures based on non-correlated signals. Let the input signal be equal to 1 with probability p, and 0 with probability $1 - p$ (Figure 8.7b). The entropy of the input signals is

$$\begin{aligned} H_{in} &= -(1-p)^2 \times \log_2(1-p)^2 - 2(1-p) \times \log_2(1-p)p - p^2 \times \log_2 p^2 \\ &= -2(1-p) \times \log_2(1-p) - 2p \times \log_2 p \;\; \mathit{bit}. \end{aligned}$$

The output of the AND function is equal to 1 with probability p^2, and equal to 0 with probability $1 - p^2$. Hence, the entropy of the output signal is

$$H_{out} = -p^2 \times \log_2 p^2 - (1-p)^2 \times \log_2(1-p)^2 \;\; \mathit{bit}.$$

The maximum value of the output entropy is equal to 1 when $p = 0.707$. Hence, the input entropy of the AND function is 0.745 *bits* (Figure 8.8). We observe that in the case of non-correlated signals, information losses are less.

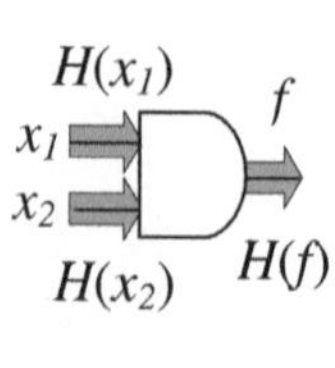

x_1	x_2	f
0	0	0
0	1	0
1	0	0
1	1	1

$f = x_1 x_2$

Method 1
The entropy of the input pattern, $p_i = 0.25$, since four combinations are possible

$$H_{in} = -4 \times 0.25 \times \log_2 0.25 = 2 \; bit/pattern$$

Entropy of the output signal

$$H_{out} = -0.25 \times \log_2 0.25 - 0.75 \times \log_2 0.75 = 0.811 \; bit/pattern$$

Loss of information

$$H_{loss} = H_{out} - H_{in} = 2.0 - 0.811 = 1.189 \; bit$$

Method 2
Entropy of the input signal ($p = 0.707$)

$$\begin{aligned} H_{in} &= -2(1-p) \times \log_2(1-p) - 2p \times \log_2 p \\ &= -2(1-0.707) \times \log_2(1-0.707) - 2 \times 0.707 \times \log_2 0.707 \\ &= 1.745 \; bit \end{aligned}$$

Output entropy

$$\begin{aligned} H_{out} &= -0.707^2 \times \log_2 0.707^2 - (1-0.707)^2 \times \log_2(1-0.707)^2 \\ &= 0.804 \; bit \end{aligned}$$

Loss of information

$$H_{loss} = H_{out} - H_{in} = 1.745 - 0.804 = 0.941 \; bit$$

FIGURE 8.8
Information measures of AND functions of two variables.

Useful properties. In Table 8.2, the entropies of elementary switching functions are grouped and calculated. We observe that

- ▶ A large group of the functions is characterized by the same output entropy $H(f) = 0.81$, conditional entropies $H(f|x_1) = 0.5$ and $H(f|x_2) = 0.5$, mutual information $I(f;x_1) = 0.31$ and $I(f;x_2) = 0.31$.
- ▶ EXOR and EQUIVALENCE functions have maximum values of entropies $H(f) = H(f|x_1) = H(f|x_2) = 1$.
- ▶ The entropy measures for a constant function, logical 1 and 0, are equal to 0.

In addition, a single-input single-output function (gate) does not lose information. Any many-input single-output logic function always results in a loss of information. An n-input n-output logic function is a lossless if and only if for each input combination there is exactly one output combination, and viceversa. This function is called *reversible*, and its fundamental properties are utilized in reversible computing (see Chapter 35 for references).

TABLE 8.1
Information measures of elementary switching functions of two variables.

Function	Information estimations
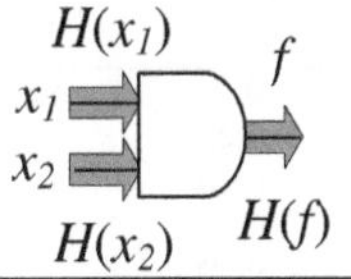 $H(x_1)$, x_1, x_2, $H(x_2)$, f, $H(f)$	$f = x_1x_2$ $H(f) = -p_{\|f=0} \cdot \log_2 p_{\|f=0} - p_{\|f=1} \cdot \log_2 p_{\|f=1}$ $p(f) = 0.5 \cdot 0.5 = 0.25$ $H(f) = -0.25 \cdot \log_2 0.25 - 0.75 \cdot \log_2 0.75 = 0.8113\ bit$
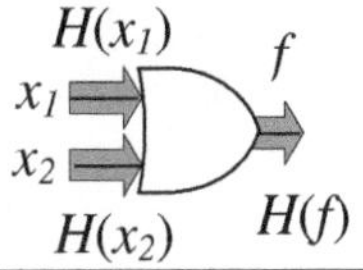 $H(x_1)$, x_1, x_2, $H(x_2)$, f, $H(f)$	$f = x_1 \vee x_2$ $H(f) = -p_{\|f=0} \cdot \log_2 p_{\|f=0} - p_{\|f=1} \cdot \log_2 p_{\|f=1}$ $p(f) = 1 - (1 - 0.5) \cdot (1 - 0.5) = 0.75$ $H(f) = -0.75 \cdot \log_2 0.75 - 0.25 \cdot \log_2 0.25 = 0.8113\ bit$
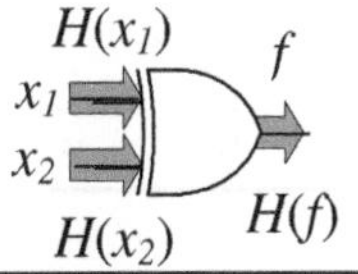 $H(x_1)$, x_1, x_2, $H(x_2)$, f, $H(f)$	$f = x_1 \oplus x_2$ $H(f) = -p_{\|f=0} \cdot \log_2 \|f = 0 - p_{\|f=1} \cdot \log_2 p_{\|f=1}$ $p(f) = 0.5 \cdot 0.5 + 0.5 \cdot 0.5 = 0.5$ $H(f) = -0.5 \cdot \log_2 0.5 - 0.5 \cdot \log_2 0.5 = 1\ bit$
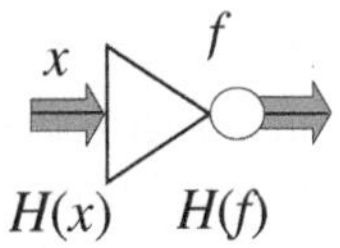 x, f, $H(x)$, $H(f)$	$f = \overline{x}$ $H(f) = p_{\|f=0} \cdot \log_2 p_{\|f=0} - p_{\|f=1} \cdot \log_2 p_{\|f=1}$ $p(f) = 1 - 0.5 = 0.5$ $H(f) = -0.5 \cdot \log_2 0.5 - 0.5 \cdot \log_2 0.25 = 1\ bit$

8.4 Information-theoretical measures and symmetry

In this section, an example of reduction of a search space using information-theoretical measures is introduced. An approach is demonstrated for symmetry detection problems.

Symmetries refer to permutations of variables (with possible complementation) that leave a logic function unchanged. Symmetries provide insights into the data structure that can be used to facilitate computations and can also serve as a guide for preserving that structure when the function is transformed, for example, into other representations. The size of a decision diagram can be reduced by using a variable order that places symmetric variables contiguously.

TABLE 8.2
Information measures of elementary switching functions of two variables.

f	Entropy $H(f)$	$H(f\|x_1)$	$H(f\|x_2)$	Mutual information $I(f;x_1)$	$H\|f;x_2$
x_1x_2	0.81	0.5	0.5	0.31	0.31
$x_1\overline{x}_2$	0.81	0.5	0.5	0.31	0.31
$\overline{x}_1x_2$	0.81	0.5	0.5	0.31	0.31
$x_1 \vee x_2$	0.81	0.5	0.5	0.31	0.31
$x_1 \uparrow x_2$	0.81	0.5	0.5	0.31	0.31
$x_1 \rightarrow x_2$	0.81	0.5	0.5	0.31	0.31
$x_2 \rightarrow x_1$	0.81	0.5	0.5	0.31	0.31
$x_1 \mid x_2$	0.81	0.5	0.5	0.31	0.31
$x_1 \oplus x_2$	1	1	1	0	0
$x_1 \sim x_2$	1	1	1	0	0
x_1	1	0	1	1	0
x_2	1	0	0	0	1
$\overline{x}_1$	1	1	0	0	1
$\overline{x}_2$	1	0	1	1	0
const 1	0	0	0	0	0
const 0	0	0	0	0	0

The search space can be reduced by utilizing the information-theoretical properties of symmetric functions. The important feature of an information-theoretical approach is that there are no losses of symmetric variables through the reduction of a search space.

After reduction of a search space, an exact method for recognizing symmetries can be used, for example, a method based on logic differences (Figure 8.9).

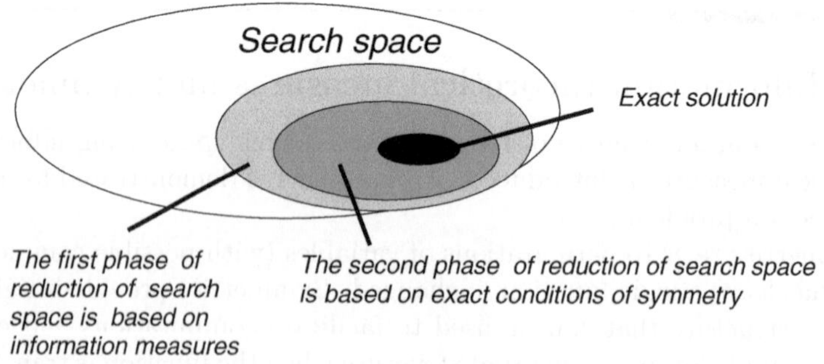

FIGURE 8.9
Search space reduction by a two phase strategy.

8.4.1 Entropy of distribution of values of logic function

The distribution of the values of logic function f given $x_i = a$, denoted by $\Delta_{x_i=a}$, is a set

$$\{\Delta_{\left|\substack{f=0\\x_i=a}\right.}, \Delta_{\left|\substack{f=1\\x_i=a}\right.}, \ldots, \Delta_{\left|\substack{f=m-1\\x_i=a}\right.}\},$$

where $\Delta_{\left|\substack{f=b\\x_i=a}\right.}$ is the number of terms for which $f = b$ given $x_i = a$; $a, b \in (0, \ldots, m-1)$. The distribution of fixed order is denoted as $\Delta^*_{x_i=a}$.

Example 8.5 *In Figure 8.10, the distribution is given of the value of a ternary function*

$$f = x_1^0 x_3^0 \vee x_2^0 x_3^0 \vee 2x_1^1 x_2^1.$$

For instance, for $x_2 = 0$ function takes value 0 six times, value 1 three times, and value 3 zero times.

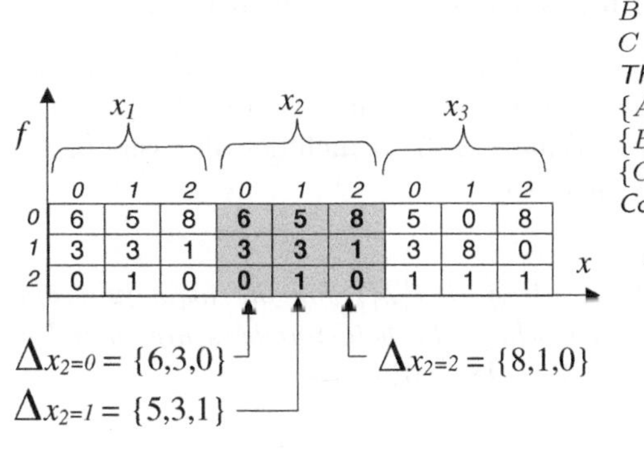

$A = \{6, 3, 0\}$
$B = \{5, 3, 1\}$
$C = \{8, 1, 0\}$
The permutation $\alpha\{\Delta_{x_1}\}$
$\{A; B; C\}, \{A; C; B\},$
$\{B; A; C\}, \{B; C; A\},$
$\{C; A; B\}, \{C; B; A\}$
Conditional entropy

$$\begin{aligned} H(f|x_1) = {} & {}^{6}/_{27} \cdot log_3 2 \\ & + {}^{3}/_{27} \cdot log_3 1 \\ & + {}^{0}/_{27} + {}^{5}/_{27} \cdot log_3 {}^{5}/_{3} \\ & + {}^{3}/_{27} \cdot log_3 1 \\ & + {}^{1}/_{27} \cdot log_3 {}^{1}/_{3} \\ & + {}^{8}/_{27} \cdot log_3 {}^{8}/_{3} \\ & + {}^{1}/_{27} \cdot log_3 {}^{1}/_{3} + {}^{0}/_{27} \end{aligned}$$

FIGURE 8.10
Distribution of variables of the function (Examples 8.5, 8.6, and 8.7).

Permutation over the set. Let $\alpha\{\Delta_{x_i}\}$ and $\alpha\{\Delta_{x_i=j}\}$ mean permutation over the sets Δ_{x_i} and $\Delta_{x_i=j}$ respectively.

Example 8.6 *Figure 8.10 explains the permutation of the values of logic function f with respect to variable x_1. In particular, the distribution for $x_1 = 0$ be $\Delta_{x_1=0} = \{6, 3, 0\}$, so*

$$\alpha\{\Delta_{x_1=0}\} \in \{6,3,0\}; \{6,0,3\}; \{3,6,0\}; \{3,0,6\}; \{0,6,3\}; \{0,3,6\}$$

Conditional entropy and distribution. The conditional entropy of an m-valued function f with respect to variable x_i is given by

$$H(f|x_i) = -\sum_{k=0}^{m}\sum_{l=0}^{m} p_{\big|\substack{f=k\\x_i=l}} (\log_m p_{\big|\substack{f=k\\x_i=l}} - \log_m p_{|x_i=l}). \tag{8.6}$$

The cofactors of probabilities (the number of assignments) in the conditional entropy $H(f|x_i)$ are elements within the permutation $\alpha\{\Delta_{x_i}\}$. This implies that the entropy includes all assignments f and x_i, and also $\alpha\{\Delta_{x_i}\}$.

Example 8.7 *The computing of conditional entropy is shown in Figure 8.10.*

8.4.2 Condition of symmetry over the distribution

The cofactors of probabilities (the number of assignments) in the conditional entropy $H(f|x_i)$ are the elements within the permutation $\alpha\{\Delta_{x_i}\}$. It is implied from this that the entropy (Equation 27.2) includes all assignments f and x_i, and so $\alpha\{\Delta_{x_i}\}$ includes them too.

Example 8.8 *Computing $H(f|x_1)$ for the logic function is given in Figure 8.10, where the bold numbers are included in $\Delta_{x_1} = \{\{6,3,0\}, \{5,3,1\}, \{8,1,0\}\}$.*

The corollary below establishes relationships between $\frac{\partial f}{\partial x_i}$ and Δ_{x_i}.

Corollary 8.1 *If a function f is symmetric in $\{x_i, x_j\}$, i.e., $\frac{\partial f}{\partial x_i} = \frac{\partial f}{\partial x_j}$, then there exists an equality between the distributions of fixed order for both variables:*

$$\Delta^*_{x_i} = \Delta^*_{x_j}. \tag{8.7}$$

The proof is given in [44].

Example 8.9 *The switching function*

$$f = \overline{x}_2x_3\overline{x}_4 \vee \overline{x}_1x_2\overline{x}_3 \vee x_1\overline{x}_2\overline{x}_3 \vee \overline{x}_1\overline{x}_2x_3 \vee x_1x_2x_3\overline{x}_4$$

and quaternary function

$$f \;=\; x_3^0 \vee\; x_2^0 \vee\; 3x_1^3x_2^2x_3^2$$

are symmetric in $\{x_1, x_2, x_3\}$ *and* $\{x_2, x_3\}$ *respectively. The entropies and distributions for these functions are as follows:*

$$\begin{cases} H(f|x_1) = H(f|x_2) = H(f|x_3) = 0,977; \\ H(f|x_4) = 0,883; \end{cases}$$

$$\Downarrow$$

$$\begin{cases} \Delta_{x_1} = \{\{4,4\};\{5,3\}\}; \\ \Delta^*_{x_2} = \Delta^*_{x_3} = \{\{5,3\};\{4,4\}\}; \\ \Delta_{x_4=1} = \{\{3,5\};\{2,6\}\}; \end{cases}$$

$$\begin{cases} H(f|x_2) = H(f|x_3) = 0,530; \\ H(f|x_1) = 0,343; \end{cases}$$

$$\Downarrow$$

$$\begin{cases} \Delta_{x_1} = \{\{9,7,0,0\};\{9,7,0,0\};\{9,7,0,0\}; \\ \qquad\{8,7,0,1\}\}; \\ \Delta^*_{x_2} = \Delta^*_{x_3} = \{\{0,16,0,0\};\{12,4,0,0\}; \\ \qquad\{11,4,0,1\};\{12,4,0,0\}\}; \end{cases}$$

Note, that $\Delta^*_{x_2} = \Delta^*_{x_3}$ *implies* $\frac{\partial f}{\partial x_2} = \frac{\partial f}{\partial x_3}$, *as shown in Figure 8.11.*

Corollary 8.2 *If logic function* f *is symmetric in* $\{x_i, x_j\}$, *the conditional entropy for both variables is equal.*

Proof: The symmetry $x_i \sim x_j$ implies that $\Delta_{x_i=a} = \Delta_{x_j=a}$. It means that $H(f|x_i) = H(f|x_j)$.

Theorems 8.1 and 8.2 guarantee that we can perform the first stage to detect the symmetry without losing the symmetric variables.

Theorem 8.1 *The necessary condition for a function* f *to be symmetric in* $\{x_i, x_j\}$ *is the equality:*

$$H(f|x_i) = H(f|x_j). \tag{8.8}$$

The proof follows from the definition of symmetry and relationship between the logic difference $\partial f/\partial x_i$ and Δ_{x_i} (Equation 8.7), that implies $\alpha\{\Delta_{x_i}\} = \alpha\{\Delta_{x_j}\}$.

The condition given by Equation 8.8 is not sufficient to detect the symmetry $x_i \sim x_j$. This is because the distribution Δ^*x_i of fixed order for logic difference $\partial f/\partial \widehat{x}_i$ is a subset of the permutation $\alpha\{\Delta_{x_i}\}$, that corresponds to $H(f|x_i)$: $\Delta^*_{x_i} \subset \alpha\{\Delta_{x_i}\}$. Ergo, Equation 8.8 can be true for symmetric as well as non-symmetric variables.

If logic function f is symmetric in $\{x_i, x_j\}$, the conditional entropy for both variables is equal. This statement follows from the fact that symmetry $x_i \sim x_j$ implies that $\Delta_{x_i=a} = \Delta_{x_j=a}$. This means that $H(f|x_i) = H(f|x_j)$. Note that opposite statement is not valid, i.e., the equality of entropies does not guarantee the symmetry of variables.

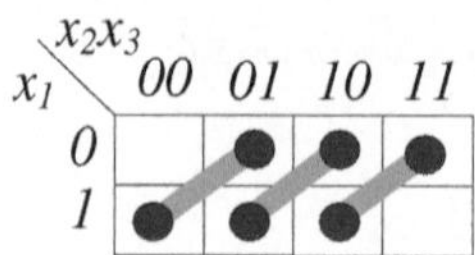

$f = \overline{x}_2x_3\overline{x}_4 \vee \overline{x}_1x_2\overline{x}_3 \vee x_1\overline{x}_2\overline{x}_3$
$\vee\ \overline{x}_1\overline{x}_2x_3 \vee x_1x_2x_3\overline{x}_4$

Conditional entropy:

$$H(f|x_1) = H(f|x_2) = H(f|x_3) = 0.977\ bit$$
$$H(f|x_4) = 0.883\ bit$$

Distribution:

$$\Delta_{x_1} = \{\{4,4\};\{5,3\}\}$$
$$\Delta^*_{x2} = \Delta^*_{x3} = \{\{5,3\};\{4,4\}\}$$
$$\Delta_{x_4=1} = \{\{3,5\};\{2,6\}\}$$

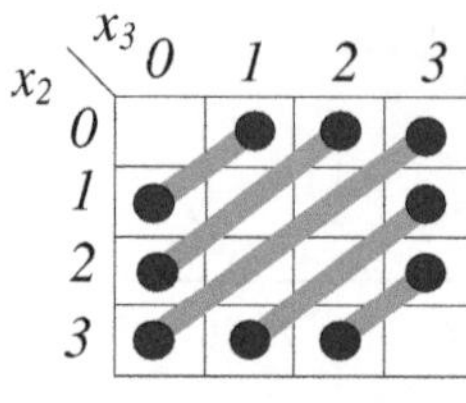

$f = x_3^0 \vee x_2^0 \vee 3x_1^3x_2^2x_3^2$

Conditional entropy:

$$H(f|x_1) = 0, .343\ bit$$
$$H(f|x_2) = H(f|x_3) = 0.530\ bit$$

Distribution:

$$\Delta_{x_1} = \{\{9,7,0,0\};\{9,7,0,0\};\{9,7,0,0\};\{8,7,0,1\}\};$$
$$\Delta^*_{x2} = \{\{0,16,0,0\};\{12,4,0,0\};\{11,4,0,1\};\{12,4,0,0\}\}$$
$$\Delta^*_{x2} = \Delta^*_{x3}$$

FIGURE 8.11
Detecting symmetries in a switching function and a quaternary function (Example 8.9).

8.5 Information and flexibility

The conditions under which an alternative function can replace a function at certain point in a circuit are called the *functional flexibility* of the point. The concept of functional flexibility can be implemented, in particular, using:

- Look-up-table-based correction,
- Sets of pairs of functions to be distinguished (SPFDs), and
- Correction based on information-theoretical method.

> *The concept of* neighborhood *is one of the possible approaches to implementing functional flexibility. Given a function f, the neighbor functions are defined as the set of functions g such that a transformation of g into f is possible through the composition of g with some correcting function.*

This concept was developed by Cheushev et al. [3, 18]. Using an information-theoretical approach, it is possible to verify not only that a network achieves the target functionality, but also that this network can be automatically *corrected* to achieve this.

8.5.1 Sets of pairs of functions to be distinguished

The concept of sets of pairs of functions to be distinguished (SPFD). This was introduced by Yamashita et al. [62] to represent the flexibility of a node in a multilevel network. It should be noted that the concept of neighborhood [3, 18] and SPFD [62] are closely related approaches from the information theoretical point of view.

The functions contained in an SPFD are all the functions which can satisfy the SPFD. An SPFD attached to a node specifies which pairs of primary input minterms can be or have to be distinguished by the node. This data is defined as the *information content* of the node since it indicates what information the node contributes to the network.

> **Example 8.10** *Figure 8.12 shows the information content of the OR gate.*

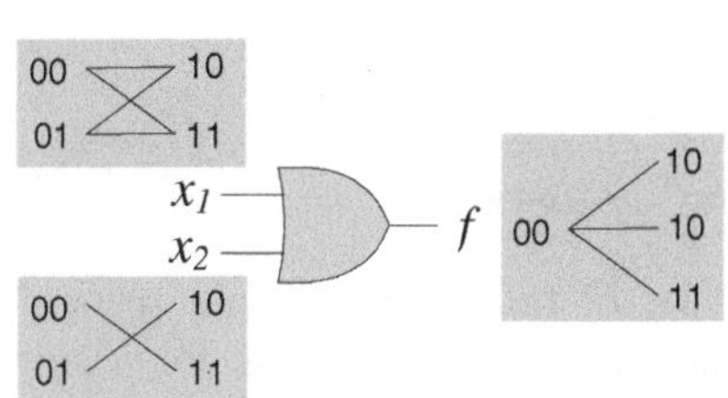

- *Input x_1 distinguishes 00 and 01 from 10 and 11*
- *Input x_2 distinguishes 00 and 10 from 01 and 11*
- *Output f distinguishes 00 from 10, 01, and 11*

$$\begin{aligned} H(f) &= -p_{|f=0} \cdot \log_2 p_{|f=0} - p_{|f=1} \cdot \log_2 p_{|f=1} \\ &= -0.75 \cdot \log_2 0.75 - 0.25 \cdot \log_2 0.25 \\ &= 0.8113 \\ p(f) &= 1 - (1 - 0.5) \cdot (1 - 0.5) = 0.75 \end{aligned}$$

FIGURE 8.12
Information flow through an OR gate (Example 8.10).

> **Example 8.11** *Given an inverter, $f = \overline{x}$:*
>
> $$\begin{aligned} H_{IN} &= H(x) = p_{|x=0} \cdot \log_2 p_{|x=0} - p_{|x=1} \cdot \log_2 p_{|x=1} = 1 \\ H_{OUT} &= H(f) = p_{|f=0} \cdot \log_2 p_{|f=0} - p_{|f=1} \cdot \log_2 p_{|f=1} \\ &= -0.5 \cdot \log_2 0.5 - 0.5 \cdot \log_2 0.5 = 1.0 \ bit \end{aligned}$$
>
> *Loss of information is*
>
> $$H_{loss} = H_{out} - H_{in} = 1 - 1 = 0 \ bit$$

An n-input n-output gate whose function is reversible does not lose information (see details in Chapter 35).

8.5.2 Concept of neighborhood of a function in terms of information

The task is to synthesize a network, with some desired characteristics, to implement a given function f. Let

- Net_i be a network obtained after the i-th iteration of an iterative synthesis strategy, and
- g be a logic function implemented by Net_i.

Verification of the evolved network is the answer to the question whether or not Net_i is a solution of the task. The answer is

$$Circuit\ solution = \begin{cases} \text{“yes”, if } g \Leftrightarrow f, \\ \text{“no”, otherwise,} \end{cases}$$

where $g \Leftrightarrow f$ means "g is equal to f". Hence, for a m-valued function f of n variables we look for a single solution in the search space of different networks implementing m^{m^n} possible functions g.

Suppose, the class $T(f)$ of functions in the neighborhood of f, such that any function $g \in T(f)$ may be transformed into f is known. Thus,

$$Circuit\ solution = \begin{cases} \text{“yes”, if } g \in T(f), \\ \text{“no”, otherwise.} \end{cases}$$

Assuming that $T(f)$ contains a finite set of functions, we can recognize them to be candidates for a solution, instead of having to look for a single solution.

Network correction with a constant. Let a and b be values of the functions f and g correspondingly. Finite and non-empty sets of all possible values of f and g are denoted by A and B, respectively. In our task, f and g take values from a restricted subsets $A' \subseteq A$ and $B' \subseteq B$.

Given the function f, the class $E(f)$ is the set of functions such that for any $g \in E(f)$, there exists a mapping $\varphi : B' \rightarrow A'$,

$$(g = b) \longrightarrow (f = a), \tag{8.9}$$

where $b \in B', a \in A'$, $|B'| \geq |A'|$. If φ is a mapping defined by (8.9), then no different values of f correspond to equal values of g:

$$(g(u_i) \neq g(u_j)) \longrightarrow (f(u_i) \neq f(u_j)),$$

where u_i, u_j denote assignments of variables in f and g. If there exists a mapping φ defined by (8.9), this mapping is a logic function.

Let φ denote a "*correcting function*". If such a function exists, then having g we can correct it with φ to obtain f. The *neighborhood* of f is then the set $\{g|\varphi(g) = f\}$.

Example 8.12 *Condition (8.9) holds for switching functions* $f = [01100111]$ *and* $g = [10011000]$ *with the correcting function* $\varphi(g) = \overline{g} = f$. *The truth vector of* φ *is* $[10]$.

In more detail:

- If $|B'| = |A'|$ and condition (8.9) holds, there exists the *bijective* mapping $B' \leftrightarrow A'$ such that $(g = b) \longleftrightarrow (f = a)$. A particular case of bijective mappings is the *identical* mapping, such that $B' = A'$ and $g \Leftrightarrow f$.
- If $|B'| > |A'|$ and condition (8.9) holds, the mapping $B' \rightarrow A'$ is a *surjection.*

So, $E(f)$ is the class of logic functions such that for any $g \in E(f)$ there exist a bijective mapping $g \leftrightarrow f$, if these functions take the same number of values, or a surjective mapping $g \rightarrow f$, if f takes less values than g.

Example 8.13 *Consider the ternary adder with outputs* $f_{sum} = [012120201]$ *and* $f_{carry} = [000001011]$. *Given the logic function* $g_1 = [012120201]$, $g_2 = [120201012]$, $g_3 = [112110200]$, *there exist the following identical, bijective, and surjective mappings:*

$$\begin{aligned} g_1 \Leftrightarrow f_{sum}: &\quad \varphi = [012], \\ g_2 \leftrightarrow f_{sum}: &\quad \varphi = [201], \\ g_3 \rightarrow f_{carry}: &\quad \varphi = [100]. \end{aligned}$$

Flexibility of restriction. We can conclude that

- If $|A'| = |B'| = m$, then the class $E(f)$ includes $m!$ functions.
- If an m-valued function f takes less than m possible values, i.e., $|A'| < m$, then mappings of g_i and g_j $(g_i, g_j \in E(f))$ into f can be the same.

Information-theory conditions of $g \in E(f)$. The following group of information-theory conditions describes the fact that $g \in E(f)$:

$$\begin{aligned} H(g) \geq H(f),\ H(f,g) = H(g), \\ H(f|g) = 0,\ \ I(g;f) = H(f). \end{aligned}$$

If $|A'| = |B'|$ and, therefore, only the bijective mapping $g \leftrightarrow f$ can exist, the following conditions hold:

$$\begin{aligned} H(g) &= H(f), \\ H(f,g) &= H(g) = H(f), \\ H(f|g) &= H(g|f) = 0, \\ I(g;f) &= H(f) = H(g). \end{aligned}$$

The equality $H(f|g) = 0$ is the necessary and sufficient condition for g to belong to the class $E(f)$. If $H(f|g) = 0$, then there exists a logic function φ such that $f = \varphi(g)$.

The relationship between our concept of neighborhood of f and the concept of $SPFDs$:

- If $H(f|g) = 0$ then $SPFD(f) \subseteq SPFD(g)$.
- If $(H(f|g) = 0) \wedge (H(g|f) = 0)$ then $SPFD(f) = SPFD(g)$.

8.6 Further reading

The problem of decision making in the presence of uncertainty is recognized as being of great importance within the field of logic nanoIC design. Many methods rely on the use of numerical information to handle imperfections.

Historical remarks. A physically based concept used in estimation is that of *entropy*, which originates from the second law of thermodynamics. In 1850, Rudolf Clausius, introduced entropy as a quantity that remains constant in the absence of heat dissipation. Entropy has since been interpreted as the amount of disorder in the system. Boltzmann (1984) derived the continuous form of entropy from kinetic theory applied to gases. Entropy is a measure of some property of a system or process that has inherent uncertainty. The entropy function gives the degree of randomness affecting a random variable.

A hundred years later, in 1948 [39], Shannon suggested a measure to represent the information by a numerical value, nowadays known as *Shannon entropy*. Since then, the term "uncertainty" is interchangable with the term "entropy." Consider an information source of symbols $s_1, s_2, \ldots, s_q$ with probabilities $p_1, p_2, \ldots, p_q$ respectively. For any discrete probability distribution there is the value of the *entropy function* $H = \sum_{i=1}^{q} p_i log(1/p_i)$. The function H of the probability distribution p_i measures the amount of uncertainty, or information the distribution contains, or information we get from the outcome of a circuit, or the outcome of some experiment. In designing an experiment, the entropy function is usually maximized. To implement this strategy, the experiment must be designed suitably, i.e., under controlled probabilities of outcomes. Shannon's noiseless coding theorem shows that by using a suitably large extension of the source, the average length of the encoded message can be brought as close to the entropy of the source as desired.

Shannon decomposition. In 1938, Shannon introduced a method for the decomposition of switching functions [38] known as *Shannon expansion*. In state-of-the-art of decision diagram technique, Shannon expansion of a switching function f with respect to a variable x_i is used in the form $f = \overline{x}_i f_0 \vee x_i f_1$, where $f_0 = f_{x_i=0}$ and $f_1 = f_{x_i=1}$.

Shannon entropy. Shannon suggested a measure to represent the information in numerical values, denoted as the *Shannon entropy* [39].

The Shannon information theory has been developed for many applications in circuit design. The latest characterization of a computing system as a communication system is consistent with the von Neumann concept of a computer. The bit strings of information are understood as messages to be communicated from a messenger to a receiver. Each message i is an event that has a certain probability of occurrence p_i with respect to its inputs. The measure of information produced when one event is chosen from the set of N events is the entropy of message i: $-\sum_{i \in N} p_i \log p_i$.

State-of-the-art decision diagram techniques. In state-of-the-art decision diagram techniques, the information theoretical notation of a Shannon expansion of a switching function f with respect to a variable x is used in the form $H^S(f|x) = p_{|x=0} \cdot H(f_{x=0}) + p_{|x=1} \cdot H(f_{x=1})$, where $H(f_{x=0})$ and $H(f_{x=1})$ is the entropy of function f given $x = 0$ and $x = 1$ respectively.

Fundamentals of information theory. The amount of randomness in a probability distribution is measured by its entropy (or information). In a fundamental sense, the concept of information proposed by Shannon captures only the case when unlimited computing power is available. However, computational cost may play a central role. This and other aspects of information theory can be found in [3, 16, 20, 24].

Applications in logic design. Useful results can be found in the book *Artificial Intelligence in Logic Design* edited by S.N. Yanushkevich, Kluwer Academic Publishers, 2004, that includes nine papers on the fundamentals of logic function manipulation based on the artificial intelligence paradigm, evolutionary circuit design, information measures in circuit design, and logic design of nanodevices.

Testing. The analysis is based upon a model where all signals are assumed to have certain statistical properties. The dynamic flavor of entropy has been studied in many papers to express testability (observability and controllability) measures for gate level circuits. For example, Agraval has shown that the probability of fault detection can be maximized by choosing test patterns that maximize the information at output [1]. The problem of the construction of sequential fault location for permanent faults has been considered by Varshney et al. [46].

Power dissipation. Existing techniques for power estimation at gate and circuit levels can be divided in dynamic and static. These techniques rely on probabilistic information in the input stream. The average switching activity per node (gate) is the main parameter that needs to be correctly determined. These and related problems are the focus many researchers.

For example, in [14, 33], it is demonstrated that the average switching activity in a circuit can be calculated using either entropy or information energy averages.

Finite state machines. Most of the algorithms for minimization of state assignments in finite state machines target reduced average switching per transition, i.e., average Hamming distance between states. Several papers have used entropy based models to solve the above problem. In particular, Tyagi's paper [45] provides theoretical lower bounds on the average Hamming distance per transition for finite state machines based on information-theoretical methods.

Evolutionary circuit design. There have already been some approaches to evolutionary circuit design. The main idea is that an evolutionary strategy would inevitably explore a much richer set of possibilities in the design spaces that are beyond the scope of traditional methods. In [2, 3, 18, 21, 22] the evolutionary strategy and information theoretical measures were used in circuit design.

Functional decomposition using information relationship measures was studied by Rawski et al. [37].

Information engine, computational work, and complexity. A deep and comprehensive analysis of computing systems' information engines has been done by Watanabe [83]. The relationship between function complexity and entropy is conjectured by Cook and Flynn [14]. The complexity of a switching function is expressed by the cost of implementing the function as a combinational network.

Hellerman has proposed so-called *logic entropy* [18]. Computation is considered as a process that reduces the disorder (or entropy) in the space of solutions while finding a result. The number of decisions required to find one correct answer in the space of solutions has been defined as *entropy of computation*, or *logic entropy* calculated as $log\frac{S}{A}$, where s is the number of solutions, A is the number of answers. This definition is consistent with the Shannon entropy provided that the space of solutions is all possible messages (bit strings) of a given length. The answer is one of the messages, so the entropy is the numbers of bits required to specify the correct answer. The term *logical entropy* owes its name to the fact that it depends on the number of logic operations required to perform the computation. In the beginning of the computation, the entropy (disorder) is maximum, at the end of computation the entropy is reduced to zero.

The other form of entropy is *spatial entropy*, and it is relevant to mapping the computation onto a domain where data travels over a physical distance. The data communication process is a process of removal of spatial entropy, while performing logical operations is aimed at removal of logical entropy (disorder). The spatial entropy of a system is a measure of the effort needed to bring data from the input location to the output locations. The removal of the spatial entropy corresponds to reduction of the distance between the input and the output.

Other applications. In machine learning, information theory has been recognized as a useful criterion [29]. To classify objects from knowledge of a training set of examples whose classes are previously known, a decision tree rule induction method known as the ID3 algorithm was introduced by Quinlan [30]. The method is based on recursive partitioning of the sample space and defines classes structurally by using decision trees.

A number of improved algorithms exist that use general to specific learning in order to build simple knowledge based systems by inducing decision trees from a set of examples [31, 32], and the method of quantitative information of logical expressions developed by Zhong and Ohsuga [50].

Detection of symmetry. Shannon [37] has shown that partially symmetric switching functions may be realized with considerably fewer components than most functions. McCluskey [25] studied group invariance of Boolean functions. Epstein [15] studied the implementation of symmetric functions over the library of AND, OR, and NOT gates. Born and Scidmore [5] have shown that any switching function can be represented by a symmetric function. Arnold and Harrison [4] developed the concept of canonical symmetric functions.

Pandey and Bryant [27] used symmetry for verification of transistor-level circuits by symbolic trajectory evaluation. Symmetries in circuits are classified as structural symmetries (from circuit structure) and data symmetries (data values in the system). Lee SC and Lee ET [19] investigated symmetric properties of multivalued functions.

The size of a decision diagram can be reduced by using a variable order that places symmetric variables contiguously, for example, using sifting procedures. Symmetries can be utilized to improve the efficiency of functional equivalence checking. For example, Scholl et al. [35] used symmetries in BDD minimization.

Flexibility. Sentovich and Brand [71] classified a flexibility that is used in logic design including the sets of pairs of functions to be distinguished (SPFDs). The method of SPFDs originally introduced by Yamashita et al., [48, 62], have been further developed, in particular, by Brayton [6] and Sinha et al. [47, 43].

References

[1] Agraval V. An information theoretic approach to digital fault testing. *IEEE Transactions on Computers*, 30(8):582–587, 1981.

[2] Aguirre AH and Coello CA. Evolutionary synthesis of logic circuits using information theory. In Yanushkevich SN, Ed., *Artificial Intelligence in Logic Design.* Kluwer, Dordrecht, pp. 285–311, 2004.

[3] Ash RB. *Information Theory.* John Wiley and Sons, New York, 1967.

[4] Arnold RF and Harrison MA. Algebraic properties of symmetric and partially symmetric Boolean functions. *IEEE Transactions on Electronic Circuits*, pp. 244–251, June, 1963.

[5] Born RC and Scidmore AK. Transformation of switching functions to completely symmetric switching functions. *IEEE Transactions on Computers*, 17(6):596–599, 1968.

[6] Brayton R. Understanding SPFDs: a new method for specifying flexibility. In Notes of the International Workshop on Logic Synthesis, Tahoe City, CA, May, 1997.

[7] Butler JT, Dueck GW, Shmerko VP, and Yanushkevich SN. On the number of generators of transeunt triangles, *Discrete Applied Mathematics*, 108:309–316, 2001.

[8] Butler JT, Dueck GW, Shmerko VP, and Yanushkevich SN. Comments on SYMPATHY: fast exact minimization of fixed polarity Reed-Muller expansion for symmetric functions. *IEEE Transactions on Computer-Aided Design of Integrated Circuits and Systems*, 19(11):1386–1388, 2000.

[9] Butler JT and Sasao T. On the properties of multiple-valued functions that are symmetric in both variables and labels. In *Proceedings of the IEEE 28th International Symposium on Multiple-Valued Logic*, pp. 83–88, 1998.

[10] Butler JT, Herscovici D, Sasao T, and Barton R. Average and worst case number of nodes in binary decision diagrams of symmetric multiple-valued functions. *IEEE Transactions on Computers*, 46(4):491–494, 1997.

[11] Cheushev VA, Yanushkevich SN, Moraga C, and Shmerko VP. Flexibility in Logic Design. An Approach Based on Information Theory Methods, *Research Report*, Forschungsbericht 741, University of Dortmund, Grant DAAD A/00/01908, Germany, 2000.

[12] Cheushev VA, Yanushkevich SN, Shmerko VP, Moraga C, and Kolodziejczyk J. Information theory method for flexible network synthesis. In *Proceedings of the IEEE 31st International Symposium on Multiple-Valued Logic*, pp. 201–206, 2001.

[13] Cheushev VA, Shmerko VP, Simovici D, and Yanushkevich SN. Functional entropy and decision trees. In *Proceedings of the IEEE 28th International Symposium on Multiple-Valued Logic*, pp. 357–362, 1998.

[14] Cook RW and Flynn MJ. Logical network cost and entropy. *IEEE Transactions on Computers*, 22(9):823-826, 1973.

[15] Epstein G. Synthesis of electronic circuits for symmetric functions. *IRE Transactions on Electronic Computers*, pp. 57–60, March 1958.

[16] Hamming RW. *Coding and Information Theory.* Prentice-Hall, New York, 1980.

[17] Hartmann CRP, Varshney PK, Mehrotra KG, and Gerberich CL. Application of information theory to the construction of efficient decision trees. *IEEE Transactions on Information Theory*, 28(5):565–577, 1982.

[18] Hellerman L. A measure of computation work. *IEEE Transactions on Computers*, 21(5):439–446, 1972.

[19] Lee SC and Lee ET. On multivalued symmetric functions. *IEEE Transactions on Computers*, pp. 312–317, March, 1972.

[20] Lo H, Spiller T, and Popescu S. *Introduction to Quantum Computation and Information.* World Scientific, Hackensack, New York, 1998.

[21] Łuba T, Moraga C, Yanushkevich SN, Shmerko VP, and Kolodziejczyk J. Application of design style in evolutionary multi-level networks synthesis. In *Proceedings of the IEEE Symposium on Digital System Design*, pp. 156–163, Maastricht, Netherlands, 2000.

[22] Luba T, Moraga C, Yanushkevich SN, Opoka M, and Shmerko VP. Evolutionary multi-level network synthesis in given design style. In *Proceeding of the IEEE 30th International Symposium on Multiple-Valued Logic*, pp. 253–258, 2000.

[23] Marculescu D, Marculesku R, and Pedram M. Information theoretic measures for power analysis. *IEEE Transactions on Computer Aided Design of Integrated Circuits and Systems*, 15(6):599–610, 1996.

[24] Martin NFG and England JW, *Mathematical Theory of Entropy.* Addison-Wesley, Reading, MA, 1981.

[25] McCluskey EJ Yr. Detection of group invariance or total symmetry of a Boolean functions. *The Bell System Technical Journal*, pp. 1445–1453, Nov., 1956.

[26] Miller DM and Muranaka N.. Multiple-valued decision diagrams with symmetric variable nodes. In *Proceedings of the IEEE International Symposium on Multiple-Valued Logic*, pp. 242–247, 1996.

[27] Pandey M and Bryant RE. Exploiting symmetry when verifying transistor-level circuits by symbolic trajectory evaluation. *IEEE Transactions on Computer-Aided Design of Integrated Circuits and Systems*, 18(7):918–935, 1999.

[28] Pomeranz I and Reddy SM. On determining symmetries in inputs of logic circuits. *Computer - Aided Design of Integrated Circuits and Systems*, 13(11):1478–1433, 1994.

[29] Principe JC, Fisher III JW, and Xu D. Information theoretic learning. In Haykin S, Ed., *Unsupervised Adaptive Filtering*, John Wiley and Sons, New York, 2000.

[30] Quinlan JR. Induction of decision trees. In *Machine Learning*, Vol. 1, pp. 81–106, Kluwer, Dordrecht, 1986.

[31] Quinlan JR. Probabilistic decision trees. In Kockatoft Y and Michalshi R, Eds., *Machine Learning, Vol. 3, An AI Approach*, Kluwer, Dordrecht, pp. 140–152, 1990.

[32] Quinlan JR. Improved use of continuos attributes in C4.5. *Journal of Artificial Intelligence Research*, 4:77–90, 1996.

[33] Ramprasad S, Shanbhag NR, and Hajj IN. Information-theoretic bounds on average signal transition activity. *IEEE Transactions on Very Large Scale Integration (VLSI) Systems*, 7(3):359–368, 1999.

[34] Rawski M, Józwiak L, and Łuba T. Functional decomposition with an efficient input support selection for sub-functions based on information relationship measures. *Journal of Systems Architecture*, 47:137–155, 2001.

[35] Scholl C, Moller D, Molitor P, and Drechsler R. BDD minimization using symmetries. *IEEE Transactions on Computer-Aided Design of Integrated Circuits and Systems*, 18(2):81–100, 1999.

[36] Sentovich E and Brand D. Flexibility in logic. In Hassoun S and Sasao T, Eds., Brayton RK, consulting Ed. *Logic Synthesis and Verification*, Kluwer, Dordrecht, pp. 65–88, 2002.

[37] Shannon CE. The synthesis of two-terminal switching circuits. *Bell Sys. Tech. Journal*, 28:59–98, 1949.

[38] Shannon C. A Symbolic analysis of relay and switching circuits. *Transactions AIEE*, 57:713–723, 1938.

[39] Shannon C. A Mathematical theory of communication. *Bell Systems Technical Journal*, 27:379–423, 623–656, 1948.

[40] Shmerko VP, Popel DV, Stanković RS, Cheushev VA, and Yanushkevich SN. Entropy based algorithm for 4-valued functions minimization. In *Proceedings of the IEEE 30th International Symposium on Multiple-Valued Logic*, pp. 265–270, 2000.

[41] Shmerko VP, Popel DV, Stanković RS, Cheushev VA, and Yanushkevich SN. Information theoretical approach to minimization of AND/EXOR expressions of switching functions. In *Proceedings of the IEEE International Conference on Telecommunications*, pp. 444–451, Yugoslavia, 1999.

[42] Sinha S, Mishchenko A, and Brayton RK. Topologically constrained logic synthesis. In *Proceedings of the International Conference on Computer-Aided Design*, pp. 679–686, 2002.

[43] Sinha S, Khatri S, Brayton RK, and Sangiovanni-Vincentelli AL. Binary and Multi-valued SPFD-based wire removal in PLA network. In *Proceedings of the International Conference on Computer-Aided Design*, pp. 494–503, 2000.

[44] Tomaszewska A, Dziurzanski P, Yanushkevich SN, and Shmerko VP. Two-Phase Exact Detection of Symmetries, In *Proceedings of the IEEE 31st International Symposium Multiple-Valued Logic*, pp. 213–219, 2001.

[45] Tyagi A. Entropic bounds of FSM switching. *IEEE Transactions on Very Large Scale Integration (VLSI) Systems*, 5(4):456–464, 1997.

[46] Varshney P, Hartmann C, and De Faria J. Application of information theory to sequential fault diagnosis. *IEEE Transactions on Computers*, 31:164–170, 1982.

[47] Watanabe H. A basic theory of information network. *IEICE Transactions Fundamentals*, E76-A(3):265–276, 1993.

[48] Yamashita S, Sawada H, and Nagoya A. A new method to express functional permissibilities for LUT-based FPGAs and its applications. In *Proceedings of the International Conference on Computer-Aided Design*, pp. 254–261, 1996.

[49] Yamashita S, Sawada H, and Nagoya A. SPFD: a method to express functional flexibility. *IEEE Transactions on Computer-Aided Design of Integrated Circuits and Systems*, 19(8):840–849, 2000.

[50] Zhong N and Ohsuga S. On information of logical expression and knowledge refinement. *Transactions of Information Processing Society of Japan*, 38(4):687–697, 1997.

9

Event - Driven Analysis

This chapter revisits the field of logic design, called *Boolean differential calculus*, that analyzes behavior of computing structures in terms of change. The operators of the boolean differential calculus are employed for analysis of behavior of logic circuit in terms of change, in particular, for sensitivity and observability analysis, testability, etc.

An expansion of a switching function, which includes Boolean differences and is called, therefore, logic Taylor expansion, is an alternative model for representation and calculation of switching functions in Reed–Muller domain.

9.1 Introduction

The notation of elementary change in a system is serviceable in the study of "static" and "dynamic" changes in a circuit, caused by an "event" (e.g., a fault on the line of a circuit). This analysis is event-driven and the mathematical tool for it is Boolean differential calculus. It is used to estimate transition density, measured in events per second (see Chapter 21). In this chapter, we discuss:

- The definition of a mathematical tool for detection of an "event" (change) in a binary system, a Boolean difference of a switching function;
- Differential operators for analysis of the properties and behavior of switching functions; and
- Data structures and techniques for computing differential operators.

Differential operators, of which the basic one is the Boolean difference, provide an opportunity for analysis of circuit properties, such as flexibility (ability to be modified without compromising functionality), symmetry, monotony, i.e., detection of properties that are prerequisites for optimization. It provides the opportunity for analysis of event-driven circuit behavior (consequences of "events," sensitivity to signal changes, testability, etc.)

Summarizing, event-driven analysis has found application in

- Sensitivity and observability evaluation,
- Test pattern generation,
- Symmetry recognition,
- Power dissipation estimation, and
- Verification.

The Boolean difference is a certain analog of the Taylor cofactor of an algebraic function. The analog of Taylor expansion on switching theory is the Reed–Muller expansion, as well as arithmetic and Walsh forms.

Thus, the Boolean difference can be utilized to calculate Reed–Muller, arithmetical and Walsh coefficients. On the other hand, Taylor expansion gives a useful interpretation of spectral technique because of the structure of each spectral coefficient in terms of change.

9.2 Formal definition of change in a binary system

In switching theory, definition of the elementary change in a binary system is introduced through the Boolean difference. Other operators based on this primary notation have been developed, including its extension to multivalued logic, called logic derivative.

9.2.1 Detection of change

Detection of a change in a binary system. A signal in a binary system is represented by two logical levels, 0 and 1. Let us formulate the task as a detection of the change in this signal. The simplest solution is to deploy an EXOR operation, modulo 2 sum of the signal s_{i-i} before an "event" and the signal s_i after the "event" (e.g., a faulty signal), i.e., $s_{i-i} \oplus s_i$.

Example 9.1 *For the signal depicted in Figure 9.1, four possible combinations of the logical values or signals 0 and 1 are analyzed.*

It follows from this example that if not change itself but direction of change is the matter, then two logical values 0 and 1 can characterize the behavior of the logic signal $s_i \in \{0, 1\}$ in terms of change, where 0 means any change of a signal, and 1 indicates that one of two possible changes has occurred, $0 \rightarrow 1$ or $1 \rightarrow 0$.

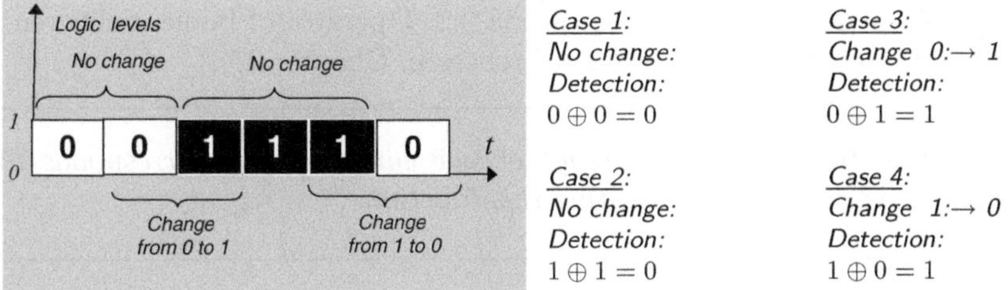

FIGURE 9.1
The change of a binary signal and its detection (Example 9.1).

Detection of change in a switching function. Let the i-th input of a switching function changed from the value x_i to the opposite value, $\overline{x}_i$. This causes the circuit output to be changed from the initial value. Note that values $f(x_i)$ and $f(\overline{x}_i)$ are not necessarily different. The simplest way to recognize whether or not they are different is to find a difference between $f(x_i)$ and $f(\overline{x}_i)$.

Model of single change: Boolean difference. The Boolean difference of a switching function f of n variables with respect to a variable x_i is defined by the equation

$$\frac{\partial f}{\partial x_i} = \underbrace{f(x_1, \ldots, x_i, \ldots, x_n)}_{Initial\ function} \oplus \underbrace{f(x_1, \ldots, \overline{x}_i, \ldots, x_n)}_{Function\ with\ complemented\ x_i} \qquad (9.1)$$

It follows from the definition of Boolean difference that

$$\frac{\partial f}{\partial x_i} = \underbrace{f(x_1, \ldots, 0, \ldots, x_n)}_{x_i\ is\ replaced\ with\ 0} \oplus \underbrace{f(x_1, \ldots, 1, ..., x_n)}_{x_i\ is\ replaced\ with\ 1} = f_{x_i=0} \oplus f_{x_i=1}. \qquad (9.2)$$

Therefore, the simplest (but optimal) algorithm to calculate the Boolean difference of a switching function with respect to a variable x_i includes two steps:

(a) Replace x_i in the switching function with 0 to get a cofactor $f_{x_i=0}$; similarly, replacement of x_i with 1 yields $f_{x_i=1}$, and
(b) Find modulo 2 sum of the two cofactors.

These manipulations can be implemented on BDDs. For example, EXOR of two sub-graphs, corresponding to $f_{x_i=0}$ and $f_{x_i=1}$, results in a BDD representing $\frac{\partial f}{\partial x_i}$, in accordance with Equation 9.2. Operator of Boolean difference is included in CUDD package (see references in Chapter 27).

> *Computing the Boolean difference of switching function corresponds to manipulation on BDD representing this function.*

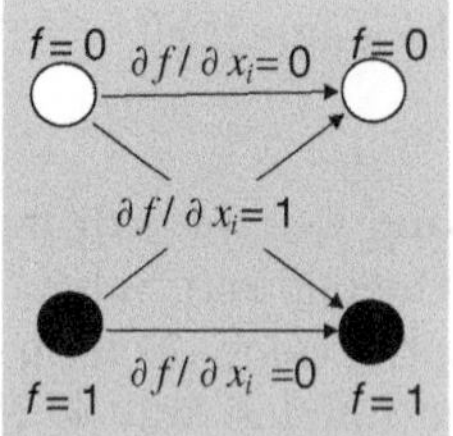

Behavior of switching function	$\frac{\partial f}{\partial x_i}$
if $(f_{x_i=0} = 0)$ *and* $(f_{x_i=1} = 0)$ *then*	0
if $(f_{x_i=0} = 0)$ *and* $(f_{x_i=1} = 1)$ *then*	1
if $(f_{x_i=0} = 1)$ *and* $(f_{x_i=1} = 0)$ *then*	1
if $(f_{x_i=0} = 1)$ *and* $(f_{x_i=1} = 1)$ *then*	0

FIGURE 9.2
The formal description of change by Boolean difference.

Figure 9.2 gives an interpretation of Boolean difference (Equation 9.1).

> **Example 9.2** *There are four combinations of possible changes of the output function* $f = x_1 \vee x_2$ *with respect to input* x_1 (x_2)*. The Boolean differences of a switching function* f *with respect to* x_1 *and* x_2 *are calculated in Figures 9.3. Figure 9.4 illustrates deriving a BDD representing the Boolean difference* $\frac{\partial f}{\partial x_1}$ *using BDDs of functions* $f(x_1, x_2)$ *and* $f(\overline{x}_1, x_2)$ *(a), and* $f_{x_1=0}$ *and* $f_{x_1=1}$ *(b).*

The Boolean difference (Equation 9.1) possesses the following properties:

- ► The Boolean difference is a switching function calculated by the Exclusive OR (EXOR) operation of the primary function and the function derived by complementing variable x_i; otherwise, it can also be calculated as EXOR of co-factors $f_{x_i=0}$ and $f_{x_i=1}$.
- ► The Boolean difference is a switching function of $n-1$ variables $x_1, x_2, \ldots, x_{i-1}, x_{i+1}, \ldots, x_n$, i.e., it does not depend on variable x_i.
- ► The value of the Boolean difference reflects the fact of local change of the switching function f with respect to changing the i-th variable x_i: the Boolean difference is equal to 0 when such change occurs, and it is equal to 1 otherwise.

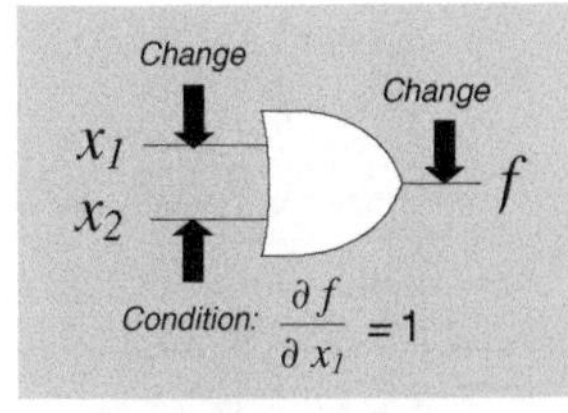

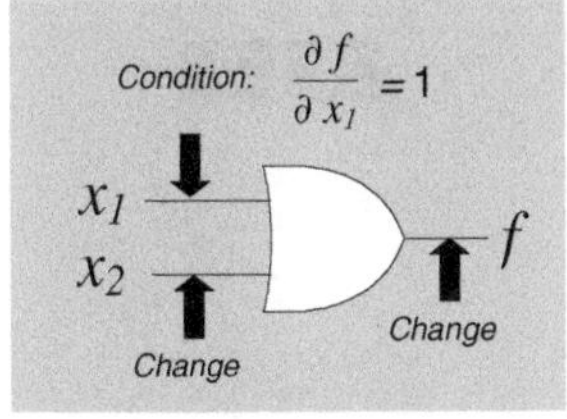

Boolean difference with respect to x_1

$$\frac{\partial f}{\partial x_1} = f_{x_1} \oplus f_{\overline{x}_1}$$
$$= (x_1 \vee x_2) \oplus (\overline{x}_1 \vee x_2),$$
$$\frac{\partial f}{\partial x_1} = f_{x_1=0} \oplus f_{\overline{x}_1=1}$$
$$= (0 \vee x_2) \oplus (1 \vee x_2) = \overline{x}_2$$

Boolean difference with respect to x_2

$$\frac{\partial f}{\partial x_2} = (x_1 \vee x_2) \oplus (x_1 \vee \overline{x}_2)$$
$$= (x_1 \vee 0) \oplus (x_1 \vee 1) = \overline{x}_1$$

FIGURE 9.3
Computing the Boolean difference $\frac{\partial f}{\partial x_1}$ for a two-input OR gate (Example 9.2).

9.2.2 Matrix model of change

Symbolic manipulation using the rules above are costly in terms of time complexity. Hence, efficient algorithms are needed to compute models based on differential operators. In this section, matrix methods of computing are introduced. These methods are useful in different design styles, in particular, in massive parallel computing on parallel-pipelining arrays such as cellular arrays or systolic arrays.

Boolean difference of a switching function f with respect to a variable x_i is defined by Equation 9.1. Let 2×2 matrix $\tilde{D}_2$ be

$$\tilde{D}_2 = \begin{bmatrix} 0 & 1 \\ 1 & 0 \end{bmatrix}$$

Let us form the 2×2 matrix D_2 by the rule

$$D_2 = I_2 \oplus \tilde{D}_2 = \begin{bmatrix} 1 & 0 \\ 0 & 1 \end{bmatrix} \oplus \begin{bmatrix} 0 & 1 \\ 1 & 0 \end{bmatrix} = \begin{bmatrix} 1 \oplus 0 & 0 \oplus 1 \\ 0 \oplus 1 & 1 \oplus 1 \end{bmatrix} = \begin{bmatrix} 1 & 1 \\ 1 & 1 \end{bmatrix},$$

where I_2 is the identity matrix.

The matrix form of a Boolean difference (Equation 9.1) with respect to the i-th variable x_i of a switching function f of n variable given by truth vector **F** is defined as

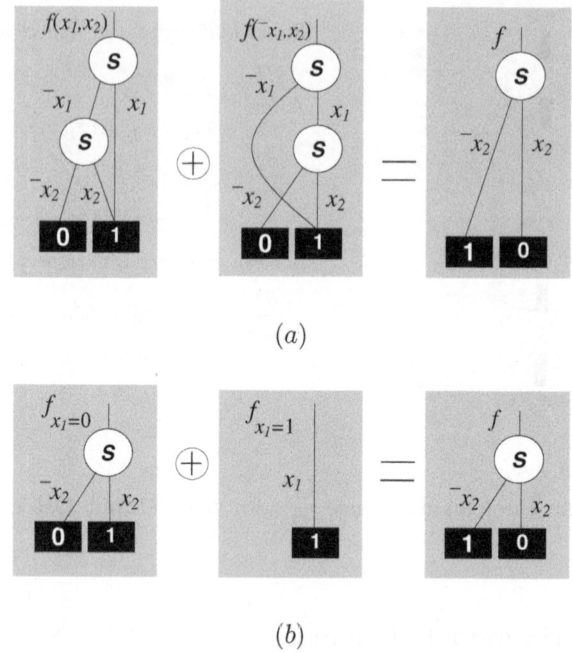

FIGURE 9.4
Computing Boolean differences for a two-input OR gate by Equation 9.1 (a) and Equation 9.2 (b) (Example 9.2).

$$\frac{\partial \mathbf{F}}{\partial x_i} = D^i_{2^n}\mathbf{F}, \tag{9.3}$$

where $2^n \times 2^n$ matrix $D^i_{2^n}$ is called a *Boolean differential matrix* generated by the rule

$$D^i_{2^n} = I_{2^{i-1}} \otimes \begin{bmatrix} 1 & 1 \\ 1 & 1 \end{bmatrix} \otimes I_{2^{n-i}}, \tag{9.4}$$

$I_{2^{i-1}}$, $I_{2^{n-i}}$ are the identity matrices.

Example 9.3 *Matrices $D^1_{2^2}$ and $D^1_{2^3}$ are constructed by Equation 9.4 as follows:*

$$D^{(1)}_{2^2} = I_{2^{1-1}} \otimes D_2 \otimes I_{2^{2-1}}$$

$$= 1 \otimes \begin{bmatrix} 1 & 1 \\ 1 & 1 \end{bmatrix} \otimes \begin{bmatrix} 1 & 0 \\ 0 & 1 \end{bmatrix} = \begin{bmatrix} I_2 & I_2 \\ I_2 & I_2 \end{bmatrix} = \begin{bmatrix} 1 & & 1 & \\ & 1 & & 1 \\ 1 & & 1 & \\ & 1 & & 1 \end{bmatrix};$$

$$D^{(1)}_{2^3} = I_{2^{1-1}} \otimes D_2 \otimes I_{2^{3-1}}$$

$$= 1 \otimes \begin{bmatrix} 1 & 1 \\ 1 & 1 \end{bmatrix} \otimes \begin{bmatrix} 1 & & & \\ & 1 & & \\ & & 1 & \\ & & & 1 \end{bmatrix} = \begin{bmatrix} I_{2^2} & I_{2^2} \\ I_{2^2} & I_{2^2} \end{bmatrix} = \begin{bmatrix} 1 & & & & 1 & & & \\ & 1 & & & & 1 & & \\ & & 1 & & & & 1 & \\ & & & 1 & & & & 1 \\ 1 & & & & 1 & & & \\ & 1 & & & & 1 & & \\ & & 1 & & & & 1 & \\ & & & 1 & & & & 1 \end{bmatrix}$$

Example 9.4 *Figure 9.5 illustrates computing the Boolean differences with respect to variable x_1, x_2, and x_3 by multiplication of the truth vector $\mathbf{F} = [f(0) \;\; f(1) \ldots \; f(7)]^T$ and the corresponding matrix $D^i_{2^n}$ for $n = 3$ and $i = 1, 2, 3$, using Equation 9.3. Let $f = x_1x_2 \vee x_3$ and $\mathbf{F} = [01010111]^T$. Then*

$$\frac{\partial \mathbf{F}}{\partial x_1} = [00100010]^T, \quad \frac{\partial f}{\partial x_1} = \overline{x}_3(\overline{x}_1x_2 \vee x_1x_2),$$
$$\frac{\partial \mathbf{F}}{\partial x_2} = [00001010]^T, \quad \frac{\partial f}{\partial x_2} = \overline{x}_3(x_1\overline{x}_2 \vee x_1x_2).$$

The Boolean difference (Equation 9.1) has a number of limitations, in particular: it cannot recognize the direction of change and cannot recognize the change in a function while changing a group of variables. This is the reason to extend the class of differential operators.

9.2.3 Model for simultaneous change

Consider the model of change with respect to simultaneously changed values of input signals. This model is called the Boolean difference with respect to *vector of variables.* Given a switching function f, the Boolean difference of n variables $x_1 \ldots x_n$ with respect to the vector of k variables $x_{i_1}, \ldots, x_{i_k}$, $i_1, \ldots, i_n \in \{1, \ldots, n\}$, is defined as follows

$$\frac{\partial f}{\partial(x_{i_1}, x_{i_2}, \ldots, x_{i_k})} = \overbrace{f(x_1, \ldots, x_{i_1}, x_{i_2}, \ldots, x_{i_k}, \ldots, x_n)}^{Initial\ function} \oplus \underbrace{f(x_1, \ldots, \overline{x}_{i_1}, \overline{x}_{i_2}, \ldots, \overline{x}_{i_k}, \ldots, x_n)}_{Function\ while\ \overline{x}_{i_1}, \overline{x}_{i_2}, \ldots, \overline{x}_{i_k}} \qquad (9.5)$$

$$\frac{\partial \mathbf{F}}{\partial x_1} = (I_{2^{1-1}} \otimes \begin{bmatrix} 1 & 1 \\ 1 & 1 \end{bmatrix} \otimes I_{2^{3-1}})\mathbf{F}$$

$$= \begin{bmatrix} 1 & & & & 1 & & & \\ & 1 & & & & 1 & & \\ & & 1 & & & & 1 & \\ & & & 1 & & & & 1 \\ 1 & & & & 1 & & & \\ & 1 & & & & 1 & & \\ & & 1 & & & & 1 & \\ & & & 1 & & & & 1 \end{bmatrix} \begin{bmatrix} f(0) \\ f(1) \\ f(2) \\ f(3) \\ f(4) \\ f(5) \\ f(6) \\ f(7) \end{bmatrix}$$

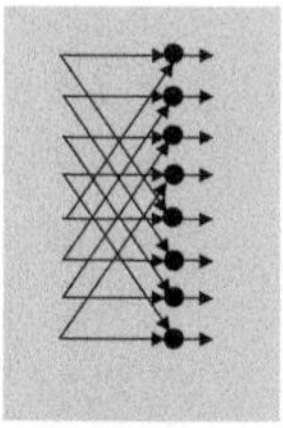

$$\frac{\partial \mathbf{F}}{\partial x_2} = (I_{2^{2-1}} \otimes \begin{bmatrix} 1 & 1 \\ 1 & 1 \end{bmatrix} \otimes I_{2^{3-2}})\mathbf{F}$$

$$= \begin{bmatrix} 1 & & 1 & & & & & \\ & 1 & & 1 & & & & \\ 1 & & 1 & & & & & \\ & 1 & & 1 & & & & \\ & & & & 1 & & 1 & \\ & & & & & 1 & & 1 \\ & & & & 1 & & 1 & \\ & & & & & 1 & & 1 \end{bmatrix} \begin{bmatrix} f(0) \\ f(1) \\ f(2) \\ f(3) \\ f(4) \\ f(5) \\ f(6) \\ f(7) \end{bmatrix}$$

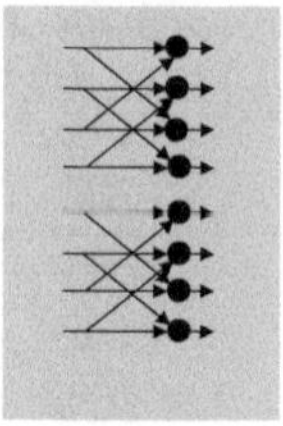

$$\frac{\partial \mathbf{F}}{\partial x_3} = (I_{2^{3-1}} \otimes \begin{bmatrix} 1 & 1 \\ 1 & 1 \end{bmatrix} \otimes I_{2^{3-3}})\mathbf{F}$$

$$= \begin{bmatrix} 1 & 1 & & & & & & \\ 1 & 1 & & & & & & \\ & & 1 & 1 & & & & \\ & & 1 & 1 & & & & \\ & & & & 1 & 1 & & \\ & & & & 1 & 1 & & \\ & & & & & & 1 & 1 \\ & & & & & & 1 & 1 \end{bmatrix} \begin{bmatrix} f(0) \\ f(1) \\ f(2) \\ f(3) \\ f(4) \\ f(5) \\ f(6) \\ f(7) \end{bmatrix}$$

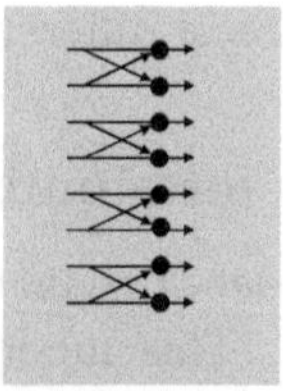

FIGURE 9.5

Matrix based computing of the Boolean differences and corresponding data flowgraphs of the algorithm; a node flowgraph implements EXOR operation (Example 9.4).

Given $k = 2$, it implies from Equation 9.5 that

$$\frac{\partial f}{\partial(x_i, x_j)} = f(x_1, \ldots, x_i, x_j, \ldots, x_n) \oplus f(x_1, \ldots, \overline{x}_i, \overline{x}_j, \ldots, x_n) \quad (9.6)$$

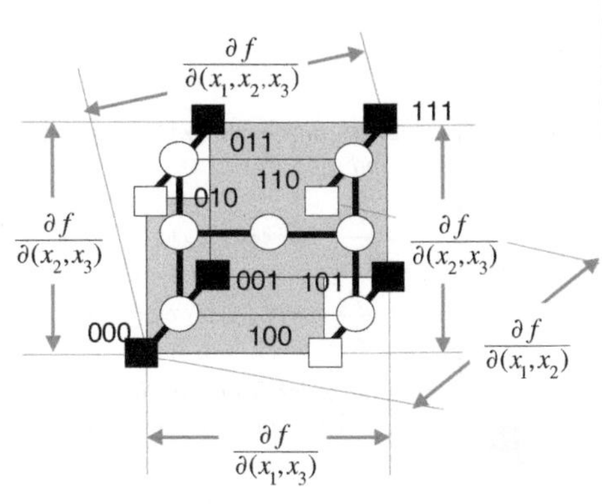

Boolean difference $\frac{\partial f}{\partial(x_1,x_2)}$ *with respect to vector of variables* (x_1, x_2)

$$\begin{aligned}
\frac{\partial f}{\partial(x_1, x_2)} &= f(x_1, x_2, x_3) \oplus f(\overline{x}_1, \overline{x}_2, x_3) \\
&= (x_1x_2 \vee x_3) \oplus (\overline{x}_1\overline{x}_2 \vee x_3) \\
&= x_1x_2 \oplus x_3 \oplus x_1x_2x_3 \\
&\oplus \overline{x}_1\overline{x}_2 \oplus x_3 \oplus \overline{x}_1\overline{x}_2x_3 \\
&= (x_1x_2 \oplus \overline{x}_1\overline{x}_2)\overline{x}_3 \\
&= x_1x_2\overline{x}_3 \vee \overline{x}_1\overline{x}_2\overline{x}_3
\end{aligned}$$

FIGURE 9.6
Interpretation of a Boolean difference with respect to a vector of variables by $\mathcal{N}$-hypercube (Example 9.6).

Example 9.5 *To calculate the Boolean difference of the switching function* $f = x_1x_2 \vee x_3$ *with respect to a vector of variables, one can apply Equation 9.6. Figure 9.6 illustrates this calculation and also shows the Boolean differences with respect to other vectors of variables of the given function using* $\mathcal{N}$*-hypercube.*

One can observe from the above example that the Boolean difference with respect to a vector of variables depends on the variables included in this vector.

9.2.4 Model of multiple change: *k*-ordered Boolean differences

Multiple, or k-ordered, Boolean difference is defined as

$$\frac{\partial^k f}{\partial x_{i_1} \partial x_{i_2} \ldots \partial x_{i_k}} = \underbrace{\frac{\partial}{\partial x_{i_1}} \left(\frac{\partial}{\partial x_{i_2}} \left(\cdots \frac{\partial f}{\partial x_{i_k}} \right) \cdots \right)}_{Either\ way}. \quad (9.7)$$

It follows from Equation 9.7 that

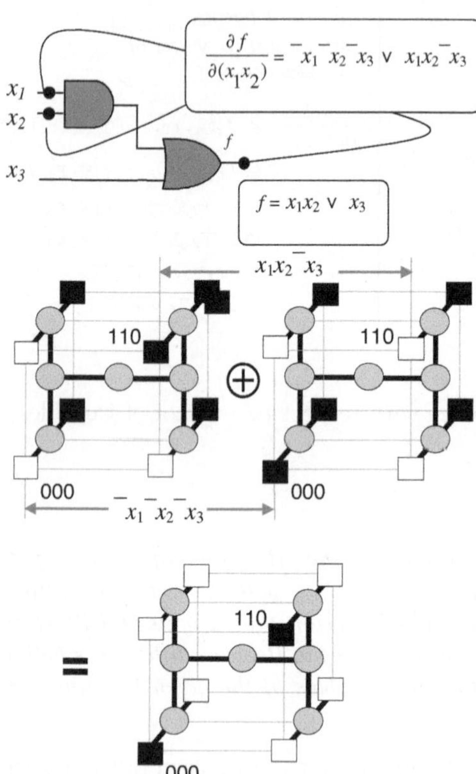

Boolean difference with respect to x_1 and x_2

$$\frac{\partial f}{\partial x_1} = x_2 \oplus x_2x_3 = x_2\overline{x}_3$$

$$\frac{\partial f}{\partial x_2} = x_1 \oplus x_1x_3 = x_1\overline{x}_3$$

Second order Boolean difference with respect to x_1 and x_2

$$\begin{aligned}\frac{\partial^2 f}{\partial x_1 \partial x_2} &= \frac{\partial}{\partial x_1}\left(\frac{\partial f}{\partial x_2}\right)\\ &= \frac{\partial}{\partial x_1}(x_1 \oplus x_1x_3)\\ &= (0 \oplus 0x_3) \oplus (1 \oplus 1x_3)\\ &= 1 \oplus x_3 = \overline{x}_3\end{aligned}$$

Boolean difference with respect to vector of variables (x_1, x_2)

$$\begin{aligned}\frac{\partial f}{\partial(x_1, x_2)} &= \frac{\partial f}{\partial x_1} \oplus \frac{\partial f}{\partial x_2} \oplus \frac{\partial^2 f}{\partial x_1 \partial x_2}\\ &= x_2\overline{x}_3 \oplus x_1\overline{x}_3 \oplus \overline{x}_3\\ &= \overline{(x_1 \oplus x_2)}\overline{x}_3\\ &= x_1x_2\overline{x}_3 \vee \overline{x}_1\overline{x}_2\overline{x}_3\end{aligned}$$

FIGURE 9.7
Measuring the input sensitivity of a circuit in the case of simultaneous change of input signals (Example 9.6).

- High-order differences can be obtained from single-order differences;
- The order of calculation of the Boolean differences does not influence the result.

Let $k = 2$, then the second order Boolean difference with respect to variables x_i and x_j will be

$$\frac{\partial^2 f}{\partial x_i \partial x_j} = \underbrace{\frac{\partial}{\partial x_i}\left(\frac{\partial}{\partial x_j}\right) = \frac{\partial}{\partial x_j}\left(\frac{\partial}{\partial x_i}\right)}_{Either\ way}. \tag{9.8}$$

Note that the multiple Boolean difference with respect to $x_i, ..., x_j$ does not depend on these variables.

There is a relationship between the second order Boolean difference (Equation 9.8) and Boolean difference with respect to vector of two variables (Equation 9.6)

$$\begin{cases} \frac{\partial f}{\partial(x_i,x_j)} = \frac{\partial f}{\partial x_i} \oplus \frac{\partial f}{\partial x_j} \oplus \frac{\partial^2 f}{\partial x_i \partial x_j} \\ \frac{\partial^2 f}{\partial x_i \partial x_j} = \frac{\partial f}{\partial x_i} \oplus \frac{\partial f}{\partial x_j} \oplus \frac{\partial f}{\partial(x_i,x_j)} \end{cases} \tag{9.9}$$

This relationship for two variables can be generalized for $k \leq n$ variables, i.e., between multiple or k-ordered Boolean difference (Equation 9.7) and the Boolean difference with respect to a vector of k variables (Equation 9.5).

Example 9.6 *Calculation of 2-ordered Boolean difference of the switching function $f = x_1x_2 \vee x_3$ with respect to variables x_1, x_2 and the vector of variables (x_1, x_2), is shown in Figure 9.7. To calculate Boolean difference $\frac{\partial f}{\partial(x_1,x_2)}$, Equation 9.9 was utilized.*

Matrix form of a multiple Boolean difference Given the truth vector $\mathbf{F}$ of a switching function f of n variables $x_1...x_n$, its Boolean difference with respect to k variables $x_{i_1}, ..., x_{i_k}$ is defined by

$$\frac{\partial^k \mathbf{F}}{\partial x_{i_1} \partial x_{i_2} ... \partial x_{i_k}} = \prod_{p \in \{i_1 ... i_k\}} D_{2^n}^{(p)} \mathbf{F} \quad over\ GF(2) \tag{9.10}$$

Example 9.7 *The data flowgraph of the calculation of the second-order Boolean difference is given in Figure 9.8. Here, Equation 9.10 is used.*

9.2.5 Boolean difference with respect to a vector of variables in matrix form

The matrix technique can be used for computing of Boolean differences with respect to a vector of k variables, and thus, multiple Boolean differences.

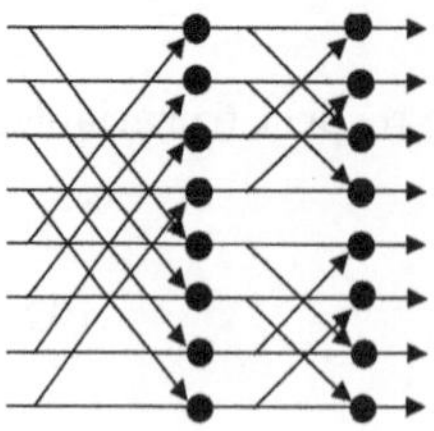

$$
\begin{aligned}
\frac{\partial^2 \mathbf{F}}{\partial x_1 \partial x_2} &= \prod_{p=1}^{2} D^{(p)} \mathbf{F} \\
&= D_{2^3}^{(2)} D_{2^3}^{(1)} \mathbf{F} \\
&= \begin{bmatrix} 1 & & & & 1 & & & \\ & 1 & & & & 1 & & \\ & & 1 & & & & 1 & \\ & & & 1 & & & & 1 \\ 1 & & & & 1 & & & \\ & 1 & & & & 1 & & \\ & & 1 & & & & 1 & \\ & & & 1 & & & & 1 \end{bmatrix}
\begin{bmatrix} 1 & & 1 & & & & & \\ & 1 & & 1 & & & & \\ 1 & & 1 & & & & & \\ & 1 & & 1 & & & & \\ & & & & 1 & & 1 & \\ & & & & & 1 & & 1 \\ & & & & 1 & & 1 & \\ & & & & & 1 & & 1 \end{bmatrix}
\begin{bmatrix} 0 \\ 1 \\ 0 \\ 1 \\ 0 \\ 1 \\ 1 \\ 1 \end{bmatrix}
= \begin{bmatrix} 1 \\ 0 \\ 0 \\ 0 \\ 1 \\ 0 \\ 0 \\ 0 \end{bmatrix}
\end{aligned}
$$

FIGURE 9.8
Data flowgraphs for computing multiple Boolean difference $\frac{\partial^2 \mathbf{F}}{\partial x_1 \partial x_2}$ (Example 9.7).

Matrix form of Boolean difference with respect to a vector of k variables $x_{i_1}, ..., x_{i_k}$, $i_1, ..., i_n \in \{1, ..., n\}$ (Equation 9.5) is defined as

$$
\frac{\partial \mathbf{F}}{\partial (x_{i_1}, ..., x_{i_k})} = \sum_{j=1}^{2^k - 1} (D_{2^n}^{(i_1)})^{j_1} \ldots (D_{2^n}^{(i_n)})^{j_n} \ \mathbf{F} \quad over\ GF(2), \quad (9.11)
$$

where

$$
\begin{aligned}
(D_{2^n}^{(i_p)})^0 &= \mathbf{I}_{2^n}, \\
(D_{2^n}^{(i_p)})^1 &= D_{2^n}^{(i_p)},
\end{aligned}
$$

$p = 1, 2, \ldots,\ k$, and $j_1 j_2 \ldots\ j_n$ is a binary representation of j. Given $n = 2$, Equation 9.11 implies:

$$
\begin{aligned}
\frac{\partial \mathbf{F}}{\partial (x_i, x_j)} &= D_{2^n}^{(i)} \mathbf{F} \oplus D_{2^n}^{(j)} \mathbf{F} \oplus D_{2^n}^{(i)} D_{2^n}^{(j)} \mathbf{F} \\
&= \frac{\partial \mathbf{F}}{\partial x_i} \oplus \frac{\partial \mathbf{F}}{\partial x_j} \oplus \frac{\partial^2 \mathbf{F}}{\partial x_i \partial x_j}
\end{aligned}
$$

9.2.6 Symmetric properties of Boolean difference

By inspection of Equation 9.1, one can observe the symmetry in the computation, that is:

$$
\frac{\partial f_{x_i = 0}}{\partial x_i} = \frac{\partial f_{x_i = 1}}{\partial x_i}.
$$

The signal graph of the computation has a symmetrical structure known as "butterfly" (in signal processing). The graph input is the truth vector **F** of the given switching function f, and the result is the truth vector of the Boolean difference.

Example 9.8 *Figure 9.9 illustrates the data flowgraphs whose input is the truth column vector* **F** *of an initial function* f*, output is truth column vector of the Boolean differences* $\frac{\partial \mathbf{F}}{\partial x_i}$*, i=1,2,3, where EXOR operation is implemented in the nodes. This corresponds to matric transform (Equation 9.3).*

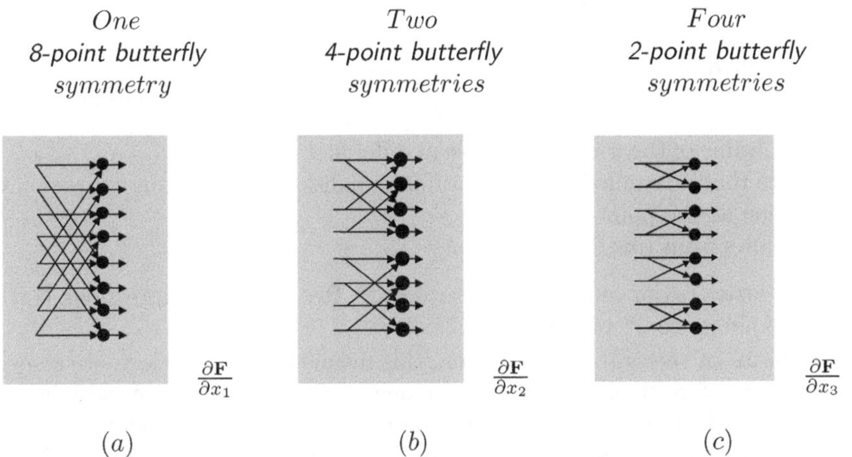

FIGURE 9.9
Illustration of symmetric properties of Boolean difference by data flowgraphs for switching function of three variables: with respect to the first variable (a), the second variable (b), and the third variable (c) (Example 9.8).

We observe from the above example that:

- The Boolean difference is symmetric with respect to x_i;
- Symmetries are represented by the "butterfly" configuration of the data flowgraphs.

Note that symmetries of the multiple Boolean differences are composed of the Boolean differences with respect to variables.

9.3 Generating Reed–Muller expressions with logic Taylor series

The Boolean difference is relevant to Reed–Muller expansion of a switching function. The *logic* Taylor series for a switching function f of n variables at the point $c \in 0, 1, \ldots, 2^n - 1$ is defined as

$$f = \bigoplus_{i=0}^{2^n-1} f_i^{(c)} \underbrace{(x_1 \oplus c_1)^{i_1} \ldots (x_n \oplus c_n)^{i_n}}_{i-th\ product},$$

where $c_1, c_2, \ldots, c_n$ and $i_1, i_2, \ldots, i_n$ are the binary representations of c and i respectively, and the i-th coefficient is defined as

$$f_i^{(c)}(d) = \left. \frac{\partial^n f(c)}{\partial x_1^{i_1} \partial x_2^{i_2} \ldots \partial x_n^{i_n}} \right|_{d=c} \quad \text{and} \quad \partial x_i^{i_j} = \begin{cases} 1, & i_j = 0 \\ \partial x_j, & i_j = 1 \end{cases}$$

that is a value of the n-ordered Boolean difference of f where $x_1 = c_1, \ldots, x_n = c_n$. Note that c is called a *polarity* of an expansion, i.e., it is an expansion of a function at the point c.

It follows from this definition that

- ► The logic Taylor expansion generates 2^n Reed–Muller expressions corresponding to 2^n polarities;
- ► In terms of spectral interpretation, this means that expressions are a spectrum of a Boolean function in one of 2^n polarities. A variable x_j is 0-polarized if it enters into the expansion uncomplemented, and 1-polarized otherwise. The components of the logic Taylor series are Boolean differences.

Details on computing Reed–Muller and other spectra via Taylor-like expansions and by using decision diagrams are given in Chapter 39.

9.4 Computing Boolean differences on decision diagrams

In this section, a techniques for computation of Boolean differences using a decision tree, diagram, and an $\mathcal{N}$-hypercube is introduced. There are two approaches:

The first approach is based on interpretation of a decision tree, or diagram and $\mathcal{N}$-hypercube whose nodes implement Shannon expansion. Calculation of Boolean differences is accomplished by manipulation of the data structure (tree or hypercube);

The second approach is oriented at the Davio tree and the corresponding $\mathcal{N}$-hypercube structure. The right branches of the Davio tree correspond to particular Boolean Differences, thus, no further manipulation on the tree is required to find the values of Boolean differences.

> *The nodes of a level of the Davio tree implement iteration of the fast Fourier transform, that is, an n-iteration matrix transform. Thus, an n-level Davio tree is associated with n iterations of the fast calculation of a corresponding Reed–Muller or Kronecker form, since a Reed–Muller form becomes a Taylor expansion of the coefficients that are interpreted as the values of Boolean differences of various order.*

For instance, a set of 2^n Reed–Muller trees representing fixed polarity Reed–Muller forms include values of all Boolean differences.

9.4.1 Boolean difference and $\mathcal{N}$-hypercube

A Boolean difference can be computed on a binary decision tree or $\mathcal{N}$-hypercube. The examples below introduce this technique.

Example 9.9 *The binary decision tree and $\mathcal{N}$-hypercube in Figure 9.10 represent the switching function $x_1x_2 \vee x_3$. To analyze the behavior of this function, let us detect the changes as follows.*

- *Boolean difference with respect to variable x_1 is $\frac{\partial f}{\partial x_1} = x_2\overline{x}_3$. The logic equation $x_2\overline{x}_3 = 1$ yields the solution $x_2x_3 = 10$. This specifies the conditions to detect the changes at x_1: when $x_2x_3 = 10$, a change at x_1 causes a change at f. This can be seen on the decision tree and on the $\mathcal{N}$-hypercube (Figure 9.10a).*
- *Boolean difference with respect to variable x_2 is $\frac{\partial f}{\partial x_2} = x_1\overline{x}_3$. The logic equation $x_1\overline{x}_3 = 1$ specifies the condition of observation as a change at f while changing x_i: $x_1x_3 = 10$(Figure 9.10b).*
- *Boolean difference with respect to variable x_3 is $\frac{\partial f}{\partial x_3} = \overline{x_1x_2}$. The logic equation $\overline{x_1x_2} = 1$ determines the condition: $x_1x_2 = \{00, 01, 10\}$ (Figure 9.10c).*

9.4.2 Boolean difference, Davio tree, and $\mathcal{N}$-hypercube

Here, we show how to use

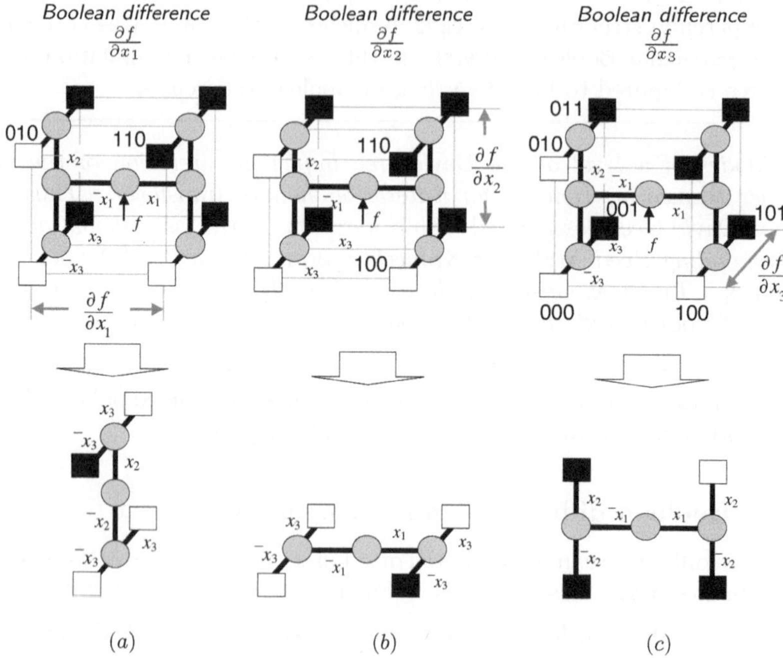

FIGURE 9.10
Interpretation of Boolean differences by $\mathcal{N}$-hypercube: Boolean difference with respect to x_1 (a), x_2 (b), and x_3 (c) (Example 9.9).

- The Davio tree, and
- The $\mathcal{N}$-hypercube which implements positive Davio expansion in the nodes

to compute Boolean differences.

The Davio tree is AND-EXOR tree, its nodes implement positive or negative Davio expansion. Let us rewrite positive Davio expansion in the form

$$f = f_{x_i=0} \oplus x_i\ (f_{x_i=0} \oplus f_{x_i=1}) = \underbrace{f_{x_i=0}}_{Left\ branch} \oplus \underbrace{x_i\ \frac{\partial f}{\partial x_i}}_{Right\ branch}$$

It follows from this form that:

- Branches of the Davio tree carry information about Boolean differences;
- The values of terminal nodes correspond to coefficients of logic Taylor expansion, that is the positive Davio tree includes values of all single and

multiple Boolean differences given a variable assignment $x_1x_2\ldots x_n = 00\ldots0$. This assignment corresponds to the calculation of Reed–Muller expansion of polarity 0, so in the Davio tree, positive Davio expansion is implemented at each node. It should be noted that any polarity can be represented by the corresponding Davio tree (with positive and negative expansion at the nodes).

- ► Representation of a switching function in terms of change is a unique representation; it means that the corresponding decision diagram is canonical.

The Davio tree can be embedded in an $\mathcal{N}$-hypercube, and the above-mentioned properties are valid for that data structure as well. In addition, the $\mathcal{N}$-hypercube enables computing of the Reed–Muller coefficients/Boolean differences, assuming that the processing is organized using parallel-pipelined, or systolic processing (see Chapter 4 for details).

Example 9.10 *Figure 9.11 shows a Davio tree and corresponding $\mathcal{N}$-hypercube for an arbitrary switching function of two and three variables. The values of Boolean differences are shown for the variable assignment $x_1x_2x_3 = 000$.*

Example 9.11 *Let $f = x_1 \vee x_2$. The values of Boolean differences given assignment $x_1x_2 = \{00\}$ are:*

$$f(00) = 0, \quad \frac{\partial f(00)}{\partial x_1} = 1, \quad \frac{\partial f(00)}{\partial x_2} = 1, \quad \textit{and} \quad \frac{\partial^2 f(00)}{\partial x_1 \partial x_2} = 1.$$

They correspond to the terminal nodes of the Davio tree and $\mathcal{N}$-hypercube (Figure 9.12).

One can conclude from Example 9.11 that data structure in the form of a positive Davio tree carries information about

- ► Reed–Muller representation of switching functions, and
- ► Representation of switching functions in terms of change.

It is not necessary to implement further calculation since the Davio that represent a Reed–Muller expansion of fixed polarity c contains values of single and multiple Boolean differences given the variable assignment $c_1c_2...c_n$.

The edges and values in terminal nodes of a Davio tree and $\mathcal{N}$-hypercube carry information about the behavior of a switching function. For example, let us compare the decision trees in Figure 9.10 and Figure 9.12. They demonstrate the relationship of Boolean differences and logic Taylor expansion. Moreover, manipulation of a decision tree can be interpreted in terms of change: reduction of the decision tree to a decision diagram leads to minimization of Reed–Muller expression and can be used as a behavioral model of this function in terms of change.

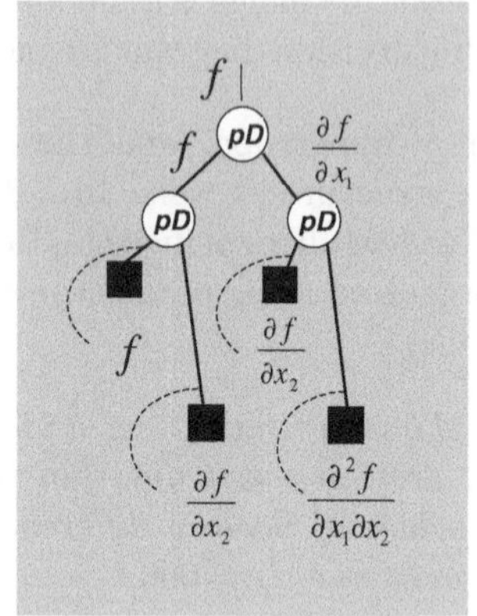

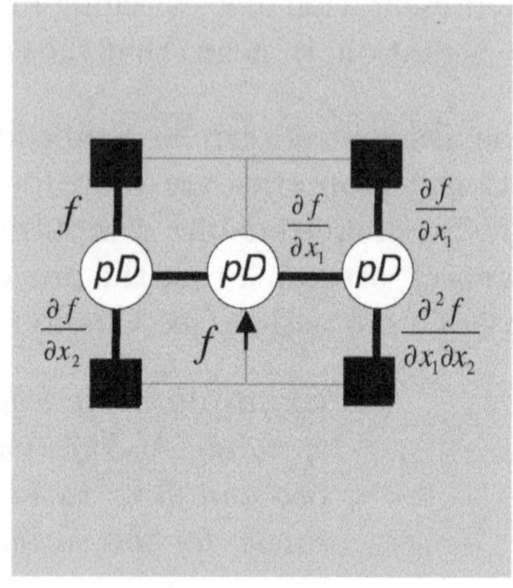

(*a*)

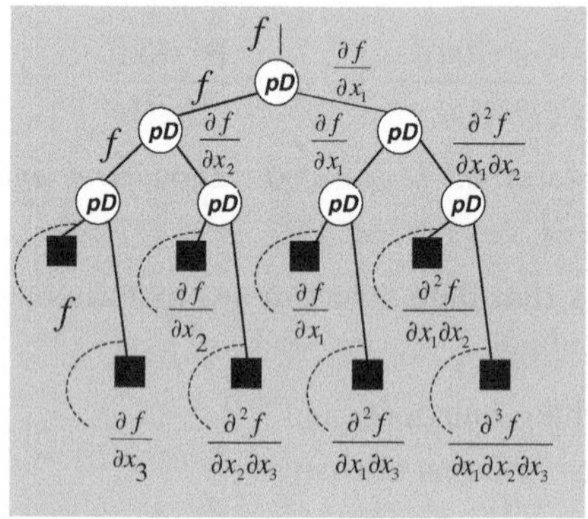

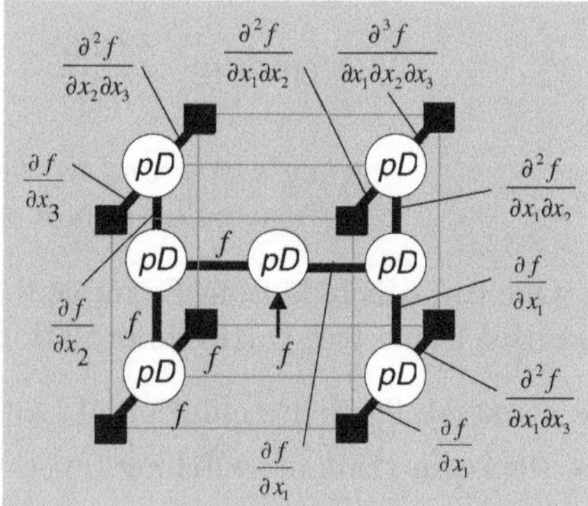

(*b*)

FIGURE 9.11
Computing Boolean differences on a Davio tree and a $\mathcal{N}$-hypercube for a switching function of two (a) and three (b) variables (Example 9.10).

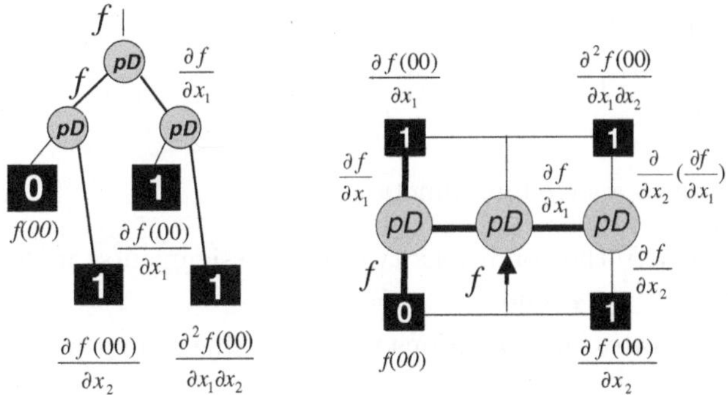

FIGURE 9.12
Computing Boolean differences of the switching function $f = x_1 \vee x_2$ (Example 9.11).

9.5 Models of logic networks in terms of change

In this section, we consider the simplest behavior models of combinational circuits in terms of change. The problem is formulated as follows: given a multiinput multioutput combinational circuit, analyze its behavior in terms of change.

9.5.1 Event-driven analysis of switching function properties: dependence, sensitivity, and fault detection

Consider a binary system with n inputs and, for simplicity's sake, with one output. Suppose that input signals are changed *independently*. The problem is to analyze the behavior of this system in terms of change. Boolean difference (Equation 9.1) has the ability to detect dependence of a function f on a variable x_i, i.e., the sensitivity of switching function f to change at x_i. Formally, the unconditional independence/dependence of output on the input x_i can be detected as follows:

► Switching function f is *unconditionally independent* of i-th variable x_i if $\frac{\partial f}{\partial x_i} \equiv 0$. This is because

$$if \quad \frac{\partial f}{\partial x_i} = 0, \quad then \quad f(x_1, \ldots, x_i, \ldots, x_n) = f(x_1, \ldots, \overline{x}_i, \ldots, x_n).$$

► Switching function f is *unconditionally dependent* on i-th variable x_i if

$\frac{\partial f}{\partial x_i} \equiv 1$. This is because

$$if \ \ \frac{\partial f}{\partial x_i} = 1, \ \ then \ \ f(x_1, \ldots, x_i, \ldots, x_n) \neq f(x_1, \ldots, \overline{x}_i, \ldots, x_n).$$

Therefore, given a switching function,

(*a*) $\frac{\partial f}{\partial x_i} = 0$ specify the conditions (variables assignments) under which f is independent of x_i; and

(*b*) $\frac{\partial f}{\partial x_i} = 1$ generates the conditions under which f is dependent on x_i.

Behavior of elementary functions. Here we show that the differential operators considered above allow us to effectively extract information about the behavior of a circuit. Table 9.1 summarizes results of analysis of change in elementary gates using three types of Boolean differential operators: Boolean difference with respect to a variable, vector of variables, and multiple Boolean difference.

TABLE 9.1
Boolean differences of two-input switching function functions.

	$\frac{\partial f}{\partial x_1}$	$\frac{\partial f}{\partial x_2}$	$\frac{\partial f}{\partial (x_1, x_2)}$	$\frac{\partial^2 f}{\partial x_1 \partial x_2}$
AND	x_2	x_1	$x_1 \sim x_2$	1
OR	$\overline{x}_2$	$\overline{x}_1$	$x_1 \sim x_2$	1
EXOR	1	1	0	0
NOR	$\overline{x}_2$	$\overline{x}_1$	$x_1 \sim x_2$	1
NAND	x_2	x_1	$x_1 \sim x_2$	1

Fault detection. Consider the simplest case of application of Boolean differences to fault detection.

Let us analyze what happens if a fault has occurred in a line (connection) that transmits binary signals in a circuit. *Stuck-at-0* or *stuck-at-1* is a fault type that causes a wire to be stuck-at-zero or one respectively. The conditions to observe the fault at input x_i and its transportation to output are described by the Boolean equation $\frac{\partial f}{\partial x_i} = 1$. Solutions to the equations $x_i \frac{\partial f}{\partial x_i} = 1$ and $\overline{x}_i \frac{\partial f}{\partial x_i} = 1$ specify the tests for detecting both *stuck-at-0* and *stuck-at-1* faults.

Example 9.12 *Figure 9.13 shows the switching function in the form of truth vector* **F** *and conditions to detect stuck-at-0 and stuck-at-1 faults. The test patterns to detect the fault are shown as well.*

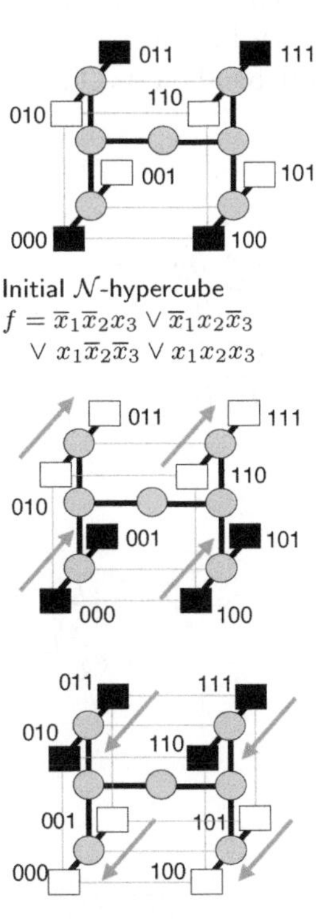

$x_1x_2x_3$	**F**	$x_3\frac{\partial \mathbf{F}}{\partial x_3}$	$\overline{x}_3\frac{\partial \mathbf{F}}{\partial x_3}$
0 0 0	0	0	1
0 0 1	1	1	0
0 1 0	1	0	1
0 1 1	0	1	0
1 0 0	1	0	1
1 0 1	0	1	0
1 1 0	0	0	1
1 1 1	1	1	0

Stuck-at-0

Stuck-at-0 at x_3 causes each value $f|_{x_3=1}$ to be changed to $f|_{x_3=0}$

Equation to find test to detect stuck-at-0 faults:

$$x_3\frac{\partial \mathbf{F}}{\partial x_3} = 1$$

Solution: tests $\{001, 011, 101, 111\}$

Stuck-at-1

Stuck-at-1 at x_3 causes each value $f|_{x_3=0}$ to be changed to $f|_{x_3=1}$

Equation to find test to detect stuck-at-1 faults:

$$\overline{x}_3\frac{\partial \mathbf{F}}{\partial x_3} = 1$$

Solution: tests $\{000, 010, 100, 110\}$

FIGURE 9.13
Deriving the tests to detect *stuck-at-0* and *stuck-at-1* faults (Example 9.12).

9.5.2 Useful rules

Rule 1. A complement of a switching function f does not change the Boolean difference with respect to variable x_i:

$$\frac{\partial \overline{f}}{\partial x_i} = \frac{\partial f}{\partial x_i}.$$

For instance, the tests for AND and NAND gates are identical.

Rule 2. Given a constant function c, $\partial c/\partial x_i = 0$.

Rule 3. (Operations with a constant.) Let c be a constant and f be a switching function. Then

$$\frac{\partial(cf)}{\partial x_i} = c\frac{\partial f}{\partial x_i}, \quad \frac{\partial(c \vee f)}{\partial x_i} = \bar{c}\frac{\partial f}{\partial x_i}, \quad and \quad \frac{\partial(c \oplus f)}{\partial x_i} = \frac{\partial f}{\partial x_i}.$$

These formulas describe situations when the constant value feeds one of the inputs of AND, OR, EXOR gate (Figure 9.14a,b,c).

Rule 4. Let f and g be switching functions that depend on x_i. Then (Figure 9.14d,e,f):

$$\frac{\partial(f \oplus g)}{\partial x_i} = \frac{\partial f}{\partial x_i} \oplus \frac{\partial g}{\partial x_i},$$

$$\frac{\partial(f \wedge g)}{\partial x_i} = f\frac{\partial g}{\partial x_i} \oplus g\frac{\partial f}{\partial x_i} \oplus \frac{\partial f}{\partial x_i}\frac{\partial g}{\partial x_i},$$

$$\frac{\partial(f \vee g)}{\partial x_i} = \bar{f}\frac{\partial g}{\partial x_i} \oplus \bar{g}\frac{\partial f}{\partial x_i} \oplus \frac{\partial f}{\partial x_i}\frac{\partial g}{\partial x_i}.$$

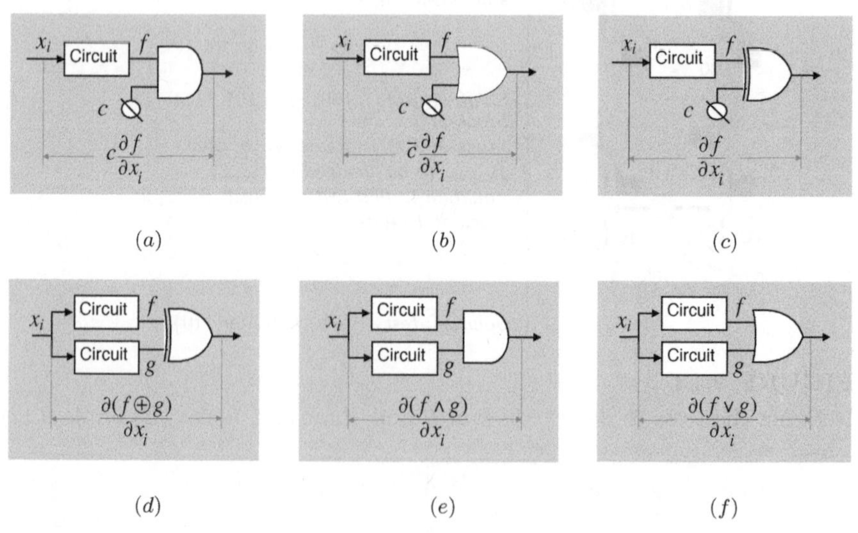

FIGURE 9.14

Boolean difference with respect to variable x_i of AND (a), OR (b), and EXOR operation (c) while the second input is a constant. Boolean difference with respect to variable x_i on logical operations of switching f of n variables x_i, $i = 1, 2, \dots, n$: EXOR operation (d), AND operation (e), and OR operation (f).

Rule 5. Multilevel circuit analysis includes computing of Boolean difference with respect to a function z.

> **Example 9.13** *In Figure 9.15, some properties of Boolean differences with respect to functions AND, OR, and EXOR of input variables x_i and x_j are shown.*

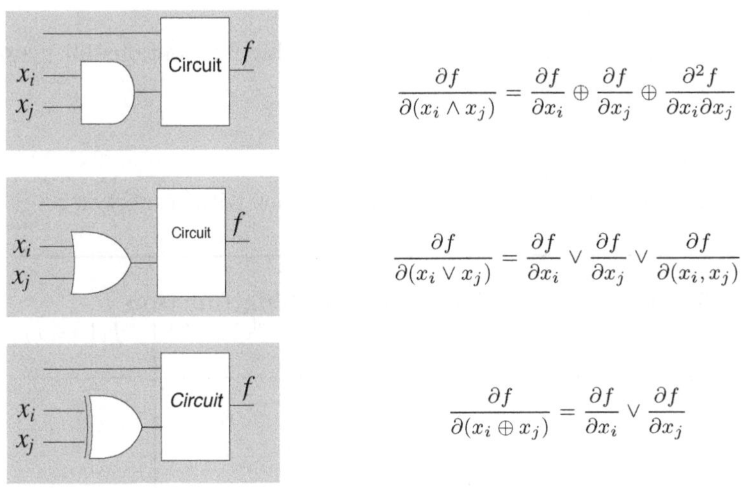

x_i
x_j
Circuit
f
$\frac{\partial f}{\partial (x_i \wedge x_j)} = \frac{\partial f}{\partial x_i} \oplus \frac{\partial f}{\partial x_j} \oplus \frac{\partial^2 f}{\partial x_i \partial x_j}$
x_i
x_j
Circuit
f
$\frac{\partial f}{\partial (x_i \vee x_j)} = \frac{\partial f}{\partial x_i} \vee \frac{\partial f}{\partial x_j} \vee \frac{\partial f}{\partial (x_i, x_j)}$
x_i
x_j
Circuit
f
$\frac{\partial f}{\partial (x_i \oplus x_j)} = \frac{\partial f}{\partial x_i} \vee \frac{\partial f}{\partial x_j}$

FIGURE 9.15
Boolean differences for AND (a), OR (b), and EXOR (c) gates with simultaneously changed inputs x_1 and x_2 (Example 9.13).

9.6 Other logic differential operators

This section revisits models that described particular changes or subsets of changes.

9.6.1 Models of directed changes in algebraic form

While the Boolean difference indicates output (function) change with respect to input (variables) changes, a directed Boolean difference investigates the direction of the changes. There are *direct* and *inverse* Boolean differences with respect to a variable for switching functions, in accordance with the direction of changes of the values of the function and variables.

The direct Boolean difference of a switching function f with respect to a variable x_i specifies the conditions of change of f and x_i along the same directions:

$$\frac{\partial f}{\partial_+ x_i} = \frac{\partial f(0 \to 1)}{\partial x_i(0 \to 1)} = \frac{\partial f(1 \to 0)}{\partial x_i(1 \to 0)}$$
$$= (\overline{x}_i \oplus f)\frac{\partial f}{\partial x_i} = \overline{f}_{x_i=0} f_{x_i=1}$$

Here $\partial_+ x_i$ denotes change $0 \longrightarrow 1$ of x_i. Table 9.2 contains all possible changes of f while changing x to $\overline{x}$.

TABLE 9.2
Truth table of the direct Boolean difference with respect to a variable.

Changing $x \to \overline{x}$	**Changing switching function** f $f(0 \to 0)$	$f(0 \to 1)$	$f(1 \to 0)$	$f(1 \to 1)$
$x(0 \to 1)$	0	1	0	0
$x(1 \to 0)$	0	0	1	0

The inverse Boolean difference of a switching function f with respect to a variable x_i defines conditions of change of f and x_i in the opposite direction

$$\frac{\partial f}{\partial_- x_i} = \frac{\partial f(0 \to 1)}{\partial x_i(1 \to 0)} = (x_i \oplus f)\frac{\partial f}{\partial x_i} = \overline{f}_{x_i=1} f_{x_i=0}$$

Table 9.3 gives more details via the primitive gates. There are four types of direct and four types of inverse Boolean differences for switching functions with respect to variable x_i.

9.6.2 Arithmetic analogs of Boolean differences and logic Taylor expansion

In this section,

- ▶ Arithmetic analogs of Boolean differences, and
- ▶ Arithmetic analogs of logic Taylor expansion

TABLE 9.3
Direct and inverse Boolean differences for the gates.

Gate	$\frac{\partial f}{\partial_+ x_1}$	$\frac{\partial f}{\partial_+ x_2}$	$\frac{\partial f}{\partial_- x_1}$	$\frac{\partial f}{\partial_- x_2}$
AND	x_2	x_1	0	0
OR	$\overline{x}_2$	$\overline{x}_1$	0	0
EXOR	$\overline{x}_2$	$\overline{x}_1$	x_2	x_1
NAND	0	0	x_2	x_1
NOR	0	0	$\overline{x}_2$	$\overline{x}_1$

are considered. An arithmetic analog of Boolean difference is called the *arithmetic difference* of a switching function. It is utilized for:

- ▶ Deriving representation of the function in arithmetic form, since arithmetic expansion is analog of logic Taylor expansion, and
- ▶ Analysis of properties of switching functions in arithmetical domain.

Arithmetic analog of Boolean difference in algebraic form is defined by equation

$$\frac{\widetilde{\partial} f}{\widetilde{\partial} x_i} = -f_{x_i} + f_{\overline{x}_i}. \tag{9.12}$$

The matrix form of the arithmetic difference (Equation 9.12) with respect to the i-th variable x_i of a switching function f of n variables given by truth vector $\mathbf{F}$ is defined as

$$\frac{\widetilde{\partial}\mathbf{F}}{\widetilde{\partial} x_i} = \widetilde{D}_{2^n}^{(i)}\ \mathbf{F}$$

where the $2^n \times 2^n$ matrix $\widetilde{D}_{2^n}$ is generated by the rule

$$\widetilde{D}_{2^n}^{(i)} = I_{2^{i-1}} \otimes \begin{bmatrix} -1 & 1 \\ 1 & -1 \end{bmatrix} \otimes I_{2^{n-i}},$$

$I_{2^{i-1}}$ and $I_{2^{n-i}}$ are the identity matrix.

A k-th-order, $k = 1, ..., n$, arithmetical difference with respect to a subset of k variables $x_{i_1}, \ldots, x_{i_k}$, is defined in algebraic and matrix form as follows

$$\frac{\widetilde{\partial}^k f}{\widetilde{\partial} x_1 \cdots \widetilde{\partial} x_k} = \frac{\widetilde{\partial}}{\widetilde{\partial} x_1}\left(\frac{\widetilde{\partial}}{\widetilde{\partial} x_2}\left(\cdots \frac{\widetilde{\partial} f}{\widetilde{\partial} x_k}\right)\cdots\right),$$

$$\frac{\widetilde{\partial}^k \mathbf{F}}{\widetilde{\partial} x_1 \cdots \widetilde{\partial} x_k} = \widetilde{D}_{2^n}^{(1)}\ \widetilde{D}_{2^n}^{(2)}\ \ldots\ \widetilde{D}_{2^n}^{(n)}\ \mathbf{F}\ \ \textit{over } GF(2).$$

Example 9.14 *The structural properties of the flowgraphs of algorithms for computing arithmetic differences with respect to all the variables are explained in Figure 9.16.*

Boolean differences

$$\frac{\partial^{(3)}\mathbf{F}}{\partial x_1 \partial x_2 \partial x_3} = \underbrace{\begin{bmatrix} 1 & & & & 1 & & & \\ & 1 & & & & 1 & & \\ & & 1 & & & & 1 & \\ & & & 1 & & & & 1 \\ 1 & & & & 1 & & & \\ & 1 & & & & 1 & & \\ & & 1 & & & & 1 & \\ & & & 1 & & & & 1 \end{bmatrix}}_{D_{2^3}^{(1)}} \underbrace{\begin{bmatrix} 1 & & 1 & & & & & \\ & 1 & & 1 & & & & \\ 1 & & 1 & & & & & \\ & 1 & & 1 & & & & \\ & & & & 1 & & 1 & \\ & & & & & 1 & & 1 \\ & & & & 1 & & 1 & \\ & & & & & 1 & & 1 \end{bmatrix}}_{D_{2^3}^{(2)}} \underbrace{\begin{bmatrix} 1 & 1 & & & & & & \\ 1 & 1 & & & & & & \\ & & 1 & 1 & & & & \\ & & 1 & 1 & & & & \\ & & & & 1 & 1 & & \\ & & & & 1 & 1 & & \\ & & & & & & 1 & 1 \\ & & & & & & 1 & 1 \end{bmatrix}}_{D_{2^3}^{(3)}} \mathbf{F}$$

Arithmetic differences

$$\frac{\widetilde{\partial}^{(3)}\mathbf{F}}{\widetilde{\partial} x_1 \widetilde{\partial} x_2 \widetilde{\partial} x_3} = \underbrace{\begin{bmatrix} -1 & & & & 1 & & & \\ & -1 & & & & 1 & & \\ & & -1 & & & & 1 & \\ & & & -1 & & & & 1 \\ 1 & & & & -1 & & & \\ & 1 & & & & -1 & & \\ & & 1 & & & & -1 & \\ & & & 1 & & & & -1 \end{bmatrix}}_{\widetilde{D}_{2^3}^{(1)}} \underbrace{\begin{bmatrix} -1 & & 1 & & & & & \\ & -1 & & 1 & & & & \\ 1 & & -1 & & & & & \\ & 1 & & -1 & & & & \\ & & & & -1 & & 1 & \\ & & & & & -1 & & 1 \\ & & & & 1 & & -1 & \\ & & & & & 1 & & -1 \end{bmatrix}}_{\widetilde{D}_{2^3}^{(2)}} \underbrace{\begin{bmatrix} -1 & 1 & & & & & & \\ 1 & -1 & & & & & & \\ & & -1 & 1 & & & & \\ & & 1 & -1 & & & & \\ & & & & -1 & 1 & & \\ & & & & 1 & -1 & & \\ & & & & & & -1 & 1 \\ & & & & & & 1 & -1 \end{bmatrix}}_{\widetilde{D}_{2^3}^{(3)}} \mathbf{F}$$

FIGURE 9.16
Matrix based computing of third-order Boolean and arithmetical differences and flowgraphs (the nodes realize EXOR and arithmetic sum respectively (Example 9.14).

An analog of the logic Taylor series for a switching function called the *arithmetical Taylor expansion* is expressed by the equation

$$P_c = \sum_{j=0}^{2^n-1} p_c^{(j)} (x_1 \oplus c_1)^{j_1} (x_2 \oplus c_2)^{j_2} \ldots (x_n \oplus c_n)^{j_n},$$

where $c_1 c_2 \ldots c_n$ and $j_1 j_2 \ldots j_n$ are the binary representations of c (polarity) and j respectively, and the j-th coefficient is defined as

$$p_c^{(j)} = \frac{\widetilde{\partial}^n f(c)}{\widetilde{\partial} x_1^{j_1} \widetilde{\partial} x_2^{j_2} \ldots \widetilde{\partial} x_n^{j_n}} \tag{9.13}$$

that is, a value of the arithmetical analog of an n-ordered Boolean difference of f given c, i.e., $x_1 = c_1$, $x_2 = c_2$, ..., $x_n = c_n$. The coefficients $p_c^{(j)}$ (Equation 39.3.1) are also called the *arithmetic spectrum* of the switching function f.

> *The arithmetic Taylor expansion produces 2^n arithmetic expressions corresponding to 2^n polarities. Similarly to multiple Boolean differences, one can calculate multiple arithmetic differences, and obtain coefficients of fixed polarity arithmetic expressions.*

9.7 Further reading

Historical remarks. Akers was the first to show that the Reed–Muller spectrum of a Boolean function is analogous to the classical MacLauren and Taylor series [1]. Fundamentals of the Boolean differential calculus have been developed in [2, 6, 23, 25]. Applications of logic differential calculus is reviewed by Bochmann et al. [3]. Generalization of Boolean differences for multivalued functions is given in Chapter 32.

Gibbs differences. An alternative difference called *Gibbs diadic differentiation* was introduced by Edward [6] based on Gibbs and Millard results [10]. Also Gibbs differences were studied in work by Butzer and Stanković [6] and Stanković [22].

Arithmetic analogs of Boolean differences. Tosić introduced arithmetical difference of switching functions [24]. The arithmetical difference was used as an analog of Taylor–Maclaurin series by Davio et al. [6]. Theoretical and applied aspects have been studied in [32].

Walsh differences. Yanushkevich [52] developed so called *Walsh differences* which are useful in a Walsh spectral representation of switching functions (see also Chapter 39).

Matrix notation of Boolean and arithmetic differences has been proposed by Yanushkevich [28, 30, 32]. This technique has been developed towards multivalued functions in [29]. Matrix based approaches to solution of Boolean and multivalued differential equations can be found in [31, 33] (see also Chapter 39). Shmerko et al. [19] proposed techniques for computing multivalued logic differences.

Analysis of dynamic systems. Bochmann [2, 4, 5] developed a theory of analysis of dynamic systems using extensions of Boolean differences.

Test pattern generation. It has been proposed by Sellers et al. [19] to use Boolean differences to find tests for switching circuits. The comprehensive approach to generate test patterns for "stuck-at" faults in a circuits was proposed by Marinos [16], and developed in many papers, in particular, by Larrabee [9].

Decomposition. Boolean differences are useful for decomposition of logic functions [6]. A comprehensive guide to application of Boolean differences in switching theory and digital design can be found in Davio et al. [6], Posthoff and Steinbach [18], and also in selected chapters of [15].

Power consumption analysis. Boolean differences was considered as a component for evaluation of transition density to measure switching activity in a circuit [7, 17] (see also Chapter 21).

Probabilistic Boolean differences. Shmulevich et al. [21] applied probabilistic Boolean differences to the study of gene regulatory networks.

References

[1] Akers SB. On a theory of Boolean functions. *Society for Industrial and Applied Mathematics*, 7(4):487–498, 1959.

[2] Bochmann D and Posthoff Ch. *Binäre Dynamische Systeme.* Akademieverlag, Berlin, 1981.

[3] Bochmann D, Yanushkevich S, Stanković R, Tosić Z, and Shmerko V. Logic differential calculus: progress, tendencies and applications. *Automation and Remote Control*, Kluwer/Plenum Publishers, 61(1):1033–1047, 2000.

[4] Bochmann D. Boolean representations of binary time series. In *Proceedings of the 5th International Workshop on Applications of the Reed–Muller Expansion in Circuit Design*, Mississippi State University, MS, pp. 139–158, 2001.

[5] Bochmann D. Boolean differences and discrete dinamics. In *Proceedings of the International Workshop on Boolean Problems*, Freiberg University, Germany, pp. 107–116, 2000.

[6] Butzer PL and Stanković RS, Eds., *Theory and Applications of Gibbs Derivatives.* Mathematical Institute, Belgrade, 1990.

[7] Chou TL and Roy K. Estimation of activity for static and domino CMOS circuits considering signal correlations and simultaneous switching. *IEEE Transactions on Computer-Aided Design of Integrated Circuits and Systems*, 15(10):1257–1265, 1996.

[8] Davio MJ, Deschamps P, and Thayse A. *Discrete and Switching Functions.* McGraw-Hill, New York, 1978.

[9] Edwards CR. The Gibbs dyadic differentiator and its relationship to the Boolean difference. *Computers and Electronic Engineering,* 5(4):335–344, 1978.

[10] Gibbs JE and Millard MS. Walsh functions as solutions of a logical differential equation. *DES Report, No.1, National Physical Laboratory* Middlesex, England, 1969.

[11] Kauffman SA. Metabolic stability and epigenesis in randomly constructed genetic nets. *Journal of Theoretical Biology,* 22:437–467, 1969.

[12] Kauffman SA. Emergent properties in random complex automata. *Physica,* D10:145-156, 1984.

[13] Kauffman SA. *The Origins of Order, Self-Organization and Selection in Evolution.* Oxford University Press, Oxford, 1993.

[14] Larrabee T. Test pattern generation using Boolean satisfiability. *IEEE Transactions on Computer-Aided Design of Integrated Circuits and Systems,* 11(1):4–15, 1992.

[15] Lee SC. *Modern Switching Theory and Digital Desing.* Prentice-Hall, New Jersey, 1978.

[16] Marinos P. Derivation of minimal complete sets of test-input sequences using Boolean differences. *IEEE Transactions on Computers*, 20(1):25–32, 1981.

[17] Najm FN. A survey of power estimation techniques in VLSI circuits. *IEEE Transactions on VLSI*, 2(4):446–455, Dec. 1994.

[18] Posthoff Ch and Steinbach B. *Logic Functions and Equations.* Springer, Heidelberg, 2004.

[19] Sellers FF, Hsiao MY, and Bearson LW. Analyzing errors with the Boolean difference. *IEEE Transactions on Computers*, 1:676–683, 1968.

[20] Shmerko VP, Yanushkevich SN, and Levashenko VG. Techniques of computing logical derivatives for MVL functions. In *Proceedings of the 26th IEEE International Symposium on Multiple-Valued Logic*, pp. 267–272, 1996.

[21] Shmulevich I, Dougherty ER, Kim S, and Zhang W. Probabilistic Boolean networks: a rule-based uncertainty model for gene regulatory networks. *Bioinformatics* 18:274–277, 2002.

[22] Stanković RS. Fast algorithm for calculation of Gibbs derivative on finite group. *Approximation Theory and its Applications,* 7(2):1–19, 1991.

[23] Thayse A and Davio M. Boolean differential calculus and its application to switching theory. *IEEE Transactions on Computers*, 22:409–420, 1973.

[24] Tosić Z. Arithmetical representation of logic functions. In *Discrete Automatics and Networks,* USSR Academy of Sciences, Publishing House "Nauka", Moscow, pp. 131–136, 1970 (In Russian).

[25] Tucker JH, Tapia MA, and Bennet AW. Boolean integral calculus for digital systems. *IEEE Transactions on Computers,* 34:78–81, 1985.

[26] Vichniac G. Simulating physics with cellular automata. *Physica D*, 10:96–115, 1984.

[27] Yanushkevich SN. Multiplicative properties of spectral Walsh coefficients of Boolean functions. *Automation and Remote Control*, Kluwer/Plenum Publishers, 64(12):1933–1947, 2003.

[28] Yanushkevich S. Arithmetical canonical expansions of Boolean and MVL functions as generalized Reed–Muller series, In *Proceedings of the IFIP WG 10.5 Workshop on Applications of the Reed–Muller Expansions in Circuit Design*, Japan, pp. 300–307, 1995.

[29] Yanushkevich SN. Systolic algorithms for arithmetic polynomial forms of k-valued functions of Boolean algebra. *Automation and Remote Control*, Kluwer/Plenum Publishers, 55(12):1812–1823, 1994.

[30] Yanushkevich SN. Development of methods of Boolean differential calculus for arithmetic logic. *Automation and Remote Control*, Kluwer/Plenum Publishers, 55(5):715–729, 1994.

[31] Yanushkevich SN. Matrix method to solve logic differential equations. *IEE Proceedings, Pt.E, Computers and Digital Technique*, 144(5):267–272, 1997.

[32] Yanushkevich SN. *Logic Differential Caluclus in Multi-Valued Logic Design.* Technical University of Szczecin Academic Publishers, Poland, 1998.

[33] Yanushkevich SN. Matrix and combinatorics solution of Boolean differential equations. *Discrete Applied Mathematics*, 117:279–292, 2001.

Part II

DECISION DIAGRAM TECHNIQUES FOR SWITCHING FUNCTIONS

- Classification
- Variable Ordering
- Decision Diagrams for Incompletely Specified Switching Functions
- Edge-Valued Decision Diagrams
- Ternary Decision Diagrams
- Spectral Decision Diagrams for Arithmetic Circuits
- Linearly Transformed Decision Diagrams
- Decision Diagrams for Arithmetic Circuits
- Word-Level Decision Diagrams
- Minimization via Decision Diagrams
- Probabilistic Techniques
- Power Consumption Analysis
- Information-Theory Measures
- Formal Verification of Circuits
- Complexity
- Programming

10

Introduction

This part includes 17 chapters on decision diagram technique for switching functions, which cover the following areas:

- ▶ Classification of decision diagrams,
- ▶ Various aspects of the construction of decision diagrams,
- ▶ Application of decision diagrams, and
- ▶ Complexity and programming of decision diagrams.

The topics also include probabilistic and information-theory measures, and various aspects of choosing a type of decision diagram based on the relationships of data structures and their requirements for implementation.

10.1 Genesis and evolution of decision diagrams

Binary decision diagrams in digital design. Binary decision diagrams (BDDs) for switching function representation date back to the late 1950s, to the paper by Lee [12] and later study by Akers [1]. With the introduction of the reduced ordered BDD (ROBDDs) as a canonical representation of switching functions by Bryant in 1986 [4], they became established as the state-of-the-art data structure for representation and manipulation of discrete functions in CAD for VLSI.

Decision tree representations of switching functions exhibit several advantages over Boolean formulas. Lee [12] showed that at most $2^n/n$ diagram nodes are required to represent any switching function of n variables, which compares very favorably with the $2^n/log_2n$ operators that may be needed by an unfactored Boolean formula. Moreover, every operator of the Boolean formula must be carried out in order to evaluate the formula, so that up to $2^n/log_2n$ operations may be performed, while a decision diagram will never require more than n variable evaluations. Thus, decision diagrams express switching functions at least as compactly as Boolean formulas and are greatly more efficient as an evaluation tool (the latter property is used, e.g., for the repeated evaluation of Boolean queries in a large database).

Basic definitions. A decision diagram is a rooted directed acyclic graph consisting of the root node, a set of non-terminal nodes and a set of terminal (constant) nodes connected with directed edges (links) (Figure 10.1). These components make up *topology* of decision trees and diagrams. The topology is characterized by a set of topological parameters such as size, number of non-terminal nodes, number of links, etc. (see details in Chapter 3). For a given switching function f, a decision diagram is designed by the reduction of the corresponding decision tree representing f.

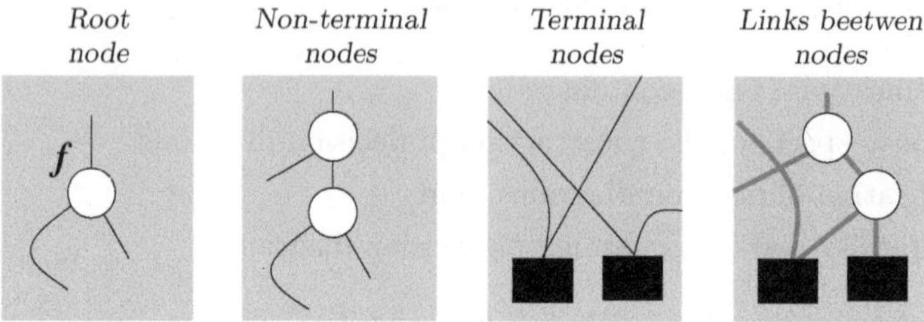

FIGURE 10.1
Components of a decision diagram.

Relationship to other data structures. Decision diagrams are graph-based data structures that represent algebraic ones. Figure 10.2 shows the relations between algebraic data structures and decision trees and diagrams for logic functions:

- ▶ The algebraic data structure can be mapped into decision trees or diagrams and vice versa.
- ▶ Manipulation of algebraic representations is based on the rules and axioms of Boolean algebra. These are aimed at simplification, factoring, composition, and decomposition. Manipulation of graphical data structures is based on the rules for reduction of nodes and links in decision trees and diagrams. There is a mapping between these representations.

Algebraic (sum-of-products, AND-EXOR, arithmetic, word-level, etc.) expressions are relevant to the corresponding graphical data structures (hypercubes and hypercube-like structures, decision trees and diagrams, etc.) For instance, a binary decision tree can be derived from a sum-of-products expression, and vice versa.

Example 10.1 *Application of the reduction rules to a decision tree of a 3-input NOR function is demonstrated in Figure 10.3.*

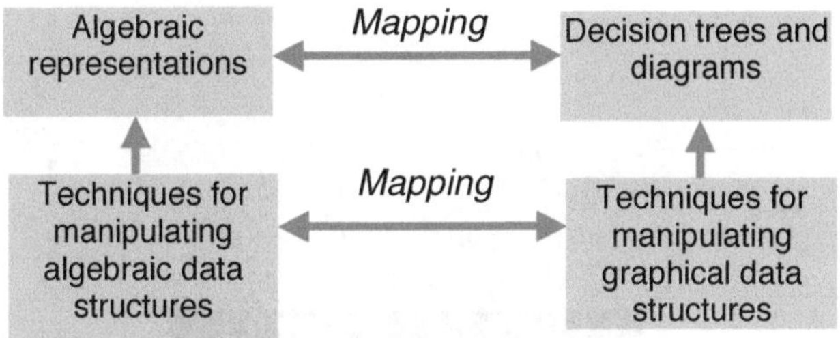

FIGURE 10.2
Relationships between algebraic representation of logic functions and decision diagrams.

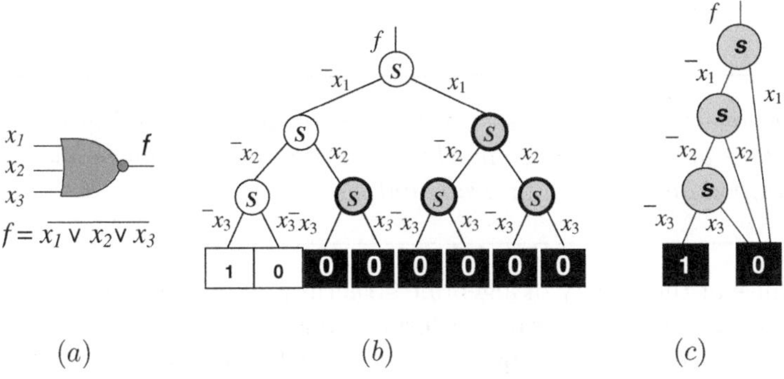

FIGURE 10.3
The 3-input NOR function (a), Shannon decision tree (b), and decision diagram with lexicographical order of variables (c) (Example 10.1).

Example 10.2 *Figure 10.4 shows the mapping of the 3-input NAND of three variables into a positive Davio tree and a Davio diagram. First, a Reed–Muller expression of polarity 0 must be derived:*

$$f = \overline{x_1 x_2 x_3} = 1 \oplus x_1 x_2 x_3.$$

Next, the coefficient vector must be derived:

$$\mathit{Coefficient\ \ vector} = [10000001]^T$$

and then the Davio tree is created and reduced to a Davio diagram.

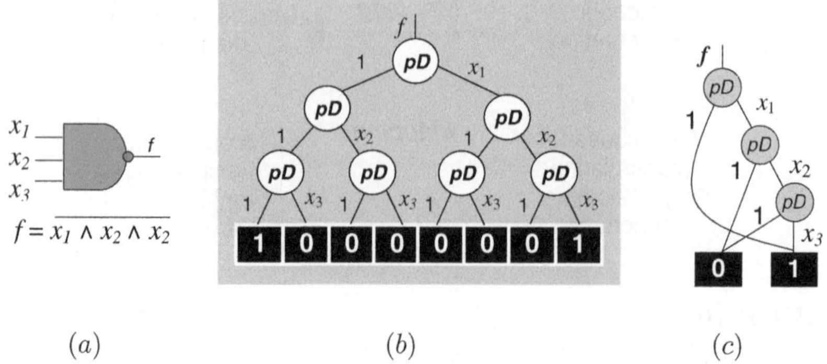

FIGURE 10.4
Mapping the switching function NAND into a Davio tree and a Davio diagram (Example 10.2).

> *If the relations between the basic rules of algebraic and graphical data structures are established, an arbitrary logic function can be mapped into a corresponding decision diagram and vice versa.*

Because of the variety of algebraic structures and techniques for their synthesis, in particular, spectral transforms, a wide variety of diagram types have been developed in the last four decades (see Chapter 13).

Example 10.3 *Figure 10.5 shows the basic principle of mapping the term $\overline{x}_1x_2$ of a sum-of-product expression and the term $1 \oplus x_2$ of a Reed–Muller expression into a graph data structure.*

10.2 Various aspects of the construction of decision diagrams

Measures on decision trees and diagrams. The implementation costs of decision trees and diagrams include:

- The tree storage cost (the sum of the storage costs of the internal nodes of the tree),
- The diagram storage cost (the sum of the storage costs of the internal nodes of the minimal diagram).

Sum-of-product algebraic representation and corresponding graphical structure

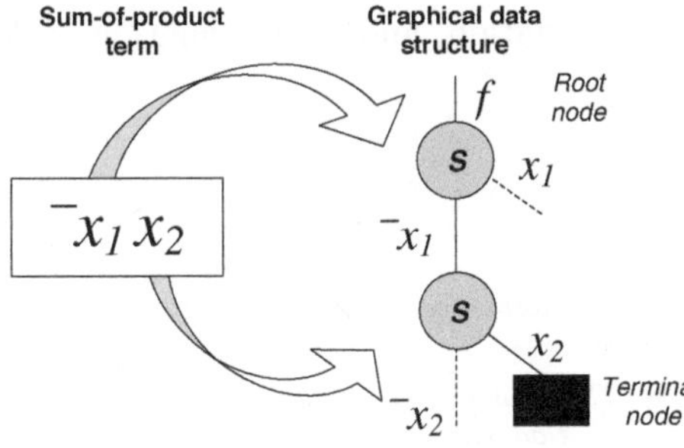

Mapping the term $\overline{x}_1 x_2$ of a sum-of-product expression into a graph data structure

- *Two nodes (rooted and intermediate) and one terminal node are needed, connected by the links of a particular topology*
- *Each node implements the Shannon expansion, the terminal node can be marked by 0 or 1*
- *If the terminal node is marked by 0, the graphical structure represents the term $\overline{x}_1 x_2 = 0$; if the terminal node is marked by 1, the graphical structure represents the term $\overline{x}_1 x_2 = 1$.*

Reed–Muller algebraic representation and corresponding graphical structure

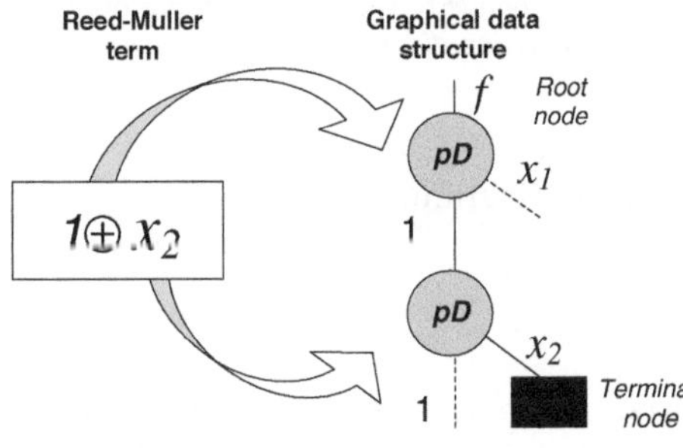

Mapping the term $1 \oplus x_2$ of a Reed–Muller expression into a graph data structure

- *Two nodes (rooted and intermediate) and the terminal node are needed*
- *Each node implements the positive Davio expansion, the terminal node can be marked by 0 or 1*
- *If the terminal node is marked by 0, the graphical structure represents the term $1 \oplus x_2 = 0$; if the terminal node is marked by 1, the graphical structure represents the term $1 \oplus x_2 = 1$.*

FIGURE 10.5
Mapping of the terms of a sum-of-product and a Reed–Muller expression into the path of a decision diagram (Example 10.3).

These can be measured on the trees and diagrams, i.e., estimated directly from their topology. The other important properties, which can be evaluated on the diagram or tree, are:

- ▶ A probability distribution that can be specified on the variables' spaces (which can be assumed uniform if not otherwise known),
- ▶ Information-theoretic measures that also use probability distributions.

These properties are considered in Chapters 20 and 24, respectively. In Figure 10.6, the information-theoretical model of decision diagrams is given. These measures are used for manipulation, optimization, and transformation of decision trees and diagrams.

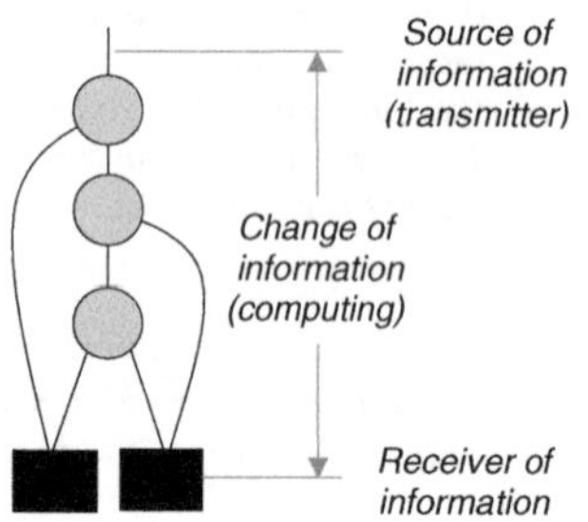

The information-theoretical model includes:

- ▶ *The source or transmitter of information, which corresponds to the input of the root node,*
- ▶ *The channel where the information is transmitted. In a decision diagram, this corresponds to processing (changing) of information by nonterminal nodes, and*
- ▶ *The receiver of information, which corresponds to the terminal nodes of a decision diagram.*

FIGURE 10.6
Information-theoretical model of a decision diagram.

Manipulation of decision diagrams. There are various techniques for the manipulation of algebraic representations of switching functions. Each of these techniques has a graphical interpretation for decision trees and diagrams. The simplest components of the techniques are called *rules* or *axioms*. Rules for manipulation of algebraic expressions of logic functions (Figure 10.2) have their equivalents in decision trees and diagrams.

> *The topology of decision trees and diagrams can be changed by the manipulation of the algebraic representation or by direct manipulation of the graphical structure.*

Example 10.4 *Manipulation of algebraic expression and graphical structure implies various topologies of decision diagrams:*

- ▶ *The topology shown in Figure 10.7a is achieved by linearization of an algebraic expression of a switching function.*
- ▶ *The topology shown in Figure 10.7b is derived by linearization of an algebraic expression and manipulation (composition) of graphical elements.*
- ▶ *The topologies shown in Figure 10.7c,d are achieved by using various representations of switching functions and their symmetric properties.*
- ▶ *The topologies shown in Figure 10.7a,b,c are classified as planar topologies of decision diagrams (they are useful in the design of networks without crossings).*

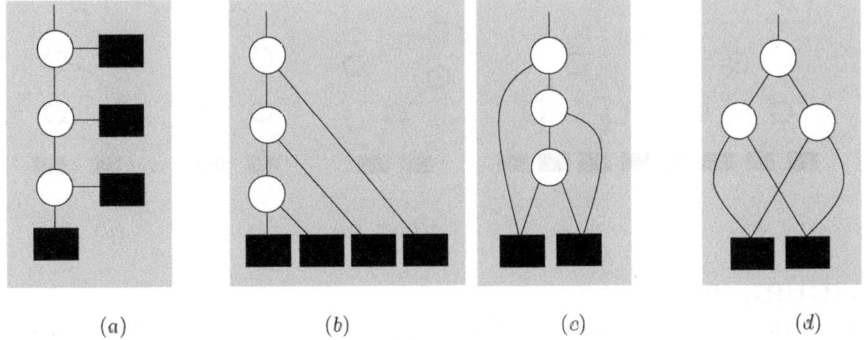

FIGURE 10.7
Various topologies of decision diagrams (Example 10.4).

Reduction of decision trees. The number of possible decision trees for a given function is in general very large (and that of possible decision diagrams even larger). A completely specified switching function of n variables can have up to

$$\prod_{i=0}^{n-1}(n-i)^{2^i}$$

distinct decision trees. This is because n choices are possible for the root, followed by $(n-1)$ choices on each of the two subtrees, etc. Several criteria have been developed in order to select an appropriate representation out of the large number of possible trees.

> *Reduction of decision trees is aimed at simplification with respect to various criteria, in particular:*
>
> - *Reduction of the number of nonterminal-nodes,*
> - *Satisfaction of topological constraints to meet the requirements of implementation,*
> - *Reduction of the length of paths from root to terminal nodes.*

Example 10.5 *The reduction of the decision tree shown in Figure 10.8a produces decision diagrams with different topological characteristics and properties (Figure 10.8b,c,d,).*

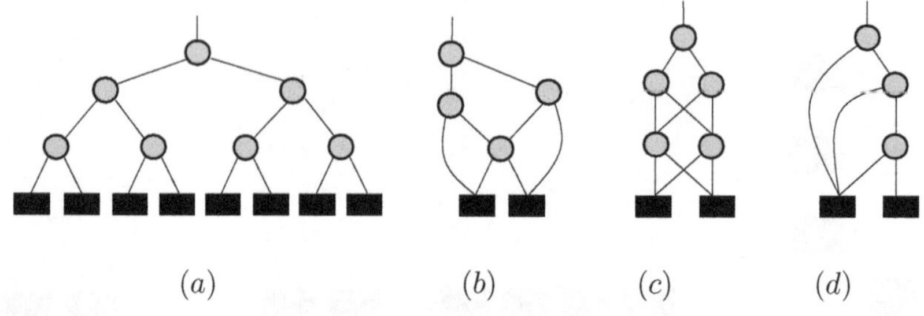

FIGURE 10.8
Reduction of decision trees produces various topologies of decision diagrams (Example 10.5).

The choice of an optimization criterion can be difficult. The measures considered above are unlikely to be simultaneously optimized; the minimization of one does not necessarily result in the minimization of the other.

In the case of binary trees and diagrams, reduced ordered binary decision diagrams (ROBDDs) provide minimal SOP representation. Detail on minimization of switching functions using ROBDD are considered in Chapter 18.

A special class of diagrams, called zero-suppressed BDD (ZBDD), are BDDs with different reduction rules:

- Eliminate all the nodes whose 1-edge points to the 0-terminal node and use subgraph of the 0-edge,
- Share an equivalent sub-graph in the same way as for ordinary BDD.

This zero-suppression rule is used for binary moment diagram (BMD) optimization, which is considered in Chapters 16 and 17.

Variable ordering. A decision diagram is ordered if the variables are encountered at most once and in the same order on every path in the decision diagram from the root to a terminal node. A BDD is a canonical representation of a switching function with respect to a fixed order of input variables. The permutation of variable order yields different BDDs for the same function. The effect of variable ordering depends on the nature of the function being represented.

A path in the decision tree consists of the edges connecting a non-terminal node with a terminal node. A complete path connects the root node with a terminal node. Decision trees and diagrams are graphical representations of functional expressions. For instance, each complete path in a binary decision tree or diagram corresponds to a term in sum-of-products expression of a function f.

Variable order is a significant factor in optimization of a decision diagram. A binary decision diagram is called *reduced* if it neither contains nodes with isomorphic sub-graphs nor nodes with both edges pointing to the same node.

A different order of variables produces decision diagrams with different numbers of nodes and, thus, different topologies. Therefore, variable ordering is an important problem in decision tree representations.

For that reason the following concepts are introduced:

- A decision tree or diagram is an *ordered* decision tree if the variables assigned to the nodes in each path appear in the same order.
- A decision tree or diagram is a *free* decision tree if the variables assigned to the nodes in each path appear in a different order.

Details of variable ordering are studied in Chapter 12.

10.3 Applications of decision diagrams

In the logic synthesis design phase of combinational circuit design, the functionality of the circuits to be designed is given in terms of a network of logical gates. An important task in logic synthesis is to optimize the representation of this circuit on the gate level with respect to several factors:

- The number of gates,

- ► The chip area,
- ► Energy consumption,
- ► Delay,
- ► Clock period, etc.

In contemporary logic synthesis, these optimization tasks are solved through manipulation on decision diagrams.

Minimization of switching functions represented by their decision diagrams is accomplished through manipulation of the diagram as described in Chapter 18. ROBDDs provide minimal SOP representation of switching functions. This representation is derived from the BDD, such that the resulting form is canonical. In the construction of a ROBDD, only the prime terms are kept (a prime term is a term not contained in any other term of the same order). A simple procedure is then used to derive a diagram optimal with respect to worst-case testing cost, expected testing cost, or storage cost. The algorithm genesis is in the relationship between Boolean formulas and optimal decision trees.

The key idea of a BDD-based minimization is that by reducing the representation, Boolean manipulation becomes much simpler computationally.

Consequently, they provide a suitable data structure for the manipulation of Boolean functions.

Sensitivity analysis and test pattern generation. Since decision diagram techniques are extended to various function manipulations, including calculation of Boolean differences (see Chapter 9), they can be employed in sensitivity analysis.

Sensitivity functions, which include Boolean difference, specify the conditions under which the circuit output is sensitive to the changing of the value on a signal line.

This is a component of automatic test generation, since the sensitivity function is included in the equation that describes the set of all tests for each single fault (usually, "stuck-at" faults are considered).

Probabilistic analysis of digital circuits. Given a switching function representing the conditions under which some event occurs, we can compute the event probability by computing the density of the function, i.e., the fraction of variable assignments for which the function yields 1. With the aid of the

Shannon expansion, the density $\rho(f)$ of a function f can be shown to satisfy the recursive equation $(f) = [\rho(f|x=0) + \rho(f|x=1)$, $\rho(1) = 1, \rho(0) = 0$. Thus, given a ROBDD representation of f, we can compute the density in linear time by traversing the graph depth first, labeling each vertex by the density of the function denoted by its subgraph (see details in Chapters 20) and 21.

ROBDDs can also be used for statistically analyzing the effects of circuit delays in a digital circuit. Consider a logic gate network in which each gate has a delay given by some probability distribution. This circuit may exhibit a range of behaviors, some of which are classified as undesirable. One simple analysis would be to treat the waveform probabilities for all signals as if they were independently distributed; then we can easily compute the behavior of each gate output according to the gate function and input waveforms, assuming that all circuit delays must be integer valued, and hence transitions occur only at discrete time points.

Thus, BDD-based symbolic analysis can be applied to systems with complex parametric variations. Although this requires simplifying the problem to consider only discrete variations, useful results can still be obtained. This approach accurately considers the effects of correlations among stochastic values.

Power consumption analysis using decision diagrams. The genesis of this application lies in the evaluation of switching activity which, in turn, depends on the input variables' "influence" on the output function, that is the switching function's structure.

Since decision diagrams explicitly reflect the function's structure and variables' "contribution," they are the perfect vehicle on which evaluation of the switching circuit activity can be performed,

Details of power consumption analysis using decision diagrams are described in Chapter 21.

Formal verification is accomplished in practical design by means of equivalence checking, which is conveniently implemented using ROBDDs. ROBDDs can be applied directly to the task of testing the equivalence of two combinational circuits. This problem arises when comparing a circuit to a network derived from the system specification or when verifying that a performed optimization has not altered the circuit functionality. Details of this application are given in Chapter 22.

Analysis of sequential circuits. In symbolic state graph methods with the state transition represented as a switching function, ROBDDs provide the nondeterministic automaton benefits. McMillan [14, 15, 16] has shown that under some conditions, the ROBDD representing the transition relation for a system grows only linearly with the number of system components, whereas the number of states grows exponentially. Notably, this property holds when the system components are connected in a linear or tree-structure.

One important application of this approach is verification of sequential digital circuits. For example, a state machine derived from the system specification can be checked for equivalence with one derived from the circuit even though it uses different state encodings.

10.4 Implementation and technologies

Complexity and programming of decision diagram. Complexity of decision diagrams for switching functions is addressed in Chapter 26. Aspects of software implementation of decision diagrams for switching functions are considered in Chapter 27.

CMOS VLSI design using decision diagram techniques. Decision diagrams are circuit models employed at many steps during synthesis and optimization of digital circuits:

- At a logic level, they are used to represent a function implemented by a circuit (technology-independent), or gate-level circuit structures (in particular, for multiplexer-based design, or timed Shannon circuits, i.e., technology dependent),
- At a physical level, they are used to represent transistor-level netlists, for verification purposes, and for direct mapping using particular techniques such as mapping of a multiplexer-based network to a physical layout of pass-transistor circuits.

Synthesis of a semiconductor microelectronic circuit follows the traditional schedule: given a functional description, an optimized network of logic gates is created, and then mapping of this network using a library of gates is accomplished. Finally, a CMOS transistor-level implementation, or circuit layout, is obtained.

At a logic level, decision diagrams are used mostly to represent a function to be implemented by a circuit. They are technology-independent models in this application; however, they are used to minimize the implementation cost.

A minimum-cost SOP can be derived from the ROBDD, and thus, a network with a minimum number of AND and OR gates. This network can then be mapped into a CMOS transistor-level circuit, optimized in terms of transistors required in total for implementation of the gates.

> *At the implementation level, decision trees and diagrams can be directly mapped into a multiplexer-based tree. The resulting circuits have few interconnections and lend themselves well to large-scale integration (for instance, binary trees form an efficient interconnection pattern). Multiplexer networks can be used as universal logic modules, thereby reducing the number of basic components needed for logic design.*

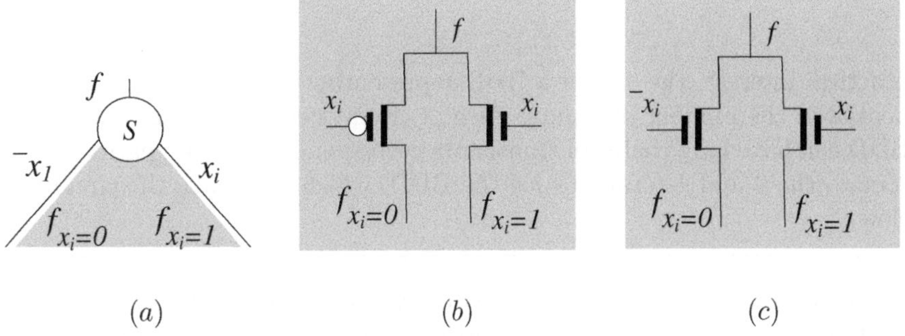

FIGURE 10.9
The BDD Shannon node (a) and its pass-transistor logic implementations (b) and (c).

Pass-transistor logic circuits can implement most functions with fewer transistors than static CMOS. This reduces the overall capacitance, resulting in faster switching times and lower power. However, level-restorers (due to the voltage drop between source and drain of pass-transistor) may increase the count of transistors.

Example 10.6 *Figure 10.10 shows the function* $f = \overline{a} \vee b\overline{c}$*, a multiplexed-based network for its implementation, and the corresponding pass-transistor logic circuit.*

Design of large pass-transistor logic circuits is, however, less developed than CMOS design. A naive BDD-based methodology for implementing pass-transistor logic circuits suffers from the drawback that for many functions of

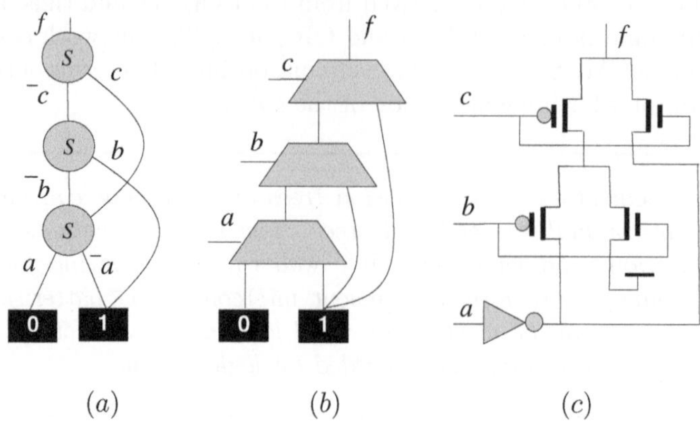

FIGURE 10.10
Interpretation of a decision diagram node in CMOS pass-transistor logic (Example 10.6).

practical interest, the size of a BDD representing the function can be exponential in the number of inputs. Also, a circuit generated from a monolithic BDD can have long chains of transistors corresponding to long paths from the root to the 0 and 1 terminals for the BDD, which can make the circuit very slow.

> *The principal advantage of BDD-based pass-transistor logic network design is that the one-to-one mapping between the BDD and the pass-transistor logic network makes the technology mapping problem very straightforward.*

As a result, we can perform circuit level optimizations by manipulating the BDD. The fact that mapping preserves the circuit structure allows us to make high-level changes that can have significant impact on area, power, and performance, but for which gains made at the high level hold at the circuit level as well. This is particularly important in the context of deep sub-micron designs, where logic level optimizations need to be driven by physical issues that depend on the circuit structure and topology. The decomposed BDD-based approach allows optimization of circuits in several ways that have no equivalent in the conventional multi-level network based synthesis flow (see details in Chapter 25).

Another solution is the use of transmission gates (a pair of transistors instead of one pass-gate), however, this also increases the count of transistors. In terms of circuit implementation oriented at universal logic blocks such as the multiplexer, decision diagrams represent an ideal candidate. The evaluation of a function then proceeds from the "leaves" (the constant values) to the "root" multiplexer; the function variables used as control signals select a unique path from the root to one leaf, and the value assigned to that leaf, propagated along the path to the output of the "root" multiplexer.

Another example of a direct mapping of the ROBDD into a CMOS circuit is the timed Shannon circuit. Since the paths from the root of a BDD to a terminal represent a set of disjointed cubes for a given variable assignment, a unique path is activated in the graph. This principle, in conjunction with a circuit mapping technique that replaces non-terminal vertices with combinational logic whose signals propagate from the top of the graph toward the terminal vertices, is implemented in timed Shannon circuits.

A significant savings in dynamic power dissipation was noted for such circuits since switching only occurs along two paths in the resulting circuit when a new set of circuit inputs are applied; first the previous path is switched off (all internal circuit nodes at logic- 1 go to logic-0) and then the new path is activated.

Note that decision diagram based techniques can be applied to verification of gate-level CMOS VLSI, and this can be accomplished at both the logic and physical level. At the logic level, a ROBDD can be created for the final optimized (with respect to a chosen criterion) circuit netlist, and then compared against a ROBDD derived from initial function description (see Chapter 22). At the physical level, a circuit schematic can be extracted from a layout and represented by a graph (diagram) model, and compared against another graph model of the schematics from which the layout was initially created.

Nanoelectronics. In nanoelectronics, the most feasible candidates for logic circuits are single-carrier electronics (low-temperature) and CMOS-molecular electronics (room-temperature). Though single-electron transistor based AND, EXOR, NOT and other gate designs have been proposed, such designs suffer from serious difficulties caused by the bilateral nature of a simple tunnel junction: separation between input and output is poor. Thus, it is difficult to construct large logic circuits using such non-unilateral circuits. The latency of the BDD corresponds to the implicit input-output presence in SET-based logic devices. In this device, quantum-dots are technologically formed using wrap-gate (WRG) structures [2].

Example 10.7 *In conventional logic design techniques, BDDs are mapped into a circuit netlist, so that each BDD node is associated with a universal element, e.g., a multiplexer. A multiplexer normally consists of two AND gates and one OR gate, in a gate-level design. Alternatively, pass-transistor logic may be used to implement a multiplexer with less transistors, although the level restoration problem causes the transistor count to increase. Figure 10.11 shows comparative designs of: multiplexer, gate-level, transistor-level, and single-electron wrap-gate nanowire device.*

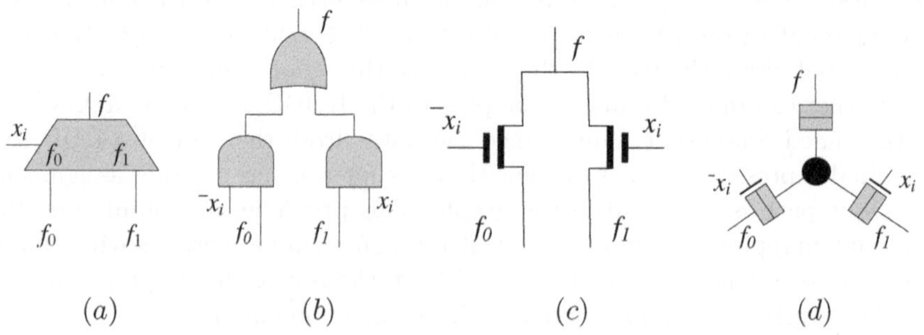

FIGURE 10.11
Implementation of Shannon node of decision diagram: multiplexer (a), gate-level (b), transistor-level (c), and single-electron wrap-gate nanowire device (d). (Example 10.7).

Those can be built on Shottky WPG or MOS WPG. The core of WPGs is the Shottky or silicon gate wrapped around a quantum wire. This branch-gate quantum-effect device is a universal multiplexer. This can be considered as a nanodevice model of a decision tree node. In a BDD-based nanodevice, 0-terminal means a connection to ground, and the 1-terminal is connected to the source of negative voltage that we apply to inject electrons into the nanowire. Multiplexer-based circuits can be built upon this principle.

Example 10.8 *The quantum dot and three tunnel junctions, two of them controlled by the wrap-gate voltage, shown schematically in Figure 10.11c, correspond to the WPG. This is used to implement a WPG circuit, as shown in Figure 10.12 for an AND gate.*

The BDD model of a circuit allows the avoidance of the interconnection problems that are present in gate-level implementation, by mapping BDDs directly to the nanowire lattice.

This approach is distinguished from the netlist-based approach (network of logic gates), and requires representation of functions as BDD. It is also the closest to implementation on SET devices (single electron tunneling) or rapid single flux quantum (SFQ) devices, in which the electron island in the SET can act as a switch. This is the reason this nonconventional (at a hardware level) representation of a logic function in terms of BDD is the basic prerequisite behind the hypercube structures considered in the further chapters of this book.

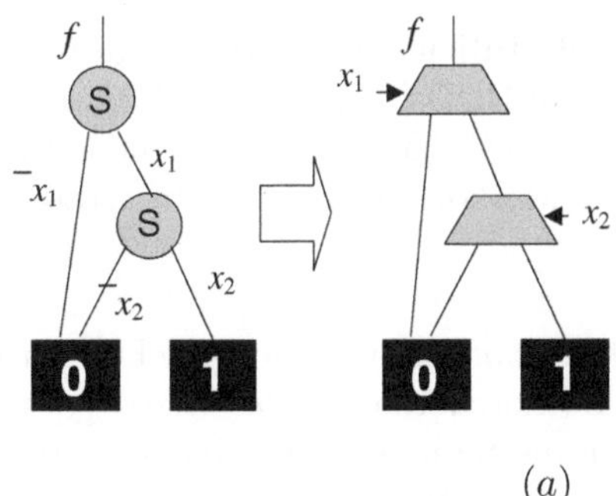

The BDD of the AND gate corresponds to a circuit formed of switches (simplest multiplexers).

(*a*)

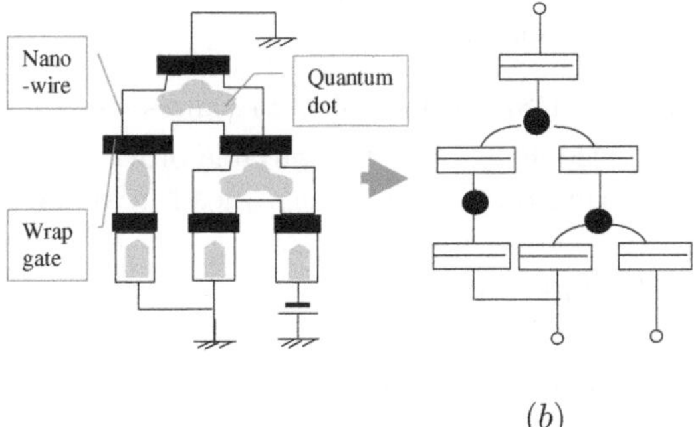

The BDD is mapped to the nanowire network, where branching of the signal (electron) is controlled by voltage applied to a wrap-gate on a nanowire.
Schematically, the network of electron junctions and quantum dots formed on the wire is shown on the right.

(*b*)

FIGURE 10.12
A demultiplexer-based single-electron circuit (Example 10.8).

10.5 Further reading

Books by Minato [17], Drechsler and Becker [8], Drechsler [9], and chapter by Yanushkevich and Shmerko [20] contain comprehensive information on decision diagram techniques in contemporary logic design.

Symbolic model checking. One of the broad BDDs applications is symbolic model checking for verification of states (see, for example, the book by McMillan [15]). Also, multiterminal BDDs presented by Clarke et.al. [7], diagrams with more than two terminal nodes, appeared to be convenient for probabilistic models. This is because probabilistic models are often represented as matrices, and multiterminal BDDs are efficient for matrix operations, including the solution of linear equations.

Computation of steady-state probabilities of Markov chains. Hatchel and Somenzi [10] applied multiterminal BDDs to the computation of steady-state probabilities of discrete-time Markov chains. It was demonstrated that large models, including discrete-time Markov chains with 10^{27} states, can be represented.

Sensitivity analysis is based on the Boolean difference (see Further reading section to Chapter 9). It should be noted that it was Akers [1] who first proposed that binary decision diagrams be used as the basis for developing tests for the switching function they represent.

Computation complexity. Wegener [19] shows that OBDDs can represent several important functions in polynomial size:

- The direct storage access function DSA or multiplexer requires size $2n+1$.
- All bits of n-bit addition need size $9n-5$ (for the variable ordering x_{n-1}, y_{n-1},...,x_0, y_0, while the ordering $x_0, y_0, ..., x_{n-1}, y_{n-1}$ leads to quadratic size ODBB.
- Conjunction or disjunction of n bits require size $n+2$.
- Parity or EXOR of n bits needs size $2n+1$.
- Arbitrary symmetric functions require at most size n^2+2.

Multiplication is the most difficult case, as lower bounds for multiplication lead for all considered BDD variants to lower bounds for squaring, the computation of the n most significant bits of the inverse of an n-bit number, and division. Wegener has reduced multiplication with read-once projections to the other mentioned arithmetic functions. Lower bounds on the computation complexity of the middle bit of multiplication leading to a $2^{n/8}$ lower bound on the OBDD size of this function is shown by Bryant [5].

One-pass synthesis. An approach to digital synthesis based on BDD mapping to multiplexer-based circuits, called one-pass synthesis, is considered in [9]. In [6], a decomposed BDD-based approach exploits some of the strengths of pass-transistor logic and is scalable.

Timed Shannon diagrams have been introduced in [11] as a synthesis approach for a low power optimization technique at the logic level. The idea of this approach was based on the assumption that overall circuit switching probabilities may be reduced in these circuits compared to traditional gate-level designs. An improvement of this principle for multi-output circuits was presented in [18], along with techniques that trade area for power reduction and a method for minimizing the overall circuit switching probability.

Self-repair and self-replacing systems. Mange et al. [13] used ROBDDs in designing self-repair and self-replacing systems based on the concept of embryonic electronics.

References

[1] Akers SB. Binary decision diagrams. *IEEE Transactions on Computers,* 27(6):509–516, 1978.

[2] Asahi N, Akazawa M, and Amemiya Y. Single-electron logic device based on the binary decision diagram. *IEEE Transactions on Electron Devices,* 44(7):1109–1116, 1997.

[3] Becker B. Testing with decision diagrams. *Integration, the VLSI Journal,* 26:5-20, 1998.

[4] Bryant RE. Graph-based algorithms for Boolean function manipulation. *IEEE Transactions on Computers*, 35(6):677–691, 1986.

[5] Bryant RE and Cen Y-A. Verification of arithmetic circuits with binary moment diagrams. In *Proceedings of the 32rd ACM/IEEE Design Automation Conference*, pp. 535–541, June 1995.

[6] Buch P, Narayan A, Newton AR, and Sangiovanni-Vincentelli AL. Logic synthesis for large pass transistor circuits. In *Proceedings of the International Conference on Computer-Aided Design,* pp. 663–670, 1997.

[7] Clarke EM, Fujita M, and Zhao X. Multi-terminal decision diagrams and hybrid decision diagrams. In *Sasao T and Fujita M, Eds, Representation of Discrete Functions*, pp. 91–108, Kluwer, Dordrecht, 1996.

[8] Drechsler R and Becker B. *Binary Decision Diagrams. Theory and Implementation.* Kluwer, Dortrecht, 1998.

[9] Drechsler R. *One-Pass Synthesis*, Kluwer, Dortrecht, 2002.

[10] Hatchel G and Somenzi F. *Logic Synthesis and Verification Algorithms.* Kluwer, Dortrecht, 1996.

[11] Lavagno L, McGeer P, Saldanha A, and Sangiovanni- Vincentelli AL. Timed Shannon circuits: A power-efficient design style and synthesis tool. In *Proceedings of the 32nd ACM/IEEE Design Automation Conference*, pp. 254-260, CA, 1995.

[12] Lee CY. Binary decision diagrams. *Bell System Technical Journal,* 38(4):985–999, 1959.

[13] Mange D, Sanchez E, Stauffer A, Tempesti G, Marchal P, and Piguet C. Embryonics: a new methodology for frdigning field-programmable gate arrays with self-repair and self-replacing properties. *IEEE Transactions on Very Large Scale Integration (VLSI) Systems*, 6(3):387–399, 1998.

[14] McMillan KL. Using unfoldings to avoid the state explosion in the verification of asynchronous circuits. In *Proceedings of the 4th Workshop on Computer-Aided Verification,* Lecture Notes in Computer Science, Vol. 663, pp. 164–174, 1992.

[15] McMillan KL. *Symbolic Model Checking.* Kluwer, Dordrecht, 1993.

[16] McMillan KL. A technique of state space search based on unfolding. *Formal Methods in System Design,* 6(1):45–65, 1995.

[17] Minato S. *Binary Decision Diagrams and Applications for VLSI CAD.* Kluwer, Dordrecht, 1996.

[18] Thornton MA, Drechsler R, and M Miller D. Multioutput timed Shannon circuits In *Proceedings of the International Symposium on VLSI*, Pittsburg, PE, pp. 47–52, April 2002.

[19] Wegener I. BDDs-design, analysis, complexity, and applications. *Discrete Applied Mathematics*, 138:229–251, 2004.

[20] Yanushkevich SN and Shmerko VP. Decision diagram technique. In *Dorf R, Editor. The Electrical Engineering Handbook*, Third Edition, CRC Press, Boca Raton, FL, 2005.

11

Classification of Decision Diagrams

Various decision diagrams have been considered in the literature. There can be enumerated almost 50 essentially different decision diagrams for several classes of discrete functions, and a few more variants and modifications of some of them (see, for example, [2] and [11]). Therefore, classification of decision diagrams currently used in the research community is important for

- Better navigation in this research space,
- Proper understanding of the decision diagrams,
- Their efficient exploitation, and
- Easy and unified interpretation of research results derived by various research groups.

11.1 Introduction

Decision diagrams are viewed as acyclic graphs consisting of non-terminal nodes and constant nodes connected by edges.

Decision diagrams are characterized by:

- *Non-terminal nodes and number of outgoing edges,*
- *Decomposition rules performed at non-terminal nodes to assign a given function f to the diagram,*
- *Constant (terminal) nodes and the values shown in them, and*
- *Edges and their labels.*

Decision diagrams can be classified with respect to any of these characteristics, which are the subject of this chapter. Table 11.1 summarizes decision diagrams and related acronyms discussed in this chapter.

TABLE 11.1
Decision diagrams and their acronyms.

Acronym	Name
ADD	Algebraic decision diagram
BDD	Binary decision diagram
BMD	Binary moment diagram
*BMD	Modified binary moment diagram
EVBDD	Edge-valued binary decision diagram
FDD	Functional decision diagram
FEVBDD	Factored edge-valued binary decision diagram
FNADD	Fourier decision diagram on finite non-Abelian groups
FNAPDD	Fourier decision diagram on finite non-Abelian groups with preprocessing
GFDD	Galois field decision diagram
RMFDD	Reed–Muller–Fourier decision diagrams with preprocessing
K*BMD	Kronecker binary moment diagram
KDD	Kronecker decision diagram
LDD	Linear decision diagram
mvMTDD	Matrix-valued multiterminal decision diagram
MTBDD	Multiterminal binary decision diagram
MD	Multiple-place diagram
MTDD	Multiterminal decision diagram
*PHDD	Power hybrid decision diagram
PKDD	Pseudo-Kronecker decision diagram
KDD	Kronecker decision diagram
WDD	Walsh decision diagram
ZBDD	Zero-suppressed binary decision diagrams

11.2 Classification of decision diagrams with respect to constant nodes

Discrete functions, defined as a mapping $f : D \to R$, where D is the domain and R the range for f, can be classified depending on the algebraic structures assumed for the domain D and the range R.

In a decision diagram, the constant nodes show either the values a function can take in R, or the values obtained after some calculation over function values depending on the decomposition rules used in definition of the decision diagram in question.

Depending on the values shown in constant nodes, decision diagrams can be classified as

- *Basic*, when constant nodes show function values,
- *Derived*, when constant nodes show values derived by performing some calculations over the function values.

In spectral interpretation, basic *diagrams represent decision diagrams where the given function f is assigned to the diagram by an identical mapping.*

Example 11.1 *Examples of decision diagrams using identical mapping are BDDs for binary-input binary-output functions, multiterminal binary decision diagrams (MTBDDs) for binary-input complex-valued output functions, and multiple-place diagrams (MDDs) for multivalued-input multivalued-output functions.*

Derived diagrams are decision diagrams where some spectral transform is performed over the function represented and then the spectrum is shown by a basic decision diagram. In this case, the constant nodes show the corresponding spectral coefficients.

Both basic and derived decision diagrams can be uniformly viewed as spectral decision diagrams, with the *identity transform* applied in the first case.

The range R for functions represented by basic decision diagrams can be different, and it may be

- The finite Galois field $GF(p)$ for logic functions, with $p = 2$ for switching functions, and $p > 2$ for multivalued functions,
- The set of integers Z for integer-valued functions,

► The field Q of rational numbers, or the field C of complex numbers, in general, for complex-valued functions.

In the first case, $GF(p)$, the corresponding decision diagrams are called bit-level diagrams, *allowing p-valued bits in the case of multi-valued functions. Otherwise, diagrams are called* word-level diagrams, *since the value of a constant node is represented by a* word *in the computer.*

The constant nodes of decision diagrams can be fuzzy values, or matrices over the field $GF(p)$, or the complex field. For instance, basic diagrams where constant nodes show fuzzy values can be viewed as *fuzzy decision diagrams.* A sub-classification of decision diagrams can be done with respect to constant nodes as

► Number-valued, and

► Matrix-valued diagrams.

The matrix-valued diagrams can be further classified into bit-level and word-level matrix-valued diagrams, depending on the entries of the matrices. Figure 11.1 shows a classification of decision diagrams with respect to constant nodes.

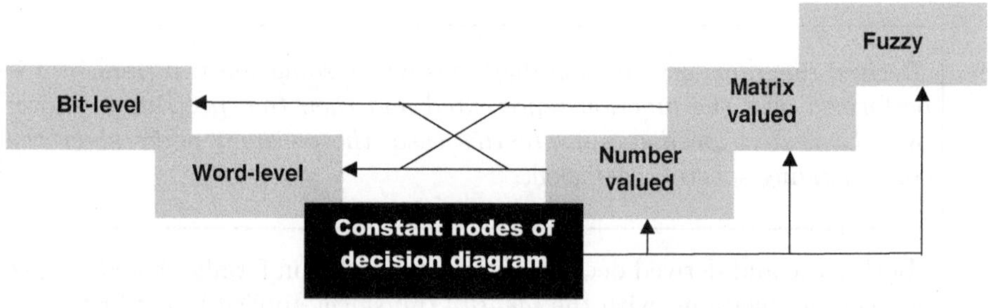

FIGURE 11.1
Classification of decision diagrams with respect to the values of constant nodes.

11.3 Classification of decision diagrams with respect to non-terminal nodes

Non-terminal nodes have outgoing edges pointing to the subtrees in the decision tree. Decision diagrams consisting of nodes with different numbers of outgoing edges are used for different classes of discrete functions.

In the spectral interpretation, functions represented by decision diagrams are viewed as functions defined on a group G, and their spectra used in definition of the diagrams are functions on the dual object Γ for G. The recursive structure of the decision trees implies that G must be decomposable as the direct product of subgroups G_i. Each variable in an n-variable function f takes values in the corresponding subgroup G_i of G. In the same way, Γ is the direct product of dual objects Γ_i of G_i.

The number of outgoing edges of non-terminal nodes is, in general, equal to the cardinality of Γ_i. In the case of Abelian groups, Γ expresses the structure of a multiplicative group isomorphic to G. It follows, that the number of outgoing edges of non-terminal nodes for decision diagrams on finite Abelian groups is equal to the values a variable can take. In the case of non-Abelian groups, the number of outgoing edges can be smaller than the number of values a variable can take.

Decision diagrams can be classified with respect of the number of outgoing edges of non-terminal nodes into decision diagrams with:

- Binary *non-terminal nodes, having two outgoing edges per node, and*
- Multiple *outgoing non-terminal nodes, with number of outgoing edges greater than two.*

From the spectral interpretation point of view, diagrams with binary non-terminal nodes are defined on finite dyadic groups, which are a direct product of n basic cyclic groups C_2. Diagrams having nodes with more than two outgoing edges are defined on groups of higher orders. For instance, *quaternary decision diagrams* are defined on a group that is the direct product of cyclic groups of order four, C_4.

If a decision tree consists of non-terminal nodes with the same number of outgoing edges, it is called a *homogeneous decision tree*, otherwise, it is *heterogeneous decision tree.* [9].

Example 11.2 *Figure 11.2 shows a homogeneous binary decision tree, a homogeneous ternary decision tree, and two heterogeneous decision trees.*

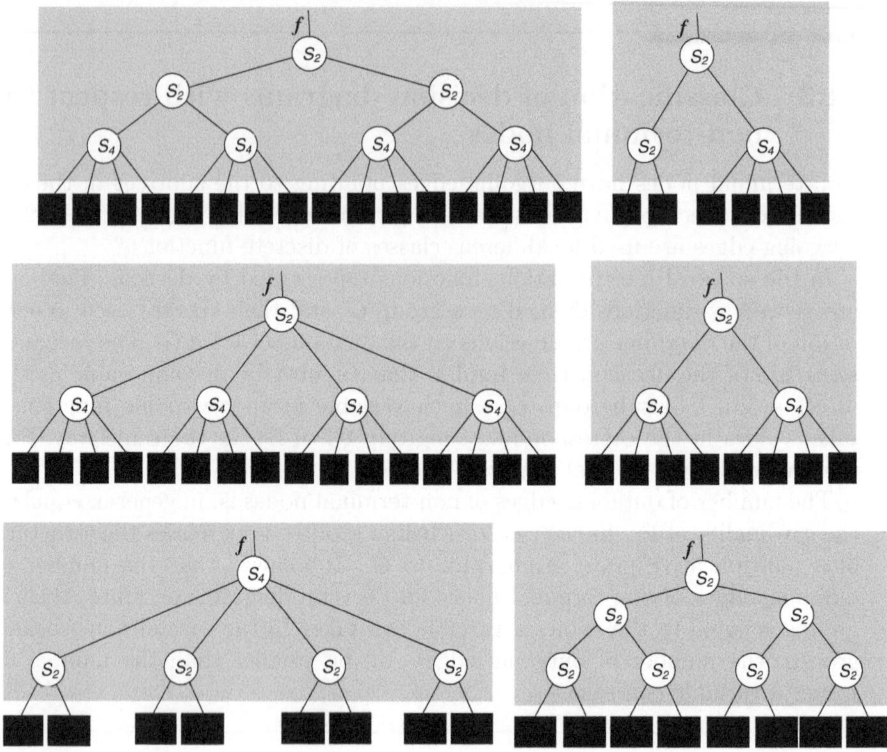

FIGURE 11.2
Decision diagrams with different numbers of outgoing edges per node.

11.4 Classification of decision diagrams with respect to decomposition rules

A given switching function f is assigned to a decision diagram by performing some decomposition rule with respect to each variable. Equivalently, the decomposition rule is formed at each node of the decision tree, which is reduced afterwards into a decision diagram. Various decision diagrams are defined with respect to different decomposition rules.

Application of the decomposition rules can be viewed as performing a spectral transform over the function to be represented before it is assigned to the basic decision tree.

Decision diagrams can be classified with respect to the decomposition rules, or equivalently, with respect to spectral transforms used in definition of particular decision diagrams. Figures 11.3 and 11.4 shows a classification of the most well known decision diagrams with respect to the spectral transforms used. Figure 11.4 refers to multiterminal diagrams (diagrams with integer-valued terminal nodes).

11.5 Classification of decision diagrams with respect to labels at the edges

In a decision diagram, we perform the decomposition of a function to be represented by using some decomposition rules, until values of constant nodes are determined. Conversely, given a decision diagram, we determine the function it represents by the inverse procedure. We start from constant nodes and traverse paths to the root node by multiplying the values of the constant nodes with labels at the edges. The labels are determined in such a way that formalized performing of the procedure inverse to the decomposition is used in assigning function f to the decision diagram.

In spectral interpretation, the product of labels at the edges determines the basis functions in terms of which a particular decision diagram is defined.

However, for some functions such decision diagrams may have exponential complexity in the number of nodes. For instance, multipliers are an example of functions having decision diagrams with an exponential complexity in the number of inputs n.

In word-level decision diagrams, there are functions requiring a large number of different values for constant nodes; however, with the same difference between values of two nodes.

Example 11.3 *For a serial four-bit adder, the vector of values for constant nodes is*

$$\mathbf{F} = [0, 1, 2, 3, 1, 2, 3, 4, 2, 3, 4, 5, 3, 4, 5, 6]^T.$$

For such functions, it appears convenient to modify the decision diagram, set the values of all constant nodes to zero, and assign additive weight coefficients to the edges. In this way edge-valued binary decision diagrams are defined.

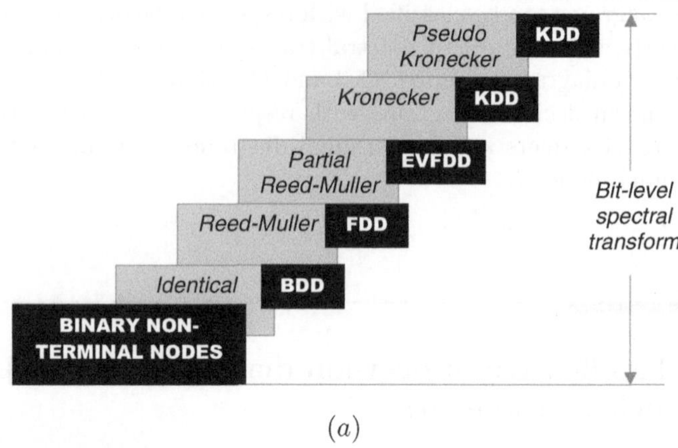

(a)

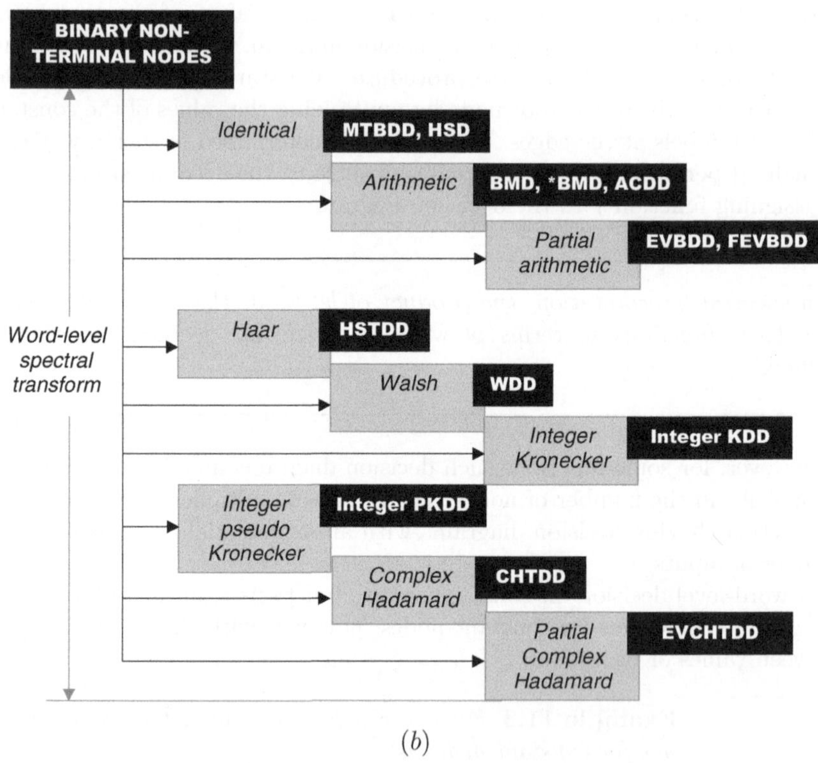

(b)

FIGURE 11.3
Classification of bit-level (a) and word-level (b) decision diagrams with respect to the decomposition rules.

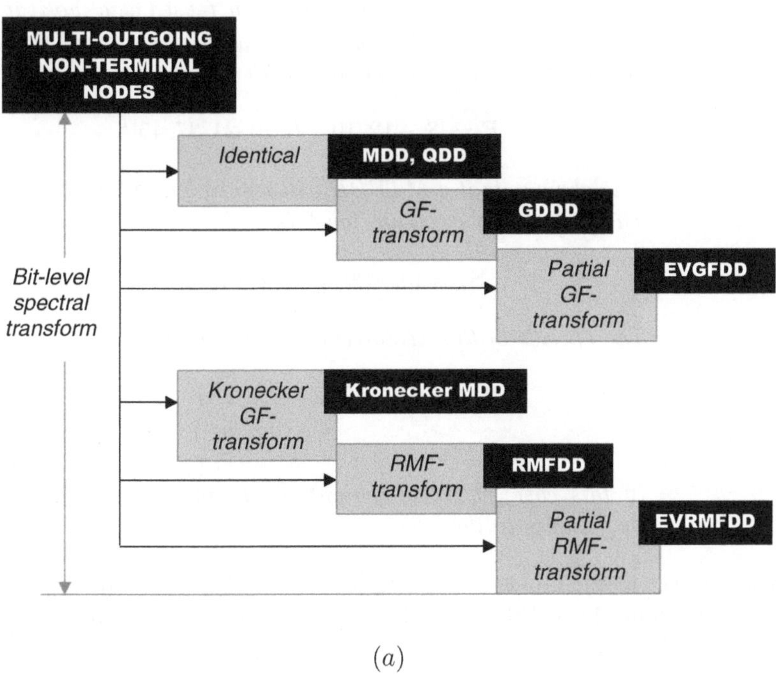

(a)

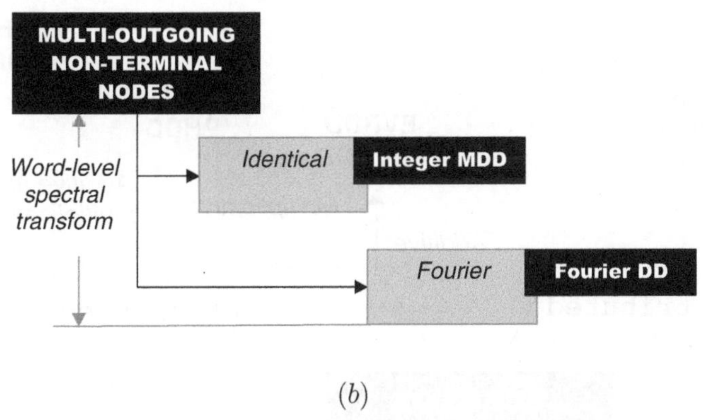

(b)

FIGURE 11.4
Classification of bit-level (a) and word-level (b) multiterminal decision diagrams with respect to decomposition rules.

Similarly, there are functions where values of constant nodes have many common factors.

Example 11.4 *A switching function f used as a motivational example for introduction of binary moment diagrams and their modification, $*$BMDs, has the truth vector*

$$\mathbf{F} = [8, -12, 10, -6, 20, 24, 37, 45]^T$$

It is obvious that a MTBDD will require many constant nodes. The arithmetic spectrum for f is

$$\mathbf{S}_f = [8, -20, 2, 4, 12, 24, 15, 0]^T.$$

The entries of this vector can be factorized as

$$2 \cdot 2 \cdot 2,\ 2 \cdot 2 \cdot (-5),\ 2 \cdot 1 \cdot 1, 2 \cdot 1 \cdot 2,\ 3 \cdot 4 \cdot 1,\ 3 \cdot 4 \cdot 2,\ 3 \cdot 1 \cdot 5,\ 3 \cdot 1 \cdot 0.$$

In this case, it is convenient to assign multiplicative weight coefficients to the edges.

With respect to the labels at the edges, decision diagrams can be classified as shown in Figure 11.5. Edge valued binary decision diagrams are considered in detail in Chapter 16.

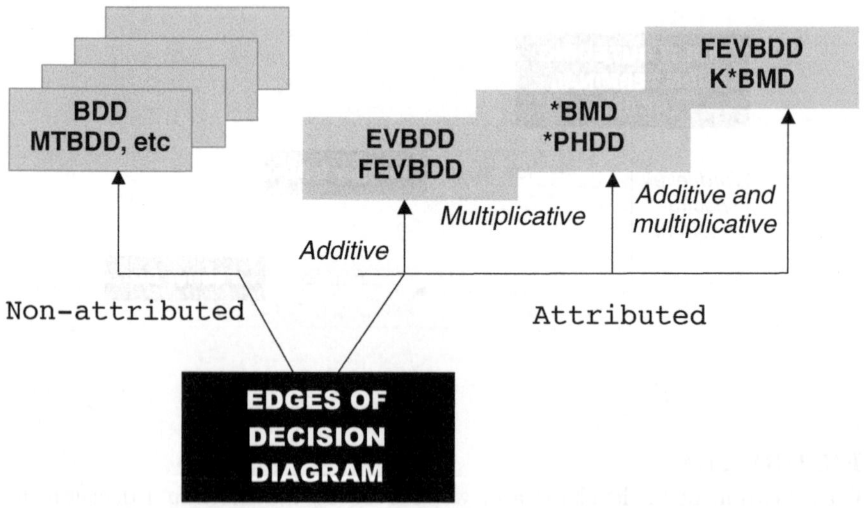

FIGURE 11.5
Classification of decision diagrams with respect to labels at the edges.

11.6 Further reading

Reviews. A comprehensive review of major types of decision diagrams is given by Bryant and Meinel [2] and Minato [5].

Classification of decision diagrams using spectral criteria was given by Stanković et al. [14, 11]. Other classifications were proposed, in particular, by Dill et al. [3] and Perkowski et al. [4].

Lattice decision diagrams. This particular class of diagrams was developed by Perkowski et al. [29] Decision diagrams of totally symmetric switching functions can be embedded into a directly. Lattice decision diagrams are based on the *Akers array* [1] which is defined as a rectangular array of identical cells, each of them being a multiplexer, where every cell obtains signals from two neighbor inputs and gives them to two neighbor outputs. All cells on a diagonal are connected to the same (control) variable. Variables have to be repeated to ensure realizability of a given or an arbitrary single-output completely specified switching function.

Perkowski et al. [29] distinguished the following types of lattice decision diagrams:

- *Ordered lattice diagrams* without repeated variables are defined as lattices with one variable on a diagonal. If the same variable appears on various levels, but only one variable in a level, than this ordered lattice diagram is a diagram with repeated variables.
- *A free lattice diagram* is a lattice in which different orders of variables in the paths are allowed.

The authors of [29] classified diagrams with respect to the type of expansion used in the nodes (by analogy with decision trees and diagrams). A lattice in which, in every level, all expansions are of the same type, or either Shannon, positive or negative Davio, is called an *ordered Kronecker lattice diagram* and *pseudo Kronecker lattice diagram.* Lindren et al. [4] studied pseudo Kronecker lattice diagrams.

Fuzzy decision diagrams were developed by Strehl et al. [6]. This is a particular class of decision diagrams.

References

[1] Akers SB. A rectangular logic array. *Transactions of IEEE.* 21:848–857, Aug. 1972

[2] Bryant RE and Meinel C. Ordered binary decision diagrams. Foundations, applications and innovations. In Hassoun S and Sasao T, Eds., Brayton RK, consulting Ed. *Logic Synthesis and Verification.* Kluwer, Dordrecht, pp. 285–307, 2002.

[3] Dill KM, Ganguly K, Safranek RJ, and Perkowski MA. A new Zhegalkin Galois logic. In *Proceedings of the 3rd International Workshop on Applications of the Reed–Muller Expansion in Circuit Design,* Oxford, UK, pp. 247–257, 1997.

[4] Lindren P, Drechsler R, and Becker B. Synthesis of pseudo Kronecker lattice diagrams. In *Proceedings of the 4-th International Workshop on Applications of the Reed–Muller Expansion in Circuit Design.* University of Victoria, BC, Canada, pp. 197–203, 1999.

[5] Minato S. *Binary Decision Diagrams and Applications for VLSI CAD.* Kluwer, Dordrecht, 1996.

[6] Strehl K, Moraga C, Temme K-H, and Stanković RS. Fuzzy decision diagrams for the representation, analysis and optimization of fuzzy rule bases. In *Proceedings of the 30th International Symposium on Multiple-Valued Logic,* pp. 127–132, 2000.

[7] Perkowski MA, Chrzanowska-Jeske M, and Xu Y. Lattice diagrams using Reed–Muller logic. In *Proceedings of the 3rd International Workshop on Applications of the Reed–Muller Expansion in Circuit Design.* Oxford, UK, pp. 85–102, 1997.

[8] Perkowski MA, Jóźwiak L, and Drechsler R. New hierarchies of AND/EXOR trees, decision diagrams, lattice diagrams, canonical forms, and regular layouts. In *Proceedings of the 3rd International Workshop on Applications of the Reed–Muller Expansion in Circuit Design.* Oxford, UK, pp. 115–142, 1997.

[9] Nagayama S and Sasao T. On the minimization of average path length for heterogeneous MDDs. In *Proceedings of the IEEE 34th International Symposium on Multiple-Valued Logic,* pp. 216–222, 2004.

[10] Stanković RS and Astola JT. *Spectral Interpretation of Decision Diagams.* Springer, Heidelrberg, 2003.

[11] Stanković RS and Sasao T. Decision diagrams for discrete functions: classiffication and uniffied interpretation. In *Proceedings of the ASP-DAC Conference,* pp. 349–446, 1998.

12

Variable Ordering in Decision Diagrams

For many problems, the complexity of a decision diagram can vary from linear to exponential depending on the variable ordering employed. As a result there has been considerable work on variable ordering, heuristic methods in particular. In this chapter, we review the concepts of the variable ordering problem and describe a number of techniques that have proven to be successful in determining good orderings.

12.1 Introduction

The variable ordering problem for decision diagrams is to find an ordering that minimizes the complexity of the decision diagram with regards to some criterion such as node count, average path length, or maximum path length.

Since there are $n!$ orderings of n variables, the problem is a complex one that can only be exactly solved for small n. Heuristic approaches must be used for problems of moderate to large size.

We first consider node count as the complexity metric. A simple example best illustrates the nature of the problem.

Example 12.1 *Consider the function:*

$$f = x_1x_4 \vee x_2x_5 \vee x_3x_6 \qquad (12.1)$$

The BDD for the variable ordering $[x_1, x_2, x_3, x_4, x_5, x_6]$ *is shown in Figure 12.1a. Figure 12.1b shows that* $[x_1, x_4, x_2, x_5, x_3, x_6]$ *yields a much simpler diagram.*

Clearly, as the number of variables is increased in an expression of the form shown in Equation 12.1, the first variable ordering leads to an increasingly complex BDD whereas the second variable ordering results in linear growth with one node per variable, the best one could hope for.

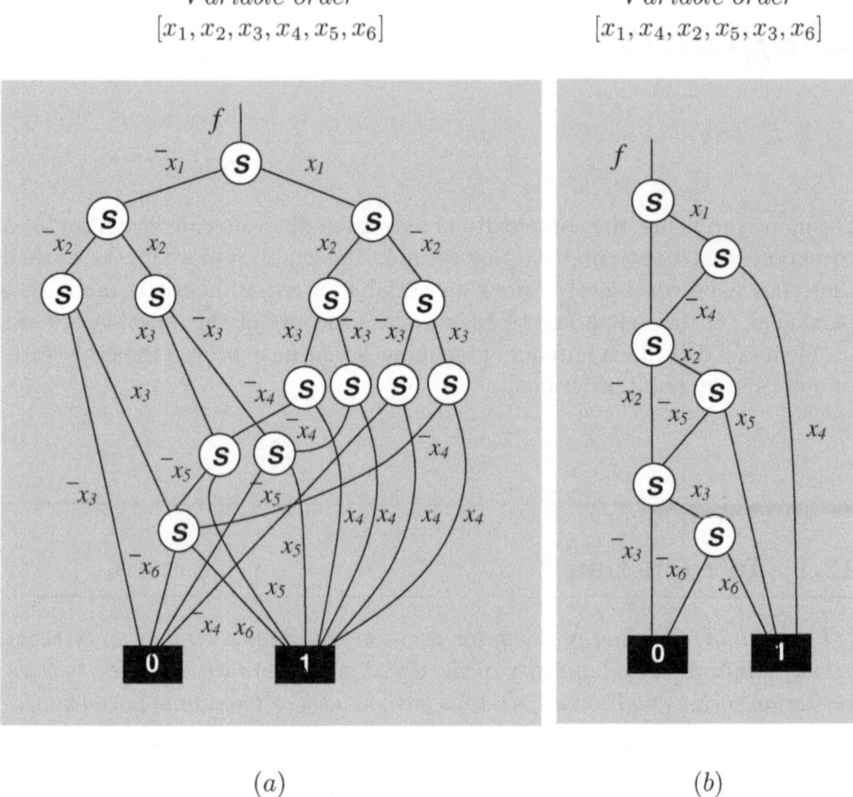

FIGURE 12.1
Decision diagrams with different variable orders (Example 12.1).

Example 12.1 illustrates two important points:

- ► The size of a decision diagram can depend critically on the variable ordering chosen.
- ► The linear growth variable ordering groups the symmetric variable pairs together. This is in fact a general principle that can be used to significant advantage in choosing a variable ordering, as will be illustrated below.

12.2 Adjacent variable interchange

As seen in Example 12.1, one can on occasion determine a good, and sometimes optimal, variable ordering by considering properties of the function or functions to be represented. However, in general this is not the case since the decision diagram is often the only representation for the functions, and properties of the functions cannot be identified while the decision diagram is being built. In practice, the problem is that we have a decision diagram built using a particular variable ordering and we want to improve that variable ordering.

The basic operation in practical variable reordering techniques, exact and heuristic, is the *interchange of two adjacent variables* in the current variable ordering. The operation is illustrated in Figure 12.2.

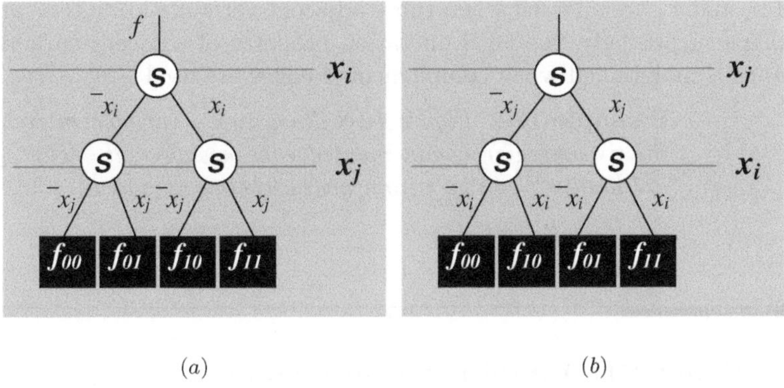

FIGURE 12.2
Adjacent variable interchange: before (a) and after(b) (without complemented edges).

Figure 12.2a shows a portion of a BDD where $f_{00}, f_{01}, f_{10}, f_{11}$ denote BDD structures representing the four co-factors. These four diagrams may share nodes but can not involve x_i or x_j.

Figure 12.2b illustrates the situation after x_i and x_j are interchanged. Note that f_{01} and f_{10} exchange positions in the two diagrams. The critical point is that the interchange of the two variables is a localized operation affecting only nodes labelled by those variables. In particular:

- Any edges pointing to the top node are not affected by the interchange since the BDD for which it is the root continues to realize the same function although the node variables are changed.

- ▶ The second level nodes in the two diagrams are roots of BDD representing different functions.
- ▶ The second level nodes in Figure 12.2a are retained if nodes higher in the BDD have edges pointing to them, otherwise they are discarded. The point is that no node above level x_i in the original diagram is affected by the exchange.
- ▶ Likewise, f_{01} and f_{10} are repositioned but their structures, including edges from above, are not changed. f_{00} and f_{11} are similarly unchanged.
- ▶ New nodes, the two second level nodes in Figure 12.2b, must be created if they do not already exist.
- ▶ Some housekeeping is required since a node originally labelled x_i is relabelled x_j.

The interchange of variables requires some adjustments to the data structures implementing the decision diagram, but as noted, the critical point that makes the interchange practical and indeed efficient, is that only nodes labelled x_i and x_j are affected when those adjacent variables are interchanged.

Perhaps surprisingly, the local operation property of adjacent variable interchange is maintained when complemented edges are used.

Example 12.2 *Figure 12.3 illustrates a variable interchange for a particular complemented edge assignment. The effects are localized to edges from nodes labelled x_i and x_j.*

12.3 Exact BDD minimization techniques

Determining an optimal variable ordering is NP-complete. A naive approach with time complexity $O(2^n \cdot n!)$ is to build the BDD for all $n!$ variable orderings step-by-step recording the ordering yielding minimal BDD size. There is a very elegant algorithm [36] that enumerates the $n!$ permutations of n objects by interchange of adjacent objects. In fact, given an initial ordering, this algorithm performs exactly $n! - 1$ interchanges to enumerate all $n!$ permutations. Hence it is possible to examine all variable orderings using only the operation of adjacent variable interchange.

Friedman and Supowit [12] have presented an exact minimization method that, while more efficient than the above, is still exponential with time complexity $O(n^2 \cdot 3^n)$ and space complexity $O(\frac{3^n}{\sqrt{n}})$. It is based upon the observation that the number of nodes labelled x_i in a BDD is independent of the order of the variables both above and below level i in the BDD. The approach uses induction and is based on the fact that if one has optimal solutions for BDD involving $s-1$ variables, optimal sizes for BDD involving s variables can be found by trying each of the s variables in turn as the top variable,

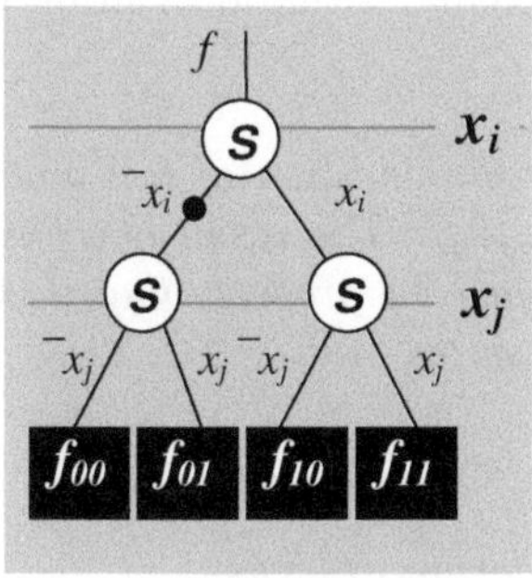

(a)

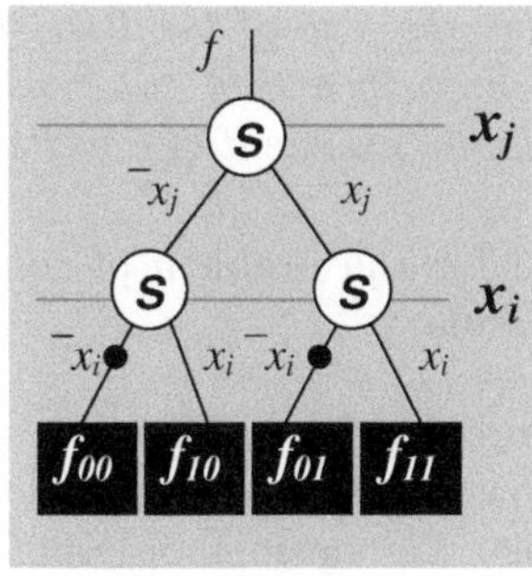

(b)

- *Enumeration of the eight cases (to achieve normalization, only edges labelled 0 can have complemented edges and the terminal node is always 1) shows locality of operation always holds.*
- *It is irrelevant if an edge leading to the top node has a complement since the function rooted by that node does not change.*

FIGURE 12.3
An adjacent variable interchange with complemented edges (Example 12.2).

and recording the minimal size and corresponding variable order found in the search. While it is of theoretical interest, this approach is not practical except for BDDs involving only a few variables.

12.4 Problem specific ordering techniques

Since finding an optimal variable ordering is in general intractable, there has been a great deal of research concerning heuristic BDD minimization techniques. These methods find a good if not optimal variable ordering. The first general approach uses properties of the function(s) under consideration and is in that sense problem specific.

A good or optimal ordering for a particular situation can on occasion be identified from the problem specification.

Example 12.3 *Consider the addition of two binary numbers* $\langle a_{n-1}, ..., a_0 \rangle$ *and* $\langle b_{n-1}, ..., b_0 \rangle$. *The result is a binary number* $\langle s_n, ..., s_0 \rangle$. *This problem has the property that it is symmetric in each* a_i, b_i *pair. It transpires that*

$$[a_{n-1}, b_{n-1}, ..., a_0, b_0]$$

is the optimal variable ordering whereas

$$[a_{n-1}, ..., a_0, b_{n-1}, ..., b_0]$$

is very poor. In particular, the size of the BDD for the first (interleaved) order grows linearly with n, *whereas the second (noninterleaved) order exhibits exponential growth.*

Many techniques have been developed for the problem of constructing a BDD from a circuit specification, in particular:

- A tree-like combinational circuit, a circuit with no fan-out, with vertex gates (AND, OR, NAND, NOR) has a BDD representation with one non-terminal node per input variable. The optimal ordering is found by a depth-first search of the circuit from the output towards the inputs. This does not hold if EXOR or EXNOR gates are introduced.
- Depth-first search approaches can be applied to circuits with fan-out. Such approaches essentially consider the fan-out free subcircuits as separate components. The performance of these methods is circuit dependent and none is in general superior to the other.
- Some approaches use multiple heuristics deciding, when a certain BDD size is reached, which heuristic is most promising for the circuit under consideration and completing the problem using that technique.

The situation becomes much more complex for multiple-output circuits. There are two basic approaches:

- Find a variable ordering for the most critical output and then insert other variables arising from the consideration of the other outputs.
- Find an ordering relative to each output and then merge the orderings.

Details can be found in [14] and [25]. While these approaches are successful in certain circumstances they have been largely superseded by dynamic variable ordering techniques because of the more general applicability of the latter.

12.5 Dynamic variable ordering

For dynamic variable ordering (DVO) techniques, construction of the BDD begins with an initial ordering. If the size of the BDD exceeds a preset limit, a DVO technique is applied to alter the variable ordering such that the BDD constructed to that point is reduced in size. DVO techniques employ the interchange of adjacent variables as discussed above.

12.5.1 Window permutation

In window permutation methods, a window of fixed size m is moved across the n variables beginning with the top m variables.

The principal features of the window permutation approach are:

- ▶ For each position of the window, all permutations of the m variables within the window are considered and the one yielding the maximum size reduction is recorded.
- ▶ This process is repeated for the $n - m + 1$ window positions after which the permutation identified as yielding the greatest size reduction across all window positions is applied.
- ▶ The complete process is repeated as long as a BDD size reduction is found. Since a permutation of variables within a window does not affect the BDD structure above or below the window (a direct result of the localized nature of adjacent variable interchange as discussed above), only window positions for which the variables covered have changed need be reconsidered.

The permutation generation algorithm from [36] can be used to efficiently examine the permutations of the window variables using adjacent variable interchange. The size of the window used affects the computation time required:

- ▶ $m = n$ is the trivial exhaustive search method. $m = 2$ is very time efficient, but the search is easily trapped in local minima;
- ▶ $m = 3$ or 4 on the other hand is still relatively efficient while yielding good results, on occasion superior to problem dependent methods.

The latter is the case because the window permutation approach addresses multiple outputs in a shared BDD together and can recognize structure across functions that may not be readily apparent to a designer.

12.5.2 Sifting

Sifting is a heuristic approach to variable reordering now used in many practical BDD packages and applications. The basic procedure operates as follows:

Sifting algorithm

Step 1 Select a variable x_i -- a simple heuristic is to choose the variable that labels the most nodes in the BDD.

Step 2 Sift x_i to the bottom of the BDD by a sequence of adjacent variable interchanges.

Step 3 Sift x_i to the top of the BDD by a sequence of adjacent variable interchanges.

Step 4 During steps (2) and (3) a record is kept of the position of x_i that yields the smallest node count in the BDD, so now sift x_i back to that position.

Step 5 Repeat steps (1) to (4) until each variable has been sifted into the *best* position. Note that once a variable is selected and sifted, it is not selected a second time.

There are $n!$ orderings of n variables. Sifting examines on the order of n^2 orderings, yet does extremely well at identifying good variable orderings. In the above procedure, each variable is shifted to its *best* position. The whole process can be iterated until there is no further improvement, an approach referred to as *sifting to convergence.*

Adjacent variable interchange is a local operation that can be performed efficiently. However, sifting can be costly due to the number of exchanges required.

Certain refinements have been added to sifting to improve its performance and are incorporated in most decision diagram implementation packages:

- *As the size of the BDD can grow substantially during the sifting process, an upper limit is commonly set. If the limit is exceeded, moving the variable further is aborted. This avoids excessively large intermediate BDD.*
- *Rather than initially moving the variable down the diagram, it can be moved to the closer of the top or bottom of the diagram and then in the other direction. This reduces the total number of adjacent variable interchanges while still trying the variable in every position.*
- *Bounding techniques are known that limit the range over which a variable must be sifted.*

Several other techniques are employed in optimized BDD packages [33].

12.6 Advanced techniques

In this section, several advances that further improve the effectiveness of sifting are reviewed.

12.6.1 Symmetric variable and group sifting

As indicated in Examples 12.1 and 12.3, and as has been identified in empirical studies, it has emerged as a general principle that symmetric variables (pairs or larger groups) should be adjacent in the variable ordering. *Symmetric variable group sifting* extends the basic sifting approach so that two variables are tested for symmetry when they become adjacent during the sifting procedure. If they are symmetric, they are grouped and sifting continues moving groups rather than individual variables. Note that symmetry is a transitive relation for totally specified functions and transitivity thus applies in this case.

Detection of symmetry between an arbitrary pair of variables from the BDD structure is an interesting problem in its own right. Since only variables adjacent in the variable ordering are tested in symmetric variable sifting, the problem is in fact considerably simpler than the general case.

Definition 12.1 *A totally-specified switching function f is* ***positive symmetric*** *in x_i and x_j if*

$$\begin{aligned} &f(x_1, x_2, ..., x_{i-1}, 0, x_{i+1}, ..., x_{j-1}, 1, x_{j+1}, ..., x_n) \\ &= f(x_1, x_2, ..., x_{i-1}, 1, x_{i+1}, ..., x_{j-1}, 0, x_{j+1}, ..., x_n) \end{aligned}$$

Definition 12.1 is the conventional definition of symmetry between a pair of function variables. In this case, the function is invariant under the interchange of x_i and x_j. The qualifier *positive* is used here to distinguish this case from *negative* symmetry (Definition 12.2). Negative symmetry means the function is invariant under the interchange of x_i and $\overline{x}_j$, or equivalently $\overline{x}_i$ and x_j. In the literature, positive and negative symmetry are also termed *equivalence* and *nonequivalence* symmetry, respectively.

Definition 12.2 *A totally-specified switching function f is* ***negative symmetric*** *in x_i and x_j if*

$$\begin{aligned} &f(x_1, x_2, ..., x_{i-1}, 0, x_{i+1}, ..., x_{j-1}, 0, x_{j+1}, ..., x_n) \\ &= f(x_1, x_2, ..., x_{i-1}, 1, x_{i+1}, ..., x_{j-1}, 1, x_{j+1}, ..., x_n) \end{aligned}$$

Figure 12.4 shows the basic tests for (a) positive and (b) negative symmetry between a pair of variables adjacent in the variable order. Note that both conditions can exist simultaneously. Also the edges shown as dangling can point to any sub-diagram including the one shown, in which case one of the

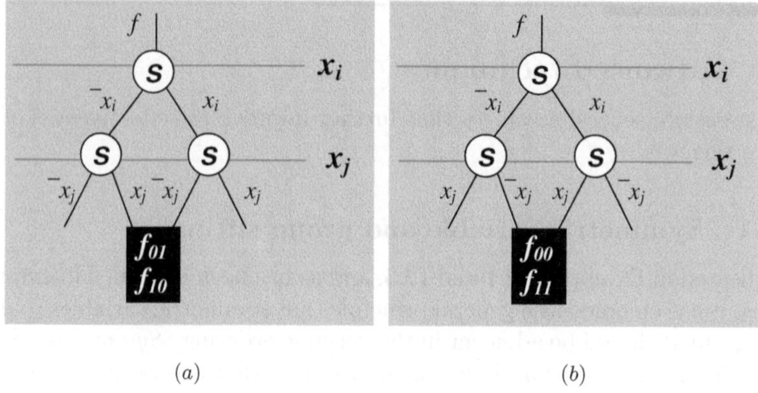

FIGURE 12.4
Symmetry conditions for adjacent variables.

nodes labelled x_j is redundant and would not in fact be present. All edges can clearly not lead to a single node as that would make all three nodes shown redundant. These tests are readily extended to include diagrams with complemented edges. Note that the numbers of complements along the two paths associated with the symmetry test must both be even, or both be odd.

The required condition must hold across the diagram, i.e., it must be satisfied for every node controlled by x_i. Hence, group sifting is based on parallel symmetry detection for all output functions in a shared decision diagram. A major advantage is that the symmetry detection is a local operation involving the nodes that are already considered in interchanging the variables. Detection of symmetry thus adds little overhead to the sifting process. Some complexity is added in that groups rather than individual variables must be interchanged but the added work is not excessive.

Two methods for aggregating variable groups are used:

- *Continuous aggregation method*: during the sifting process if two variables come together and meet the aggregation criteria, they are immediately locked into a group and remain together from that point forward.
- *Relative absolute position method*: an unsifted variable is sifted to its best position using the normal sifting method. It is then checked to see if it should be grouped with its neighbors.

The aggregation criteria used in decision diagram packages supporting group sifting typically allow for the grouping of variables based on both positive and negative symmetry and also allow for up to 10% violation in checking for symmetry across the two variables, i.e., a somewhat *relaxed* symmetry check is applied.

Many packages support aggregating variable groups based on criteria other than symmetry. For example, the user may choose to keep a set of control selection variables together in a group rather than letting them intermix with the data variables.

The computation time for functions with a large number of inputs can be quite high but it is important to note that the time penalty introduced by group sifting is small for relative absolute position method, compared to the advantage gained in reducing BDD size. The continuous aggregation method often uses less total CPU time than regular sifting. This is because groups are formed immediately so in total there are less objects to sift, albeit each object can itself be a bit more work to sift.

12.6.2 Linear sifting

In *linear sifting* use of the simple interchange of two adjacent variables x_i and x_j is replaced by the following:

Linear sifting algorithm

Step 1 Variables x_i and x_j are interchanged. Let k_1 be the number of nodes in the BDD after this interchange.

Step 2 Apply the linear transformation $x_j \leftarrow x_i \ominus x_j$. Let k_2 be the number of nodes in the BDD after this transformation.

Step 3 If $k_1 \leq k_2$ then the transformation is undone. Undoing the transformation is accomplished by simply reapplying it since it is self-inverse.

Figure 12.5 illustrates the two basic operations used in linear sifting. Figure 12.5a shows the BDD structure before transformation. Figure 12.5b shows the effect of interchanging x_i and x_j which is to interchange the subfunctions f_{01} and f_{10}, as discussed earlier. Figure 12.5c shows the effect of subsequently applying $x_j \leftarrow x_i \ominus x_j$ which is to interchange the subfunctions f_{00} and f_{01} in Figure 12.5b.

The function represented by each of the diagrams in Figure 12.5 is the same and the representation of the subfunctions f_{00} through f_{11} are not affected by the transformations. Hence linear transformation is a local operation affecting only two adjacent levels in the BDD in the same manner as discussed earlier for the interchange of variables. It is straightforward to verify that complemented edge changes are confined to those two levels as well.

Symmetry would suggest application of the transformation $x_i \leftarrow x_i \ominus x_j$ should be considered. In fact, it is, since sifting will encounter the variables in the two possible orderings. Trying both $x_j \leftarrow x_i \ominus x_j$ and $x_i \leftarrow x_i \ominus x_j$ for

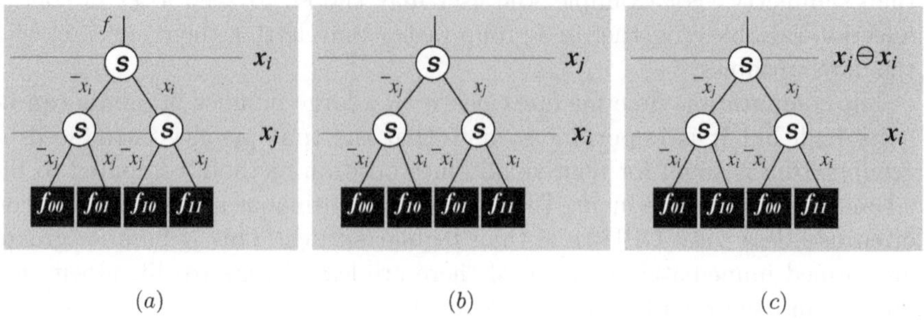

(a) (b) (c)

FIGURE 12.5
Basic operations in linear sifting.

each orientation duplicates effort.

The algorithm for linear sifting of decision diagrams was developed by Meinel et al. [26]. This algorithm is based on the class of linear transformations of variables (see details in Chapter 14). A central procedure in linear sifting implementation is the interaction matrix which shows whether there is a function in BDD which depends on variables x_i and x_j.

12.6.3 Window optimization

Linear sifting can be extended to window optimization. The difference is that while linear sifting considers a single variable in each step, in window optimization, linear transformations are considered across a window of adjacent variables.

The basic approach is:

- *First apply linear sifting to get a good initial solution.*
- *A window of size 2 is then passed across the BDD looking for liner transformations to improve the solution. Then sifting is called.*
- *If sifting improves the solution, the window is passed across the BDD again.*
- *If sifting does not improve the size, the window is increased in size by 1 and the procedure repeats.*

This continues up to a maximum window size. Naturally, the larger the window, the more time consuming is the method. Windows up to size 5 are feasible.

12.7 Path length

Logic simulation and software synthesis applications are affected as much, or more, by the path length as they are by the node count since the evaluation of a function represented by a BDD is proportional to the path length.

Definition 12.3 *The* average path length *(APL) of a BDD is the sum of the lengths of the paths for the* 2^n *assignments to the variables divided by* 2^n.

A sifting style variable reordering method can be used to optimize the APL. The key factor to the efficiency of the approach is a method for updating the value of the APL of the BDD when two adjacent variables are interchanged that involves the nodes being swapped and does not require additional traversals of the BDD.

A good initial ordering is an important starting point. One approach uses the first-order Rademacher–Walsh spectral coefficients [19]. Each first-order coefficient is a measure of correlation between a variable and the function(s). The initial variable ordering is found by arranging the variables by decreasing absolute value of the first-order coefficients. Ties are resolved by arbitrary choice.

Definition 12.4 *The* node traversal probability *is the fraction of the* 2^n *assignments of values to the variables whose path includes the node.*

The following theorem from [39] connects node traversal probability to APL.

Theorem 12.1 *The APL of a BDD is the sum of the node traversal probabilities of the non-terminal nodes of the BDD.*

Definition 12.5 *The* edge traversal probability *is the fraction of the* 2^n *assignments of values to the variables whose path includes the edge.*

When adjacent variables are swapped, the traversal probability of certain nodes is increased and the traversal probability of others is decreased. The nodes affected are those labelled by the variables being swapped and their immediate descendants. The effects are easily computed using the appropriate node and edge traversal probabilities. Since Theorem 12.1 shows the APL is just the sum of the non-terminal node traversal probabilities, the APL is readily recomputed by adding and subtracting the changes appropriately.

Given the above, the process proceeds as regular sifting except that a lower bound on APL, rather than a lower bound on node count, is used to avoid unnecessary swaps. This bound restricts how deep a variable has to be sifted down in the BDD to minimize the APL.

12.8 Reordering in BDD packages

The commonly available BDD packages support variable reordering to varying degrees. Somenzi's CUDD package [23], that is very widely used, supports a particularly large number of schemes which can be called explicitly by the application program or applied dynamically. The methods supported include:

- *Random*: Pairs of variables are randomly chosen, and swapped in the order. The swap is performed by a series of adjacent variable interchanges. The best order among those obtained by the series of swaps is retained. The number of pairs chosen for swapping equals the number of variables in the diagram.
- *Random pivot*: Same as random, but the two variables are chosen so that the first is above the variable with the largest number of nodes, and the second is below that variable. In case there are several variables tied for the maximum number of nodes, the one closest to the root is used.
- *Sifting*: This method is an implementation of Rudell's sifting algorithm [38]. There is a limit on the number of variables that will be sifted. If the diagram grows too much while moving a variable up or down, that movement is terminated before the variable has reached one end of the order. Sifting to convergence is also supported.
- *Symmetric sifting*: This implements symmetric sifting as described above and in more detail studied by Panda et al. [32]. Symmetric sifting to convergence is supported.
- *Group sifting*: Group sifting [33] and group sifting to convergence are provided.
- *Window permutation*: Window permutation [13, 20] is supported for windows of size 2, 3 and 4. Again, convergence is provided as an option.
- *Simulated annealing*: This method implements a simulated annealing approach to variable reordering. It can be very slow.
- *Genetic algorithm*: This method implements a genetic algorithm approach to variable reordering and is also potentially very slow.
- *Exact reordering*: A dynamic programming approach [18] is applied for exact reordering [12, 20, 23]. It is relatively memory efficient but is not recommended for more than 16 variables.

Another widely used suite, the BuDDy package, created by J. Lind-Nielsen [24], supports window permutation reordering, with window sizes of 2 or 3, and sifting. Reordering to convergence is supported for each of these methods. Tree-based user specified grouping of variables is also supported.

JavaBDD [21] is based on BuDDy and supports the same variable reordering options. It is written and accessible in JAVA whereas BuDDy is written in C++.

The CMU BDD library is heavily used for model checking [7]. It also supports window permutation reordering and sifting.

It should be noted that some packages, e.g. JDD [22], do not implement dynamic variable reordering since the intended domains of application for those package do not seem to benefit sufficiently to justify the associated cost.

12.9 Further reading

Variable reordering in BDD is known to be NP-complete (see, for example, papers by Bollig et al. [1, 2]. The complexity of the problem can be fully appreciated by studying the work on using BDD in the representation of multipliers, in particular, by Bryant [2], Burch [4], and Scholl and Becker [16].

The variable reordering problem has been well studied and BDD packages such as those noted above offer very robust and effective implementations. For many applications, reordering can be considered an automatic feature of the package and largely ignored by the user. Consult the documentation for the chosen decision diagram package if specific detail is required.

Exact BDD minimization results are given in [8] for a set of, not surprisingly, relatively small benchmark functions. Even at that the execution time and storage required is close to prohibitive for some of the problems. These tests were performed using the CUDD package by Somenzi [23].

Constructing a BDD from a fan-out free circuit specification is discussed in [8]. Fujita et al. [14], Malik et al. [25] and Minato et al. [27] have adapted the depth-first search approach to circuits with fan-out. Other approaches introducing a variety of heuristics are presented in [14, 25, 37].

Window permutation methods are studied by Fujita et al. [13] and Ishiura et al. [20]. Empirical results are presented in [8].

Sifting was initially introduced by Rudell [38] in 1993 and has undergone a number of refinements. It is now included in most BDD software packages [7, 21, 23]. Symmetric sifting was developed by Panda et al. [32].

Group sifting. Panda and Somenzi [33] introduced two methods for aggregating variable groups. Group sifting is a major feature of the variable reordering options CUDD package by Somenzi [23]. In [8], the use of symmetric variable groups for both window permutation and sifting have considered.

Panda and Somenzi [33] presented results for sifting with *relaxed symmetric variable groups* for a set of benchmark functions including examples with more than 600 variables. Further, results reported in [8, 33] indicate that relaxed symmetry tests outperform strict symmetry tests. Panda and Somenzi [33] also suggest extending the idea of group sifting to allow for groups to be specified based on other user identified properties, an approach adopted in various decision diagram packages. In this approach, the user specifies a tree structure that permits the specification that certain variables be grouped, and hence remain adjacent in the variable order. One can also specify that certain groups should remain adjacent.

Lower bounds for sifting. Lower bound technique for sifting was studied, in particular, by Meinel et al. [26]. During sifting, the bounds are used to abort moving a variable in a particular direction if that cannot yield an improvement over the best BDD size seen before. Heuristic adjustments relaxing the lower bound to further speedup sifting are described in [10].

Empirical results are given in [10] comparing sifting and sifting incorporating the above lower bounds. For a large set of benchmarks including the ISCAS85 functions and a number of the ISCAS89 functions, the lower bounds reduced the adjacent variable interchanges on average by 53.4% and reduced the computation time by 69.1%. Using a relaxed lower bound, Drechsler et al. were able to further reduce the sifting time by a factor of 7 over regular sifting but with only a 1% penalty in the BDD size compared to sifting with no lower bound.

Linear sifting was introduced by Meinel, Somenzi and Theobold [26] and further discussed by Günther and Drechsler [15, 16, 17]. Meinel et al. [26] have presented results for linear sifting for 77 benchmark functions, three with more than 1400 variables. The aggregate improvement over all the benchmarks is 13%.

When linear transformations are applied they must be recorded in order to later apply the appropriate mapping of input variables to BDD variables. This can be done with a second decision diagram. For example, in [16], a linear sifting results for a set of 14 benchmarks are presented. On average, the size of the decision diagram representing the linear transformations is 3.3% of the size of the linear sifted BDD.

Multivalued decision diagrams. Sifting is readily extended to MDD [9, 29]. Miller and Drechsler [16] have also extended the method of linear sifting to the multivalued case, replacing the linear exclusive-OR or equivalence function (depending on the normalization rules used) with various generalizations for multivalued logic functions. It is found that replacing exclusive-OR with sum mod-p is as effective as the alternatives tried at significantly less computational cost.

Average path length minimization has been considered in [5, 13]. In [13], Nagayama et al. introduce a sifting style variable reordering method where the minimization metric is the APL rather than the node count.

Results are given in [13] for a set of commonly used benchmark problems including the ISCAS 85 benchmarks [4]. For each circuit, each output is minimized independently and the sum of the APL over all outputs is reported. The APL is reduced on average across the benchmarks by 15%. In doing so, the number of nodes is increased sometimes dramatically. The method is quite efficient in terms of computation time.

Experimental results are also given by Nagayama et al. [13] for a set of common benchmark functions and compared to the results from an earlier path length minimization approach by Liu et al. The number of nodes is improved by 16% on average and the average path length is improved by 4% on average.

The method in [13] is effective in reducing APL quite efficiently. The fact that sifting to reduce APL often increases node count, indicates that traditional sifting to minimize node count does not guarantee a good APL. Hence, if APL is important to a particular application, it should be used as the metric for variable reordering.

Implementation: Readers interested in implementation details of decision diagram packages and reordering methods in particular are directed to the foundational paper by Brace et al. [3] and the papers by Ranjan et al. [34, 35] which presents approaches to highly effective memory utilization and to variable reordering in a breadth-first oriented decision diagram package. Details can be found for each of the mentioned BDD software packages [7, 21, 22, 23]. The programmer's manual provided with the CUDD package by Somenzi [23] is particularly informative.

Implementation details of an MDD package are discussed in [9, 28, 29].

References

[1] Bollig B, Savicky P, and Wegner I. On the improvement of variable orderings for OBDD. In *Proceedings of the IFIP Workshop on Logic and Architecture Synthesis*, pp. 71–80, 1994.

[2] Bollig B and Wegner I. Improving the variable ordering of OBDDs is NP-complete. *IEEE Transactions on Computers*, 45(9):993–1002, 1996.

[3] Brace K, Rudell R, and Bryant RE. Efficient implementation of a BDD Package. In *Proceedings of the ACM/IEEE Design Automation Conference*, pp. 40–45, 1990.

[4] Brglez F and Fujiwara H. Neutral netlist of ten combinational benchmark circuits and a target translator in FORTRAN. In *Proceedings of the International Symposium on Circuits and Systems*, pp. 663–698, 1985.

[5] Bryant RE. On the complexity of VLSI implementations and graph representations of Boolean functions with application to integer multiplication. *IEEE Transactions on Computers*, 40(2):205–213, 1991.

[6] Burch JR. Using BDDs to verify multipliers. In *Proceedings of the ACM/IEEE Design Automation Conference*, pp. 408–412, 1991.

[7] CMU BDD. http://www-2.cs.cmu.edu/ modelcheck/bdd.html, 2005.

[8] Drechsler R and Becker B. *Binary Decision Diagrams: Theory and Implementation.* Kluwer, Dordrecht, 1998.

[9] Drechsler R and Miller DM. Implementing a multiple-valued decision diagram package. In *Proceedings of the IEEE International Symposium on Multiple-Valued Logic*, pp. 52–57, 1998.

[10] Drechsler R, Günther W, and Somenzi F. Using lower bounds during dynamic BDD minimization. *RANSCAD*, 20(1):51–57, 2001.

[11] Ebendt R, Hoehme S, Günther W, and Drechsler R. Minimization of the expected path length in BDDs based on local changes. In *Proceedings of the ASP-DAC*, pp. 866–871, 2004.

[12] Friedman SJ and Supowit KJ. Finding the optimal variable ordering for binary decision diagrams. *IEEE Transactions on Computers*, 39(5):710–713, 1990.

[13] Fujita M, Matsunaga Y, and Kakuda T. On variable ordering of binary decision diagrams for the application of multi-level synthesis. In *Proceedings of the European Conference on Design Automation*, pp. 50–54, 1991.

[14] Fujita M, Fujisawa H, and Kawato N. Evaluation and improvements of Boolean comparison method based on binary decision diagrams. In *Proceedings of the International Conference on Computer-Aided Design*, pp. 2–5, 1988.

[15] Günther W and Drechsler R. Linear transformations and exact minimizations of BDDs. In *Proceedings of the Great Lakes Symposium on VLSI*, pp. 325–330, 1998.

[16] Günther W and Drechsler R. BDD minimization by linear transformations. In *Proceedings of the Conference on Advanced Computer Systems*, pp. 525–532, 1998.

[17] Günther W and Drechsler R. Minimization of BDDs using linear transformations based on evolutionary techniques. In *Proceedings of the International Symposium on Circuits and Systems*, pp. 387–390, 1999.

[18] Held M and Karp RM. A dynamic programming approach to sequencing problems. *SIAM*, 10(1):196–210, 1962.

[19] Hurst SL, Miller DM, and Muzio JC. *Spectral Techniques in Digital Logic.* Academic Press, Orlando, Florida, 1985.

[20] Ishiura N, Sawada H, and Yajima Z. Minimization of binary decision diagrams based on exchanges of variables. In *Proceedings of the International Conference on Computer-Aided Design*, pp. 472–475, 1991.

[21] JavaBDD, http://javabdd.sourceforge.net/, 2005.

[22] JDD. http://javaddlib.sourceforge.net/jdd/. 2005.

[23] Jeong SW, and Kim TS, and Somenzi F. An efficient method for optimal BDD ordering computation. In *Proceedings of the International Conference on VLSI and CAD*, 1993.

[24] Lind-Nielsen J. BuDDy. http://sourceforge.net/projects/buddy, 2005.

[25] Malik S, Wang AR, Brayton RK, and Sangiovanni-Vincentelli AL. Logic verification using binary decision diagrams in a logic synthesis environment. In *Proceedings of the International Conference on Computer-Aided Design*, 1988.

[26] Meinel C, Somenzi F, and Theobold T. Linear sifting of decision diagrams. In *Proceedings of the ACM/IEEE Design Automation Conference*, pp. 202–207, 1997.

[27] Minato S, Ishiura N, and Yajima S. Shared binary decision diagrams with attributed edges for efficient Boolean function manipulation. In *Proceedings of the ACM/IEEE Design Automation Conference*, pp. 52–57, 1990.

[28] Miller DM. Multiple-valued logic design tools. In *Proceedings of the IEEE 23rd International Symposium on Multiple-Valued Logic*, pp. 2–11, 1993.

[29] Miller DM and Drechsler R. Further improvements in implementing MVDDs. In *Proceedings of the IEEE 32nd International Symposium on Multiple-Valued Logic*, pp. 245–253, 2002.

[30] Miller DM and Drechsler R. Augmented sifting of multiple-valued decision diagrams. In *Proceedings of the IEEE 33rd International Symposium on Multiple-Valued Logic*, pp. 275–282, 2003.

[31] Nagayama S, Mishchenko A, Sasao T, and Butler JT. Minimization of average path length in BDDs by variable reordering. In *Proceedings of the International Workshop on Logic Synthesis*, pp. 28–30, 2003.

[32] Panda S, Somenzi F, and Plessier BF. Symmetry detection and dynamic variable ordering of decision diagrams. In *Proceedings of the International Conference on Computer-Aided Design*, 628—631, 1994.

[33] Panda S and Somenzi F. Who are the variables in your neighborhood. In *Proceedings of the International Conference on Computer-Aided Design*, pp. 74–77, 1995.

[34] Ranjan R, Sanghavi J, Brayton R, and Sangiovanni-Vincentelli A. High performance BDD package based on exploiting memory hierarchy. In *Proceedings of the ACM/IEEE Design Automation Conference*, pp. 635–640, 1996.

[35] Ranjan RK, Gosti W, Brayton RK, Sangiovanni-Vincenteili A. Dynamic reordering in a breadth-first manipulation based BDD package: challenges and solutions. In *Proceedings of the International Conference on Computer-Aided Design*, p. 944, 1997.

[36] Reingold EM, Nievergelt J, and Deo N. *Combinatorial Algorithms – Theory and Practice.* Prentice-Hall, 1977.

[37] Ross DE, Butler KM, Kapur R, and Mercer MR. Fast functional evaluation of candidate OBDD variable ordering. In *Proceedings of the European Conference on Design Automation*, pp. 4–9, 1991.

[38] Rudell R. Dynamic variable ordering for ordered binary decision diagrams. In *Proceedings of the International Conference on Computer-Aided Design*, pp. 42–47, 1993.

[39] Sasao T, Iguchi Y, and Matsura M. Comparison of decision diagrams for multiple-output logic functions. In *Proceedings of the International Workshop on Logic Synthesis*, pp. 379–384, 2002.

[40] Scholl C and Becker B. On the generation of multiplexor circuits for pass transistor logic. In *Proceedings of the Design and Test in Europe*, pp. 372–378, 2000.

[41] Somenzi F. CUDD: CU Decision Diagram Package. University of Colorado at Boulder, http://vlsi.colorado.edu/~fabio/CUDD/, 2005.

13

Spectral Decision Diagrams

In this chapter, the interpretation of decision trees and diagrams in terms of spectral techniques is considered. This description is called *spectral decision diagrams*. The basic statements of signal processing are as follows:

- Signals are physical conveyors of information.
- Signals are processed to extract the information in them.
- The processing is based on mathematical models of signals and transforms, including spectral techniques.
- Functions are mathematical models of signals.

13.1 The theoretical background of spectral interpretation of decision diagram techniques

Many problems can be solved by transferring the problem in the original space isomorphically to some other space that reflects particular properties of the problem. Thus, in the new space, the problem is simpler or there are at least well known solution tools. Once the problem is solved, we have to be able to return to the original space. One of the possible ways to solve the problem in the above formulation is the spectral techniques (Figure 13.1).

The theoretical background of spectral interpretation of decision diagrams is Fourier-like transform techniques (see Chapter 7).

Let $P(G)$ be a finite-dimensional vector space consisting of g elements. Each set $\Phi = \{\phi_i\}$, $i = 0, 1, \ldots, g-1$, of g linearly independent functions is a basis in $P(G)$. Thus, each function $f \in P(G)$ can be represented as a linear combination

$$f = \sum_{x \in G} s_i \phi_i(x)$$

where $s_i \in P$. No function from a set of linearly independent functions $\Phi = \{\phi_i\}$ cannot be reduced to some other function from this set by some simple linear operations over this and other functions.

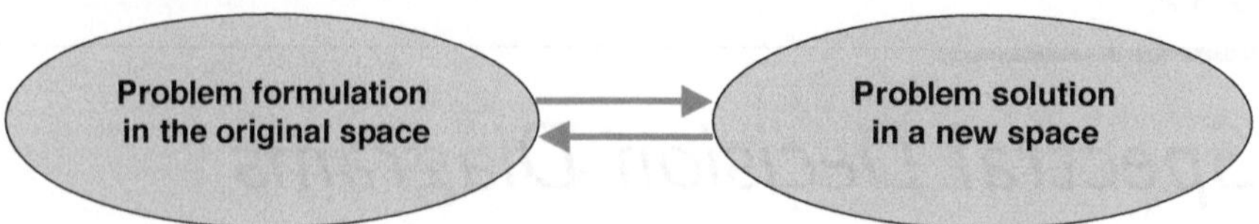

FIGURE 13.1
Transferring the problem in the original space to another space that reflects the particular properties of the problem.

Example 13.1 *Using the addition of some functions in the set $\Phi = \{\phi_i\}$, subtraction, multiplication by a constant and addition or subtraction, etc., it is impossible to reduce this set.*

13.2 Decision tree and spectrum of switching function

Let a switching function f and its decision tree be given, and let $\mathbf{Q}(n)$ be a $(2^n \times 2^n)$ non-singular matrix with elements in P. Thus, the columns of $Q(n)$ are a set of linearly independent functions. Since there are 2^n such functions, $\mathbf{Q}(n)$ determines a basis in $P(C_2^n)$.

Decision trees are graphical representations of spectral transform expansions of $f \in P(G)$ with respect to a basis Q in $P(G)$. In a decision tree, each path from the root node to the constant nodes corresponds to a basic function in Q. The constant nodes represent the Q-spectral coefficients for f.

Theorem 13.1 *A decision tree defined with respect to a basis Q represents at the same time a switching function f and the Q-spectrum of f.*

The proof follows from the relationship of decision trees and spectra, and from the existence of direct and inverse spectral transforms. This theorem yields the following statement:

Each decision tree representing a switching function f can be considered as a Shannon tree representing the Q-spectrum of f.

Example 13.2 *Figure 13.2 shows a decision tree that represents at the same time a switching function and its Q-spectrum.*

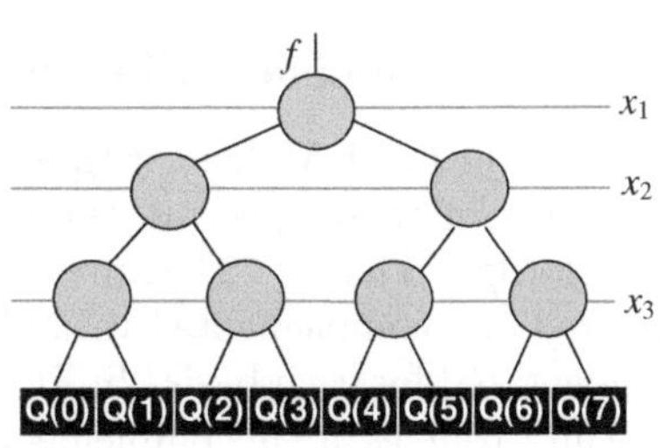

- The decision tree represents a switching function f of three variables, x_1, x_2, and x_3
- There are 8 paths from the root node to 8 terminal nodes
- Each path corresponds to a basic function in Q
- Each terminal node corresponds to a Q-spectal coefficient
- The decision tree represents a Q-spectrum of the switching function f

FIGURE 13.2
A Decision tree is a graphical representation of a spectral transform expansions of switching function f with respect to a basis Q (Example 13.2).

Decision diagram and spectrum of a switching function. Spectral interpretation of BDDs and FDDs implies definition of spectral transform decision trees (STDTs) as a concept involving different decision diagrams for particular specification of the basis Q.

Suppose that a non-singular matrix $\mathbf{Q}(n)$ is represented as the Kronecker product of n factors $\mathbf{Q}(1)$:

$$\mathbf{Q}(n) = \bigotimes_{i=1}^{n} \mathbf{Q}(1).$$

For a basis Q, the i-th basic matrix $\mathbf{Q}(1)$ defines the expansion of f with respect to the i-th variable

$$f = \mathbf{Q}^{-1}(1)\mathbf{Q}(1) \begin{bmatrix} f_0 \\ f_1 \end{bmatrix},$$

where $\mathbf{Q}^{-1}$ is the matrix inverse of $\mathbf{Q}(1)$.

Definition 13.1 *A spectral transform decision tree (STDT) is a decision tree assigned to f by the decomposition of f with respect to the basis Q.*

Definition 13.2 *Spectral transform decision diagrams (STDDs) are decision diagrams derived by the reduction of the corresponding STDTs.*

Reduction of STDTs is performed by deleting and sharing isomorphic subtrees. It is obvious that the definition of STDTs satisfies the requirements in Theorem 7.3.1 and that different decision diagrams are involved in STDTs for different specifications of the matrices $\mathbf{Q}(1)$.

13.3 Bases of spectral transforms and decision diagrams

Figure 13.3 shows three spectral representations in Reed–Muller, arithmetic, and Walsh bases of a switching function. Each of these spectral representations can be extended with respect to polarity. That is, 2^n expressions with different combinations of uncomplemented and complemented variables, where n is the number of variables, can be generated for each basis. In Figure 13.3, two particular cases are depicted where all variables are uncomplemented (positive polarity) and all variables are complemented (negative polarity).

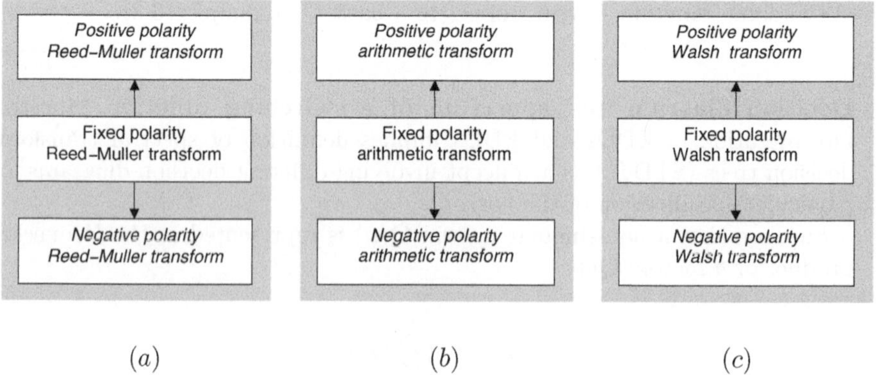

FIGURE 13.3
Fixed polarity Reed–Muller (a), arithmetic (b) and Walsh (c) transforms include positive (all variables are uncomplemented) and negative (all variables are complemented) transforms.

13.4 Fixed polarity spectral representations

Each switching function can be represented in a fixed polarity form using Reed–Muller, arithmetic, or Walsh transforms. The coefficients of the Reed–Muller, arithmetic, and Walsh expression are composed of the spectral coefficients of a Reed–Muller, arithmetic, and Walsh spectral transform, correspondingly.

Fixed polarity Reed–Muller representation. For a given positive-polarity Reed–Muller spectrum, the Reed–Muller spectrum for the polarity

$$H = (h_1; , \dots, h_n)$$

is determined by the permutation of the i-th Reed–Muller coefficient into the

$$(i_1 \oplus \overline{h}_1, \dots, i_n \oplus \overline{h}_n)$$

coefficient, since $\mathbf{R}_H(n)$ is derived from $\mathbf{R}(n)$ by the permutation of columns in which the i-th column is shifted to the position $(i_1 \oplus h_1 \dots i_n \oplus h_n)$. However, for a given switching function f of n-variables, different polarity Reed–Muller matrices can be used.

> **Example 13.3** *Figure 13.4 shows the construction of the Reed–Muller transform matrix for* $n = 3$ *and the polarity vector* $H = (0, 1, 0)$ *from the positive polarity* $H = (0, 0, 0)$ *Reed–Muller matrix* $\mathbf{R}^{(0,0,0)}(3)$ *using Hadamard ordering for product terms.*

The Reed–Muller matrix for polarity $H = (0, 0, 0)$

$$\mathbf{R}^{(0,0,0)}(3) = \left[\begin{array}{cc|cc|cc|cc} 1 & 0 & 0 & 0 & 0 & 0 & 0 & 0 \\ 1 & 1 & 0 & 0 & 0 & 0 & 0 & 0 \\ \hline 1 & 0 & 1 & 0 & 0 & 0 & 0 & 0 \\ 1 & 1 & 1 & 1 & 0 & 0 & 0 & 0 \\ \hline 1 & 0 & 0 & 0 & 1 & 0 & 0 & 0 \\ 1 & 1 & 0 & 0 & 1 & 1 & 0 & 0 \\ \hline 1 & 0 & 1 & 0 & 1 & 0 & 1 & 0 \\ 1 & 1 & 1 & 1 & 1 & 1 & 1 & 1 \end{array}\right]$$

The Reed–Muller matrix for polarity $H = (0, 1, 0)$

$$\mathbf{R}^{(0,1,0)}(3) = \left[\begin{array}{cc|cc|cc|cc} 0 & 0 & 1 & 0 & 0 & 0 & 0 & 0 \\ 0 & 0 & 1 & 1 & 0 & 0 & 0 & 0 \\ \hline 1 & 0 & 1 & 0 & 0 & 0 & 0 & 0 \\ 1 & 1 & 1 & 1 & 0 & 0 & 0 & 0 \\ \hline 0 & 0 & 1 & 0 & 0 & 0 & 1 & 0 \\ 0 & 0 & 1 & 0 & 0 & 0 & 1 & 1 \\ \hline 1 & 0 & 1 & 0 & 1 & 0 & 1 & 0 \\ 1 & 1 & 1 & 1 & 1 & 1 & 1 & 1 \end{array}\right]$$

The indices of columns in $R^{(0,1,0)}(3)$ *are defined as*

$$(i_1 \oplus h_1; i_2 \oplus h_2; i_3 \oplus h_3)$$

This implies:

$$(0; 1; 2; 3; 4; 5; 6; 7) \rightarrow (2; 3; 0; 1; 6; 7; 4; 5)$$

Given a switching function f *given by the truth-vector*

$$\mathbf{F} = [10010111]^T,$$

the Reed–Muller expansion for $H = (0, 1, 0)$ *is the expression*

$$f = x_3 \oplus \overline{x}_2 \oplus x_1 \oplus x_1 x_3 \oplus x_1 \overline{x}_2 x_3$$

FIGURE 13.4
Construction of a Reed–Muller transform matrix of polarity two from polarity zero (Example 13.3).

Table 40.1 summarizes the bases allowed and compares permitted changes in basic functions, in terms of which various AND-EXOR expressions and corresponding decision trees for functions in $GF(C_2^n)$ are defined.

TABLE 13.1
Bases of spectral transform used in the AND-EXOR expressions and related decision diagrams.

Type	Specification
Binary decision diagram, BDD	Trivial basis Fixed form of functions Fixed order of functions Kronecker product representable
Positive polarity Reed–Muller decision diagrams, PPRM	Reed–Muller basis Fixed form of functions Fixed order of functions Kronecker product representable
Fixed polarity Reed–Muller decision diagrams	Reed–Muller basis Fixed form of functions Changeable depending on levels Permuted Reed–Muller
Pseudo–Reed–Muller decision diagram	Pseudo–Reed–Muller basis Fixed form of functions Changeable order of function depending on levels Permuted and cyclic-shifted Reed—Muller
Generalized Reed–Muller decision diagram	Combined Reed–Muller basis Fixed form of functions Changeable order of functions depending on levels Combination of the permuted and Reed–Muller basis for different polarities
Kronecker decision diagram	Combined Reed–Muller and trivial bases depending on levels Changeable form of functions depending on levels Changeable order of functions depending on levels Kronecker product representable
Pseudo–Kronecker decision diagram	Pseudo–Kronecker (KRO) basis Changeable form of functions depending on levels Changeable order of functions depending on levels Permuted and cyclic shifted KRO basis

Fixed polarity arithmetic representation. The factors of the arithmetical expression are the coefficients derived from an arithmetic spectrum.

> **Example 13.4** *Figure 13.5 shows the construction of the arithmetic transform matrix for $n = 3$ and the polarity vector $H = (0, 1, 0)$ from the positive polarity $H = (0, 0, 0)$ arithmetic matrix $\mathbf{R}^{(0,0,0)}(3)$ using Hadamard ordering for product terms.*

The arithmetic matrix for polarity

$$H = (0, 0, 0)$$

$$\mathbf{A}^{(0,0,0)}(3) = \left[\begin{array}{rr|rr|rr|rr} 0 & 0 & 1 & 0 & 0 & 0 & 0 & 0 \\ 0 & 0 & -1 & 1 & 0 & 0 & 0 & 0 \\ \hline 1 & 0 & -1 & 0 & 0 & 0 & 0 & 0 \\ -1 & 1 & 1 & -1 & 0 & 0 & 0 & 0 \\ \hline 0 & 0 & -1 & 0 & 0 & 0 & 1 & 0 \\ 0 & 0 & 1 & -1 & 0 & 0 & -1 & 1 \\ \hline -1 & 0 & 1 & 0 & 1 & 0 & -1 & 0 \\ 1 & -1 & -1 & 1 & -1 & 1 & 1 & -1 \end{array}\right]$$

The arithmetic matrix for polarity

$$H = (0, 1, 0)$$

$$\mathbf{A}^{(0,1,0)}(3) = \left[\begin{array}{rr|rr|rr|rr} 0 & 0 & 1 & 0 & 0 & 0 & 0 & 0 \\ 0 & 0 & -1 & 1 & 0 & 0 & 0 & 0 \\ \hline 1 & 0 & -1 & 0 & 0 & 0 & 0 & 0 \\ -1 & 1 & 1 & -1 & 0 & 0 & 0 & 0 \\ \hline 0 & 0 & -1 & 0 & 0 & 0 & 1 & 0 \\ 0 & 0 & 1 & -1 & 0 & 0 & -1 & 1 \\ \hline -1 & 0 & 1 & 0 & 1 & 0 & -1 & 0 \\ 1 & -1 & -1 & 1 & -1 & 1 & 1 & -1 \end{array}\right]$$

The indices of columns in $A^{(0,1,0)}(3)$

$$(i_1 \oplus h_1; i_2 \oplus h_2; i_3 \oplus h_3)$$

This implies:

$$(0; 1; 2; 3; 4; 5; 6; 7) \rightarrow (2; 3; 0; 1; 6; 7; 4; 5)$$

For a switching function f given by the truth-vector

$$\mathbf{F} = [10010111]^T,$$

the arithmetic expansion for $H = (0, 1, 0)$ is the expression

$$\begin{aligned} f &= ([1 \;\; 1 - 2x_1] \\ &\otimes [1 \;\; 1 - 2\overline{x}_1] \\ &\otimes [1 \;\; 1 - 2x_1]) \\ &\quad \left(\begin{bmatrix} 1 & 1 \\ 1 & -1 \end{bmatrix} \otimes \begin{bmatrix} 1 & 1 \\ -1 & 1 \end{bmatrix} \otimes \begin{bmatrix} 1 & 1 \\ 1 & -1 \end{bmatrix} \right) \mathbf{F} \\ &= 5 - (1 - 2x_3) + (1 - 2\overline{x}_2) \\ &- (1 - 2\overline{x}_2)(1 - 2x_3) \\ &- (1 - 2x_1) + (1 - 2x_1)(1 - 2x_3) \\ &- (1 - 2\overline{x}_1)(1 - 2\overline{x}_2) \\ &- 3(1 - 2\overline{x}_1)(1 - 2\overline{x}_2)(1 - 2x_3) \end{aligned}$$

FIGURE 13.5
Construction of an arithmetic transform matrix of polarity two from polarity zero (Example 13.4).

Fixed polarity Walsh representation. The Walsh spectrum's coefficients are used to construct Walsh polynomial expressions.

> **Example 13.5** *Figure 13.6 shows the construction of the Walsh transform matrix for $n = 3$ and the polarity vector $H = (0, 1, 0)$ from the positive polarity $H = (0, 0, 0)$ Walsh matrix $\mathbf{W}^{(0,0,0)}(3)$ using Hadamard ordering for product terms.*

Note that after simplifications the Walsh expression reduces to the arithmetic polynomial of the same polarity for f. Further, if the coefficients are calculated by modulo 2, and addition and subtraction formally replaced by EXOR, this expression reduces to the FPRM for f for the considered polarity.

The Walsh matrix for polarity

$$H = (0,0,0)$$

$$\mathbf{W}^{(0,0,0)}(3) = \left[\begin{array}{cc|cc|cc|cc} 0 & 0 & 1 & 0 & 0 & 0 & 0 & 0 \\ 0 & 0 & -1 & 1 & 0 & 0 & 0 & 0 \\ \hline 1 & 0 & -1 & 0 & 0 & 0 & 0 & 0 \\ -1 & 1 & 1 & -1 & 0 & 0 & 0 & 0 \\ \hline 0 & 0 & -1 & 0 & 0 & 0 & 1 & 0 \\ 0 & 0 & 1 & -1 & 0 & 0 & -1 & 1 \\ \hline -1 & 0 & 1 & 0 & 1 & 0 & -1 & 0 \\ 1 & -1 & -1 & 1 & -1 & 1 & 1 & -1 \end{array}\right]$$

The Walsh matrix for polarity

$$H = (0,1,0)$$

$$\mathbf{W}^{(0,1,0)}(3) = \left[\begin{array}{cc|cc|cc|cc} 0 & 0 & 1 & 0 & 0 & 0 & 0 & 0 \\ 0 & 0 & -1 & 1 & 0 & 0 & 0 & 0 \\ \hline 1 & 0 & -1 & 0 & 0 & 0 & 0 & 0 \\ -1 & 1 & 1 & -1 & 0 & 0 & 0 & 0 \\ \hline 0 & 0 & -1 & 0 & 0 & 0 & 1 & 0 \\ 0 & 0 & 1 & -1 & 0 & 0 & -1 & 1 \\ \hline -1 & 0 & 1 & 0 & 1 & 0 & -1 & 0 \\ 1 & -1 & -1 & 1 & -1 & 1 & 1 & -1 \end{array}\right]$$

The indices of columns in $A^{(0,1,0)}(3)$

$$(i_1 \oplus h_1; i_2 \oplus h_2; i_3 \oplus h_3)$$

This implies:

$$(0;1;2;3;4;5;6;7) \rightarrow (2;3;0;1;6;7;4;5)$$

Given a switching function f *given by the truth-vector*

$$\mathbf{F} = [10010111]^T,$$

the Walsh expansion for $H = (0,1,0)$ *given the expression*

$$\begin{aligned} f &= ([1 \;\; 1-2x_1] \\ &\otimes [1 \;\; 1-2\overline{x}_1] \\ &\otimes [1 \;\; 1-2x_1]) \\ &\left(\begin{bmatrix} 1 & 1 \\ 1 & -1 \end{bmatrix} \otimes \begin{bmatrix} 1 & 1 \\ -1 & 1 \end{bmatrix} \otimes \begin{bmatrix} 1 & 1 \\ 1 & -1 \end{bmatrix}\right)\mathbf{F} \\ &= 5 - (1-2x_3) + (1-2\overline{x}_2) \\ &- (1-2\overline{x}_2)(1-2x_3) \\ &- (1-2x_1) + (1-2x_1)(1-2x_3) \\ &- (1-2\overline{x}_1)(1-2\overline{x}_2) \\ &- 3(1-2\overline{x}_1)(1-2\overline{x}_2)(1-2x_3) \end{aligned}$$

FIGURE 13.6
Construction of the Walsh transform matrix of polarity two (Example 13.5).

13.5 Spectral decision diagram techniques

In this section, details of construction of spectral decision diagrams are given.

TABLE 13.2
Spectral representation of switching functions.

Class	Polynomial representation	Transform matrices
Positive Davio 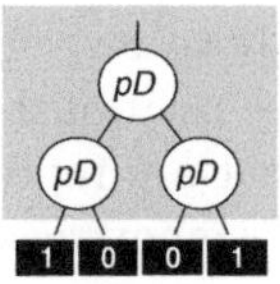	$f = \mathbf{X}_r(n)\mathbf{R}(n)\mathbf{F}$	$\mathbf{X}_r(n) = \bigotimes_{i=1}^{n} [1\ x_i]$ $\mathbf{R}(n) = \bigotimes_{i=1}^{n} \begin{bmatrix} 1 & 0 \\ 1 & 1 \end{bmatrix}$
Fixed polarity Reed–Muller 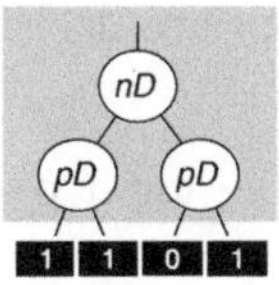	$f = \left(\bigotimes_{i=1}^{n} [1\ x_i^{h_i}]\right) \times \left(\bigotimes_{i=1}^{n} \mathbf{R}^{h_i}(1)\right)\mathbf{F}$	$\mathbf{R}_H(n) = \bigotimes_{i=1}^{n} \mathbf{R}^{h_i}(1)$ $\mathbf{R}^{h_i}(1) = \begin{cases} \begin{bmatrix} 1 & 0 \\ 1 & 1 \end{bmatrix}, & h_i = 0; \\ \begin{bmatrix} 0 & 1 \\ 1 & 1 \end{bmatrix}, & h_i = 1. \end{cases}$
Positive polarity Walsh 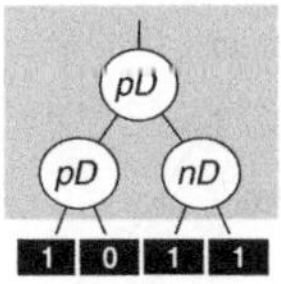	$f = 2^{-n}\mathbf{X}_w\mathbf{W}(n)\mathbf{F}$	$\mathbf{X}_w(n) = \bigotimes_{i=1}^{n} [1\ (1-2x_i]$ $\mathbf{W}(n) = \bigotimes_{i=1}^{n} \mathbf{W}(1)$ $\mathbf{W}(1) = \begin{bmatrix} 1 & 1 \\ 1 & -1 \end{bmatrix}$
Fixed polarity Walsh 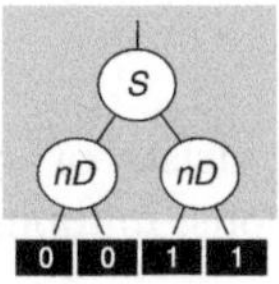	$f = \left(\bigotimes_{i=1}^{n} [1\ 1-2x_i^{h_i}]\right) \times \left(\bigotimes_{i=1}^{n} \begin{bmatrix} 1 & 1 \\ (-1)^{h_i} & (-1)^{\overline{h}_i} \end{bmatrix}\right)\mathbf{F}$	$\mathbf{W}_H(n) = \bigotimes_{i=1}^{n} \mathbf{W}^{h_i}(1)$ $\mathbf{W}^{h_i}(1) = \begin{cases} \begin{bmatrix} 1 & 1 \\ 1 & -1 \end{bmatrix}, & h_i = 0; \\ \begin{bmatrix} 1 & 1 \\ -1 & 1 \end{bmatrix}, & h_i = 1. \end{cases}$

13.5.1 Decision diagrams as relations between algebraic and graphical data structures

Figure 40.2 shows the relationship between:

- *Algebraic* representations of logic functions and the spectrum of corresponded functions,
- *Graphical data* structures of algebraic and spectral representations of logic functions.

There are a lot of non-formalized problems at each above level of relationships. Another problem is that characteristics of these relationships are strongly related to the characteristics of logic functions that are represented by decision diagrams. The above relationships are useful for many reasons, in particular,

- For classification of decision trees and diagrams,
- For the study of their properties,
- For the development of methods of choosing an appropriate decision diagram given a logic function, and
- For developing new decision diagrams that are suitable for a particular logic function.

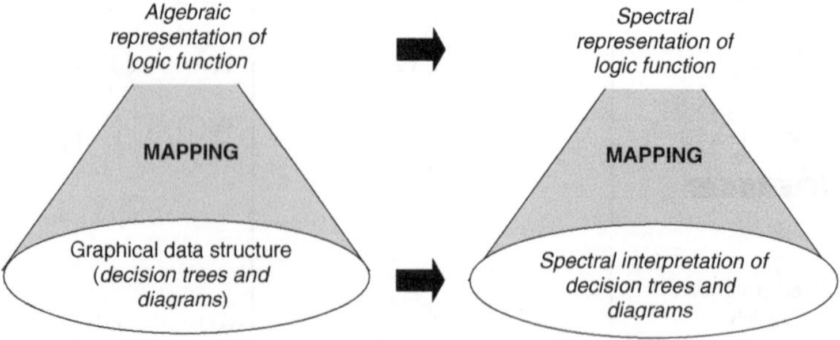

FIGURE 13.7
The spectral approach in decision diagram techniques consists of spectral interpretation of an algebraic representation of logic functions, mapping, and decision diagrams.

13.5.2 Bit-level and word-level strategy

In a bit-level decision diagram the constant nodes are logic values 0 and 1. BDDs and FDDs are particular examples of bit-level decision diagrams. A variety of bit-level decision diagrams are generated by different decomposition rules in the diagram nodes. In these decision diagrams, besides the Shannon and positive Davio decomposition, the negative Davio decomposition is used.

Algebraic decision diagrams (ADDs), and binary moment diagrams (BMDs) are examples of word-level decision diagrams derived from spectral representations (see Chapter 17 for details). A special class, linear decision diagrams (LDDs) for representation of multi-output functions (words) by linear arithmetic expressions is considered in Chapter 37.

13.6 Further reading

Fundamentals of spectral decision diagram techniques are considered in Chapter 7. Generalization for multivalued functions is given in Chapter 30. Developing new decision diagrams based on spectral techniques is considered in Chapter 40.

Hierarchies of decision trees and diagrams were studied, in particular, by Green [3], Sasao [5], Perkowski et al. [4], and Dill et al. [1]. Spectral interpretation of decision trees and diagrams is developed by Falkowski and Stanković [3], Stanković et al. [70], [47], [48], [49].

References

[1] Dill KM, Ganguly K, Safranek RJ, and Perkowski M. A new Zhegalkin Galois logic. In *Proceedings of the 3rd International Workshop on Applications of the Reed–Muller Expansion in Circuit Design*, Oxford University, UK, pp. 247–257, 1997.

[2] Falkowski BJ and Stanković RS. Spectral interpretation and applications of decision diagrams. *VLSI Design International Journal of Custom Chip Design, Simulation and Testing*, 11(2):85–105, 2000.

[3] Green DH. Families of Reed–Muller canonical forms, *International Journal of Electronics*, 2:259–280, 1991.

[4] Perkowski M, Jóźwiak L, and Drechsler R. New hierarchies of AND/EXOR trees, decision diagrams, lattice diagrams, canonical forms, and regular layouts. In *Proceedings of the 3rd International Workshop on Applications of*

the Reed–Muller Expansion in Circuit Design, Oxford University, UK, pp. 115–142, 1997.

[5] Sasao T. Representation of logic functions using EXOR operators. In Sasao T and Fujita M, Eds., *Representations of Discrete Functions*, Kluwer, Dordrecht, pp. 29–54, 1996.

[6] Stanković RS and Astola JT. *Spectral Interpretation of Decision Diagrams.* Springer, Heidelberg, 2003.

[7] Stanković RS, Sasao T, and Moraga C. Spectral transform decision diagrams. In Sasao T and Fujita M, Eds., *Representations of Discrete Functions*, Kluwer, Dordrecht, pp. 55–92, 1996.

[8] Stanković RS. *Spectral Transform Decision Diagrams in Simple Questions and Simple Answers.* NAUKA Publishers, Belgrade, Yugoslavia, 1998.

[9] Stanković RS. Non-Abelian groups in optimization of decision diagrams representations of discrete functions. *Formal Methods in System Design*, 18:209–231, 2001.

14

Linearly Transformed Decision Diagrams

An affine transform of the variables* of a switching function is aimed at the representation of a function as a superposition of:

- A system of linear switching functions and
- A residual nonlinear part of minimal complexity.

The method can be used for both bit-level and word-level polynomial expressions. In this way, the complexity of decision diagrams can be reduced. Affine transformations can also be used in dynamic reordering of variables in decision diagrams. The practical value of the affine transform of variables includes the design of fine-grain and cellar automata types of FPGAs, and programmable logic devices with EXOR gates.

14.1 Introduction

The problem of linearization can be formulated in various ways depending on the type of data structure and given criteria:

Linear topological structures. Linearization of algebraic representations of switching functions addresses the problem of word-level description of switching functions by a set of linear polynomials. These polynomials can be mapped into linear word-level decision diagrams. The problem of linearization in this formulation is aimed at construction of linear topological structures for computing switching functions and linear decision diagrams.

Reducing decision diagrams using linear algebraic transformations. Linearization can be formulated as a problem of complexity reduction in decision diagrams. The affine transform of input variables considered

*The chapter was co-authored with Dr. R. Kh. Latypov, Kazan State University, Russia.

in this chapter is one of the linearization techniques. It results in a linearly transformed decision diagram. Compared to problem of word-level linearization (Chapter 37), a linearly transformed decision diagram does not have a linear topology, but has a lower complexity in the number of nodes and edges than the original decision diagram for the same switching function.

14.1.1 Affine transforms

Affine transform of variables is a method for optimization of the various representations of switching functions.

In spectral techniques, the problem of affine transform of variables is equivalent to the problem of reduction of non-zero spectral coefficients.

Given a switching function f of n variables $x_1, \ldots, x_n$, an affine transform of variables is defined as the assignment $f_\sigma(y) = f(\sigma x)$, where σ is an affine operator over $GF(2)$ for rearrangement of input variables. The usefulness of an affine transform of variables is based on the assumption that computation of the switching function $f_\sigma(y)$ is more effective than computing f. Specifically, the problem is formulated as follows: choose an appropriate affine operator σ and construct a smaller decision diagram.

14.1.2 Linearly transformed decision diagrams

A linearly transformed decision diagram technique is based on the fundamental property of linearly dependence or independence of vectors known in linear algebra†.

Linearly transformed decision diagrams are defined by allowing linear combinations and complements of variables. This is a manipulation of variables that is more general than ordering of variables (see Chapter 12). In this way, the number of invertible affine transformations of variables is $n!$ and $2^n \prod_{i=0}^{n-1}(2^n - 2^i)$ for reordering and affine transformations of variables, respectively. Note that the second number is bigger than the first one.

Several algorithms have been proposed to determine affine transformations of variables (see "Further reading" section). Most of these algorithms exploit heuristics and, therefore, cannot guarantee optimal results. For small functions, deterministic algorithms exploit properties of switching functions in the original and spectral transform domains.

†The vectors $\mathbf{X}_1, \mathbf{X}_2, \ldots, \mathbf{X}_k$ are *linearly dependent* if there exist coefficients $a_1, a_2, \ldots, a_k$ such that $a_1\mathbf{X}_1 + a_2\mathbf{X}_2 + \cdots + a_k\mathbf{X}_k = 0$. If vectors are not linearly dependent, they are *linearly independent.* For example, vectors $\mathbf{X} = [1, 2, 3]$, $\mathbf{Y} = [2, 6, 4]$, and $\mathbf{Z} = [4, 11, 9]$ are linearly dependent because $2\mathbf{X} + 3\mathbf{Y} - 2\mathbf{Z} = 0$.

14.2 Manipulation of variables using affine transforms

In this section, a basic procedure of the method of affine transform of input variables is considered. This procedure is called *rearrangement* of variables. The rearrangement consists of several operations: reorder, complement, and EXOR of input variables of a switching function f. It results in a new switching function g.

Rearrangement of variables is performed using the following operations:

- ▶ Reordering,
- ▶ Complementing, and
- ▶ Operations over $GF(2)$.

> **Example 14.1** *Given a sequence* $\{x_1, x_2, x_3\}$, *the following rearrangement of variables can be made:*
>
> (a) *Reordering of variables* $\{x_3, x_2, x_1\}$,
> (b) *Complementing of some variable sequence* $\{\overline{x}_1, x_2, \overline{x}_3\}$,
> (c) *EXOR of variables sequence*
>
> $$\{(x_1 \oplus x_3), x_2, (x_2 \oplus x_3)\},$$
>
> (d) *Reordering with complementing and EXOR*
>
> $$\{(x_1 \oplus \overline{x}_3), \overline{x}_2, (\overline{x}_1 \oplus x_2)\}.$$

Reordering of variables. Given the vector of n variables $\mathbf{X} = [x_1, x_2, \dots, x_n]^T$ of a switching function f, reordering of variables is defined by the matrix equation

$$\mathbf{Y} = P\mathbf{X} \tag{14.1}$$

where the *permutation matrix* P has exactly one 1 in each row and in each column, and 0's elsewhere. $\mathbf{Y} = [y_1, y_2, \dots, y_n]^T$ is called a *vector of rearranged variables.*

> **Example 14.2** *Figure 14.1a shows one possible reordering of variables for the vector of variables* $\mathbf{X} = [x_1, x_2, x_3, x_4]^T$, *performed in accordance with Equation 14.1.*

Complement of variables. Let $\mathbf{X} = [x_1, x_2, \dots, x_n]^T$ be a vector of n variables of a switching function f, and $\mathbf{A} = [a_1 a_2 \dots a_n]$ be a vector with

Reodering

x_1 — $y_1 = x_4$
x_2 — $y_2 = x_2$
x_3 — $y_3 = x_1$
x_4 — $y_4 = x_3$

Given the vector of variables $\mathbf{X} = [x_1, x_2, x_3, x_4]^T$ *and the permutation matrix* P*, the vector of rearranged variables is*

$$\mathbf{Y} = P\mathbf{X} = \begin{bmatrix} 0&0&0&1 \\ 0&1&0&0 \\ 1&0&0&0 \\ 0&0&1&0 \end{bmatrix} \begin{bmatrix} x_1 \\ x_2 \\ x_3 \\ x_4 \end{bmatrix} = \begin{bmatrix} x_4 \\ x_2 \\ x_1 \\ x_3 \end{bmatrix} = \begin{bmatrix} y_1 \\ y_2 \\ y_3 \\ y_4 \end{bmatrix}$$

Complementing

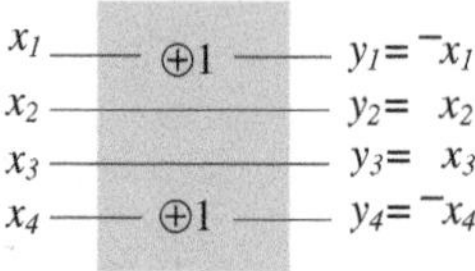

Given the polarity vector $\mathbf{A} = [1001]^T$*, the vector of rearranged variables is*

$$\mathbf{Y} = \mathbf{X} \oplus \mathbf{A} = \begin{bmatrix} x_1 \\ x_2 \\ x_3 \\ x_4 \end{bmatrix} \oplus \begin{bmatrix} 1 \\ 0 \\ 0 \\ 1 \end{bmatrix} = \begin{bmatrix} \overline{x}_1 \\ x_2 \\ x_3 \\ \overline{x}_4 \end{bmatrix} = \begin{bmatrix} y_1 \\ y_2 \\ y_3 \\ y_4 \end{bmatrix}$$

Reodering with complement

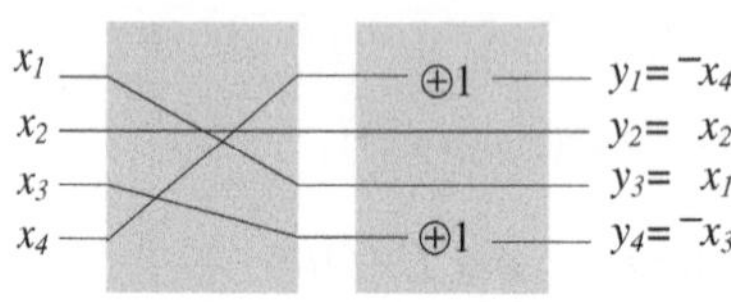

Given the permutation matrix P*, and polarity vector* $\mathbf{A} = [1001]^T$*, the vector of rearranged variables is*

$$\mathbf{Y} = P\mathbf{X} \oplus \mathbf{A}$$

$$= \begin{bmatrix} 0&0&0&1 \\ 0&1&0&0 \\ 1&0&0&0 \\ 0&0&1&0 \end{bmatrix} \begin{bmatrix} x_1 \\ x_2 \\ x_3 \\ x_4 \end{bmatrix} \oplus \begin{bmatrix} 1 \\ 0 \\ 0 \\ 1 \end{bmatrix} = \begin{bmatrix} \overline{x}_4 \\ x_2 \\ x_1 \\ \overline{x}_3 \end{bmatrix} = \begin{bmatrix} y_1 \\ y_2 \\ y_3 \\ y_4 \end{bmatrix}$$

Reodering, complementing, and operations in $GF(2)$

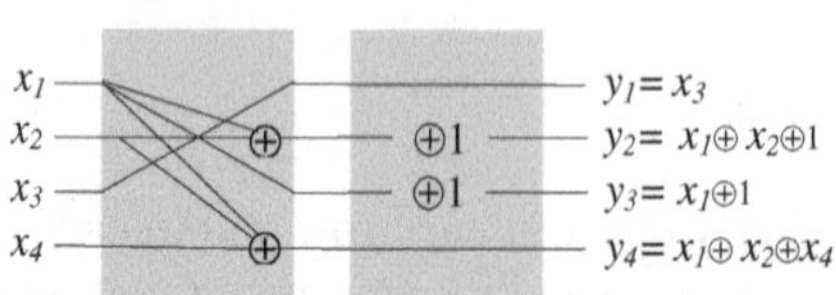

Given the transform matrix E*, and polarity vector* $\mathbf{A} = [0110]^T$*, the vector of rearranged variables is*

$$\mathbf{Y} = E\mathbf{X} \oplus \mathbf{A}$$

$$= \begin{bmatrix} 0&0&1&0 \\ 1&1&0&0 \\ 1&0&0&0 \\ 1&1&0&1 \end{bmatrix} \begin{bmatrix} x_1 \\ x_2 \\ x_3 \\ x_4 \end{bmatrix} \oplus \begin{bmatrix} 0 \\ 1 \\ 1 \\ 0 \end{bmatrix}$$

$$= \begin{bmatrix} x_3 \\ x_1 \oplus x_2 \oplus 1 \\ x_1 \oplus 1 \\ x_1 \oplus x_2 \oplus x_4 \end{bmatrix} = \begin{bmatrix} y_1 \\ y_2 \\ y_3 \\ y_4 \end{bmatrix}$$

FIGURE 14.1
Basic operations for rearrangement: permutation, complement, permutation, and permutation with an operation over Galois field $GF(2)$ (Examples 14.2,14.3, 14.4, and 14.5).

binary elements $a_i \in \{0, 1\}$, $i = 1, 2, \dots n$, called a *polarity vector*[‡]. A partial complement of input variables given vectors $\mathbf{X}$ and $\mathbf{A}$ can be described by the matrix equation

$$\mathbf{Y} = \mathbf{X} \oplus \mathbf{A}. \tag{14.2}$$

Example 14.3 *Complementing of variables given a vector* $\mathbf{X}$ *and the polarity vector* $\mathbf{A} = [1001]^T$ *using Equation 14.2 is illustrated in Figure 14.1.*

Reordering with complementing of variables. The matrix equation for reordering of variables with complementing can be defined by combining Equations 14.1 and 14.2:

$$\mathbf{Y} = P\mathbf{X} \oplus \mathbf{A}. \tag{14.3}$$

Example 14.4 *Figure 14.1 depicts a rearrangement of variables using a permutation with a complement (Equation 14.3) for a given vector* $\mathbf{X}$*, permutation matrix, and polarity vector.*

Operations in Galois field GF(2). Given a vector $\mathbf{X} = [x_1, x_2, \dots, x_n]^T$, polarity vector $\mathbf{A}$, and transform matrix E, operations over $GF(2)$ are defined by the equation

$$\mathbf{Y} = E\mathbf{X} \oplus \mathbf{A}. \tag{14.4}$$

The transform matrix E is a nonsingular matrix[§] specified for each particular example. If $\mathbf{A} = 0$ in Equation 14.4, then the affine transform is called a *linear* affine transform.

Example 14.5 *Figure 14.1 shows an affine transform (Equation 14.4) of variables given vector* $\mathbf{X}$*, a transform matrix, and a polarity vector.*

14.2.1 Direct and inverse affine transforms

There are two types of affine transforms of input variables of switching functions: *direct* and *inverse* (Figure 14.2). Given a vector of input variables $\mathbf{X}$,

---

[‡]For two Boolean vectors of the same dimension, various Boolean operations can be applied, for example, EXOR, AND, etc. Bitwise AND and EXOR result in a vector whose bits consist of the AND and EXOR of the corresponding bits of input vectors, accordingly.

[§]A matrix without an inverse is called *noninvertible* or *singular*. If a matrix has an inverse, it is called *invertible* or *nonsingular*. Matrix inverses are unique. For nonsingular matrices A and B, the following properties held: $AA^{-1} = I_n = A^{-1}A$, $(BA)^{-1} = A^{-1}B^{-1}$, $(A^{-1})^T = (A^T)^{-1}$.

transform matrix E, and polarity vector $\mathbf{A}$, the pair of affine transforms is defined by the matrix equations

$$\mathbf{Y} = E\mathbf{X} \oplus \mathbf{A} \tag{14.5}$$

$$\mathbf{X} = E^{-1}\mathbf{Y} \oplus E^{-1}\mathbf{A} \tag{14.6}$$

where E^{-1} is the inverse transform matrix.

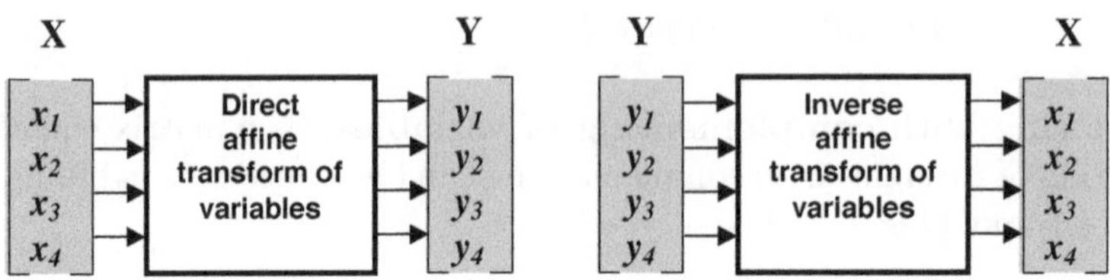

FIGURE 14.2
Direct and inverse affine transforms of input variables.

Direct affine transform of a switching function f of n variables $x_1, x_2, \ldots, x_n$, is defined by Equation 14.5 for an arbitrary nonsingular matrix E and a binary vector $\mathbf{A}$. We change function f by substituting each x_i, $i = 1, 2, \ldots, n$, with the corresponding expression specified by Equation 14.5. Applying an affine transform to the variables of a given switching function f, a new switching function g of n variables $y_1, y_2, \ldots, y_n$ can be generated.

> **Example 14.6** *Figure 14.3a illustrates a direct affine transform of the input variables for an AND function given the transform matrix E and the polarity vector $\mathbf{A} = [10]^T$.*

The inverse affine transform is defined by Equation 14.6. The inverse transform recovers an initial switching function f of n variables $x_1, x_2, \ldots, x_n$ from the affine equivalent function g of n variables $y_1, y_2, \ldots, y_n$, where the same matrix E and binary vector $\mathbf{A}$ are used as in the direct affine transform.

> **Example 14.7** *Figure 14.3b shows an inverse affine transform of variables for NOR function given the inverse transform matrix and the polarity vector $\mathbf{A} = [11]^T$. Note that computing $g(y_1, y_2)$ yields:*

$$g(\underbrace{1 \oplus x_1}_{y_1}, \underbrace{x_1 \oplus x_2}_{y_2}) = \overline{\bar{x}_1 \vee (x_1 \oplus x_2)}$$

$$= x_1\overline{(x_1 \oplus x_2)}$$

$$= x_1(\bar{x}_1 \oplus x_2) = x_1x_2.$$

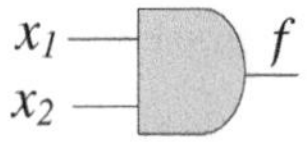

$f = x_1 x_2$

x_1	x_2	f
0	0	0
0	1	0
1	0	0
1	1	1

$$\mathbf{Y} = E\mathbf{X} \oplus \mathbf{A}$$

$$= \begin{bmatrix} 1 & 0 \\ 1 & 1 \end{bmatrix} \begin{bmatrix} x_1 \\ x_2 \end{bmatrix} \oplus \begin{bmatrix} 1 \\ 0 \end{bmatrix} = \begin{bmatrix} x_1 \\ x_1 \oplus x_2 \end{bmatrix} \oplus \begin{bmatrix} 1 \\ 0 \end{bmatrix},$$

thus

$$y_1 = 1 \oplus x_1$$
$$y_2 = x_1 \oplus x_2$$

(*a*)

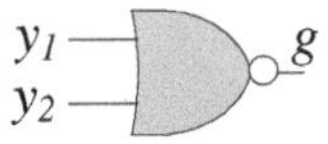

$g = \overline{y_1 \vee y_2}$

y_1	y_2	g
0	0	1
0	1	0
1	0	0
1	1	0

$$\mathbf{X} = E^{-1}\mathbf{Y} \oplus E^{-1}\mathbf{A}$$

$$= \begin{bmatrix} 1 & 0 \\ 1 & 1 \end{bmatrix} \begin{bmatrix} y_1 \\ y_2 \end{bmatrix} \oplus \begin{bmatrix} 1 & 0 \\ 1 & 1 \end{bmatrix} \begin{bmatrix} 1 \\ 0 \end{bmatrix}$$

$$= \begin{bmatrix} 1 & 0 \\ 1 & 1 \end{bmatrix} \begin{bmatrix} y_1 \\ y_2 \end{bmatrix} \oplus \begin{bmatrix} 1 \\ 1 \end{bmatrix} = \begin{bmatrix} y_1 \\ y_1 \oplus y_2 \end{bmatrix} \oplus \begin{bmatrix} 1 \\ 1 \end{bmatrix}$$

thus

$$x_1 = 1 \oplus y_1$$
$$x_2 = 1 \oplus y_1 \oplus y_2$$

(*b*)

FIGURE 14.3
Direct (a) and inverse (b) affine transforms of variables for AND and NOR functions (Examples 14.7 and 14.8).

The pair of functions f and g formed by an affine transform of variables is called an *affine equivalent.* This pair of functions is called a *linear affine equivalent* if it is formed by a linear affine transform of variables.

Example 14.8 *The switching functions* $f = x_1 x_2$ *and* $g = \overline{y_1 \vee y_2}$ *are an affine equivalent (Figure 14.3a). By analogy, AND and NOR functions are an affine equivalent (Figure 14.3b).*

14.2.2 Affine transforms of the adjacent variables

An adjacency is a useful property of algebraic and topological structures. In an undirected graph, the adjacency relation is symmetric, that is, if a vertex v is adjacent to a vertex u, then a vertex u is adjacent to a vertex v (see details in Chapter 3).

Consider linear transforms of the adjacent variables x_i and x_{i+1}. The structure of the transform matrix is defined by the 2×2 kernel matrix

$$<\texttt{Kernel matrix}> = \begin{bmatrix} a_{i,i} & a_{i,i+1} \\ a_{i+1,i} & a_{i+1,i+1} \end{bmatrix}$$

The diagonalized kernel matrix (the elements of this matrix form the diagonal of identity matrix) is called an *elementary* matrix. Any nonsingular matrix of a linear transform can be represented by a product of elementary matrices. Thus, any affine transform of variables is a sequence of affine transforms of adjacent variables. Some elementary matrices of nonidentical linear transforms given adjacent variables x_i and x_{i+1} are shown in Table 14.1.

TABLE 14.1
Nonidentical transforms.

Replacement	Matrix
$\begin{cases} x_i = x_i \oplus x_{i+1} \\ x_{i+1} = x_{i+1} \end{cases}$	$\begin{bmatrix} a_{i,i} & a_{i,i+1} \\ a_{i+1,i} & a_{i+1,i+1} \end{bmatrix} = \begin{bmatrix} 1 & 1 \\ 0 & 1 \end{bmatrix}$
$\begin{cases} x_i = x_i \\ x_{i+1} = x_{i+1} \oplus x_i \end{cases}$	$\begin{bmatrix} a_{i,i} & a_{i,i+1} \\ a_{i+1,i} & a_{i+1,i+1} \end{bmatrix} = \begin{bmatrix} 1 & 0 \\ 1 & 1 \end{bmatrix}$
$\begin{cases} x_i = x_{i+1} \\ x_{i+1} = x_i \end{cases}$	$\begin{bmatrix} a_{i,i} & a_{i,i+1} \\ a_{i+1,i} & a_{i+1,i+1} \end{bmatrix} = \begin{bmatrix} 0 & 1 \\ 1 & 0 \end{bmatrix}$

14.3 Linearly transformed decision trees and diagrams

Linearly transformed decision diagrams are constructed in two steps:

- The affine transform of variables is used to find a satisfactory (with respect to a given criterion) permutation of input variables.
- A new decision decision diagram is constructed.

Figure 14.4 shows how to perform two steps on a network of serially connected linear and nonlinear blocks¶. For an n-variable function, the complexity of the linear block increases slowly, whereas the complexity of the nonlinear block is almost always an exponential function of n. Therefore, the complexity of the linear block may be ignored.

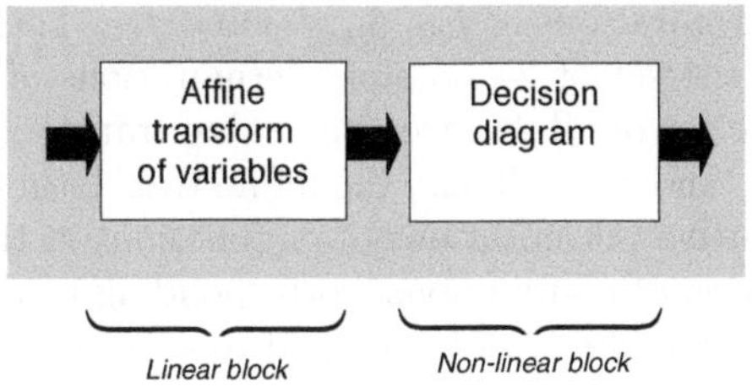

- *The linear block consists of EXOR circuits*
- *The number of 2-input gates of the linear block increases asymptotically, upperbounded by $n^2/log_2 n$ for $n \to \infty$ [11]*
- *The complexity of the nonlinear block increases exponentially as n increases*

FIGURE 14.4
The two steps of construction of linearly transformed decision diagram.

Reduction of the number of product terms in a SOP expression of a given function f corresponds to reduction of the number of paths in the decision diagram for f. Thus, the effectiveness of a linearly transformed decision diagram is defined by

- The number of paths,
- The length of the path that determines the distribution of nodes per level and the shape of the decision diagram,
- The size of the decision diagram,
- The number of different values of terminal nodes that corresponds to the number of spectral coefficients (for word-level diagrams).

Since affine transformations of variables aim at reduction of the number of products, they can also improve the topological characteristics of decision diagrams.

14.3.1 Affine transforms in decision trees

Consider two adjacent levels of a binary tree. The Shannon decomposition of a switching function f with respect to two adjacent variables x_i and x_{i+1}

¶*A linear* function is represented by the equation $a_0 \oplus a_1 x_1 \oplus \cdots \oplus a_n x_n$. A linear function is a parity function if $a_1 = a_2 = \cdots = a_n = 1$. There are 2^{n+1} linear functions of n variables. A linear function is either a self-dual or a self-antidual function. The function obtained by linear composition of linear functions or by assigning a linear function in an arbitrary variable of a linear function is also a linear switching function (see detail in Chapter 4).

results in the expression

$$f = \overline{x}_i\overline{x}_{i+1}f_{00} \vee \overline{x}_i x_{i+1}f_{01} \vee x_i\overline{x}_{i+1}f_{10} \vee x_i x_{i+1}f_{11},$$

where co-functions are defined as follows:

$$f_{00} = f(x_i = 0, x_{i+1} = 0),\ \ f_{01} = f(x_i = 0, x_{i+1} = 1),$$
$$f_{10} = f(x_i = 1, x_{i+1} = 0),\ \ f_{11} = f(x_i = 1, x_{i+1} = 1).$$

There are $n! = 4! = 24$ possible permutations of f_{00}, f_{01}, f_{10} and f_{11}. There exists one-to-one mapping between the set of 24 possible permutations of 4 co-functions $f_{00}, f_{01}, f_{10}, f_{11}$, and the set of all 24 invertible affine transforms with respect to adjacent variables x_i and x_{i+1}. Thus, the affine transform of variables can be obtained by consecutive permutations of co-functions in the levels of a decision tree. A permutation of co-functions, corresponding to the affine transform with respect to two variables, can be applied to an arbitrary level of a decision tree.

Example 14.9 *Figure 14.5 shows several transforms that produce permutations of f_{00}, f_{01}, f_{10}, and f_{11}.*

14.3.2 Affine transforms in decision diagrams

Affine transforms of variables can reduce the number of nodes in a BDD. While the ROBDD of a given function is unique, transform of variables is one of the few ways to minimize the ROBDD size.

The affine transform of variables converts **F** into a vector with a pair of adjacent equal values, which permits reduction of a non-terminal node in the corresponding linearly transformed BDD.

The following example illustrates optimization of decision diagrams by ordering and transform of variables.

Example 14.10 *Let a switching function $f = f(x_1, x_2)$ of two variables be given by the truth vector*

$$\mathbf{F} = [f(0), f(1), f(2), f(3)]^T = [f_{00}, f_{01}, f_{10}, f_{11}]^T,$$

and let $f(01) = f(10)$, i.e., $f(1) = f(2)$. Two transforms (successful and unsuccessful) of variables which result in the reduction of f are given in Figure 14.6.

That is, the function is symmetric and, therefore,

$$f = \overline{x}_1\overline{x}_2 f_{00} \oplus \overline{x}_1 x_2 f_{01} \oplus x_1\overline{x}_2 f_{01} \oplus x_{10} x_2 f_{11}$$
$$= \overline{x}_1\overline{x}_2 f_{00} \oplus (x_1 \oplus x_2) f_{01} \oplus x_1 x_2 f_{01}.$$

The BDD with the permuted variables x_2, x_1 represents the function f by the same expression, and therefore, the size of BDD is not reduced.

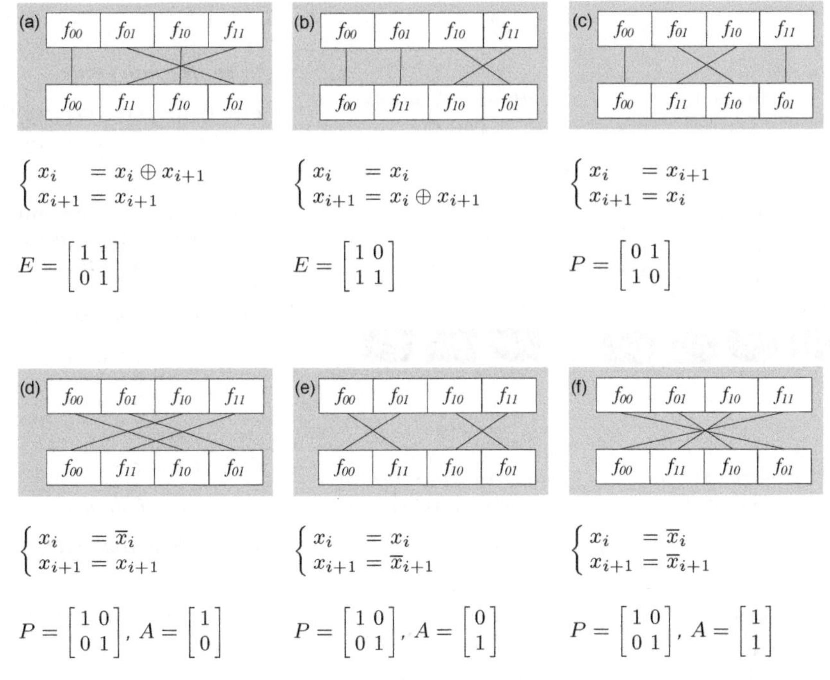

FIGURE 14.5
Arbitrary permutations of f_{00}, f_{01}, f_{10}, and f_{11} (Example 14.9).

Example 14.11 *(Continuation of Example 14.10). the first transform yields:*

$$f = \overline{y}_1\overline{y}_2 f_{00} \oplus \overline{y}_1 y_2 f_{11} \oplus y_1 f_{01},$$

which, when written in terms of primary variables, is

$$f = \overline{(x_1 \oplus x_2)}\overline{x}_2 f_{00} \oplus \overline{(x_1 \oplus x_2)} x_2 f_{11} \oplus (x_1 \oplus x_2) f_{01}.$$

Thus, the linear transform detects the appearance of the term $x_1 \oplus x_2$ in f. This term is placed in the upper level of the decision diagram, and this further leads to reduction of the diagram. Notice that the function f considered in this example is a symmetric one, i.e.,

$$f(x_1, x_2) = f(x_2, x_1)$$

for any combination of logic values for x_1 and x_2. Reodering of variables will be equally unsuccessful for any symmetric function of an arbitrary number of variables. In this case, the affine transform of variables may be useful to reduce the size of the BDD.

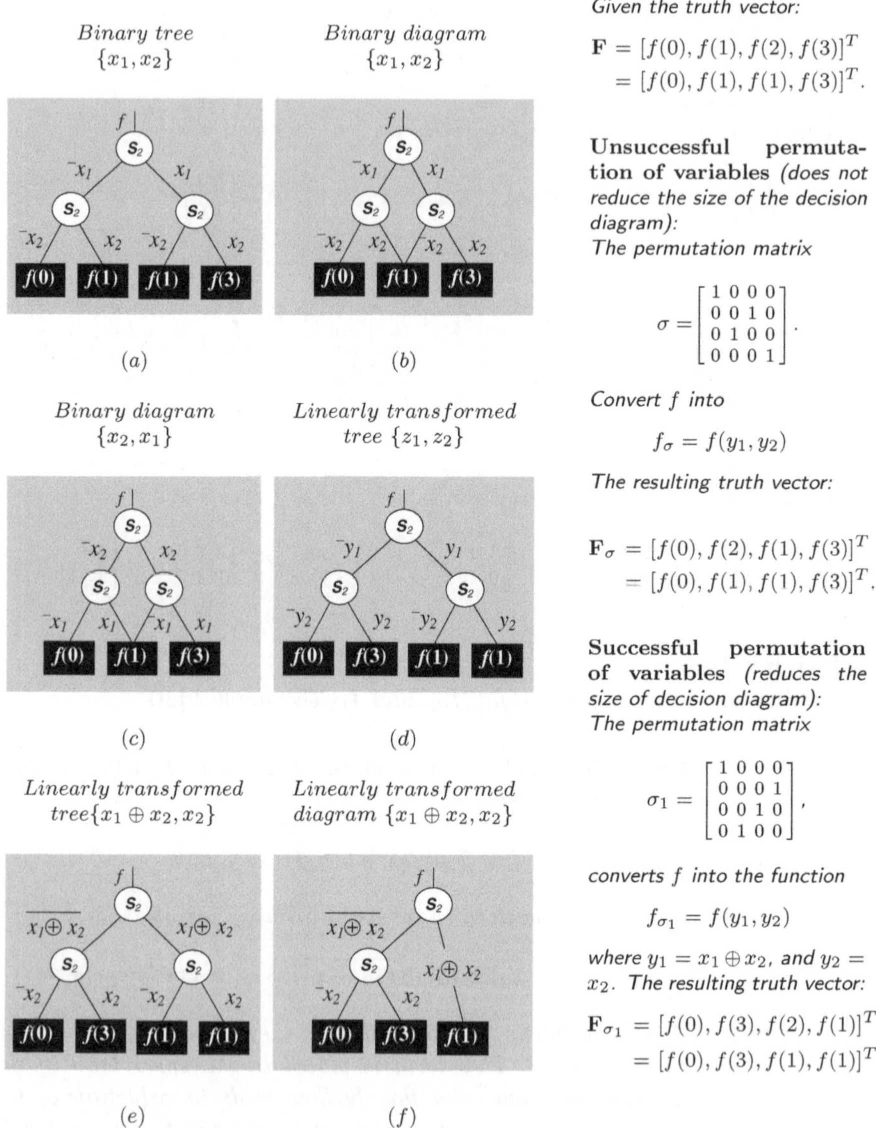

Binary tree $\{x_1, x_2\}$
Binary diagram $\{x_1, x_2\}$
(a)
(b)
Binary diagram $\{x_2, x_1\}$
Linearly transformed tree $\{z_1, z_2\}$
(c)
(d)
Linearly transformed tree$\{x_1 \oplus x_2, x_2\}$
Linearly transformed diagram $\{x_1 \oplus x_2, x_2\}$
(e)
(f)
Given the truth vector:
$$\mathbf{F} = [f(0), f(1), f(2), f(3)]^T = [f(0), f(1), f(1), f(3)]^T.$$
Unsuccessful permutation of variables *(does not reduce the size of the decision diagram):*
The permutation matrix
$$\sigma = \begin{bmatrix} 1 & 0 & 0 & 0 \\ 0 & 0 & 1 & 0 \\ 0 & 1 & 0 & 0 \\ 0 & 0 & 0 & 1 \end{bmatrix}.$$
Convert f *into*
$$f_\sigma = f(y_1, y_2)$$
The resulting truth vector:
$$\mathbf{F}_\sigma = [f(0), f(2), f(1), f(3)]^T = [f(0), f(1), f(1), f(3)]^T.$$
Successful permutation of variables *(reduces the size of decision diagram):*
The permutation matrix
$$\sigma_1 = \begin{bmatrix} 1 & 0 & 0 & 0 \\ 0 & 0 & 0 & 1 \\ 0 & 0 & 1 & 0 \\ 0 & 1 & 0 & 0 \end{bmatrix},$$
converts f *into the function*
$$f_{\sigma_1} = f(y_1, y_2)$$
where $y_1 = x_1 \oplus x_2$*, and* $y_2 = x_2$*. The resulting truth vector:*
$$\mathbf{F}_{\sigma_1} = [f(0), f(3), f(2), f(1)]^T = [f(0), f(3), f(1), f(1)]^T$$

FIGURE 14.6
Affine transform of variables for binary decision trees and diagrams (Example 14.10).

Reodering of variables in a function of n variables allows $n!$ permutations in the vector $\mathbf{F}$ of function values for f. The affine transform of variables increases the number of possible permutations in the vector of function values and, therefore, permits reduction of the sizes of BDDs in cases when reordering cannot be used.

14.4 Examples of linear transform techniques

Kolpakov and Latypov [13, 14] developed several heuristics for linearly transformed decision trees and diagrams based on affine transforms of variables. Finding an affine transform of variables given a function of a large number of variables is computationally expensive. To reduce the search space, various methods can be used which utilize properties of the binary tree such as weight, balance, or homogeneity.

Some properties of binary trees can be used as criteria for the evaluation of the constructed linearly transformed decision trees. The following characteristics are introduced below:

- The *weight* $w(f)$ of a switching function f is a number of ones in the truth table of f. If the switching functions f and g are affine equivalent, then their weights are equal, i.e., $w(f) = w(g)$.
- The *balance* of subtrees A and B is defined as a difference of their weights.
- *Homogeneity* H_f of a switching function f of n variables given a truth vector $\mathbf{F}$ is defined as the absolute value of the difference between number of 1's, n_1, and the number of 0's, n_0, i.e., $H_f = |n_1 - n_0| = |2^n - 2n_0|$.
- *Homogeneity* $H(L)$ of level L of a decision tree that represents a switching function f of n variables is defined as $H(L) = \sum_{i=0}^{2^L-1} |2^{n-L} - 2n_{0i}|$, where n_{0i} is the number of 0's in the truth table corresponding to the subtree.

Example 14.12 *Figure 14.7 illustrates the calculation of characteristics of decision trees.*

These characteristics are used for linearly transformed decision tree and diagram construction. Details on the characteristics of decision trees and manipulation are given in the "Further reading" section and in Chapters 3 and 34.

In the fragment of algorithm given below, several techniques are shown.

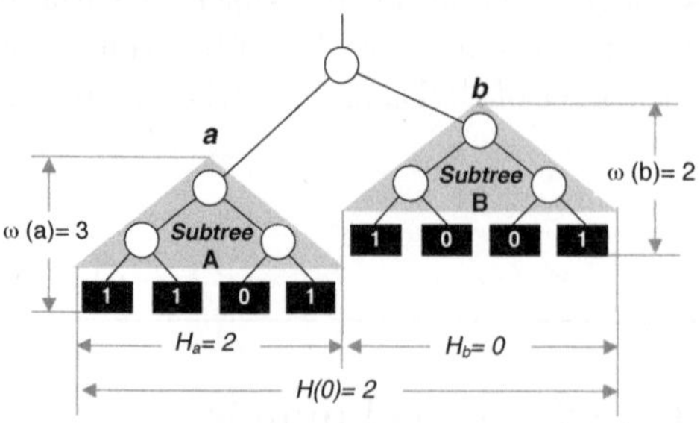

- Weights of subtrees A and B: $w(a) = 3$ and $w(b) = 2$
- Balance of subtrees A and B: $B = w(a) - w(b) = 3 - 2 = 1$
- Homogeneity of subtree A: $H = |n_1 - n_0| = |3 - 1| = 2$
- Homogeneity of subtree B: $H = |n_1 - n_0| = |2 - 2| = 0$
- Homogeneity of a level: $L = 0, H(0) = 2 + 0 = 2$

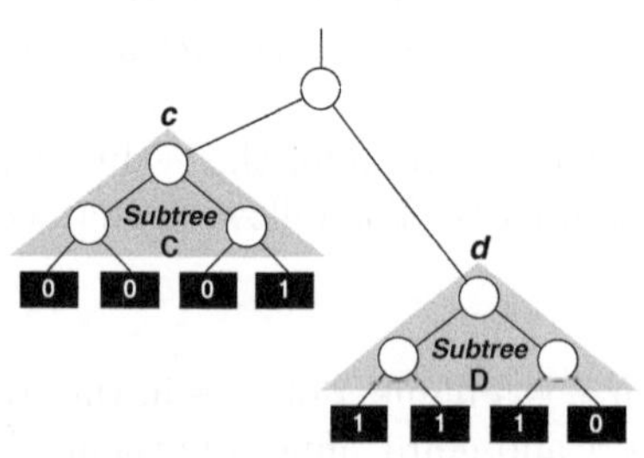

- Weights of subtrees C and D: $w(c) = 1$ and $w(d) = 3$
- Balance of subtrees C and D: $B = w(c) - w(d) = 1 - 3 = -2$
- Homogeneity of subtree C: $H = |n_1 - n_0| = |1 - 3| = 2$
- Homogeneity of subtree D: $H = |n_1 - n_0| = |3 - 1| = 2$
- Homogeneity of a level: $L = 0, H(0) = 2 + 2 = 4$

FIGURE 14.7
Computing the weights, balances, and the homogeneity of subtrees (Example 14.12).

Input data: Decision diagram.

Step 1 Set level $L = 1$. Rearrange functions f_0, f_1, f_2, f_3, in level $L1$ with respect to criterion of weight $w(f_0) \leq w(f_1) \leq w(f_2) \leq w(f_3)$.

Step 2 : Set level $L = L + 1$.

- Calculate the homogeneity of the current level, $HL(current)$.
- Using permutations of subsets of sub-functions of this level, calculate $HL(new) > HL(current)$ and go back to level $L - 1$.
- If there are no $HL(new) > HL(current)$ for any permutation, go to the next level $L = L + 2$.

Step 3 Stop if L is equal to the number of levels in the decision tree. Construct the matrix of affine transform.

Output data Linearly transformed decision diagram and affine transform.

Example 14.13 *Consider a switching function of four variables, given by its truth vector* $\mathbf{F} = [0101110001110001]^T$. *Figure 14.8 shows the initial BDD of 7 nodes. The total homogeneity is*

$$H(total) = H(0) + H(1) + H(2) + H(3) = 0 + 8 + 8 + 12 = 28.$$

The final BDD is of 5 nodes. Figures 14.9 and 14.10 illustrate the details of computing. The final result is the affine equivalent switching function $\mathbf{F} = [0000010100111111]^T$.

14.5 Further reading

Linearly transformed decision diagrams technique. The size of decision diagrams is very sensitive to variable ordering. There are many techniques to minimize the size of decision diagrams. Rudell [21] developed a reordering algorithm called *sifting* (see details in Chapter 12). Linearly independent transformations have been studied, in particular, by Bern et al. [2], Fujita et al. [9], Perkowski et al. [22, 23, 24], Meinel et al. [16], and Rahardia and Falkowski [20]. Notice that linear transform of variables is a classical method for optimization of various representations of switching functions (see, for example, the paper by Nechiporuk [19]).

A linear transform with respect to the variables can be represented by an $(n \times n)$ matrix over $GF(2)$, and the required space can be neglected compared to the space required to store a BDD or a linearly transformed BDD. The latter can be considered as a bottleneck for applications of linearly transformed BDDs, although there are heuristic algorithms to determine a suitable linear combination of variables.

In the algorithm proposed by Meinel et al. [15, 16], the set of variables is splited into subsets of adjacent variables. The sifting method is used for variable ordering in decision diagrams. The authors utilized the fact that invertible linear transformations can be generated by a set of elementary linear transformations. The number of invertible linear transformations of variables is $\prod_{i=0}^{n-1}(2^n - 2^i)$ for reordering (if polarity vector $\mathbf{A} = 0$ in Equation 14.4). The affine transform for variables defined by Equation 14.4, $\mathbf{Y} = E\mathbf{X} \oplus \mathbf{A}$, where $\mathbf{A}$ is a polarity vector, and E is a transform matrix, generates $2^n \prod_{i=0}^{n-1}(2^n - 2^i)$ invertible affine transformations i.e., 2^n times greater compared with the number of invertible linear transformations of variables.

The algorithms for efficient manipulations of linearly transformed BDDs are implemented by Günther et al. [10] as an extension of the CUDD package by Somenzi [27] for further support of linearly transformed BDDs. Karpovsky et al. [11] discussed the applications of total autocorrelation functions to reduce

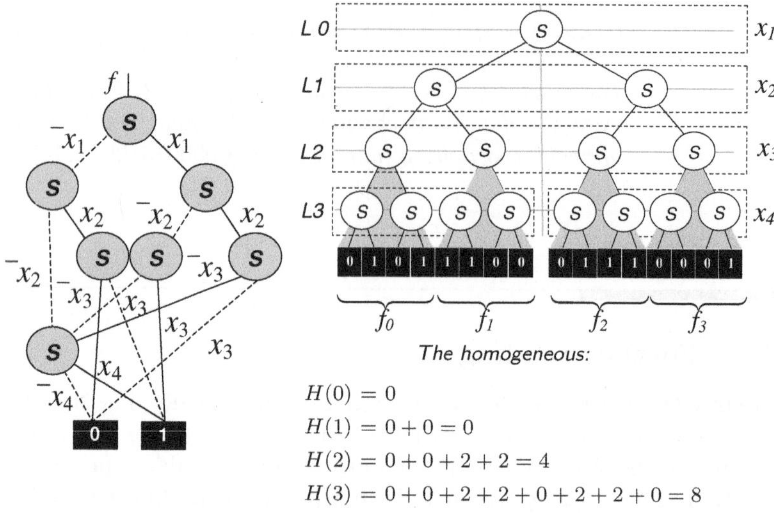

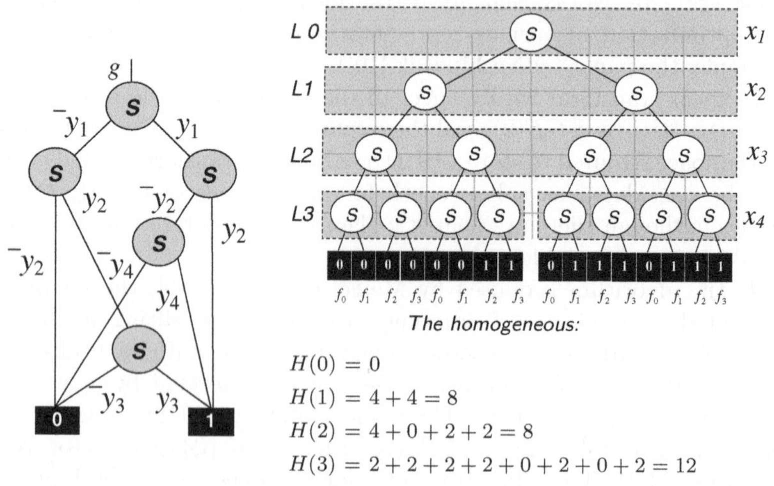

FIGURE 14.8
Original ROBDD for the switching function $f(x_1, x_2, x_3, x_4)$, its decision tree, and the linearly transformed ROBDD and its decision tree after affine transform of variables for the switching function $g(y_1, y_2, y_3, y_4)$ (Example 14.13).

Step 1 : Level 1

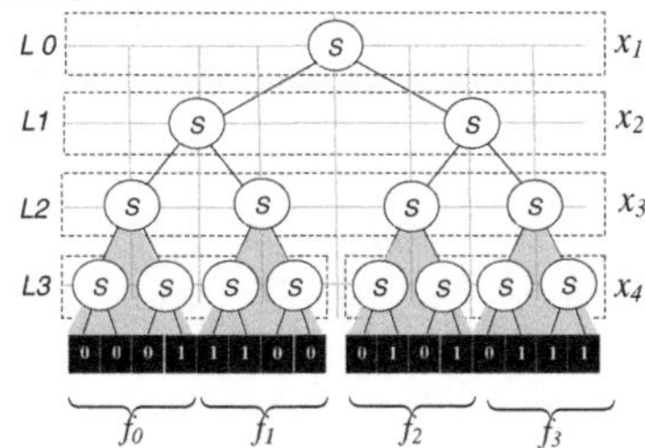

Reorder co-functions at the level $L1$ using permutation f_3, f_1, f_0, f_2. Apply a weight criteria $w(f_3) \leq w(f_1) \leq w(f_0) \leq w(f_2)$. Calculate the homogeneity and affine transforms:

$$H(1) = 2 + 2 = 4, \;\; H(2) = 4, \;\; H(3) = 8$$

$$\begin{bmatrix} x_1 \\ x_2 \\ x_3 \\ x_4 \end{bmatrix} = \begin{bmatrix} 1 \oplus z_1 \oplus z_2 \\ 1 \oplus z_1 \\ z_3 \\ z_4 \end{bmatrix} \text{ and } \begin{bmatrix} z_1 \\ z_2 \\ z_3 \\ z_4 \end{bmatrix} = \begin{bmatrix} 1 \oplus x_2 \\ x_1 \oplus x_2 \\ x_3 \\ x_4 \end{bmatrix}$$

Step 2 : Level 2

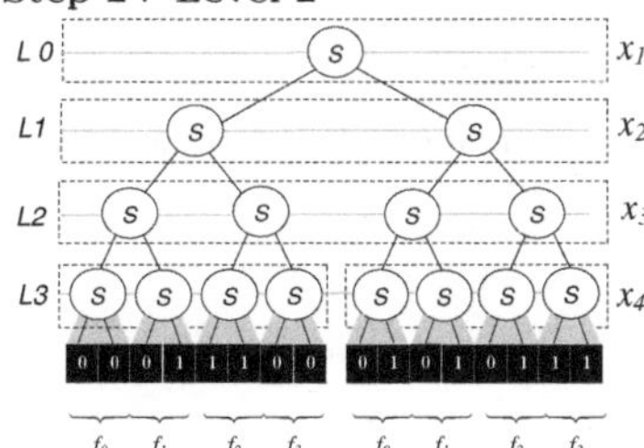

Calculate the homogeneity:

$$H(2) = 2 + 0 + 0 + 2 = 4$$

Step 3 : Level 2

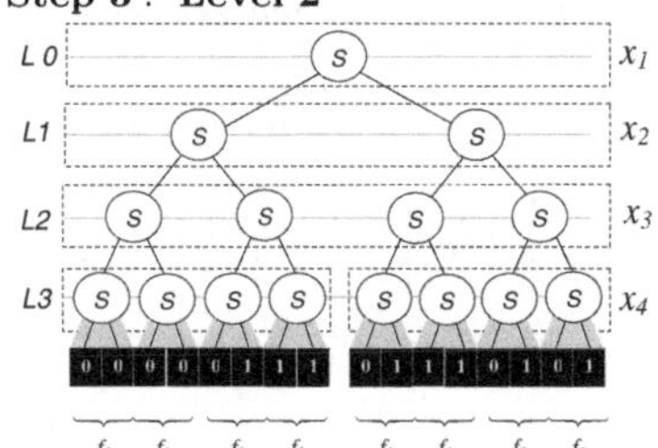

Reorder co-functions at the level $L2$ using the permutation f_0, f_3, f_2, f_1. Calculate the homogeneity and affine transforms:

$$H(2) = 4 + 2 + 2 + 0 = 8$$

$$\begin{bmatrix} z_1 \\ z_2 \\ z_3 \\ z_4 \end{bmatrix} = \begin{bmatrix} u_1 \\ u_3 \\ u_2 \oplus u_3 \\ u_4 \end{bmatrix} \text{ and } \begin{bmatrix} u_1 \\ u_2 \\ u_3 \\ u_4 \end{bmatrix} = \begin{bmatrix} z_1 \\ z_1 \oplus z_2 \\ z_3 \\ z_4 \end{bmatrix}$$

Step 4 : Level 1

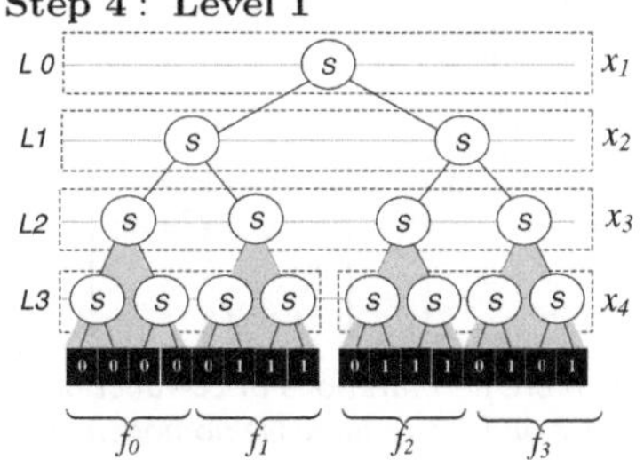

Go to level $L1$

FIGURE 14.9

Initial tree and the first three steps of the algorithm for affine transform of variables given the ROBDD in Figure 14.8a (Example 14.13).

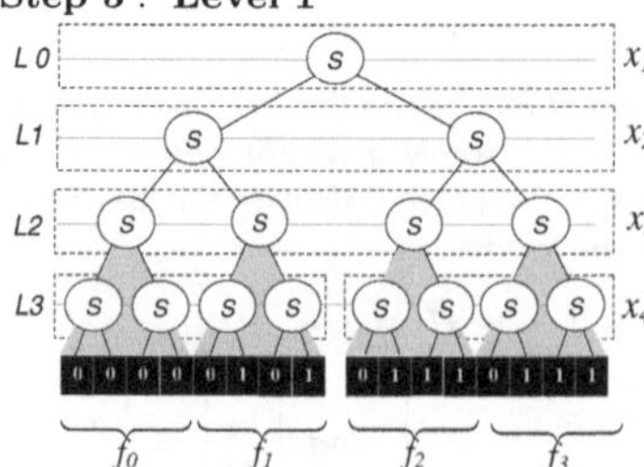

Reorder co-functions using the permutation f_0, f_3, f_2, f_1*. Apply a weight criterion* $w(f_0) \leq w(f_3) \leq w(f_2) \leq w(f_3)$*. Calculate the affine transforms:*

$$\begin{bmatrix} u_1 \\ u_2 \\ u_3 \\ u_4 \end{bmatrix} = \begin{bmatrix} v_1 \oplus v_2 \\ v_2 \\ v_3 \\ u_4 \end{bmatrix} \quad \text{and} \quad \begin{bmatrix} v_1 \\ v_2 \\ v_3 \\ v_4 \end{bmatrix} = \begin{bmatrix} u_1 \oplus u_2 \\ u_2 \\ u_3 \\ u_4 \end{bmatrix}$$

Step 6 : Level 2

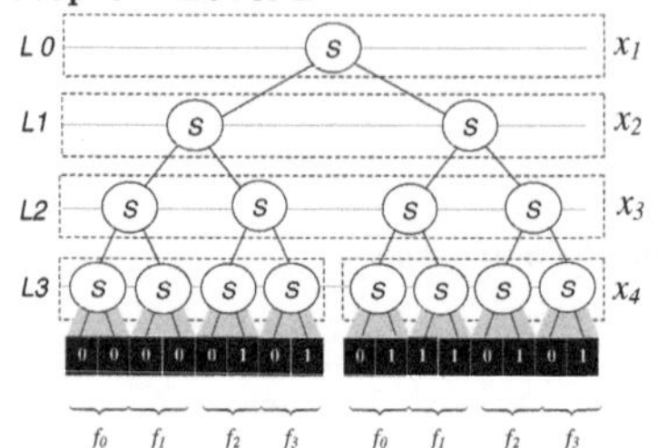

Calculate the homogeneity

$$H(2) = 4 + 0 + 2 + 2 = 8$$

There is no permutation of co-functions that could change $H(2)$

Step 7 : Level 3

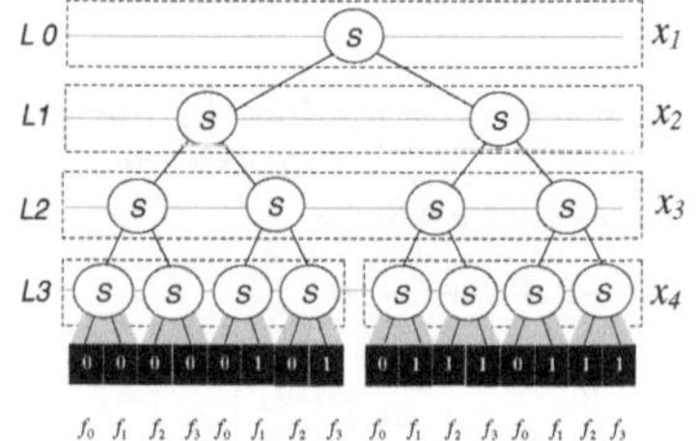

Reorder co-functions using the permutation f_0, f_2, f_1, f_3*. Calculate the homogeneity and affine transforms*

$$H(3) = 2 + 2 + 0 + 0 + 0 + 2 + 0 + 2 = 8$$

$$\begin{bmatrix} v_1 \\ v_2 \\ v_3 \\ v_4 \end{bmatrix} = \begin{bmatrix} y_1 \\ y_2 \\ y_3 \\ y_4 \end{bmatrix} \quad \text{and} \quad \begin{bmatrix} y_1 \\ y_2 \\ y_3 \\ y_4 \end{bmatrix} = \begin{bmatrix} v_1 \\ v_2 \\ v_4 \\ v_3 \end{bmatrix}$$

Step 8: Level 3

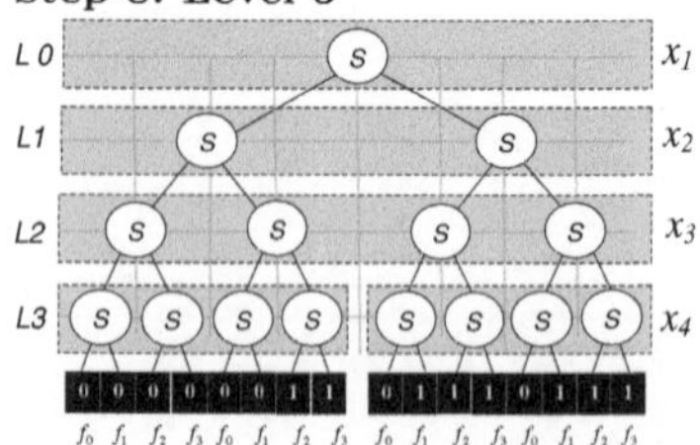

The final affine transform

$$\begin{bmatrix} y_1 \\ y_2 \\ y_3 \\ y_4 \end{bmatrix} = \begin{bmatrix} v_1 \\ v_2 \\ v_4 \\ v_3 \end{bmatrix} = \begin{bmatrix} u_1 \oplus u_2 \\ u_2 \\ u_4 \\ u_3 \end{bmatrix} = \begin{bmatrix} z_1 \oplus z_2 \oplus z_3 \\ z_1 \oplus z_2 \\ z_4 \\ z_2 \end{bmatrix} = \begin{bmatrix} 1 \oplus x_1 \oplus x_3 \\ x_1 \oplus x_2 \oplus x_3 \\ x_4 \\ x_1 \oplus x_2 \end{bmatrix}$$

There are no more permutations of co-functions at the levels $L2$ *and* $L3$ *to increase homogeneity as well as in the level* $L1$ *to apply weight criteria.*

FIGURE 14.10
Continuation of the algorithm for affine transform of variables given the ROBDD in Figure 14.8a. Final ROBDD is given in Figure 14.8b (Continuation Example 14.13).

the size of decision diagrams. The maximum values of the total autocorrelation function can be used as a measure to determine the number of pairs of isomorphic subtrees rooted at the nodes at the same level in the decision diagram. Using this property, the authors of [11] developed a method for minimization of the width of decision diagrams level by level, providing at each level a maximum number of isomorphic subtrees. Under this assumption, the presented method can be considered as a deterministic method for reduction of the size of decision diagrams.

Stanković and Astola [29] discussed an application of the linear transform of input variables for optimization of adders. Kolpakov and Latypov [13, 14] studied the effectiveness of heuristic strategy for optimization of decision diagrams using affine transforms of variables.

Tree-based algorithms. Topological characteristics of trees are important in affine transform of variables. There are various definitions of a balanced tree. For example, a tree can be called the balanced tree if there exists a bipartition of the vertices into two equal-sized sets and no edges are entirely within one of the parts. Embedding trees into balanced trees is aimed at minimizing the average dilation. Wagner [31] developed an algorithm for embedding trees into balanced trees.

The problem of linear transform of variables in terms of signal processing. Stanković and Astola [29] formulated the linearization problem as follows. For a given $f : C_2^n \rightarrow C_2$, find a nonsingular over $GF(2)$ $(n \times n)$ matrix σ such that $f(x) = \nu(\sigma x)$, and $\mu(\nu) \rightarrow max_{\sigma}$. A solution of this problem is using the Wiener-Khintchine theorem. The complexity of solving the linearization problem for a given f does not exceed $O(n2^n)$ and can be further reduced assuming a compact description of f is given.

Autocorrelation and spectrum of switching functions. Autocorrelation is a useful concept in spectral methods for analysis and synthesis of networks realizing logic functions. For a given n-variable switching function f, the autocorrelation function B_f is defined as

$$B_f(\tau) = \sum_{x=0}^{2^n-1} f(x) f(x \oplus \tau),$$

where $\tau \in \{0, \ldots, 2^n - 1\}$ is a Hamming distance. The product term $f(x)f(x \oplus \tau)$ takes the value of 1 or 0 depending on whether both $f(x)$ and $f(x)f(x \oplus \tau)$ are 1 or one of them is zero, respectively. The Winer–Khintchine theorem states a relationship between the autocorrelation function and Walsh (Fourier) coefficients.The autocorrelation function is invariant to the shift operator $\oplus$ in terms of which B_f is defined. Autocorrelation is useful in applications where we are interested in the equality of some function values, and not in their magnitude. This approach was utilized in [29]. Stanković et al. [30] compared four methods for calculation of the autocorrelation functions using decision diagrams.

Daemen and Rijmen [8] used the correlation properties of switching functions in developing advanced encryption technique. Linear cryptanalysis exploits large correlations between switching functions. The correlation between two switching functions ranges between -1 and 1. If the correlation is different from zero, the switching functions are *correlated.* If the correlation is equal to 1, the switching functions are equal, and if the correlation is equal to -1, the switching functions are each other's complements.

Various results of linear cryptanalysis can be useful in linearly transformed decision diagram construction, in particular:

- ▶ The composition properties of the Walsh–Hadamard spectrum of switching functions derived as convolution of the corresponding spectra. For example, the spectrum of EXOR expressions can be computed by *dyadic shifts.*
- ▶ Properties of correlation matrices and their applications in permutations of switching functions. For example, the following statement is used: a transform of switching function is invertible iff it has an invertible correlation matrix.
- ▶ Cross-correlation (autocorrelation) including the Winer–Khintchine theorem for switching functions.

Autocorrelation for multivalued functions. A generalization of autocorrelation to multioutput p-valued functions of m-variables is straightforward: $f(z) = \{f^{(i)}(z^{(0)}, \ldots, z^{(n-1)})\}$, $i = 0, \ldots, k-1$, $z^{(i)} \in \{0, \ldots, p-1\}$, a system of characteristic functions is defined as $f_r^{(i)} = 1$ for $f^{(i)}(z) = r$, and $f_r^{(i)} = 0$ if $f^{(i)}(z) \neq r$. The total autocorrelation function is defined as

$$B_f(\tau) = \sum_{r=0}^{p-1} \sum_{i=0}^{k-1} B_r^{(i)}(\tau).$$

The same definition applies to integer valued-functions on groups of order p^n, thus it can be used for integer-valued equivalent functions for a multioutput switching function. Further generalizations to functions on arbitrary finite Abelian groups are also straightforward. In this case, the Wiener-Khinchin theorem is defined with respect to Fourier transforms defined in terms of group characters. This study can be found in the paper by Stanković and Astola [29].

Linear transform of variables for word-level expressions. Stanković and Astola [29] used a linear transform of variables for word-level expressions. For a k-output switching function $f^{(i)}$, $i = 0, \ldots, k-1$, of n variables the total autocorrelation function is defined as the sum of autocorrelation functions of each output: $B_f(\tau) = \sum_{i=0}^{k-1} B_{f^{(i)}}(\tau)$. Note that for any $\tau \neq 0$, $B_f \leq B_f(0)$. The set $G_I(f)$ of all values for τ such that $B_f(\tau) = B_f(0) = \sum_{i=0}^{k-1} \sum_{x=0}^{2^n-1} f^{(i)}(x)$ is a group with respect to the EXOR operation.

Linearization of word-level expressions. Linear (affine) transform of input variables is aimed at the improvement of characteristics of decision diagrams (size, shape, and paths). In such an approach, the embedding properties of decision diagrams cannot be controlled and, in general, the resulting decision diagram is not linear.

Linearization of word-level expressions results in linear word-level decision diagrams [1, 28, 32]. An arbitrary logic function can be represented by a set of linear decision diagrams, which can be embedded into arbitrary topological structure (see details in Chapters 34 and 37).

Planar decision diagrams. Karpovsky et al. used linear transformations in construction of planar decision diagrams [12]. Networks without crossings are useful in FPGAs design because crossings produce considerable delays. Planar decision diagrams result in planar networks. It was shown in [12] that:

- ▶ Linearly transformed BDDs constructed by Walsh coefficients have sizes comparable to the sizes of BDDs for the optimal order of variables,
- ▶ For many functions planar linearly transformed BDDs are smaller than planar BDDs (see, for example, the results by Sasao and Butler [25, 26]).

Other techniques. Symmetry is a useful property of switching functions that often addresses the problem of linearization (see, for example, results by Butler et al. [3, 2, 5]). Butler and Sasao [6], and Nagayama et al. [56, 18] proposed to use the average path length in decision diagram techniques. To minimize the average path length, swapping technique of adjacent variables is used, similar to permutations in the linear transform of variables. Ciriani [7] proposed to describe switching functions by so-called *pseudocubes* using affine spaces.

References

[1] Antonenko V, Ivanov A, and Shmerko V. Linear arithmetical forms of k-valued logic functions and their realisation on systolic arrays. *Automation and Remote Control*, Kluwer/Plenum Publishers, 56(3):419–432, Pt.2, 1995.

[2] Bern J, Meinel C, and Slobodova A. Efficient OBDD-based Boolean manipulation in CAD beyond current limits. In *Proceedings of the Design Automation Conference*, pp. 408–413, 1995.

[3] Butler JT, Nowlin JL, and Sasao T. Planarity in ROMDD's of multiple-valued symmetric functions. In *Proceedings of the IEEE 26th International Symposium on Multiple-Valued Logic*, pp. 236–241, 1996.

[4] Butler JT, Dueck G, Shmerko VP, and Yanushkevich SN. Comments on SYMPATHY: fast exact minimization of fixed polarity Reed-Muller expansion for symmetric functions. *IEEE Transactions on Computer-Aided Design of Integrated Circuits and Systems*, 19(11):1386–1388, 2000.

[5] Butler JT, Dueck G, Shmerko VP, and Yanushkevich SN. On the number of generators of transeunt triangles. *Discrete Applied Mathematics*, 108:309–316, 2001.

[6] Butler JT and Sasao T. On the average path length in decision diagrams of multiple-valued functions. In *Proceedings of the IEEE 33rd International Symposium on Multiple-Valued Logic*, pp. 383–390, 2003.

[7] Ciriani V. Synthesis of SPP three-level logic networks using affine spaces. *IEEE Transactions on Computer-Aided Design of Integrated Circuits And Systems*, 22(10):1310–1323, 2003.

[8] Daemen J and Rijmen V. *The Design of Rijndael: AES – The Advanced Encryption Standard.* Springer, Heidelberg, 2002.

[9] Fujita M, Kukimoto Y, and Brayton RK. BDD minimization by truth table permutation. In *Proceedings of the International Symposium on Circuits and Systems*, 4:596–599, 1996.

[10] Günther W and Drechsler R. Efficient manipulation algorithms for linearly transformed BDDs. In *Proceedings of the 4th International Workshop on Applications of Reed–Muller Expansion in Circuit Design*, Victoria, Canada, pp. 225–232, 1999.

[11] Karpovsky MG, Stanković RS, and Astola JT. Reduction of sizes of decision diagrams by autocorrelation functions. *IEEE Transactions on Computers*, 52(5):592–606, 2003.

[12] Karpovsky MG, Stanković RS, and Astola JT. Construction of linearly transformed planar BDD by Walsh coefficients. In *Proceedings of the International Symposium on Circuits and Systems*, Vol. 4, pp. 517–520, 2004.

[13] Kolpakov AV and Latypov RKh. Minimization of BDDs using liner transformation of variables. In *Proceedings of the 5th International Workshop on Applications of Reed–Muller Expansion in Circuit Design*, Mississipi State University, pp. 57–65, 2001.

[14] Kolpakov AV and Latypov RKh. Approximate minimization algorithm for binary decision diagrams using linear transformation of variables. *Automation and Remote Control*, (Kluwer/Plenum Publishers), Special Issue on Arithmetical Logic in Control Systems, 65(6):893–912, 2004.

[15] Meinel Ch and Tehobald T. *Algorithms and Data Structures in VLSI Design.* Springer, Heidelberg, 1998.

[16] Meinel Ch, Somenzi F, and Tehobald T. Linear sifting of decision diagrams and its application in synthesis. *IEEE Transactions on Computer-Aided Design of Integrated Circuits and Systems*, 19(5):521–533, 2000.

[17] Nagayama S and Sasao T. On the minimization of average path length for heterogeneous MDDs. In *Proceedings of the IEEE 34th International Symposium on Multiple-Valued Logic*, pp. 216–222, 2004.

[18] Nagayama S, Mishchenko A, Sasao T, and Butler JT. Exact and heuristic minimization of the average path length in decision diagrams. *International Journal on Multi-Valued Logic and Soft Computing*, in press.??????

[19] Nechiporuk EI. On the synthesis of networks using linear transformations of variables. *Proceedings of USSR Academy of Sciences*, 123(4):610–612, 1958.

[20] Rahardia S and Falkowski BJ. Fast linearly independent arithmetic expansion, *IEEE Transactions on Computers,* 48(9):991–999, 1999.

[21] Rudell R. Dynamic variable ordering for ordered binary decision diagrams. In *Proceedings of the IEEE Conference on Computer Aided Design*, Santa Clara, CA, pp. 42–47, 1993.

[22] Perkowski MA. A fundamental theorem for EXOR circuits. In *Proceedings of the IFIP WG 10.5 Workshop on Applications of the Reed–Muller Expression in Circuit Design*, Hamburg, Germany, pp. 52–60, 1993.

[23] Perkowski M, Sarabi A, and Beyl F. Fundamental theorems and families of forms for binary and multiple-valued linearly independent logic. In *Proceedings of the IFIP WG 10.5 International Workshop on Applications of the Reed–Muller Expansions in Circuit Design*, pp. 288–299, Japan, 1995.

[24] Perkowski MA, Jóźwiak L, Drechsler R, and Falkowski B. Ordered and shared, linearly-independent, variable-pair decision diagrams. In *Proceedings of the International Conference on Information, Communications and Signal Processing*, Singapore, vol. 1, pp. 261–265, 1997.

[25] Sasao T and Butler JT. Planar multiple-valued decision diagrams. In *Proceedings of the IEEE 25th International Symposium on Multiple-Valued Logic*, 1995.

[26] Sasao T and Butler JT. Planar decision diagrams for multiple-valued functions. *International Journal on Multiple-Valued Logic,* 1:39–64, 1996.

[27] Somenzi F. CUDD: CU Decision Diagram Package. University of Colorado at Boulder, http://vlsi.colorado.edu/~fabio/CUDD/, 2001.

[28] Shmerko VP. Malyugin's theorems: a new concept in logical control, VLSI design, and data structures for new technologies. *Automation and Remote Control*, Plenum/Kluwer Publishers, Special Issue on Arithmetical Logic in Control Systems, 65(6):893–912, 2004.

[29] Stanković RS and Astola JT. Some remarks on linear transform of variables in representation of adders by word-level expressions and spectral transform decision diagrams. In *Proceedings of the IEEE 32th International Symposium on Multiple-Valued Logic*, pp. 116–122, 2002.

[30] Stanković RS, Bhattacharaya M, and Astola JT. Calculation of dyadic autocorrelation through decision diagrams. In *Proceedings of the European Conference on Circuit Theory and Design*, pp. 337–340, 2001.

[31] Wagner AS. Embedding all binary trees in the hypercube. *Journal of Parallel and Distributed Computing,* 18:33–43, 1993.

[32] Yanushkevich SN, Shmerko VP, and Dziurzanski P. Linearity of word-level models: new understanding. In *Proceedings of the IEEE/ACM 11th International Workshop on Logic and Synthesis,* pp. 67–72, New Orleans, LA, 2002.

15

Decision Diagrams for Arithmetic Circuits

In this chapter, different arithmetical circuits such as adders and multipliers are revisited, and bit-level and word-level decision diagrams for representation of adders and multipliers are considered.

15.1 Introduction

In arithmetical circuits, representation of many bits at once, that is, word-level data, are used. Therefore, data structures that can represent both bit-level and word-level functions are appropriate. Many types of diagrams beneficial for representation of arithmetic circuits have been proposed, such as:

- Arithmetic decision diagrams (ADDs),
- Edge-valued binary decision diagrams (EVBDDs),
- Multiterminal binary decision diagrams (MTBDDs),
- Binary moment diagrams (BMDs) and hybrid decision diagrams (HDDs), and
- Hybrid spectral transform decision diagrams (HSTDDs).

For example, MTBDDs, BMDs, and HDDs were introduced first for verification of adders and multipliers, and BMDs and HDDs have been shown to implement integer multiplication efficiently (see Chapters 16, 17 and 22).

15.2 Binary adders

The operands of addition are the *addend* and *augend*. The addend is added to the augend to form the sum. In most arithmetic circuits, the augmented operand (the augend) is replaced by the sum, whereas the addend is unchanged. High speed adders are not only for addition but also for subtraction, multiplication and division. The speed of a digital processor depends heavily on the speed of adders. The adders add vectors of bits and the principal problem is to speed up the carry signal.

An *n-bit binary adder* is a combinational circuit that has two n-bit inputs $A = a_{n-1}, \dots a_0$ and $B = b_{n-1}, \dots b_0$ representing the operands A and B, respectively, and n-bit output $S = s_{n-1}, \dots s_0$, and performs binary addition of the input operands. Additional input and output signals, *carry-in* C_i and *carry-out* C_{i+1} are used to implement module-based architecture, i.e., the design of larger adders.

15.2.1 Full-adder

A logic network that performs addition at a single bit position is the generic cell used not only to perform addition but also arithmetic multiplication, division, and filtering operations.

Example 15.1 *The truth table for this network (1-bit adder) is given in Figure 15.1a. Such a network is referred to as a binary* full-adder *(FA). The optimized functions of the full adder outputs, sum* S_i *and* C_{i+1} *are illustrated in Figure 15.1b. The OR and two-level AND-OR combination circuit to implement* S_i *and* C_{i+1} *correspondingly are shown in Figure 15.1c, and an alternative circuit formed of two half-adders (HAs), the subcircuits that compute* p_i *and* g_i*, is given in Figure 15.1d.*

Example 15.2 *Figure 15.2 illustrates the ROBDD, a graph-based representation of the two-output function of the full adder. In this graph, Shannon expansion (S) is used to represent switching functions* s_i *and* c_i *by decision diagrams. Then, the nodes of the minimized decision diagrams, ROBDD, are represented by multiplexers (MUX), so that a multiplexer-based full adder is built (Figure 15.2b).*

15.2.2 Ripple-carry adder

Ripple-carry adder is a multilevel network designed by the connection of full-adders (Figure 15.3). For the multilevel adder, the total time required is calculated as the delay from carry-in C_i to carry-out C_{i+1}.

Depending on the position at which a carry signal has been generated, the propagation time can be variable. In the best case, when there is no carry generation, the addition time will only take into account the time to propagate the carry signal. With a ripple-carry adder, if the input bits A_i and B_i are

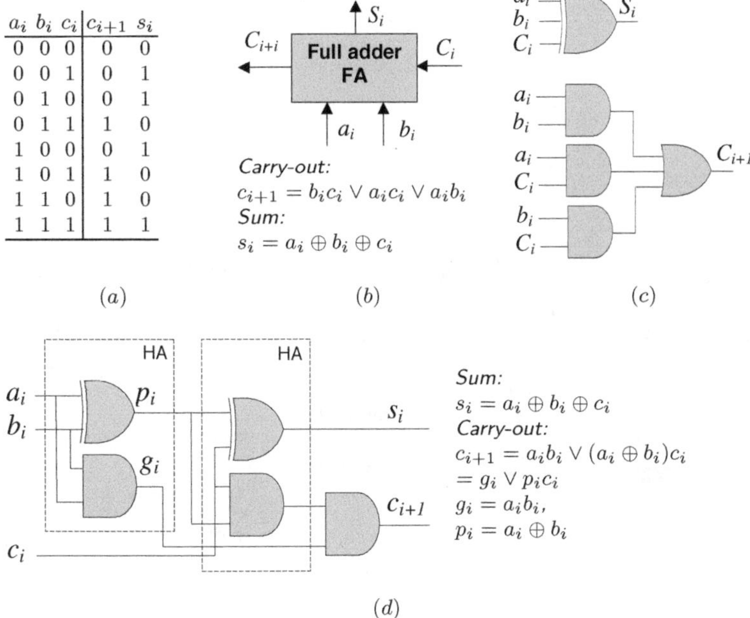

a_i	b_i	c_i	c_{i+1}	s_i
0	0	0	0	0
0	0	1	0	1
0	1	0	0	1
0	1	1	1	0
1	0	0	0	1
1	0	1	1	0
1	1	0	1	0
1	1	1	1	1

FIGURE 15.1
Full adder: the truth table (a), the formal description (b), the logic network over library of AND, OR and EXOR gates (c) and the half adder (HA) based design (d).

different for all positions i, then the carry signal is propagated at all positions (thus never generated), and the addition is completed when the carry signal has propagated through the whole adder. In this case, the ripple-carry adder is as slow as it is large. Actually, ripple-carry adders are fast only for some configurations of the input words, where carry signals are generated at some positions. They can be divided into blocks, where a special circuit detects quickly if all the bits to be added are different. These *carry skip adders* take advantage of both the generation and the propagation of the carry signal.

15.2.3 Binary-coded-decimal format

Let two decimal digits (operands) be denoted by the binary codes $A = a_3a_2a_1a_0$ and $B = b_3b_2b_1b_0$. Note that 4-bit A and B cannot take values larger than 9 (e.g. 1010,1011,1100,1101,1110, and 1111). A carry-in and carry-out are denoted by C_0 and C_3. The data output (sum) is $S = s_3s_2s_1s_0$. This sum must be corrected. These cases are shown in the Table in Figure 15.4, along with the circuit of the BCD adder that consists of the binary

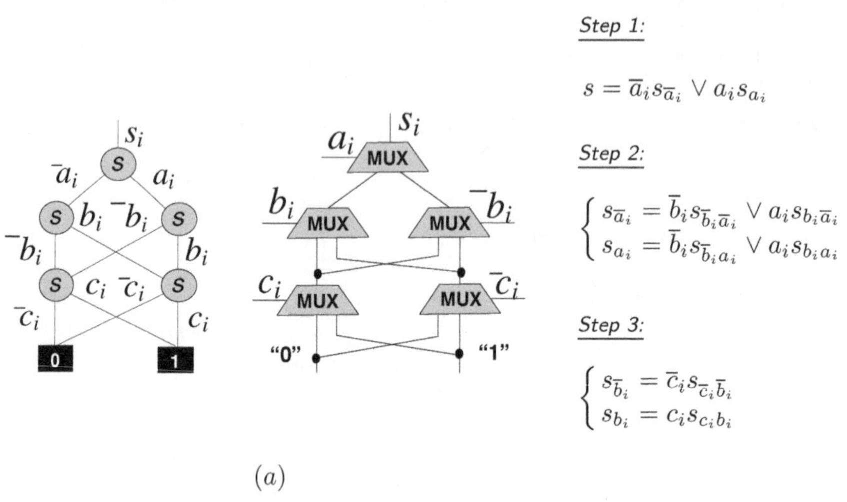

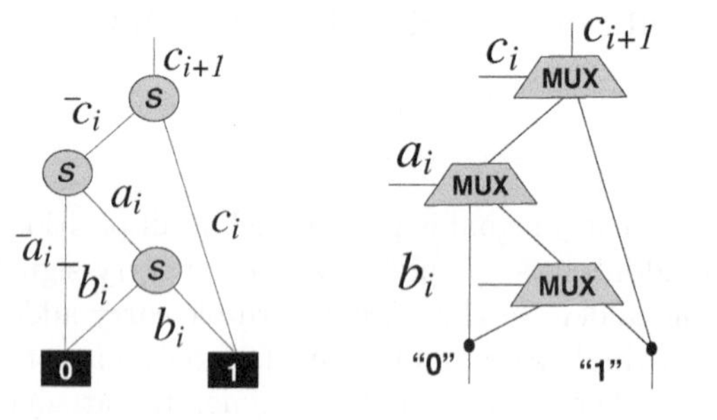

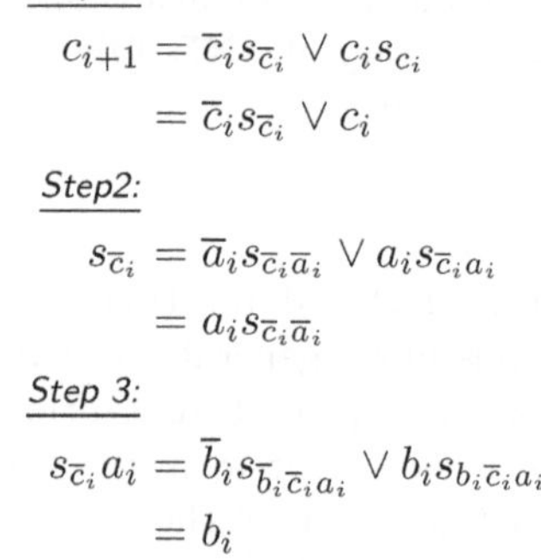

(b)

FIGURE 15.2
Multiplexer-based synthesis of full adder using ROBDDs of functions sum s_i (a) and carry c_i (b).

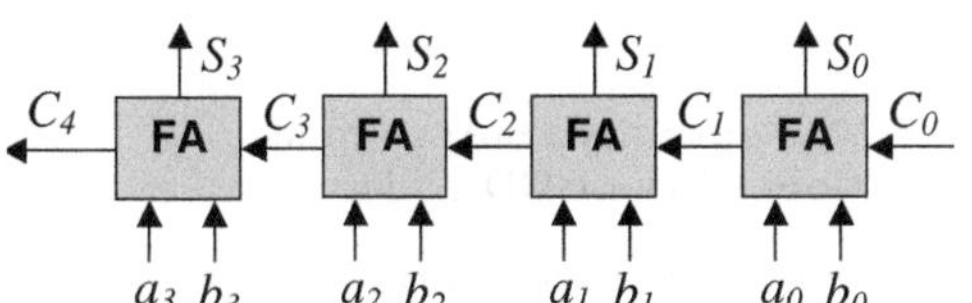

- *Consists of four full adders*
- *Adds two 4-bit unsigned binary numbers and a carry input*
- *Produces 4-bit sum and carry output*

FIGURE 15.3
Ripple-carry 4-bit adder.

adder and the correction circuit. The correction circuit implements addition of number 0110 in the indicated cases. The combinational circuit is implemented with the optimal representation of the function c_{out}, z_3, z_2, z_1, and z_0. Alternatively, MUX based design can be used: formally, the circuit can be represented by a ROBDD (Figure 15.5).

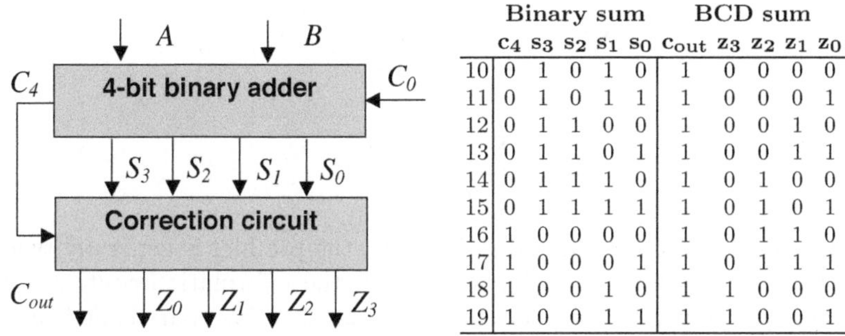

	Binary sum					BCD sum				
	c_4	s_3	s_2	s_1	s_0	c_{out}	z_3	z_2	z_1	z_0
10	0	1	0	1	0	1	0	0	0	0
11	0	1	0	1	1	1	0	0	0	1
12	0	1	1	0	0	1	0	0	1	0
13	0	1	1	0	1	1	0	0	1	1
14	0	1	1	1	0	1	0	1	0	0
15	0	1	1	1	1	1	0	1	0	1
16	1	0	0	0	0	1	0	1	1	0
17	1	0	0	0	1	1	0	1	1	1
18	1	0	0	1	0	1	1	0	0	0
19	1	0	0	1	1	1	1	0	0	1

FIGURE 15.4
Decade of BCD adder and the part of a truth table where correction is needed.

15.3 Multipliers

A combinational multiplier for positive integers is used in floating-point processors and signal processing applications. An $n \times m$ bits combinational multiplier is a combinational circuit that produces the multiplication $0 \leq A \times B \leq (2^n - 1)(2^m - 1)$ (product) of two integer numbers: $0 \leq A \leq 2^n - 1$ (multipli-

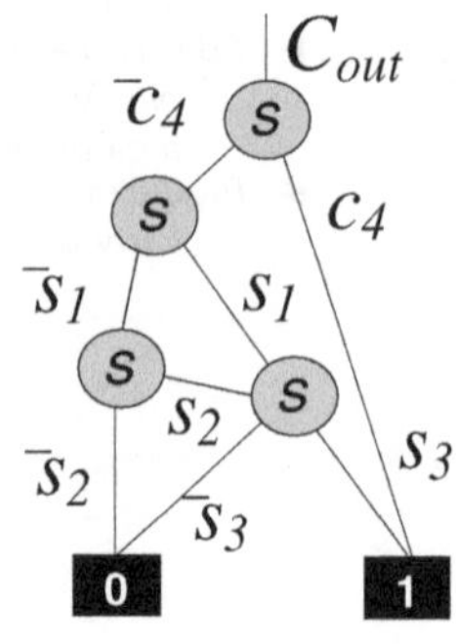

The optimized output functions of the BCD adder:

$$c_{out} = c_4 \vee s_1 s_3 \vee s_2 s_3$$
$$z_0 = s_0$$
$$z_1 = s_1 \overline{s}_3 \overline{c}_4 \vee \overline{s}_1 s_2 s_3 \vee \overline{s}_1 c_4$$
$$z_2 = c_4 \vee s_1 s_2 \vee s_2 \overline{s}_3$$
$$z_3 = s_1 c_4 \vee s_3 c_4 \vee \overline{s}_1 \overline{s}_2 s_3$$

The ROBDD represents the logic function c_{out}

FIGURE 15.5
Decision diagram of the correction function of the BCD adder.

cand) and $0 \leq B \leq 2^m - 1$ (multiplier):

$$A = \sum_{i=0}^{m-1} a_i 2^i, \quad B = \sum_{j=0}^{n-1} b_j 2^j, \quad m \geq n$$

$$A \times B = \sum_{h=0}^{m+n-2} \left(\sum_{i+j=h} a_i b_j \right) 2^h$$

When the operands are interpreted as integers, the product is generally twice the length of the operands in order to preserve the information content*. It is possible to decompose multipliers in two parts. The first part is dedicated to the generation of partial products, and the second one collects and adds them†.

The simplest multiplication can be viewed as repeated shifts and adds (one adder, a shift register, and a small amount of control logic). The disadvantage is that it is slow. One fairly simple improvement to this is to form the matrix of partial products in parallel, and then use a 2-dimensional array of full adders to sum the rows of partial products. This 8×6 structure shown in Figure 15.6 is known as an *array multiplier*. The multiplier consists of $m - 1 = 5$

*Multiplication can be considered as a series of repeated additions. This repeated addition method that is suggested by the arithmetic definition is so slow that it is almost always replaced by an algorithm that makes use of positional number representation.

†As for adders, it is possible to enhance the intrinsic performances of multipliers. Acting in the generation part, the Booth (or modified Booth) algorithm is often used because it reduces the number of partial products (see "Further Reading" section. The collection of the partial products can then be made using a regular array, a Wallace tree, or a binary tree.

n = 8-bit carry-ripple adders and m = 6 arrays of n AND gates. The delay of the multiplier is defined as the critical path equal to the sum of the delays of the buffer circuit connecting input signals and AND gates, the delay of AND gate, and the delays of the adders.

The advantage of the array multiplier is that it is a regular structure and a local interconnect: each cell is connected only to its neighbors. The disadvantage is that the worst case delay path goes from the upper left corner diagonally down to the lower right corner and then across the ripple carry adder, i.e., the delay is linearly proportional to the operand size. The method that can be employed to decrease the delay of the array multiplier is to replace the ripple carry adder with a carry-lookahead adder[‡].

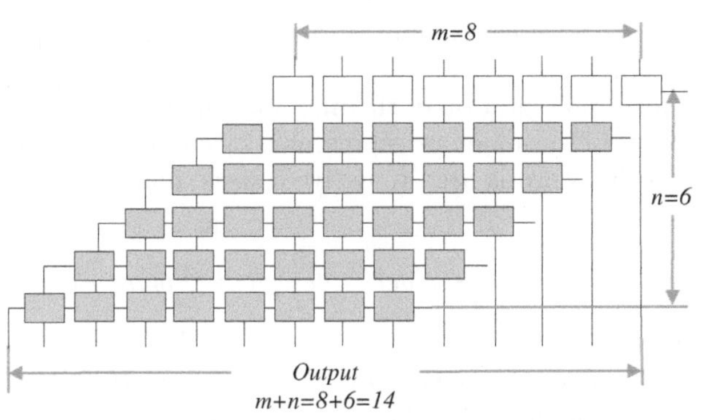

$$
\begin{array}{cccccccccccccc}
 & & & & & & a_7b_0 & a_6b_0 & a_5b_0 & a_4b_0 & a_3b_0 & a_2b_0 & a_1b_0 & a_0b_0 \\
 & & & & & a_7b_1 & a_6b_1 & a_5b_1 & a_4b_1 & a_3b_1 & a_2b_1 & a_1b_1 & a_0b_1 & \\
 & & & & a_7b_2 & a_6b_2 & a_5b_2 & a_4b_2 & a_3b_2 & a_2b_2 & a_1b_2 & a_0b_2 & & \\
 & & & a_7b_3 & a_6b_3 & a_5b_3 & a_4b_3 & a_3b_3 & a_2b_3 & a_1b_3 & a_0b_3 & & & \\
 & & a_7b_4 & a_6b_4 & a_5b_4 & a_4b_4 & a_3b_4 & a_2b_4 & a_1b_4 & a_0b_4 & & & & \\
 & a_7b_5 & a_6b_5 & a_5b_5 & a_4b_5 & a_3b_5 & a_2b_5 & a_1b_5 & a_0b_5 & & & & &
\end{array}
$$

Architecture of an $n \times m$ bit multiplier:
$m - 1 = 5$ $n = 8$-bit carry-ripple adders;
m arrays of n 2-input AND gates

Formal description

$$A \times B = A(\sum_{i=0}^{m-1} b_i 2^i) = \sum_{i=0}^{m-1} A b_i 2^i$$

The multiplication is performed by adding the integers $Ab_i 2^i$. Because b_i is either 0 or 1, we get

$$Ab_i = \begin{cases} 0, & if\ b_i = 0; \\ A, & if\ b_i = 1. \end{cases}$$

FIGURE 15.6
An 8 × 6 multiplier: topology of architecture and multiplication scheme.

Efficient BDDs for multiplier representation can be derived based on the redundant binary adder tree [7] (see "Further reading" section).

[‡]Another approach to collection of the partial products is based on the so called *Wallace tree multiplier*. The number of operations occurring in multiplication is at most $O(mn)$.

15.4 Word-level decision diagrams for arithmetic circuits

Arithmetic circuits are multi-input multi-output structures since they operate with words, that is, binary representations of decimal numbers. Naturally, some word-level diagrams (see Chapters 6, 16, and 17 for details) can be more compact data structures for arithmetic circuits compared against BDDs. This is because the BDD data structure may grow exponentially with the size of the circuit. This causes difficulties of verification of arithmetical circuits (adders and multipliers, as shown in Chapter 22).

15.4.1 Spectral diagrams for representation of arithmetical circuits

Arithmetic, Walsh, and Haar representations exhibit efficient polynomial representation of functions of arithmetical circuits, and so do the corresponding spectral diagrams (see Chapter 7).

Consider a Haar form for representation of the multioutput function of the binary adders. Many outputs of the adder are represented by an integer-valued truth vector, or word, and this word is used to derive a spectrum.

Example 15.3 *A two-bit adder is described (Figure 15.7):*

- *By the truth vector* **F**.
- *By the Haar spectrum* $\mathbf{H}_f$*; the corresponding HSTDD is given in Figure 15.7a. Note that any permutation of the function values may destroy the constant subvectors.*
- *By the Haar spectrum* $\mathbf{H}_f$ *after the linear transformation of variables. The corresponding HSTDD is given in Figure 15.7b. In this HSTDD the number of non-zero coefficients is reduced from 15 to 7.*

This implies that:

- The HSTDD is the simplest of these representations and
- Linear transform for variables increases complexity, since reordering of the function values destroys sequences of identical values.

Some types of spectral diagrams derived from word-level representation of multi-output functions of the multiplier are very compact compared with BDDs.

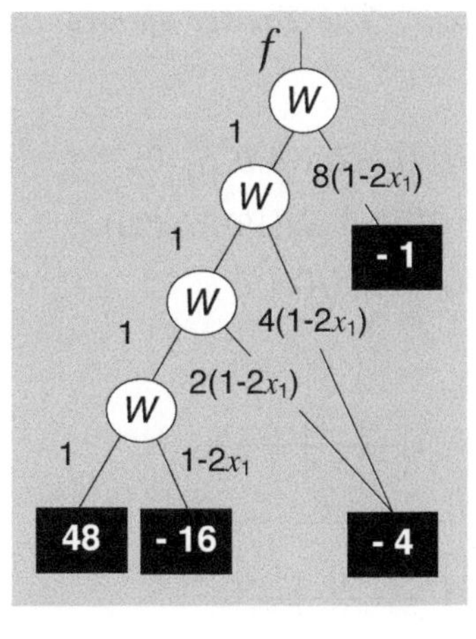

(*a*)

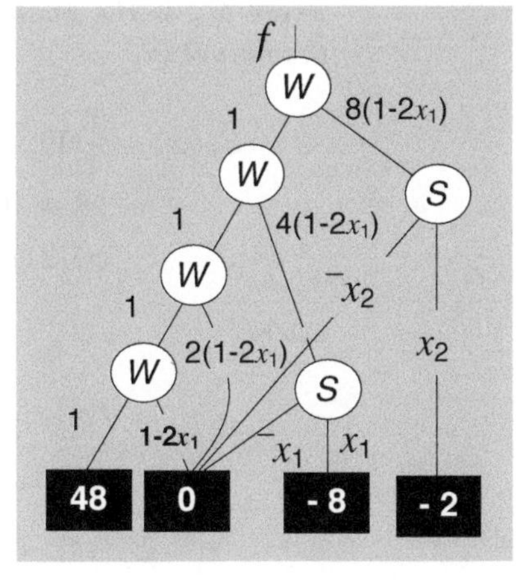

(*b*)

Truth vector — *Haar spectrum* — *Haar spectrum after linear transformation of variables*

$$\mathbf{F} = \begin{bmatrix} 0 \\ 1 \\ 2 \\ 3 \\ 1 \\ 2 \\ 3 \\ 4 \\ 2 \\ 3 \\ 4 \\ 5 \\ 3 \\ 4 \\ 5 \\ 6 \end{bmatrix} \qquad \mathbf{H}_f = \begin{bmatrix} 48 \\ -16 \\ -4 \\ -4, \\ -4 \\ -4 \\ -4 \\ -4 \\ -1 \\ -1 \\ -1 \\ -1 \\ -1 \\ -1 \\ -1 \\ -1 \end{bmatrix} \qquad \mathbf{H}_f = \begin{bmatrix} 48 \\ 0 \\ 0 \\ 0 \\ -8 \\ 0 \\ -8 \\ 0 \\ -2 \\ -2 \\ 0 \\ 0 \\ -2 \\ -2 \\ 0 \\ 0 \end{bmatrix}$$

FIGURE 15.7
Two-bit binary adder representation using HSTDD (a) and HSTDD with linearly transformed variables (b) (Example 15.3).

Example 15.4 *The outputs of the 3-bit multiplier are represented in a word-level format. Word-level description is considered as a function on the finite non-Abelian group*

$$G_{64} = Q_2 \times Q_2,$$

where Q_2 is the quaternion group. The Fourier spectrum of the output is

$$\begin{aligned} S_f(w) = \frac{1}{4}[&49, -7, 28, 0, S_f(4), -7, 1, 4, 0, S_f(9), \\ -&28, 4, 16, 0, S_f(14), 0, 0, 0, S_f(19), S_f(20), \\ &S_f(21), S_f(22), S_f(23), S_f(24)]^T, \end{aligned}$$

where

$$S_f(4) = S_f(20) = -7\left[\begin{array}{c|c} 1-i & -1+i \\ \hline -1+i & 1+i \end{array}\right],$$

$$S_f(9) = S_f(21) = \left[\begin{array}{c|c} 1-i & -1+i \\ \hline -1+i & 1+i \end{array}\right],$$

$$S_f(14) = S_f(22) = 4\left[\begin{array}{c|c} 1-i & -1+i \\ \hline -1+i & 1+i \end{array}\right],$$

$$S_f(19) = S_f(23) = \left[\begin{array}{c|c} 0 & 0 \\ \hline 0 & 0 \end{array}\right],$$

$$S_f(24) = 2\left[\begin{array}{r|r|r|r} -i & -1+i & 1 & -1-i \\ \hline -1+i & 1 & 1+i & i \\ \hline i & 1-i & 1 & -1-i \\ \hline -1+i & -1 & 1+i & i \end{array}\right].$$

Three-bit multiplier representation by MTDD on

$$G_{64=C_4 \times C_4 \times C_4}$$

and matrix-valued FNADD on

$$G_{64=Q_2 \times Q_2}$$

are shown in Figure 15.8a and 15.8b correspondingly.

Details for this example are given in [70]. In [70], the FNADD with number-valued constant nodes for the 3-bit multiplier for the decomposition $G_{64} = Q_2 \times Q_2$ was designed. In this FNADD, matrix-valued nodes in mvFNADD are represented by MTBDDs.

15.4.2 Representation of adders by linear decision diagrams

Details on designing linear decision diagrams (LDDs), a special case of word-level diagrams, are given in Chapter 37. In this section, a particular example of utilizing LDDs for representation of multilevel and two-level implementation

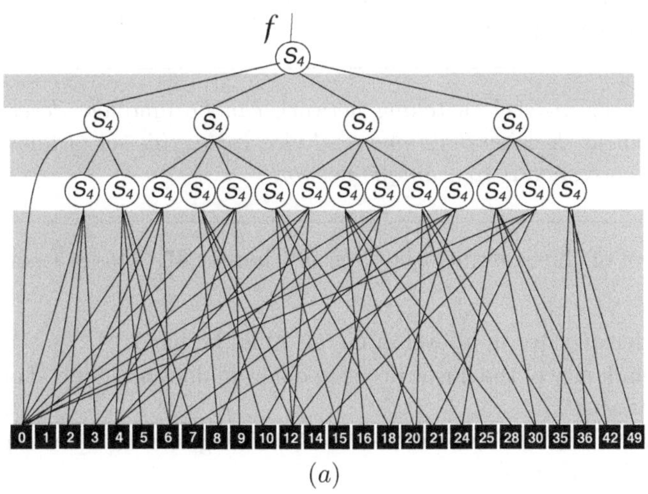

(a)

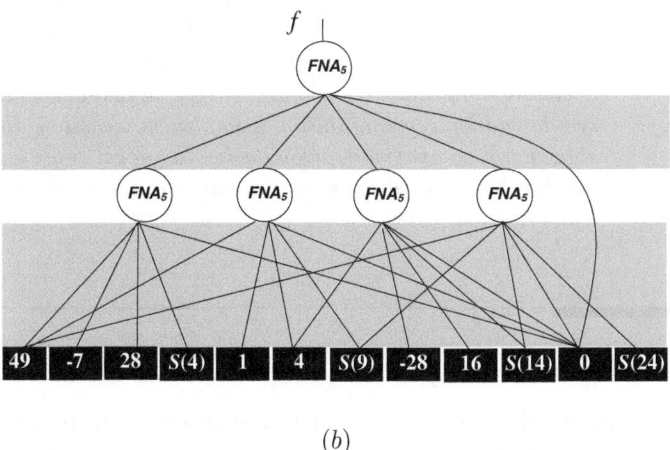

(b)

FIGURE 15.8
Three-bit multiplier representation by MTDD on $G_{64=C_4 \times C_4 \times C_4}$ (a) and matrix-valued FNADD on $G_{64=Q_2 \times Q_2}$ (Example 15.4).

of the binary adder is considered.

An arbitrary r-level combinational circuit is represented by r linear decision diagrams, i.e., for each level of a circuit a linear diagram is designed. The complexity of this representation is $O(G)$, where G is the number of gates in the circuit. The outputs of this model are calculated by transmission data through this set of diagrams.

> *An arbitrary m-level switching network can be uniquely described by a set of m linear decision diagrams, and vice versa; this set of linear decision diagrams corresponds to a unique network.*

The proof of this statement is given in Chapter 37. This statement implies that:

- ▶ The order of gates in a level of circuit must be fixed.
- ▶ The complexity of the linear decision diagram does not depend on the order of variables.
- ▶ Data transmission between linear decision diagrams representing the levels must be provided.

Example 15.5 *Figure 15.9 illustrates a set of linear decision diagrams that represents the three-level circuit depicted in Figure 15.1(d). This is an example of the appropriately chosen output labels (MSB and LSB) which yield the simplest linear decision diagrams. In Figure 15.10, a two-level (PLA-type) two-bit adder representation using linear decision diagrams is shown for comparison. Obviously, two-level representation is not the best choice for deriving linear decision diagram from.*

15.5 Further reading

Overview. An excellent overview of digital arithmetic can be found in [6]. Conversion between digital systems employed in the decision diagram techniques was studied in [11].

Word-level decision diagram.

ADDs were introduced by Bahar et al. [1]. Lai and Sastry [10] proved that EVBDDs are efficient for circuit representation and verification (see Chapter 22). MTBDDs were introduced by Clarke et al. [7] for manipulation of bit-wise representation of many functions at once (the terminal nodes of the MTBDDs are integer valued, but otherwise they are like BDDs).

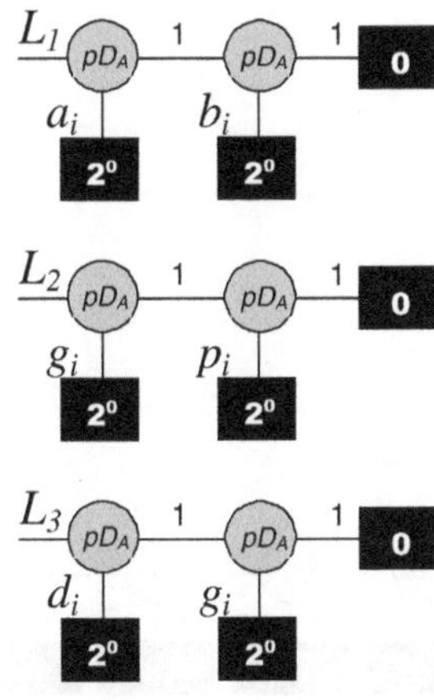

Level L1:

$$L_1 = a_i + b_i$$
$$g_i = \Xi^1\{L_1\}$$
$$p_i = \Xi^0\{L_1\}$$

Level L12:

$$L_2 = g_i + p_i,$$
$$c_{i+1} = \Xi^1\{L_2\}$$
$$s_i = \Xi^0\{L_2\}$$

Level L3:

$$L_3 = d_i + g_i,$$
$$c_{i+1} = \Xi^1\{L_3\}$$

FIGURE 15.9
Multilevel two-bit adder representation using linear decision diagrams (Example 15.5); each linear decision diagram describes the linear form L_i; the masking operators for each level output are shown as well (Example 15.5).

Bryant and Chen [4] proposed binary moment diagrams (BMDs) for the design and verification of arithmetic circuits. These diagrams utilize a decomposition called *moment decomposition*, that is $f = f_{\overline{x}} + x(f_x - f_{\overline{x}})$. This is an arithmetic analog of positive Davio decomposition, useful for representation of arithmetical forms of a switching function.

Hybrid decision diagrams. Clarke et al. [4] introduced HDDs, a trade-off between MTBDDs and BMDs, to represent the integer valued function in the task of verification of arithmetic circuits. An extension of BMDs is *BMDs, which adopt the concept of edge weights. For instance, *BMDs have been used to verify multipliers [5]. HSTDDs have been considered by Stankovic and Astola in [70].

Residue BDDs have been proposed by Kimura [8]. Residue BDDs avoid the node explosion problem of BDDs, and were used to verify large arithmetic circuits. Binary-to-residue and residue-to-binary conversion were considered by Vinnakota [15] and Wang [16].

Other implementations of arithmetic functions. Harata et al. [7] deployed the binary adder tree to design a high-speed multiplier. Sinha and Srimani [12] employed systolic processors to implement binary multiplication.

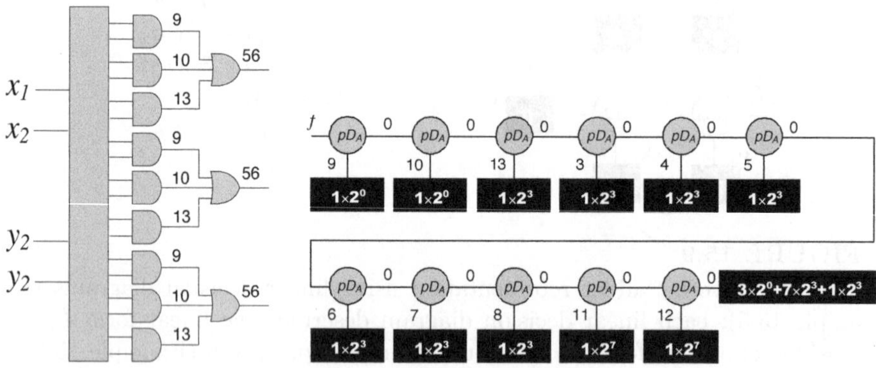

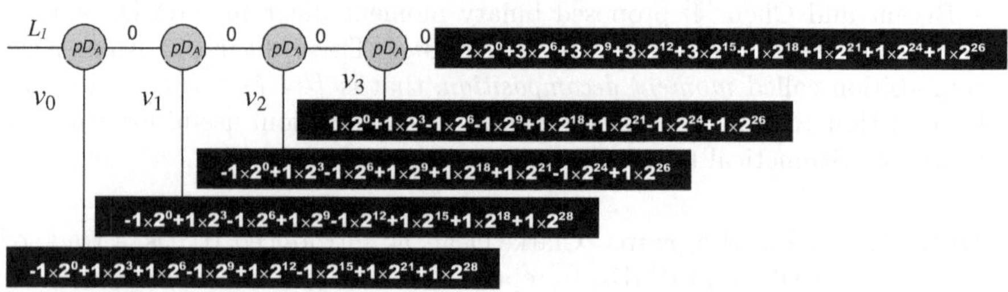

FIGURE 15.10

Two-level two-bit adder representation using linear decision diagrams (Example 15.5).

Cellular neural network was proposed by Siu and Brick [13] for implementation of arithmetic functions.

References

[1] Bahar RI, Frohm EA, Gaona CM, Hachtel GD, Macii E, Pardo A, and Somenzi F. Algebraic decision diagrams and their application. In *Proceedings of the International Conference on CAD*, pp. 188-191, 1993.

[2] Bryant RE and Chen YA. Verification of arithmetic circuits with binary moment diagrams. In *Proceedings of the 32rd ACM/IEEE Design Automation Conference*, pp. 535–541, 1995.

[3] Clarke E, Fujita M, McGeer P, McMillan KL, Yang J, and Zhao X. Multi terminal binary decision diagrams: An efficient data structure for matrix representation. In *Proceedings of the International Workshop on Logic Synthesis*, pp. P6a:1-15, 1993.

[4] Clarke EM, Fujita M, and Zhao X. Hybrid decision diagrams – overcoming the limitations of MTBDDs and BMDs. In *Proceedings of the International Conference on CAD*, pp. 159-163, 1995.

[5] Drechsler R, Becker B, and Ruppertz S. K*BMD – efficient data structure for verification. In *Proceedings of the European Design and Test Conference*, pp. 2–8, 1996.

[6] Ercegovac MD and Lang T. *Digital Arithmetic.* Morgan Kaufmann, 2003.

[7] Harata Y, Nakamura Y, Nagase H, Takigawa M, and Takagi N. A high-speed multiplier using a redundant binary adder tree. *IEEE Journal of Solid States Circuits*, 22(1):28–34, 1987.

[8] Kimura S. Residue BDD and its application to the verification of arithmetic circuits. In *Proceedings of the 32nd Design Automation Conference*, pp. 542–545, 1995.

[9] Kinoshita E and Lee Ki-Ja. A residue arithmetic extension for reliable scientific computation. *IEEE Transactions on Computers*, 46(2):129-138, 1997.

[10] Lai YT and Sastry S. Edge-valued binary decision diagrams for multi-level hierarchical verification. In *Proceedings of the DAC*, pp. 668–613, 1992.

[11] Phatak DS and Koren I. Hybrid signeddigit number systems: a unified framework for redundant number representations with bounded carry propagation chains. *IEEE Transactions on Computers,* Special issue on Computer Arithmetic TC43, (8):880–891, 1994.

[12] Sinha BP, and Srimani PK. Fast parallel algorithms for binary multiplication and their implementation on systolic architectures. *IEEE Transactions on Computers*, 38(3):424–431, 1989.

[13] Siu KY and Bruck J. Neural computation of arithmetic functions. *Proceedings IEEE*, 78:1669–1675, 1990.

[14] Stanković RS and Astola JT. *Spectral Interpretation of Decision Diagrams.* Springer, Heidelberg, 2003.

[15] Vinnakota B, and Rao VVB. Fast conversion technique for binary-residue number systems. *IEEE Transactions on Circuits and Systems - I*, 41(12):927–929, December 1994.

[16] Wang Y. Residue-to-binary converters based on new Chinese remainder theorems. *IEEE Transactions on Circuits and Systems - II*, pp. 197–206, March 2000.

16

Edge - Valued Decision Diagrams

In this chapter, edge-valued decision diagrams are discussed.

Edge-valued decision diagrams for switching functions are related to the arithmetic transform and represent a given function f in the form of the algebraic polynomials.

Edge-valued diagrams are mostly used for representation of multiple-output switching functions and, therefore, are convenient for representation of arithmetic functions such as adders, multipliers, and similar devices.

16.1 Introduction

A rationale for introduction of edge-valued decision diagrams and the basic principles upon which they are based can be found in the following considerations:

- ► For many functions often met in practice as, for instance, arithmetic circuits, BDDs have exponential complexity. For instance, to represent dividers by BDDs is a hard task and requires a considerable memory. Various edge-valued decision diagrams have been introduced to overcome this problem.
- ► BDDs and shared BDDs for multi-output functions have a large size when a function takes many different values, or equal values are distributed in such a way that they do not produce isomorphic subtrees. However, in such cases it may be that these values contain many common factors. This feature is exploited to define a particular class of decision diagrams having some attributes assigned to the edges.
- ► Similarly, in such functions, it may be better to set the values of all constant nodes, which would be many, to zero, thus representing them by a single constant node showing the value 0, and instead of that, assign an appropriately determined weighting coefficient to the edges.

In this way, several edge-valued decision diagrams have been introduced and the theory extended to multivalued functions.

16.2 Terminology and abbreviations of edge-valued decision trees and diagrams

Using a spectral approach, edge-valued decision diagrams can be classified with respect to non-attributed and attributed edges (Figure 16.1). The weighting coefficients of the attributed edges can be distinguished with respect to rules of manipulation of coefficients.

In Table 16.1, a specification of various edge-valued decision trees and diagrams is given with respect to node values and edge weights.

Edge-valued binary decision trees (EVBDTs) and diagrams (EVBDDs) are defined as a modification of BDDs by using additive weighting coefficient at the edges. Edge-valued decision diagrams belong to the class of diagrams with *attributed edges.* These diagrams are characterized by the following features:

- ▶ Constant nodes do not show function values or values of Fourier-like spectrums as in other decision diagrams, but:
  - (*a*) Are reduced to zero, or
  - (*b*) Are selected as some common factors in certain Fourier-like spectrum,
  - (*c*) Represent values of coefficients in partial Fourier-like transforms.
- ▶ To correct values of constant nodes, some weighting coefficients are assigned to the edges, relating these values to the values of the functions represented.

Factored edge-valued binary decision trees (FEVBDTs) and diagrams (FEVBDDs) are defined as another modification of BDDs by using both additive and multiplicative weighting coefficients at the edges.

Binary moment decision trees (*BMTs) and diagram (*BMDs) are defined as an edge-valued version of binary moment diagrams (BMDs) by exploiting common factors in the arithmetic spectrum for the functions represented.

Kronecker binary moment decision trees (K*BMTs) and diagrams (K*BMDs) are defined as a further generalization of the methods used in previously mentioned diagrams, by allowing different decomposition rules for nodes.

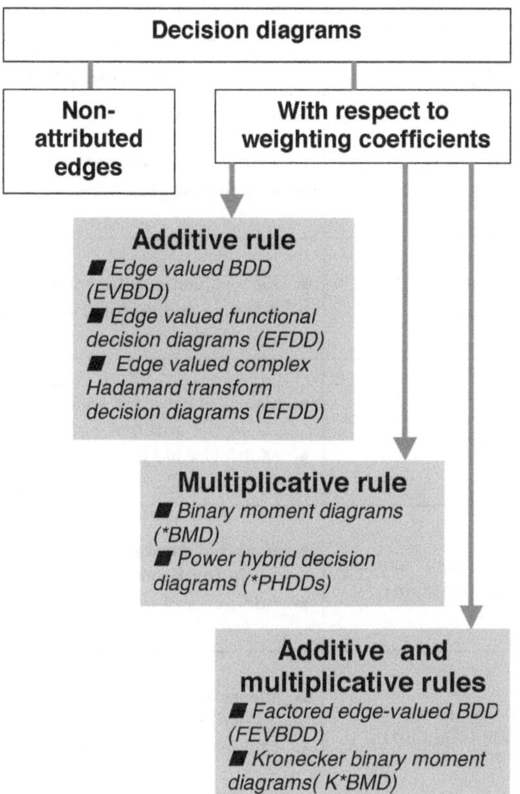

FIGURE 16.1
Classification of edge-valued decision diagrams based on a spectral approach.

Multiplicative power hybrid decision diagrams (*PHDDs) are defined as another particular extension of the idea of exploiting common factors in the Kronecker spectra of functions to be represented.

TABLE 16.1
Specifications of edge-valued decision trees and diagrams.

Type	Specification
Edge-valued decision tree (diagram)	▶ Zero-valued constant nodes ▶ Additive weights
Factored edge-valued decision tree (diagram)	▶ Zero-valued constant nodes ▶ Additive and multiplicative weights
*Binary moment decision tree (diagram)	▶ Values of constant nodes are common factors in the arithmetic spectrum ▶ Multiplicative weights
Kronecker *binary moment decision tree (diagram)	▶ Values of constant nodes are common factors in the Kronecker spectra ▶ Additive and multiplicative weights
*Multiplicative power hybrid decision tree (diagram)	▶ Values of constant nodes are derived from Kronecker spectra ▶ Multiplicative weights

Edge-valued binary decision diagrams, EVBDDs, are decision diagrams in which weighting coefficients are assigned to the edges in the decision diagram to achieve compact representations.

In EVBDDs:

- *One constant node and its value is set to zero irrespective of the represented switching function f.*
- *Procedures to assign f to the decision tree, and to determine f from this tree, are related to the edges and weighting coefficients at the edges, not to the nodes.*

With respect to the structure of weights, decision diagrams with

- *Additive weights*; examples of diagrams with this structure of weights are EVBDDs and FEVBDDs.
- *Multiplicative weights*; examples of diagrams with this structure of weights are *BMDs.
- Both additive and multiplicative weights; examples of diagrams with this structure of weights are FEVBDDs and K*BMDs.

16.3 Spectral interpretation of edge-valued decision diagrams

First, edge-valued binary decision diagrams, EVBDDs, [7], [9] were introduced to improve the efficiency of representation of multiple-output switching functions for which MTBDDs are of exponential size. For example, such functions describe outputs of two-variable function generators, n-bit adders, n-bit multipliers, and some other arithmetic circuits.

Spectral interpretation of edge-valued decision diagrams is based on partial spectral transforms.

Edge-valued decision diagrams on dyadic groups. In a group-theoretic approach, these decision diagrams are decision diagrams on dyadic groups. Thus, they consist of nodes with two outgoing edges. Spectral interpretation relates them to some spectral transforms and *partial* spectral transforms for functions on dyadic groups into the complex field C (see details in Chapter 13).

Partial spectral transforms. Consider a spectral transform whose transform matrix can be represented by the Kronecker product of some basic transform matrices. Thus, the transform matrix is

$$\mathbf{K}(n) = \bigotimes_{i=1}^{n} \mathbf{K}_i(1),$$

where $\mathbf{K}_i(1)$ are basic transform matrices and n is the number of variables. From the theory of fast Fourier transform (FFT) algorithms, such a matrix $\mathbf{K}(n)$ can be factorized into the ordinary matrix product of sparse n matrices, which can be also represented as the Kronecker product. Therefore,

$$\mathbf{K}(n) = \prod_{i=1}^{n} \mathbf{C}_i(n),$$

where

$$\mathbf{C}_i(n) = \bigotimes_{j=1}^{n} \mathbf{D}_j(1),$$

and

$$\mathbf{D}_j = \begin{cases} \mathbf{K}_i(1), \text{ if } j = i, \\ \mathbf{I}(1), \quad \text{if } j \neq i, \end{cases}$$

with the identity matrix $\mathbf{I}(1)$.

In the FFT algorithm based on this factorization, each of the matrices $\mathbf{C}_i(n)$ performs the transform with respect to the ith variable. Therefore, these matrices are transform matrices that define the partial Kronecker transforms related to the considered Kronecker transform.

Function of a non-terminal node. In EVBDDs, a non-terminal node represents the switching function

$$f = x_i(v_i + f_l) + (1 - x_i)f_r,$$

where f_l, f_r denote the left and right outgoing edge of a node in EVBDTs, and v_i is the weight associated to the left edge. The weight assigned to f_r is always equal to zero, since that provides EVBDDs for a fixed order of variables that are canonical representations for a switching function f.

The decomposition function in EVBDDs can be written as

$$f = x_i v_i + f_0 + x_i(f_1 - f_0),$$

to point out the similarity to the arithmetic transform. Therefore, the nodes in EVBDDs are integer counterparts of the positive Davio nodes.

Example 16.1 *Figure 16.2 shows the EVBDT for a switching function f of three variables ($n = 3$). The left and right outgoing edges correspond to f_1, and f_0, respectively. It follows that all the calculations in EVBDDs are done from the right to the left.*

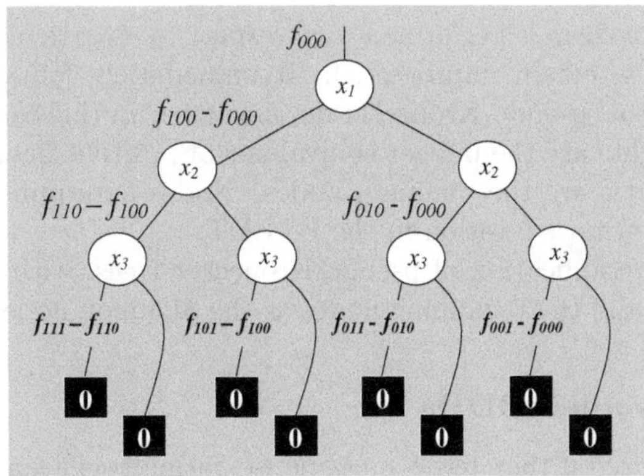

A non-terminal node

$f = x_i(v_i + f_l) + (1 - x_i)f_r$

The weights at the edges

$$
\begin{aligned}
S_{1f}(4) &= f(100) - f(000),\\
S_{2f}(2) &= f(010) - f(000),\\
S_{2f}(6) &= f(110) - f(100),\\
S_{3f}(1) &= f(001) - f(000),\\
S_{3f}(3) &= f(011) - f(010),\\
S_{3f}(5) &= f(101) - f(100),\\
S_{3f}(7) &= f(111) - f(110)
\end{aligned}
$$

FIGURE 16.2
An edge valued binary decision tree, EVBDT, for a switching function f of three variables ($n = 3$) and the partial arithmetic transform coefficients (Examples 16.1 and 16.2).

The following remark can be derived by inspection of the weighting coefficients in EVBDDs and a comparison with partial arithmetic transform coefficients.

In EVBDDs, the weights at the edges at the i-th level are equal to certain partial arithmetic transform coefficients with respect to the i-th variable.

Example 16.2 *In Figure 16.2, the weights at the edges in EVBDT are equal to $S_{1f}(4) = f(100) - f(000)$. The weights at the level i are values of spectral coefficients in the partial arithmetic spectrum with respect to the i-th variable. Involved are coefficients where the i-th bit in binary representation of the coefficient index changes from 0 to 1, but are written from right to left.*

These weights in EVBDT shown in Figure 16.2 are viewed as *partial arithmetic transform* coefficients. A particular primary product of switching variables corresponds to each coefficient corresponds in terms of which the arithmetic transform is defined. Therefore, when weights are multiplied by the corresponding primary products, it follows that an EVBDT represents the function f in the form of the arithmetic polynomial for f.

16.3.1 Relationship of EVBDDs and Kronecker decision diagrams

In a partial transform, the transform is performed with respect to a particular variable, while other variables remain unprocessed. It immediately follows that an EVBDT corresponds to a pseudo-Kronecker decision tree. In this tree, nodes determined by the weights are the integer equivalents of positive Davio nodes, and the remaining nodes are the Shannon nodes. Nodes determined by weights are nodes at the leftmost position in the EVBDT.

Therefore, EVBDTs are a modification of pseudo-Kronecker trees, with n integer positive Davio nodes, and the remaining nodes as the Shannon nodes.

16.3.2 Factored edge-valued BDDs

In EVBDTs, nodes can be shared if they have the same weighting coefficients assigned to the edges. In EVBDTs for some functions there are many nodes at the last level above the constant nodes. *Factored EVBDDs* (FEVBDD) are a further modification of EVBDDs introduced by attempting to increase the possibility of sharing nodes at this level. In FEVBDTs, the weights assigned to the edges of the nodes at the last level are normalized to 1, and the corresponding multiplicative weights are assigned to the edges at the upper levels. Again it is attempted to also share these weights among different nodes whenever possible.

Function of a non-terminal node. In FEVBDDs, the non-terminal nodes represent the function

$$f = x_i(v_i + factor_i f_1) + (1 - x_i)f_0,$$

where $factor_i$ is the multiplicative weight at the edge corresponding to f_l.

Example 16.3 *A FEVBDT has the same form as an EVBDT, and the difference is just in the weights assigned to the edges. In FEVBDTs, besides additive weights determined in the same way as in EVBDTs, the multiplicative weights derived from the normalization of the additive weights at the edges towards the constant nodes are allowed.*

16.4 Binary moment diagrams

BMDs have been introduced by Bryant and Chen [1] and belong to the class of multiterminal decision diagrams for representation of arithmetic spectra. Hence, they are word-level diagrams. More details on BMDs are given in Chapter 17.

16.4.1 Binary moment decision diagrams –*BMDs

Binary moment decision diagrams (*BMDs) are an edge-valued version of BMDs. In *BMDs, weights at the edges are factors in the arithmetic transform coefficients. They can be determined during the calculation of arithmetic transforms over BDDs for a switching function f, step by step by using the fast Fourier-like algorithm adapted for implementation over BDDs.

Function of a non-terminal node. The decomposition used in *BMDs is the modification of that in BMDs and is given by

$$f = w_i(f_0 + x_i(f_1 - f_0)),$$

where w_i are the weighting coefficients. The number of factors allowed in the factorization of the arithmetic transform coefficients is equal to or less than the number of levels in the decision tree, i.e., to the number of variables in switching function f.

Example 16.4 *Figure 16.3 shows the *BMD for f given its arithmetic spectrum:*

$$\mathbf{A}_f = [8, -20, 2, 4, 12, 24, 15, 0]^T$$

These values can be factored as

$$\overbrace{2 \cdot 2 \cdot 2}^{8},\ \overbrace{2 \cdot 2 \cdot (-5)}^{-20},\ \overbrace{2 \cdot 1 \cdot 1}^{2},\ \overbrace{2 \cdot 1 \cdot 2}^{4},\ \overbrace{3 \cdot 4 \cdot 1}^{12}, \overbrace{3 \cdot 4 \cdot 2}^{24},\ \overbrace{3 \cdot 1 \cdot 5}^{15},\ \overbrace{3 \cdot 1 \cdot 0}^{0}$$

*To produce a *BMD for f the first and second factors are moved to the first and the second level in the BMD. The third factors are kept as the values of constant nodes.*

16.4.2 Arithmetic transform related decision diagrams

Spectral interpretation shows that the arithmetic transform is used in several decision diagrams to decompose f and assign it to a decision tree.

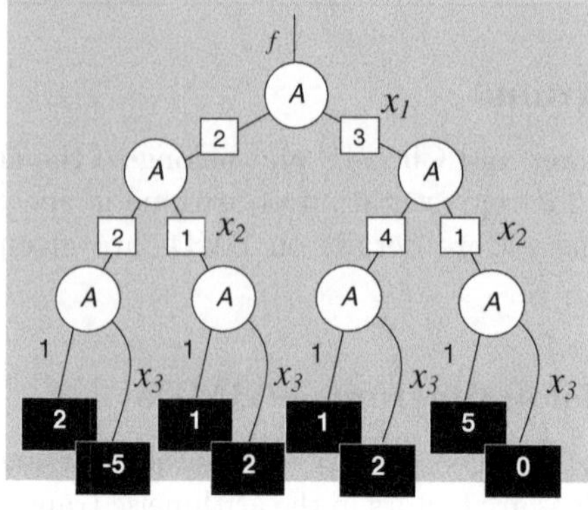

The arithmetic spectrum
$\mathbf{A}_f = [8, -20, 2, 4, 12, 24, 15, 0]^T$.
Factored representation

$$\begin{aligned} 8 &= 2 \cdot 2 \cdot 2, \\ -20 &= 2 \cdot 2 \cdot (-5), \\ 2 &= 2 \cdot 1 \cdot 1, 2 \cdot 1 \cdot 2, \\ 4 &= 3 \cdot 4 \cdot 1, \\ 12 &= 3 \cdot 4 \cdot 2, \\ 24 &= 3 \cdot 1 \cdot 5, \\ 15 &= 3 \cdot 1 \cdot 0 \end{aligned}$$

FIGURE 16.3
*BMD for switching function f (Example 16.4).

*ACDDs, BMDs, EVBDDs, FEVBDDs, and *BMDs represent switching function f in the form of the arithmetic expression for f. These decision diagrams can be regarded as* arithmetic transform related decision diagrams.

These diagrams represent an integer counterpart of the Reed–Muller related decision diagrams, which are a subset of AND-EXOR related decision diagrams.

16.4.3 Kronecker binary moment diagrams

Kronecker diagrams are introduced with the main reason being to increase the number of possible decomposition rules, which results in an increased number of different decision diagrams to represent a given function f. The same reason motivated the introduction of edge-valued versions of Kronecker BMDs, Kronecker *BMDs (K*BMDs).

In a K*BMD, a Kronecker spectrum is determined for switching function f. Then, the spectral coefficients are factored and some factors assigned to the edges as in *BMDs.

16.5 Further reading

Edge-valued diagrams. First, edge-valued binary decision diagrams, EVBDDs, [7], [9] were introduced to improve the efficiency of representation of multiple-output switching functions for which MTBDDs are of exponential size. For example, such functions describe outputs of two-variable function generators, n-bit adders, n-bit multipliers, and some other arithmetic circuits.

Factored edge-valued diagrams. Factored EVBDDs have been introduced by Vrudhula et al. [27] to further improve reduction possibilities in EVBDDs, by increasing the number of isomorphic subtrees at the last level in the diagram. It was shown in [6] that FEVBDDs are EVBDDs with normalized weights in the edges towards the constant nodes. The additive weights are taken as 1 and multiplicative weights as 0.

Binary moment diagrams. BMDs are introduced by Bryant and Chen [1] in order to efficiently represent arithmetic circuits for their verification efficiently in terms of space and time. *BMDs are edge-valued version of BMDs. The difference with EVBDDs is in the order of the decompositions applied in the definition of these decision diagrams. It was shown in [4, 5] that, similarly to EVBDDs, the constant nodes can be set to zero and both additive and multiplicative weights introduced. Similarly to FEVBDDs, the normalization of the weights can be performed to get K*BMDs with integer valued weights.

Kronecker binary moment diagrams. Kronecker *BMDs, K*BMDs, [3] are derived by replacing the role of arithmetic transform in *BMDs by the Kronecker transforms [6].

Multiplicative power hybrid diagrams. Unlike other decision diagrams, multiplicative power hybrid decision diagrams (*PHDD) are capable of representing floating point arithmetics [2]. They are introduced to improve the performance of decision diagrams useful in circuit verification and similar applications.

References

[1] Bryant RE and Chen YA, Verification of arithmetic functions with binary moment diagrams. In *Proceedings of the CMU-CS-94-160*, 1994.

[2] Chen YA, and Bryant RE. An efficient graph representation for arithmetic circuit verification. *IEEE Transactions on CAD*, 20(12):1443–1445, 2001.

[3] Drechsler R and Becker B. OKFDDs-algorithms, applications and extensions. in Sasao T and Fujita M, Eds., *Representations of Discrete Functions*, pp. 163–190, Kluwer, Dordrecht, 1996.

[4] Drechsler R, Becker B, and Rupertz S. K*BMD: a data structure for verification. In *Proceedings of the European Design and Test Conference*, pp. 2–8, 1996.

[5] Derchsler R, Becker B, and Rupertz S. K*BMDs: a verification data structure. *IEEE Design and Test of Computers*, pp. 51–59, 1997.

[6] Drechsler B, Stanković RS, and Sasao T, Spectral transforms and word-level decision diagrams. In *Proceedings of the SASIMI-97*, pp. 39–44, 1997.

[7] Lai YT and Sastry S. Edge-valued binary decision diagrams for multi-level hierarchical verification. In *Proceedings of the DAC*, pp. 668–613, 1992.

[8] Stanković RS and Astola JT. *Spectral Interpretation of Decision Diagrams.* Springer, Heidelberg, 2003.

[9] Vrudhula SBK, Lai YT, and Pedram M. Efficient computation of the probability and Reed–Muller spectra of Boolean functions using edge-valued binary decision diagrams. In Sasao T and Fujita M, Eds., *Proceedings of the IFIP WG 10.5 Workshop on Applications of the Reed-Muller Expansion in Circuit Design*, pp. 62–69, Makuhari, Chiba, Japan, 1995.

[10] Vrudhula SBK, Pedram M, and Lai Y-T, Edge valued binary decision diagrams. In Sasao T and Fujita M, Eds., *Representations of Discrete Functions*, pp. 109–132, Kluwer, Dordrecht, 1996.

17

Word-Level Decision Diagrams

In this chapter, a class of word-level decision diagrams is introduced. This class represents integer-valued, rational, or complex-valued functions according to classification of decision diagrams with respect to terminal nodes (Chapter 11).

From this class, the following decision diagrams and trees are considered:

- Multiterminal binary decision diagrams,
- Binary moment decision diagrams,
- Kronecker binary decision diagrams, and
- Haar spectral diagrams.

17.1 Introduction

In binary decision diagrams, BDDs, values of constant nodes are logic values 0 and 1. Therefore, BDDs are an example of bit-level decision diagrams.

There are different classes of discrete functions, and switching binary-valued functions are a particular class of them. For the representation of various classes of discrete functions, the definition of BDDs has to be extended and generalized to be adapted to the domain and the range of functions which should be represented. In this chapter, we consider decision diagrams for representation of functions $f : C_2^n \to P$, where

$$C_2^n = \times_{i=1}^n C_2, \quad C_2 = (\{0,1\}, \oplus),$$

and P can be the field of complex numbers C, the field of rational numbers Q, some finite (Galois) field of order larger than two, or the set of integers Z. We denote the space of such functions by $P(C_2^n)$, and specify P for particular classes of functions.

> *Word-level decision diagrams are decision diagrams in which the constant nodes are elements of either finite (Galois) fields, integers, or complex numbers.*

Therefore, word-level decision diagrams permit representation of integers or in general, complex-valued functions. Multiple-output switching functions

$$f_{q-1} * f_{q-2} * \cdots * f_0$$

are represented by their integer-valued equivalents by using the mapping

$$f_z = \sum_{i=0}^{q-1} 2^i f_i,$$

and then represented by the word-level decision diagrams.

In Table 17.1, specification of various decision trees and diagrams is given.

17.2 Multiterminal decision trees and diagrams

The optimization of the decision diagram representation of multi-output functions with respect to the number of non-terminal nodes can be achieved by using MTBDDs and shared MTBDDs. Further optimization can be achieved by the pairing of outputs in shared MTBDDs.

Compared with generalizations of spectral transforms, word-level decision diagrams are derived from bit-level decision diagrams by changing the field $GF(2)$ where constant nodes take their values with C, in the same way as the arithmetic transform is derived from the Reed–Muller transform.

> *Word-level diagrams are integer counterparts of bit-level decision diagrams or AND-EXOR related decision diagrams.*

MTBDTs are a generalization of binary decision trees by allowing integers, or complex numbers, at constant nodes. Thus, MTBDDs represent functions in $C(C_2^n)$. A node of the MTDD is identical to a node in a BDD (Shannon expansion), except that it is implemented for a string of bits.

Example 17.1 *In Figure 17.1, the nodes of MTBDTs are shown. Figure 17.2 shows an MTBDT for functions of three variables.*

TABLE 17.1
Specification of a sample of word-level decision trees and diagrams.

Type	**Specification**
Multiterminal, MTBDT and MTBDD	Generalization of binary decision trees by allowing integers, or complex numbers, at terminal (constant) nodes
ADD	Generalization of FDD by allowing integers at terminal nodes
Binary moment, BMT and BMD	Derived by the reduction of binary moment trees by using the ZBDD rules
Linear word-level, LDD	Subclass of BMD
Kronecker, KDT and KDD	Generalization of Davio trees and diagrams
Walsh, WDT and WDD	MTBDT and MTBDD of Walsh spectrum
Haar spectral, HST and HSD	MTBDT and MTBDD of Haar spectrum with edge values
Haar spectral transform, HSTDT and HSTDD	MTBDT and MTBDD of Haar spectrum

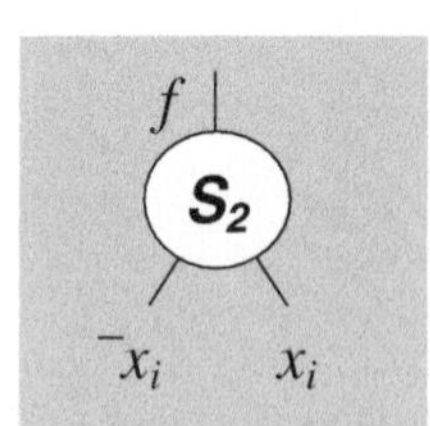

Algebraic description

$$f = \overline{x}_i f_0 \vee x_i f_1$$
$$f_0 = f_{x_i=0}$$
$$f_1 = f_{x_i=1}$$

Matrix description

$$f = [\, \overline{x}_i \; x_i \,] \begin{bmatrix} 1 & 0 \\ 0 & 1 \end{bmatrix} \begin{bmatrix} f_0 \\ f_1 \end{bmatrix}$$
$$= [\overline{x}_i \; x_i] \begin{bmatrix} f_0 \\ f_1 \end{bmatrix}$$

FIGURE 17.1
Node function in a multiterminal binary decision diagram (Example 17.1).

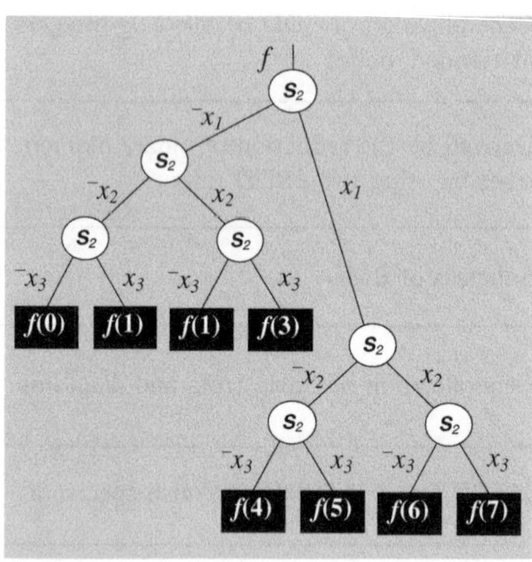

Switching functions of three variables $n = 3$ given by the vector of function values

$$\mathbf{F} = [f(0), f(1), \ldots, f(7)]^T$$

The constant nodes show function values for f in the corresponding points specified by minterms generated as products of the labels at the edges in the decision tree.

FIGURE 17.2
Multiterminal binary decision diagram for switching function of three variables (Example 17.1).

17.3 Spectral interpretation of word-level decision diagrams

Knowing spectral interpretation of BDDs, the spectral interpretation of MTBDDs is obvious. In spectral interpretation, MTBDTs are defined in terms of the identical mapping in $C(C_2^n)$. Thus, they are integer counterparts of BDDs, since they are defined with respect to the same mapping over different fields, $GF(2)$ for BDDs and C for MTBDDs. That is,

- ► The MTBDD of the Reed–Muller transform represents arithmetic analog of Davio diagram, also called Kronecker decision diagram (KDD)
- ► The MTBDD of arithmetic transform represents BMD, and
- ► The MTBDD of Walsh transform represents Walsh diagram.

An advantage of the spectral interpretation of decision diagrams is the possibility to define new decision diagrams by the decomposition of the switching function f with respect to different spectral transforms.

17.4 Binary moment trees and diagrams

Binary moment trees (BMTs). The following consideration concerns extensions of other bit-level decision diagrams into word-level counterparts as well as some modifications and further generalizations of decision diagrams.

> **Example 17.2** *In Figure 17.3, a node of MTBDTs of the arithmetic transform are shown. This is also a node of BMT and BMDa node of BMD, which right branch implements so-called* binary moment*:*
>
> $$f_2 = f_1 - f_0.$$
>
> *Figure 17.4 shows binary moment tree for the switching function f in Example 17.1.*

Binary moment decision diagrams. In BDD for some functions, the number of non-terminal nodes is larger than the number of BDDs for their arithmetic spectra. In this case, it is more effective to represent such functions by algebraic decision diagrams (ADDs), also called *arithmetic spectral transform decision diagrams* (ACDDs). ADDs are defined by using the same decomposition rule as in BMTs, but the reduction is performed by BDD reduction rules. In both BMDs and ADDs:

- ► A given switching function f is assigned to the decision tree by the decomposition in terms of the arithmetic transform which is the integer counterpart of the Reed–Muller transform.
- ► The constant nodes represent the arithmetic spectrum for switching function f.

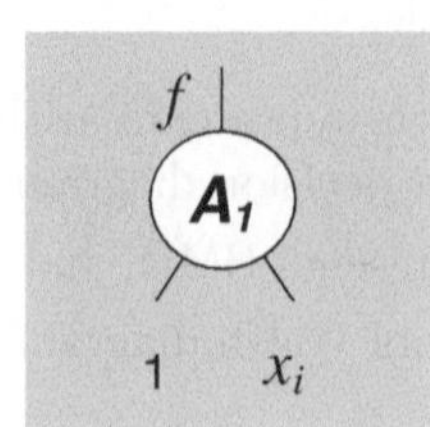

Algebraic description

$$f = f_0 + x_i f_2$$
$$f_0 = f_{x_i=0}$$
$$f_2 = f_{x_i=1} - f_{x_i=0}$$

Matrix description

$$f = [\,1\ x_i\,] \begin{bmatrix} 1 & 0 \\ -1 & 1 \end{bmatrix} \begin{bmatrix} f_0 \\ f_1 \end{bmatrix}$$
$$= [1\ x_i] \begin{bmatrix} f_0 \\ f_1 - f_0 \end{bmatrix}$$

FIGURE 17.3
Node function in a moment tree (Example 17.2).

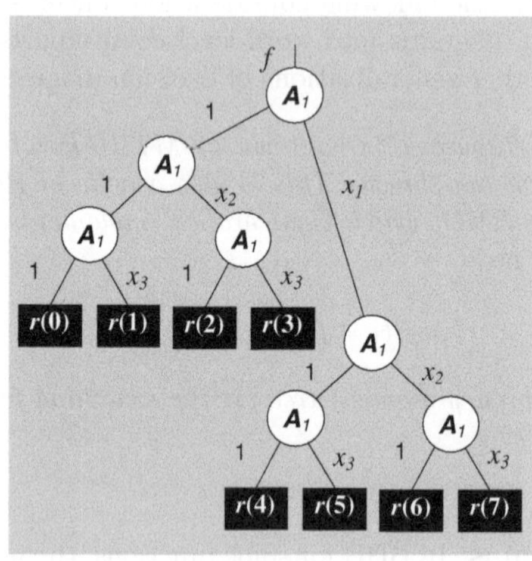

Binary moment tree for switching function f of three variables ($n = 3$) given by the vector of function values

$$\mathbf{F} = [f(0), f(1), \ldots, f(7)]^T$$

The constant nodes $r(i)$, $i = 0, 1, \ldots 7$ show values of arithmetic spectral transform coefficients for f:

$$\begin{aligned} f &= r(0) \\ &+ r(1)x_3 \\ &+ r(2)x_2 \\ &+ r(3)x_2x_3 \\ &+ r(4)x_1 \\ &+ r(5)x_1x_3 \\ &+ r(6)x_1x_2 \\ &+ r(7)x_1x_2x_3 \end{aligned}$$

FIGURE 17.4
Binary moment tree (Example 17.2).

BMDs and ADDs are integer counterparts of bit-level functional decision diagrams (FDDs) that are defined in terms of the Reed–Muller transform. BMDs and ADDs differ in the reduction rules. BMDs use zero-suppressed reduction rules, while ADDs use generalized BDDs rules.

ADDs are derived from the corresponding decision trees by using the BDD reduction rules. However, since the integer Davio nodes are used, the reduction of these trees can be performed by a using the zero-suppressed reduction rules as in the case of their bit-level equivalents, the Reed–Muller decision diagrams. In this way *binary moment diagrams* (BMDs) are defined [1]. Therefore, in BMDs, constant nodes show values of the arithmetic transform coefficients.

BMDs are derived by the reduction of BMTs by using the zero-suppressed binary decision diagrams (ZBDDs) rules.

> **Example 17.3** *Consider a two-output function of a half-adder (see Chapter 15). Its arithmetical polynomial is* $f = x_1 + x_2$. *Figure 17.5 shows various word-level decision diagrams for representation* f.

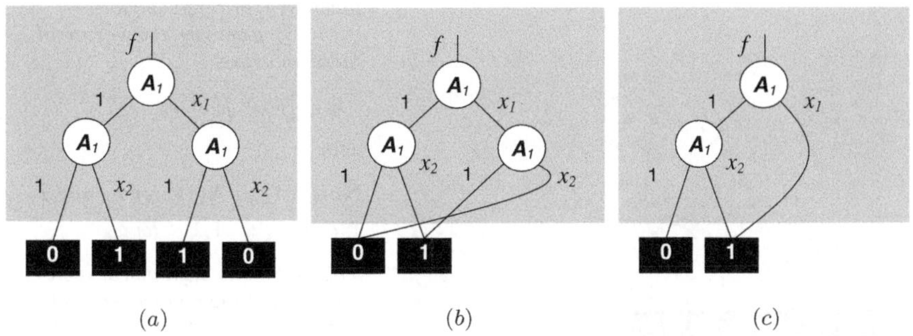

FIGURE 17.5
BMT (a), ADD (b), and BMD (c) for two-output switching function of a half-adder (Examples 17.3 and 17.4).

Linear decision diagrams. A subset of BMDs are linear decision diagrams (LDDs).

> **Example 17.4** *The BMD in Figure 17.5c is also LDD provided that each terminal node has only one ingoing edge (terminal node "1" is duplicated to avoid multiple ingoing edges).*

LDDs are word-level diagrams which correspond to arithmetic representation of multioutput functions. More details and examples on LDDs are given in Chapter 37.

Kronecker binary moment trees, KBMTs, are integer counterparts of Kronecker decision trees, KDTs, for switching binary-valued functions. They are defined by freely choosing decomposition matrices for nodes at each level in the decision tree from the set

$$KI = \left\{ \mathbf{I}(1) = \begin{bmatrix} 1 & 0 \\ 0 & 1 \end{bmatrix}, \; \mathbf{A}(1) = \begin{bmatrix} 1 & 0 \\ -1 & 1 \end{bmatrix}, \; \overline{\mathbf{A}}(1) = \begin{bmatrix} 0 & 1 \\ 1 & -1 \end{bmatrix} \right\}.$$

Example 17.5 *Figure 17.6 shows KBMT for a three-variable function f in Example 17.1. The constant nodes show values of spectral coefficients with respect to the basis specified by the decomposition rules assigned to the nodes in the decision tree.*

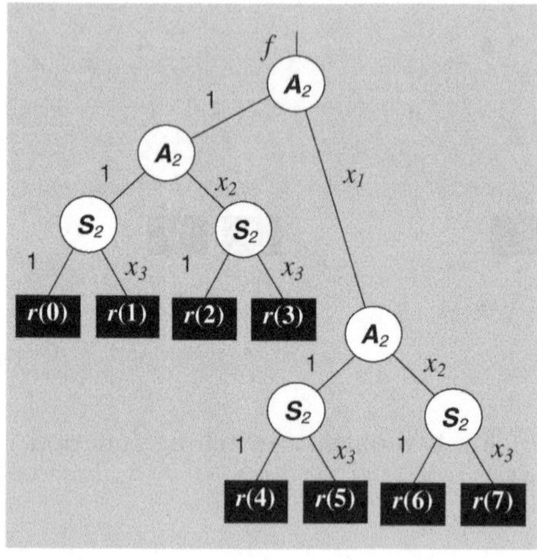

Walsh decision tree for switching function f of three variables ($n = 3$) given by the vector of function values

$$\mathbf{F} = [f(0), f(1), \ldots, f(7)]^T$$

with

Node A_2 : $\mathbf{A}(1)$ *for* x_1 *and* x_2

Node S_2 : $\mathbf{I}(1)$ *for* x_3

This KBMT is a graphical representation of the Fourier series-like expansion of f with respect to the basis:

$\varphi_0 = \overline{x}_3, \;\; \varphi_1 = x_3, \;\; \varphi_2 = x_2\overline{x}_3,$
$\varphi_3 = x_2x_3, \;\; \varphi_4 = x_1\overline{x}_3,$
$\varphi_5 = x_1x_3, \;\; \varphi_6 = x_1x_2\overline{x}_3,$
$\varphi_7 = x_1x_2x_3,$

where the variables x_i take integer values 0 and 1, and $\overline{x}$ is considered as $1 - x$.

FIGURE 17.6
Kronecker binary moment tree (Example 17.5).

Walsh spectral diagrams. The MTDD of Walsh transform is called a *Walsh decision diagrams.*

Example 17.6 *In Figure 17.7, the nodes of Walsh decision diagrams are shown. If we replace the arithmetic transform with the Walsh transform, we define the Walsh decision diagrams, WDDs. In representation of switching functions by WDDs, advantages are taken from properties of the Walsh spectrum of switching functions. For example, if the function values* $(0,1)$ *are encoded by* $(1,-1)$*, then the constant nodes in WDDs for* n*-variable switching functions are even integers in the set* $\{-2^n, \ldots, 2^n\}$.

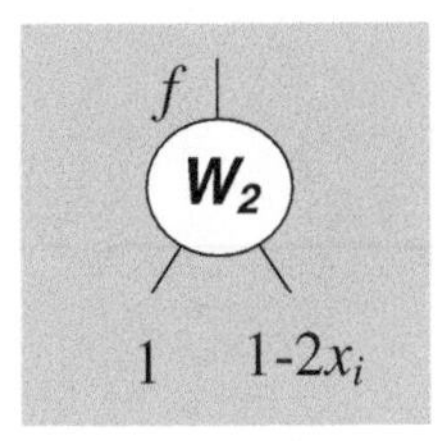

Algebraic description

$$f = f_1 + (1-2x_i)f_0$$
$$f_0 = f_{x_i=0}$$
$$f_1 = f_{x_i=1}$$

Matrix description

$$f = [\,\overline{x}_i\ x_i\,]\begin{bmatrix} 1 & 1 \\ -1 & 1 \end{bmatrix}\begin{bmatrix} f_0 \\ f_1 \end{bmatrix}$$
$$= [\overline{x}_i\ x_i]\begin{bmatrix} f_0 + f_1 \\ f_1 - f_0 \end{bmatrix}$$
$$= \begin{bmatrix} 1-2x_i \\ 1 \end{bmatrix}\begin{bmatrix} f_0 \\ f_1 \end{bmatrix}$$

FIGURE 17.7
Node function in a Walsh tree (Example 17.6).

Example 17.7 *Figure 17.8 shows WDT for switching function* f *in Example 17.1. In WDTs, constant nodes show the Walsh coefficients for the represented switching function* f*. WDDs are derived from Walsh decision trees by using the generalized BDD reduction rules.*

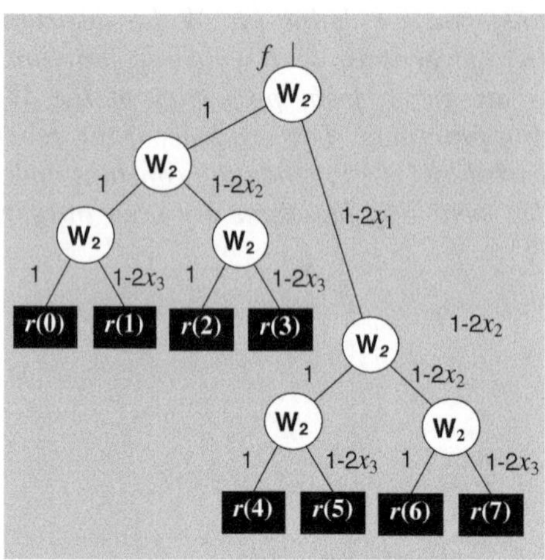

Walsh decision tree for switching function f of three variables ($n = 3$) given by the vector of function values

$$\mathbf{F} = [f(0), f(1), \ldots, f(7)]^T$$

The Walsh expression is as follows:

$$\begin{aligned} f &= r(0) \\ &+ r(1)(1-2x_3) \\ &+ r(2)(1-2x_2) \\ &+ r(3)(1-2x_2)(1-2x_3) \\ &+ r(4)(1-2x_1) \\ &+ r(5)(1-2x_1)(1-2x_3) \\ &+ r(6)(1-2x_1)(1-2x_2) \\ &+ r(7)(1-2x_1)(1-2x_2)(1-2x_3) \end{aligned}$$

FIGURE 17.8
Walsh decision tree (Example 17.7).

17.5 Haar spectral diagrams

Haar spectral diagrams, HSDs, are defined as an edge-valued modification of multiterminal binary decision diagrams, MTBDDs, to represent the Haar spectrum for a given switching function f.

HSDs are defined by an analogy to the edge-valued binary decision diagrams (EVBDDs) and their relationship to the arithmetic transform. In HSDs:

- ▶ The Haar coefficients are assigned to the incoming edge of the root node and the right outgoing edges of nodes in the MTBDDs for f.
- ▶ The assignment of the Haar coefficients to the MTBDDs for f is performed through a correspondence among the Haar coefficients and Haar functions in terms of which they are determined.
- ▶ A coefficient is situated at the edge ending the subpath consisting of edges whose labels are used to describe the corresponding Haar function in terms of binary variables.

In HSDs, nodes are the Shannon nodes, thus the same as MTBDD nodes. Therefore:

- ► From HSD for f, we can read f as from the MTBDD(f), and
- ► We can read the Haar coefficients as from the MTBDD for the Haar spectrum S_f for f.
- ► From MTBDD(S_f), however, we cannot read f in terms of the Haar coefficients and Haar functions.

If f is a switching function in coding (1, -1), the values of constant nodes in the MTBDD for f can be determined by the values of spectral coefficients assigned to the edges of nodes corresponding to x_n. This is possible, since from the definition of the Haar matrix, the processing of nodes at this level consists of subtraction of the function values at the neighboring points. However, that property cannot be used for integer or complex-valued functions to which the Haar transform can be also applied.

Example 17.8 *Figure 17.9 shows an example of the Haar spectral tree (HST) for* $n = 3$.

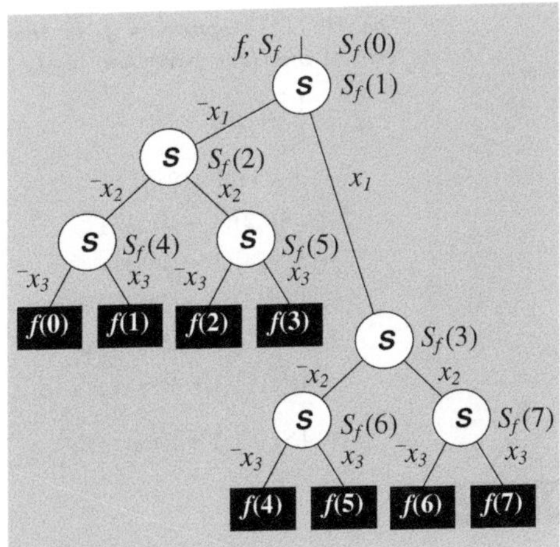

The relation of the correspondence among

- ► *The Haar functions,*
- ► *Labels at the edges in the multiterminal binary decision tree, and*
- ► *The Haar coefficients:*

$f(0)$	$har(0,x) = 1$	$S_f(0)$
$f(1)$	$har(1,x) = (1-2x_1)$	$S_f(1)$
$f(2)$	$har(2,x) = (1-2x_2)\overline{x}_1$	$S_f(2)$
$f(3)$	$har(3,x) = (1-2x_2)x_1$	$S_f(3)$
$f(4)$	$har(4,x) = (1-2x_3)\overline{x}_1\overline{x}_2$	$S_f(4)$
$f(5)$	$har(5,x) = (1-2x_3)\overline{x}_1 x_2$	$S_f(5)$
$f(6)$	$har(6,x) = (1-2x_3)x_1\overline{x}_2$	$S_f(6)$
$f(7)$	$har(7,x) = (1-2x_3)x_1 x_2$	$S_f(7)$

FIGURE 17.9
The Haar spectral tree (Example 17.8).

17.6 Haar spectral transform decision trees

> *The Haar spectral transform decision trees, HSTDTs, are defined as the graphic representation of the Haar expression for f:*
>
> - *Each path from the root node to a constant node corresponds to a Haar function $har(w, x)$.*
> - *The constant nodes show the values of Haar coefficients.*

Example 17.9 *Figure 17.10 shows the HSTDT for $n = 3$ defined by using the non-normalized Haar transform.*

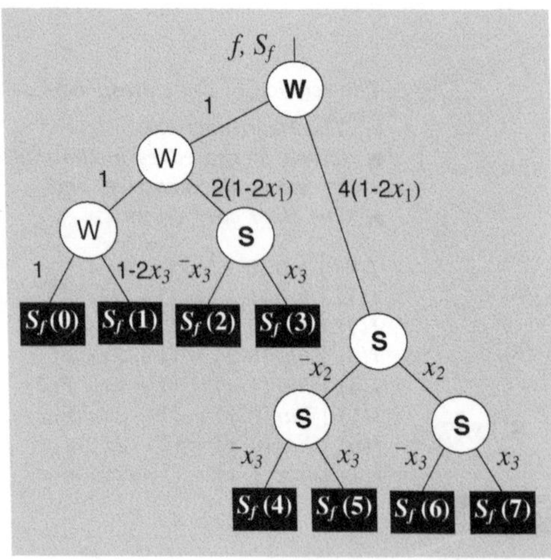

The HSTDT represents f in the form of the Haar expression for f:

$$
\begin{aligned}
f = \frac{1}{8}\Big[& S_f(0) \\
& + S_f(1)(1-2x_1) \\
& + 2S_f(2)(1-2x_2)\overline{x}_1 \\
& + 2S_f(3)(1-2x_2)x_1 \\
& + 4Sf(4)(1-2x_3)\overline{x}_2\overline{x}_1 \\
& + 4S_f(5)(1-2x_3)\overline{x}_2 x_1 \\
& + 4S_f(6)(1-2x_3)x_2\overline{x}_1 \\
& + 4S_f(7)(1-2x_3)x_2 x_1\Big]
\end{aligned}
$$

FIGURE 17.10
The Haar spectral transform decision trees (Example 17.9).

17.7 Other decision diagrams

Many other decision diagrams can be defined by using different spectral transforms to decompose switching function f and assign it to a decision tree. Such examples are

- Complex Hadamard transform decision diagrams (CHTDDs),
- Hybrid decision diagrams (HDDs).

References can be found in "Further reading" section.

17.8 Further reading

The main applications of word-level decision diagrams are in the area of verification of logic networks. While verification using bit-level diagrams requires multiple traversing, word-level diagrams offer simpler verification procedures [1, 5]. The other argument in favour of word-level decision diagrams is their compactness in terms of topology (number of paths and nodes).

The complexity of the verification procedure depends on the number of paths in the decision diagram. In this respect, Haar spectral transform decision diagram (HSTDD) [19], in particular, have a useful property in that the number of paths in these diagrams can be reduced by the application of the algorithm for reduction of the number of non-zero coefficients in the Haar spectrum.

Spectral interpretation of decision diagrams. In Stanković et al. [17], the MTBDTs are defined in terms of the identical mapping in $C(C_2^n)$, i.e., as an integer counterparts of BDDs.

Binary moment diagrams (BMDs) in Bryant and Chen [1] are derived by the reduction of BMTs by using the ZBDD rules. Multiterminal binary decision diagrams (MTBDTs) were studied by Clarke et al. [2, 3, 4].

Kronecker binary moment trees. Some results on KBMTs one can find, in particular, in [5, 6].

Arithmetic transform decision diagrams (ACDDs) are discussed in [17]. The BMDs and ACDDs are integer counterpart of bit-level functional decision diagrams (FDDs) (see, for example, the paper by Kebschull et al. [11]) that are defined in terms of the Reed–Muller transform. A subclass of BMDs, linear decision diagrams (LDDs) introduced by Yanushkevich et al. [23] and Shmerko [16], is discussed in Chapter 37.

Walsh decision diagrams (WDDs) were studied by Stanković et al. [17].

Haar spectral diagrams (HSDs) have been introduced by Hansen and Sekine [9]. HSDs were defined as an edge-valued modification of MTBDDs introduced by Clarke and Zhao [3]. The HSDs were also defined by an analogy to the EVBDDs by Lai et al. [7]. Stanković showed their relationship to the arithmetic transform [18, 18]. Stanković et al. [21, 73] showed that the Haar spectral transform decision trees (HSTDTs) are defined as the graphic representation of the Haar expression for switching function f. Stanković and Astola [19] defined a HSTDT by using the non-normalized Haar transform.

Complex Hadamard transform decision diagrams are developed by Falkowski and Rahardja [8]. Hybrid decision diagrams (HDDs) were introduced by Clarke et al. [4].

The optimization of decision diagram representations of multi-output functions with respect to the number of non-terminal nodes can be achieved by using multiterminal binary decision diagrams (MTBDDs) and shared MTBDDs (SMTBDDs). This approach was studied by Sasao and Butler [15]. Sasao and Babu [10] proposed further optimization that was achieved by pairing of outputs in SMTBDDs.

Function expansions. Exploiting of function expansions defined with respect to arbitrary bases to construct decision diagrams has been suggested by Perkowski [13] and Sasao [14].

References

[1] Bryant RE and Chen YA. Verification of arithmetic functions with binary moment diagrams. In *Proceedings of the CMU-CS-94-160*, 1994.

[2] Clarke EM, McMillan KL, Zhao X, Fujita M, and Yang J. Spectral transforms for large Boolean functions with application to technology mapping. In *Proceedings of the DAC)*, 1993.

[3] Clarke EM, McMillan KL, Zhao X, and Fujita M. Spectral transforms for extremely large Boolean functions. In Kebschull U, Schubert E and Rosenstiel W, Eds., *Proceedings of the IFIP WG 10.5 International Workshop on Applications of the Reed–Muller Expansions in Circuit Design*, Hamburg, Germany, pp. 86–90, 1993.

[4] Clarke EM, Fujita M, and Zhao X. Multiterminal decision diagrams and hybrid decision diagrams. In [14], pp. 93–108.

[5] Drechsler R. *Formal Verification of Circuits.* Kluwer, Dordrecht, 2000.

[6] Drechsler R, Sarabi A, Theobald M, Becker B, and Perkowski MA. Efficient representations and manipulation of switching functions based on ordered Kronecker functional decision diagrams. In *Proceedings of the DAC*, pp. 415–419, 1994.

[7] Egiazarian K, Astola JT, Stanković RS, and Stanković M. Construction of compact word-level representations of multiple-output switching functions by wavelet packets. In *Proceedings of the 6th. International Symposium on Representations and Methodology for Future Computing Technologies*, Trier, Germany, pp. 77–84, 2003.

[8] Falkowski BJ and Rahardja S. Complex spectral decision diagrams. In Proceedings of the IEEE 26th International Symposium on Multiple-Valued Logic, pp. 255–260, 1996.

[9] Hansen JP and Sekine M. Decision diagrams based techniques for the Haar wavelet transform. In *Proceedings of the IEEE International Conference on Information, Communications and Signal Processing,* Singapore, vol. 1, pp. 59–63, 1997.

[10] Hasan-Babu HMd and Sasao T. A method to represent multiple-output switching functions by using binary decision diagrams. In *Proceedings of the SASIMI'96*, pp. 212–217, 1996.

[11] Kebschull U, Schubert E, and Rosenstiel W. Multilevel logic synthesis based on functional decision diagrams. In *Proceedings of the EDAC*, pp. 43–47, 1992.

[12] Lai YF, Pedram M, and Vrudhula SBK. EVBDD-based algorithms for integer linear programming, spectral transformation, and functional decomposition. *IEEE Transactions on Computer-Aided Design of Integrated Circuits and Systems*, 13(8):959–975, 1994.

[13] Perkowski MA. The generalized orthonormal expansions of functions with multiple-valued inputs and some of its applications. In *Proceedings of the IEEE 22nd International Symposium on Multiple-Valued Logic*, pp. 442–450, 1992.

[14] Sasao T. A transformation of multiple-valued input two-valued output functions and its application to simplification of exclusive-OR sum-of-products expressions. In *Proceedings of the IEEE International Symposium on Multiple-Valued Logic*, pp. 270–279, 1991.

[15] Sasao T and Butler JT. A method to represent multiple-output functions by using multi-valued decision diagrams. In *Proceedings of the IEEE 26th International Symposium on Multiple-Valued Logic*, pp. 248–254, 1996.

[16] Shmerko VP. Malyugin's theorems: a new concept in logical control, VLSI design, and data structures for new technologies. *Automation and Remote Control*, Plenum/Kluwer Publishers, Special Issue on Arithmetical Logic in Control Systems, 65(6):893–912, 2004.

[17] Stanković RS. Some remarks about spectral transform interpretation of MTB-DDs and EVBDDs. In *Proceedings of the ASP-DAC'95*, pp. 385–390, 1995.

[18] Stanković RS. *Spectral Transform Decision Diagrams in Simple Questions and Simple Answers.* Nauka, Belgrade, 1998.

[19] Stanković RS and Astola JT. *Spectral Interpretation of Decision Digrams.* Kluwer, Dordrecht, 2003.

[20] Stanković RS, Sasao T, and Moraga C. Spectral transform decision diagrams. In Sasao T and Fujita M, Eds., *Representations of Discrete Functions*, pp. 55–92, Kluwer, Dordrecht, 1996.

[21] Stanković RS, Stanković M, Astola JT, and Egiazarian K. Karpovsky's old Haar spectrum theorem in a new light. In *Proceedings of the 1st International Workshop on Spectral Techniques and Logic Design for Future Digital Systems*, Tampere, Finland, TICSP Series #10, pp. 95–102, Dec. 2000.

[22] Stanković RS, Stanković M, Astola JT, and Egiazarian K. Haar spectral transform decision diagrams with exact algorithm for minimization of the number of paths. In *Proceedings of the 4th International Workshop on Boolean Problems*, Freiberg, Germany, pp. 113–129, 2000.

[23] Yanushkevich SN, Shmerko VP, and Dziurzanski P. Linearity of word-level models: new understanding. In *Proceedings of the IEEE/ACM 11th International Workshop on Logic and Synthesis*, pp. 67–72, New Orleans, LA, 2002.

18

Minimization via Decision Diagrams

In this chapter, the minimization of switching functions using decision diagrams is considered. Decision diagram minimization is accomplished through the reduction of nodes and isomorphic sub-graphs. These operations lead to different effects for different diagrams. For example, for some switching functions a free decision diagram provides good effects in terms of node reduction (in a free decision diagram each variable is encountered at most once on each path from the root to a terminal vertex). In some cases, to achieve minimization, the variables must be encountered in the same order on each path in the diagram from the root to a terminal vertex (ordered decision diagram).

18.1 Introduction

Reducing a switching function is called *simplification*, or *minimization*. Minimization can be accomplished through manipulation on cubes derived from a truth table, or through symbolic manipulation of logic expressions (SOP, ESOP, etc.). Decision trees can be derived from a functional equation or truth table. They provide a canonical representation of functions in graphical form, so that, for a fixed order of variables, there is a bijection between a switching function and a decision diagram. Therefore, minimization can be implemented on decision trees by reducing them to optimal decision diagrams.

The complete decision tree is a canonical data structure, just as a ROBDD is. The ROBDD is constructed by reducing a decision tree. The ROBDD represents an optimal AND-OR expression of the switching function in sum-of-products form (SOP). Other types of diagrams considered in this chapter are exclusive-sum-of-products (ESOP) and arithmetic forms.

18.2 Terminology

A literal is the appearance of a variable or its complement. In determining the complexity of an expression, one of the measures is the number of literals. Each appearance of a variable is counted.

> **Example 18.1** *There are six literals in the expression* $x_1x_2\overline{x}_3 \vee \overline{x}_1\overline{x}_2 \vee x_4$.

A sum term is one or more literals connected by an OR operation.

> **Example 18.2** $x_1 \vee \overline{x}_2 \vee x_3$ *and* $\overline{x}_1$ *are sum terms.*

A product term is one or more literals connected by an AND operation.

> **Example 18.3** $x_1x_2\overline{x}_3$ *and* x_4 *are product terms.*

A minterm is a product term that includes all variables of the function, either complemented or uncomplemented.

> **Example 18.4** *For a switching function of three variables* x_1, x_2 *and* x_3, *there are two minterms* $x_1x_2\overline{x}_3$ *and* $\overline{x}_1x_2x_3$ *in the expression* $x_1x_2\overline{x}_3 \vee \overline{x}_1\overline{x}_2 \vee \overline{x}_1x_2x_3$.

A sum-of-products (SOP) expression is one or more product terms connected by OR operations.

> **Example 18.5** *The expression* $x_1x_2\overline{x}_3 \vee \overline{x}_1\overline{x}_2 \vee x_4$ *is a SOP.*

A canonical sum is a sum of minterms.

> **Example 18.6** *The expression* $f = x_1x_2\overline{x}_3 \vee \overline{x}_1x_2x_3$ *is a canonical sum and* $f = \overline{x}_2 \vee \overline{x}_1x_3$ *is not.*

A minimum sum-of-products expression is a SOP expression that has the fewest number of product terms.

A maxterm is a sum term that includes all variables of the function, either complemented or uncomplemented.

> **Example 18.7** *For a switching function of three variables* x_1, x_2 *and* x_3, *there are two maxterms,* $x_1 \vee x_2 \vee \overline{x}_3$ *and* $\overline{x}_1 \vee x_2 \vee x_3$, *in the expression* $(x_1 \vee x_2 \vee \overline{x}_3)(\overline{x}_1 \vee \overline{x}_2)(\overline{x}_1 \vee x_2 \vee x_3)$.

A product of sums (POS) is one or more sum terms connected by AND operations.

> **Example 18.8** *The expression* $f = (x_2 \vee \overline{x}_3)(x_1 \vee \overline{x}_3)(\overline{x}_1 \vee x_2 \vee x_3)$ *is a product of sums.*

A canonical product is a product of maxterms.

> **Example 18.9** *The expression* $f = (x_1 \vee x_2 \vee \overline{x}_3)(\overline{x}_1 \vee x_2 \vee x_3)$ *is a canonical product given a switching function of three variables* x_1, x_2 *and* x_3.

A minimum product of sums expression is a POS expression that has the least number of sum terms.

A minimum representation of a switching function is the expression with the least number of terms, and, among that with the same number of terms, those with the least number of literals. A given switching function can be reduced to a minimum sum of products form, or to a minimum product of sums form. They may both have the same number of terms and literals or either may have fewer than the other.

Some useful rules for the manipulation of AND-OR and OR-AND expressions are given below:

- $xy \vee x\overline{y} = x$,
- $(x \vee y)(x \vee \overline{y}) = x$
- $x \vee \overline{x}y = x \vee y$
- $x(\overline{y} \vee y) = xy$
- $x \vee x = x$
- $xx = x$
- $x \vee xy = x$
- $x(x \vee y) = x$
- $x \vee 0 = x$
- $x \vee 1 = 1$
- $x0 = 0$
- $x1 = x$

Manipulation of AND-OR expressions to reduce them can be done with binary decision trees, which represent SOPs.

> **Example 18.10** *The first two rules,* $xy \vee x\overline{y} = x$ *and* $(x \vee y)(x \vee \overline{y}) = x$, *are called* absorbtion *rules and are useful for the minimization of SOP expressions. The absorbtion corresponds to a node reduction in binary decision trees and diagrams (Figure 18.1).*

An implicant of a switching function is a product term that can be used in a sum of products expression for that function. The function is 1 whenever the implicant is 1. An implicant is any product term in an SOP expression. A *prime implicant* is a product term that cannot be further reduced.

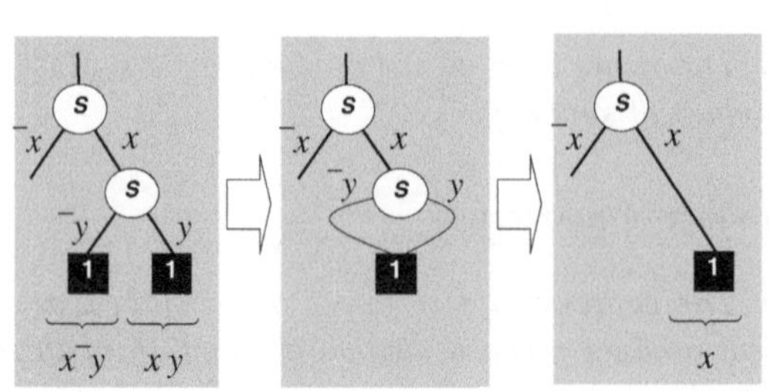

Algebraic form

$$f = xy \vee x\overline{y}$$
$$= x(y \vee \overline{y}) = x$$

Graphical form

(a) *Terminal nodes* $1 \cdot x\overline{y}$ *and* $1 \cdot xy$ *are juncted,*
(b) *Edges* $\overline{y}$ *and* y *are merged,*
(c) *The lower node* S *is eliminated*

FIGURE 18.1
Absorbtion rules of Boolean algebra in algebraic form and in terms of decision diagrams (Example 18.10).

A prime implicant is an implicant that is not fully contained in any one other implicant, i.e., if any literal is removed from that term, it is no longer an implicant.

An essential prime implicant. If a minterm of a switching function is included in only one prime implicant, that implicant is called an *essential* prime implicant.

Example 18.11 *In the expression*

$$f = \overline{x}_1x_2x_3 \vee x_1\overline{x}_2x_3 \vee x_1x_2\overline{x}_3 \vee x_1x_2x_3$$
$$= x_2x_3 \vee x_1x_3 \vee x_1x_2$$

none of the implicants x_2x_3, x_1x_3, *and* x_1x_2 *can be reduced further without violating the truth table of* f*; they are prime implicants. Neither* x_2x_3, x_1x_3, *nor* x_1x_2 *can be taken away from* f *without changing its truth table; each is an essential prime implicant. In terms of formal logic, each product term is a sufficient condition, and all product terms are the necessary conditions.*

Implication. The switching function f_1 implies the switching function f_2 if there is no assignment of values to those variables that makes f_1 equal to 1 and f_2 equal to 0.

Example 18.12 *Switching function* $f_1 = x_1x_2 \vee x_2x_3$ *implies switching function* $f_1 = x_1x_2 \vee x_2x_3 \vee \overline{x}_1x_3$.

Irredundant expression is defined as an expression in sum-of-products form such that (*a*) Every product term is a prime implicant, and

(*b*) No product term may be eliminated from the expression without changing the function.

Similarly, in product-of-sums form an irredundant expression can be defined as an expression such that every sum term is a prime implicant and no sum term can be eliminated from the expression without changing the switching function.

The consensus theorem states that $xy \vee \overline{x}z \vee yz = xy \vee \overline{x}z$. This theorem is used to eliminate redundant terms from a switching function expression.

> **Example 18.13** *An interpretation of the consensus theorem illustrated by a BDD is given in Figure 18.2.*

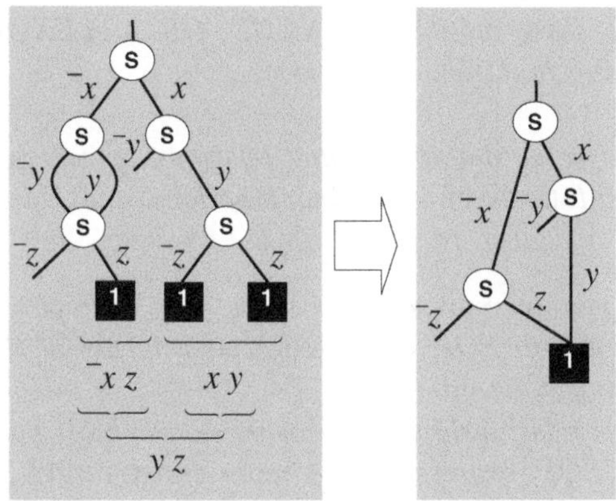

Algebraic form

$f = xy \vee \overline{x}z \vee yz = xy \vee \overline{x}z$

Graphical form

(*a*) *Terminal nodes* $1 \cdot xy\overline{z}$ *are juncted to* $1 \cdot xy$ *and lower right* S *node is eliminated*

(*b*) *Lower left* S *node is eliminated*

FIGURE 18.2
Interpretation of the consensus theorem in terms of decision diagrams (Example 18.13).

Don't care conditions. In some systems, the values of the output are specified for only some of the input conditions. Such functions are referred to as *incompletely specified* functions. For the remaining input conditions, it does not matter what the output is, i.e., we don't care. In the truth table, don't cares are indicated by an **x**. In real systems, don't cares occur; for example, there may be some input combinations that never occur (see details in Chapter 19).

18.3 Minimization using hypercubes

The relationship between variables and switching functions can be expressed in terms of hypergraphs. Polynomial expressions for a switching function, such as sum-of-products and AND-EXOR forms, can be expressed in terms of singular hypercubes. Cubes, or n-tuples, correspond to the terms in polynomial expressions. The corresponding representation of a switching function of n variables is the n-dimensional hypercube such that:

- The vertices of the hypercube denote the minterms, thus, the hypercube is a collection of minterms;
- The number of minterms is a power of two, 2^m, for some $m \leq n$;
- The number of edges in a hypercube is $3 \cdot s^{n-1}$.

Example 18.14 *In Figure 18.3, techniques for the minimal representation of switching functions based on cubes and various types of decision diagrams are shown. For demonstration of these techniques, three-input AND, NAND, OR, and EXOR elementary switching functions are chosen.*

Example 18.15 *Useful details of the relationship between cube-based representations of switching functions and Davio decision diagrams are given in Figure 18.4:*

- *In terms of algebraic representations, these two graphical structures are both SOP and Reed-Muller forms of the same switching function.*
- *There are direct relationships between these graphical data structures. SOP expressions be represented in AND-EXOR form and vice versa.*

Note that cube-based descriptions are useful for manipulation. Decision diagram are used in manipulation and can be implemented into hardware directly.

Relationship of data structures. Tabular (truth tables) and graphical (hypercubes, decision trees and diagrams) representations of switching functions have a one-to-one relationship.

Example 18.16 *Relationship of schematic, algebraic, truth table, hypercube, decision tree, decision diagram, and spatial ($\mathcal{N}$ – hypercube) representations for the switching function $f = x_1\overline{x}_2 \vee \overline{x}_3$ are given in Figure 18.5.*

Let $x_j^{i_j}$ be a literal of a Boolean variable x_j such that $x_j^{i_j} = \overline{x}_j$ if $i_j = 0$, and $x_j^{i_j} = x_j$ if $i_j = 1$. A product of literals $x_1^{i_1}x_2^{i_2}\dots x_n^{i_n}$ is called a *product term.*

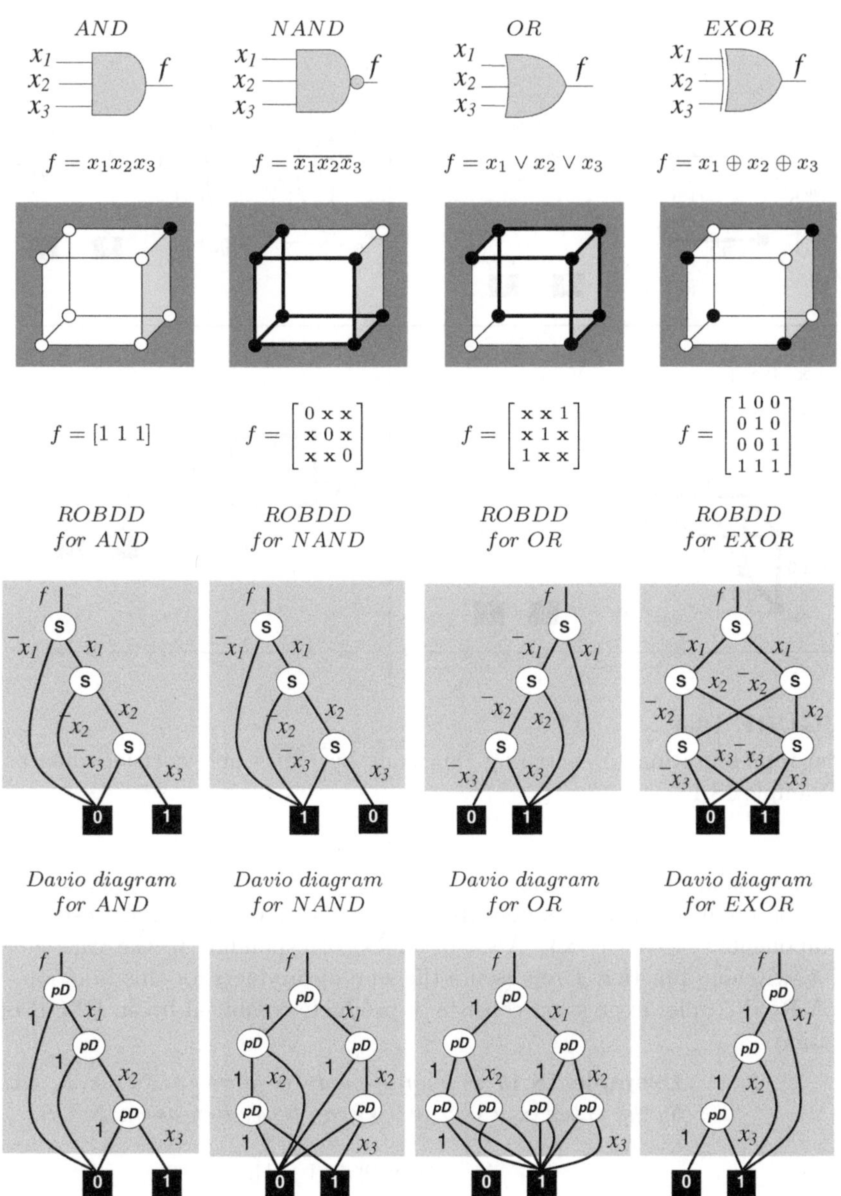

FIGURE 18.3
Representation of AND, NAND, OR and EXOR 3-input gates by cubes and decision diagrams.

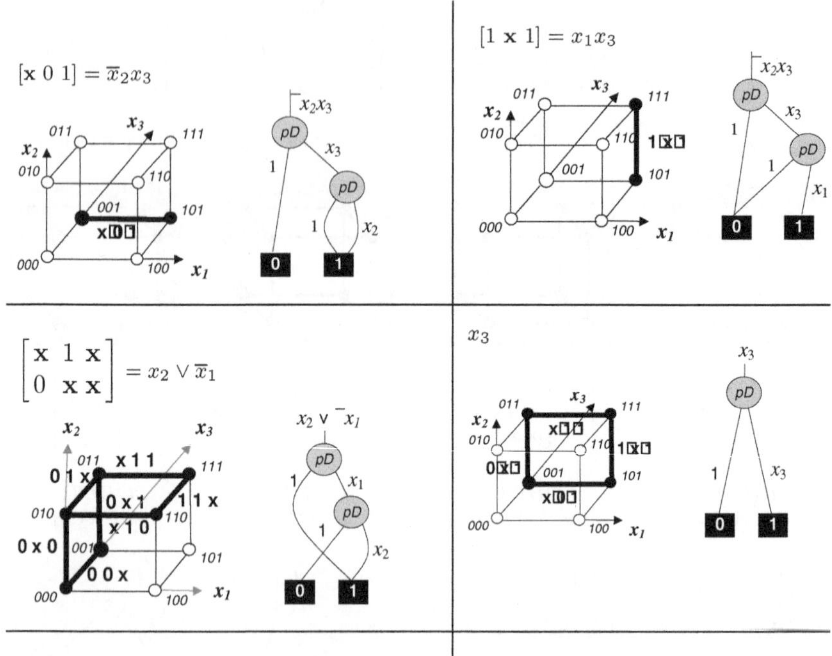

FIGURE 18.4
The representation of switching functions by cubes and decision diagrams (Example 18.15).

If the variable x_j is not present in a cube, $i_j = \mathbf{x}$ (don't care), i.e., $x_j^{\mathbf{x}} = 1$. In cube notation, a term is described by a cube that is a ternary n-tuple of components $i_j \in \{0,\ 1,\ \mathbf{x}\}$. A set of cubes corresponding to the true values of a switching function f represents the sum-of-products for this function.

A Reed–Muller expression consists of products combined by an EXOR operation.

Example 18.17 *A sum-of-products form $f = \overline{x}_3 \vee x_1\overline{x}_2$ given by the cubes* $[\mathbf{x}\ \mathbf{x}\ 0] \vee [1\ 0\ \mathbf{x}]$ *can be written as an ESOP,*

$$[\mathbf{x}\ \mathbf{x}\ 0] \oplus [1\ 0\ 1],$$

that is, $\overline{x}_3 \oplus x_1\overline{x}_2 x_3$. The different cubes arise because of the different operations between the products in the expressions.

Thus, the manipulation of the cubes involves OR, AND and EXOR operations applied to the appropriate literals following the rules given in Figure 18.6.

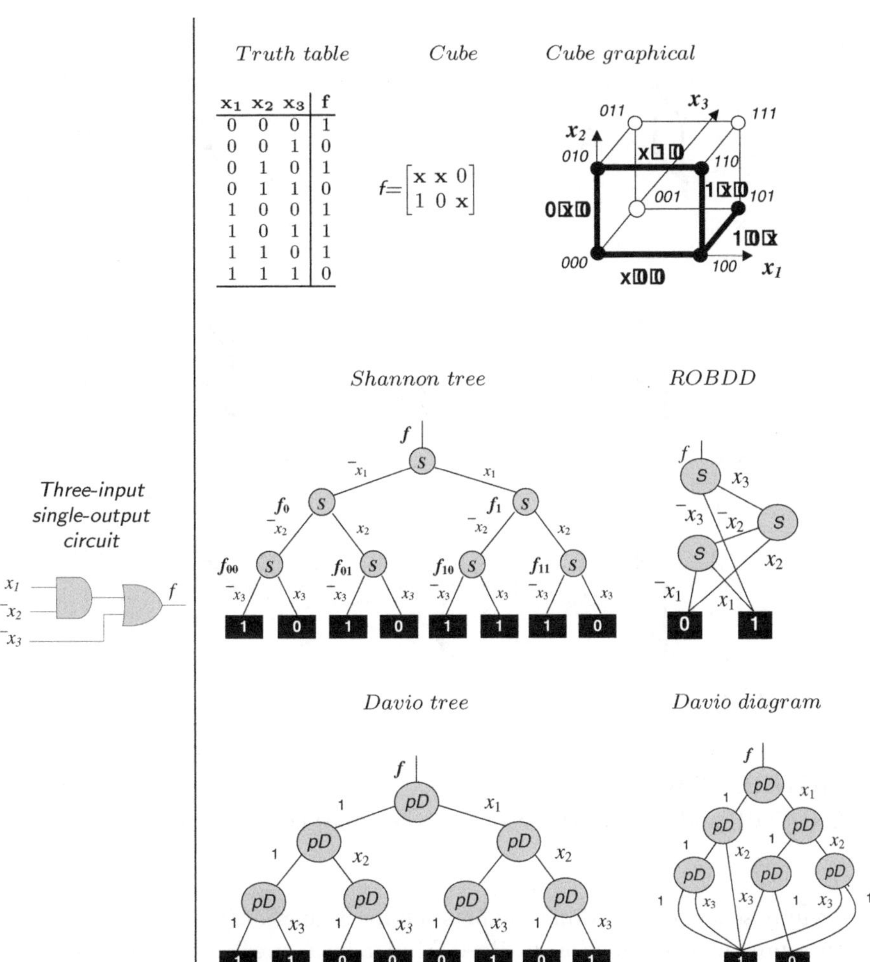

FIGURE 18.5
Various data structures for the minimization of a three-input single-output circuit: cube-based, Shannon decision tree, ROBDD, and Davio tree and diagram (Example 18.16).

Example 18.18 *Given the cubes* [1 1 **x**] *and* [1 0 **x**], *the AND, OR, and EXOR operations on these cubes are shown in Figure 18.6.*

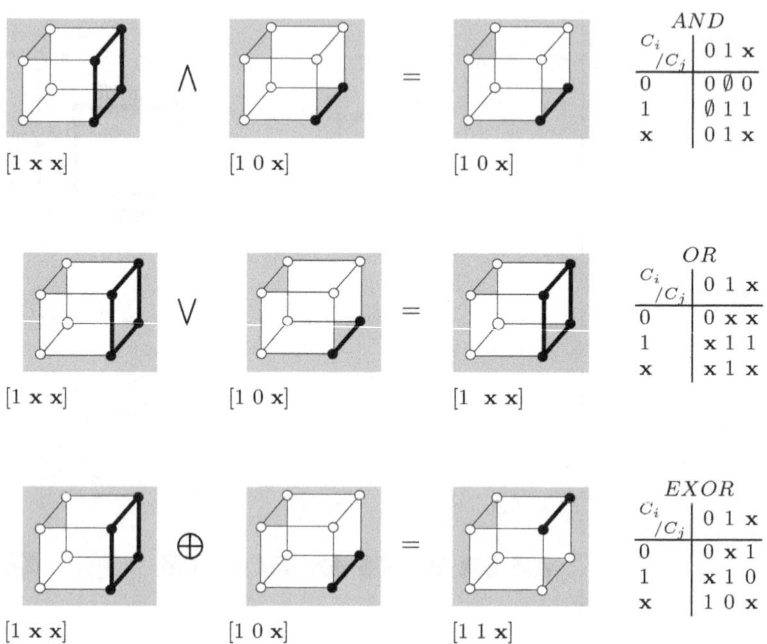

FIGURE 18.6
AND, OR, and EXOR operations on the cubes (Example 18.18).

Suppose a sum-of-products expression for a function f is given by its cubes. To represent this function in Reed–Muller form, we have to generate cubes based on the equation

$$x \vee y = x \oplus y \oplus xy$$

that can be written in cube notation as

$$[C_1] \vee [C_2] = [C_1] \oplus [C_2] \oplus [C_1][C_2]. \tag{18.1}$$

Example 18.19 *Consider a switching function given in a sum-of-products form by four cubes,*

$$f = [\mathbf{x}\ 1\ 0\ 1] \vee [1\ 0\ 0\ \mathbf{x}] \vee [0\ \mathbf{x}\ \mathbf{x}\ 0] \vee [\mathbf{x}\ \mathbf{x}\ 1\ 0].$$

To find its ESOP expression, an OR must be replaced by an EXOR operation and AND must be computed for each cube distinguished in only one literal. The cube representation is as follows:

$$f = [\mathbf{x}\ 1\ 0\ 1] \oplus [1\ 0\ 0\ \mathbf{x}] \oplus [0\ \mathbf{x}\ \mathbf{x}\ 0] \oplus [\mathbf{x}\ \mathbf{x}\ 1\ 0] \oplus [0\ \mathbf{x}\ 1\ 0],$$

that is,

$$f = x_2\overline{x}_3x_4 \oplus x_1\overline{x}_2\overline{x}_3 \oplus \overline{x}_1\overline{x}_4 \oplus x_3\overline{x}_4 \oplus \overline{x}_1x_3\overline{x}_4.$$

Note that ESOP is a *mixed* polarity form where a variable may be used in both complemented and uncomplemented form.

18.4 Minimization of AND-EXOR expressions

Simplification of AND-EXOR expressions. To simplify AND-EXOR expressions, the following rules are applied:

- $x \oplus 0 = x$ and $x \oplus 1 = \overline{x}$
- $x \oplus x = 0$ and $x \oplus \overline{x} = 1$
- $\overline{x \oplus y} = x \oplus \overline{y} = \overline{x} \oplus y$
- $xy \oplus \overline{x}y = y$
- $\overline{x}y \oplus y = xy$
- $xy \oplus y = \overline{x}y$
- $0 \oplus 0 = 0$
- $0 \oplus 1 = 1$
- $1 \oplus 1 = 0$

Example 18.20 *Interpretations of the rules* $1 \oplus 0 = 1$ *and* $0 \oplus 0 = 0$ *in forms of positive and negative Davio trees reduction is given in Figure 18.7.*

Each node in a Davio tree of a switching function f corresponds to the Davio decomposition of the function with respect to each variable x_i. There exist:

The positive Davio expansion

$$f = f_0 \oplus x_i f_2, \tag{18.2}$$

where $f_0 = f_{x_i=0}$ and $f_2 = f_{x_i=1} \oplus f_{x_i=0}$, and
The negative Davio expansion

$$f = \overline{x}_1 f_2 \oplus f_1, \tag{18.3}$$

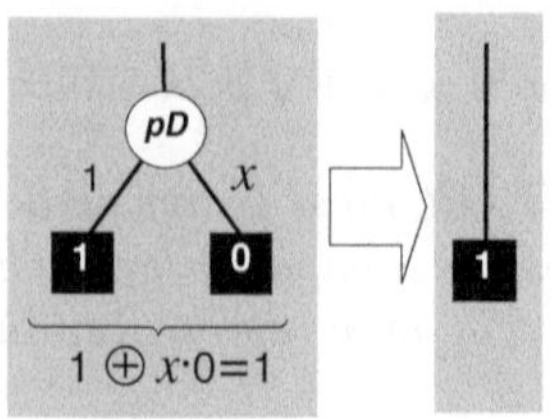

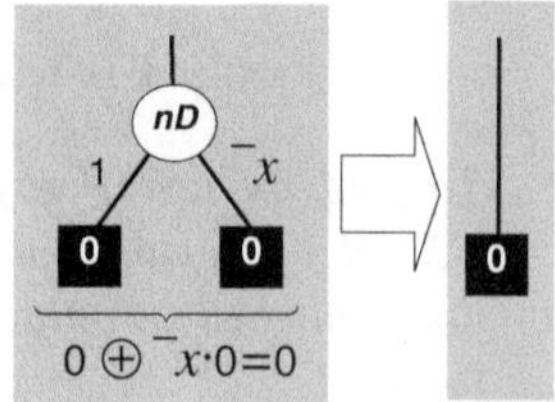

FIGURE 18.7
Interpretation of the simplest AND-EXOR rules in terms of decision diagrams (Example 18.20).

where $f_1 = f_{x_i=1}$.

Positive and negative Davio decompositions are labeled as pD and nD respectively (Figure 18.8). A binary decision tree that corresponds to the fixed positive polarity Reed-Muller canonical representation of a switching function is called a positive *Davio tree.* A Davio decision diagram can be derived from a Davio tree.

In matrix notation, the switching function f of the Davio node is a function of a single variable x_i given by the truth vector $\mathbf{F} = [\ f(0)\ f(1)\ ]^T = [\ f_0\ f_1\ ]^T$ and is defined as

$$
\begin{aligned}
f &= [\ \overline{x}_i\ x_i\ ] \begin{bmatrix} 1 & 0 \\ 1 & 1 \end{bmatrix} \begin{bmatrix} f_0 \\ f_1 \end{bmatrix} = [\ \overline{x}_i\ x_i\ ] \begin{bmatrix} f_0 \\ f_1 \end{bmatrix} \\
&= \overline{x}_i f_0 \oplus x_i f_1 = (1 \oplus x_i) f_0 \oplus x_i f_1 = f_0 \oplus x_i f_2 \\
&= [1\ x_i] \begin{bmatrix} f_0 \\ f_1 \end{bmatrix}.
\end{aligned}
$$

The recursive application of the positive Davio expansion to the function f given by the truth-vector $\mathbf{F} = [f(0)\ f(1) \dots f(2^n - 1)]^T$ can be expressed in matrix notation as follows

$$
f = \widehat{\mathbf{X}}\ \mathbf{R}_{2^n}\ \mathbf{F}, \tag{18.4}
$$

where

$$
\widehat{\mathbf{X}} = \bigotimes_{i=1}^{n} [\ 1\ x_i\ ], \quad \mathbf{R}_{2^n} = \bigotimes_{i=1}^{n} \mathbf{R}_2, \quad \mathbf{R}_2 = \begin{bmatrix} 1 & 0 \\ 1 & 1 \end{bmatrix},
$$

and $\otimes$ denotes the Kronecker product.

In decision trees and diagrams for representing AND-EXOR expressions, Davio decomposition in nodes is used, and that is why they are often called *Davio* decision trees and diagrams.

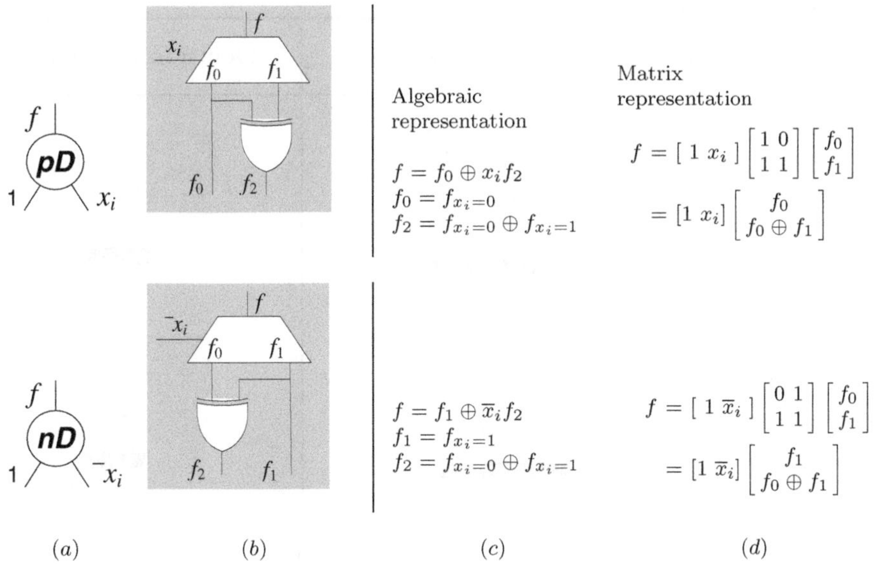

FIGURE 18.8
The node of a Davio tree (a), MUX-based implementation (b), algebraic (c), and matrix (d) descriptions.

The Davio diagram is derived from the Davio tree by deleting redundant nodes, and by sharing equivalent subgraphs.

The rules below produce the reduced Davio diagram (see details in Chapter 4).

Example 18.21 *Application of the reduction rules to the three-variable NAND function is demonstrated in Figure 18.9.*

Types of decision trees using AND-EXOR operations are given in Table 18.1. Recall that a Shannon decision tree is constructed by applying the Shannon expansion recursively to a switching function. This type of tree corresponds to the sum-of-products expression of the function.

TABLE 18.1
The effects of minimization of AND-EXOR expressions on various types of decision diagrams.

Class	Decision tree	Decision diagram
Shannon	$\overline{x} \vee \overline{y} = \overline{x}\,\overline{y} \vee \overline{x}y \vee x\overline{y}$ 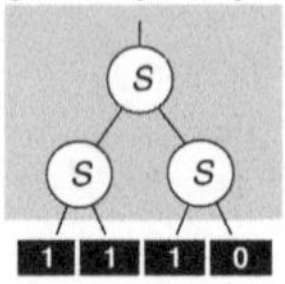	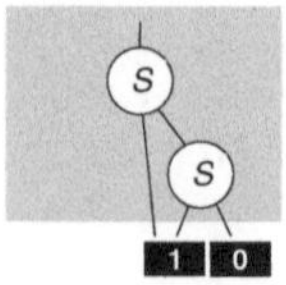
Positive (negative) Davio	$1 \oplus xy$ 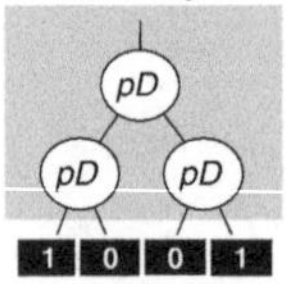	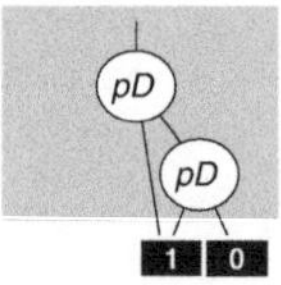
Reed–Muller (RM)	$1 \oplus y \oplus \overline{x}y$ 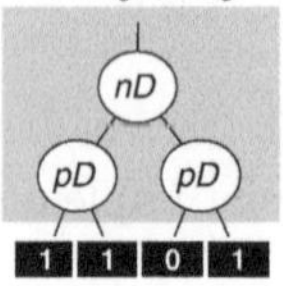	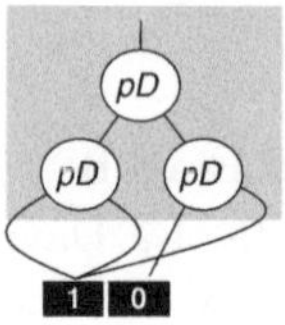
Pseudo Reed–Muller (PSRM)	$1 \oplus x \oplus x\overline{y}$ 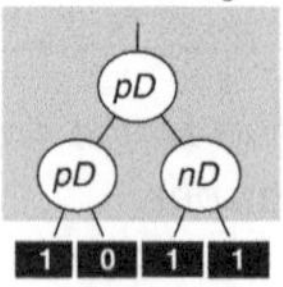	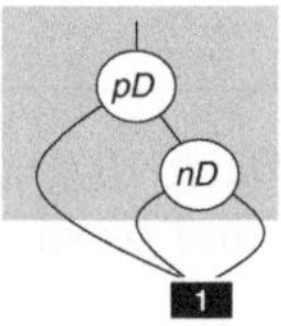
Kronecker (KRO)	$\overline{x} \oplus x\overline{y}$ 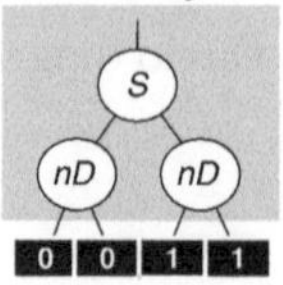	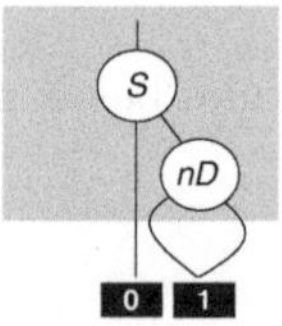
Pseudo Kronecker (PSKRO)	$1 \oplus x \oplus \overline{y} \oplus \overline{x}y$ 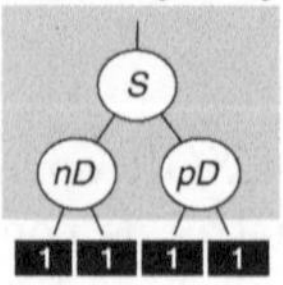	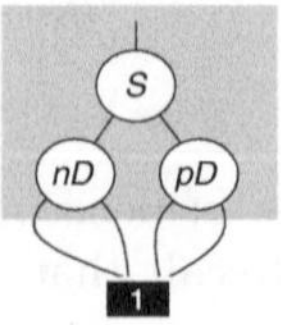

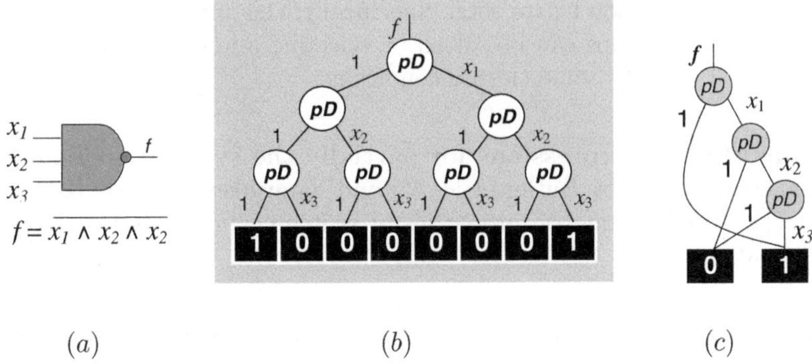

FIGURE 18.9
The three-variable NAND function, its Davio tree, and its reduced Davio decision diagram (Example 18.21).

18.5 Further reading

Simplification of Boolean expressions. Classical methods of the simplification of switching functions can be found in textbooks, in particular, by Givone [6], Roth [11], and Brown and Vranesic [1].

Morreale's algorithm for generating prime-irredundant cubes [7] is useful in decision diagram techniques (see Chapter 24). Let g_0 and g_1 denote covers of subfunctions for left and middle edges, respectively. Then we remove the covered cubes from f_0 and f_1, to produce f_0'' and f_1''. The subfunction f_d is obtained as the product of f_0'' and f_1'' and consists of the cubes that do not depend on the variable used in the considered node (Table 24.5).

TABLE 18.2
Morreale's operators.

f'_0, f'_1: f_1, f_0 \ f_0, f_1	0	1	d	f''_0, f''_1: g_i \ f_i	0	1	d	f_d: f_1'' \ f_0''	0	1	d
0	0	1	d	0	0	1	d	0	0	0	0
1	0	d	d	1	-	d	d	1	0	1	1
d	0	d	d					d	0	1	d

Minimization of ESOP expressions. Riese and Besslich [10] developed an algorithm for the minimization of incompletely specified ESOPs expressions of switching functions. They also used topological representation of ESOP by hypercubes. Oliveira et al. [8] studied the problem of BDD exact minimization in the presence of don't care sets. Specifically, the authors show that BDD minimization problems can be solved by selecting a minimum sized cover for a graph that satisfies some closure conditions.

Optimal FPRM expressions for partially the symmetric logic functions were studied by Butler et al. [2] and Yanushkevich et al. [17]. Scholl et al. [12] used symmetric properties of switching functions in BDD-based minimization.

The information-theoretical approach can be used in minimization of various representations of logic functions [3], [18], [5], [7], [11]. For example, in [68, 14, 67], Shmerko et al. discussed minimization of Reed–Muller expressions. Details on information-theoretical approach can be found in Chapters 8 and 24.

References

[1] Brown S and Vranesic Z. *Fundamentals of Digital Logic with VHDL Design.* McGraw-Hill, New York, 2000.

[2] Butler J, Dueck G, Shmerko V, and Yanushkevich S. Comments on SYMPATHY: fast exact minimization of fixed polarity Reed–Muller expansion for symmetric functions. *IEEE Transactions on Computer-Aided Design of Integrated Circuits and Systems*, 19(11):1386–1388, 2000.

[3] Cheushev VA, Yanushkevich SN, Moraga C, and Shmerko VP. Flexibility in logic design. An approach based on information theory methods. *Research Report 741*, Forschungsbericht, University of Dortmund, Germany, 2000.

[4] Cheushev VA, Yanushkevich SN, Shmerko VP, Moraga C, and Kolodziejczyk J. Remarks on circuit verification through the evolutionary circuit design. In *Proceedings of the IEEE 31st International Symposium on Multiple-Valued Logic*, pp. 201–206, 2001.

[5] Cheushev VA, Shmerko VP, Simovici D, and Yanushkevich SN. Functional entropy and decision trees. In *Proceedings of the IEEE 28th International Symposium on Multiple-Valued Logic*, pp. 357–362, 1998.

[6] Givone DD. *Digital Principles and Design.* McGraw-Hill, New York, 2003.

[7] Kabakcioglu AM, Varshney PK, and Hartman CRP. Application of information theory to switching function minimization. *IEE Proceedings*, Pt E, 137(5):389–393, 1990.

[8] Oliveira AL, Carloni LP, Villa T, and Sangiovanni-Vincentelli AL. Exact minimization of binary decision diagrams using implicit techniques. *IEEE Transactions on Computers*, 47(11):1282–1296, 1998.

[9] Morreale E. Recursive operators for prime implicant and irredundant normal form determination. *IEEE Transactions on Computer*, 19(6):504–509, 1970.

[10] Riese MW and Besslich PhW. Low-complexity synthesis of incompletely specified multiple-output mod-2 sums. *IEE Proceedings*, Pt E, 139(4):355–362, 1992.

[11] Roth CH Jr. *Fundamentals of Logic Design.* Thomson, Brooks/Cole, Belmont, CA, 2004.

[12] Scholl C, Miller D, Molitor P, and Drechsler R. BDD minimization using symmetries. *In IEEE Transactions on CAD of Integrated Circuits and Systems*, 18(2):81–100, 1999.

[13] Shmerko VP, Popel DV, Stanković RS, Cheushev VA, and Yanushkevich SN. Entropy based algorithm for 4-valued functions minimization. In *Proceedings of the IEEE 30th International Symposium on Multiple-Valued Logic*, pp. 265–270, 2000.

[14] Shmerko VP, Popel DV, Stanković RS, Cheushev VA, and Yanushkevich SN. AND/EXOR minimization of switching functions based on information theoretic approach. In *Facta Universitatis*, Series Electronics and Energetic, 13(1):11–25, 2000.

[15] Shmerko VP, Popel DV, Stanković RS, Cheushev VA, and Yanushkevich SN. Information theoretical approach to minimization of AND/EXOR expressions of switching functions. In *Proceedings of the IEEE International Conference on Telecommunications*, pp. 444–451, Yugoslavia, 1999.

[16] Yanushkevich SN, Shmerko VP, Dziurzanski P, Stanković RS, and Popel DV. Experimental verification of the entropy based method for minimization of switching functions on pseudo-ternary decision trees. In *Proceedings of the IEEE International Conference on Telecommunications in Modern Satellite, Cable and Broadcasting Services*, pp. 452–459, Yugoslavia, 1999.

[17] Yanushkevich S, Butler J, Dueck G, and Shmerko V. Experiments on FPRM expressions for partially symmetric logic functions. In *Proceedings of the IEEE 30th International Symposium on Multiple-Valued Logic*, pp. 141–146, 2000.

19

Decision Diagrams for Incompletely Specified Functions

An optimal decision diagram structure improves the performance and flexibility of many applications in logic design, circuit synthesis, and engineering where the specification is either uncertain or incomplete. This chapter* considers the problem of building various types of decision diagrams for representation and minimization of incompletely specified functions. It also defines the incompletely specified decision diagrams along with techniques for their manipulation, such as redefinition of unspecified values, "don't care about don't cares," fusion, and other operations.

19.1 Introduction

Incompletely specified logic functions are commonly used for representation of problems with uncertain or incomplete specifications. In incompletely specified functions, unlike completely specified ones, values are assigned to a function for only a proper subset of combinations of values of variables.

Two types of *don't care* conditions are distinguished:

- Some input states may never occur and the output states are irrelevant or cannot be observed.
- For some input states, the corresponding output states need not be specified.

In a synthesized network, every input state must correspond to a certain output state, i.e., a network must be completely deterministic. Unspecified values, called *don't cares*, are often used while designing a network. The aim of this process is to achieve additional flexibility in manipulation of network components or formal descriptions.

*This chapter was co-authored with Dr. Denis V. Popel, Baker University, Kansas

Some problems can be classified as *weakly specified* considering the initial specification of functions on a restricted number of combinations. In fact, many practical problems in engineering require up to 200 input variables with an enormous number of *don't cares.* Problems with an unusually high percentage of *don't care* values of switching functions occur also in circuit design when using VHDL compilers.

Though decision diagrams proved to be a practical tool for symbolic verification and logic function manipulation, they are not always efficient in dealing with incompletely specified functions, especially weakly specified ones. Moreover, the problem of designing an optimal decision diagram for incompletely specified data is NP-complete and only heuristic approaches are of practical use. As an alternative, new graph structures and novel principles might be exploited and validated for the representation and manipulation of incompletely specified functions. This chapter revisits decision diagram techniques for the representation and manipulation of incompletely specified functions.

19.2 Representation of incompletely specified functions

An *incompletely specified function* f is a mapping from the set $\{0,1\}^n$ into the set $\{0,1,d\}$, where d is an undefined value of the function called a *don't care.* That is, a function with *don't cares* is the relation where certain combinations of its variables cannot occur. The truth table of the function f does not generate output values for every possible combination of input values.

More formally, the function f can be represented by the sets:

$$\begin{aligned} \mathcal{C}_i &= \{\epsilon \in \{0,1\}^n | f(\epsilon) = i,\ i = 0,1\} \text{ and} \\ \mathcal{DC} &= \{\epsilon \in \{0,1\}^n | f(\epsilon) = d\}, \end{aligned}$$

where ϵ is a cube, and $k = \sum_{i=0}^{m-1} |\mathcal{C}_i|$ is the number of cubes ϵ of the function f. A cube ϵ is labeled with a decimal value $j \in (0, 2^n - 1)$. It is obvious that, given a completely specified function f, $\mathcal{DC} = \oslash$.

> **Example 19.1** *The Karnaugh map along with the set and symbolic descriptions for an incompletely specified function f of four variables is given in Figure 19.1a.*

There are two basic approaches to handling *don't cares* for incompletely specified functions:

- The first approach is based on redefining unspecified values of the function, that is, assigning 0's or 1's to the *don't care* values for the sake of optimization of the representation form.
- The second approach utilizes the idea "*don't care about don't cares*" which stands for excluding unspecified values from further analysis.

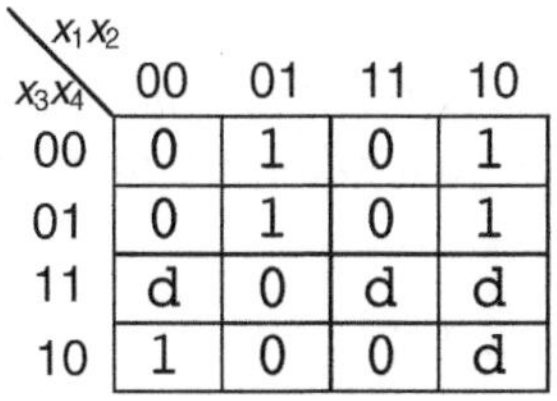

x_3x_4 \ x_1x_2	00	01	11	10
00	0	1	0	1
01	0	1	0	1
11	d	0	d	d
10	1	0	0	d

An incompletely specified function is given by:

$$\mathcal{C}_0 = \{0, 1, 6, 7, 12, 13, 14\}$$
$$\mathcal{C}_1 = \{2, 4, 5, 8, 9\}$$
$$\mathcal{DC} = \{3, 10, 11, 15\}$$

(*a*)

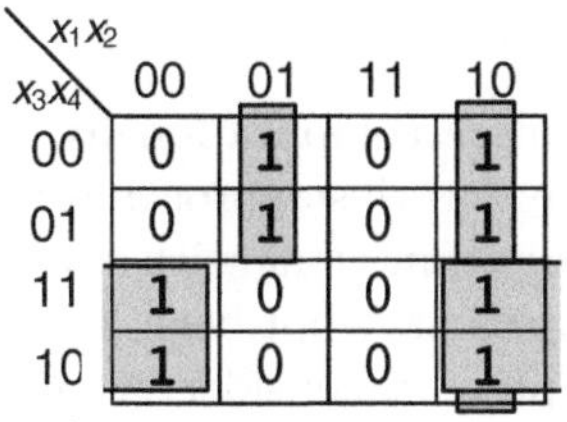

x_3x_4 \ x_1x_2	00	01	11	10
00	0	1	0	1
01	0	1	0	1
11	1	0	0	1
10	1	0	0	1

- *Assign the value 0 to all don't cares:*

$$f = x_1'x_2x_3' + x_1x_2'x_3' + x_1'x_2'x_3x_4'.$$

- *Assign 1 to all don't cares:*

$$f = x_1'x_2x_3' + x_1x_3x_4 + x_1x_2' + x_2'x_3$$

- *Assign 1 to don't cares $\mathcal{DC} = \{3, 10, 11\}$, and define $\mathcal{DC} = \{15\}$ with 0:*

$$f = x_1'x_2x_3' + x_1x_2' + x_2'x_3$$

(*b*)

FIGURE 19.1
An incompletely specified function given by different representation forms (Examples 19.1 and 19.2).

There are various representation forms used to describe the incompletely specified functions which are characterized by a different assignment of binary values to don't cares. Finding the assignment that leads to the smallest representation form is known to be NP-complete. Therefore, the exact techniques are computationally expensive, and practical approaches are based on certain heuristics.

The major obstacle in the representation of incompletely specified functions is the size explosion problem through redefining *don't cares*, which commonly raises questions about suggested heuristics and their practical applicability.

Example 19.2 *(Continuation of Example 19.1). Consider three different assignments of values to the* don't cares*:*

- *Assigning the value 0 to all* don't cares,
- *Assigning 1 to all* don't cares,
- *Assigning 1 to* don't cares $\mathcal{DC} = \{3, 10, 11\}$, *and 0 to* $\mathcal{DC} = \{15\}$.

The last choice of values leads to the simplest solution (Figure 19.1b).

The "*don't care about don't cares*" approach considers only the sets $\mathcal{C}$ to specify the function f and manipulate the cubes.

> **Example 19.3** *(Continuation of Example 19.1) The "don't care about don't cares" approach leaves the following care sets for future analysis:*
>
> $$\mathcal{C}_0 = \{0, 1, 6, 7, 12, 13, 14\}$$
> $$\mathcal{C}_1 = \{2, 4, 5, 8, 9\}$$
>
> *Further manipulation techniques depend on the problem and the final representation of the function.*

The "*don't care about don't cares*" approach is more computationally attractive than the redefinition of the *don't cares.* Note that this approach is not limited to single-output functions, it can be applied to functions with several outputs. However, decision diagram related techniques have not been as well developed as the redefinition approach. In this Chapter, we consider so-called *incompletely specified decision diagrams* based on the "*don't care about don't cares*" principle.

19.3 Decision diagrams for incompletely specified logic functions

> *Incompletely specified functions can be represented using BDDs or ternary decision diagrams (TDDs).*

Spectral decision diagrams are appropriate if the task is to find an optimal spectral form for incompletely specified functions. The same approach can be deployed for incompletely specified multiple-valued functions.

Given a variable ordering, the ROBDD representation of a completely specified function is unique. For an incompletely specified function, however, many ROBDDs can be used to represent the function, each associated with a different assignment of *don't cares* to binary values. The problem called *redefinition* is formulated as finding an assignment of don't cares that yields a small ROBDD representation.

19.3.1 Safe BDD minimization

The redefinition-based heuristics for deriving minimal decision diagram for incompletely specified functions are aimed at maximizing the instances of node sharing or sibling-substitution during the minimization process. BDD nodes become shared if the reassignment of *don't cares* makes their associated functions identical. Sibling-substitution is a special case of node sharing where a child of a BDD node is replaced by the other child. Sibling-substitution leads to fewer nodes because a parent and its two children are replaced by the child when the two children are made identical.

Most of the redefinition based approaches (see Further Reading) utilize so-called *sibling substitution.* If this substitution is performed on nodes that will not cause an increase in BDD size, this is called *safe* BDD minimization.

Let us consider the safe minimization in detail assuming that an incompletely specified function is given by a pair of completely specified functions $[f, c]$, where f is a cover of the incompletely specified function and c denotes the care-function, that is, the combination of the sets C_0 and C_1.

Definition 19.1 *A set of cubes f is a cover of the original function g if $C_0(g) \subseteq C_0(f)$ and $C_1(g) \subseteq C_1(f)$.*

Two BDDs $[F, C]$ are built for the functions $[f, c]$. The safe minimization algorithm consists of two phases:

- The *mark-edge* phase handles the preprocessing of the original BDDs $[F, C]$ identifying nodes for which applying sibling-substitution does not increase overall BDD size.
- In the second phase, called *build-result,* sibling-substitution is selectively applied to the nodes identified in the first phase.

Example 19.4 *Let an incompletely specified function $g = g(x_1, x_2, x_3)$, represented by the sets*

$$\begin{aligned} \mathcal{C}_0 &= \{2\}, \\ \mathcal{C}_1 &= \{5, 7\}, \\ \mathcal{DC} &= \{0, 1, 3, 4, 6\}, \end{aligned}$$

be given by two decision diagrams F and C (Figure 19.2(a) and (b) respectively). By traversing the diagrams, the sets of care values are

$$\begin{aligned} \mathcal{C}_0(F) &= \{0, 1, 2, 4, 6\}, \\ \mathcal{C}_1(F) &= \{3, 5, 7\}, \\ \mathcal{C}_0(C) &= \{2, 3\}, \\ \mathcal{C}_1(C) &= \{0, 1, 4, 5, 6, 7\} \end{aligned}$$

The result of the first edge-marking phase is shown in Figure 19.2c. The second phase, build-result, produces the diagram in Figure 19.2d.

The two basic phases of the safe BDD minimization are outlined below.

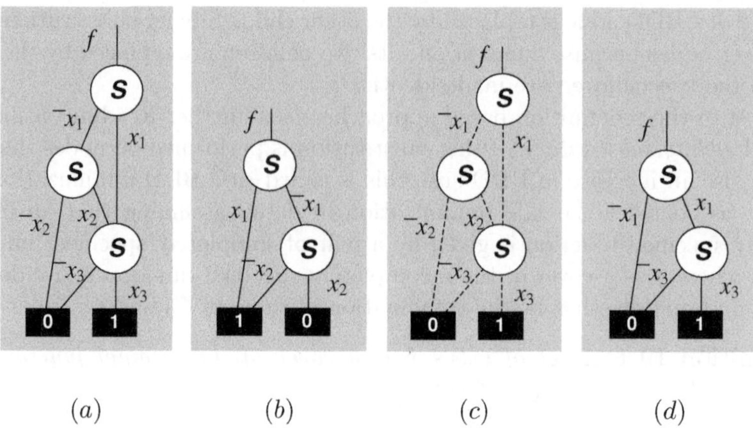

FIGURE 19.2
Safe BDD minimization of the incompletely specified function: (a) the first diagram; (b) the second diagram; (c) edge-marked result, and (d) minimized diagram (Example 19.4).

Mark-edges recursive algorithm:

> **Step 1.** Compare nodes from F and C for being a leaf and with 0 respectively. If it is not a leaf-0, continue with Steps 2-4.
>
> **Step 2.** Get the top variable x from F and C which will be used as a substitution.
>
> **Step 3.** For non-terminal nodes and the left sub-BDD, mark the edge and call the recursive function passing $F_{x'}$ and $C_{x'}$.
>
> **Step 4.** For non-terminal nodes and the right sub-BDD, mark the edge and call the recursive function passing F_x and C_x.

Build-result recursive algorithm:

> **Step 1.** Compare nodes from F for being a leaf. If it is not a leaf, continue with Steps 2-4.
>
> **Step 2.** Get the top variable x from F.
>
> **Step 3.** If the left sub-BDD is marked and the right sub-BDD is not, call the recursive function passing F_x.
>
> **Step 4.** If the left sub-BDD is not marked and the right sub-BDD is, call the recursive function passing $F_{x'}$.

Recalling that a BDD minimization using don't cares is safe if the minimized BDD is guaranteed to be smaller than the original BDD, the two-phase compaction algorithm is considered to be safe [14]. The result of minimization is produced by replacing some nodes with one of their descendants. This is safe because it ensures that no node will be split. This property can be deducted from the structure of the build-result algorithm. It creates one node for each node it visits and visits each node at most once. Specifically, nodes that are not reachable from the root by a path of marked edges are not visited by the build-result algorithm and, therefore, are not included in the minimized BDD.

19.3.2 Kleene function and ternary decision diagrams

A Kleene function K is $\mathbf{T}^n \to \mathbf{T}$ over the variable set $X = \{x_1, \cdots, x_n\}$, where $\mathbf{T}$=$\{0,1,d\}$, n is the number of variables, and d denotes unknown input or output values. The Kleene function represents the behavior of a logic function in the presence of unknown values. For a given two-valued logic function, the Kleene function is unique.

Ternary decision diagrams (TDDs) are similar to BDDs, except that each non-terminal node has three successors (see details in Chapter 23). Having three outgoing edges, it is possible to represent switching functions with unspecified (third) values. The ternary structures implement Kleene functions.

Generally, a TDD needs less space than a pair of BDDs. Moreover, BDD pairs cannot handle the unknown input or *don't care* output directly, while a TDD can. Comparing to BDDs, TDDs have one more terminal node, and one more edge for non-terminal nodes. TDDs can be reduced the same way as BDDs, and a TDD usually refers to the reduced one. To obtain a TDD from a BDD, the following transformations are needed.

- Expansion: expand the BDD into a decision tree.
- Alignment: apply the alignment operation recursively to create an unspecified edge:

$$\texttt{Alignment}(x, y) = \begin{cases} x \ if \ x = y \\ d \ if \ x \neq y. \end{cases}$$

- Reduction: reduce the ternary decision tree to a directed acyclic graph.

A TDD is canonical as well, so all the properties for canonical forms still apply. An *abbreviated TDD* is a TDD without unspecified edges. A *purged TDD* is a combination of abbreviated and full TDDs. Because the unspecified edge could be calculated with the alignment operation at any time, if it is done earlier, then the unspecified branch is not needed. A purged TDD only calculats the unspecified branch when it is necessary. This method could save time and space significantly. However, with the size explosion problem, TDDs are often too large to build.

> **Example 19.5** *The identity function for a variable x in Kleenean logic is given in Figure 19.3a. Figure 19.3b depicts an example of the full TDD for an incompletely specified function $f(x_1, x_2, x_3)$. Correspondingly, the abbreviated TDD is given in Figure 19.3c. The resulting expression is $f = x_1 x_2 \vee x_3$.*

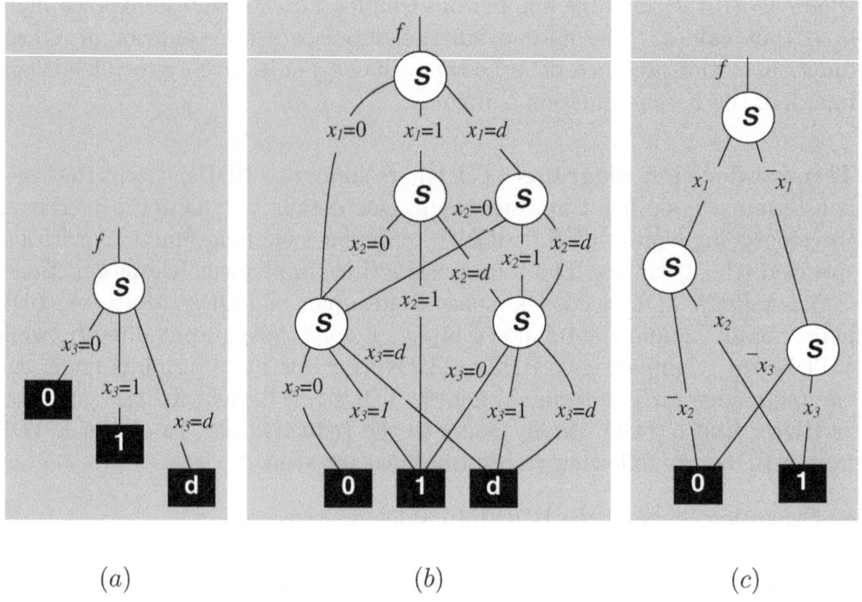

(a) (b) (c)

FIGURE 19.3
Ternary decision diagram minimization: for one variable (a), a full diagram (b), and an abbreviated diagram (c) (Example 19.5).

There are several functional extensions to TDDs for incompletely specified functions. These extensions include a class of Kronecker ternary decision diagrams, diagrams that have two kinds of nodes: binary nodes as in Kronecker decision diagrams, and ternary nodes as in TDDs. For certain functions, Kronecker ternary decision diagrams are smaller and more efficient for logic optimization and technology mapping than TDDs.

19.3.3 Manipulation of TDDs

The following changes are made to the original BDD algorithms to enable manipulation with unspecified values:

Reduce. The *reduce operator* is used to generate reduced TDDs from TDDs according to the reduction rules. All the nodes in a TDD are labelled with integers, then the nodes with the same label are combined. We use a bottom-up method to label the nodes one by one. A non-terminal node cannot be labelled until all its branches have been completely labelled. $id(n)$ denotes the label of the node n, $0(n)$ and $1(n)$ represent the corresponding successor of the node n. The labelling method has the following sequence of steps:

Step 1. If $id(0(n)) = id(1(n))$, then $id(n) = id(0(n))$.

Step 2. If there exists a node m with $id(0(m)) = id(0(n))$ and $id(1(m)) = id(0(n))$, then $id(n) = id(m)$. Otherwise, we assign the next unused label to $id(n)$.

Restrict. The *restrict operator* computes a new TDD with the same variable ordering, but restricts certain variables to a given value. We use $f(x = t)$ to denote a formula obtained by replacing all the occurrences of x in f by t. The TDD for $f(x = t)$ is constructed by forcing all the edges pointed to the node associated with x to point to the root of its proper edge instead. The reduce operation has to be executed afterwards.

Example 19.6 *An incompletely specified function is given by its TDD as shown in Figure 19.4a. The second step of the reduce operation eliminates one node with common successors from the second level of the diagram minimizing the number of nodes and their interconnections (Figure 19.4b). Using the obtained diagram and assuming that $x_2 = d$, the restrict operation forces all outgoing edges from the first level to point to terminal nodes as shown in Figure 19.4c.*

19.4 Incompletely specified decision diagrams

The deficiencies of existing heuristic algorithms, which redefine don't cares and manipulate variable order, can be avoided using different principles and an extension of the traditional graph structure. In this section, we give a definition of an incompletely specified decision diagram using the idea "*don't care about don't cares.*"

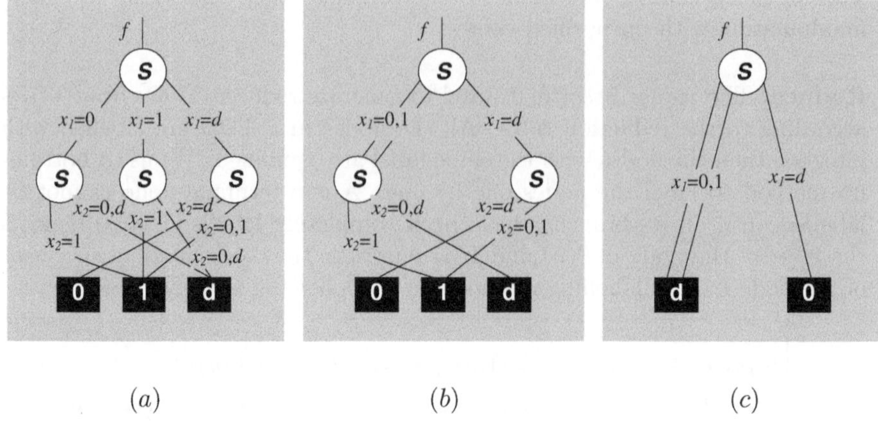

(a) (b) (c)

FIGURE 19.4
Ternary decision diagram operations: (a) original TDD, (b) result of reduce, and (c) result of restrict for $x_2 = d$ (Example 19.6).

19.4.1 Definition of the incompletely specified decision diagram

An incompletely specified decision diagram Γ is a directed acyclic graph with a node set in which each non-terminal node u has at least one successor.

Example 19.7 *An example of an incompletely specified decision diagram is shown in Figure 19.6a. Thus, the corresponding function is specified on three combinations:*

$$\{x_1^1 x_2^0 x_3^2\},\ \ \{x_1^2 x_2^0 x_3^0\},\ \ \{x_1^2 x_2^0 x_3^1\}.$$

The order of variables in the diagram is $\prec x_2\ x_1\ x_3 \succ$.

The properties of incompletely specified decision diagrams are defined similarly to those for traditional decision diagrams.

19.4.2 Manipulation of incompletely specified decision diagrams

There are two basic operations on incompletely specified decision diagrams:

▶ Variable reordering and

▶ Minimization.

These operations support the dynamic essence of incompletely specified decision diagrams: if the specification of incompletely specified function is updated, the structure of the diagram will be modified instantaneously.

Variable reordering. Many heuristics have been proposed for finding a good variable ordering, but it is evident that none of them guarantees that the solution is optimal. In this approach, we have chosen to use sifting, because of its dynamic nature. The basic idea of the sifting algorithm is to select the best position for one variable assuming that the relative order of all others remains the same. This process is repeated for all variables, starting with variables situated in the level with the largest number of nodes. dynamic variable ordering is integrated into the algorithm outlined below. The method consists of the following steps:

Step 1. `The levels are sorted according to their size. The largest level is considered first.`

Step 2. `For each variable:`

`2.1 The variable is exchanged with its successor variable until it is the last variable in the ordering.`

`2.2 The variable is exchanged with its predecessor until it is the topmost variable.`

`2.3 The variable is moved back to the closest position which has led to the minimal size of the BDD or MDD, respectively.`

Details are given in Chapter 12.

Example 19.8 *Consider the incompletely specified decision diagram given in Figure 19.5a:*

▶ *The dynamic reordering exchanges variables at the first and second levels, assigning x_2 to the first level and x_1 to the second level.*

▶ *The reordering results in a diagram shown in Figure 19.5b. This diagram has fewer nodes and interconnections, and will be more preferable as the result of optimization.*

Minimization. The compactness of incompletely specified decision diagrams is guaranteed by two rules:

▶ Merging that shares equal functions, and

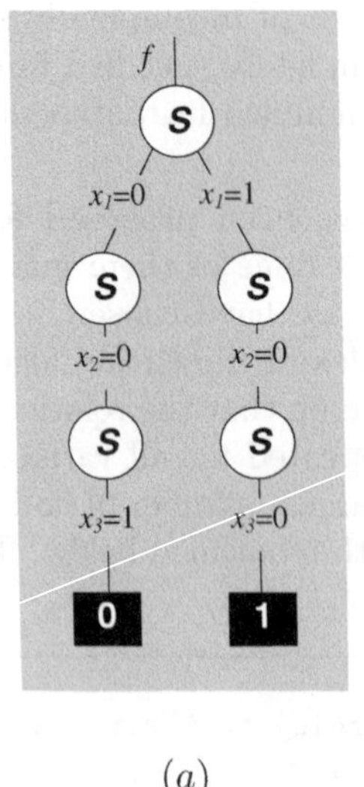

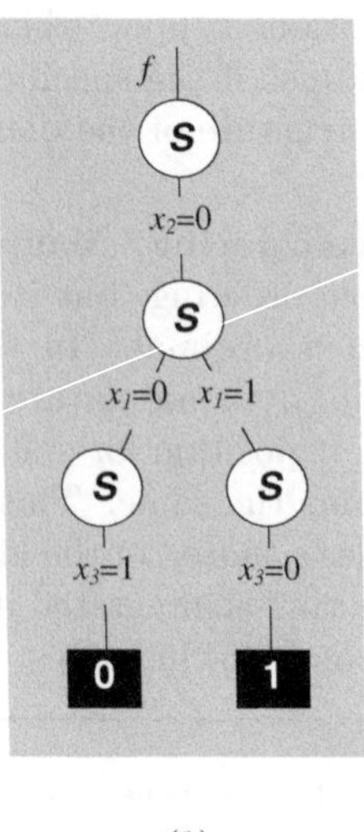

(a) (b)

FIGURE 19.5
Incompletely specified decision diagrams: (a) before the dynamic reordering, (b) after the reordering is completed (Example 19.8).

- Deletion that deletes a node where all r children are equal.

These rules are formulated similarly to the ones for completely specified functions except in the case of nodes with fewer successors than the radix of variables.

A post-processing operation is required. It is called *minimization* and serves to eliminate nodes where defined children are equal. It can be considered as a transition from an incompletely specified decision diagram into a traditional decision diagram. Before considering the minimization algorithm, let us introduce some more definitions.

A chain. Let Γ be a linear incompletely specified decision diagram composed of

- Non-terminal nodes associated with each variable x of the function f, and
- Terminal nodes with the value $f(\epsilon)$.

This represents a cube ϵ as a conjunction of a set of variable values: $\epsilon \Rightarrow \Gamma$. Note that don't cares are excluded from consideration.

Example 19.9 *A chain for the cube* $\{x_2^0 x_1^2 x_3^1\}$ *from the set* C_0 *is depicted in Figure 19.6(b).*

The variable ordering in the chain Γ is adjusted to the variable ordering in the existing diagram according to the output of dynamic variable reordering. To add the chain Γ into the current incompletely specified decision diagram, we merge graph structures through a *fusion* operation defined below.

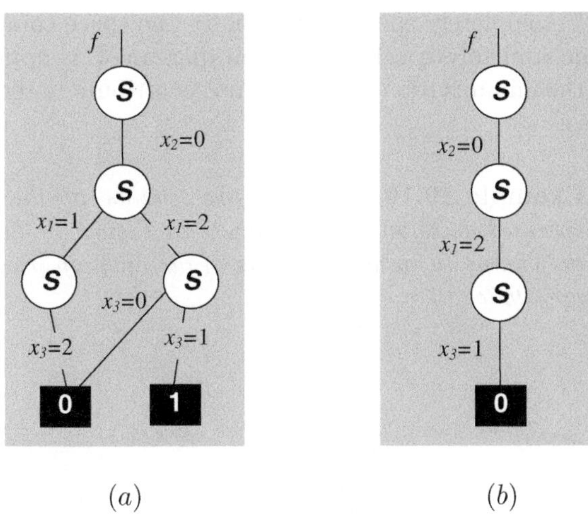

FIGURE 19.6
Incompletely specified decision diagram (a) and its chain for the cube $\{x_2^0 x_1^0 x_3^1\}$ (b) (Examples 19.7 and 19.9).

A fusion operation $\coprod$ unites two incompletely specified decision diagrams Γ_1 and Γ_2 with an identical variable ordering:

$$\Gamma = \Gamma_1 \coprod \Gamma_2.$$

In the following, we consider the design of the incompletely specified decision diagram Γ for the given incompletely specified multivalued function f as a sequence of fusion operations on k chains γ:

$$\Gamma = \coprod_k \gamma_k.$$

This process can be described iteratively as a sequence of snapshots: $\Gamma_{t+1} = \Gamma_t \coprod \gamma_{t+1}$, where $t = 1, \ldots, k$ and $\Gamma_1 = \gamma_1$.

The computational complexity of the fusion operation $\coprod$ on the current incompletely specified decision diagram Γ and the chain γ is in $O(n)$, and the computational complexity of the design of the incompletely specified decision diagram Γ is $O(k \cdot n)$. This makes it feasible for even weakly specified functions. The design of the incompletely specified decision diagram Γ has a space complexity approximated by $k \cdot n$.

The fusion operation $\coprod$ reduces the space complexity, which can be finally approximated by $k{\cdot}(n - \log_r k + 1) - 1$. The latter formula leads to the conclusion that the usability of this method is restricted to weakly specified functions, where k is significantly less than r^n: $k \ll r^n$. For the function f where $k = r^n$ (completely specified function), the space complexity of the design of the incompletely specified decision diagram Γ is approximated by r^n. Obviously, the minimization stage does not contribute to reduction of the number of nodes.

Example 19.10 *Consider two chains of the incompletely specified decision diagram given in Figure 19.7a,b. Fusion of the chains (a) and (b) results in the diagram presented in Figure 19.7c.*

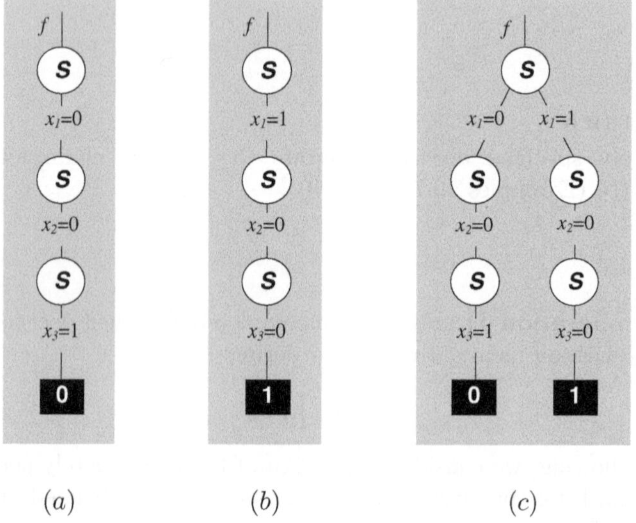

FIGURE 19.7
Fusion operation: (a) the first diagram, (b) the second diagram, and (c) the result of fusion (Example 19.10).

A sketch of the minimization algorithm for the function f given on the sets $\mathcal{C}$ is shown below:

Step 1. `Initialize the variable ordering lexicographically:` $x_1 \prec x_2 \prec \ldots \prec x_n$ `(`x_1 `is the topmost variable), and the incompletely specified decision diagram` $\Gamma = \oslash$.

Step 2. `Consider a cube` ϵ `from the given set` $\mathcal{C}$: $\epsilon \in \mathcal{C}$. `Build a corresponding chain` γ `applying current variable ordering:` $\epsilon \Rightarrow \gamma$.

Step 3. `Merge the chain` γ `with the existing incompletely specified decision diagram` Γ `applying the fusion operation:` $\Gamma = \Gamma \coprod \gamma$.

Step 4. `Reorder the obtained incompletely specified decision diagram` Γ`, applying sifting.`

Step 5. `Perform the compaction of the incompletely specified decision diagram` Γ `to eliminate nodes that share equal functions, and parent nodes with equal children.`

Step 6. `If there are other cubes, i.e.,` $\mathcal{C} \neq \oslash$`, go to Step 2. Otherwise, terminate the algorithm and do the minimization of the incompletely specified decision diagram` Γ.

Step 7. `(Postprocessing) Perform the minimization of the incompletely specified decision diagram` Γ `to eliminate nodes with` *`don't cares`*.

Example 19.11 *Let us consider the following incompletely specified 3-valued function* $f = f(x_1, x_2, x_3)$ *given on three combinations of variable values:*

$$\mathcal{C}_0 = \{x_1^2 x_2^0 x_3^1\}$$
$$\mathcal{C}_1 = \{x_1^1 x_2^0 x_3^2, x_1^2 x_2^0 x_3^0\}.$$

The iterative process of constructing an incompletely specified Multivalued decision diagram and its minimization is illustrated in Figure 19.8:

- *The first step combines two chains* $x_1^1 x_2^0 x_3^2$ *and* $x_1^2 x_2^0 x_3^0$ *by applying the fusion operation.*
- *The post-processing reordering (sifting) swaps variables assigned to the first and second levels. The resulting diagram is combined next with the remaining chain* $x_1^2 x_2^0 x_3^1$.
- *The final reordering operation does not change the order of variables in the diagram.*

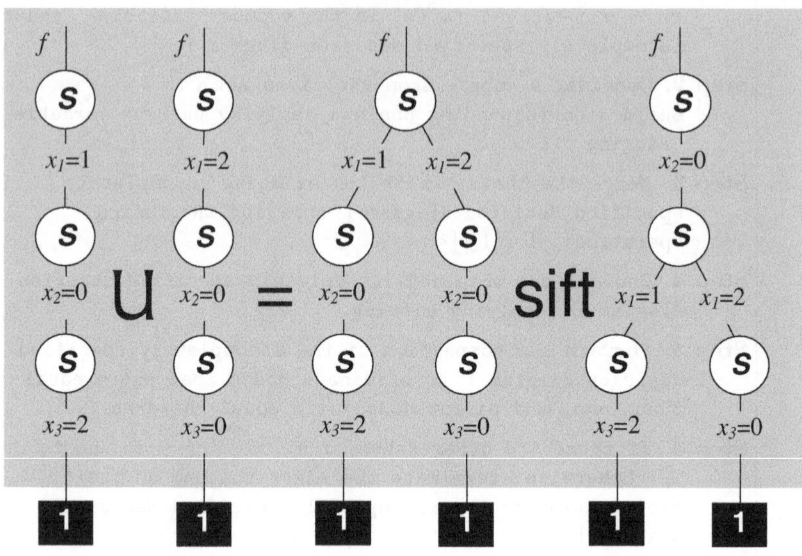

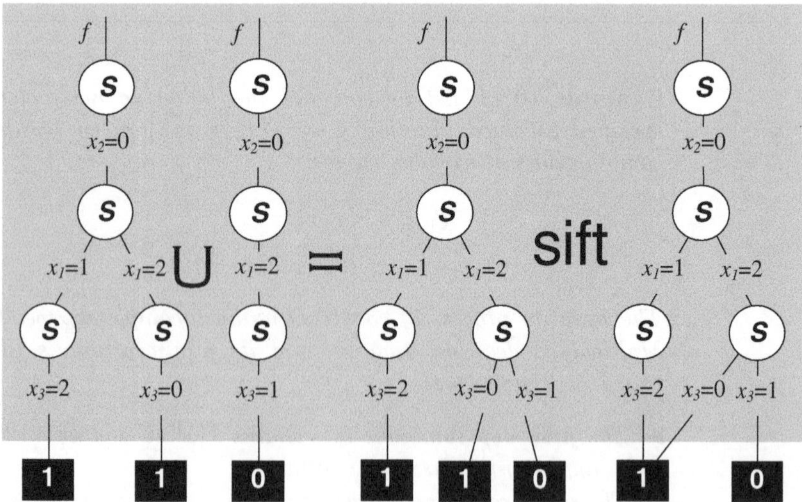

FIGURE 19.8
Multiple-valued decision diagram minimization for the function f (Example 19.11): Step 1 is forming the diagram from two chains $x_1^1 x_2^0 x_3^2$ and $x_1^2 x_2^0 x_3^0$, and Step 2 is adding the chain $x_1^2 x_2^0 x_3^1$.

19.5 Further reading

The fundamentals of incompletely specified functions, their nature and applicability in electronics are introduced, in particular, by Brayton et al. [4], Green [10], Varma and Trachtenberg [37], and Smith [34]. Decision diagram representations of incompletely specified functions as well as manipulation techniques are used in a variety of settings including micro- and nanoelectronic design, for example, by Iguchi et al. [16], Bergamaschi et al. [2], and Hong et al. [13]. Brayton et al. [5] studied different varieties of don't cares occuring in Boolean networks. *Structural* don't cares are created by the internal structure of a Boolean network. A single function has no structural don't cares because it has no internal structure. There are two types of structural don't cares: *satisfiability* don't cares occur when an intermediate variable value is inconsistent with its function inputs, and *observability* don't cares occur when an intermediate variable's value doesn't affect the network's primary outputs. Sentovich and Brand [30] studied don't care conditions as a form of flexibility.

The two basic approaches to handling *don't cares* of an incompletely specified function have been proposed so far:

- Finding the assignment to the *don't cares*, also called *redefining* the unspecified values of the function, for the sake of deriving the smallest diagram. This principle has been extensively exploited for several years, in particular, by Hong et al. [13], Oliveira et al. [23], and Sauerhoff and Wegener [29]. Finding the assignment that leads to the smallest diagram is known to be NP-complete [29] and exact techniques [23] are typically too computationally expensive. Therefore, heuristic algorithms have been developed to address this minimization problem (see, for example, the work by Shiple et al. [32]). Zilic and Vranesic [42] considered the minimization of incompletely specified functions in Reed-Muller form as polynomial interpolation problem over finite field.
- The second approach utilizes the idea called "*don't care about don't cares*". This approach was initially developed by Zakrevskij [40]. Based on this approach, a so called *stair-case* algorithm was developed by Zakrevskij et al. [41], Holowinski et al. [11, 12], and Yanushkevich [39]. Popel et al. [27] introduced the concept of incompletely specified decision diagrams and appropriate algorithms for diagram minimization. This methodology was extended to multivalued functions in [26].

There are multiple heuristics suggested to minimize BDDs/ROBDDs. Thus, a framework of sibling-substitution-based heuristics was proposed by Shiple et al. [32]. These heuristics, specifically *restrict* and *constrain*, outperform others in terms of both run-time and resulting BDD size. Another method of assigning binary values to *don't cares*, by traversing the BDD structure from top to bottom, was outlined by Chang et al. [7]. Being computationally complex, it makes sub-BDDs shared, which results in overall diagram reduction. More recently, restrict and constrain heuristics were adjusted to minimize BDDs safely [14]. The idea of the *safe* BDD minimization is to perform sibling-substitution only on nodes that will not cause increase in BDD size.

Various aspects of decision diagram representation and optimization for incompletely specified data have been extensively studied in the following areas:

- Decomposition of functions using decision diagrams (see, for example, the paper by Mishchenko et al. [22], and Chapter 25),
- Minimization of logic networks (see, for example, results by Jiang and Brayton [18]), and
- Synthesis of finite state machines (see, for example, detail in work by Oliveira and Edwards [24]).

Many applications can take advantage of optimizing the diagram size for incompletely specified functions:

Verification of FSM. Symbolic reachability analysis using decision diagrams was introduced in [9], and demonstrated superior results for large FSMs against explicit state techniques that process one state at a time. The decision diagram based techniques minimize the transition relation of an FSM with respect to unreachable states [9, 32]. The performance of such an analysis can be improved via diagram minimization using *don't cares.*

Synthesis of FPGAs. The selection of minimum size representations is also important in logic synthesis applications where decision diagrams are used to derive gave-level solutions, e.g. Shannon circuits and multiplexer-based *Field Programmable Gate Arrays* (FPGAs). Some FPGA techniques work from decision diagram representations to map circuits to multiplexer-based FPGAs [20, 33]. For incompletely specified circuits, the output of decision diagram minimization can advance to smaller circuit implementation.

Data mining. The majority of data mining problems are given in functional form. Decision diagram representations serve as a tool for decision making processes, where the minimum diagram size can lead to faster decision making (see, for example, the results by Popel and Hakeem [27]). In many cases in inductive learning, applications that use decision diagrams as the representation scheme, the accuracy of the inferred hypothesis is strongly dependent on the complexity of the result.

Logic function evaluation with unknown inputs. In logic simulation, it is advantageous to keep a selected subset of inputs as unknowns. For example, if some inputs do not affect the outputs, these inputs can be retained as unknown. The initialization of logic networks also assumes unknown inputs. For such cases, conventional logic simulators take too much time. Also, they often produce imprecise results for the networks with reconvergence. Cycle-based logic simulators based on BDDs are faster than conventional ones. However, to evaluate a logic function in the presence of unknown inputs, we have to evaluate the functions where the unknown values d are replaced by 0 and 1, in all possible combinations. Thus, the simulation time increases exponentially with the number of unknown inputs.

The Kleenean strong logic is used to represent incompletely specified functions in the form of TDDs [19]. Several TDD manipulation techniques were adopted from BDDs by Jennings [17].

Iguchi et. al. [16] suggested that to evaluate the output values precisely, the simulator needs to use a regular ternary logic function (RT function) for the given two-valued logic function, and a Kleene TDD (Kleene ternary decision diagram). This produces precise values in $O(n)$ computation time. Unfortunately, a Kleene TDD requires $O(3^n/n)$ memory storage to represent an n-variable function. To make logic simulation faster, a special hardware can be used such as a look-up-table (LUT), cascades, and LUT rings. It has been shown that the double-rail realizations of RT functions are efficient and smaller than the straightforward RAM realizations.

Sets of pairs of functions to be distinguished (SPFDs) introduced by Yamashita et al. [38] are used to express the flexibility and represent incompletely specified functions (see detail in Chapter 8). SPFDs provide more freedom in manipulation of incompletely specified functions and computing the flexibility at a node in a Boolean network.

References

[1] Allen M and Zilberstein Z. Automated conversion and simplification of plan representations. In *Proceedins of the International Conference on Autonomous Agents and Multiagent Systems*, pp. 1272–1273, 2004.

[2] Bergamaschi R, Brand D, Stok L, Berkelaar M, and Prakash S. Efficient use of large don't cares in high-level and logic synthesis. In *Proceedins of the IEEE/ACM International Conference on Computer Aided Design*, pp. 272–278, 1995.

[3] Brace K, Rudell R, and Bryant R. Efficient implementation of a BDD package. In *Proceedins of the IEEE/ACM International Design Automation Conference*, pp. 40–45, 1990.

[4] Brayton RK, Hachtel GD, McMullen C, and Sangiovanni-Vincentelli AL. *Logic Minimization Algorithms for VLSI Synthesis.* Kluwer, Dordrecht, 1984.

[5] Brayton RK, Hachtel GD, and Sangiovanni-Vincentelli AL. Multilevel logic synthesis. *Proceedings of the IEEE*, 78(2):264–300, 1990.

[6] Bryant R. Binary decision diagrams and beyond: enabling techniques for formal verification. In *Proceedins of the International Conference on Computer-Aided Design*, pp. 236–243, 1995.

[7] Chang S, Marek-Sadowska M, and Hwang T. Technology mapping for TLU FPGA's based on decomposition of binary decision diagrams. *IEEE Transactions on Computer-Aided Design of Integrated Circuits and Systems*, 15(10):1226–1248, 1996.

[8] Clarke E, Fujita F, McGeer P, McMillan K, Yang J, and Zhao X. Multi terminal binary decision diagrams: an efficient data structure for matrix representation. In *Proceedings of the International Workshop on Logic Synthesis*, pp. P6a:1–15, 1993.

[9] Coudert O, Berthet C, and Madre J-C. Verification of synchronous sequential machines based on symbolic execution. In *Proceedings of the Workshop on Automatic Verification Methods for Finite State Systems, LN407 in Computer Science*, pp. 365–373, 1989.

[10] Green DH. Reed–Muller expansions of incompletely specified functions. *IEE Proceedings*, Pt. E. 134(5):228–236, 1987.

[11] Holowinski G and Yanushkevich SN. Fast heuristic minimization of MVL functions in generalized Reed-Muller domain. In *Proceedings of the 3rd International Conference on Applications of Computer Systems*, Poland, pp. 58–64, 1996.

[12] Holowinski G, Malecki K, Dueck GW, Shmerko VP, and Yanushkevich SN. Development of Zakrevskij's minimization strategy towards arithmetical polynomial domain. In *Proceedings of the International Workshop on Boolean Problems*, Germany, pp. 101–108, 1998.

[13] Hong H, Beerel P, Burch J, and McMillan K. Safe BDD minimization using don't cares. In *Proceedings of the IEEE/ACM International Design Automation Conference*, pp. 208–213, 1997.

[14] Hong H, Beerel P, Burch J, and McMillan K. Sibling-substitution-based BDD minimization using don't cares. *IEEE Transactions on Computer-Aided Design of Integrated Circuits and Systems*, 19(1):44–55, 2000.

[15] Hong Y, Beerel P, Lavagno L, and Sentovich E. Don't care-based BDD minimization for embedded software. In *Proceedings of the Design Automation Conference*, pp. 506–509, 1998.

[16] Iguchi Y, Sasao T, and Matsuura M. A method to evaluate logic functions in the presence of unknown inputs using LUT cascades. In *Proceedings of the IEEE 34th International Symposium on Multiple-Valued Logic*, pp. 302–308, 2004.

[17] Jennings G. Symbolic incompletely specified functions for correct evaluation in the presence of indeterminate input values. In *Proceedings of the Annual Hawaii International Conference on System Sciences*, pp. 23–31, 1995.

[18] Jiang Y and Brayton Y. Don't cares and multi-valued logic network minimization. In *Proceedings of the IEEE/ACM International Conference on Computer-Aided Design*, pp. 520–525, 2000.

[19] Kleene S. *Introduction to Metamathematics.* Van Nostrand, Princeton, New York, 1964.

[20] Lee G. Logic synthesis for cellular architecture FPGA using BDDs. In *Proceedings of the IEEE International ASP Design Automation Conference*, pp. 253–258, 1997.

[21] Lindgren P. Improved computational methods and lazy evaluation of the ordered ternary decision diagrams. In *Proceedins of Asia and South Pacific Design Automation Conference*, pp. 379–384, 1995.

[22] Mishchenko A, Files C, Perkowski M, Steinbach B, and Dorotska C. Implicit algorithms for multi-valued input support minimization. In *Proceedins of the International Workshop on Boolean Problems*, pp. 9–20, 2000.

[23] Oliveira A, Carloni L, Villa T, and Sangiovanni-Vincentelli A. Exact minimization of Boolean decision diagrams using implicit techniques. Technical report, University of California, CA, 1996.

[24] Oliveira A and Edwards S. Limits of exact algorithms for inference of minimum size finite state machines. In *Proceedings of the Algorithmic Learning Theory Workshop*, 1160:59–66, 1996.

[25] Perkowski M, Chrzanowska-Jeske M, Sarabi A, and Schafer I. Multi-level logic synthesis based on Kronecker decision diagrams and Boolean ternary decision diagrams for incompletely specified functions. *Proceedings VLSI*, 3(3-4):301–313, 1995.

[26] Popel D and Drechsler R. Efficient minimization of multiple-valued decision diagrams for incompletely specified functions. In *Proceedins of the IEEE 33rd International Symposium on Multiple-Valued Logic*, pp. 241–246, 2003.

[27] Popel D and Hakeem N. Multiple-valued logic in decision making and knowledge discovery. Technical report, Baker University, KS, 2002.

[28] Reif J and Sun Z. Movement planning in the presence of flows. *Lecture Notes in Computer Science*, 2125:450–461, 2000.

[29] Sauerhoff M and Wegener I. On the complexity of minimizing the OBDD size for incompletely specified functions. *IEEE Transactions on Computer-Aided Design of Integrated Circuits and Systems*, 15(11):1435–1437, 1996.

[30] Sentovich E and Brand D. Flexibility in logic. In Hassoun S and Sasao T, Eds., Brayton RK, consulting Ed. *Logic Synthesis and Verification*, Kluwer, Dordrecht, pp. 65–88, 2002.

[31] Shmerko VP, Holowinski G, Song N, Dill K, Yanushkevich SN, and Perkowski M. High-quality minimization of multi-valued input binary output exclusive-or sum of product expressions for strongly unspecified multi-output functions. In *Proceedings of the International Conference on Applications of Computer Systems*, Poland, pp. 248–255, 1997

[32] Shiple T, Hojati R, Sangiovanni-Vincentelli A, and Brayton R. Heuristic minimization of BDDs using don't cares. In *Proceedings of the IEEE/ACM International Design Automation Conference*, pp. 225–231, 1994.

[33] Sinha S and Brayton R. Implementation and use of SPFDs in optimizing Boolean networks. In *Proceedings of the IEEE/ACM International Conference on Computer-Aided Design*, pp. 103–110, 1998.

[34] Smith DR. Complexity of partially defined combinational switching functions. *IEEE Transactions on Computers,* 20(2):204–208, 1971.

[35] Srinivasan A, Kam T, Malik S, and Brayton R. Algorithms for discrete function manipulation. In *Proceedins of the IEEE/ACM International Conference on Computer-Aided Design*, pp. 92–95, 1990.

[36] Wang C, Hachtel G, and Somenzi F. The compositional far side of image computation. In *Proceedings of the International Conference on Computer-Aided Design*, 2003.

[37] Varma D and Trachtenberg E. Computational of Reed–Muller expansions of incompletely specified Boolean functions from reduced representations. *IEE Proceedings,* Pt. E, 138(2):85–92, 1991.

[38] Yamashita S, Sawada H, and Nagoya A. SPFD: a method to express functional flexibility. *IEEE Transactions on Computer-Aided Design of Integrated Circuits and Systems*, 19(8):840–849, 2000.

[39] Yanushkevich SN. *Logic Differential Calculus in Multi-Valued Logic Design.* Technical University of Szczecin Academic Publisher, Szczecin, Poland, 1998.

[40] Zakrevskij A. Optimizing polynomial implementation of incompletely specified Boolean functions. In *Proceedins of the Workshop on Application of the Reed-Muller Expansions in Circuit Design*, pp. 250–256, 1995.

[41] Zakrevskij AD, Yanushkevich SN, and Jaroszewicz S. Minimization of Reed–Muller expressions for system of incompletely specified MVL functions. In *Proceedins of the 3rd International Conference on Methods and Models in Automation and Robotics*, pp. 1085–1090, Miedzyzdroje, Poland, Sept. 1996.

[42] Zilic Z and Vranesic Z. A multiple valued Reed–Muller transform for incompletely specified functions. *IEEE Transactions on Computers*, 44(8):1012–1020, 1995.

20

Probabilistic Decision Diagram Techniques

In this chapter, we present an efficient method for computing output probability of a function directly from its BDD representation. The method is applicable for functions in a shared BDD representation and accommodates complemented edges. Computation of individual spectral coefficients is presented as one application of this method.

20.1 Introduction

The concept of output probability is formally defined as follows.

Definition 20.1 *The* output probability *of a switching function is a real-valued quantity that specifies the probability the function will assume the value 1 given the probabilities that each of the input variables are assigned the value 1.*

Example 20.1 *Assuming all inputs have equal probability of being 0 or 1:*

- *The output probability of a two-input AND gate is 1/4,*
- *The output probability of a three-input OR gate is 7/8, and*
- *Two-input exclusive-OR gate has output probability 2/4.*

Example 20.2 *Computing the output probability for even a simple function is not always straightforward since in general the terms in a sum-of-products expression are not disjoint. For example, switching function*

$$f = x_1x_2 \vee x_3x_4 \vee x_5x_6$$

can be verified to be 1 for 37 minterms so

$$< \texttt{The output probability} >= \frac{37}{64} = 0.578125.$$

Note that result of Example 20.2 is not readily seen from the minimal sum-of-products expression.

Output probability is applicable in various applications such as analyzing the random testability of combinational circuits, low-power design and the computation of spectral coefficients. The interest here is how output probability can be computed directly on the BDD representation of switching functions.

Every path in a BDD from the top (root) node to the terminal node 1 represents a term in a disjoint sum-of-products representation of the function represented by the BDD. In other words, every such path represents a set of minterms for which the function evaluates to 1, and those sets are pairwise disjoint. The disjoint property is a direct consequence of the separation inherent in the Shannon expansion in each nonterminal node.

A naive approach to computing the output probability is to traverse the BDD computing the output weight for each path from the root node to terminal node 1. This weight is $2^{(n-m)}$ where m is the number of the n function input variables that appear on the path. Since the paths are disjoint, the output probability is the sum of these weights divided by 2^n.

Below, we will show that the output probability can be computed as the BDD is constructed, thereby avoiding a separate traversal. Further, while the naive method implicitly assumes all input variables have equal probability (1/2) of being 0 or 1, the methods discussed here permit different input probabilities.

Computation of spectral coefficients is briefly discussed as one application of output probabilities. Readers interested in other applications should consult the literature cited for further reading at the end of the Chapter.

20.2 BDD methods for computing output probability

Let a switching function f be represented by a binary decision diagram. The node in this diagram assigned with the i-th variable, x_i, and with outgoing edges corresponding to $f_{x_i=0}$ and $f_{x_i=1}$, computes the Shannon expansion

$$f = \overline{x}_i f_0 \vee x_i f_1$$

where $f_0 = f_{x_i=0}$ and $f_1 = f_{x_i=1}$.

This equation in arithmetic form is the following:

$$
\begin{aligned}
f &= \overline{x}_i f_0 + x_i f_1 - \overbrace{(\overline{x}_i f_0)(x_i f_1)}^{Equal\ to\ 0)} \\
&= \overline{x}_i f_0 + x_i f_1.
\end{aligned}
$$

That is, the probability at the node's root can be calculated using the equation:

$$
\begin{aligned}
p(f) &= p(\overline{x}_i f_0) + p(x_i f_1) \\
&= p(\overline{x}_i)p(f_0) + p(x_i)p(f_1) \\
&= p(x_i = 0)p(f_{x_i=0}) + p(x_i = 1)p(f_{x_i=1})
\end{aligned}
$$

because of independency of co-factors $\overline{x}_i$ and f_0 and x_i and f_1.

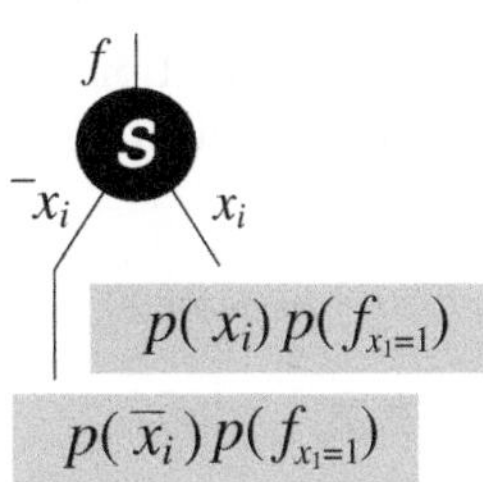

A decision diagram represents a switching function f of n random and independent variables $x_1, \ldots, x_i, \ldots, x_n$.
The i-th node of the Shannon expansion propagate probabilities of variables

$$
\begin{aligned}
p(f) &= p(\overline{x}_i)p(f_0) + p(x_i)p(f_1), \\
f_0 &= f_{x_i=1} \\
f_1 &= f_{x_i=0}
\end{aligned}
$$

FIGURE 20.1
Probability propagation through the Shannon node in decision diagram.

The output probability of a circuit can be evaluated on BDD using top-down or bottom-up approach.

20.2.1 A top-down approach

Thornton and Nair [11] presented a top-down approach to computing the output probability of a function represented as a BDD. Assuming equal probability of 0 and 1 for each input, the algorithm is given in Figure 20.2.

Example 20.3 gives a simple example of applying this algorithm. One can see drawbacks to this approach:

- ▶ The algorithm uses a top-down approach so the complete BDD must be constructed and then a separate traversal made to compute the probabilities.

```
Step 1. Assign a probability of 1 to the BDD's root node.
Step 2. Proceeding through the nodes in top to bottom
    breadth-first order:

    (a) The probability assigned to each of the two outgoing
        edges from a node is set to 1/2 of the probability
        assigned to that node.
    (b) The probability assigned to a node is the sum of the
        probabilities assigned to edges that point to that
        node.

Step 3. The output probability of the function is the
    probability assigned to the terminal 1 node in the BDD.
```

FIGURE 20.2
Top-down algorithm to computing the probability of a function represented as a BDD.

- ► The probability assigned to a node does not represent the output probability of the function represented by the BDD rooted by that node.
- ► As a result, in a shared BDD representation of a set of functions, recomputation is required for the shared nodes since they will in general be assigned different probabilities for each function.

Example 20.3 *The BDD for switching function*

$$f(x_1, x_2, x_3) = x_1 \vee \overline{x}_2\overline{x}_3$$

is shown in Figure 20.3. The probabilities shown for each node are as computed by Thornton and Nair's top-down algorithm.

20.2.2 A bottom-up approach

A better approach is to assign each node in the BDD the actual output probability for the function represented by the BDD for which that node is the root. This means the output probability for the function represented by a BDD is assigned to the root and not the terminal 1 node.

Observe the following:

- ► Since each nonterminal node in the BDD implements a Shannon expansion, the probability assigned to each node is 1/2 of the sum of the probabilities assigned to its two children nodes (the nodes the outgoing edges point to). This is assuming the probability of each input being 0 or 1 is equal.
- ► The terminal nodes represent constant functions and clearly the terminal 1 has probability 1.0 and the terminal 0 has probability 0.0.

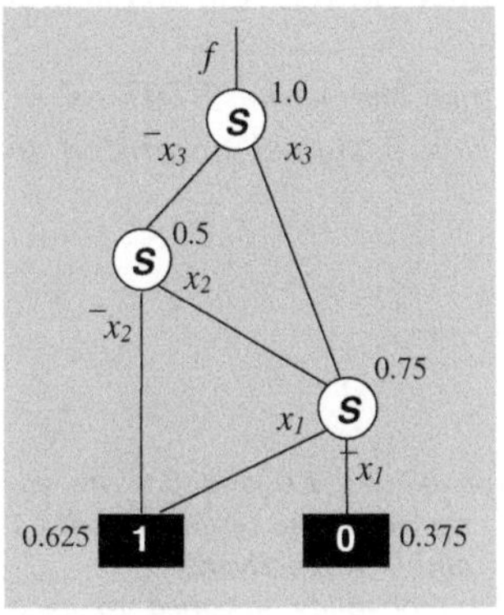

- *Probabilities are shown for nodes only.*
- *The probability on each edge is 0.5 of the probability of its parent node.*
- *The output probability for f is the probability assigned to the terminal node labelled 1.*

FIGURE 20.3
Application of bottom-up probability assignment algorithm (Example 20.3).

- The probability assigned to a node is fixed and, in particular, is the same when used multiple times in a shared BDD.
- The above implies node probabilities should be assigned using a bottom-up traversal. In fact, the probability of a node can be assigned when a node is created. This is indeed a bottom-up approach but a separate traversal is not required.
- The computation is done as the BDD is formed whereas the top-down approach requires construction of the complete BDD before any probabilities can be assigned.

The top-down algorithm for computing probability is given in Figure 20.4.

Step 1. Assign a probability of 1.0 to the terminal node 1 and a probability of 0.0 to the terminal node 0.

Step 2. Whenever a new nonterminal node (ν) is created assign it the probability $p = 1/2(p_0 + p_1)$ where p_0 is the probability assigned to the node the 0-edge of ν points to, and p_1 is the probability assigned to the node the 1-edge of ν points to.

FIGURE 20.4
The bottom-up algorithm for computing output probability for the function represented by the BDD.

Example 20.4 illustrate the application of the bottom-up algorithm. Note the approach is totally distinct from the top-down approach and the probabilities assigned to a node by the two methods have no relation to each other.

Example 20.4 *Figure 20.5 shows the same BDD as in Example 20.3 with node probabilities assigned according to the bottom-up algorithm.*

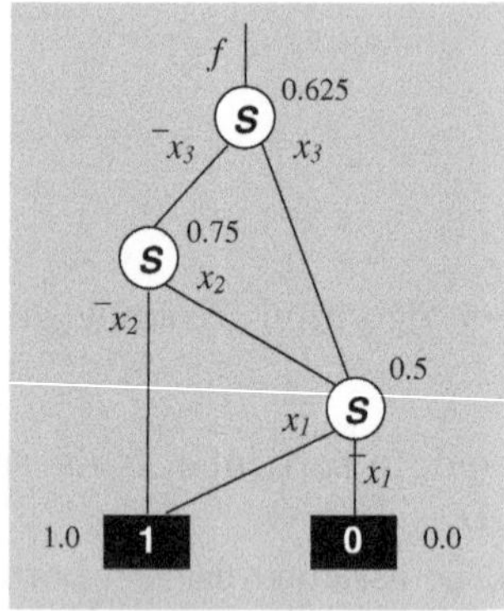

- *Probabilities 1.0 and 0.0 are assigned to the terminal nodes 1 and 0, respectively.*
- *The probability assigned to each nonterminal is 1/2 the sum of the probabilities assigned to its children.*
- *Probabiliteis are assigned as nodes are created and a separate traversal is not required.*

FIGURE 20.5
Application of bottom-up probability assignment algorithm (Example 20.4).

Both of the above approaches are easily adapted to take complemented edges into account.

For the bottom-up method, an edge points to a node with a particular (fixed) probability p which is the output probability for a particular subfunction. If the edge has a complement, it actually points to the complement of that subfunction which clearly has output probability $1-p$.

Example 20.5 *Figure 20.6 shows a BDD with complemented edges for the function*

$$f = x_1x_2 \vee x_3x_4 \vee x_5x_6.$$

As seen in Example 20.2 this function is 1 for 37 of 64 input patterns yielding and output probability of 0.578125 as is shown to be computed by the bottom-up algorithm.

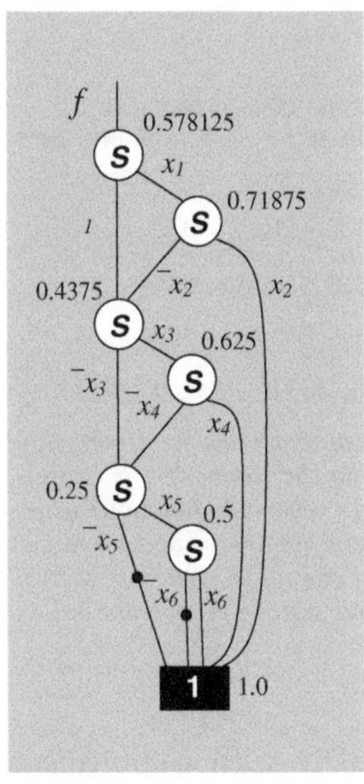

- *The terminal node is assigned probability 1.0*
- *The output probabilities for each nonterminal node in order from bottom up are:*

x_6: $\frac{1}{2}(1-1.0)+\frac{1}{2}(1.0)=0.5$

x_5: $\frac{1}{2}(1-1.0)+\frac{1}{2}(0.5)=0.25$

x_4: $\frac{1}{2}(1-0.25)+\frac{1}{2}(1.0)=0.625$

x_3: $\frac{1}{2}(0.25)+\frac{1}{2}(0.625)=0.4375$

x_2: $\frac{1}{2}(0.4375)+\frac{1}{2}(1.0)=0.71875$

x_1: $\frac{1}{2}(0.4375)+\frac{1}{2}(0.71875)=0.578125$

FIGURE 20.6
Bottom-up probability assignment for a BDD with complemented edges (Example 20.5).

Example 20.6 *Figure 20.7 shows a shared BDD with complemented edges for a 3-input, 3-output function that represents a simple cyclic increment, i.e.,*

$$f(000)=001, f(001)=010, ..., f(111)=000.$$

This a reversible function and by definition each output probability is 0.5 (see Chapter 35).

The above has been presented assuming every input variable has equal (1/2) probability of being 0 or 1. The approaches are easily modified to allow for unequal probabilities. In the bottom-up approach in particular, this simply means different weights are used when computing the probability associated with a node based on the probabilities assigned to its two children. If the probabilities vary, recomputation is required.

20.2.3 Symbolic computation

There is of course the possibility to adapt the algorithms above to allow one or more of the input probabilities to be symbolically represented. Let p_i denote the probability that input $x_i = 1$. The probability that $x_i = 0$ is $1 - p_i$. Figure 20.8 illustrates the symbolic computation for the BDD in Example

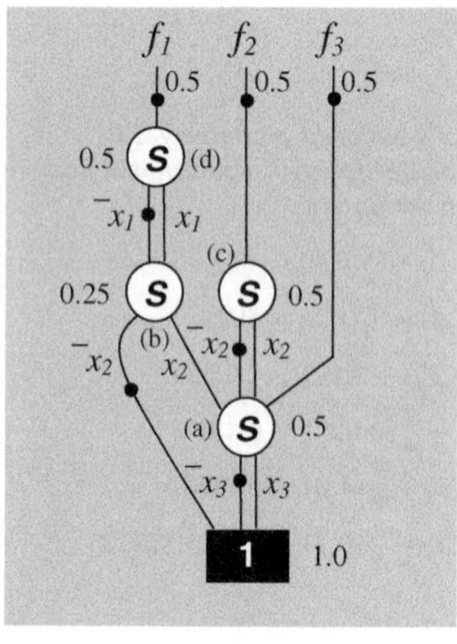

- *The terminal node is assigned probability 1.0.*
- *The output probabilities for each node are computed as follows:*

  a: $\frac{1}{2}(1-1.0)+\frac{1}{2}(1.0)=0.5$

  b: $\frac{1}{2}(1-1.0)+\frac{1}{2}(0.5)=0.25$

  c: $\frac{1}{2}(1-0.5)+\frac{1}{2}(0.5)=0.5$

  d: $\frac{1}{2}(1-0.25)+\frac{1}{2}(0.25)=0.5$
- *The function output probabilities are 1 minus the probability labeling the top nodes. It is peculiar to this case (a reversible function) where the output patterns are unique and are in fact a permutation of the input patterns, that all three outputs have output probability 0.5.*

FIGURE 20.7
Bottom-up probability assignment for a shared BDD with complemented edges (Example 20.6).

20.3 treating all variables as symbolic.

20.3 Cross-correlation of functions

In this section, we again assume the probability of an input being 0 or 1 is equal (1/2). We consider the computation of the cross-correlation of two functions f and f_c as defined in [4]. f_c is termed the *constituent* function.

By choosing appropriate constituent functions one can compute Rademacher–Walsh, Haar, Reed–Muller, and other spectral coefficients [3]. We use the definition of a generalized spectral coefficient as defined in [11] given by

$$\begin{aligned} S_f(f_c) &= 2^n - 2N_d \\ &= 2N_s - 2^n, \end{aligned} \tag{20.1}$$

where n is the number of input variables, which is the same for both functions, N_s is the number of input assignments for which the functions yield the same value, and N_d is the number of input assignments for which the functions yield different values.

Let $P(f)$ denote the output probability. Clearly,

$$N_s = 2^n \times (P(f \cdot f_c) + P(\overline{f} \cdot \overline{f}_c))$$

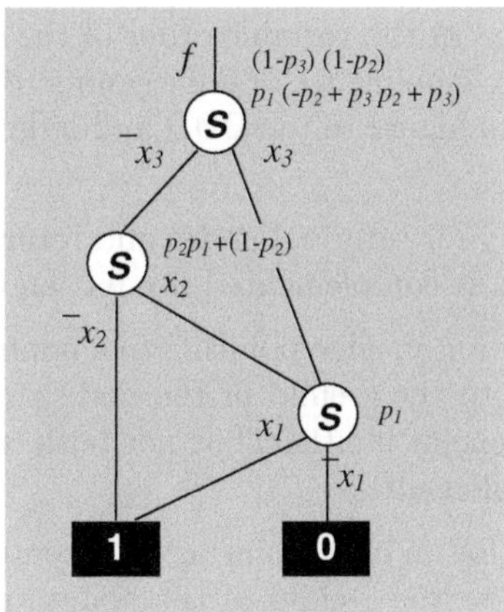

FIGURE 20.8
Symbolic computation for BDD in Example 20.3.

since $P(f \cdot f_c)$ is the probability both functions are 1 and $P(\overline{f} \cdot \overline{f}_c)$ is the probability both functions are 0. Hence,

$$S_f(f_c) = 2^n \cdot (2 \cdot [P(f \cdot f_c) + P(\overline{f} \cdot \overline{f}_c)] - 1). \tag{20.2}$$

Computing the spectral coefficient directly as formulated in Equation 20.2 requires BDD for f, f_c, $\overline{f}$, and $\overline{f}_c$. The first two are known. Their complements could be computed using Bryant's *Apply* operation [5] or is in fact a trivial operation if complemented edges are used. Two logical AND of functions are required that can also be computed using *Apply*. The appropriate probabilities can then be computed as described in the previous section, and finally the value of the spectral coefficient can be computed using Equation 20.2.

To simply the computation, we first observe that

$$N_d = 2^n \cdot [P(f) + P(f_c) - 2P(f \cdot f_c)],$$

which follows from the above and will be recognized as a standard probabilistic computation of the number of minterms in the exclusive-OR of the two functions, which is the number of assignments for which the functions differ. Substitution into Equation 20.1 yields

$$S_f(f_c) = 2^n \cdot [1 - 2P(f) - 2P(f_c) + 4P(f \cdot f_c)] \tag{20.3}$$

The computational process implied in Equation 20.3 can be further simplified since it is possible to compute $P(f \cdot f_c)$ without having to construct the BDD for $f \cdot f_c$. This is accomplished by a recursive algorithm which allows shared nodes and complemented edges. The algorithm uses the following notation:

index(e) is the position in the variable order of the variable associated with the node that edge e points to, or 0 if e points to a terminal node. Nodes of higher index are higher in the BDD and index value n corresponds to the root node.

value(e) is defined only for terminal nodes and returns the value of the node e points to, which is complemented if e has an edge complement.

zero(e) is defined only for an edge pointing to a nonterminal node. It returns an edge pointing to the 0-child of the node e points to. The returned edge has a complement if either, but not both, of e and the 0-child edge has an edge complement.

one(e) is defined only for an edge pointing to a nonterminal node. It returns an edge pointing to the 1-child of the node e points to. The returned edge has a complement if e has an edge complement.

prob(e) returns the probability assigned to the node e points to, or 1 minus that value if e has an edge complement. Note that the probability of a terminal node is 1.0 if it is logic 1 and 0.0 if the node is logic 0.

The algorithm is initially invoked with e_1 as the edge pointing to the root of the BDD for f and e_2 as the edge pointing to the root of the BDD for f_c. It returns the value of $P(f \cdot f_c)$.

20.3.1 Recursive AND probability algorithm

Input: Two edges e_1 and e_2. If $index(e_2) > index(e_1)$, e_1 and e_2 are swapped. This reduces the number of cases that must be considered and is applicable since AND is a commutative operation.

Output: Output probability of the function that is the AND of the functions represented by the BDD e_1 and e_2 point to.

The first of the following rules that is applicable is applied:

Rule 1. If e_1 and e_2 point to the same node, then the result is 0 if one but not both have edge complements and $prob(e_1)$ otherwise.

Rule 2. If $index(e_2) = 0$, then if $value(e_2) = 0$ the result is 0, otherwise the result is $prob(e_1)$.

Rule 3. If $index(e_1) = index(e_2)$, then the result is 1/2 the sum of the output probabilities found by recursively applying the algorithm to $[zero(e_1),$ $zero(e_2)]$ and $[one(e_1),\ one(e_2)]$.

Rule 4. If $index(e_1) > index(e_2)$, then the result is 1/2 the sum of the output probabilities found by recursively applying the algorithm to $[zero(e_1),\ e_2]$ and $[one(e_1),\ e_2]$.

This algorithm is fairly straightforward and is essentially an adaptation of Bryant's *Apply*. We note the following.

Case 1 is the situation where both edges point to the same node. In that case, the answer is immediate and the only issue to consider is the effect of edge complements.

Case 2 applies when e_2 points to a terminal node and e_1 points to a nonterminal node. The result is immediate depending only on the value of the node e_2 points to.

In case 3, both the edges point to two distinct nodes controlled by the same variable. In this case two recursions are required operating on the 0-children and the 1-children.

Case 4 is similar to case 3, except the recursions only descend the edges from the node pointed to by e_1 since the BDD e_2 is independent of the control variable at the top of the BDD pointed to by e_1.

Example 20.7 *Consider two switching functions*

$$f = \overline{x}_1 \vee x_2 x_3$$
$$f_c = x_1 x_2$$

Figure 20.9 shows a shared BDD representation for these two functions with complemented edges. A trace of applying the recursive AND probability algorithm is shown in Figure 20.10. The computed probability is 0.125 which is as expected since

$$f \cdot f_c = x_1 x_2 x_3.$$

Example 20.8 *(Continuation of Example 20.7). Each tree node shows an instance of the recursive procedure with the parameters and applicable algorithm case (in brackets) in the left cell. The parameters are shown as the node rather than the edge that points to it for clarity. A bar over a node label indicates the negation of the function represented by the BDD rooted by that node. The result is shown in the right cell of each node. It computed upon returning from the right branch out of the node. The tree nodes are numbered in the order they are visited in the depth-first recursion. The probability values in the right cell of each tree node are computed as shown as each phase of the recursion completes.*

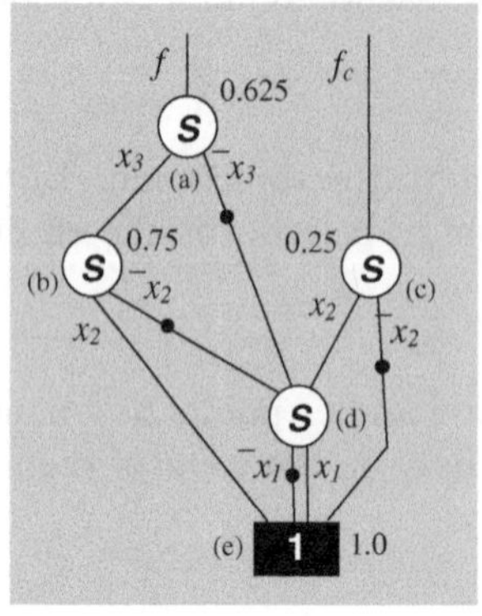

- *Probabilities shown for each note are computed using the bottom-up algorithm.*
- *The node labels (a), etc., are for reference in Figure 20.10.*

FIGURE 20.9
Shared BDD with complemented edges for f and f_c in Example 20.7.

Even this rather small example illustrates the algorithm operates correctly for shared BDD and for edge complements. Note that the recursion does not reach pairs of terminal nodes since once one parameter becomes a terminal, the result is immediate. What this example does not illustrate is that the same computation can be encountered more than once during the recursion. This is easily handle using a compute table technique that records the results of recent computations for subsequent use without recomputation.

20.3.2 Modification of an algorithm

The above algorithm is readily modified for other operations such as the OR or exclusive-OR of two switching functions. The difference is in the terminal cases, the recursive structure is the same. Indeed, it is straightforward to implement a single procedure with the terminal cases table driven selected by the operation being performed.

20.3.3 Computing spectral coefficients

As noted earlier, the methods described above can be used to compute individual spectral coefficients using equation

$$S_f(f_c) = 2^n[1 - 2P(f) - 2P(f_c) + 4P(f \cdot f_c)].$$

The desired coefficient is found by setting f_c appropriately.

> **Example 20.9** *If $f_c = x_i$, the value of $S_f(f_c)$ is the ith first-order Rademacher–Walsh coefficient. By setting f_c to an appropriate exclusive-OR function, higher order Rademacher–Walsh coefficients can be similarly computed.*

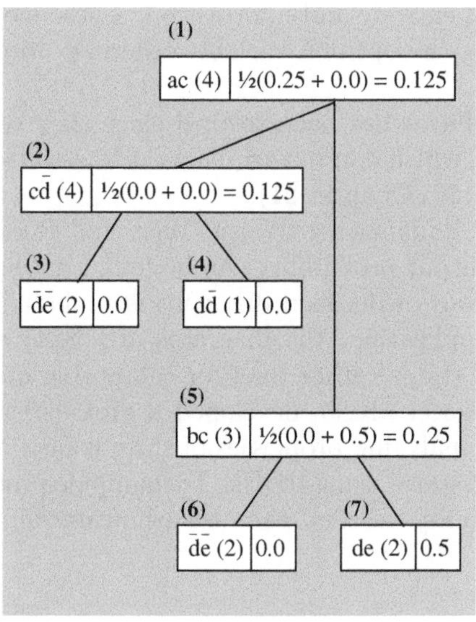

Node (1) *is the initial call to the procedure: the value of 0,125 computed after processing both descendants is the final probability for the AND of f and f_c.*

Node (2) *is reached from tree node (1) by following the 0-edge from BDD node a.*

Node (3) *is a terminal case since the second parameter identifies the constant value 0 so the probability of the AND is 0.0.*

Node (4) *is another terminal case since both edges point to the same node. One is complemented so the AND is 0 and the probability is 0.0.*

Node (5) *is reached from tree node (1) by following the 1-edge from BDD node a. BDD nodes b and c are both controlled by x_2 so algorithm case 3 applies.*

Node (6) *is reached by following the 0-edges from BDD nodes b and c together. Again the second parameter points to 0 so the resulting probability is 0.0.*

Node (7) *is reached by following the 1-edges from BDD nodes b and c together. In this instance, the second parameter points to logic 1 so the result of the AND is the function represented by the other parameter which in this case has probability 0.5 (it is the function x_1).*

FIGURE 20.10
Trace of the application of the recursive AND probability algorithm for Example 20.7.

Haar spectral coefficients can be computed in a manner analogous to computation of Rademacher–Walsh coefficients. For Reed–Muller coefficients, the procedure is the same with the addition that the resulting value is the value computed using the methods above modulo 2. Note that $P(f_c)$ is fixed for a given spectral coefficient.

Example 20.10 *It is 1/2 for all Rademacher–Walsh coefficients except the 0-order coefficient for which it is 1. $P(f \cdot f_c)$ is computed using the recursive procedure above avoiding the construction of the BDD for the AND of the two functions.*

20.4 Further reading

Output probability was initially proposed by Parker and McCluskey [6] to evaluate the effectiveness of random testing in combinational logic circuits. Output probabilities are also calculated on decision diagrams for information-theoretical measures such as functional entropy and information. These measures are used as heuristics for finding the optimal variable ordering. More details are given in Chapters 8 and 24.

Recently, interest in output probabilities has been revived since they can be used to estimate switching activity, which can in turn be used to estimate overall power dissipation [2, 7, 8] (see also Chapter 21).

More detail on the computation of Rademacher–Walsh, Haar and Reed–Muller spectral coefficients [3] using output probability can be found in [9].

One of recently developed applications of decision diagrams is probabilistic verification of systems. In this application, the diagrams are deployed mostly for manipulation of the sets of states rather than for computing output probabilities, though. Kwiatkowska et al. [5] developed a probabilistic model checker PRISM which involves analyzing properties of state transition systems and the manipulation of sets of states using BDDs. To manipule probability transition matrices and probability vectors, they used multiterminal BDD (MTBDD).

References

[1] Bryant RE. Graph-based algorithms for boolean function manipulation. *IEEE Transactions on Computers*, 35(8):677–691, 1986.

[2] Ghosh A, Devadas S, Keutzer K, and White J. Estimation of average switching activity in combinational and sequential circuits. In *Proceedings of the IEEE/ACM Design Automation Conference*, pp. 253–259, 1992.

[3] Hurst SL, Miller DM, and Muzio JC. *Spectral Techniques in Digital Logic*, Academic Press, 1985.

[4] Karpovsky MG. *Finite Orthogonal Series in the Design of Digital Devices*, Wiley, 1976.

[5] Kwiatkowska M, Norman G, and Parker D. PRISM: Probabilistic symbolic model checker. In *Proceedings of the TOOLS 2002, Lecture Notes in Computer Science*, 2324, pp. 200–204, 2002.

[6] Parker KP and McCluskey EJ. Analysis of logic circuits with faults using input probabilities. *IEEE Transactions on Computers*, 24:573–578, 1975.

[7] Pedram M. Power minimization in IC design. *ACM Transactions on Design Automation of Electronic Systems*, 1(1): 3–56, 1996.

[8] Sasao T. *Logic Synthesis and Optimization*, Kluwer, Dordrecht, 1993.

[9] Thornton MA, Miller DM, and Drechsler R. *Spectral Techniques in VLSI CAD*, Kluwer, Dordrecht, 2001.

[10] Shen A and Devadas S. Probabilistic manipulation of Boolean functions using free Boolean diagrams. *IEEE Transactions on Computer-Aided Design of Integrated Circuits and Systems*, 14(1):87–94, 1995.

[11] Thornton MA and Nair VSS. Efficient calculation of spectral coefficients and their applications. *IEEE Transactions on Computer-Aided Design of Integrated Circuits and Systems*, 14:1328-=1341, 1995.

21

Power Consumption Analysis using Decision Diagrams

One of the factors that influence power dissipation in a circuit is its *switching activity*. The latter is relevant to the circuit's logic through the *variable's activity*. This, in its turn, can be easily evaluated using decision trees or diagrams, since the variable's activity is employed as a criterion to find the optimal decision tree or diagram. This chapter focuses on these analytical techniques for switching activity estimation based on decision diagrams. In the application, decision trees are associated with the probability distribution specified on the (input) variables' space. Therefore, a decision diagram associated with the probabilities which describe variable activity can be used to evaluate switching activity for power dissipation estimation.

21.1 Introduction

Power dissipation in digital CMOS circuits is caused by various sources: the leakage current, which is primarily determined by the fabrication technology, the standby current, the short-circuit current, and the capacitance current. Evaluation of power dissipation at the logic level is based on the following assumptions.

The dominant source of power dissipation in CMOS circuits is the charging and discharging of the node capacitances, so the average power consumption in a circuit is calculated by the following equation:

$$Power = \frac{1}{2} V_{DD}^2 f_{clk} \sum_i C_i E_i, \tag{21.1}$$

where

- V_{DD}^2 is the supply voltage,
- f_{clk} is the clock frequency (or more generally, the average periodicity of data arrivals),

- C_i is the sum of all input capacitances of the transistors that are driven by signal i, and
- E_i is the *transitions probability* at signal i, also called *switching activity.*

All these parameters except for E_i are physical parameters, and only E_i can be evaluated in both register-transfer and gate-level description analytically. It is, in fact, the average number of output transitions per $1/f_{clk}$ time. Since dynamic power dissipation takes place when the output of a CMOS gate makes a $0 \to 1$ or a $1 \to 0$ transition, E_i is the probability that there is a $0 \to 1$ or a $1 \to 0$ transition on signal i from one clock cycle to the next:

$$E_i = p(0 \to 1) + p(1 \to 0) \tag{21.2}$$

The product of E_i and f_{clk}, which is the number of transitions per second (how many transitions each arrival will generate), is referred to as the *transition density.*

Switching activity at the output of a logic gate is dependent on the switching function of the gate itself. This is because the logic function of a gate determines the probability that the present value of the gate output is different from its previous value. Note that the probability of a signal is the probability that this signal is high (that is, it takes value 1). Parker's and McCluskey's [22] polynomial method generates a polynomial that represents the probability that the gate output is a 1. This method and how to calculate the circuit output probability is considered in detail in Chapter 20.

Example 21.1 *The probabilities of the signal at the outputs of the AND, OR and NOT gates are calculated using the following rules:*

$$\begin{aligned} p(x_1 \wedge x_2) &= p(x_1)p(x_2) \\ p(x_1 \vee x_2) &= p(x_1) + p(x_2) - p(x_1)p(x_2) \\ p(\overline{x}) &= 1 - p(x) \end{aligned}$$

It should be noted that these expressions are exactly the arithmetical polynomial forms considered in Chapter 13 and Chapter 17, substituting probabilities instead of variables. These output probabilities are used below to calculate transition probabilities.

Transition probabilities E_i of signals for any given logic representation can be estimated under a general delay model first introduced by Cirit [5] using global BDD. An alternative approach is to compute probability polynomial propagation. This can be accomplished using arithmetical decision diagrams, in particular, ADDs or BMDs.

21.2 Switching activity

Probability polynomials are used to determine signal transition probabilities, that is, switching activity.

Zero-delay model assumes zero-delay of the signal at a gate. It also assumes that the probabilities of the gate's inputs are independent, that is, that no spatial correlation is present. Let the signal takes value i_{t_1} at a time t_1, and at the next time t_2 the signal takes value i_{t_2}. The switching activity E_i of signal i in a zero delay model is represented by:

$$E_i = \sum p(i_{t_1} \to i_{t_2}), \quad i_{t_1} \neq i_{t_2}$$

that turns into Equation 21.2 given a switching circuit.

Example 21.2 *Assuming that the input signals are uncorrelated, switching activity at the output f of a (static) two-input NAND gate is derived from the analysis of the the truth table containing transitions (Figure 21.1).*

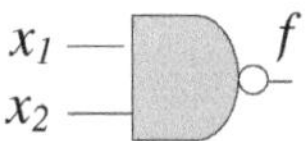

x_1	x_2	f
$0 \to 0$	$0 \to 0$	$1 \to 1$
$0 \to 0$	$0 \to 1$	$1 \to 1$
$0 \to 0$	$1 \to 0$	$1 \to 1$
$0 \to 0$	$1 \to 1$	$1 \to 1$
$0 \to 1$	$0 \to 0$	$1 \to 1$
$0 \to 1$	$0 \to 1$	$1 \to 0$
$0 \to 1$	$1 \to 0$	$1 \to 1$
$0 \to 1$	$1 \to 1$	$1 \to 0$
$1 \to 0$	$0 \to 0$	$1 \to 1$
$1 \to 0$	$0 \to 1$	$1 \to 1$
$1 \to 0$	$1 \to 0$	$0 \to 1$
$1 \to 0$	$1 \to 1$	$0 \to 1$
$1 \to 1$	$0 \to 0$	$1 \to 1$
$1 \to 1$	$0 \to 1$	$1 \to 0$
$1 \to 1$	$1 \to 0$	$0 \to 1$
$1 \to 1$	$1 \to 1$	$0 \to 0$

There are three $0 \to 1$ and three $1 \to 0$ transitions in the truth table, therefore,

$$p(0 \to 1) = 3/16$$
$$p(1 \to 0) = 3/16$$
$$E_f = p(0 \to 1) + p(1 \to 0) = 3/8$$

The same is true for AND, OR and NOR gates, while at the output of a two-input XOR gate it is $1/2$. Switching activity at the output of a k-input AND, OR, NAND or NOR gate approaches

$$E_f = \frac{1}{2}(k-1)$$

for large k, whereas that for a k-input XOR gate remains at $1/2$.

FIGURE 21.1
Switching activity at the output of the NAND gate (Example 21.3).

Note, that the transition $i_{t_1} \rightarrow i_{t_2}$ means change of a signal from i_{t_1} to i_{t_2} with zero-delay, that is,

$$p(i_{t_1} \rightarrow i_{t_2}) = p(i_{t_1})p(i_{t_2})$$

This leads to the following equations given a switching circuit:

$$p(0 \rightarrow 1) = p(i = 0)p(i = 1)$$
$$p(1 \rightarrow 0) = p(i = 1)p(i = 0)$$

Thus, E_i is expressed as below:

$$\begin{aligned} E_i &= p(i = 0)p(i = 1) + p(i = 1)p(i = 0) \\ &= 2p(i = 0)p(i = 1) = 2p(i)(1 - p(i)) \end{aligned}$$

where $p(i = 0)$ and $p(i = 1)$ are the probabilities that the signal value at the node i is zero and one, respectively, and $p(i)$ is a transition ($0 \rightarrow 1$ or $1 \rightarrow 0$) probability.

Probability polynomials. Given a switching function f, let us express $p_{f=0}$ and $p_{f=1}$ in terms of probability polynomial for the Shannon expansion:

$$p(f) = p(\overline{x}_i)p(f_0) + p(x_i)p(f_1)$$

Thus,

$$\begin{aligned} p(f = 0) &= p(\overline{x}_i)p(\overline{f}_0) + p(x_i)p(\overline{f}_1) \\ &= (1 - p(x_i))(1 - p(f_1)) + p(x_i)(1 - p(f_1)), \\ p(f = 1) &= p(\overline{x}_i)p(f_0) + p(x_i)p(f_1) \\ &= (1 - p(x_i))p(f_0) + p(x_i)p(f_1) \end{aligned}$$

and, therefore,

$$\begin{aligned} p(0 \rightarrow 1) &= p(f = 0)p(f = 1) \\ &= ((1 - p(x_i))(1 - p(f_0)) + p(x_i)(1 - p(f_1))) \\ &\quad \times ((1 - p(x_i))p(f_0) + p(x_i)p(f_1)) \\ p(1 \rightarrow 0) &= p(f = 1)p(f = 0) \\ &= ((1 - p(x_i))p(f_1) + p(x_i)p(f_1)) \\ &\quad \times ((1 - p(x_i))(1 - p(f_0)) + p(x_i)(1 - p(f_1))) \end{aligned}$$

General-delay model. To handle gate delays, a generalization of the Parker-McCluskey method, called the *general-delay model*, is required. This models leads directly to an exact power estimation algorithm, which requires us to sum up the values of appropriate probability polynomials to obtain the average switching activity at any gate in the circuit. At each output there will be a waveform of polynomial groups, termed a polynomial waveform, where each group represents the conditions at the gate output at a particular instant in time.

In the general-delay model, E_i can be interpreted as the probability that a power-consuming transition will occur during a single data period. This switching activity at the output of a gate depends not only on the switching activities at the inputs and the logic function of the gate, but also on the signal correlation of the input signals. In particular, dependencies among the gate inputs can be described by:

- Spatial correlations (internal signals may be correlated because of fanout signals; primary inputs are usually considered spatially uncorrelated),
- Temporal correlations (next value of a signal can depend on its current value), and
- Mixed spatial and temporal correlations

The general delay model takes into account spatial correlation (reconvergent fanouts). Reconvergent nodes are those that receive inputs from two signal paths that fanout from some other circuit node. For networks with reconvergent fanout, computing of switching activity is more challenging because internal signals may become strongly correlated.

In sequential circuits, in addition to the above correlations, states and state lines can be temporally correlated.

Example 21.3 *Consider a two-input NAND gate. Possible spatial and temporal correlation of its inputs are illustrated in Figure 21.2.*

21.3 Decision diagrams for power consumption estimation

Transition probabilities E_i of signals for any given logic representation can be estimated under a general delay model using global BDD.

This requires a BDD evaluation for each primary output of the symbolic network because the symbolic network may contain internal correlations that were never present in the original network. An alternative approach is to compute probability polynomial propagation. This can be accomplished using arithmetical decision diagrams, in particular, BMDs.

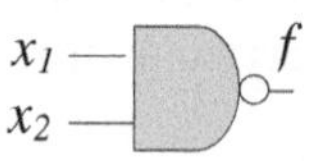

x_1	x_2	f
$0 \rightarrow 0$	$1 \rightarrow 1$	$1 \rightarrow 1$
$0 \rightarrow 1$	$1 \rightarrow 0$	$1 \rightarrow 1$
$0 \rightarrow 1$	$1 \rightarrow 1$	$1 \rightarrow 0$
$1 \rightarrow 0$	$0 \rightarrow 1$	$1 \rightarrow 1$
$1 \rightarrow 0$	$1 \rightarrow 1$	$0 \rightarrow 1$
$1 \rightarrow 1$	$0 \rightarrow 0$	$1 \rightarrow 1$
$1 \rightarrow 1$	$0 \rightarrow 1$	$1 \rightarrow 0$
$1 \rightarrow 1$	$1 \rightarrow 0$	$1 \rightarrow 1$
$1 \rightarrow 1$	$1 \rightarrow 1$	$0 \rightarrow 0$

Spatial correlations between gate inputs

Assuming that it is known that patterns 00 cannot be applied to the gate inputs x_1, x_2, and that the other patterns are equally likely, then $E_f = 4/9$.

Temporal correlations between gate inputs

Assuming that it is known that every 0 applied to input x_1 is immediately followed by a 1, then $E_f = 1/2$.

FIGURE 21.2
The spatial and temporal dependencies among the NAND gate's inputs (Example 21.3).

21.3.1 BDD for switching activity evaluation

Given a switching circuit, the signal probability at the output of a node is calculated by first building an OBDD corresponding to the global function of the node (i.e., the function of the node in terms of the circuit inputs) and then performing a postorder traversal of the OBDD.

The signal probability at a BDD node. As shown in Chapter 20, the signal probability at a BDD node with outgoing edges corresponding to f_0 and f_1 can be calculated using the equation

$$p_{node} = p(x_i = 0)p(f_0) + p(x_i = 1)p(f_1)$$

where $p(f_0) = p(f_{x_i=0})$ and $p(f_1) = p(f_{x_i=1})$

The first step in calculating information measures is to determine the probabilities of a switching function and their sub-functions.

A depth-first-traversal of the BDD, with a post-order evaluation of probability at every node can be used for calculation of the root probability (bottom-up approach, as described in Chapter 20). This can be implemented using the "scan" function of the CUDD package.

The total power dissipated in the circuit. After calculation of output probability and switching probability at each node, the total power dissipated in the circuit can be evaluated. The recursive traversal algorithm is described below. It makes only one pass through all nodes in the BDD. Since $p(f = 1)$ can be calculated from the probability of the left branch $p(f_0)$ and the right branch $p(f_1)$, each node only needs to be visited once. If a node is required in multiple paths, the value of that node and its children will remain the same so they do not need to be recalculated. A root node in the BDD is an output of the circuit described by the BDD. A BDD can have many output functions. The algorithm starts at a root node and is used for all root nodes in the circuit.

```
BDD Recursive Traversal Algorithm:

For each output of the circuit call calculateProbability()
Parameters:  Node, the number of negative arcs in the Path
Return Value:  Node probability

calculateProbability() {
Step 1. if Node = terminal one node
      Return 1
      if Node = terminal zero node
      Return 0
      if Node has been visited before
      Return previously calculated value of p(f = 1)
Step 2. Call calculateProbability() for the one branch of this
      node Call calculateProbability() for the zero branch of
      this node
Step 3. Calculate p(f = 1) for this Node using the equation
```

$$p(f = 1) = p(x = 0)p(f_0 = 0) + p(x = 1)p(f_1 = 1)$$

```
Step 4. Calculate switching probability for this Node using the
      equation
```

$$\begin{aligned} p_{sw} &= 2(p(x = 0)p(f_0 = 0) + p(x = 1)p(f_1 = 0)) \\ &\times (p(x = 0)p(f_0 = 1) + p(x = 1)p(f_1 = 1)) \end{aligned}$$

```
      (assume p(x = 0) = p(x = 1) = 0.5)
Step 5. Create a new NodeCalculation object for this Node and
      Store it in the correct index in the Node Data storage
      structure
      Return p(f = 1).
}
```

The runtime of the algorithm is based on the number of nodes in the ROBDD, which in turn is dependent on the number of input variables to the logic function. The number of output nodes does not affect the time spent traversing the ROBDD.

The capacitive load C_i can be estimated for a mapped BDD node (its fanout). The resulting circuit is implemented by mapping BDD nodes to multiplexer circuits implemented using CMOS transmission gates and static inverters. Similar BDD mapping methods based on pass transistor logic circuits can also be used.

The total power lost in the circuit is calculated from Equation 21.1 by summing the power lost through each node. To determine the capacitance at a node the circuit will be realized as a network of multiplexers. The number of transistors in the multiplexer determines capacitance and the capacitance value of a transistor is consistent with that used in the SIS power calculations.

$$C = fanout \times 0.01pF + transistors \times 0.005pF$$

21.3.2 Information measures on BDDs and switching activity

This method is based upon calculation of the switching probability for each node through calculation of the entropy at each node and using the constraint that the average switching activity is upper bounded by half of its entropy. Chapter 24 described information and entropy measures on decision diagrams. In this chapter, this approach is employed. The proposed heuristic is based on the assumption that a BDD's variable order, chosen on the criterion of minimal entropy, leads to less switching activity as well.

The entropy of a function f is calculated as follows:

$$H(f) = -\sum_{a=0}^{1} p(f = a) \cdot \log p(f = a) \tag{21.3}$$

This calculation involves the algorithm described above for calculating the output probabilities. The conditional entropy $H(f|x)$ of the function f with respect to the variable x can be simplified using the statement below:

$$H(f|x) = p(x = 0)H(f|x = 0) + p(x = 1)H(f|x = 1)$$

This means that the entropy of all sub-functions is involved in the calculation of the conditional entropy. This requires the calculation of conditional probability.

This algorithm calculates conditional and joint probabilities for computing conditional entropy according to Equation 21.3. Thus, for the joint probability $p(f = 1, x = 1)$ it is necessary to set $p(x = 1) = 1$ and $p(x = 0) = 0$ before BDD traversal. This approach allows us to develop a technique for calculating the whole range of probabilities using only one BDD traversal.

The criterion for choosing a decomposition variable x for the arbitrary level of BDD is that the conditional entropy of the function with respect to this variable has to be minimal:

$$H(f|x) = min(H(f|x_i)|i = 1, ..., n).$$

An algorithm for generating a new variable order for a BDD is a recursive heuristic presented below:

```
                Algorithm for variable reordering:

Input:  BDD for the given function f
Output:  New order of the variables
Reorder() {
Step 1. for (∀ level:  level = 1...n)
        for (∀x of the level {≥ level)
        Calculate p(f = 1|x = 0) and p(f = 1|x = 1);

Step 2. Calculate H(f|x_i);

Step 3. Reorder variables according to the equation:

                  H(f|x_{j1}) ≤ ... ≤ H(f|x_{jn});

Step 4. Rebuild BDD using x_{j1} for the current level;
}
```

In this way, the variables are arranged in order from the the most significant (top level) to the least significant (bottom level) taking into consideration the heuristic criterion of entropy to find the best ordering of variables and, at the same time, evaluate the variables' "activity".

Example 21.4 *Given a circuit which implements the switching function*

$$f = (\overline{x_3 \vee x_2}) \vee x_1$$

with truth vector $\mathbf{F} = [10001111]^T$*, we have to find conditional entropies with respect to the variables* x_1*,* x_2 *and* x_3*. The initial BDD with the variable order* $< x_3x_2x_1 >$ *and the probability and conditional entropy calculation is shown in Figure 21.3.2. According to the considered algorithm, the best orders of variables in BDD for the function* f *are* $< x_1x_2x_3 >$ *and* $< x_1x_3x_2 >$*. Indeed, the power dissipated by circuits with varying orders of variables is as follows:*

$$\{x_1x_2x_3\} \ \ or \ \ \{x_1x_3x_2\} \rightarrow Power = 31.87$$
$$\{x_2x_1x_3\} \ \ or \ \ \{x_3x_1x_2\} \rightarrow Power = 45.0$$
$$\{x_3x_2x_1\} \ \ or \ \ \{x_2x_3x_1\} \rightarrow Power = 37.50,$$

that is, the power consumption for the best case is 1.5 times smaller than for the worst case, taking into account the same size of BDDs for both cases.

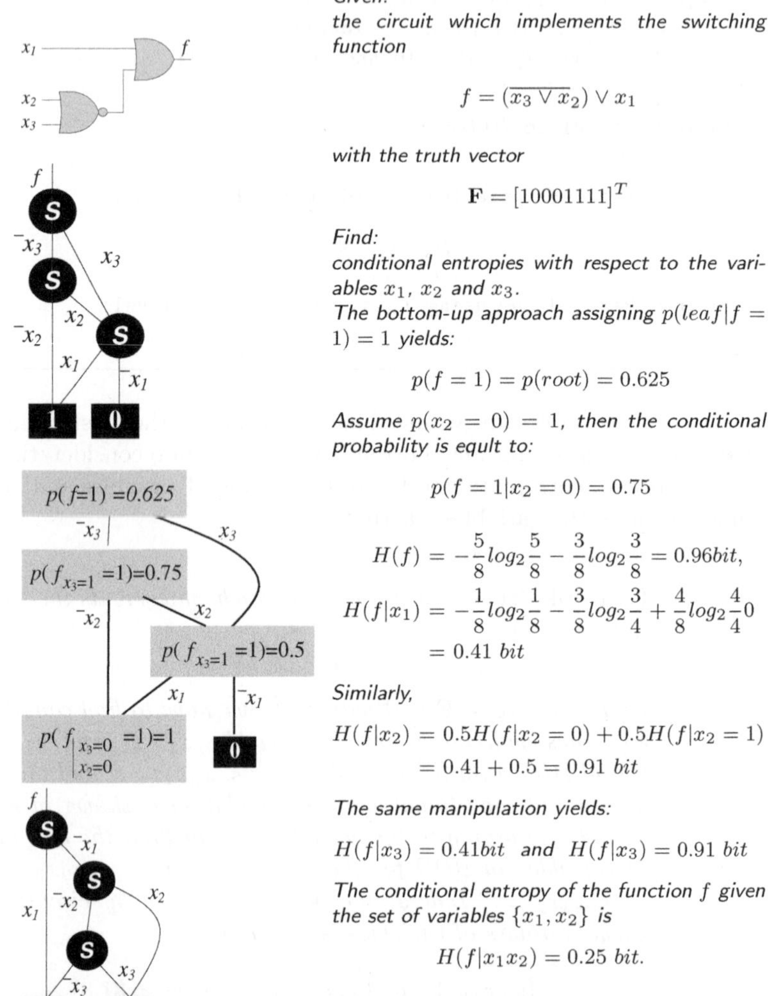

Given:
the circuit which implements the switching function
$f = (\overline{x_3 \vee x_2}) \vee x_1$
with the truth vector
$\mathbf{F} = [10001111]^T$
Find:
conditional entropies with respect to the variables x_1, x_2 and x_3.
The bottom-up approach assigning $p(leaf|f = 1) = 1$ yields:
$p(f = 1) = p(root) = 0.625$
Assume $p(x_2 = 0) = 1$, then the conditional probability is eqult to:
$p(f = 1|x_2 = 0) = 0.75$
$H(f) = -\frac{5}{8}log_2\frac{5}{8} - \frac{3}{8}log_2\frac{3}{8} = 0.96bit,$
$H(f|x_1) = -\frac{1}{8}log_2\frac{1}{8} - \frac{3}{8}log_2\frac{3}{4} + \frac{4}{8}log_2\frac{4}{4}0$
$= 0.41\ bit$
Similarly,
$H(f|x_2) = 0.5H(f|x_2 = 0) + 0.5H(f|x_2 = 1)$
$= 0.41 + 0.5 = 0.91\ bit$
The same manipulation yields:
$H(f|x_3) = 0.41bit$ and $H(f|x_3) = 0.91\ bit$
The conditional entropy of the function f given the set of variables $\{x_1, x_2\}$ is
$H(f|x_1x_2) = 0.25\ bit.$
x_1
x_2
x_3
f
S
$\bar{x}_3$
x_3
$\bar{x}_2$
x_2
x_1
$\bar{x}_1$
1
0
$p(f=1) = 0.625$
$p(f_{x_3=1} = 1) = 0.75$
$p(f_{x_3=1} = 1) = 0.5$
$p(f|_{x_3=0, x_2=0} = 1) = 1$

FIGURE 21.3
Computing entropies of a switching function on a BDD.

21.3.3 Other decision diagrams for switching activity evaluation

The Parker-McCluskey method was generalized by Najm to work with transition probabilities [19]. In this extension, each input x_i has four probability values corresponding to the input staying low, making a rising transition, making a falling transition, and staying high. Thus, for each gate, there are four polynomials

$$P_{00}, \; P_{01}, \; P_{10}, \; \text{and} \; P_{11},$$

corresponding to the probability that the gate stays low, makes a rising transition, makes a falling transition, or stays high, respectively. That is,

$$\begin{aligned} &P_{00} \text{ corresponds to } P(0 \rightarrow 0), \\ &P_{01} \text{ corresponds to } P(0 \rightarrow 1), \\ &P_{10} \text{ corresponds to } P(1 \rightarrow 0), \text{ and} \\ &P_{11} \text{ corresponds to } P(1 \rightarrow 1). \end{aligned}$$

These four polynomials are different for each gate.

Example 21.5 *Given the two inputs x_1 and x_2 and the probability variables denoted by $x_i^{00}, x_i^{01}, x_i^{10}, x_i^{11}, i = 1, 2,$ corresponding to the input staying low, making a rising transition, a falling transition and staying high, the probability polynomials for NAND gate are given below:*

$$\begin{aligned} P_{00} &= x_1^{11} x_2^{11} \\ P_{01} &= x_1^{10} x_2^{10} \vee x_1^{10} x_2^{11} \vee x_1^{11} x_2^{10} \\ P_{10} &= x_1^{01} x_2^{01} \vee x_1^{010} x_2^{11} \vee x_1^{11} x_2^{01} \\ P_{11} &= x_1^{00} x_2^{00} \vee x_1^{00} x_2^{01} \vee x_1^{00} x_2^{10} \vee x_1^{00} x_2^{11} \\ &\quad \vee x_1^{01} x_2^{00} \vee x_1^{01} x_2^{10} \vee x_1^{10} x_2^{00} \vee x_1^{10} x_2^{01} \vee x_1^{11} x_2^{00}. \end{aligned}$$

These probability polynomials are used in the *polynomial simulation* method [18] for switching activity computation under a general delay. For that purpose, the global BDD is utilized: a BDD for each output of the network is created, considering the transition probability at primary inputs as BDD variables. A probability 0.25 is normally used for all primary input events described by the probability variables $x_i^{00}, x_i^{01}, x_i^{10}, x_i^{11}$.

The polynomial simulation method which requires global BDD suffers from exponential complexity. This can be alleviated by using some approximation, for example, restricting analysis of switching activity using the *limited depth reconvergent path* [6]. The general-delay model takes signal correlation up to a certain depth in terms of logic levels. That is, reconvergent paths are considered only if their length is, at most, the chosen depth. The exact switching activity can be estimated if the chosen depth is equal to the total number of levels in the circuit, and the method turns into a zero-delay model, which takes into account all internal signal correlations.

To manipulate probability polynomials, a *switching probability binary moment diagram* (switching probability BMD) has been used by Ferreira et al. [10].

Example 21.6 *Given the probability polynomial P_{01} for a NAND gate, a switching probability BMD can be derived for its representation and manipulation. This BMD is shown in Figure 21.3.3. The reduced polynomial, given some input transition probabilities, is presented in Figure 21.3.3 as well.*

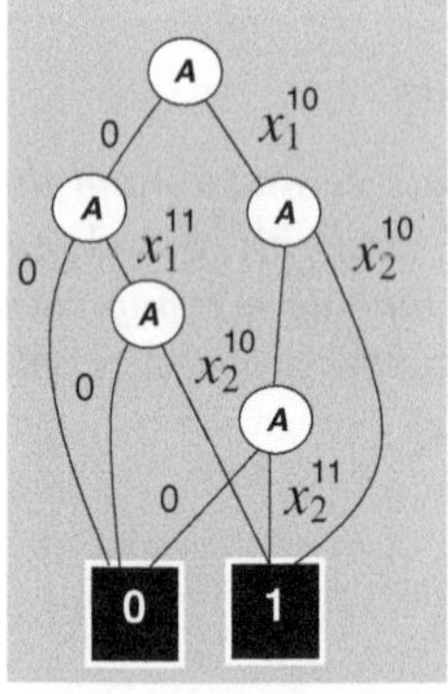

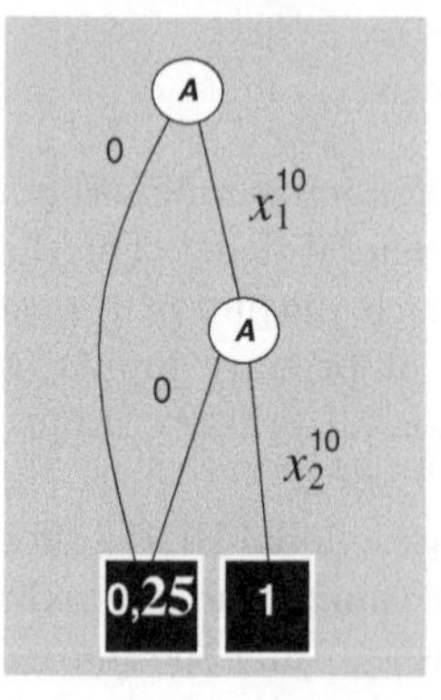

Given:
a NAND gate
and the probability polynomial

$$P_{01} = x_1^{10}x_2^{10} \vee x_1^{10}x_2^{11} \vee x_1^{11}x_2^{10}$$

The BMD consists of five binary moment nodes. Assume that some probability variables can be substituted by their probability values:

$$x_2^{11} = 0.25,\ x_2 1^{11} = 0.25$$

This is possible because the variables outside the converged path will not generate spatial correlations.
The polynomial's complexity can be reduced to

$$P_{01} = x_1^{10}x_2^{10} \vee 0.25x_1^{10} \vee 0.25x_2^{10}$$

The corresponding BMD consists of two binary moment nodes

FIGURE 21.4
Switching probability BMDs for representation of probability polynomials.

Furthermore, Ferreira et al. [11] combined BMD and zero-suppressed techniques using zero-suppressed decision diagram (ZBDD) to calculate probability polynomials. All the described techniques produce quite accurate estimates of switching activity and, therefore, power dissipation in the circuits at gate level.

21.4 Further reading

Comprehensive overview of low-power techniques can be found in surveys by Najm [20] and Pedram [23].

Information-theoretical approach. Entropy characterizes the uncertainty of a sequence of applied vectors and thus, intuitively, is related to switching activity [15, 16]. Indeed, it is shown by Ramprasad et al. [25] that, under the temporal independence assumption, the average switching activity of a bit is upper-bounded by one half of its entropy. The average entropy per circuit line is calculated and used as an estimate of the average switching activity per signal line.

In Marculescu et al. [16, 17], the temporal correlation between values of some signal x in two successive clock cycles is modeled by a time-homogeneous Markov chain. The various transition probabilities can be computed exactly using the OBDD representation of the logic function of x in terms of the circuit inputs.

Decision diagram techniques. Binary Decision Diagrams(BDD) for signal probability were first proposed by Chakravarti et al [3].

In this method, the signal probability at the output is calculated by building OBDDs corresponding to the function of the node in terms of circuit inputs, and then performing a post-order traversal of the OBDD.

The above methods failed to take into account the gate delay, ignoring the power dissipation due to hazards and glitches. Ghosh et al. [12] estimated average power dissipated in combinational and sequential circuits, using a general delay formula. They use symbolic simulation to produce a set of switching functions that represent the condition for switching at different time points for each gate. From the input switching rate, probability of each gate switching at any point in time is calculated. The sum of switching activity in the entire circuit over all the time points for all the gates corresponding to a clock cycle is calculated. The major disadvantage of this method is that for medium to large circuits the symbolic formulae become too large to build.

Marculescu et al. [16] developed a method using local OBDD constructions. This work has been extended to handle highly correlated input streams using the notions of conditional independence and isotropy of signals [17].

Buch et al. [1] computed the power dissipation of each node by the estimated switching activity and the node fanout, without considering any temporal signal correlation. This technique can also be used with BDDs using complemented edges, so that the local switching probabilities can be calculated during variable exchange operations on BDDs with complemented edges.

This technique was developed by Drechsler et al. [8, 9], Lindgren et al. [13], and Popel [24] toward BDD mapped circuits (that is, circuit is obtained by mapping BDD nodes to pass-transistor logic circuits) using temporal correlation.

The concept of a probability waveform is introduced in Burch et al. [2]. This waveform consists of an event list, that is, a sequence of transition edges or events over time from the initial steady state to the final steady state where each event is annotated with an occurrence probability. The probability waveform of a node is a compact representation of the set of all possible logical waveforms at that node. Given these waveforms, it is straightforward to calculate the switching activity.

Najm [19] proposed the *transition density* function to measure switching activity in a circuit. To measure switching activity, *transition density*, $D(y)$, can be used:

$$D(y) = \sum_{i=1}^{n} P(\frac{\partial y}{\partial x_i}) D(x_i),$$

where $P(\frac{\partial y}{\partial x_i})$ is the probability of the Boolean difference of y with respect to x_i.

Najm developed an algorithm based on the Boolean difference operation to propagate the transition densities from circuit inputs throughout the circuit.

Tsui et al. [26] proposed a tagged probabilistic simulation approach. In this method logic waveforms at a node are broken into four groups, each group being characterized by its steady state values. Each group is then combined into a probability waveform with an appropriate steady-state tag. Given the tagged probability waveforms at the input of node n, it is possible to compute tagged probability waveforms at the output. The correlation between probability waveforms at inputs is approximated by the correlation between the steady state values of these lines, which is calculated by describing the node function in terms of some set of intermediate variables in the circuit. This method does not take into account the slew in the waveforms. Higher order spatial correlations are also not modelled in this method.

The accuracy of the transition density propagation equation can be improved by using higher-order Boolean difference terms as shown by Chou et al. [4]. A major source of error is the assumption that x_i are independent.

Other decision diagrams for switching activity evaluation. Ferreira et al. [10] estimated power consumption by calculating probability polynomials on ZBDDs. Ferreira and Trullemanns [11] further combined BMD and ZBDD techniques to calculate probability polynomials. They introduced switching probability BMDs, which are BMDs with manipulations based on ZBDD algorithms.

References

[1] Buch P, Narayan A, Newton AR, and Sangiovanni-Vincentelli AL. Logic synthesis for large pass-transistor circuits. In *Proceedings of the International Conference on CAD*, 1993, pp. 663–670.

[2] Burch R, Najm FN, Yang P, and Trick T. A Monte Carlo approach for power estimation. *IEEE Transactions on VLSI Systems*, 1(1):63–71, 1993.

[3] Chakravarti S. On the complexity of using BDDs for the synthesis and analysis of Boolean circuits. In *Proceedings of the 27th Annual Conference on Communication, Control and Computing*, pp. 730-739, 1989.

[4] Chou TL and Roy K. Statistical estimation of sequential circuit activity. In *Proceedings of the IEEE International Conference on Computer Aided Design*, pp. 34–37, 1995.

[5] Cirit M. Estimating Dynamic Power Consumption of CMOS circuits, In *Proceedings of the International Conference on Computer-Aided Design*, pp. 534–537, Nov. 1987.

[6] Costa J, Monteiro J, and Devadas S. Switching activity estimation using limited depth reconvergent path analysis. In *Proceedings of the International Symposium on Low Power Electronics and Design*, pp. 184-189, Aug. 1997.

[7] Devadas S, Keutzer K, and White J. Estimation of power dissipation in CMOS combinational circuits using Boolean function manipulation. *IEEE Transactions on Computer Aided Design of Integrated Circuits Systems*, 11(3):373–383, 1992.

[8] Drechsler R, Kerttu M, Lindgren P, and Thornton M. Low power optimization techniques for BDD mapped circuits. In *Proceedings of the Conference on Asia South Pacific Design Automation*, Yokohama, Japan, pp. 615–621, 2001.

[9] Drechsler R, Kerttu M, Lindgren P, and Thornton M. Low power optimization techniques for BDD mapped circuits using temporal correlation. In *Proceedings of the International Workshop on System-on-Chip for Real Time Applications*, Banff, Canada, pp. 400–409, July 2002.

[10] Ferreira R and Trullemans A-M. BDD variants for probability polynomials. In *Proceedings of the International Conference MALOPD'99*, Moscow, pp. 12–19, Sept. 1999.

[11] Ferreira R, Trullemans A-M, Costa J, and Monteiro J. Probabilistic bottom-up RTL power estimation. In *Proceedings of the First International Symposium on Quality of Electronic Design*, pp. 439–443, 2000.

[12] Ghosh A, Devadas S, Keutzer K, and White J, Estimation of Average Switching Activity in Combinational and Sequential Circuits. In *Proceedings of the 29th Design Automation Conference*, Anaheim, CA, pp. 253-259, June 1992.

[13] Lindgren P, Kerttu M, Thornton M, and Drechsler R. Low Power Optimization Technique for BDD Mapped Circuits. In *Proceedings of the IEEE/IEICE/ACM Asia South Pacific Design Automation Conference*, pp. 615–621, Jan. 2001.

[14] Marculescu D, Marculesku R, and Pedram M. Information theoretic measures for power analysis. *IEEE Transactions on Computer Aided Design of Integrated Circuits and Systems*, 15(6):599–610, 1996.

[15] Marculescu R, Marculesku D, and Pedram M. Sequence compaction for power estimation: theory and practice. *IEEE Transactions on Computer Aided Design of Integrated Circuits and Systems*, 18(7):973–993, 1999.

[16] Marculescu R, Marculescu D, and Pedram M. Logic level power estimation considering spatiotemporal correlations. In *Proceedings of the IEEE International Conference on Computer Aided Design*, pp. 294–299, 1994.

[17] Marculescu R, Marculescu D, and Pedram M. Efficient power estimation for highly correlated input streams. In *Proceedings of the 32nd Design Automation Conference*, pp. 628-634, 1995.

[18] Monteiro J, Devada S, Ghosh A, Keutzer K, and White J. Estimation of average switching activity in combinational logic circuits using symbolic simulation. *IEEE Transactions on Computer Aided Design of Integrated Circuits and Systems*, 16(1):121–127, 1997.

[19] Najm F. Transition density: a new measure of activity in digital circuits. *IEEE Transactions on Computer Aided Design of Integrated Circuits and Systems*, 12(2):310–323, 1993.

[20] Najm F. A survey of power estimation techniques in VLSI. *IEEE Transactions on VLSI Systems,* 2(4):446-455, 1994.

[21] Nemani M. and Najm F. Towards a high-level power estimation capability. *IEEE Transactions on Computer Aided Design of Integrated Circuits and Systems*, 15(6):588-598, 1996.

[22] Parker KP and McCluskey J. Probabilistic treatment of general combinational networks. *IEEE Transactions on Computers*, 24(6):668–670, 1975.

[23] Pedram M. Power minimization in IC design: principles and applications. *ACM Transactions on Design Automation of Electronic Systems (TODAES)*, 1(1):3–56, 1996.

[24] Popel DV. Synthesis of Low-Power Digital Circuits Derived from Binary Decision Diagrams. In *Proceedings of the IEEE European Conference on Circuit Theory and Design*, vol. 3, pp. 317–320, 2001.

[25] Ramprasad S, Shanbhag NR, and Hajj IN. Information-theoretic bounds on average signal transition activity. *IEEE Transactions on Very Large Scale Integration (VLSI) Systems*, 7(3):359–368, 1999.

[26] Tsui CY, Monteiro J, Pedram M, Devadas S, Despain AM, and Lin B. Power estimation in sequential logic circuits. *IEEE Transactions on VLSI Systems*, 3(3):404–416, 1995.

22

Formal Verification of Circuits

Verification is an intrinsic part of logic synthesis. Its most advanced techniques are reasoning methods such as Boolean satisfiability (SAT) and binary decision diagrams. The scenario of verification based on binary decision diagrams is the following: let the specifications of a circuit to be implemented be given in terms of a function f, a verified circuit realizing f and a new realization which is claimed to realize f. This is aimed at verifying that the realization and specification are equivalent, i.e., the input-output behavior of these has to be proved to be equal.

22.1 Introduction

In logic synthesis of combinational circuits, the functionality of the circuits to be designed is given in terms of a network of logical gates. An important task in logic synthesis is to optimize the representation of this circuit on the gate level with respect to several factors. These optimization criteria include the number of gates, the chip area, energy consumption, delay, clock period, etc.

First, the internal circuit representation must be generated from a given netlist of gates. This process is called *symbolic simulation*. The representations of the literals, i.e., the representation of the functions being computed in the input nodes, must be constructed. Then, in topological order, the representations of the functions being computed in the individual gates are determined depending on the functions of the corresponding predecessor gates. This is made possible by applying the Boolean operation performed in the gate to the representation of the inputs of the gate.

Example 22.1 *Consider the problem of transforming the circuit to a smaller circuit which only consists of NAND gates. The initial circuit and resulting circuit are functionally equivalent, i.e., they compute the same functions. To guarantee that the functionality of a circuit has not been modified through the synthesis and optimization process, a verification task has to be solved, for example, for small dimensions by formal description and simplification (minimization) of formulas.*

In the above example, the representation of the original circuit is interpreted as a specification. The representation of the optimized circuit is considered as an implementation.

It has to be proven formally that the specification and implementation are functionally equivalent, i.e., that both compute exactly the same switching function.

22.2 Verification of combinational circuits

Practical design verification means validating that an implemented gate-level design matches its desired behavior as specified at the register-transfer level. Formal verification of digital circuits demands that a mathematical proof of correctness implicitly covers all possible input patterns (stimuli). It must, on the other hand, avoid the explicit enumeration of an exponential number of input patterns. This is accomplished in practical design by means of equivalence checking. This is not the same as formal verification, which is proving that a design has a desirable property:

- Correctness is defined as the equivalence of two designs, and
- Equivalence checking is usually localized by finding the correspondence between latches, i.e., checking if they have the same next-state function.

Equivalence checking for combinational switching networks can be reformulated as follows:

Given two networks, check if their corresponding outputs are equal for all possible input patterns.

The following statement holds:

Two logic circuits are equivalent if and only if the canonical representations of their output functions are the same.

Example 22.2 *A complete binary tree and a ROBDD are canonical forms. Thus, two logic circuits are equivalent if their ROBDDs are isomorphic.*

In practice, subfunctions of two circuits (called above specification and implementation) are transformed into BDD by simulating the circuits gate by gate, normally in a depth-first manner. The BDDs to be checked for equivalence must be transformed into a ROBDD.

> **Example 22.3** *Figure 22.1 shows the two simplest circuits: an AND gate and a circuit consisting of three NOT gates and one AND gate. To check their equivalence, a ROBDD for the first circuit (AND gate) is derived. Derivation of a ROBDD for the second circuit requires two steps: generation of the ROBDD of the AND gate with inverted inputs, and finding the complement of this ROBDD, that is, inverting the terminal node values. By inspection, it can be seen that the ROBDD of the AND gate and the last derived ROBDD are isomorphic, and, therefore, both circuits are functionally equivalent.*

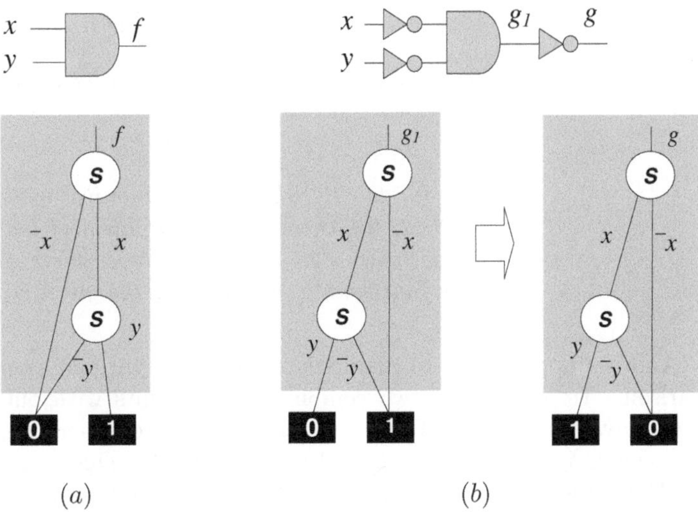

FIGURE 22.1
Two simple circuits and the derivation of their ROBDDs (Example 22.3).

The ROBDD of the circuit can be found by applying a recursive algorithm that synthesizes a ROBDD for each gate by the symbolic manipulation of the ROBDDs of its inputs based on the functionality of this gate. This is done for both circuits to be verified for equivalence. Next, the check for isomorphism of two ROBDDs is performed.

Example 22.4 *Consider the two circuits shown in Figure 22.2a, d.*
For the first circuit *(Figure 22.2a), a ROBDD of the inverter is used to construct the ROBDD of the AND gate (function $x_1\overline{x}_3$); in the same manner the ROBDD of the other AND gate is created (function x_2x_3), Figure 22.2b. Next, their two AND functions are considered to be the inputs of the OR gate, which forms the function*

$$f_1 = x_1\overline{x}_3 \vee x_2x_3.$$

This manipulation results in the final ROBDD for f_1, depicted in Figure 22.3.
For the second circuit *(Figure 22.4a), the ROBDDs for both AND gates in the first level of the circuit are derived first (Figure 22.4b). Next, recursive derivation of ROBDDs for an OR gate (function $x_1 \vee x_2x_3$) is shown in Figure 22.4c, and then the ROBDD of an EXOR gate is constructed by the manipulation of the rule for that function,*

$$(x_1 \vee x_2x_3) \oplus x_1x_3,$$

Figure 22.5. Note that the variable ordering is fixed to

$$x_1 < x_2 < x_3$$

in order for the ROBDD to be canonical with respect to this order. Next, the check for isomorphism of the ROBDDs presented in Figure 22.3 and 22.5 proves their equivalence and, thus, the functional equivalence of the given circuits.

Another approach is to manipulate ROBDDs without isomorphism comparison. To verify that two combinational circuits with outputs F and G are equivalent, the OBDD for $f \sim g$ is constructed, where f and g represent the switching functions for F and G, respectively. Due to the canonicity of OBDD's, the two circuits implement the same switching function if and only if the resulting OBDD is identical to the terminal 1.

Example 22.5 *Let us derive the ROBDD for the function $f \sim g$ where f and g are the functions considered in Example 22.3. Figure 22.6 illustrates calculation on ROBDDs of f and $\overline{g}$ using the equation $f \sim g = \overline{f \oplus g} = f \oplus \overline{g}$. Since the resulting diagram is a constant terminal "1", both circuits are considered to be equivalent.*

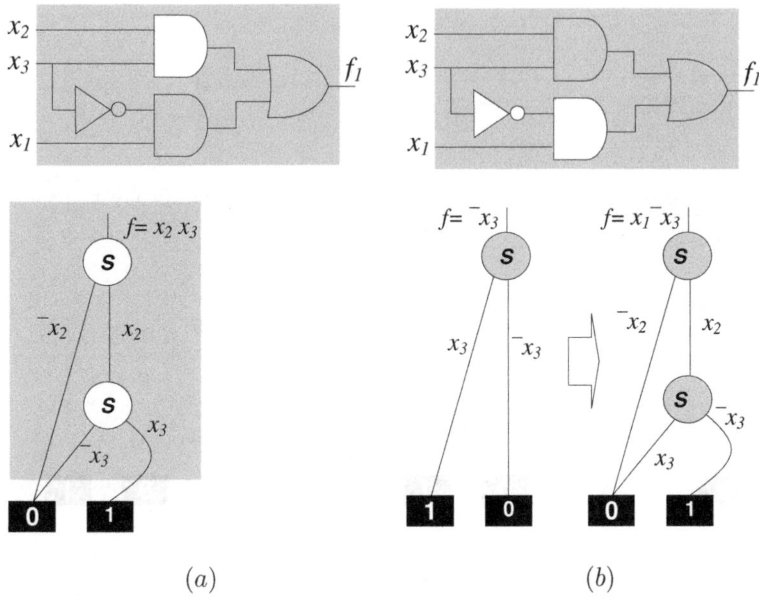

FIGURE 22.2
The circuit and ROBDDs of the gate AND x_2x_3, gate NOT $\overline{x}_3$, and the other gate AND $x_1\overline{x}_3$ (Example 22.4).

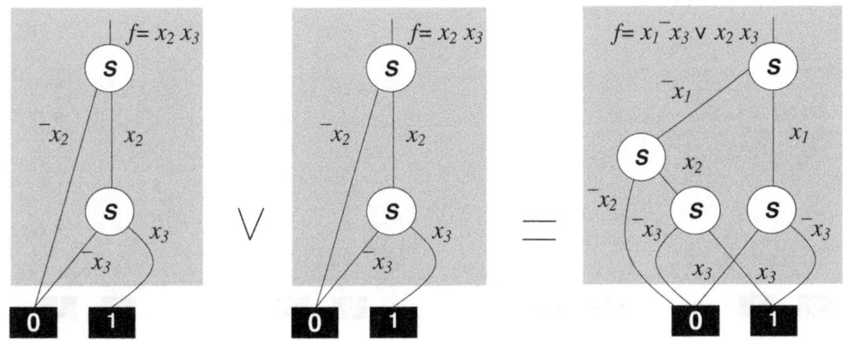

FIGURE 22.3
Construction of the ROBDD for the switching function $f_1 = x_1\overline{x}_3 \vee x_2x_3$ (Example 22.4).

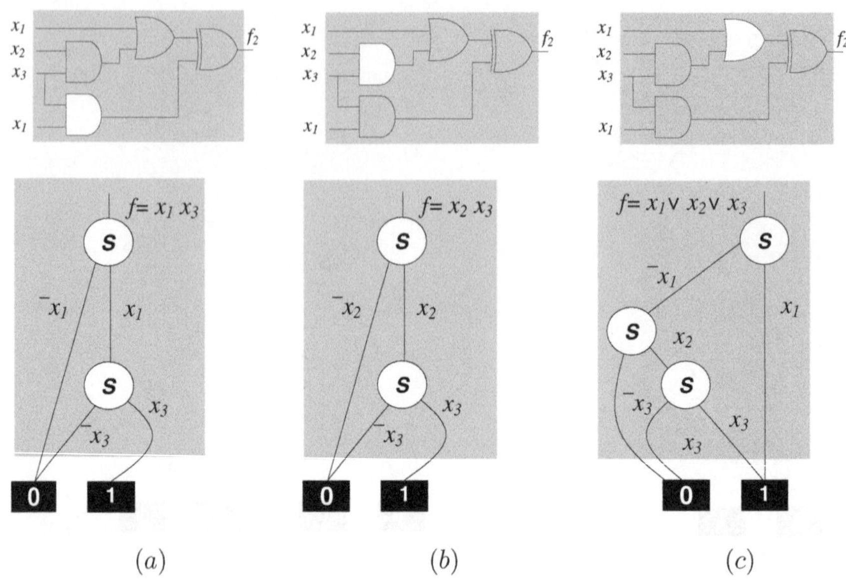

FIGURE 22.4
The circuit and the ROBDD of its two AND gates (a) and (b), and OR gate (function $x_1 \vee x_2x_3$) (c) (Example 22.4).

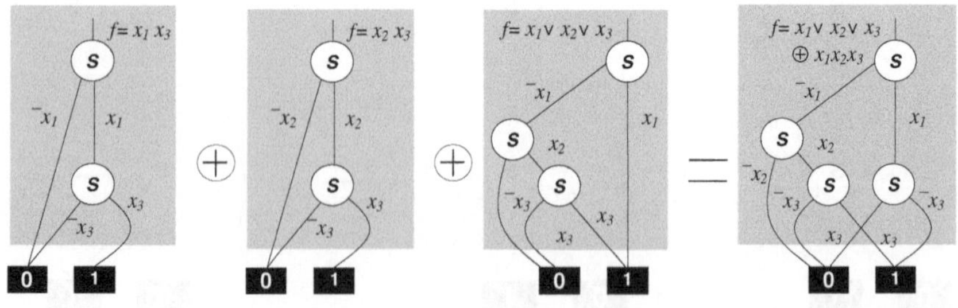

FIGURE 22.5
Construction of the ROBDD for the function $(x_1 \vee x_2x_3) \oplus x_1x_3$ (Example 22.4).

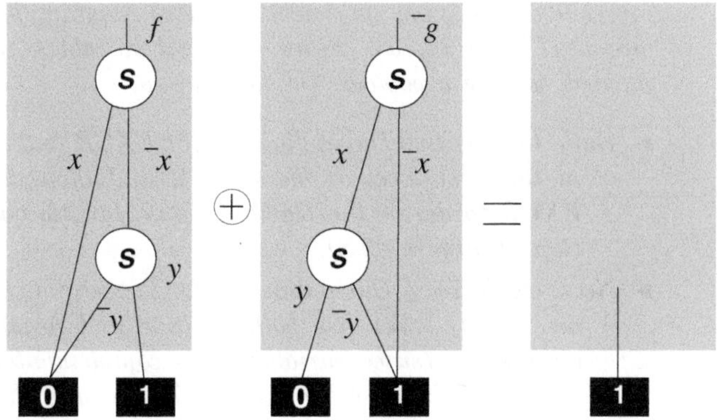

FIGURE 22.6
Calculation on ROBDDs of f and $\overline{g}$ (Example 22.5).

22.3 Verification using other types of diagrams

Other types of diagrams that are also canonical can be employed for equivalence checking.

Ordered functional decision diagrams (OFDDs) derived from Davio trees, are also canonical, and can be used for verification of the circuits assuming their OFDD are isomorphic under the same variable order. This is also true for ordered Kronecker functional decision diagrams (OKFDDs), which are derived from Kronecker decision trees.

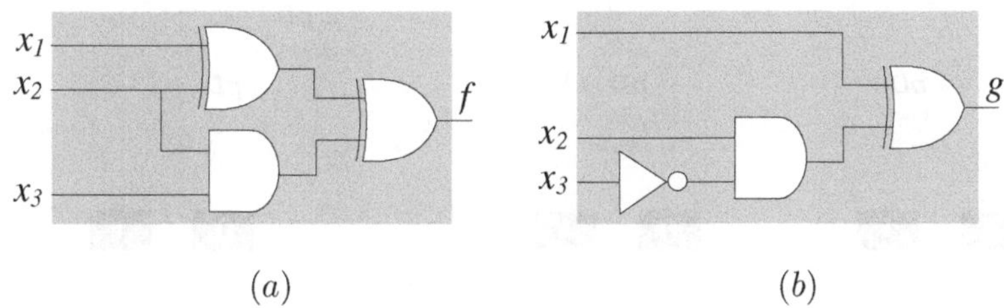

FIGURE 22.7
Two three-input single-output circuits (Example 22.6).

Example 22.6 *Figure 22.7 shows two circuits to be checked for equivalence. Let us derive a Davio diagram for the first circuit (Figure 22.7a). Since no input variables are complemented, we use a positive Davio diagram:*

- *Build the positive Davio diagram for EXOR and AND gates in the first level of the circuit, and then derive their EXOR to derive the Davio diagram for the output function f (Figure 22.8).*
- *Next, we derive a Davio diagram for the second circuit (Figure 22.7b). We use both positive and negative Davio expansions (since variable x_3 is complemented) to construct the diagram for the AND gate, and then we find the EXOR of this diagram and the diagram for x_1; the resulting Davio diagram is presented in Figure 22.9.*

Since the node of the negative Davio expansion is equivalent to the node of the positive Davio expansion with complemented terminal nodes (Figure 22.10), we can conclude that the resulting Davio diagrams in Figures 22.8 and 22.9 are isomorphic, and, thus, the circuits given in Figure 22.7 are equivalent.

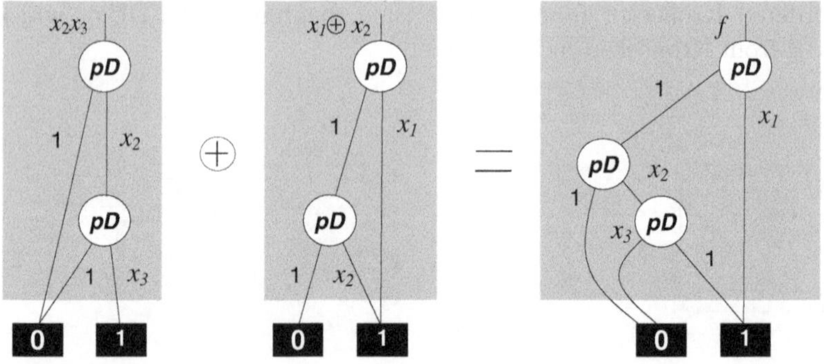

FIGURE 22.8
Derivation of the positive Davio diagram for the circuit given in Figure 22.7a (Example 22.6).

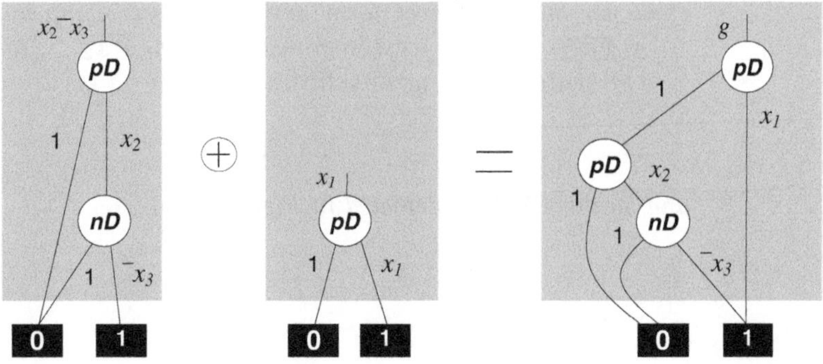

FIGURE 22.9
Derivation of the positive Davio diagram for the circuit given in Figure 22.7b (Example 22.6).

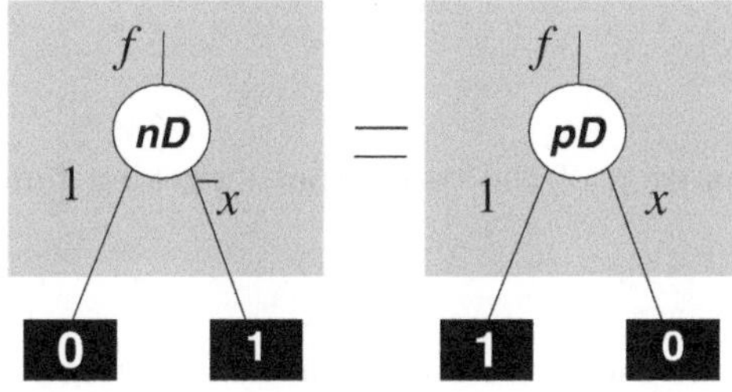

FIGURE 22.10
Relationship between positive and negative Davio expansion in terms of Davio diagrams (Example 22.6).

22.3.1 Word-level BDDs

Some compact representations, that can resolve the size explosion problem that ROBDDs face for some practical circuits such as arithmetic functions (adders and multipliers) have been employed for verification. These are, for example, Edge-Valued Decision Diagrams (EVBDDs) which represent integer-valued functions.

EVBDDs are usually much more compact than OBDDs when used to describe an arithmetic function (for example, 129 nodes are required to represent a 64-bit adder). In EVBDD, there is only one terminal node, constant 0. A non-terminal node is assigned with a quadruple $< x, q, r, v >$, where x is a variable and a node label, q and r are two subgraphs rooted at x, and v is a label assigned to the left edge of x.

Example 22.7 *Consider a two-output switching function given by the word-level form with the most significant bit,* $f_1 = \overline{x_1 x_2}$*, and the least significant bit,* $\overline{x_1 \oplus x_2}$*. This can be represented by the arithmetical expression*

$$3 - x_1 - x_2$$

and by the EVBDD depicted in Figure 22.11.

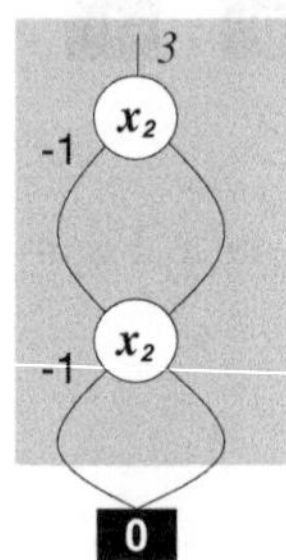

The two-output switching function

$$f_1 = \overline{x_1 \oplus x_2}$$
$$f_2 = \overline{x_1 x_2}$$

is specified by the arithmetic expression

$$f = 3 - x_1 - x_2$$

FIGURE 22.11
EVBDD implementation of the two-output switching function given in Example 22.7.

Verification is performed on EVBDDs in the following way: the equivalence of a multioutput function representing the circuits, and its arithmetical expression, can be checked for equivalence through constructing the EVBDD. The example below illustrates this approach.

Example 22.8 *Consider a two-output switching function* $x_1 \vee x_2$*,* $x_1 \oplus x_2$*. EVBDD for this function can be created in several iterations (Figure 22.12). Its arithmetical expression implies an EVBDD in another way (Figure 22.13). The resulting EVBDDs are isomorphic; therefore, the equivalence of the function in the form of a circuit graph or netlist, and its specification in the form of arithmetical expression, is proven.*

Another canonical representation of switching and integer-valued functions given by their word-level specifications are multiterminal BDDs (MTBDDs), binary moment diagrams (BMDs), and K*BMD. The nodes of word-level diagrams implement an arithmetical analog of positive Davio expansion, called a *linear moment*:

$$f = f|_{x=0} + x(-f|_{x=0} + f|_{x=1})$$

Verification using these diagrams is based on the following methodology: one must prove that there is a correspondence between the logic circuit rep-

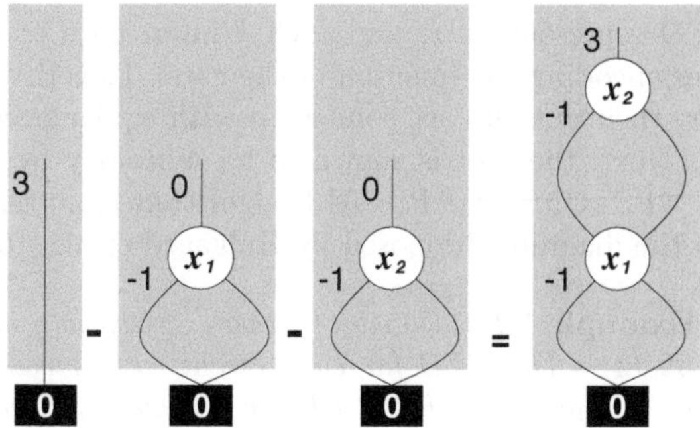

FIGURE 22.12
Construction of the EVBDD representing the arithmetical expression $f = 3 - x_1 - x_2$ (Example 22.8).

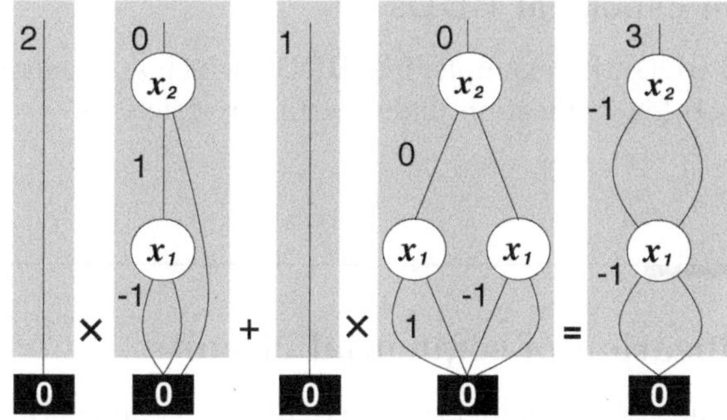

FIGURE 22.13
Construction of the EVBDD of the two-output function f from the EVBDDs of the two switching functions $f_1 = \overline{x_1 \oplus x_2}$ and $f_2 = \overline{x_1 x_2}$ by the manipulation $f = 2f_2 + f_1$ (Example 22.8).

resented by a truth table, and the specification represented by a word-level expression of the function. This is accomplished by checking the match between the circuit output interpreted as a word, and the specification when applied to word-level interpretations of the circuit inputs.

22.3.2 Boolean expression diagrams

A Boolean expression diagram (BED) is an extension of a OBDD that allows any combinational circuit to be represented in linear space. A BED can be constructed directly from the circuit graph; thus, it reflects the structure of the circuit. The BED for each circuit is derived by representing each input

x_i with the BED representing x_i and each k-input gate by a tree of $k-1$ operator vertices encoding the function of the gate. To verify the equivalence of these circuits, their outputs are connected with equivalence $\equiv$ (also called bi-implication). Next, the root is shown to be tautology by constructing an OBDD using a UP__ONE or UP__ALL algorithm (that is, moving one or all variables up the diagram) proposed by Hulgaard et al. [16]

> **Example 22.9** *Consider two switching circuits (Figure 22.14). The BED for each circuit is derived by representing each input x_i with the BED representing x_i and each 2-input gate by one operator vertex. The outputs are connected with equivalence $\equiv$. Next, the root is shown to be equal to a "constant 1" node by constructing an OBDD using a UP_ONE or UP_ALL algorithm.*

22.3.3 Non-canonical BDDs

Non-canonical BDDs, for example, free BDDs, which are generally not canonical, can be used for verification under certain restrictions.

22.4 Verification of sequential circuits

While the outputs of a combinational circuit are completely determined by the circuit input, the outputs of a sequential circuit additionally depend on values computed in the past. The outputs of the sequential circuit can depend both on the current inputs and on the values in the memory elements. Therefore, sequential circuits can be seen as an interconnection of combinational logic gates and registers. The behavior of such circuits is described by finite state machines. Hence, the verification of sequential circuits requires checking functional equivalence for all possible input sequences, that is, every input sequence must produce the same output sequence in both circuits: one that specifies a functional behavior, and the other one an optimized implementation.

The solution lies in the derivation of finite state machines from both the circuits, and then an *equivalence test* for this pair must be performed.

The equivalence test of finite state machines has been investigated in computer science for many years. Traditional explicit representation of state sets, e.g., in the form of lists, is not appropriate for large systems since the total number of all states is huge. Therefore, simple simulation on a subset of this large number of states cannot provide a complete proof of the correctness of the sequential circuits.

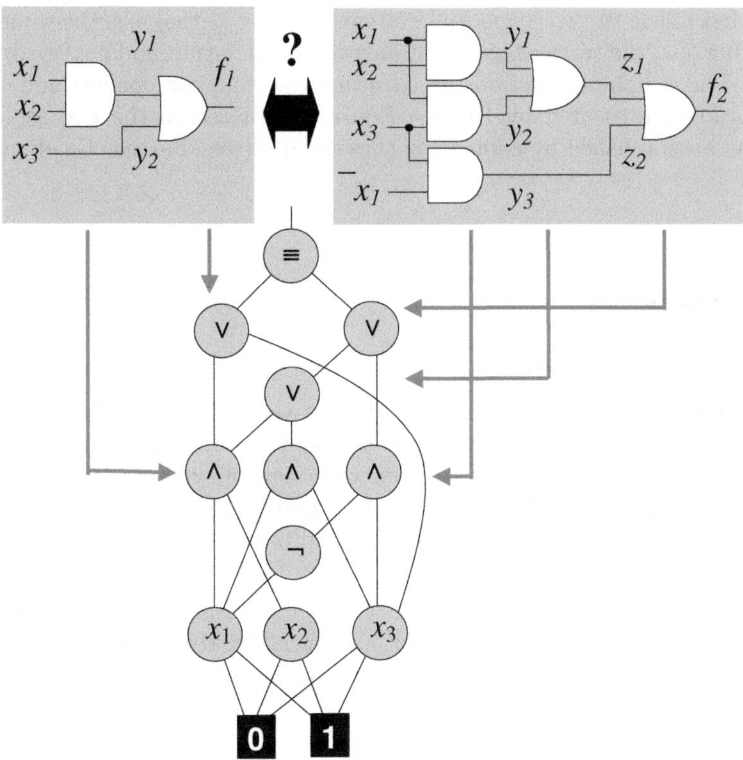

FIGURE 22.14
Verification of two switching circuits using BED.

The contemporary approaches are based on an *implicit* set representation. The problem is formulated as follows. Let a circuit contain a combinational part that implements a switching function in n variables, and a register. Let a subset $S \in \{0,1\}^n$ contain all n-tuples of register assignments that can be reached by arbitrary input sequences of length k. S is a set, and χ_S is the *characteristic function* of S,

$$\chi_S(x_1, \ldots, x_n) = 1 \Leftrightarrow (x_1, \ldots, x_n) \in S.$$

Hence, set representation and set operation can be completely reduced to the manipulation of switching functions. If a compact representation of switching functions is provided, the implicit set descriptions remain small as well. It can be proven that the equivalence test of two finite state machines M_1 and M_2 can be reduced to the manipulation of switching functions by means of implicit set representation.

Thus, ROBDD based equivalence checking methods can be applied to verify isomorphism of two sequential systems, as long as they use the same state encoding, i.e., the two systems must have identical output and next-state functions. The switching functions representing the state components and output signals are represented by BDDs. Equivalence between the two sequential circuits is established by comparing these BDDs (see "Further Reading" Section).

22.5 Further reading

ROBDD based verification methodology was first described by Bryant [2]. The advantage of ROBDD technique is that it is functional (and canonical) data structure, rather than structural. Overview of symbolic model checking techniques can be found in the book by McMillan [22]. Ordered functional decision diagrams (OFDDs) were introduced in [11].

Boolean expression diagrams. Structural equivalence employs so-called Boolean expression diagrams (BED) introduced by Hulgaard et al. [16], and AND/INVERTER graphs first considered by Jeong et al.[18].

Special types of BDDs were introduced to perform verification effectively for functions given by word-level specification. Among those functions are arithmetical and linear ones that cannot be represented using ROBDDs because of the size explosion problem. These are edge-valued binary decision diagrams (EVBDDs) introduced by Lai et al. [20], multiterminal BDDs (MTBDDs) proposed by Clarke et al. [7]. Binary moment diagrams (BMDs) introduced by Bryant and Chen [4], and K*BMD [12].

Some alternative approaches to formal verification of digital circuits include, in particular, probabilistic verification. The probabilistic verification approach was studied by Jain et al. [17]. In this method, every minterm of a function f is converted into an integer value under some random integer assignment p to the input variables.

Residue decision diagrams. Another approach is residue verification [26]. The method is based on residue arithmetic and Chinese reminder theorem. Verification is performed by interpreting the outputs of the circuits as integers and verifying the residues of the outputs with respect to a set of moduli (chosen by using Chinese reminder theorem). The technique builds the residue algebraic decision diagram (ADD) of the variables describing outputs in the multiplier, and composes the circuit from the outputs to the inputs into the residue. The residue ADD is checked against the specification.

Sequential circuits equivalence checking. The idea of applying ROBDD for sequential circuit equivalence checking was proposed by Bryant [3]: given two sequential circuits using the same state encoding, their equivalence can be established by showing combinational equivalence for the implementation of their next state and output functions. Suppose the next-state and output functions of two sequential circuits to be verified are specified in a procedural hardware description language (HDL). The switching functions for the state components and output signals are extracted from the HDL description, via symbolic execution. These functions are represented by BDDs. Equivalence between the two sequential circuits is established by comparing these BDDs. Touati et al. [27], and Burch et al. [5], used symbolic techniques based on BDDs to store and manipulate the characteristic functions of the sets of states. Alternatively, ROBDD can be derived from the vector of the next-state functions without use of a relational representation, as proposed by Coudert et al. [9]. Similarly, Pixley et al. [24] deployed BDD-based techniques to find synchronizing sequences for sequential circuits.

Hu et al. [15] applied implicitly conjoined BDDs to the characteristic functions of the sets of states, in order to decompose some of the BDDs to prevent size explosion and apply a simplification procedure to pairs of BDDs. It is based on the observation that one BDD defines a "don't care" set for a second BDD with which it is conjoined (see details in Chapter 19). McMillan [23] proposed a conjunctively decomposed representation that also provides canonicity. Ravi and Somenzi [25] proposed to reduce the size of intermediate BDDs by replacing them with smaller BDDs of dense subsets, which leads to state space traversal in a combination of breadth-first and depth-first exploration. Hu and Dill's technique is based on the elimination of state variables that can be expressed as functions of other state variables [14].

Software implementation. Example of software implementation of the above mentioned techniques are EVER package by Hu et al. [13], package SMV by McMillan [21], and Murφ system by Dill [10]. To support word-level model checking, hybrid decision diagrams (HDDs) were used by Clarke et al. [8] as the underlying data structure, which permits model checking of properties involving words. BDDs are also employed in VIS package developed by Brayton et al. [1]. VIS integrates model checking with other verification techniques such as combinational and sequential equivalence checking.

Yet another approach is to implement equivalence checking between a register-transfer level specification and a gate- or transistor-level implementation, as implemented by Kuehlmann et al. in VERITY [19]. The specified and implemented designs are described as a hierarchy of cells. Leaf nodes in the hierarchy are verified via combinational equivalence checking. For that, a functional extractor computes switching functions (represented as BDDs) for the next-state and output functions of the implementation circuit, which are then compared with the corresponding functions obtained from the register-transfer level specification.

References

[1] Brayton RK, Hatchel GD, Sangiovanni-Vincentelli A, Somenzi F, Aziz A, Cheng S-T, Edwards SA, Khatri SP, Kukimoto Y, Pardo A, Qadeer S, Ranjan RK, Sarwary S, Shiple TR, Swamy G, and Villa T. VIS. In *Proceedings of the 1st International Conference on Formal Methods in Computer-Aided Design*, Srivas M and Camilleri A, Eds., Lecture Notes in Computer Science, Vol. 1166, pp. 248-256, Springer, Heidelberg, 1996.

[2] Bryant RE. Graph-based algorithms for Boolean function manipulation. *IEEE Transactions on Computers*, 35(6):677–691, 1986.

[3] Bryant RE. Symbolic Boolean manipulation with ordered binary decision diagrams. *ACM Computing Surveys*, 24:293–318, 1992.

[4] Bryant RE and Cen Y-A. Verification of arithmetic circuits with binary moment diagrams. In *Proceedings of the 32rd ACM/IEEE Design Automation Conference*, pp. 535–541, June 1995.

[5] Burch J, Clarke E, Long D, McMillan K, and Dill D. Symbolic model checking for sequential circuit verification. *IEEE Transactions on Computer-Aided Design of Integrated Circuits and Systems*, 13(4):401-424, 1994.

[6] Cabodi G, Camurati P, and Quer S. Improving the efficiency of BDD-based operators by means of partitioning. *IEEE Transactions on Computer-Aided Design of Integrated Circuits and Systems*, 18(5):545–556, 1999.

[7] Clarke E, McMillan KL, Zhao X, Fujita M, and Yang JC-H. Spectral transforms for large Boolean functions with application to technology mapping. In *Proceedings of the 30th ACM/IEEE Design Automation Conference*, Dallas, TX, pp. 54–60, 1993.

[8] Clarke EM, Khaira M, and Zhao X. Word level model checking: avoiding the Pentium FDIV error. In *Proceedings of the 33rd Conference on Design Automation*, Las Vegas, NV, pp. 645-648, June 1996.

[9] Coudert O, Berthet Ch, and Madre JCh. Verification of sequential machines usign Boolean functional vectors. In *Proceedings of the IMEC-IFIP International Workshop on Applied Formal Methods for Correct VLSI Design*, pp. 111–128, Nov. 1989.

[10] Dill DL. The Murφ verification system. In *Proceedings of the 8th International Conference on Computer-Aided Verification*, Lecture Notes in Computer Science, vol. 1102, pp. 390-393, Springer, 1996.

[11] Drechsler R, Sarabi A, Theobald M, Becker B, and Perkowski MA. Efficient representation and manipulatiomn of switching functions based on or-

dered Kronecker functional decision diagrams. In *Proceedings of the 34th ACM/IEEE Design Automation Conference*, pp. 415–419, 1994.

[12] Drechsler R, Becker B, and Ruppertz S. K*BMD – efficient data structure for verification. In *Proceedings of the European Design and Test Conference*, pp. 2–8, 1996.

[13] Hu AJ, Dill DL, Drexler AJ, and Yang CH. Higher-level specification and verification with BDDs. In *Proceedings of the 4th International Conference on Computer-Aided Verification, Lecture Notes in Computer Science*, vol. 663, Springer, pp. 82-95, 1992.

[14] Hu AJ and Dill DL. Reducing BDD size by exploiting functional dependencies. In *Proceedings of the 30th International Conference on Design Automation*, Dunlop AE, Ed., pp. 266-271, Dallas, TX, June 1993.

[15] Hu AJ, York G, and Dill DL. New techniques for efficient verification with implicitly conjoined BDDs. In *Proceedings of the 31st Annual Conference on Design Automation*, Lorenzetti M, Ed., pp. 276-282, San Diego, CA, June 1994.

[16] Hulgaard H, Williams PW, and Andersen HR. Equivalence checking of combinational circuits using Boolean expression diagrams. *IEEE Transactions on Computer-Aided Design of Integrated Circuits and Systems*, 18(7):903–917, 1999.

[17] Jain J, Narayan A, Fyjita M, and Sangiovanni-Vincentelli A. Formal verification of combinational circuits. *In Proceedings of the 10th International Conference on VLSI Design*, pp. 218–225, 1997.

[18] Jeong SW, Plessier B, Hatchel G, and Somenzi F. Extended BDD's: trading off canonicity for structure in verification algorithms. *Digest of Technical Papers of the IEEE International Conference of Computer-Aided Design*, pp. 464–467, IEEE, Nov. 1991.

[19] Kuehlmann A, Srinivasan A, and Lapotin DP. Verity - a formal verification program for custom CMOS circuits. *IBM Journal of Research and Development*, 39(1/2):149-165, January/March 1995.

[20] Lai Y-T and Sastry S. Edge-valued binary decision diagrams for multi-level hierarchical verification. In *Proceedings of the 29th ACM/IEEE Design Automation Conference*, Anaheim, CA, pp. 608–613, 1992.

[21] McMillan KL. Symbolic model checking: an approach to the state explosion problem. *Ph.D. Dissertation. School of Computer Science, Carnegie Mellon University*, Pittsburgh, PA, 1992.

[22] McMillan KL. *Symbolic Model Checking.* Kluwer, Dordrecht, 1993.

[23] McMillan KL. A conjunctively decomposed boolean representation for symbolic model checking. In *Proceedings of the 8th International Conference on Computer-Aided Verification*, pp. 13-25, Lecture Notes in Computer Science, vol. 1102. Springer, 1996.

[24] Pixley C, Jeong S-W, and Hatchel GD. Exact calculation of synchronizing sequences based on binary decision diagrams. *IEEE Transactions on Computer-Aided Desing of Integrated Circuits*, 13(8):1024-1034, 1994.

[25] Ravi K and Somenzi F. High-density reachability analysis. In *Proceedings of the IEEE/ACM International Conference on Computer-Aided Design*, Rudell R, Ed., pp. 154-158, San Jose, CA, Nov. 1995.

[26] Ravi K, Pardo A, Hatckel G, and Somenzi F. Modular verification of multipliers. *Formal Methods in Compuer-Aided Design*, Lecture Notes on Computer Science, Vol. 1196, pp. 49–63, Springer, Heidelberg, 1996.

[27] Touati HJ, Savoj H, Lin B, Brayton RK, and Sangiobvanni-Vincentelli A. Implicit state enumeration for finite state machines using BDD's. *Digest of Technical Papers of the IEEE International Conference of Computer-Aided Design*, pp. 130–133, IEEE, Nov. 1990.

23

Ternary Decision Diagrams

In this chapter, ternary decision trees and diagrams for the representation of functions of binary-valued variables are presented. They should not be mixed with ternary decision trees and diagrams to represent functions of three-valued variables.

The usage of a ternary diagram to represent functions of binary-valued variables is not intended to provide the most compact representations, since some redundancy may be expected. Therefore, there are other reasons to discuss ternary decision diagrams for representing switching functions. Due to redundancy, the information content of a ternary decision diagram is considerably increased compared to that of a binary decision diagram for a given switching function f. Thanks to that, various ternary decision diagrams provide solutions for several problems, for instance, representation of all the prime implicants, logic simulation in the presence of unknown inputs, optimization of AND-EXOR expressions, etc.

23.1 Terminology and abbreviations

AND, OR, EXOR ternary decision trees and diagrams are defined as generalizations of BDDs, by allowing the third outgoing edge for each nonterminal node. The first two edges point to the cofactors f_0 and f_1 the same as in the BDDs. However, the third edge points to the

$$f_0 * f_1,$$

where $*$ is an operation over cofactors f_0 and f_1. In AND, OR, and EXOR ternary decision diagrams the operation $*$ is defined as the logic AND, OR, and EXOR, respectively.

Kleene ternary decision trees and diagrams are defined as ternary decision diagrams, where the operation for the third edge is the Kleene alignment operation defined below.

Arithmetic ternary decision trees and diagrams (ATDDs) are defined as a modification of EXOR ternary decision diagrams where the logic EXOR is replaced by the addition to the set of integers.

Arithmetic (spectral) transform ternary decision trees and diagrams (AC-TDDs) are defined as another modification of EXOR ternary decision diagram, however, in this case, the EXOR is replaced by the subtraction in the set of integers.

Bit-level ternary decision trees and diagrams are defined as ternary decision diagrams where the values of constant nodes show logic values 0 and 1 or undetermined values as in the case of Kleene ternary decision diagrams.

Word-level ternary decision trees and diagrams are defined as ternary decision diagrams where the values of constant nodes are integers.

Table 23.1 provides specification of various ternary decision trees and diagrams.

TABLE 23.1
Specification of a sample of ternary decision trees and diagrams.

Type	Specification
EXOR ternary decision tree (diagram)	A super Shannon expansion rule is applied.
Bit-level ternary decision tree (diagram)	The concept of the extended truth-vector resulting in the AND, OR, $EXOR$, and Kleene ternary decision trees for the switching function.
Word-level ternary decision trees (diagrams)	Include arithmetic and arithmetic spectral transform ternary decision trees (diagrams).
Arithmetic (word-level) ternary decision tree (diagram)	A generalization of EXOR ternary decision diagrams with the interpretation of EXOR as the addition modulo 2. The third outgoing edge of a node in the arithmetic ternary decision diagram points to the value $f_2 = f_0 + f_1$, where the addition ($+$) is the addition of integers in the complex field.
Arithmetic transform (word-level) ternary decision tree (diagram)	Arithmetic super Shannon expansion is used. This is an integer-valued counterpart of the super Shannon expansion rule used in definition of EXOR ternary decision diagrams.

23.2 Definition of ternary decision trees

Ternary decision trees are generalizations of binary decision trees derived by allowing three outgoing edges from nodes to which binary-valued decision variables are assigned.

Node of a ternary decision tree. In a ternary decision tree for a given switching function f of n binary-valued variables, the outgoing edges of a node at the i-th level (Figure 23.1) point to the subfunctions

$$\begin{aligned} f_0 &= f_{x_i=0}, \\ f_1 &= f_{x_i=1}, \\ f_2 &= f_0 * f_1, \end{aligned}$$

where $*$ denotes a particular operation performed over co-factors f_0 and f_1 of f. The symbol S_3 relates to the Shannon nodes, which are used in binary decision trees, and the index shows the number of outgoing edges.

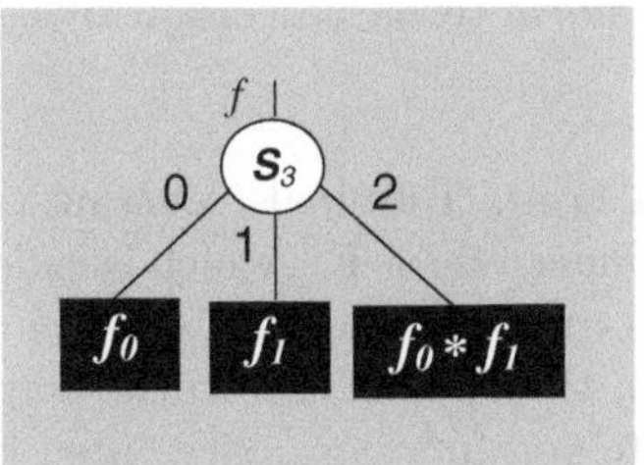

Alignment $x \odot y$

$\odot$	0	1	u
0	0	u	u
1	u	1	u
u	u	u	u

FIGURE 23.1
Node S_3 of a ternary decision diagram (the index shows the number of outgoing edges).

Assignments of the operation. Similar to binary decision diagrams, there are

- *Bit-level* and
- *Word-level* ternary decision diagrams.

These various diagrams are defined by using different assignments of the operation $*$ (Table 24.14). For instance:

TABLE 23.2
Various ternary decision trees with different assignments of the operation.

Type of tree	Assignment
AND ternary decision tree	$*$ = AND
OR ternary decision tree	$*$ = OR
EXOR ternary decision tree	$*$ = EXOR
Kleene ternary decision tree	$* = \odot$ (Figure 23.1)
Arithmetic ternary decision tree	$*$ = <Addition>
Arithmetic spectral transform decision tree	$*$ = <Subtraction>

- ▶ Assignments of logic operation AND, OR, and EXOR produce AND, OR, and EXOR ternary decision trees, which are examples of bit-level ternary decision trees.
- ▶ When $*$ is the addition in the field of real numbers, this tree is word-level, or *arithmetic* ternary decision trees.
- ▶ When $*$ is the subtraction in the field of real numbers, this word-level tree is called an *arithmetic (spectral) transform* ternary decision tree.

Terminal nodes of a ternary decision tree. Terminal (constant) nodes in a ternary decision tree represent a 3^n-element vector $\mathbf{F}_e$, which is called the *extended truth-vector* [14], since its entries are the function values and also the values determined by the application of the operation $*$ to the entries of the 2^n-element truth-vector of a given switching function represented by the ternary decision tree.

Matrix notation of a ternary decision tree. In matrix notation,

- ▶ The extended truth-vector $\mathbf{F}_e$ can be determined as the product of a $3^n \times 2^n$ matrix.
- ▶ Due to the recursive structure of decision trees, this $3^n \times 2^n$ matrix can be expressed as the Kronecker product of the basic 3×2 transform matrix $\mathbf{E}(1)$ suitably defined depending on the operation $*$ and the range for the function values of the functions represented.
- ▶ Calculation of extended truth-vector $\mathbf{F}_e$ from the truth-vector $\mathbf{F}$ of the switching function represented can be performed by a fast Fourier-like algorithm, which can also be implemented over the binary decision diagram for $\mathbf{F}$.

The expansion rules used in ternary decision trees and determination of extended truth-vector $\mathbf{F}_e$ will be illustrated by the examples of EXOR ternary decision trees, arithmetic ternary decision trees, and AC-ternary decision trees.

Ternary decision diagrams are derived from the corresponding ternary decision trees by reduction similar to that used in BDDs. The reduction consists in sharing isomorphic subtrees and deleting the redundant information from the tree.

23.3 EXOR ternary decision trees

Similarly to in binary decision trees, a switching function f is assigned to an EXOR ternary decision tree by an expansion rule, which in matrix notation can be written as

$$f = \begin{bmatrix} x_i & \overline{x}_i & 1 \end{bmatrix} \mathbf{E}(1) \begin{bmatrix} f_0 \\ f_1 \end{bmatrix},$$

where $\mathbf{E}(1) = \begin{bmatrix} 1 & 0 \\ 0 & 1 \\ 1 & 1 \end{bmatrix}$, and all the calculations are performed modulo 2, i.e., by using EXOR as the addition and AND as the multiplication.

This expansion rule is called the *super Shannon* expansion rule [19] by analogy to the Shannon expansion rule used in binary decision trees and referring to the extended truth-vectors.

Example 23.1 *Figure 23.2 shows an EXOR-ternary decision tree for the function ($n = 2$)*

$$f = (\begin{bmatrix} x_1 & \overline{x}_1 & 1 \end{bmatrix} \otimes \begin{bmatrix} x_2 & \overline{x}_2 & 1 \end{bmatrix})(\mathbf{E}(1) \otimes \mathbf{E}(1))\mathbf{F}.$$

The coefficients q_i, $i = 0, \ldots, 7$, are called the *extended Reed–Muller coefficients* [19].

23.4 Bit-level ternary decision trees

The concept of the extended truth-vector for EXOR ternary decision trees, as well as the way of calculating it, can be applied to the vector of values of

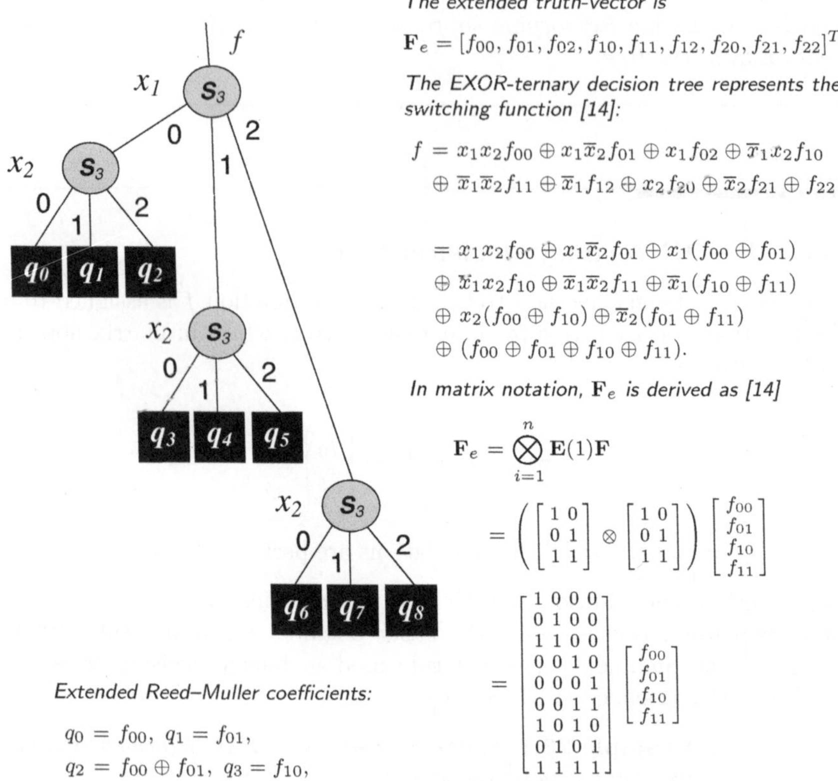

f
x_1
S_3
0
1
2
x_2
S_3
0
1
2
q_0 q_1 q_2
x_2
S_3
0
1
2
q_3 q_4 q_5
x_2
S_3
0
1
2
q_6 q_7 q_8
The extended truth-vector is
$\mathbf{F}_e = [f_{00}, f_{01}, f_{02}, f_{10}, f_{11}, f_{12}, f_{20}, f_{21}, f_{22}]^T$.
The EXOR-ternary decision tree represents the switching function [14]:
$f = x_1 x_2 f_{00} \oplus x_1 \overline{x}_2 f_{01} \oplus x_1 f_{02} \oplus \overline{x}_1 x_2 f_{10}$
$\oplus \overline{x}_1 \overline{x}_2 f_{11} \oplus \overline{x}_1 f_{12} \oplus x_2 f_{20} \oplus \overline{x}_2 f_{21} \oplus f_{22}$
$= x_1 x_2 f_{00} \oplus x_1 \overline{x}_2 f_{01} \oplus x_1 (f_{00} \oplus f_{01})$
$\oplus \overline{x}_1 x_2 f_{10} \oplus \overline{x}_1 \overline{x}_2 f_{11} \oplus \overline{x}_1 (f_{10} \oplus f_{11})$
$\oplus x_2 (f_{00} \oplus f_{10}) \oplus \overline{x}_2 (f_{01} \oplus f_{11})$
$\oplus (f_{00} \oplus f_{01} \oplus f_{10} \oplus f_{11})$.
In matrix notation, $\mathbf{F}_e$ is derived as [14]
$\mathbf{F}_e = \bigotimes_{i=1}^{n} \mathbf{E}(1)\mathbf{F}$
$= \left(\begin{bmatrix} 1 & 0 \\ 0 & 1 \\ 1 & 1 \end{bmatrix} \otimes \begin{bmatrix} 1 & 0 \\ 0 & 1 \\ 1 & 1 \end{bmatrix} \right) \begin{bmatrix} f_{00} \\ f_{01} \\ f_{10} \\ f_{11} \end{bmatrix}$
$= \begin{bmatrix} 1 & 0 & 0 & 0 \\ 0 & 1 & 0 & 0 \\ 1 & 1 & 0 & 0 \\ 0 & 0 & 1 & 0 \\ 0 & 0 & 0 & 1 \\ 0 & 0 & 1 & 1 \\ 1 & 0 & 1 & 0 \\ 0 & 1 & 0 & 1 \\ 1 & 1 & 1 & 1 \end{bmatrix} \begin{bmatrix} f_{00} \\ f_{01} \\ f_{10} \\ f_{11} \end{bmatrix}$
$= [q_0, q_1, q_2, q_3, q_4, q_5, q_6, q_7, q_8]^T$
Extended Reed–Muller coefficients:
$q_0 = f_{00},\ q_1 = f_{01},$
$q_2 = f_{00} \oplus f_{01},\ q_3 = f_{10},$
$q_4 = f_{11},\ q_5 = f_{10} \oplus f_{11},$
$q_6 = f_{00} \oplus f_{10}, q_7 = f_{01} \oplus f_{11},$
$q_8 = f_{00} \oplus f_{01} \oplus f_{10} \oplus f_{11}$

FIGURE 23.2
EXOR-ternary decision tree for $n = 2$ (Example 23.1).

constant nodes in other ternary decision trees, provided calculations are made in terms of the corresponding operations [14], [19].

> **Example 23.2** *Figure 23.3 shows the AND, OR, EXOR, and Kleene ternary decision trees for the switching function* $f(x_1, x_2, x_3) = x_1\overline{x}_2 \vee x_2x_3$.

23.5 Word-level ternary decision trees

> Arithmetic-ternary decision diagrams *are defined as a word-level generalization of EXOR ternary decision diagrams with the interpretation of EXOR as the addition modulo 2.*

Therefore, the third outgoing edge of a node in the arithmetic ternary decision diagram points to the value

$$f_2 = f_0 + f_1$$

where the addition represents the addition of integers in the complex field C.

ATDDs are used in functional decomposition and detection of prime implicants and prime implicates [12] since from an arithmetic ternary decision diagram, we can read f and all the prime implicants for f. The expansion rule that is recursively applied to variables of the given function determines an arithmetic ternary decision diagram.

Definition 23.1 *The ATDDs for a given switching f are determined by the recursive application of the expansion rule*

$$f = \frac{1}{2}((-1)^{x_i} + (-1)^{\overline{x}_i} f_1 + 1 \cdot (f_0 + f_2)).$$

In matrix notation, this expansion rule is [14]

$$f = \left[(-1)^{x_i} \; (-1)^{\overline{x}_i} \; 1\right] \mathbf{E}_i(1) \begin{bmatrix} f_0 \\ f_1 \end{bmatrix}.$$

ATDDs are defined through:

- ► A combination of expansion rules used in *algebraic* decision diagrams (ADDs) (see Chapter 17), and
- ► Labels at the edges used in *Walsh* decision diagrams (see details in Chapter 40).

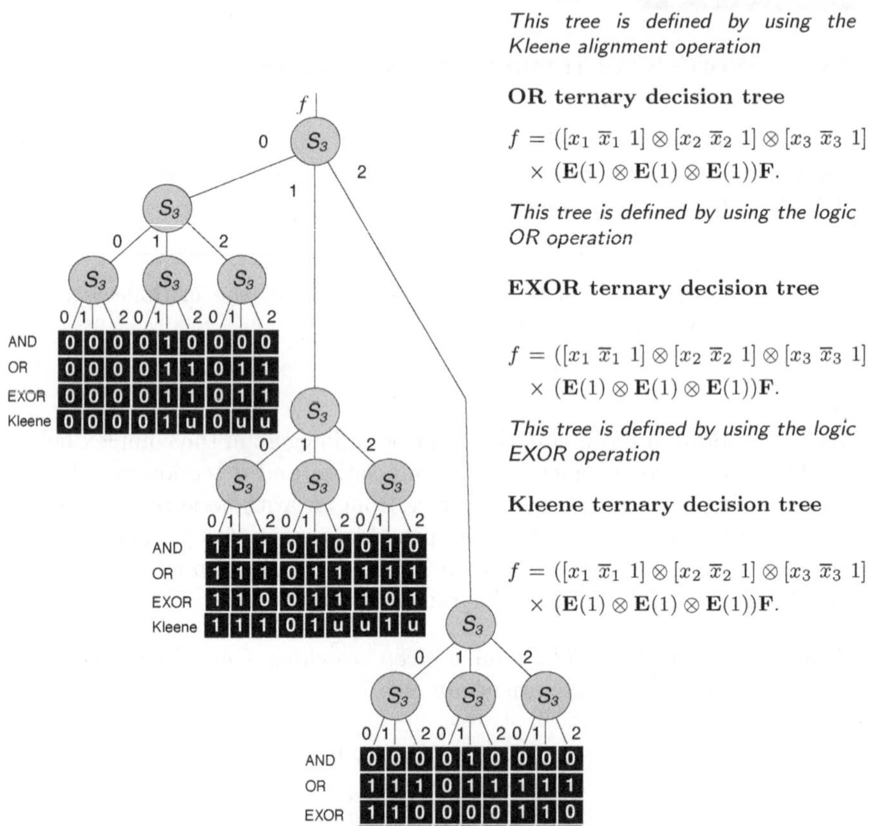
AND ternary decision tree
$f = ([x_1\ \overline{x}_1\ 1] \otimes [x_2\ \overline{x}_2\ 1] \otimes [x_3\ \overline{x}_3\ 1] \times (\mathbf{E}(1) \otimes \mathbf{E}(1) \otimes \mathbf{E}(1))\mathbf{F}.$
This tree is defined by using the Kleene alignment operation
OR ternary decision tree
$f = ([x_1\ \overline{x}_1\ 1] \otimes [x_2\ \overline{x}_2\ 1] \otimes [x_3\ \overline{x}_3\ 1] \times (\mathbf{E}(1) \otimes \mathbf{E}(1) \otimes \mathbf{E}(1))\mathbf{F}.$
This tree is defined by using the logic OR operation
EXOR ternary decision tree
$f = ([x_1\ \overline{x}_1\ 1] \otimes [x_2\ \overline{x}_2\ 1] \otimes [x_3\ \overline{x}_3\ 1] \times (\mathbf{E}(1) \otimes \mathbf{E}(1) \otimes \mathbf{E}(1))\mathbf{F}.$
This tree is defined by using the logic EXOR operation
Kleene ternary decision tree
$f = ([x_1\ \overline{x}_1\ 1] \otimes [x_2\ \overline{x}_2\ 1] \otimes [x_3\ \overline{x}_3\ 1] \times (\mathbf{E}(1) \otimes \mathbf{E}(1) \otimes \mathbf{E}(1))\mathbf{F}.$
f
S_3
0 1 2
AND 0 0 0 0 1 0 0 0 0
OR 0 0 0 0 1 1 0 1 1
EXOR 0 0 0 0 1 1 0 1 1
Kleene 0 0 0 0 1 u 0 u u
AND 1 1 1 0 1 0 0 1 0
OR 1 1 1 0 1 1 1 1 1
EXOR 1 1 0 0 1 1 1 0 1
Kleene 1 1 1 0 1 u u 1 u
AND 0 0 0 0 1 0 0 0 0
OR 1 1 1 0 1 1 1 1 1
EXOR 1 1 0 0 0 0 1 1 0
Kleene u u u 0 1 u u u u

FIGURE 23.3
AND, EXOR, and Kleene-ternary decision tree of $f = x_1\overline{x}_2 \vee x_2x_3$ (Example 23.2).

23.6 Arithmetic transform ternary decision diagrams

Arithmetic transform ternary decision diagrams (AC-TDDs), another example of word-level ternary decision diagrams, are defined by the interpretation of EXOR as the subtraction modulo 2 [16].

Definition 23.2 *The arithmetic transform ternary decision diagram for a given switching function f is derived by the recursive application of the expansion rule*

$$f = x_i f_0 + \overline{x}_i f_1 + (-1)^{\overline{x}_i}(-f_0 + f_1), \qquad (23.1)$$

to all the variables in f.

> *Expansion 23.1 is called the* arithmetic super Shannon *expansion rule, since it is an integer-valued counterpart of the super Shannon expansion rule used in definition of EXOR ternary decision diagrams.*

The multiplicative factor $(-1)^{\overline{x}_i}$ takes into account the change of the sign of the arithmetic coefficients calculated with respect to the *a*rithmetic positive Davio and *a*rithmetic negative Davio expansion rules. Thus, it is denoted as the sign alternating factor [14].

> *An AC-TDD represents f and the arithmetic transform coefficients for f for all possible different polarities.*

From an AC-ternary decision diagram, we can read f and all the possible polynomial expressions for f in the vector space of complex valued functions of n binary valued variables, $C(C_2^n)$, since the coefficients in these expressions are taken in the set of function values and the set of coefficients in all fixed-polarity arithmetic expressions for f.

Therefore, AC-TDDs are used in the determination of various arithmetic polynomial expressions over C, compact representation of multi-output switching functions, and exact minimization of fixed-polarity arithmetic expressions.

23.7 The relationships between arithmetic transform ternary decision diagrams and other decision trees

The relationships between of AC-ternary decision trees, AC-TDDs, and other decision trees can be specified as follows. Denote by R_w the set of decision trees consisting of MTBDTs, ACDTs, and all possible word-level KDTs, and PKDTs. Then, the AC-ternary decision tree is the power set $P(R_w)$ of R_w. Thus, it is the union of all these word-level decision trees.

Example 23.3 *Figure 23.4 shows examples of four binary decision trees contained in an EXOR ternary decision tree for $n = 2$. These trees are specified as follows:*

- ► *Binary decision tree* : $\{q_0, q_1 q_3, q_4\}$,
- ► *Positive polarity Reed–Muller tree* : $\{q_0, q_2, q_6, q_8\}$,
- ► *FPRM(pE,nD)* : $\{q_1, q_2, q_7, q_8\}$,
- ► *Kronecker tree KDT(S,pD)* : $\{q_0, q_2, q_3, q_5.\}$

which shows the constant nodes in the paths from the root node to the constant nodes in these trees.

Many other decision trees with two outgoing edges per node can be determined from this EXOR-ternary decision tree [15].

23.8 Ternary decision diagrams and differential operators for switching functions

Ternary decision diagrams are defined in terms of some operations over the cofactors f_0 and f_1 that determine the subfunctions rooted at the nodes where the third outgoing edges of nodes point. If these are operations in terms of which some differential operators are defined, the relationships between ternary decision diagrams and differential operators is straightforward.

Example 23.4 *In EXOR-ternary decision diagrams, the elements of* $\mathbf{F}_e$ *for* f *are the function values of* f *and the values of all its Boolean differences, since the operation assigned to the third outgoing edge is EXOR, in terms of which the Boolean difference is defined.*

Recursive application of the decomposition rules to build ternary decision diagrams corresponds to the recursive calculation of logic differential operators of different orders and with respect to different subsets of variables.

The same statement applies to word-level diagrams with respect to partial Gibbs derivatives which can be viewed as integer counterparts of the Boolean difference (see Chapters 9 and 39).

The relationships between AC-ternary decision diagrams and partial dyadic Gibbs derivatives are similar to those of EXOR ternary decision diagrams and

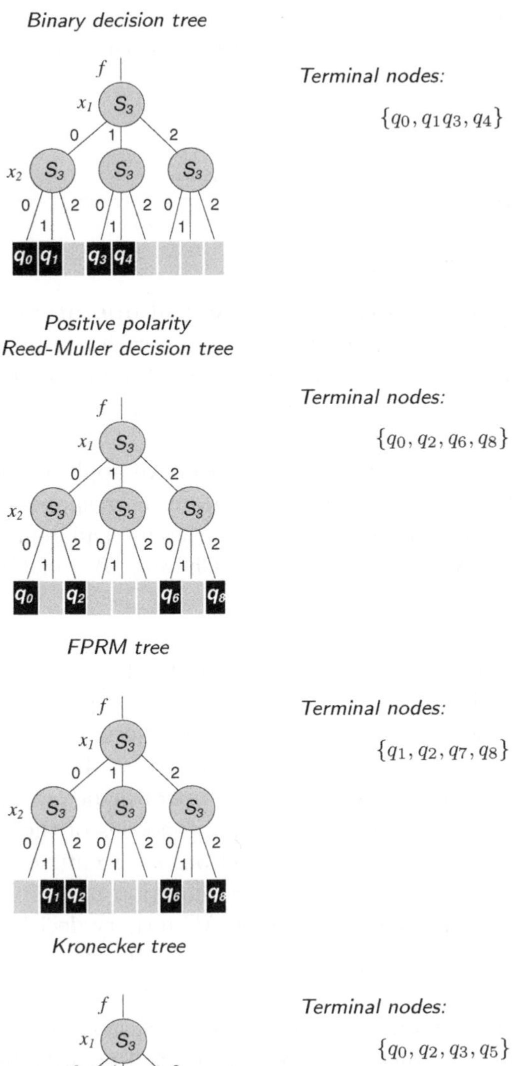

FIGURE 23.4
Example of four binary decision trees contained in an EXOR-ternary decision tree for $n = 2$ (Example 23.3).

the Boolean differences. The values of partial dyadic Gibbs derivatives are elements of the extended arithmetic transform spectrum, thus, they appear as the values of constant nodes in AC-ternary decision diagrams determined by the corresponding n-tuples.

23.9 Further reading

Ternary decision diagrams and differential operators. Due to the redundancy introduced by representing a function with a truth-vector of 2^n elements by an EXOR-ternary decision diagram with an extended truth-vector of 3^n elements, the information content of EXOR ternary decision trees is rather larger than that of BDDs [15].

Stanković and Astola [14] proved that recursive application of the decomposition rules to build ternary decision diagrams corresponds to the recursive calculation for functions of binary-valued variables. Differential operators include, in particular, Boolean differences (Chapter 9) and Gibbs differences (derivatives).

The Gibbs derivatives were first defined for functions on finite dyadic groups [3], since their introduction was motivated by problems of differentiation of Walsh functions. This derivative is, therefore, denoted as the dyadic Gibbs derivative or the logic Gibbs derivative [1, 2, 4]. In this setting, the logic Gibbs derivative is defined as a linear operator possessing properties of differential operators corresponding to the properties of the Newton-Leibniz derivative, and having discrete Walsh functions as eigenfunctions. This property permits many generalizations of the definition of Gibbs derivatives on various groups. The relationships between AC-ternary decision diagrams and partial dyadic Gibbs derivatives were studied in [16, 18].

Minimization via ternary decision trees. Sasao and Izuhara [14] proposed an algorithm for exact minimization of fixed-polarity Reed–Muller expressions by using EXOR ternary decision diagrams. The same algorithm can be extended to the minimization of fixed-polarity arithmetic expressions through AC-ternary decision diagrams.

Ternary decision diagrams. Stanković and Sasao [19] proposed the concept of the extended truth-vector in ternary decision diagrams. Sasao [11, 12] showed that arithmetic ternary decision diagrams can be used in functional decomposition and detection of prime implicants. A detailed discussion for EXOR ternary decision trees is given in work by Sasao and Izuhara [14]. For other examples of ternary decision trees see the book by Stanković and Astola

[14]. *Free ternary decision diagrams* in which the ordering of variables may be different for each path from the root node to the constant nodes have been considered in papers by Higuchi and Kameyama [5], [7], [8]. Kleene ternary decision diagrams have been used for logic simulation in the presence of unknown input variables (see, for example, the paper by Jennings [6]). Various canonical and non-canonical forms of ternary decision diagrams have been proposed by Perkowski et al. [10]. Arithmetic transform ternary decision diagrams were introduced by Stanković [16]. Details on these diagrams can be also found in [14].

Using EXOR-ternary decision diagrams in minimization of switching functions. Papakonstantinou [9] discussed minimization of EXOR-sum-of-product (ESOP) expressions with up to four variables by using EXOR ternary decision diagrams.

Information-theoretical measures. A ternary decision diagram is a redundant data structure. It is explained in [15] that the information content of a ternary decision diagram is considerably increased compared to that of a binary decision diagram for a given switching function. The redundancy of a ternary decision diagram can be efficiently used in searching, for example, prime implicants and logic simulation with uncertainty. Yanushkevich et al. [11] used an information-theoretical approach for minimization of switching functions on ternary decision trees (details are given in Chapter 24).

References

[1] Edwards CR. The generalized dyadic differentiator and its application to 2-valued functions defined on an n-space. In *Proceedings IEE, Comput. and Digit. Techn.*, 1(4):137–142, 1978.

[2] Edwards CR. The Gibbs dyadic differentiator and its relationship to the Boolean difference. *Comput. and Elect. Engng.*, 5:335–344, 1978.

[3] Gibbs JE. Walsh spectrometry, a form of spectral analysis, well suited to binary digital computation. *NPL DES Repts.*, National Physical Lab., Teddington, Middlesex, England, 1967.

[4] Gibbs JE. Local and global views of differentiation. In Butzer PL and Stanković RS, Eds., *Theory of Gibbs Derivatives and Applications.* Matematički Institut, Beograd, pp. 1–19, 1990.

[5] Higuchi T and Kameyama M. Ternary logic system based on T-gate. In *Proceedings of the IEEE 5th International Symposium on Multiple-Valued Logic*, pp. 290–304, 1975.

[6] Jennings G. Symbolic incompletely specified functions for correct evaluation in the presence of indeterminate input values. In *Proceedings of the 28th Hawaii International Conference on System Sciences*, Vol. 1, pp. 23–31, 1995.

[7] Kameyama M and Higuchi T. Synthesis of multiple-valued logic based on tree-type universal logic module. *IEEE Transactions on Computers*, 26:1297–1302, 1977.

[8] Kameyama M and Higuchi T. Synthesis of optimal T-gate networks in multiple-valued logic. In *Proceedings of the IEEE 9th International Symposium on Multiple-Valued Logic*, pp. 190–195, 1979.

[9] Papakonstantinou G. Minimization of modulo-2 sum of products. *IEEE Transactions on Computers*, 28:163–167, 1979.

[10] Perkowski MA, Chrzanowska-Jeske M, Sarabi A, and Schafer I. Multi-level logic synthesis based on Kronecker decision diagrams and Boolean ternary decision diagrams for incompletely specified functions. *VLSI Des.* (Switzerland), 3(3–4):301–313, 1995.

[11] Sasao T. Ternary decision diagrams. In *Proceedings of the 28th International Symposium on Multiple-Valued Logic*, pp. 241–250, 1997.

[12] Sasao T. Arithmetic ternary decision diagrams and their applications. In *Proceedings of the 4th International Workshop on Applications of Reed–Muller Expansion in Circuit Design*, Victoria, B.C., Canada, pp. 149–155, 1999.

[13] Sasao T and Fujita M, Eds. *Representations of Discrete Functions.* Kluwer, Dordrecht, 1996.

[14] Sasao T and Izuhara F. Exact minimization of FPRMs using multi-terminal EXOR-TDDs. In [14], pp. 191–210.

[15] Stanković RS. Information content of ternary decision diagrams. *Automation and Remote Control,* Kluwer/Plenum Publishers, 63(4):666–681, 2002.

[16] Stanković RS. Word-level ternary decision diagrams and arithmetic expressions. In *Proccedings of the Workshop on Applications of Reed–Muller Expressions in Circuit Design*, Mississippi State University, Starkville, pp. 34–50, 2001.

[17] Stanković RS and Astola JT. *Spectral Interpretation of Decision Diagrams.* Springer, Heidelberg, 2003.

[18] Stanković RS and Astola J. Relationships between logic derivatives and ternary decision diagrams. In *Proceedings of the 5th International Workshop on Boolean Problems*, Freiberg, Germany, pp. 53–60, 2002.

[19] Stanković RS and Sasao T. Spectral interpretation of TDDs. In *Proceedings of the 17th Workshop on Synthesis and System Integration of Mixed Technologies*, Osaka, Japan, 45–50, 1997.

[20] Stanković RS, Sasao T, and Moraga C. Spectral transform decision diagrams, In [14], pp. 55–92.

[21] Yanushkevich SN, Shmerko VP, Dziurzanski P, Stanković RS, and Popel DV. Experimental verification of the entropy based method for minimization of switching functions on pseudo-ternary decision trees. In *Proceedings of the IEEE International Conference on Telecommunications in Modern Satellite, Cable and Broadcasting Services*, Yugoslavia, pp. 452–459, 1999.

24

Information - Theoretical Measures in Decision Diagrams

24.1 Introduction

Entropy based strategies are effective for tasks that can be formalized as the conversion of decision tables into decision trees. Decision tables are used in computer aided design to specify which action must be taken for any condition in a set of conditions.

24.2 Information-theoretical measures

In this section, the following basic measures on decision trees and diagrams are considered:

- The entropy,
- Mutual entropy,
- Conditional entropy, and
- Information.

Let $A = \{a_1, a_2, \ldots, a_n\}$ be a complete set of events with the probability distribution

$$\{p(a_1), p(a_2), \ldots, p(a_n)\}.$$

The *entropy* of the finite field A is given by

$$H(A) = -\sum_{i=1}^{n} p(a_i) \cdot \log p(a_i) \tag{24.1}$$

where the logarithm is base 2. The entropy can never be negative, i.e., $\log p(a_i) \leq 0$, and thus $H(A) \geq 0$. The entropy is zero if and only if A

contains one event only (see details on information-theoretical measures in Chapter 8).

Example 24.1 *For a switching function that takes the value 1 with the probability p_1 and the value 0 with the probability p_0, the entropy is minimally*

$$H(A) = 0 \ \ bit/pattern$$

when $p_0 = 0$ or $p_0 = 1$, and the entropy reaches its maximum

$$H(A) = 1 \ \ bit/pattern$$

when $p_0 = p_1 = 0.5$.

Example 24.2 *The entropy for the completely specified switching function f of three variables given by the truth vector $\mathbf{F} = [10111110]^T$ is calculated as follows:*

$$\begin{aligned} H(f) &= -0.25 \cdot \log 0.25 - 0.75 \cdot \log 0.75 \\ &= 0.8113 \ bit/pattern. \end{aligned}$$

The entropies of variables:

$$\begin{aligned} H(x_1) = H(x_2) &= H(x_3) \\ &= -0.5 \cdot \log 0.5 - 0.5 \cdot \log 0.5 \\ &= 1 \ bit/pattern. \end{aligned}$$

i.e., uncertainty in the variables of this function is the same.

Let A and B be finite fields of events with the probability distributions

$$\begin{aligned} &\{p(a_i)\}, \quad i = 1, 2, \ldots, n \text{ and} \\ &\{p(b_j)\}, \quad j = 1, 2, \ldots, m \end{aligned}$$

respectively. The *conditional entropy* of the finite field A with respect to B is defined by

$$H(A|B) = -\sum_{i=1}^{n}\sum_{j=1}^{m} p(a_i, b_j) \cdot \log \frac{p(a_i, b_j)}{p(b_j)} \tag{24.2}$$

Example 24.3 *(Continuation of Example 24.2.) The conditional entropy of switching function f with respect to the variable x_1, x_2 and x_3 is (Figure 24.1):*

$$H(f|x_1) = 0.8113\ bit/pattern$$
$$H(f|x_2) = 0.8113\ bit/pattern$$
$$H(f|x_3) = 0.5\ bit/pattern$$

The variable x_3 conveys more information than the variables x_1 and x_2.

$x_1x_2x_3$	f
000	1
001	0
010	1
011	1
100	1
101	1
110	1
111	0

The conditional entropy of switching function f with respect to the variable x_1:

$$\begin{aligned} H(f|x_1) &= -0.125 \cdot \log \frac{0.125}{0,5} - 0.125 \cdot \log \frac{0.125}{0.5} \\ &- 0.375 \cdot \log \frac{0.375}{0.5} - 0.375 \cdot \log \frac{0.375}{0.5} \\ &= 0.8113\ bit/pattern \\ H(f|x_2) &= -0.125 \cdot \log \frac{0.125}{0,5} - 0.125 \cdot \log \frac{0.125}{0.5} \\ &- 0.375 \cdot \log \frac{0.375}{0.5} - 0.375 \cdot \log \frac{0.375}{0.5} \\ &= 0.8113 bit/pattern, \\ H(f|x_3) &= -0 \cdot \log \frac{0}{0.5} - 0.25 \cdot \log \frac{0.25}{0.5} \\ &- 0.5 \cdot \log \frac{0.5}{0.5} - 0.25 \cdot \log \frac{0.25}{0.5} \\ &= 0.5\ bit/pattern \end{aligned}$$

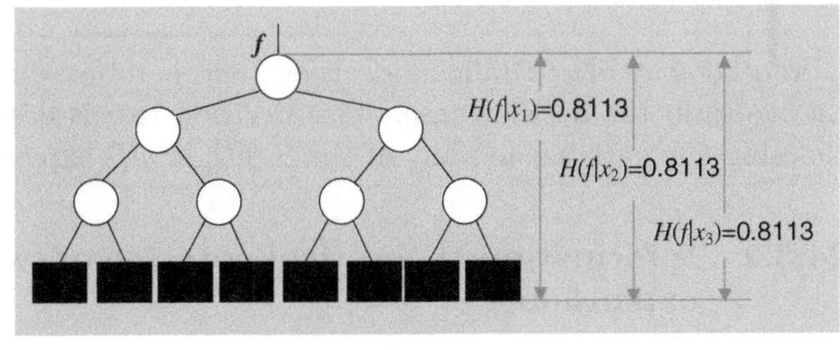

FIGURE 24.1
The conditional entropy of a completely specified switching function f with respect to the variables x_1, x_2, and x_3 (Example 24.3).

24.3 Information-theoretical measures in decision trees

In this section, we address the design of decision trees with nodes of three types: Shannon (S) positive Davio (pD) and negative Davio (nD) based on information theoretical approaches. An approach revolves around choosing the "best" variable and the "best" expansion type with respect to this variable for any node of the decision tree in terms of information measures. This means that:

> *In any step of the decision making on the tree, we have an opportunity to choose both the variable and the type of expansion based on the criterion of minimum entropy.*

An entropy-based optimization strategy can be thought of as generation of *optimal paths* in a decision tree, with respect to the minimum entropy criterion.

24.3.1 Decision tree induction

The best known use of entropy and information measures on decision trees is the induction of decision trees (ID3) algorithm for optimization.

Example 24.4 *Figure 24.2 illustrates the calculation of entropy on a decision tree.*

> Free *binary decision trees are derived by permitting the permutation of the order of variables in a subtree independently of the order of variables in the other subtrees related to the same nonterminal node.*

Another way of generalizing decision trees is to use different expansions at the nodes in the decision tree. This decision tree is designed by arbitrarily choosing any variable and any of the S, pD, or nD expansions for each node.

24.3.2 Information-theoretical notation of switching function expansion

In the top-down decision tree design two information measures are used:

$$\begin{aligned} < \texttt{Conditional entropy} > &= H(f|Tree), \\ < \texttt{Mutual information} > &= I(f;Tree). \end{aligned}$$

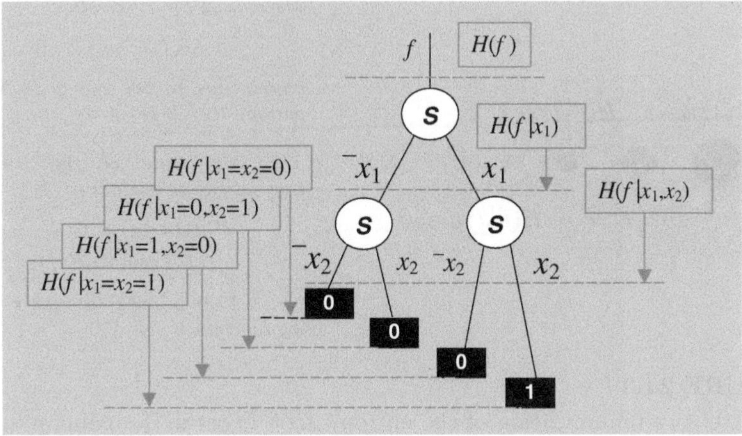

FIGURE 24.2
Measurement of entropy on a decision tree (Example 24.4).

The *initial state* of this process is characterized by the maximum value for the conditional entropy

$$H(f|Tree) = H(f).$$

Nodes are recursively attached to the decision tree by using the top-down strategy. In this strategy the entropy $H(f|Tree)$ of the function is reduced, and the information $I(f;Tree)$ increases, since the variables convey the information about the function. Each *intermediate state* can be described in terms of entropy by the equation

$$I(f;Tree) = H(f) - H(f|Tree). \tag{24.3}$$

The goal of such an optimization is to maximize the information $I(f;Tree)$ that corresponds to the minimization of entropy $H(f|Tree)$, in each step of the decision tree design.

The final state of the decision tree is characterized by $H(f|Tree) = 0$ and $I(f;Tree) = H(f)$, i.e., $Tree$ represents the switching function f (Figure 24.3).

The decision tree design process is a recursive decomposition of a switching function. A step of this recursive decomposition corresponds to the expansion of a switching function f with respect to the variable x. Assume that the

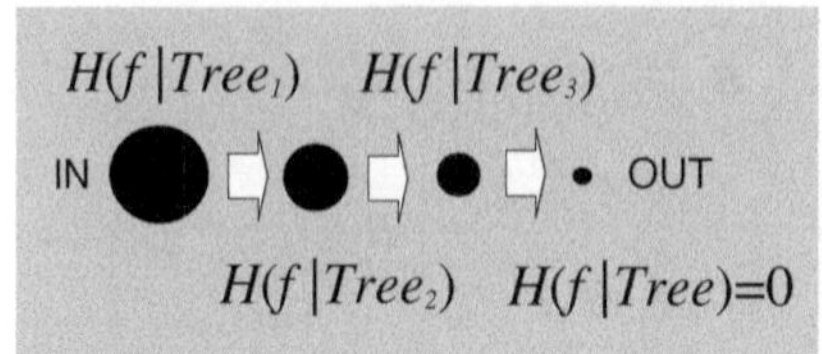

Maximizing the information

$$I(f; Tree_i)$$

corresponds to the minimization of entropy $H(f|Tree_i)$.

The final state of the decision tree design corresponds to

$$H(f|Tree_3) = 0$$
$$I(f; Tree) = H(f),$$

i.e., $< Tree >$ *represents the switching function* f

FIGURE 24.3
Four steps to minimization of the entropy $H(f|Tree)$ in designing a decision tree for reduction of uncertainty.

variable x in f conveys information that is, in some sense, the rate of influence of the input variable on the output value for f.

The initial state of the expansion $\omega \in \{S, pD, nD\}$ can be characterized by the entropy $H(f)$ of f, and the final state by the conditional entropy $H^{\omega}(f|x)$. The ω expansion of the switching function f with respect to the variable x is described in terms of entropy as follows

$$I^{\omega}(f; x) = H(f) - H^{\omega}(f|x). \tag{24.4}$$

A formal criterion for completing the sub-tree design is $H^{\omega}(f|x) = 0$.

Information notation of S expansion

The designed decision tree based on the S expansion is mapped into a sum-of-products expression as follows: a leaf with the logic value 0 is mapped into $f = 0$, and with the logic value 1 into $f = 1$; a nonterminal node is mapped into $f = \overline{x} \cdot f_{x=0} \vee x \cdot f_{x=1}$ (Figure 24.4). The information measure of S expansion for a switching function f with respect to the variable x is represented by the equation

$$H^S(f|x) = p_{|x=0} \cdot H(f_{x=0}) + p_{|x=1} \cdot H(f_{x=1}). \tag{24.5}$$

The information measure of S expansion is equal to the conditional entropy $H(f|x)$:

$$H^S(f|x) = H(f|x). \tag{24.6}$$

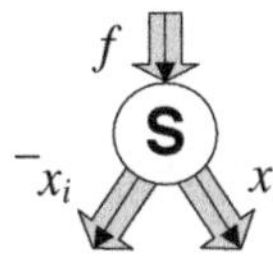

Shannon expansion

$$f = \overline{x} \cdot f_{x=0} \oplus x \cdot f_{x=1}$$

Information-theoretical notation

$$H^S(f|x) = \underbrace{p_{|x=0} H(f_{x=0})}_{Left\ leaf} + \underbrace{p_{|x=1} H(f_{x=1})}_{Right\ leaf}$$

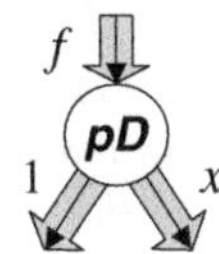

Positive Davio expansion

$$f = f_{x=0} \oplus x \cdot (f_{x=0} \oplus f_{x=1})$$

Information-theoretical notation

$$H^{pD}(f|x) = \underbrace{p_{|x=0} H(f_{x=0})}_{Left\ leaf} + \underbrace{p_{|x=1} H(f_{x=0} \oplus f_{x=1})}_{Right\ leaf}$$

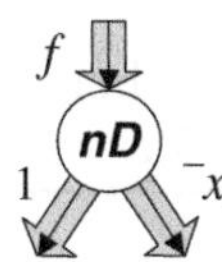

Negative Davio expansion

$$f = f_{x=1} \oplus \overline{x} \cdot (f_{x=0} \oplus f_{x=1})$$

Information-theoretical notation

$$H^{nD}(f|x) = \underbrace{p_{|x=1} H(f_{x=1})}_{Left\ leaf} + \underbrace{p_{|x=0} H(f_{x=0} \oplus f_{x=1})}_{Right\ leaf}$$

FIGURE 24.4
Shannon and Davio expansions and their information measures for a switching function.

Information notation of pD and nD expansion

The information measure of a pD expansion of a switching function f with respect to the variable x is represented by

$$H^{pD}(f|x) = p_{|x=0} \cdot H(f_{x=0}) + p_{|x=1} \cdot H(f_{x=0} \oplus f_{x=1}). \tag{24.7}$$

The information measure of the nD expansion of a switching function f with respect to the variable x is

$$H^{nD}(f|x) = p_{|x=1} \cdot H(f_{x=1}) + p_{|x=0} \cdot H(f_{x=0} \oplus f_{x=1}). \tag{24.8}$$

Note, that, since

$$H^S(f|x) = H(f|x) \text{ and } I^S(f;x) = I(f;x)$$

the mutual information of f and x is

$$I(f;x) = H(f) - H(f|x).$$

This is because

$$\begin{aligned} H(f|x) &= p_{|x=0} \cdot H(f_{x=0}) + p_{|x=1} \cdot H(f_{x=1}) + p_{|x=1} \cdot H(f_{x=0} \oplus f_{x=1}) \\ &\quad - p_{|x=1} \cdot H(f_{x=0} \oplus f_{x=1}) \\ &= H^{pD}(f|x) + p_{|x=1} \cdot (H(f_{x=1}) - H(f_{x=0} \oplus f_{x=1})). \end{aligned}$$

Thus

$$I(f;x) = H(f) - H^{pD}(f|x) - p_{|x=1} \cdot (H(f_{x=1}) - H(f_{x=0} \oplus f_{x=1})),$$
$$I(f;x) = H(f) - H^{nD}(f|x) - p_{|x=0} \cdot (H(f_{x=0}) - H(f_{x=0} \oplus f_{x=1})).$$

Example 24.5 *Given a switching function of three variables, calculate the entropy of Shannon and Davio expansions with respect to all variables. The results are summarized in Table 24.1. We observe that the minimal value of the information theoretical measure corresponds to Shannon expansion with respect to the variable x_2.*

TABLE 24.1
Choosing the type of expansion for a switching function of three variables (Example 24.5).

	Shannon $H^S(f_1\|x)$	Positive Davio $H^{pD}(f_1\|x)$	Negative Davio $H^{nD}(f_1\|x)$
x_1	0.88	0.95	0.88
x_2	**0.67**	0.88	0.75
x_3	0.98	0.98	0.95

24.3.3 Optimization of variable ordering in a decision tree

The entropy based design of an optimal decision tree can be described as the optimal (with respect to the information criterion) node selection process. A path in the decision tree starts from a node and finishes in a terminal node. Each path corresponds to a term in the final expression for f.

The criterion for choosing the decomposition variable x and the expansion type $\omega \in \{S, pD, nD\}$ is that the conditional entropy of the function with respect to this variable has to be minimum:

$$H^{\omega}(f|x) \to MIN.$$

The entropy based algorithm for the minimization of AND/EXOR expressions is introduced in the example below. In this algorithm, the ordering restriction is relaxed. This means that

- ▶ Each variable appears once in each path, and
- ▶ The orderings of variables along the paths may be different.

Example 24.6 *The design of an AND/EXOR decision tree for the hidden weighted bit function is given in Figure 24.5. The order of variables in the tree is evaluated based on a measure of entropy of the switching function f with respect to variables x_1 x_2, and x_3. According to the criterion of minimum entropy, x_1 is assigned to the root. The other assignments are shown in Figure 24.5. The quasi-optimal Reed–Muller expression corresponding to this tree is:*

$$f = x_2x_3 \oplus x_1\overline{x}_3.$$

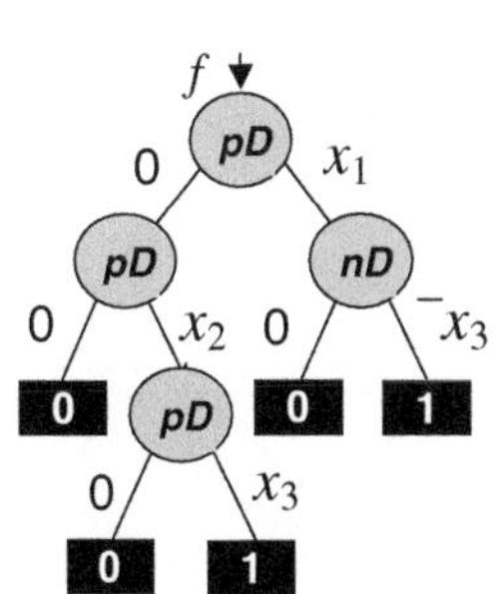

x_1	x_2	x_3	f
0	0	0	0
0	0	1	0
0	1	0	0
0	1	1	1
1	0	0	1
1	0	1	0
1	1	0	1
1	1	1	1

Step 1. *Choose the variable x_1 and pD expansion for the root node*

$H^{pD}(f|x_1) = 0.91$ *bit*
Decision: select the $f_0 = f_{x_1=0}$

Step 2. *Choose the variable x_2 and pD expansion for the next node*

$H^{pD}(f|x_2) = 0.5$ *bit*
$f_0 = f_{x_2=0} = 0$
Decision: select
$f_1 = f_{x_2=0} \oplus f_{x_2=1}$

Step 3. *Select pD expansion for the variable x_3*

$f_0 = f_{x_3=0} = 0$ *and*
$f_1 = f_{x_3=0} \oplus f_{x_3=1} = 1$
Decision: select
$f_0 = f_{x_1=0} \oplus f_{x_1=1}$

Step 4. *Choose the variable x_3 and select nD expansion*

$f_0 = f_{x_3=1} = 0$ *and*
$f_1 = f_{x_3=0} \oplus f_{x_3=1} = 1$

FIGURE 24.5
AND/EXOR decision tree design (Example 24.6).

Example 24.7 *Consider the design of the Shannon tree given a sum-of-products expression and a hidden weighted bit function. The Shannon tree is shown in Figure 24.6.*

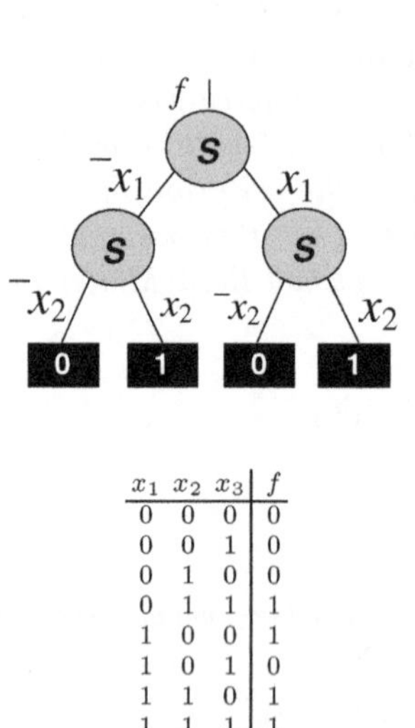

x_1	x_2	x_3	f
0	0	0	0
0	0	1	0
0	1	0	0
0	1	1	1
1	0	0	1
1	0	1	0
1	1	0	1
1	1	1	1

Entropy of the function

$$\begin{aligned} H(f) = & -{}^1/_2 \cdot \log {}^1/_2 \\ & -{}^1/_2 \cdot \log {}^1/_2 \\ & = 1 \ bit/pattern \end{aligned}$$

Conditional entropy of f with respect to the variable x_1

$$\begin{aligned} H(f|x_1) = & -{}^3/_8 \cdot \log {}^3/_8 \\ & -{}^1/_8 \cdot \log {}^1/_8 \\ & -{}^1/_8 \cdot \log {}^1/_8 \\ & -{}^3/_8 \cdot \log {}^3/_8 \\ & = 0.81 \ bit/pattern \end{aligned}$$

Sum-of-products expression
$f = \overline{x}_3 \cdot x_1 \vee x_3 \cdot x_2$

FIGURE 24.6
Shannon decision tree design (Example 24.7).

24.3.4 Information measures in the $\mathcal{N}$-hypercube

It has been shown that information-theoretical measures for logic networks can be evaluated using decision trees. In this section we focus on the details of information measures in an $\mathcal{N}$-hypercube based on information measures in decision trees.

Example 24.8 *Figure 24.7 illustrates the calculation of entropy of an $\mathcal{N}$-hypercube. Starting with the root, where entropy is maximal, we approach variables in sequence. Approaching x_1 reveals information about this variable, etc. Approaching terminal nodes means that the entropy becomes 0.*

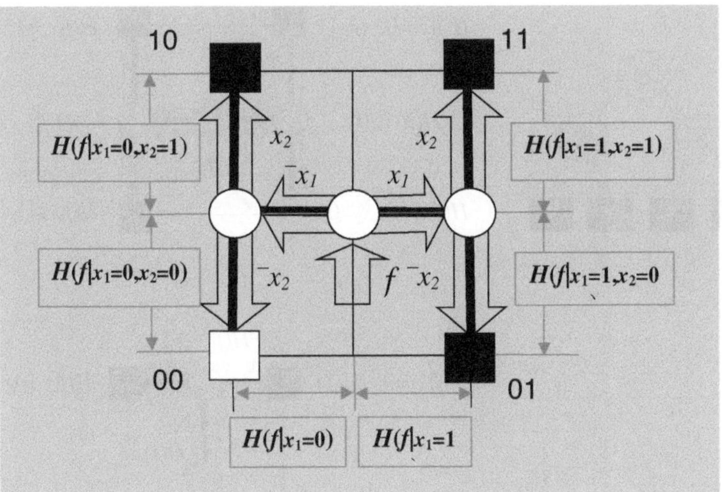

FIGURE 24.7
Measure of entropy on an $\mathcal{N}$-hypercube (Example 24.8).

Example 24.9 *Let the order of variables of a two variable switching function be* $\{x_1, x_2\}$. *In Figure 24.8a, the corresponding decision tree and* $\mathcal{N}$*-hypercube are depicted. Let us change the order of variables:* $\{x_2, x_1\}$. *It follows from Figure 24.8b that it is necessary to reallocate terminal values in the decision tree, but we do not need to do so in the* $\mathcal{N}$*-hypercube.*

The calculation of information measures of two-variable switching functions is given in Table 24.2 and Table 24.3. Here we suppose that input patterns are generated with equal probabilities. An alternative approach is based on the calculation of input and output entropy assuming that input patterns are generated with different probabilities.

Example 24.10 *The calculation of information and entropy on the tree for a switching function given by the truth table* [0 1 0 1] *is illustrated in Figure 24.9.*

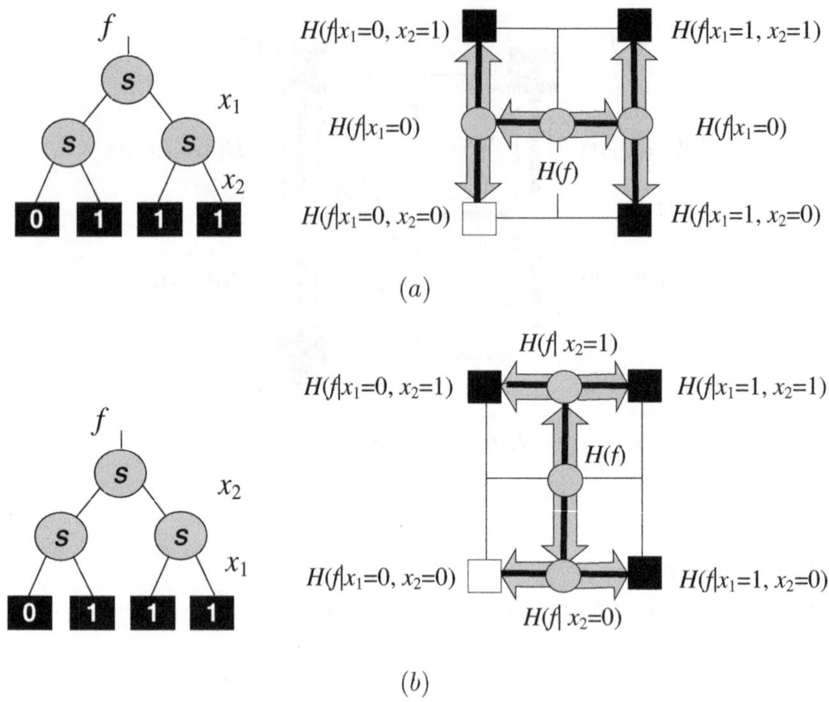

FIGURE 24.8
Order of variables in a decision tree and rotation of an $\mathcal{N}$-hypercube (Example 24.9).

24.4 Information-theoretical measures in multivalued functions

Information-theoretical measures can be applied to multivalued logic functions. The focus is S, pD, and nD expansions and decision tree design in terms of entropy and information.

Information-theoretical measures can be applied to m-valued functions. For calculation, the logarithm base m is applied, e.g., $\log_3$ for ternary function, $\log_4$ for quaternary function, etc. The example below demonstrates the technique for computing information-theoretical characteristics for the function given by a truth table.

TABLE 24.2
Information measures of elementary switching functions of two variables.

Function	Information estimations
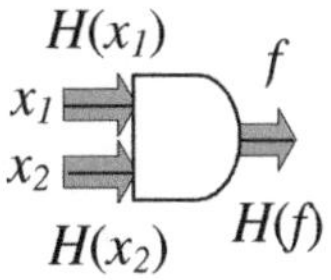 $H(x_1)$, x_1, x_2, $H(x_2)$, f, $H(f)$	$f = x_1 x_2$ $H(f) = -0.25 \cdot \log_2 0.25 - 0.75 \cdot \log_2 0.75$ $= 0.8113$ *bit*
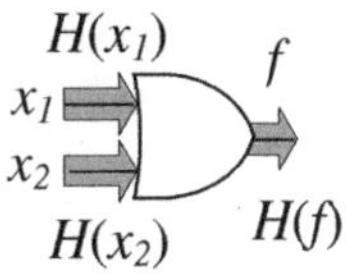 $H(x_1)$, x_1, x_2, $H(x_2)$, f, $H(f)$	$f = x_1 \vee x_2$ $H(f) = -0.75 \cdot \log_2 0.75 - 0.25 \cdot \log_2 0.25$ $= 0.8113$ *bit*
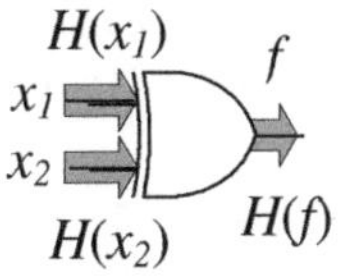 $H(x_1)$, x_1, x_2, $H(x_2)$, f, $H(f)$	$f = x_1 \oplus x_2$ $H(f) = -0.5 \cdot \log_2 0.5 - 0.5 \cdot \log_2 0.5$ $= 1$ *bit*
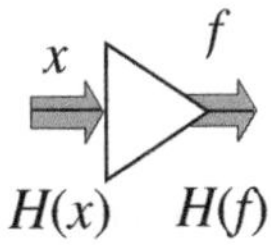 x, $H(x)$, f, $H(f)$	$f = \overline{x}$ $H(f) = -0.5 \cdot \log_2 0.5 - 0.5 \cdot \log_2 0.25$ $= 1$ *bit*

Example 24.11 *Computing entropy, conditional entropy, and mutual information for a 4-valued function f given its truth column vector* $[0000\ 0231\ 0213\ 0321]^T$ *are shown in Figure 24.10.*

24.4.1 Information notation of S expansion

The information measures of Shannon expansion in $GF(4)$

$$f = J_0(x) + xJ_1(x) + x^2 J_2(x) + x^3 J_3(x)$$

are given in Table 24.4 (first row), where $J_i(x)$, $i = 0, \ldots, k-1$, are the characteristic functions, denoted by $J_i(x) = 1$, if $x = i$ and $J_i(x) = 0$, otherwise. The average Shannon entropy is equal to conditional entropy $H(f|x)$ of

TABLE 24.3
Information measures of elementary switching functions of two variables.

Tree and $\mathcal{N}$-hypercube		Information estimates
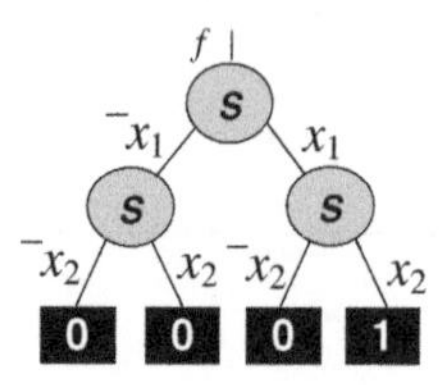	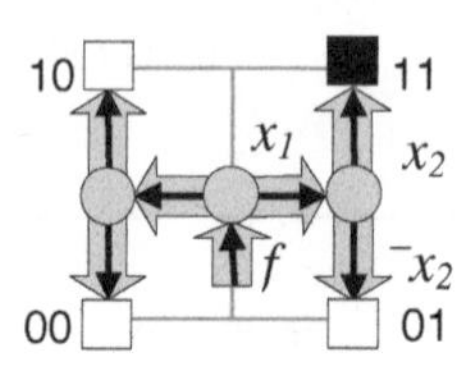	$f = x_1x_2$ $H(f) = -\ p_{\|f=0} \cdot \log_2 p_{\|f=0}$ $-\ p_{\|f=1} \cdot \log_2 p_{\|f=1}$ $p_f = 0.5 \cdot 0.5$ $= 0.25$ $H(f) = -0.25 \cdot \log_2 0.25$ $-0.75 \cdot \log_2 0.75$ $= 0.8113\ bit$
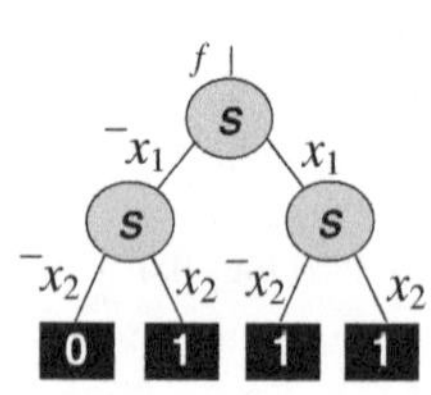	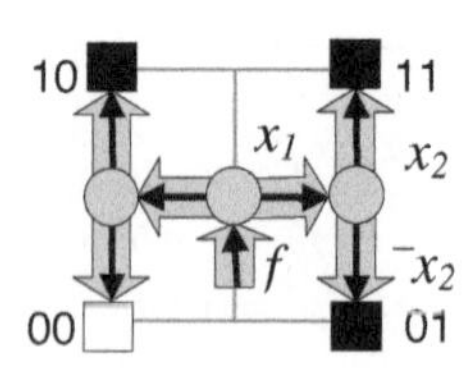	$f = x_1 \vee x_2$ $H(f) = -p_{\|f=0} \cdot \log_2 p_{\|f=0}$ $-p_{\|f=1} \cdot \log_2 p_{\|f=1}$ $p_f = 1 - (1-0.5) \cdot (1-0.5)$ $= 0.75$ $H(f) = -0.75 \cdot \log_2 0.75$ $-0.25 \cdot \log_2 0.25$ $= 0.8113\ bit$
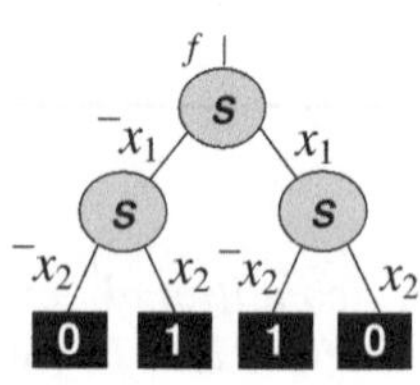	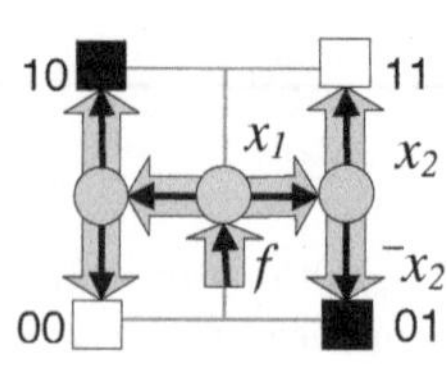	$f = x_1 \oplus x_2$ $H(f) = -p_{\|f=0} \cdot \log_2 p_{\|f=0}$ $-p_{\|f=1} \cdot \log_2 p_{\|f=1}$ $p_f = 0.5 \cdot 0.5 + 0.5 \cdot 0.5$ $= 0.5$ $H(f) = -0.5 \cdot \log_2 0.5$ $-0.5 \cdot \log_2 0.5$ $= 1\ bit$
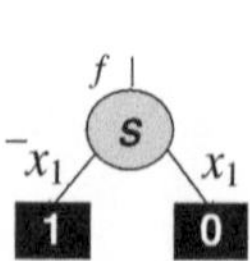	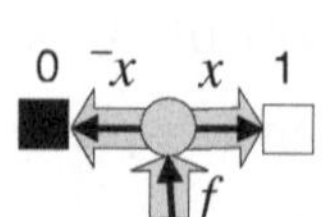	$f = \overline{x}$ $H(f) = p_{\|f=0} \cdot \log_2 p_{\|f=0}$ $-p_{\|f=1} \cdot \log_2 p_{\|f=1}$ $p_f = 1 - 0.5$ $= 0.5$ $H(f) = -0.5 \cdot \log_2 0.5$ $-0.5 \cdot \log_2 0.5$ $= 1\ bit$

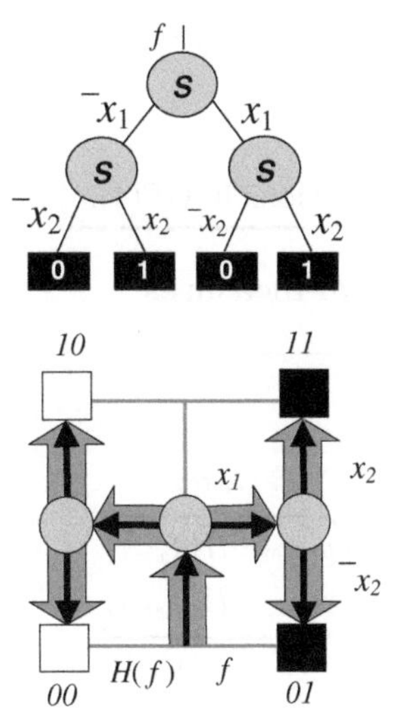

Input entropy

$$
\begin{aligned}
H_{in} &= -\,2(1-p) \times \log_2(1-p) \\
&\quad -\,2p \times \log_2 p \\
&= -\,2(1-0.707) \times \log_2(1-0.707) \\
&\quad -\,2 \times 0.707 \times \log_2 0.707 \\
&= 1.745 \; bit
\end{aligned}
$$

Output entropy

$$
\begin{aligned}
H_{out} &= -\,0.707^2 \times \log_2 0.707^2 \\
&\quad -\,(1-0.707)^2 \times \log_2(1-0.707)^2 \\
&= 0.804 \; bit
\end{aligned}
$$

Loss of information

$$
\begin{aligned}
H_{loss} &= H_{out} - H_{in} \\
&= 1.745 - 0.804 = 0.941 \; bit
\end{aligned}
$$

FIGURE 24.9
Information measures on an $\mathcal{N}$-hypercube (Example 24.10).

function f with respect to x:

$$H^S(f|x) = H(f|x). \tag{24.9}$$

This is because the terminal nodes of Shannon expansion represent the values of the function. The information notation of Shannon expansion is specified as follows:

$$I^S(f;x) = I(f;x). \tag{24.10}$$

Since $H^S(f|x) = H(f|x)$ and $I^S(f;x) = I(f;x)$, then for S expansion, we can write $I(f;x) = H(f) - H(f|x)$.

24.4.2 Information notations of *pD* and *nD* expansion

Consider $m = 4$. Then the positive and negative Davio expansions of a 4-valued logic function are defined as:
Positive Davio expansion pD

$$
\begin{aligned}
f &= f_{x=0} + x(f_{x=1} + 3f_{x=2} + 2f_{x=3}) \\
&\quad + x^2(f_{x=1} + 2f_{x=2} + 3f_{x=3}) \\
&\quad + x^3(f_{x=0} + f_{x=1} + f_{x=2} + f_{x=3}) \\
&= f_{x=0} + xf_1 + x^2 f_2 + x^3 f_3
\end{aligned}
$$

TABLE 24.4
Information measures of Shannon and Davio expansions in GF(4).

Type	Information theoretical measures
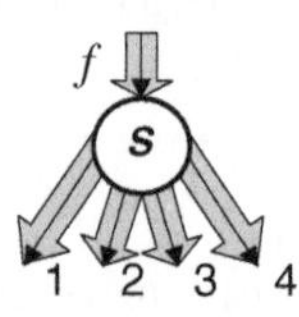	$H^S(f\|x) = \overbrace{p_{\|x=0} \cdot H(f_{x=0})}^{Leaf\ 1} + \overbrace{p_{\|x=1} \cdot H(f_{x=1})}^{Leaf\ 2} + \underbrace{p_{\|x=2} \cdot H(f_{x=2})}_{Leaf\ 3} + \underbrace{p_{\|x=3} \cdot H(f_{x=3})}_{Leaf\ 4}$
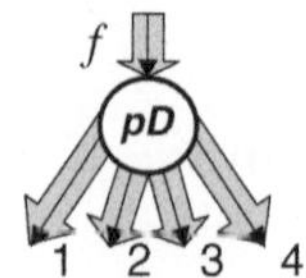	$H^{pD}(f\|x) = p_{\|x\neq 0} \cdot \frac{H(f_1)+H(f_2)+H(f_3)}{3} + p_{\|x=0} \cdot H(f_{x=0})$
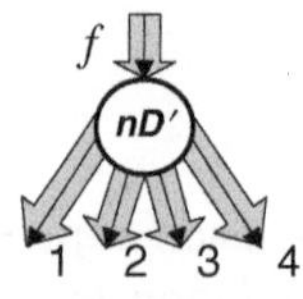	$H^{nD'}(f\|x) = p_{\|x\neq 1} \cdot \frac{H(f_0)+H(f_2)+H(f_3)}{3} + p_{\|x=1} \cdot H(f_{x=1})$
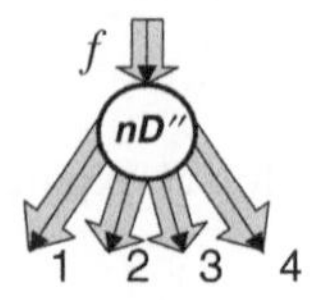	$H^{nD''}(f\|x) = p_{\|x\neq 2} \cdot \frac{H(f_0)+H(f_1)+H(f_3)}{3} + p_{\|x=2} \cdot H(f_{x=2})$
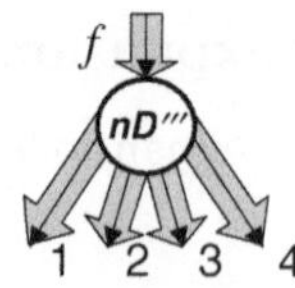	$H^{nD'''}(f\|x) = p_{\|x\neq 3} \cdot \frac{H(f_0)+H(f_1)+H(f_2)}{3} + p_{\|x=3} \cdot H(f_{x=3})$

x_1	x_2	f
0	0	0
0	1	0
0	2	0
0	3	0
1	0	0
1	1	3
1	2	1
1	3	0
2	0	2
2	1	0
2	2	1
2	3	3
3	0	0
3	1	3
3	2	2
3	3	1

The probabilities of the logic function values are

$$p_{|f=0} = {}^7/_{16}, \quad p_{|f=1} = p_{|f=2} = p_{|f=3} = {}^3/_{16}$$

The entropy of the logic function f is

$$\begin{aligned} H(f) &= -{}^7/_{16} \cdot \log_2 {}^7/_{16} - 3 \cdot {}^3/_{16} \cdot \log_2 {}^3/_{16} \\ &= 1.88 \ bit \end{aligned}$$

The conditional entropy of the logic function f with respect to the variable x_1 is

$$\begin{aligned} H(f|x_1) &= -{}^4/_{16} \cdot \log_2 1 - 12 \cdot {}^1/_{16} \cdot \log_2 {}^1/_4 \\ &= 1.5 \ bit. \end{aligned}$$

The conditional entropy with respect to variable x_2 is

$$H(f|x_2) = 1.25 \ bit$$

The mutual information for the logic function f and the variables x_1 and x_2 is

$$\begin{aligned} I(f; x_1) &= 0.38 \ bit \\ I(f; x_2) &= 0.63 \ bit \end{aligned}$$

FIGURE 24.10
Information-theoretical measures of a 4-valued function (Example 24.11).

Negative Davio expansion nD'

$$f = f_{x=1} + \widehat{x} f_0 + \widehat{x}^2 f_2 + \widehat{x}^3 f_3$$

Negative Davio expansion nD''

$$f = f_{x=2} + \widehat{\widehat{x}} f_0 + \widehat{\widehat{x}}^2 f_1 + \widehat{\widehat{x}}^3 f_3$$

Negative Davio expansion nD'''

$$f = f_{x=3} + \widehat{\widehat{\widehat{x}}} f_0 + \widehat{\widehat{\widehat{x}}}^2 f_1 + \widehat{\widehat{\widehat{x}}}^3 f_2$$

The entropy associated with the above expansions is shown in Table 24.4.

24.4.3 Information criterion for decision tree design

The main properties of the information measure are

- The recursive character of S, pD, and nD expansions and their generalization for the 4-valued case, and
- The possibility of choosing a decomposition variable and expansion type based on the information measure.

Decision tree design can be interpreted as an optimized (with respect to information criterion) node selection process. The criterion for choosing the decomposition variable x and expansion type $\omega \in \{S, pD, nD\}$ is that the conditional entropy of the logic function given a variable x_i has to be minimal

$$H^{\omega}(f|x) = MIN(H^{\omega_j}(f|x_i) \mid \forall \; pairs \; (x_i, \omega_j)) \tag{24.11}$$

In the algorithm, the ordering restriction is relaxed. This means that (i) each variable appears once in each path and (ii) the order of variables along with each path may be different.

Example 24.12 *Consider the design of a decision tree for the logic function f from Example 24.11.*

Step 1. Choose variable x_2 and 4–pD expansion for root node, because the minimal entropy is $H^{pD}(f|x_2) = 0.75$ bit. Functions $f_0 = f_{x_2=0}$ and $f_3 = f_{x_2=0} + f_{x_2=1} + f_{x_2=2} + f_{x_2=3}$ both take logic value 0. Select the function

$$f_1 = f_{x_2=1} + 3f_{x_2=2} + 2f_{x_2=3}.$$

Step 2. Choose variable x_1 and pD expansion for the next node. The successors are constant: $f_0 = 0$, $f_1 = 0$, $f_2 = 3$, and $f_3 = 1$. Select the function

$$f_2 = f_{x_2=1} + 2f_{x_2=2} + 3f_{x_2=3}.$$

Step 3. Select $H^{nD'}$ expansion for variable x_1. The successors are constant: $f_0 = 0$, $f_1 = 1$, $f_2 = 0$, and $f_3 = 1$.

The decision tree obtained is shown in Figure 24.11(a). The corresponding Reed–Muller expression is

$$f = 3 \cdot x_2 \cdot x_1^2 + x_2 \cdot x_1^3 + x_2^2 \cdot {}^{1-}x_1 + x_2^2 \cdot {}^{3-}x_1.$$

By analogy, the logic expression corresponding to the decision tree shown in Figure 24.11(b), is

$$f = 2 \cdot x_2 \cdot J_1(x_1) + 3 \cdot x_2 \cdot J_2(x_1) + 2 \cdot x_2^2 \cdot J_2(x_1) + 3 \cdot x_2^2 \cdot J_3(x_1).$$

24.4.4 Remarks on information-theoretical measures in decision diagrams

Given a completely specified switching function

$$f: \; p(x = 0) = p(x = 1) = \frac{1}{2}.$$

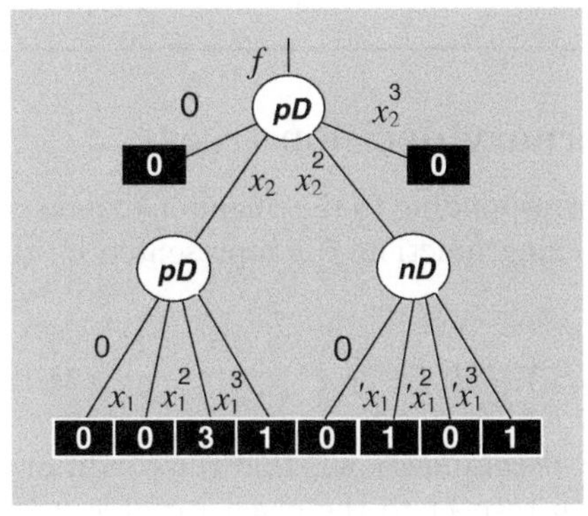

(a)

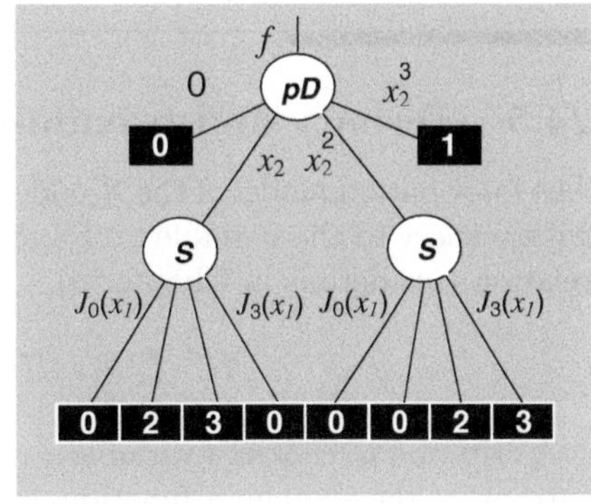

(b)

FIGURE 24.11
Decision tree design (Example 24.12).

Since each node of the BDD is an instance of the Shannon expansion, a probability assignment algorithm in a top-down fashion works as follows:

$$p(f) = \frac{1}{2}p(f_{x=0}) + \frac{1}{2}p(f_{x=1}).$$

Another approach is the following. Consider $p(leaf|f = 1) = 1$ and $p(leaf|f = 0) = 0$, and output probability $p(f = 1) = p(root)$ and $p(f = 0) = 1 - p(root)$. Application of the recursive strategy:

$$p(node) = p(x = 0)p(edgel) + p(x = 1)p(edger)$$

provides calculation of conditional and joint probabilities for computing conditional entropy. To find the joint probability $p(f = 1, x = 1)$, it is necessary to set $p(x = 1) = 1$ and $p(x = 0) = 0$ before BDD traversal. This allows the calculation of the whole range of probabilities using only one BDD traversal.

Example 24.13 *Given the switching function* $f = x_3 \vee x_2 \vee x_1$ *by its truth vector* $\mathbf{F} = [10001111]$*:*

$$\begin{aligned} H(f) &= -5/8 \cdot log_2(5/8) - 3/8 \cdot log_2(3/8) = 0.96 \; bit, \\ H(f|x_1) &= -1/8 \cdot log_2(1/4) - 3/8 \cdot log_2(3/4) - 4/8 \cdot log_2(4/4) - 0 \\ &= 0.41 \; bit. \end{aligned}$$

Similarly, $H(f|x_2) = 0.91$ *bit,* $H(f|x_3) = 0.91$ *bit. A top-down approach of assigning* $p(leaf|f = 1) = 1$ *yields* $p(f = 1) = p(root) = 0.625$ *(BDD with three nodes). The result of setting* $p(x_2 = 0) = 1$ *is a conditional probability* $p(f_{x_2=0}) = 0.75$.

24.5 Ternary and pseudo-ternary decision trees

The information model of the S-node corresponding to the Shannon expansion with respect to the variable x of a switching function f is represented by the relation (see details in Chapter 8)

$$H^S(f|x) = H(f_{x=0}) + H(f_{x=1}). \tag{24.12}$$

A criterion to choose a variable x in S-decomposition is that the conditional entropy of the switching function f with respect to the given variable x has to be minimum.

Ternary decision trees are a generalization of the binary decision trees derived by permitting nodes with three outgoing edges (see details in Chapter 23). In ternary decision trees derived as a generalization of the binary decision trees, it is assumed that

- The first two outgoing edges point to the cofactors of f determined as in the Shannon decomposition for f, and
- The third outgoing edge points to a subfunction $f_0 \# f_1$, where # denotes a binary operation.

By choosing different operations, different ternary decision trees are defined. Examples are AND-, OR-, EXOR-ternary decision trees, and Kleene-ternary decision trees, defined by using the logic operations AND, OR, EXOR, and the Kleene operation, respectively. In these ternary decision trees, the correspondence between a ternary decision tree and a switching function f is defined as follows:

(*a*) For a terminal node v, $v = \begin{cases} f_v = 1, & \text{if v=1;} \\ f_v = 0, & \text{if v=0.} \end{cases}$

(*b*) For a nonterminal node

$$f_v = \overline{x} \cdot f_0 \vee x \cdot f_1 \vee 1 \cdot f_2,$$

where subfunctions f_0, f_1, and f_2 are computed as follows:

$$\begin{aligned} f_0 &= f_{x=0} \\ f_1 &= f_{x=1} \\ f_2 &= f_0 \# f_1 \end{aligned}$$

the symbol # denotes any of the operations AND, OR, EXOR, or the Kleene operator.

Example 24.14 *The EXOR ternary decision tree based on an entropy based technique for the switching function f given by truth vector $\mathbf{F} = [10111110]^T$ is constructed as shown in Figure 24.12. Construction is started with variable x_3, because it follows from Example 24.3 that the order of decomposed variables is $\{x_3, x_2, x_1\}$.*

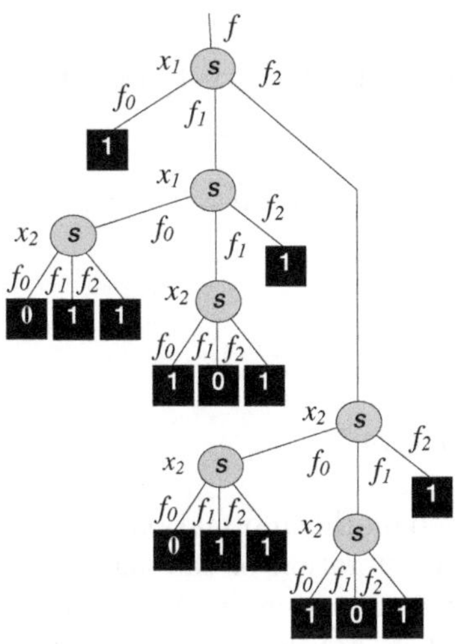

The first edge corresponds to $x_3 = 0$,
$f_0 = \{000, 010, 100, 110\}$.
The second edge corresponds to $x_3 = 1$,
$f_1 = \{001, 011, 101, 111\}$.
The third edge is formed as
$f_2 = f_0 \oplus f_1$.

The result of building the decision tree yields the following AND/EXOR form:

$$f = \overline{x}_3 \oplus x_1x_3 \oplus x_2x_3$$

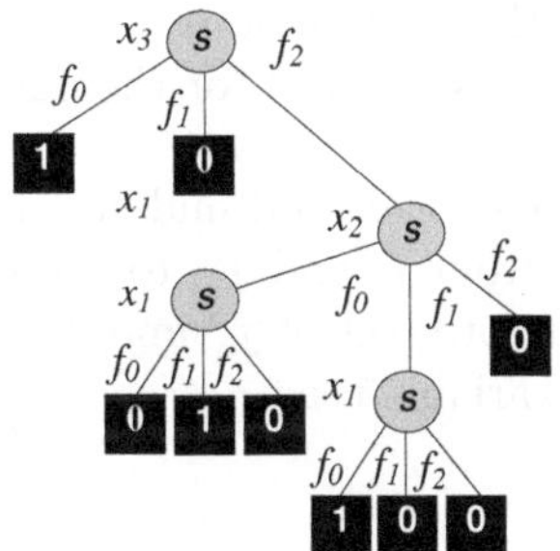

FIGURE 24.12
Example of entropy based decision tree design for a free ternary decision tree with final result $f = \overline{x}_3 \oplus x_1x_3 \oplus x_2x_3$ (a) and free pseudo ternary decision tree with resulting function $f = \overline{x}_3 \vee \overline{x}_1x_2 \vee x_1\overline{x}_2$ (b) given a switching function of 3-variables (Example 24.14).

A pseudo ternary decision tree is a decision tree with the third outgoing edge of the node pointing to the subfunction in the form of *Morreale's* operator defined in Table 24.5 [7]. In this table,

- ▶ The left part shows how to obtain the subfunctions for the first f'_0 and the second edge f'_1.
- ▶ The central part shows how to remove the covered assignments to obtain f''_0 and f''_1.
- ▶ The right part shows how to obtain the subfunction of the third edge f_d for a node.

The correspondence between an AND/OR pseudo ternary decision tree and a switching function is defined as follows:

(*a*) For a terminal node v, $f_v = 1$ if $v = 1$ and $f_v = 0$ if $v = 0$,

(*b*) For a nonterminal node

$$f_v = \overline{x} \cdot f_0 \vee x \cdot f_1 \vee f_d,$$

where f_d is computed as shown in Table 24.5.

Consider a technique for computation based on the information estimations of Morreale's algorithm. The results of each iteration of the algorithm are denoted by f_0, f_1, f''_0, f''_1. In each node, we choose one variable for the S-expansion, then we determine the cubes that do not depend on this variable. In the next step, we eliminate these cubes from the subfunctions f_0 and f_1. The results of these operations give f'_0 for the left edge and f'_1 for the middle edge.

Let g_0 and g_1 denote covers of subfunctions for the left and middle edges, respectively. Then we remove the covered cubes from f_0 and f_1, to produce f''_0 and f''_1. The subfunction f_d is obtained as the product of f''_0 and f''_1 and consists of the cubes that do not depend on the variable used.

TABLE 24.5
Morreale's operators.

f'_0, f'_1:	$f_1, f_0 \backslash f_0, f_1$	0	1	d	f''_0, f''_1:	$g_i \backslash f_i$	0	1	d	f_d:	$f''_1 \backslash f''_0$	0	1	d
	0	0	1	d		0	0	1	d		0	0	0	0
	1	0	d	d		1	-	d	d		1	0	1	1
	d	0	d	d							d	0	1	d

Example 24.15 *In Figure 24.13, the minimization of a switching function by using the entropy based on Morreale's algorithm.*

$x_1x_2x_3$	f
000	1
001	0
010	1
011	1
100	1
101	1
110	1
111	0

Step 1 *Choose a variable with the smallest conditional entropy* $\min\{H(f|x)\}$. *For the first node,*
$H(f|x_1) = 0,8113$ bit/pattern
$H(f|x_2) = 0,8113$ bit/pattern
$H(f|x_3) = 0,5$ bit/pattern.
The variable x_3 is selected.

Step 2 *Create two outgoing edges: the first, f_0, corresponds to all the assignments of the values to variables for which $x_3 = 0$, the second, f_1, corresponds to the assignments where $x_3 = 1$:*
$f_0 = \{000, 010, 100, 110\}$,
$f_1 = \{001, 011, 101, 111\}$.

Step 3 *Create the third edge f_d. Change the values of functions f_0 and f_1, into f_0' and f_1', respectively, as it is defined in Table 24.5a. The results of this step are two subfunctions without cubes which do not depend on the variable selected in the first step.*

Step 4 *Steps 1, 2, and 3 are repeated for the next nodes, until a leaf node is achieved. If the leaf node takes the value 1, change the values of the functions f_0' or f_1' into f_0'' or f_1'' respectively, by deleting the covered assignments as shown in Table 24.5a. In our case, the second level of the DT contains two leaf nodes – the first one with the value 1 (g_0) and the next one with the value 0 (g_1). The first terminal node corresponds to $\overline{x}_3$, so it covers all the assignments in $f_0 = \{000, 010, 100, 110\}$.*

Step 5 *For each node whose outgoing edges are ended by the leaf nodes, create the edge f_d. It includes all the assignments not yet covered by the assignments for edges f_0 and f_1. In our case, these are the assignments covered by the variable $\overline{x}_3$: i.e., {001,011,101,111} (the variable that corresponds to the third edge is omitted, as shown in Table 24.5c.*

FIGURE 24.13
Minimization of a switching function by using the entropy based on Morreale's algorithm (Example 24.15).

24.6 Entropy-based minimization using pseudo-ternary trees

In this chapter, an algorithm for sum-of-products minimization of switching functions based on information estimates in pseudo-ternary decision trees is introduced.

Example 24.16 *Let a switching function of three variables be given by the truth vector shown in Figure 24.13. Figure 24.14 illustrates the calculation of conditional entropies in Morreale's minimization algorithm. The optimal path corresponds to the following calculated entropies:*

$$\{H(f|x_3), H(f|x_2), H(f|x_1)\}$$

Example 24.17 *Figure 24.15 illustrates the entropy based method for a free pseudo-ternary decision tree for finding a minimal SOP expression using the free pseudo-ternary tree. The resulting minimal SOP is*

$$f = \overline{x}_3 \vee \overline{x}_1 x_2 \vee x_1 \overline{x}_2$$

(a) Initial function

$x_1x_2x_3$	f	$x_1x_2x_3$	f
000	1	100	1
001	0	101	1
010	1	110	1
011	1	111	0

(a) Calculate the conditional entropy

$$H^S(f|x_i)$$

by using (24.12) for each variable x_i, $i = 1, 2, 3$, to chose a variable for the optimal path. It follows from Example 24.3 that the variable x_3 is chosen first.

(b) 1st node, 1st and 2nd edges

x_1	x_2	f_0	f_1	f_0'	f_1'
0	0	1	0	1	0
0	1	1	1	d	d
1	0	1	1	d	d
1	1	1	0	1	0

(b) Form the subfunctions f_0 and f_1 for the Shannon decomposition of f, i.e.,

$$f_0 = f_{x_3=0}$$
$$f_1 = f_{x_3=1}$$

Then compute subfunctions f_0' and f_1' with respect to the rules given in Table 24.5a.

(c) 1st node

x_1	x_2	f_0	g_0	f_0''	f_1	g_1	f_1''	f_d
0	0	1	1	d	0	0	0	0
0	1	1	1	d	1	0	1	1
1	0	1	1	d	1	0	1	1
1	1	1	1	d	0	0	0	0

(c) Perform the algorithm for subtrees with the roots $f = f_0$ and $f = f_1$. Let g_0 be the cover of f_0 and g_1 be the cover of f_1. To specify subfunctions f_0'' and f_1'' we have to use the rules from Table 24.5b. At the moment we have enough information to determine the third edge, f_d, of this decision tree node (Table 24.5c).

(d) The 2nd node

x_2	f_0	f_0'	g_0	f_0''	f_1	f_1'	g_1	f_1''	f_d
0	0	0	0	0	1	1	1	d	0
1	1	1	1	d	0	0	0	0	0

(d) Perform the algorithm for $f = f_d$. The conditional entropies are

$$H^S(f|x_1) = H^S(f|x_2) = 1$$

so we have to select any of these variables to add to the optimal path and create a node. Let us choose x_1. Then we obtain the following assignments to the edge (d), so that f_{d0} (e) and f_{d1} (f).

(e) The 3rd node

f_0	f_0'	g_0	f_0''	f_1	f_1'	g_1	f_1''	f_d
0	0	0	0	1	1	1	d	0

(f) The 4th node

f_0	f_0'	g_0	f_0''	f_1	f_1'	g_1	f_1''	f_d
1	1	1	d	0	0	0	0	0

FIGURE 24.14
Entropy based minimization (Example 24.16).

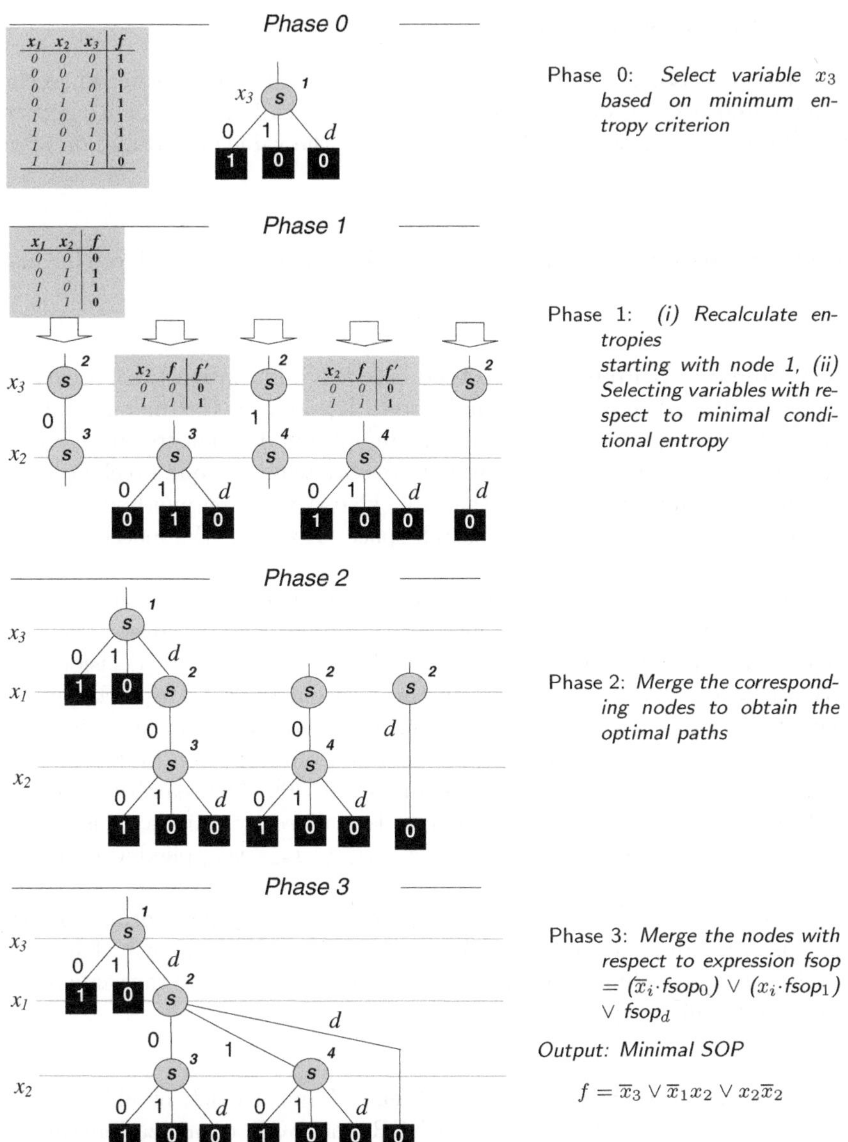

FIGURE 24.15
Entropy based, improved Morreale algorithm for minimization of completely specified switching function of three variables on free pseudo-ternary decision tree (Example 24.17).

24.7 Further reading

Historical remark. The decomposition of switching functions known as classical Shannon expansion was introduced in 1938*. In 1948, Shannon suggested information entropy as a measure to represent the value of information in numerical values.†

A technique of information-theoretical measures is introduced in Chapter 8.

Designing an optimal decision tree based on information-theoretical measures was studied by Goodman and Smyth [2], and by Hartmann et al. [3]. There are some approaches to designing an optimal decision tree from a given decision table [2]. Among them, we note the branch-and-bound technique and dynamic programming methods that are exhaustive search methods. Decomposition allows us to reduce the search space. However, the methods do not guarantee the optimal solution.

Hartmann et al. [3] applied the information theoretic criterion to convert decision tables with don't cares into near-optimal decision trees. The truth table of a logic function is considered as a special case of a decision table with variables replacing the tests in the decision table. The technique for convert the decision table into a decision tree and its optimization has been applied for switching and multivalued function minimization by Kabakcioglu et al. [4], Lloris et al. [5] and Lloris-Ruiz et al. [6]. The search for the best implicants to appear in the final solution is realized by using a decision tree, so that Shannon expansion formula $\overline{x} \cdot f_{x=0} \oplus x \cdot f_{x=1}$ is applied at each node in the decision tree.

Cheushev et al. [1] gave a justification for the use of the information measure for switching and multivalued logic functions. It was shown that in the simplest understanding, the information theory methods give the same results as the probabilistic approach [9].

The design of ternary decision trees. Yanushkevich et al. [11] and Shmerko et al. [67] studied AND/EXOR and SOP minimization through ternary decision trees [8]. The nodes of such a decision tree have three outputs and the third edge assigns don't care values with respect to Morreale's operator [7] for generating the prime-irredundant cubes.

---

* *Trans. AIEE, Vol. 27, 1938, 713-723*

† *Bell Syst. Tech. J., Vol.27, 1948, 379-423, 623-656.*

References

[1] Cheushev V, Shmerko V, Simovici D, and Yanushkevich S. Functional entropy and decision trees. In *Proceedings of the IEEE 28th International Symposium on Multiple-Valued Logic*, pp. 357–362, 1998.

[2] Goodman RM and Smyth P. Decision tree design from a communication theory standpoint. *IEEE Transactions on Information Theory*, 34(5):979–994, 1988.

[3] Hartmann CRP, Varshney PK, Mehrotra KG, and Gerberich CL. Application of information theory to the construction of efficient decision trees. *IEEE Transactions on Information Theory*, 28(5):565–577, 1982.

[4] Kabakcioglu AM, Varshney PK, and Hartman CRP. Application of information theory to switching function minimization. *IEE Proceedings, Pt E*, 137:389–393, 1990.

[5] Lloris L, Gomez JF, and Roman R. Using decision trees for the minimization of multiple-valued functions. *International Journal on Electronics*, 75(6):1035–1041, 1993.

[6] Lloris-Ruiz A, Gomez-Lopera JF, and Roman-Roldan R. Entropic minimization of multiple-valued functions. In *Proceedings of the IEEE 23rd International Symposium on Multiple-Valued Logic*, pp. 24–28, 1993.

[7] Morreale E. Recursive operators for prime implicant and irredundant normal form determination. *IEEE Transactions on Computer*, 19(6):504–509, 1970.

[8] Sasao T. Ternary decision diagrams and their applications. Chapter 12, pp. 269–292. In Sasao T and Fujita M, Eds., *Representations of Discrete Functions*. Kluwer, Dordrecht. 1995.

[9] Shen A and Devadas S. Probabilistic manipulation of Boolean functions using free Boolean diagrams. *IEEE Transactions on Computer-Aided Design of Integrated Circuits and Systems*, 14(1):87-94, 1995.

[10] Shmerko VP, Popel DV, Stanković RS, Cheushev VA, and Yanushkevich SN. Information theoretical approach to minimization of AND/EXOR expressions of switching functions. In *Proceedings of the IEEE International Conference on Telecommunications*, pp. 444–451, Yugoslavia, 1999.

[11] Yanushkevich SN, Shmerko VP, Dziurzanski P, Stanković RS, and Popel DV. Experimental verification of the entropy based method for minimization of switching functions on pseudo-ternary decision trees. In *Proceedings of the IEEE International Conference on Telecommunications in Modern Satellite, Cable and Broadcasting Services*, pp. 452–459, Yugoslavia, 1999.

25

Decomposition Using Decision Diagrams

The number of inputs of gates in a combinational circuit is restricted for technical reasons. Therefore, only simple switching functions can be implemented by a two level circuit and all larger functions require a multilevel circuit. In this chapter*, we will review some concepts of the design of multilevel circuits using decomposition methods based on binary decision diagrams (BDDs).

25.1 Introduction

Generally, there are two basic strategies to synthesize a circuit, i.e., to find a circuit structure given a switching function:

- The *covering strategy* is to find a cover for the function by selected cubes. The background for this strategy is sum-of-products (SOP) and product-of-sums representations of switching functions (see details in Chapters 2 and 18). In this approach, the given function is covered by a minimal number of prime implicants.
- The second strategy is called *decomposition.* Decomposition methods are more difficult than covering methods. Depending on the required computation power, the practical application starts in the last two decades, much later than covering methods.

BDD-based decomposition methods can be classified into:

- Methods which detect decomposition properties of a function using BDD operations, and
- Methods which utilize BDD data structure for decomposition.

*This chapter was co-authored with Dr. B. Steinbach, Technical University of Freiberg, Germany.

25.1.1 Covering strategy

Assume that the variables of a function are available in both polarities and there is no restriction on the number of gate inputs. In this case, a two-level circuit can be built directly from the minimized normal form.

Example 25.1 *A switching function*

$$f = x_1x_2x_3 \vee \overline{x}_1\overline{x}_2x_3 \vee \overline{x}_1\overline{x}_2x_4 \vee x_1x_2x_4 \vee \overline{x}_3\overline{x}_4x_5x_6$$

is a minimal sum-of-product form, since it includes only prime implicants (products). This means that it is impossible to remove a variable from the product without the function's change.

Note that these products cover the function so that the sum-of-product form is minimal. The two-level circuit structure is shown in Figure 25.1. There is a direct mapping of Example 25.1 into an AND-OR circuit of Figure 25.1a or a NAND-NAND circuit of Figure 25.1b, respectively. It can be observed from the two level circuit structure of Example 25.1 that:

- ► There is a direct mapping between the minimized SOP and the two-level circuits.
- ► The number of gate inputs in two-level representation is not restricted to two inputs, in general.

In practice, the number of gate inputs must be limited. For the rest of this chapter we assume that each gate must have no more than two inputs. To meet this requirement, a decomposition using a tree-like multilevel circuit may be utilized.

Example 25.2 *A simple way to transform a two-level circuit into a multilevel circuit is to replace a gate having more than two inputs with a decomposition tree of two-input gates of the same type (Figure 25.2).*

Once this simple decomposition is applied to each gate of a two-level circuit, a multilevel circuit is created. The internal structure of such a circuit is a tree. Only the input variables are reused. This circuit can be simplified if several gates of the same type are controlled by the same inputs. In this case such a gate can be reused for the associated subfunctions. Obviously, this optimization approach depends on the distribution of the variables to the gates while creating the basic decomposition trees.

Example 25.3 *Figure 25.3 shows the optimized multilevel circuit obtained by decomposition of the function given in Example 25.1.*

A multilevel circuit is not necessary a result of minimization of a two-level circuit. The reason is that the criteria of the primary minimization does not necessarily meet the requirements of a minimal circuit. Thus, an alternative strategy for design of multilevel circuits should be developed.

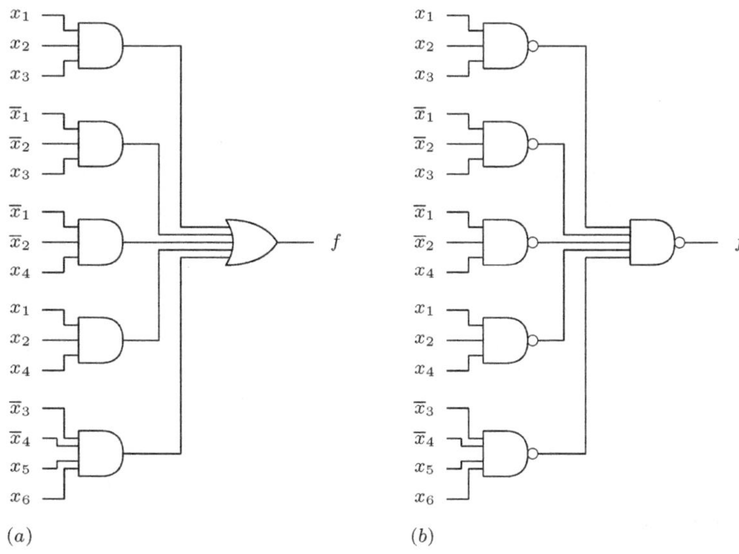

FIGURE 25.1
Two-level circuit: AND-OR (a) and NAND-NAND (b) (Example 25.1).

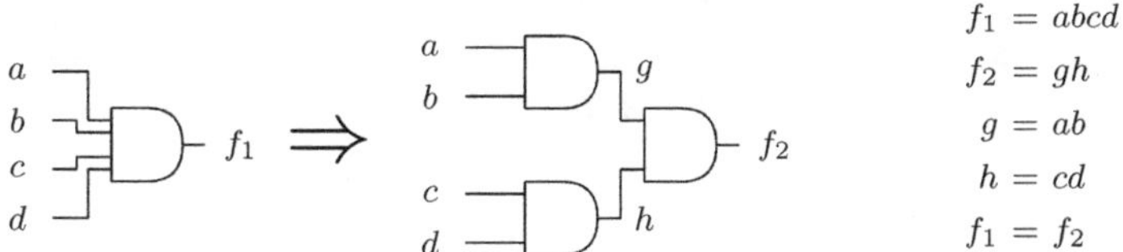

FIGURE 25.2
Decomposition tree: a four-input AND gate is transformed into a tree of three two-input AND gates (Example 25.2).

25.1.2 Decomposition strategy

A complementary synthesis strategy is the *functional decomposition.* The main idea of this decomposition is to split a given switching function into two or more simpler functions which can be combined in a simple circuit. This strategy can be applied recursively and, thus, will result in a multilevel circuit.

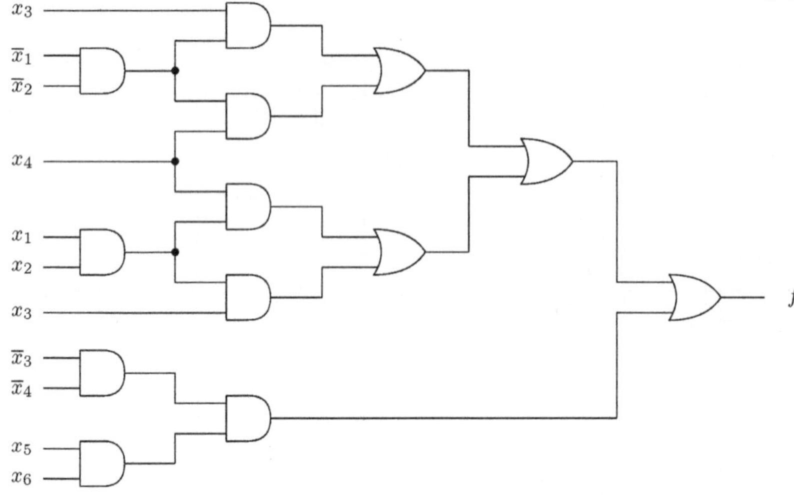

FIGURE 25.3
Multilevel circuit created from a minimized two-level circuit (Example 25.3).

Example 25.4 *Figure 25.4 shows a possible circuit structure which implements the switching function given Example 25.1. The comparison of the Figures 25.3 and 25.4 implies that:*

- *The number of gates correlates with the chip area and is reduced from 13 gates to 6 gates, and*
- *This circuit is faster than in Figure 25.3, since the decomposition reduces the path length form 5 to 3.*

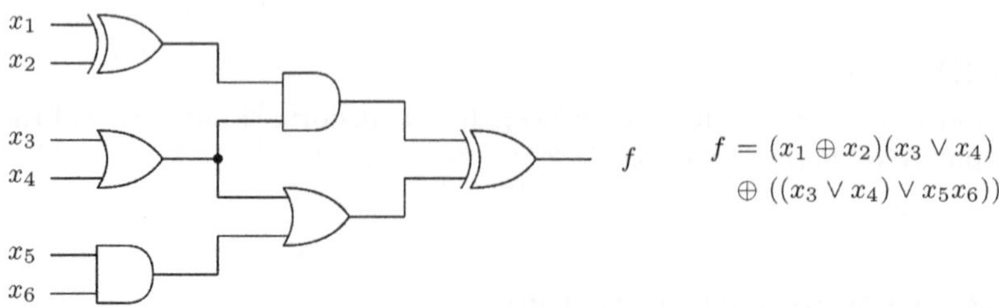

FIGURE 25.4
A multilevel circuit synthesized by the functional decomposition (Example 25.4).

Functional decomposition is much more powerful for the design of multilevel circuit than covering. Admittedly, it is more difficult than covering methods.

25.2 Decomposition types

The most general functional decomposition of a switching function $f(\underline{x})$ is expressed as follows:

$$f(\underline{x}) = f_c\ \overbrace{(f_{d1}(\underline{x}), \ldots, f_{dk}(\underline{x})),}^{Decomposition\ functions} \quad (25.1)$$

where $f_{d1}(\underline{x})$, ...,f_{dk} are k independent decomposition functions $f_{d1}(\underline{x})$,..., $f_{dk}(\underline{x})$ and $f_c(\underline{f_d(\underline{x})})$ is a single composition function (Figure 25.5).

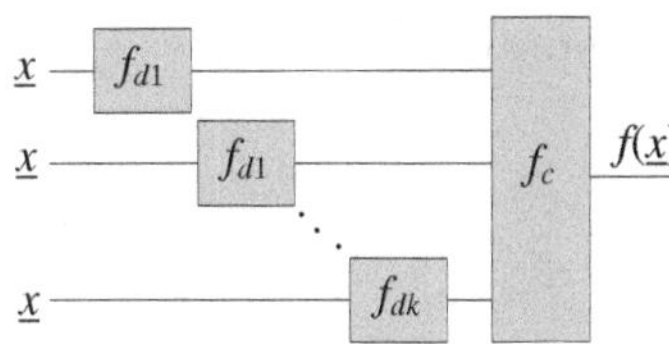

▶ *The given function $f(\underline{x})$ is a composition of f_c of k decomposition functions*

▶ *Decomposition functions $f_{di}(\underline{x})$ depend on all variables so that this most general decomposition does not simplify the associated circuit structure*

FIGURE 25.5
Circuit structure corresponding to the most general decomposition.

There are several approaches to restricting the most general decomposition (Equation 25.1) so that both the decomposition functions $f_{di}(\underline{x})$ and the composition function f_c are simpler than the given function $f(\underline{x})$.

25.2.1 Shannon decomposition

Shannon decomposition is a restricted type of decomposition specified by equation Equation 25.1. It consists of three special decomposition components. Two of them depend on the variables $\underline{x}\backslash x_i$ and the third one is equal to x_i.

$$f(\underline{x}) = f_c(f_{d1}(\underline{x}_0), f_{d2}(\underline{x}_0), x_i) \quad (25.2)$$

where the functions and sets of variables are specified in Figure 25.6. The Shannon decomposition is used to create the nodes of a BDD. Thus, the associated decomposition structure can be build directly from the BDD, as shown in Figure 25.6,

▶ The Shannon decomposition is implemented using a multiplexer,

- The Shannon decomposition exists for each switching function with respect to each variable,
- Both decomposition functions $f_{d1}(\underline{x}_0)$ and $f_{d2}(\underline{x}_0)$, in this case the cofactors, are simpler than the given function, and
- The simplification of the decomposition functions is small because the number of variables is reduced only by one.

Details are given in Chapters 2, 3, and 4.

FIGURE 25.6
Circuit structure and formulas of Shannon decomposition.

25.2.2 Davio decomposition

By changing the composition function f_c of Equation 25.2 we can define two further decomposition types, called *Davio decomposition*. The decomposition and the composition function are defined as the *positive* Davio decomposition (Figure 25.7a). Alternatively, the *negative* Davio decomposition is used in Figure 25.7b. The effect of simplification of the functions due to Davio decomposition is similar to that achieved by the Shannon decomposition (see details in Chapter 4).

25.2.3 Ashenhurst decomposition

A powerful decomposition was suggested by *Ashenhurst* [1]. We consider below

- *Disjoint* and
- *Undisjoint*

In this decomposition only one decomposition function is used, as defined by Equation 25.4.

Disjoint Ashenhurst decomposition. In disjoint Ashenhurst decomposition (Figure 25.8), the decomposition function depends on the bound subset

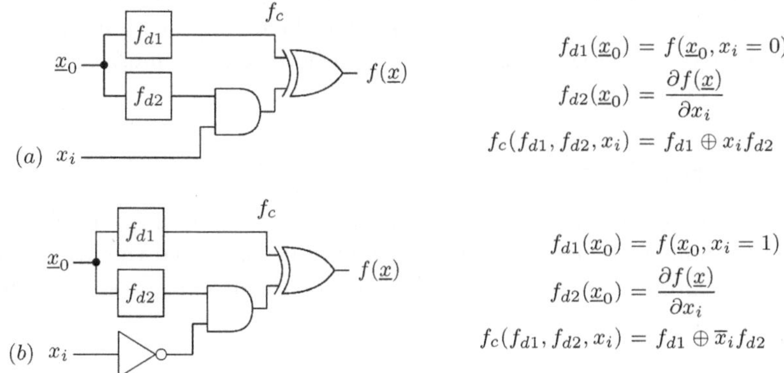

FIGURE 25.7
Circuit structures of positive (a) and negative (b) Davio decomposition.

$\underline{x}_1 \subset \underline{x}$ of variables. The remaining free subset of variables $\underline{x}_0$ does not overlap with the bound set:

$$\underline{x}_0 = \underline{x} \backslash \underline{x}_1 \tag{25.3}$$

The free variables are directly used by the composition function

$$f(\underline{x}) = f(\underline{x}_1, \underline{x}_0) = f_c(f_{d1}(\underline{x}_1), \underline{x}_0) \tag{25.4}$$

Disjoint Ashenhurst decomposition

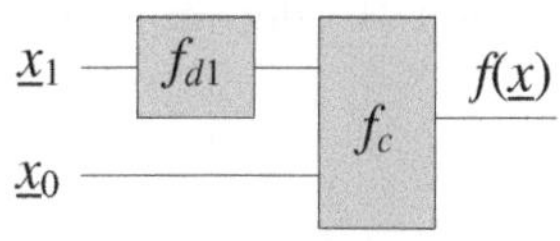

Both the decomposition function $f_{d1}(\underline{x}_1)$ and the composition function $f_c(f_{d1}, \underline{x}_0)$ are simpler than the given function, if the number of variables in both subsets are approximately equal.

Undisjoint Ashenhurst decomposition

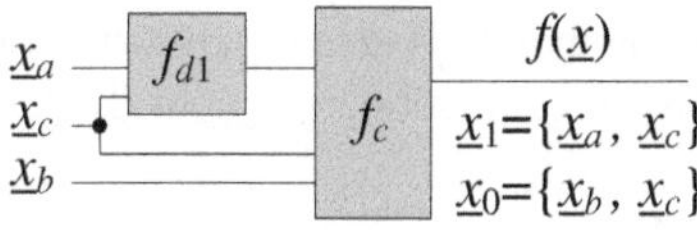

The more variables are commonly used in both sets of variables the less is the benefit of the decomposition.

FIGURE 25.8
Circuit structure of disjoint and undisjoint Ashenhurst decompositions.

Unfortunately, such a simple decomposition exists only for a few functions. Further restrictions are required, in particular, for undisjoint decomposition.

Undisjoint Ashenhurst decomposition. The decomposition is *undisjoint* (Figure 25.8) if common variables are allowed in the bound set $\underline{x}_1$ and the free set $\underline{x}_0$. The more variables commonly used in both sets the more undisjoint decomposition exist.

25.2.4 Curtis decomposition

A slightly modified decomposition was suggested by Curtis [14]. In order to restrict the number of commonly used variables, he suggested the utilization of several decomposition functions, where all of them depend on the same bound set of variables, $\underline{x}_1$,

$$f(\underline{x}) = f(\underline{x}_1, \underline{x}_0) = f_c(f_{d1}(\underline{x}_1), ..., f_{dk}(\underline{x}_1), \underline{x}_0). \tag{25.5}$$

In disjoint Curtis decomposition, in addition to Equation 25.5, the condition of the disjoint subsets of variables holds (Equation 25.3). If the free and the bound set of variables overlap, we have an undisjoint Curtis decomposition.

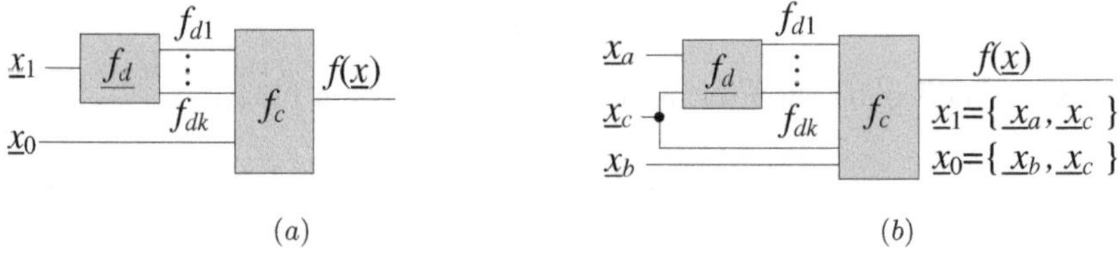

FIGURE 25.9
Circuit structure of the Curtis disjoint (a) and undisjoint (b) decomposition.

25.2.5 Bi-decomposition

A quite different family of decomposition approaches is *bi-decomposition*. Let the set of variables $\underline{x}$ be divided into three disjoint subsets $\underline{x}_a$, $\underline{x}_b$, and $\underline{x}_c$:

- The sets $\underline{x}_a$ and $\underline{x}_b$, which are not empty, and
- The set $\underline{x}_c$, called a *common* set.

A bi-decomposition of a switching function $f(\underline{x}_a, \underline{x}_b, \underline{x}_c)$ with respect to the composition function $f_c(f_{d1}, f_{d2})$ and the dedicated sets $\underline{x}_a$ and $\underline{x}_b$ is a pair of decomposition functions

$$\langle f_{d1} = g(\underline{x}_a, \underline{x}_c), f_{d2} = h(\underline{x}_b, \underline{x}_c)\rangle ,$$

such that

$$f(\underline{x}_a, \underline{x}_b, \underline{x}_c) = f_c(g(\underline{x}_a, \underline{x}_c), h(\underline{x}_b, \underline{x}_c)). \tag{25.6}$$

If the common set $\underline{x}_c$ is empty, the decomposition is called disjoint. The composition function $f_c(f_{d1}, f_{d2})$ determines the output gate of such a decomposition. It can be an OR-gate, an AND-gate, or an EXOR-gate. The associated decompositions are called *OR-bi-decomposition*, *AND-bi-decomposition*, and *EXOR-bi-decomposition*, respectively (Figure 25.10).

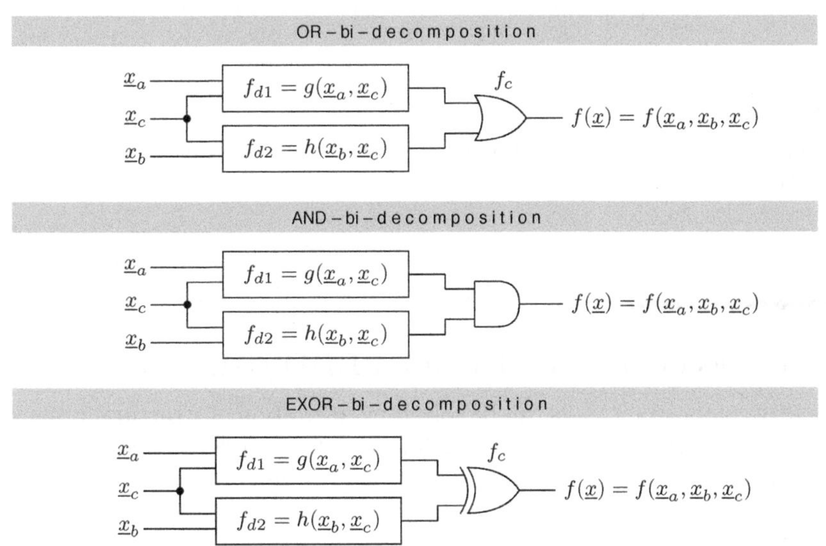

FIGURE 25.10
Circuit structure for OR-bi-decomposition, AND-bi-decomposition, and EXOR-bi-decomposition.

For completeness of the bi-decomposition approach, a *weak bi-decomposition* was introduced:

$$f(\underline{x}_a, \underline{x}_c) = f_c(g(\underline{x}_a, \underline{x}_c), h(\underline{x}_c)). \tag{25.7}$$

By inspection of Equations 25.6 and 25.7, one can find the difference between the bi-decomposition and the weak bi-decomposition: the dedicated set $\underline{x}_b$ is empty in the weak bi-decomposition.

The weak EXOR-bi-decomposition is possible for each function but does not guarantee the simplification. Figure 25.11 shows the circuit structure for the weak OR-bi-decomposition and AND-bi-decomposition.

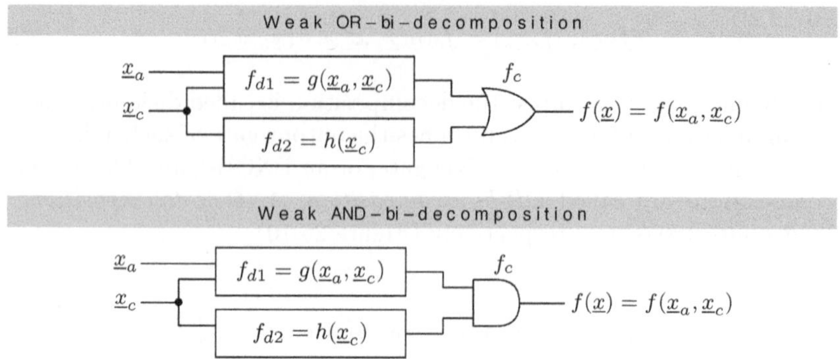

FIGURE 25.11
Circuit structure of a weak OR-bi-decomposition and weak AND-bi-decomposition.

25.3 Decomposition based on BDD structure

Some decomposition tasks can be solved by utilization of structural properties of a BDD. In this section we use such properties in order to decompose completely specified functions using the decomposition types considered above.

25.3.1 Multiplexer-based circuits

Each node of a ROBDD visualizes a Shannon decomposition. The associated switching element of such a node is a multiplexer. Thus the BDD can be mapped directly into a multiplexer-based circuit.

Example 25.5 *Consider the function:*

$$f = \overline{x}_1 x_2 x_3 \vee x_1 \overline{x}_2 x_3 \vee x_1 x_2 \overline{x}_3.$$

Figure 25.12a shows the associated BDD. The multiplexer decomposition is applied to the root node x_1 of the BDD, the decomposition functions are defined by the nodes of the cofactors. Figures 25.12b and (c) show the BDDs of these decomposition functions. The decomposition functions do not depend on the control variable x_1 of the multiplexer. In this example, each decomposition function (Figure 25.12e) represents a simple gate. Figure 25.12d shows the corresponding circuit.

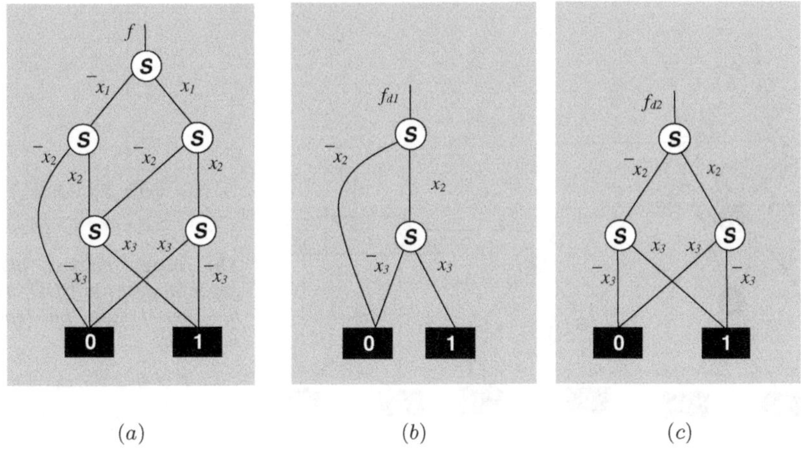

(a) (b) (c)

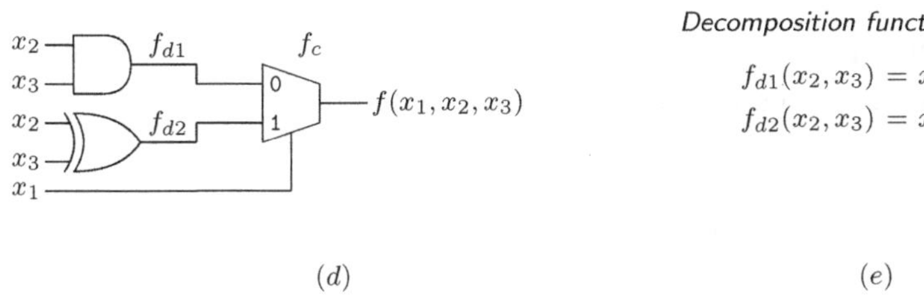

(d) (e)

FIGURE 25.12
Decomposition by a multiplexer: BDD of the function to be decomposed (a), BDD of f_{d1} (b), BBD of f_{d2} (c), circuit (d), and decomposition functions (e) (Example 25.5).

A BDD can be decomposed as

$$f = \overline{g} f_1 \vee g f_2,$$

where f_1 and f_2 are the functions implemented by two subBDDs, and g is obtained by redirecting root f_1 to 0, and root f_2 to 1. This is a special case of a disjoint Ashenhurst decomposition where a multiplexer on the output of the composition function f_c is controlled by the decomposition function $f_{d1} = g$.

> **Example 25.6** *Given a BDD that implements a switching function f (Figure 25.13a), the result of a functional MUX-based decomposition is shown in Figure 25.13b.*

25.3.2 Disjoint Ashenhurst decomposition

Given a pattern on the free set $\underline{x}_0$ of a disjoint Ashenhurst decomposition of Equation 25.4, one of four functions can be observed at the output:

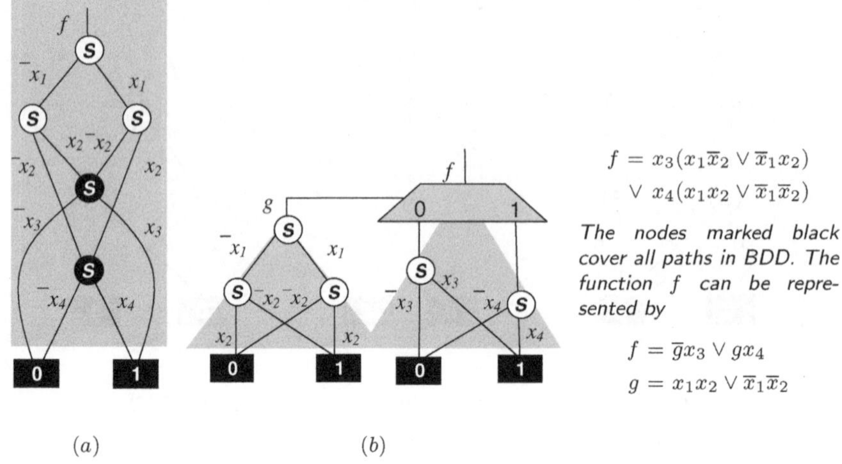

f = x3(x1x̄2 ∨ x̄1x2)
∨ x4(x1x2 ∨ x̄1x̄2)
The nodes marked black cover all paths in BDD. The function f can be represented by
f = ḡx3 ∨ gx4
g = x1x2 ∨ x̄1x̄2
(a)
(b)

FIGURE 25.13
Functional MUX decomposition of the BDD (Example 25.6).

- $f(\underline{x}_0 = \underline{c}_{01}, \underline{x}_1) = f_{d1}(\underline{x}_1)$
- $f(\underline{x}_0 = \underline{c}_{02}, \underline{x}_1) = \overline{f_{d1}(\underline{x}_1)}$
- $f(\underline{x}_0 = \underline{c}_{03}, \underline{x}_1) = 0$
- $f(\underline{x}_0 = \underline{c}_{04}, \underline{x}_1) = 1$

This property can be checked easily in a BDD which uses complemented edges. A disjoint Ashenhurst decomposition exists for a given function if a single node occurs in one level of the BDD with complemented edges. Such a BDD can be found as a result of a sifting algorithm (see Chapter 12). The nodes above this single node specify the free set. The single node is the root node of the decomposition function f_{d1}. For a disjoint Ashenhurst decomposition it is additionally allowed that edges from the free set to the leaves of the BDD are cut.

Example 25.7 *Consider the function:*

$$f = x_1\overline{x}_2 \vee x_1x_3 \vee x_1\overline{x}_4 \vee \overline{x}_2\overline{x}_3x_4.$$

Figure 25.14a shows the associated BDD that includes one complemented edge. The BDD includes only one node on the level x_3. A cut on the single node divides the variable into the free set on top and the bound set below the cut. The decomposition function f_{d1} is used in the composition function directly in the case $\overline{x}_1\overline{x}_2$ and negated in the case x_1x_2 (Figure 25.14b). Figure 25.14c shows the circuit structure of the disjoint Ashenhurst decomposition.

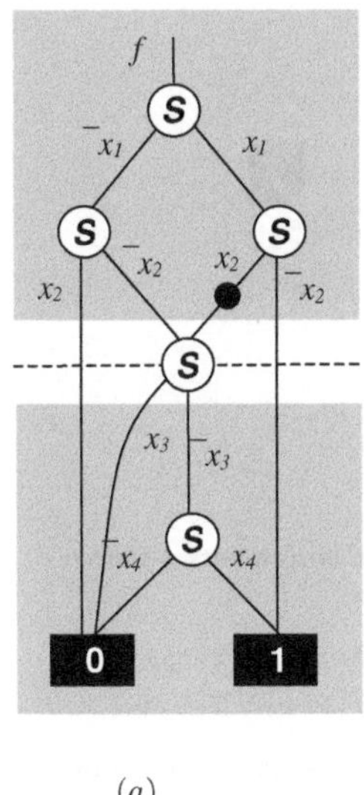

Decomposition functions:

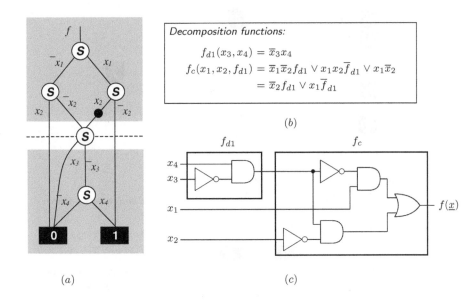

$$f_{d1}(x_3, x_4) = \overline{x}_3 x_4$$

$$f_c(x_1, x_2, f_{d1}) = \overline{x}_1\overline{x}_2 f_{d1} \vee x_1 x_2 \overline{f}_{d1} \vee x_1\overline{x}_2$$

$$= \overline{x}_2 f_{d1} \vee x_1 \overline{f}_{d1}$$

(*b*)

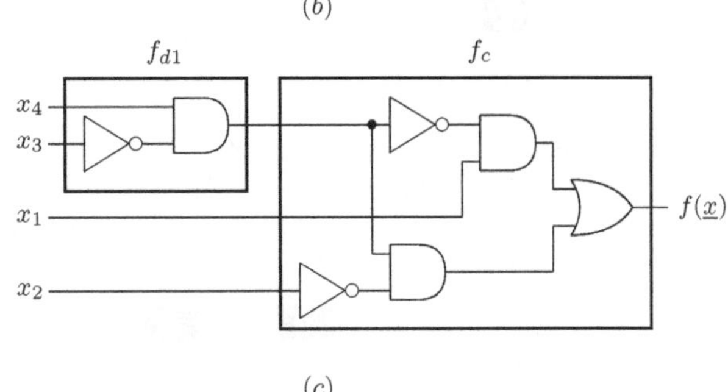

FIGURE 25.14
Disjoint Ashenhurst decomposition: BDD of a switching function with a complemented edge (a), decomposition function f_{d1} and composition function f_c (b), and circuit structure (c) (Example 25.7).

Alternatively, an edge cut can be used. This requires the exchange of the free set with the bound set in variable order in BDD. In a traditional approach, this order of variables is chosen based on a decomposition chart. Given this variable order, a BDD is generated from the circuit, and then a cut set is selected to partition the variables into two sets.

> **Example 25.8** *Figure 25.15 illustrates disjoint Ashenhurst decomposition using a BDD with variable order starting from a bound set: $(x_1 \rightarrow x_2)$ followed by the free set: $(x_3 \rightarrow x_4)$. The cut on edges is implemented.*

25.3.3 Curtis decomposition

The Curtis decomposition generalizes the edge cut approach of the Ashenhurst decomposition. Assume that there are k decomposition functions $f_{di}(\underline{x}_1)$, $i = 1, 2, \ldots, k$, in a disjoint Curtis decomposition. These decomposition functions can select at most 2^k different composition functions $f_{cj}(\underline{x}_0)$. To find a Curtis decomposition, we move the variables of the bound set $\underline{x}_1$ on top of the BDD and the variables of the free set $\underline{x}_0$ below the cutting line of the BDD. A Curtis decomposition with k decomposition function, $f_{di}(\underline{x}_1)$, exists, if no more than

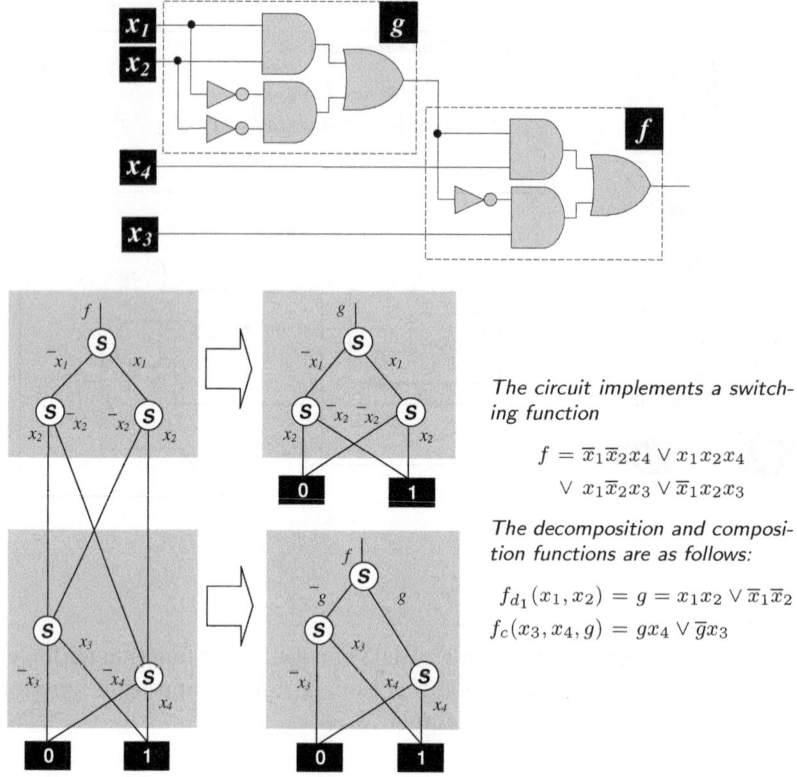

x_1
x_2
x_4
x_3
g
f
The circuit implements a switching function
$f = \overline{x}_1\overline{x}_2x_4 \vee x_1x_2x_4 \vee x_1\overline{x}_2x_3 \vee \overline{x}_1x_2x_3$
The decomposition and composition functions are as follows:
$f_{d_1}(x_1, x_2) = g = x_1x_2 \vee \overline{x}_1\overline{x}_2$
$f_c(x_3, x_4, g) = gx_4 \vee \overline{g}x_3$

FIGURE 25.15
Disjoint Ashenhurst decomposition: a circuit, the ROBDD of f, and ROBDD of the decomposition g and composition functions $f = f_c$ of the BDD (Example 25.8).

2^k different nodes in lower part of the BDD are reached directly through the edges from the upper part. The decomposition effect can be achieved if the number of variables in the bound set is lager than k. Using a sifting algorithm, the appropriated distribution of the variables between the bound set and the free set may be found.

Example 25.9 *Consider the function:*

$$f = \overline{x}_1x_2x_3x_4\overline{x}_5 \vee \overline{x}_1\overline{x}_4(\overline{x}_2 \vee \overline{x}_3) \vee x_1x_2x_3\overline{x}_4$$
$$\vee x_1x_4(\overline{x}_2 \vee \overline{x}_3) \vee x_1x_5 \vee \overline{x}_4x_5.$$

Figure 25.16 shows all BDDs of the disjoint Curtis decomposition, and Figure 25.17 depicts the associated functions and circuit structure.

In Figure 25.16:

- The dashed line separates the bound set (x_1, x_2, x_3) on top of the BDD from the free set (x_4, x_5) below this line. Four nodes of the lower part of the BDD are reached directly by the edges of the upper part (Figure 25.16a).
- Figure 25.17b shows the circuit structure of the calculated Curtis decomposition. The upper part of the BDD can be substituted by a decomposition tree of the Curtis decomposition function f_{d1} and f_{d2}. The BDD describes the composition function $f_c(f_{d1}, f_{d2}, x_4, x_5)$. The selected decomposition tree fixes the required values below the cutting line for the composition function.
- The labels in Figure 25.16b indicate those values, which are used to find the decomposition function f_{d1} as shown in Figure 25.16c and the decomposition function f_{d2} as shown in Figure 25.16d, respectively.

In Example 25.9, the decomposition function f_{d1} is very simple. Another coding may change the complexity of the decomposition function f_{d1} and f_{d2}.

25.3.4 Bi-decomposition

Simple cases of the bi-decomposition can be found using the structure of BDDs. Karplus [21] introduced the concept of 1-dominator and 0-dominator which allows the discovery of a completely specified function:

- Disjoint OR-bi-decomposition or
- Disjoint AND-bi-decomposition.

Example 25.10 *Figure 25.18 illustrates Karplus's approach for bi-decomposition given the switching functions*

$$f = x_1\overline{x}_2 \vee x_3x_4$$
$$f = (x_1 \vee x_2)(\overline{x}_3 \vee x_4).$$

Ravi et al. [36] developed a weak AND-bi-decomposition using BDD approximation. They defined a BDD approximation as a construction of a new BDD smaller in size, which implements a simpler switching function depending on the same variables. In this way the decomposition function $f_{d1}(\underline{x})$ of a weak bi-decomposition is found.

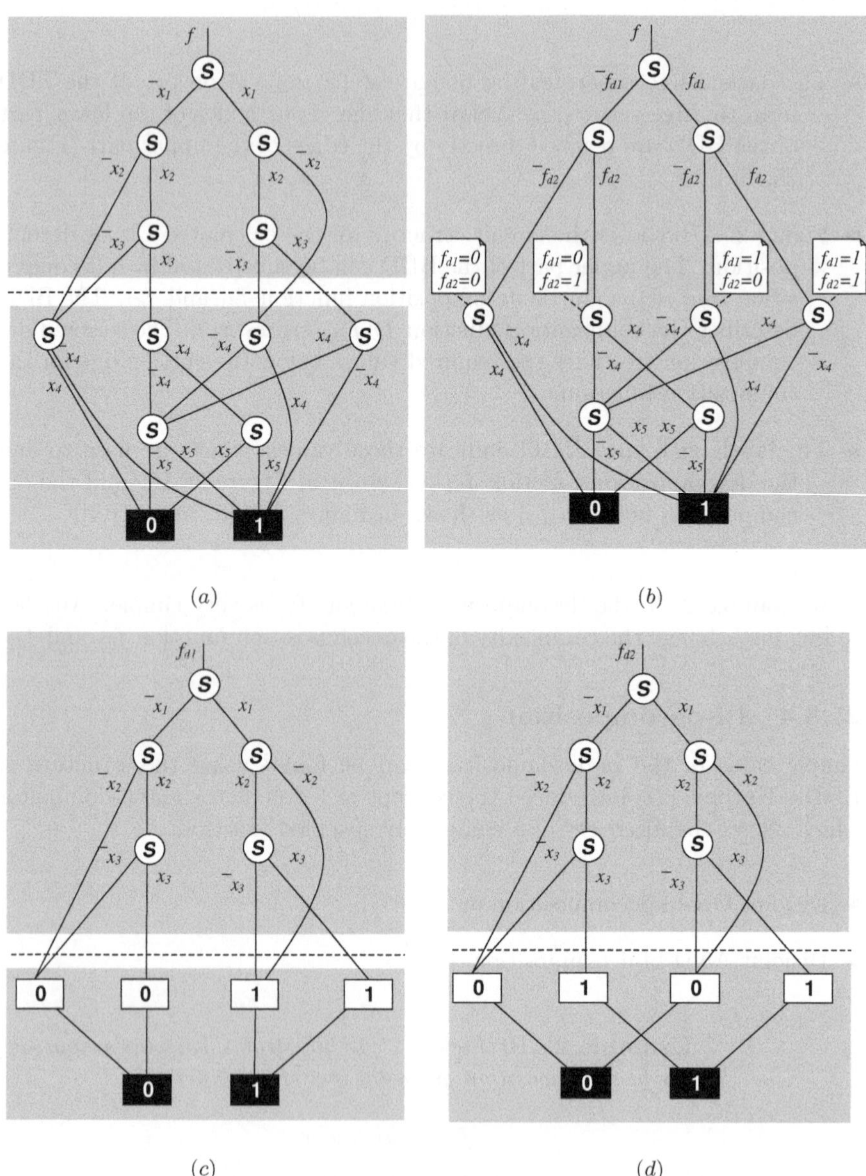

FIGURE 25.16
BDDs of the Curtis decomposition: switching function f to be decomposed (a), composition function f_c (b), decomposition function f_{d1} (c), and decomposition function f_{d2} (d), (Example 25.9).

$$f_c(f_{d1}, f_{d2}, x_4, x_5) = \overline{f}_{d1}\overline{f}_{d2}\overline{x}_4 \vee \overline{f}_{d1}f_{d2}(x_4 \oplus x_5) \vee f_{d1}\overline{f}_{d2}x_4 \vee f_{d1}f_{d2}\overline{x}_4$$
$$f_{d1}(x_1, x_2, x_3) = x_1$$
$$f_{d2}(x_1, x_2, x_3) = x_1\overline{x}_2 \vee x_2x_3$$

(*a*)

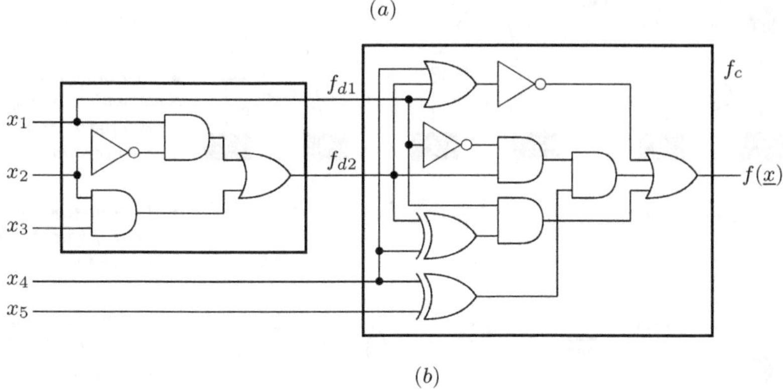

(*b*)

FIGURE 25.17
Curtis decomposition: decomposition functions f_{d1}, f_{d2}, composition function f_c (a), and the circuit (b) (Example 25.9).

Example 25.11 *Given the BDD of the switching function,*

$$f = \overline{x}_1x_2\overline{x}_3 \vee x_1x_2\overline{x}_3 \vee x_1\overline{x}_2x_3,$$

find the BDD of f_{d1} and f_{d2} such that $f = f_{d1} \wedge f_{d2}$. In Figure 25.19, the approximation is calculated by remapping the node marked black into the constant 1. The BDD for f_{d1} represents function $f_{d1} = x_1x_3 \vee x_2$. The BDD for f_{d2} is constructed as $f \subseteq f_{d2} \subseteq f_{d2} \vee \overline{f}_{d1}$, and represents the function $f_{d2} = \overline{x}_2 \vee \overline{x}_3$. The function f_{d2} depends on less variables than the given function, so that it is a decomposition function f_{d2} of a weak AND-bi-decomposition. Hence,

$$f = f_{d1} \wedge f_{d2} = (x_1x_3 \vee x_2) \wedge (\overline{x}_2 \vee \overline{x}_3).$$

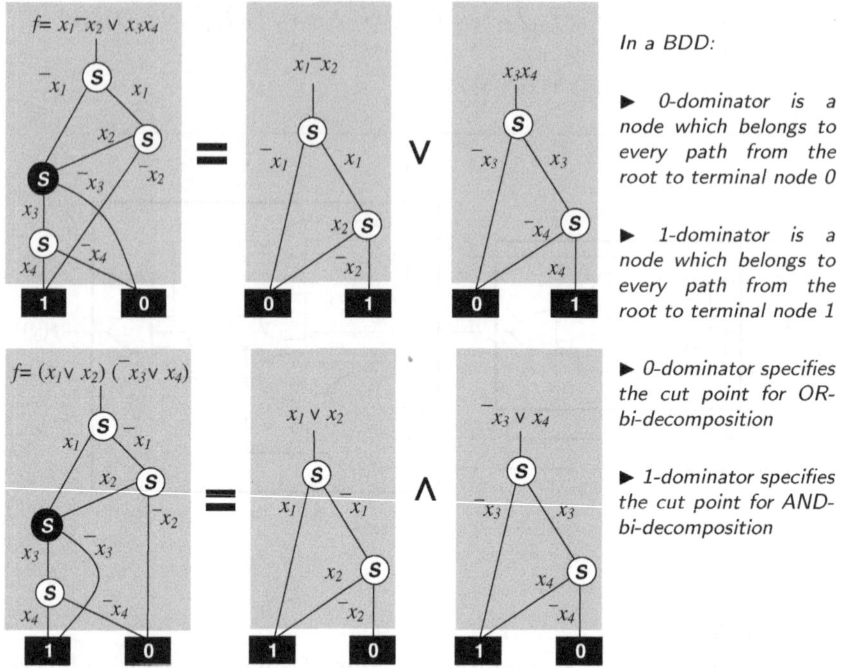

f= x1¯x2 ∨ x3x4
x1¯x2
x3x4
In a BDD:
▶ 0-dominator is a node which belongs to every path from the root to terminal node 0
▶ 1-dominator is a node which belongs to every path from the root to terminal node 1
▶ 0-dominator specifies the cut point for OR-bi-decomposition
▶ 1-dominator specifies the cut point for AND-bi-decomposition
f= (x1∨ x2) (¯x3∨ x4)
x1 ∨ x2
¯x3 ∨ x4

FIGURE 25.18
OR-bi-decomposition and AND-bi-decomposition based on the Karplus approach (Example 25.10).

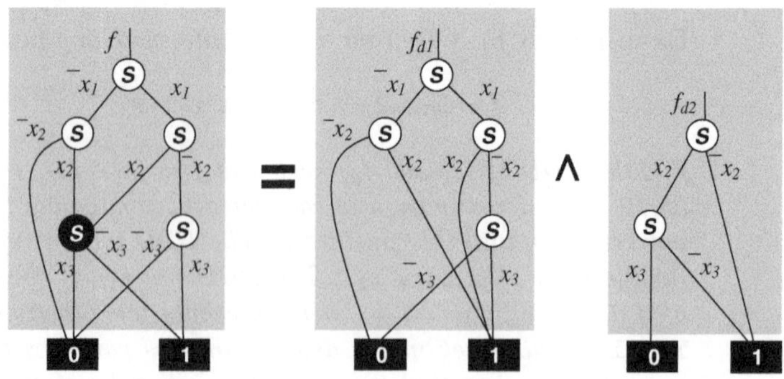

FIGURE 25.19
Weak AND-bi-decomposition by approximation (Example 25.11).

25.4 Decomposition based on BDD operations

It is a property of the switching function itself whether one or another decomposition with respect to selected sets of variables exists. As shown in the previous sections, some of these properties are directly relevant to manipulation of the BDD of the function to be decomposed. Given an incompletely specified switching function, its decomposition properties can be detected using the BDD derived from implicitly calculated function.

25.4.1 Incompletely specified functions

An incompletely specified function is defined as below:

- ▶ It defines the function values only for a subset of the input pattern set.
- ▶ The function values for the don't care set (DC-set) can be chosen without restriction.
- ▶ There are $2^{|DC|}$ allowed completely specified functions if the DC-set of the incompletely specified function includes the $|DC|$ pattern. Thus, the chance to find a decomposable function increases for incompletely specified functions.
- ▶ The $2^{|DC|}$ allowed completely specified functions have the algebraic structure of a *lattice*. This means that there is a smallest and a largest function in the lattice, and both the conjunction and the disjunction of any pair of these functions result in a function of the lattice.

An incompletely specified function $f(\underline{x})$ is described as follows:

- ▶ The ON-set is denoted as $f_q(\underline{x})$.
- ▶ The OFF-set is denoted as $f_r(\underline{x})$.
- ▶ Sometimes the DC-set is denoted as $f_\varphi(\underline{x})$.
- ▶ The pair of functions is denoted as $< f_q(\underline{x}), f_r(\underline{x}) >$.

Completely specified functions can be expressed with the same pair of functions, where $f_r(\underline{x}) = \overline{f_q(\underline{x})}$. Consequently, it is not necessary to distinguish between completely and incompletely specified functions inside of the decomposition procedure. Even if we start with a completely specified function to be decomposed, an incompletely specified function may occur as a decomposition function. It is not necessary to distinguish between the don't cares coming from the given function and the don't cares created in the decomposition process.

In the following subsection we explain the bi-decomposition of incompletely specified functions. All required calculations can be executed using BDDs. We describe the BDD based calculation of necessary operations of the Boolean differential calculus and explain the formulas to check for several types of bi-decomposition together with the formulas for calculation of the decomposition functions. Using the same example function of the Curtis decomposition, the benefit of the bi-decomposition approach will be demonstrated.

25.4.2 BBD based calculation of the m-fold minimum and m-fold maximum

Calculations in the field of bi-decomposition involve two operations of the Boolean differential calculus:

- The *minimum* operator

$$\min_{x_i} f(\underline{x}) = f(x_1, \ldots, x_i, \ldots, x_n) \wedge f(x_1, \ldots, \overline{x}_i, \ldots, x_n) \quad (25.8)$$

- The *maximum* operator

$$\max_{x_i} f(\underline{x}) = f(x_1, \ldots, x_i, \ldots, x_n) \vee f(x_1, \ldots, \overline{x}_i, \ldots, x_n) \quad (25.9)$$

The names of these derivate operations reflect the following property

$$\min_{x_i} f(\underline{x}) \leq f(\underline{x}) \leq \max_{x_i} f(\underline{x}).$$

Both the minimum and the maximum do not depend on the variable x_i.

The derivate operations minimum and maximum can be calculated sequentially with respect to a set of variables. We divide the variables of the function into two disjoint sets of Boolean variables

$$\underline{x}_0 = (x_1, x_2, ..., x_m) \text{ and } \underline{x}_1 = (x_{m+1}, x_{m+2}, ..., x_n)$$

respectively. The *m-fold minimum* of Equation 25.10 and the *m-fold maximum* of Equation 25.11 are further derivative operations of the Boolean differential calculus

$$\min_{\underline{x}_0}{}^m f(\underline{x}_0, \underline{x}_1) = \min_{x_m} \left(\ldots \left(\min_{x_2} \left(\min_{x_1} f(\underline{x}_0, \underline{x}_1) \right) \right) \ldots \right) \quad (25.10)$$

$$\max_{\underline{x}_0}{}^m f(\underline{x}_0, \underline{x}_1) = \max_{x_m} \left(\ldots \left(\max_{x_2} \left(\max_{x_1} f(\underline{x}_0, \underline{x}_1) \right) \right) \ldots \right) \quad (25.11)$$

The result of both m-fold derivative operations does not depend on the sets of variables $\underline{x}_0$.

The m-fold minimum is equal to 1 for such subspaces $\underline{x}_1 = const$ where the function $f(\underline{x}_0, \underline{x}_1 = const)$ is equal to 1 for all patterns of the remaining variables $\underline{x}_0$. This property can be comprehended as a projection of the function values 0 to the subspaces defined by $\underline{x}_1 = const$. The more variables in the subset $\underline{x}_0$ the smaller is the m-fold minimum,

$$\min_{(\underline{x}_0, x_i)}{}^m f(\underline{x}_0, x_i, \underline{x}_1) \leq \min_{\underline{x}_0}{}^{m-1} f(\underline{x}_0, x_i, \underline{x}_1) \leq f(\underline{x}_0, x_i, \underline{x}_1).$$

The m-fold minimum can be calculated by the following algorithm using BDDs.

Algorithm for computing m-fold minimum

Step 1. `Move the variables of the subset` $\underline{x}_0$ `to the bottom of the BDD by means of a sifting algorithm.`

Step 2. `Remove all nodes of the subset` $\underline{x}_0$ `and redirect the cut edges to the terminal node` 0.

Step 3. `Apply the reduction rules in order to simplify the BDD.`

Example 25.12 *Figure 25.20 depicts this algorithm for Equation 25.12 with respect to variables* (x_3, x_4). *Step one is omitted because the variables are already ordered as required.*

$$f(x_1, x_2, x_3, x_4) = x_1x_2 \vee x_1x_3 \vee x_2x_3\overline{x}_4 \tag{25.12}$$

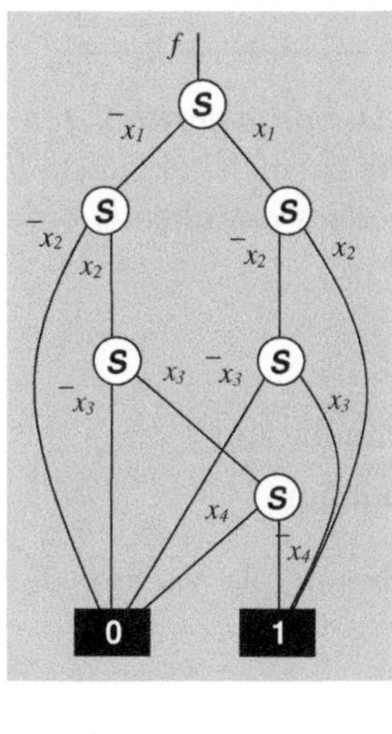

(a)

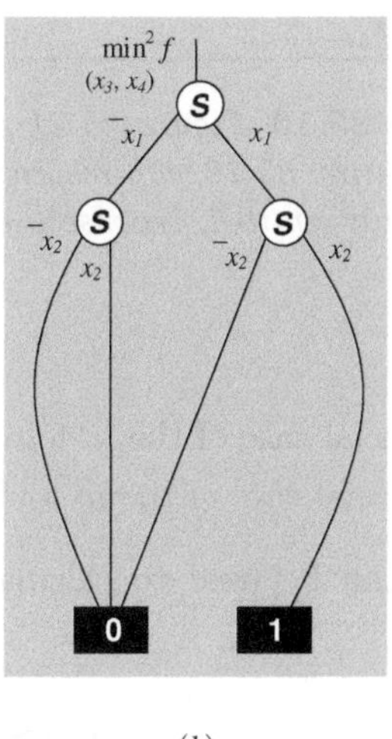

(b)

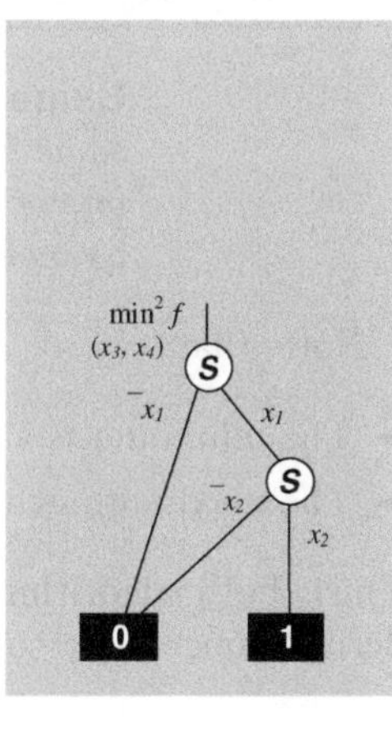

(c)

FIGURE 25.20
Calculation of the 2-fold minimum of a switching function f with respect to the variables (x_3, x_4): BDD of the given function $f(x_1, x_2, x_3, x_4)$ (a), BDD of $\min_{(x_3,x_4)}{}^m f(x_1, x_2, x_3, x_4)$ after step two of the algorithm (b), final BDD of $\min_{(x_3,x_4)}{}^m f(x_1, x_2, x_3, x_4) = x_1x_2$ (c) (Example 25.12).

The m-fold maximum is equal to 1 for such subspaces $\underline{x}_1 = const$ where at least one function value of the function $f(\underline{x}_0, \underline{x}_1 = const)$ is equal to 1. In the opposite of the m-fold minimum, this property can be comprehended as a projection of the function values 1 in the subspaces defined by $\underline{x}_1 = const$. The more variables in the subset $\underline{x}_0$ the larger is the m-fold maximum

$$f(\underline{x}_0, x_i, \underline{x}_1) \leq \max_{\underline{x}_0}{}^{m-1} f(\underline{x}_0, x_i, \underline{x}_1) \leq \max_{(\underline{x}_0, x_i)}{}^{m} f(\underline{x}_0, x_i, \underline{x}_1). \qquad (25.13)$$

The m-fold maximum can be calculated by the following algorithm using BDDs. In comparison to the algorithm of the m-fold minimum, the new terminal node of cut edges is changed to 1.

Algorithm for computing m-fold maximum

Step 1. Move the variables of the subset $\underline{x}_0$ to the bottom of the BDD by means of a sifting algorithm.

Step 2. Remove all nodes of the subset $\underline{x}_0$ and redirect the cut edges to the terminal node 1.

Step 3. Apply the normal reduction rules in order to simplify the BDD.

Example 25.13 *Figure 25.21 depicts this algorithm for the same Equation 25.12 with respect to the variables (x_3, x_4). Step one is again omitted, because the variables are already ordered as required.*

Note, that:

- The minimum is a special case of the m-fold minimum and
- The maximum is a special case of the m-fold maximum.

Thus, both algorithms can be used to calculate the simple version of the derivate operation, too.

25.4.3 OR-bi-decomposition and AND-bi-decomposition

OR-bi-decomposition. The DC-set of an incompletely specified function $< f_q, f_r >$ is specified as follows: at least for one of the created completely specified functions there exists an OR-bi-decomposition with respect to the dedicated sets $\underline{x}_a$ and $\underline{x}_b$ if and only if the condition given in Table 25.1 (first line) holds.

This equation prohibits that a 1-value of the given function is covered by the 0-projections in both $\underline{x}_a$ and $\underline{x}_b$ directions.

TABLE 25.1
Bi-decomposition characteristics.

Characteristic	Formal notation
	OR-bi-decomposition
Condition of existing	$f_q(\underline{x}_a, \underline{x}_b, \underline{x}_c) \wedge \max^m_{\underline{x}_a} f_r(\underline{x}_a, \underline{x}_b, \underline{x}_c) \wedge \max^m_{\underline{x}_b} f_r(\underline{x}_a, \underline{x}_b, \underline{x}_c) = 0$
Largest lattice of g-functions	$g_q(\underline{x}_a, \underline{x}_c) = \max^m_{\underline{x}_b} (f_q(\underline{x}_a, \underline{x}_b, \underline{x}_c) \wedge \max^m_{\underline{x}_a} f_r(\underline{x}_a, \underline{x}_b, \underline{x}_c))$ $g_r(\underline{x}_a, \underline{x}_c) = \max^m_{\underline{x}_b} f_r(\underline{x}_a, \underline{x}_b, \underline{x}_c)$
Lattice of the remaining h-functions using $g(\underline{x}_a, \underline{x}_c)$	$h_q(\underline{x}_b, \underline{x}_c) = \max^m_{\underline{x}_a} \left(f_q(\underline{x}_a, \underline{x}_b, \underline{x}_c) \wedge \overline{g(\underline{x}_a, \underline{x}_c)}\right)$ $h_r(\underline{x}_b, \underline{x}_c) = \max^m_{\underline{x}_a} f_r(\underline{x}_a, \underline{x}_b, \underline{x}_c)$
	AND-bi-decomposition
Condition of existing	$f_r(\underline{x}_a, \underline{x}_b, \underline{x}_c) \wedge \max^m_{\underline{x}_a} f_q(\underline{x}_a, \underline{x}_b, \underline{x}_c) \wedge \max^m_{\underline{x}_b} f_q(\underline{x}_a, \underline{x}_b, \underline{x}_c) = 0$
Largest lattice of g-functions	$g_q(\underline{x}_a, \underline{x}_c) = \max^m_{\underline{x}_b} f_q(\underline{x}_a, \underline{x}_b, \underline{x}_c)$ $g_r(\underline{x}_a, \underline{x}_c) = \max^m_{\underline{x}_b} (f_r(\underline{x}_a, \underline{x}_b, \underline{x}_c) \wedge \max^m_{\underline{x}_a} f_q(\underline{x}_a, \underline{x}_b, \underline{x}_c))$
Lattice of the remaining h-functions using $g(\underline{x}_a, \underline{x}_c)$	$h_q(\underline{x}_b, \underline{x}_c) = \max^m_{\underline{x}_a} f_q(\underline{x}_a, \underline{x}_b, \underline{x}_c)$ $h_r(\underline{x}_b, \underline{x}_c) = \max^m_{\underline{x}_a} (f_r(\underline{x}_a, \underline{x}_b, \underline{x}_c) \wedge g(\underline{x}_a, \underline{x}_c))$
	EXOR-bi-decomposition
Derivative of the lattice (f_q, f_r) with respect to a	$f_q^a(\underline{x}_b, \underline{x}_c) = \max_a f_q(a, \underline{x}_b, \underline{x}_c) \wedge \max_a f_r(a, \underline{x}_b, \underline{x}_c)$ $f_r^a(\underline{x}_b, \underline{x}_c) = \min_a f_q(a, \underline{x}_b, \underline{x}_c) \vee \min_a f_r(a, \underline{x}_b, \underline{x}_c)$
Condition of existing	$\max^m_{\underline{x}_b} f_q^a(\underline{x}_b, \underline{x}_c) \wedge f_r^a(\underline{x}_b, \underline{x}_c) = 0$
Simple possible g-function	$g(a, \underline{x}_c) = a \wedge \max^m_{\underline{x}_b} f_q^a(\underline{x}_b, \underline{x}_c)$
Lattice of the remaining h-functions using $g(a, \underline{x}_c)$	$h_q(\underline{x}_b, \underline{x}_c) = \max_a \left(\overline{g(a, \underline{x}_c)} \wedge f_q(a, \underline{x}_b, \underline{x}_c) \vee g(a, \underline{x}_c) \wedge f_r(a, \underline{x}_b, \underline{x}_c)\right)$ $h_r(\underline{x}_b, \underline{x}_c) = \max_a \left(\overline{g(a, \underline{x}_c)} \wedge f_r(a, \underline{x}_b, \underline{x}_c) \vee g(a, \underline{x}_c) \wedge f_q(a, \underline{x}_b, \underline{x}_c)\right)$

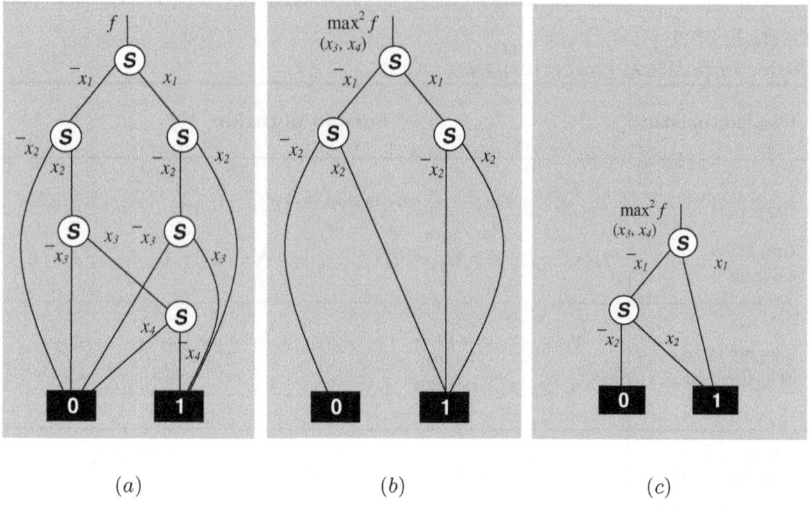

(a) (b) (c)

FIGURE 25.21
Calculation of the 2-fold maximum of $f(x_1, x_2, x_3, x_4)$ of Equation 25.12 with respect to the variables (x_3, x_4): BDD of the given function $f(x_1, x_2, x_3, x_4)$ (a), BDD of $\max_{(x_3,x_4)}{}^m f(x_1, x_2, x_3, x_4)$ after step two of the algorithm (b), and final BDD of $\max_{(x_3,x_4)}{}^m f(x_1, x_2, x_3, x_4) = x_1 \vee x_2$ (c) (Example 25.13).

If an OR-bi-decomposition exists, several pairs of decomposition functions can be calculated. The largest possible incompletely specified function that covers all allowed decomposition functions $f_{d1} = g(\underline{x}_a, \underline{x}_c)$ is the largest lattice specified by equation in the second line of Table 25.1.

Assume the function $g(\underline{x}_a, \underline{x}_c)$ is selected in a recursive bi-decomposition of the incompletely specified decomposition function $< g_q, g_r >$. Using this function $g(\underline{x}_a, \underline{x}_c)$ the remaining allowed decomposition functions $f_{d2} = h(\underline{x}_b, \underline{x}_c)$ are covered by an incompletely specified function $< h_q, h_r >$ defined by the equations in the third line of Table 25.1. The incompletely specified function $< h_q, h_r >$ can be recursively decomposed.

The AND-bi-decomposition is a dual decomposition to the OR-bi-decomposition. The DC-set of an incompletely specified function $< f_q, f_r >$ is specified as follows: at least for one of the created completely specified functions there exists an AND-bi-decomposition with respect to the dedicated sets $\underline{x}_a$ and $\underline{x}_b$ if the conditions given in the fourth line of Table 25.1 hold. This equation prohibits that a 0-value of the given function is covered by the 1-projections in both the $\underline{x}_a$ direction and the $\underline{x}_b$ direction.

If an AND-bi-decomposition exists, several pairs of decomposition functions can be calculated. The equations in the fifth line of Table 25.1 specify an incompletely specified function that covers all allowed decomposition functions

$f_{d1} = g(\underline{x}_a, \underline{x}_c)$.

Assume that the function $g(\underline{x}_a, \underline{x}_c)$ is selected in a recursive bi-decomposition of the incompletely specified decomposition function $< g_q, g_r >$. Using this function $g(\underline{x}_a, \underline{x}_c)$ the remaining allowed decomposition functions $f_{d2} = h(\underline{x}_b, \underline{x}_c)$ are covered by the incompletely specified function $< h_q, h_r >$, shown in the sixth line of Table 25.1.

25.4.4 EXOR-bi-decomposition

Consider algorithms that can verify if an incompletely specified function has an EXOR-bi-decomposition with respect to any dedicated sets $\underline{x}_a$ and $\underline{x}_b$. Let us restrict the consideration to the case that the dedicated set $\underline{x}_a$ includes only one single variable: $\underline{x}_a = (a)$. The incompletely specified function has the mark functions f_q^a and f_r^a is defined by the equations given in the seventh line of Table 25.1. This lattice covers the derivatives of the incompletely specified functions $< f_q, f_r >$ with respect to the variable a.

The DC-set of an incompletely specified function $< f_q, f_r >$ is defined as follows. At least for one of the created completely specified functions there exists an EXOR-bi-decomposition with respect to the dedicated sets a and $\underline{x}_b$ if and only if the conditions given in the eighth line of Table 25.1 hold. This equation prohibits that different values to the above introduced derivative occur in the same subspace $\underline{x}_c = const$.

If an EXOR-bi-decomposition exists, several pairs of decomposition functions can be calculated. All possible decomposition functions $f_{d1} = g(a, \underline{x}_c)$ cannot be expressed by a single incompletely specified function. For that reason, one simple completely specified function $f_{d1} = g(a, \underline{x}_c)$ is specified by the equation in the ninth line of Table 25.1. This equation selects the function $g(a) = a$ for subspaces, where $g(a)$ depends on the variable a, and for the remaining subspaces $g(a) = 0$ is selected. Using the above function $g(a, \underline{x}_c)$, all allowed decomposition functions $f_{d2} = h(\underline{x}_b, \underline{x}_c)$ are covered by the incompletely specified function $< h_q, h_r >$, shown in the last line of Table 25.1. In these equations, an EXOR operation between $g(a, \underline{x}_c)$ or $\overline{g(a, \underline{x}_c)}$ and all completely specified functions covered by $< f_q, f_r >$ are calculated implicitly.

Example 25.14 *Consider the function from Example 25.9 already used to demonstrate the Curtis decomposition. Figure 25.16a shows the associated BDD. As a result of initial analysis, an EXOR-bi-decomposition with the dedicated sets* $\underline{x}_a = x_1$ *and* $\underline{x}_b = x_2$ *is found. Both dedicated sets are extended in further detailed analysis to* $\underline{x}_a = (x_1, x_4)$ *and* $\underline{x}_b = (x_2, x_3)$*, so that the remaining common set* $\underline{x}_c = x_5$*. Starting with a completely specified function, the decomposition functions of a EXOR-bi-decomposition are completely specified, too. Figure 25.22 shows further details of the decomposition and the synthesized multilevel circuit structure.*

This example emphasizes the benefit of the bi-decomposition in terms of area, as well as delay and power consumption.

Phase 1
EXOR-bi-decomposition of f:
A possible pair of decomposition functions is:

$$f_{d1_0} = g(x_1, x_4, x_5) = \overline{x}_1\overline{x}_4 \vee x_1 x_4 \vee x_1 x_5$$
$$f_{d2_0} = h(x_2, x_3, x_5) = x_2 x_3 \overline{x}_5.$$

Phase 2
OR-bi-decomposition of f_{d1_0}:
An OR-bi-decomposition exists with respect to the dedicated sets $\underline{x}_a = x_4$ and $\underline{x}_b = x_5$. The pair of the simplest completely specified decomposition functions is

$$f_{d1_1} = g(x_1, x_4) = x_1 \oplus \overline{x}_4$$
$$f_{d2_1} = h(x_1, x_5) = x_1 x_5.$$

The recursive decomposition is terminated because each of these decomposition functions depends on two variables.

Phase 3
AND-bi-decomposition of f_{d2_0}:
Several disjoint AND-bi-decompositions exists for f_{d2_0}, in particular:

$$f_{d1_2} = g(x_2) = x_2,$$
$$f_{d2_2} = h(x_3, x_5) = x_3 \overline{x}_5.$$

The recursive decomposition is terminated because each of decomposition functions does not depend on more than two variables

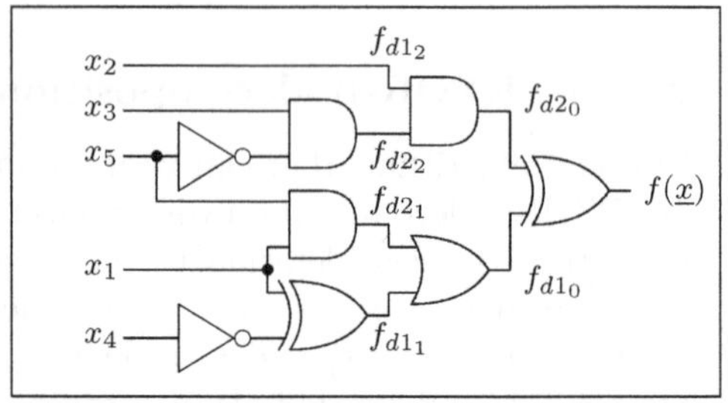

This circuit is a result of the bi-decomposition, and includes:

- *8 gates, and*
- *4 levels.*

The circuit in Figure 25.17 implements the same function and requires

- *13 gates, and*
- *7 levels.*

If we also restrict to two inputs gates, that means we split two 3-input-OR-gates each into two 2-input-OR-gates, the Curtis decomposition needs

- *15 gates and*
- *8 levels.*

The circuit in this example could be synthesized completely using the three strong types of bi-decomposition. Therefore, no weak bi-decomposition was applied.

FIGURE 25.22
Circuit structure of the bi-decomposition (Example 25.14).

25.4.5 Weak OR- and AND-bi-decomposition

Each bi-decomposition must create simpler decomposition functions. The above strong types of bi-decomposition solve this task by the reduction of the number of variables. In weak bi-decompositions the simplification is reached

TABLE 25.2
Weak bi-decomposition characteristics.

Characteristic	Formal notation
	Weak OR-bi-decomposition
Condition of existing	$f_q(\underline{x}_a, \underline{x}_c) \wedge \overline{\max_{\underline{x}_a}^m f_r(\underline{x}_a, \underline{x}_c)} \neq 0$
Largest lattice of g-functions	$g_q(\underline{x}_a, \underline{x}_c) = f_q(\underline{x}_a, \underline{x}_c) \wedge \max_{\underline{x}_a}^k f_r(\underline{x}_a, \underline{x}_c)$, $g_r(\underline{x}_a, \underline{x}_c) = f_r(\underline{x}_a, \underline{x}_c)$
Lattice of the remaining h-functions using $g(\underline{x}_a, \underline{x}_c)$	$h_q(\underline{x}_c) = \max_{\underline{x}_a}^m \left(f_q(\underline{x}_a, \underline{x}_c) \wedge \overline{g(\underline{x}_a, \underline{x}_c)} \right)$ $h_r(\underline{x}_c) = \max_{\underline{x}_a}^m f_r(\underline{x}_a, \underline{x}_c)$
	Weak AND-bi-decomposition
Condition of existing	$f_r(\underline{x}_a, \underline{x}_c) \wedge \overline{\max_{\underline{x}_a}^m f_q(\underline{x}_a, \underline{x}_c))} \neq 0$
Largest lattice of g-functions	$g_q(\underline{x}_a, \underline{x}_c) = f_q(\underline{x}_a, \underline{x}_c)$, $g_r(\underline{x}_a, \underline{x}_c) = f_r(\underline{x}_a, \underline{x}_c) \wedge \max_{\underline{x}_a}^m f_q(\underline{x}_a, \underline{x}_c)$
Lattice of the remaining h-functions using $g(\underline{x}_a, \underline{x}_c)$	$h_q(\underline{x}_c) = \max_{\underline{x}_a}^m f_q(\underline{x}_a, \underline{x}_c)$ $h_r(\underline{x}_c) = \max_{\underline{x}_a}^m \left(f_r(\underline{x}_a, \underline{x}_c) \wedge g(\underline{x}_a, \underline{x}_c) \right)$

additionally by increasing the DC-set, which extents the freedom for selection of the next bi-decomposition. For each logic function exists at least one of the strong types or one of the weak types of bi-decomposition. Therefore, the multilevel synthesis can be done completely by the bi-decomposition approach.

Weak OR-bi-decomposition. The DC-set of an incompletely specified function $< f_q, f_r >$ can be specified such that at least for one of the created completely specified function exist an weak OR-bi-decomposition with respect to the dedicated set $\underline{x}_a$ if and only if conditions equation given in the first line of Table 25.2 holds.

The DC-set of the decomposition function f_{d1} with the mark functions g_q and g_r is lager than the DC-set of the given incompletely specified function $< f_q, f_r >$ because the ON-set function g_q is restricted. The largest lattice is specified in the second line of Table 25.2.

The comparison of these equations with the associated g-equations of the OR-bi-decomposition shows that in the case of the weak OR-bi-decomposition the $|\underline{x}_b|$-fold maximum must be removed. Assume the function $g(\underline{x}_a, \underline{x}_c)$ is selected in a recursive bi-decomposition of the incompletely specified decomposition function $< g_q, g_r >$. Using this function $g(\underline{x}_a, \underline{x}_c)$ the remaining allowed decomposition functions f_{d2} are covered by the incompletely specified function $< h_q, h_r >$, or the lattice of the remaining h-function (the third line in Table 25.2). These equations comply with the associated h-equations of the OR-bi-decomposition. The $|\underline{x}_a| - fold$ maximum restricts the number of independent variables in the decomposition function f_{d2}.

Weak AND-bi-decomposition. The weak AND-bi-decomposition is a counterpart to the weak OR-bi-decomposition. The DC-set of an incompletely specified function, $< f_q, f_r >$, can be specified such that at least for one of the created completely specified function there exist a weak AND-bi-decomposition with respect to the dedicated set $\underline{x}_a$ if and only if the conditions given in the fourth line of Table 25.2 holds.

If a weak AND-bi-decomposition exists, the DC-set of a decomposition function f_{d1} with the mark functions g_q and g_r called the largest lattice (the fifth line in Table 25.2) can be lager than the DC-set of the given incompletely specified function $< f_q, f_r >$ because the OFF-set function g_r is restricted.

Assume the function $g(\underline{x}_a, \underline{x}_c)$ is selected in a recursive bi-decomposition of the incompletely specified decomposition function $< g_q, g_r >$. Using this function $g(\underline{x}_a, \underline{x}_c)$ the remaining allowed decomposition functions f_{d2} are covered by incompletely specified function $< h_q, h_r >$ (the last line in Table 25.2). The $|\underline{x}_a|$-fold maximum restricts the number of independent variables in the decomposition function f_{d2}.

25.5 Further reading

Algebraic division. Decomposition of a logic function based on manipulation of algebraic expressions is called the algebraic division. A given function f is divided by the divisor expression d such that $f = q \cdot d + r$. The expression q denotes the quotient and r is called remainder expression. Brayton and McMullen [8] proposed a fundamental theorem about how common algebraic divisors can be found using kernels. On this basis several approaches to finding kernel intersections were suggested, in particular, in [8], [9]. A more general approach developed by Rudell [40] is the rectangle covering that also find kernels. Fujita et al. [17] proposed a factoring algorithm to apply the algebraic division recursively. Some heuristics to improve the factoring algorithm was developed by Hachtel and Somenzi [18]. Vasudevamurthy and

Rajski [61] proposed to reduce the exponential effort by double-cube kernel extraction.

BDD based approaches for the Ashenhurst decomposition. Ashenhurst [1] used a decomposition chart in order to find a decomposition. Roth and Karp [39] considered functional decomposition in contrast to the above mentioned algebraic decomposition. Minato and DeMicheli [28] studied the relationship between these types of decomposition. The application of BDDs for this task was suggested by Lai et al. [24]. The decompositon of the function was realized by a cut of the BDD. Generalizations of this approach for incompletely specified functions, non-disjoint decompositions, recursive decompositions and implementation technology oriented decompositions was proposed by Bertacco and Damiani [2], Chang et al. [11], Matsunaga and Fujita [26], Sasao [42], and Stanion and Sechen [50].

BDD based approaches for the bi-decomposition. A review of decomposition methods using different function representations is given by Perkowski and Grygiel [33]. Karplus [21] developed the concept of 1-dominator and 0-dominator which allows direct bi-decomposition using a BDD. A generalization of this approach for EXOR-bi-decomposition was suggested by Yang et al. [64]. The method by Sasao and Butler [43] prefers disjoint bi-decomposition and is restricted to few variables in the common set. Stanković [49] introduced EXOR-TDDs showed that decomposition or bi-decomposition with respect to r variables can be obtained by building an EXOR-TDD for a matrix-valued function of $(n - r)$ variables.

Bi-decomposition was suggested for the first time by Böhlau [6]. Concerning the distribution of the variables into three groups the term grouping was used. Particular problems of the EXOR bi-decompositon were studied by Steinbach and Wereszczynski [58]. Dresig [16] proposed the reusing of given circuit structures in the checks for bi-decompositions. Both functional and structural properties are used by Steinbach and Hesse [51] in order to calculate bi-decomposition of extremely large circuits. A comprehensive representation of the bi-decomposition approach is given by Steinbach and Lang in [52] and by Posthoff and Steinbach in [35]. The theoretical background of bi-decomposition is Boolean differential calculus developed, in particular, by Bochmann and Posthoff [4], Bochmann and Steinbach [5], and Posthoff and Steinbach [35].

Completeness of the bi-decomposition approach. Posthoff and Steinbach [35], and Steinbach and Lang [52] proved that an arbitrary switching function can be decomposed completely using the five bi-decomposition types: OR-bi-decomposition, AND-bi-decomposition, EXOR-bi-decomposition, weak OR-bi-decomposition, and weak AND-bi-decomposition. This proof based on the previous result by Steinbach and Le [54] of weak bi-decompositions and on cognitions, that in the case that no weak OR-bi-decomposition and no weak AND-bi-decomposition exists the function can be decomposed by an EXOR-bi-decomposition.

Testability of multilevel combinational circuits. The testability is property of the structure of a multilevel combinational circuit. The bi-decomposition approach allows different decomposition structures. Some of them are not completely testable. Circuit structures, designed by means of the decomposition formulas of this chapter, are 100% testable regarding all stuck-at fault on all inputs, all outputs, all internal gate-inputs and all internal gate-outputs. It was shown by Steinbach and Stöckert [57] that the test pattern can be calculated easily during the synthesis process. Selected intermediate synthesis results must be accumulated so that the test pattern generation increases the time for synthesis approximately by 10%, only. Steinbach and Zhang [59] generalized this approach for large circuits having a global block structure.

Selection of the variable for the bi-decomposition. Steinbach et al. [52], [56] developed an algorithm for an optimal selection of dedicated sets of variables of the bi-decomposition. These algorithms lead to balanced circuit structures of small depth. Yamashita et al. [62] suggested a search tree approach for this task.

Experimental results of the bi-decomposition approach. Mishchenko et al. [30] reported experimental results on bi-decomposition. The implemented program BI-DECOMP based on the BDD package CUDD by Somenzi [48] and outperforms both SIS by Sentovich et. al. [46], and BDS by Yang and Ciesielski [64] in terms of delay. In case of EXOR-intensive circuits the program BI-DECOMP outperforms both SIS and BDS in terms of area, too.

Decomposition of multivalued logic functions. Muzio and Miller [31] generalize the decomposition of Boolean functions to the case of ternary switching functions. Three years later the same authors [27] suggested the two-place decomposition where a many-valued function is reexpressed as a composition of three functions, two of which have at most two arguments. Tokmen [60] studied the case of disjoint decompositions. Sasao [41] used an OR-AND-OR PLA structure for Boolean functions where subsets of Boolean variables are regarded as multiple-valued variable. Reischer and Simovici [38] suggested a more general approach for decomposition of multiple-valued functions and relational databases. Łuba [25] and Selvaraj et al. [44] applied set covering algorithm for the decomposition of multiple-valued functions. Perkowski et al. [34] reported an approach to Curtis like decomposition of multiple-valued functions using a generalized graph coloring method. Steinbach et al. [55] applied the method of successive value removal to the bi-decomposition of multiple-valued functions. Lang and Steinbach [52] suggested a more powerfully approach for bi-decomposition of multiple-valued functions which uses for fast calculations *binary-encoded multiple-valued decision diagrams* (BEMDD) suggested by Mishchenko et al. [29]. This approach was generalized by Steinbach and Lang [53], using the MIN-MAX multi-decomposition additionally, each multiple-valued function can be expressed by a multilevel circuit of MIN- and MAX-gates.

Information-theoretical approach in decomposition. Hartmanis and Stearns [19] developed the theory of partitions and proposed measures of information relationship in decomposition. These relationships and measures were used by Jóźwiak [20] and Rawski et al. [37] in further study of input support minimization, parallel decomposition and serial functional decompositions. Using Shannon information-theory approach in decomposition is discussed in Chapters 8 and 24.

Flexibility in decomposition. A *flexibility* in decomposition can be defined as conditions for replacing a function by an in alternative function. Details of using a flexibility in logic design can be found in the paper by Sentovich [71]. Yamashita et al. [78] developed method called *sets of pairs of functions to be distinguished* (SPFD) to to represent a flexibility in network optimization. Cheushev et al. [12] used information-theory approach in flexible network synthesis. The SPFD concept was extended by Sinha et al. [47]. Relationship of SPDF and Shannon information-theory approach is discussed in Chapter 8.

Spectral approach in decomposition. In spectral interpretation of decision diagrams, the use of different decomposition rules is equivalently the use of different spectral transforms (see details in Chapters 7 and 13).

References

[1] Ashenhurst RL. The decomposition of switching functions. In *Proceedings of the International Symposium on the Theory of Switching Functions*, pp. 74–116, 1957.

[2] Bertacco V and Damiani M. The disjunctive decomposition of logic functions. In *Proceedings of the Computer-Aided Design Conference*, pp. 78–82, 1997.

[3] Bochmann D, Dresig F, and Steinbach B. A new decomposition method for multilevel circuit design. In *Proceedings of the Euro-DAC*, pp. 374–377, 1991.

[4] Bochmann D and Posthoff C. *Binäre dynamische Systeme.* Oldenbourg Verlag, München, Germany, 1981.

[5] Bochmann D and Steinbach B. *Logikentwurf mit XBOOLE. Verlag Technik,* Berlin, Germany, 1991.

[6] Böhlau P. Eine Dekompositionsstrategie für den Logikentwurf auf der Basis funktionstypischer Eigenschaften. *Dissertation thesis,* Technical University Karl-Marx-Stadt, Germany, 1987.

[7] Brayton RK, McMullen C, Hachtel GD, and Sangiovanni-Vincentelli AL. *Logic Minimization Algorithms for VLSI Synthesis.* Kluwer, Dordrecht, 1984.

[8] Brayton RK and McMullen C. The decomposition and factorization of Boolean expressions. In *Proceedings of the IEEE International Symposium on Circuits and Systems,* pp. 49–54, 1982.

[9] Brayton RK, Rudell RL, Sangiovanni-Vincentelli AL, and Wang AR. MIS: a multiple-level logic optimization system. *IEEE Transactions on Computer-Aided Design of Integrated Circuits and Systems,* 6(6):1062–1081, 1987.

[10] Cabodi G, Camurati P, and Quer S. Improved reachability analysis of large finite state machines. In *Proceedings of the International Conference Computer-Aided Design,* Santa-Clara, CA, pp. 354–360, 1996.

[11] Chang SC, Marek-Sadowska M, and Hwang T. Technology mapping for TLUFPGAs based on decomposition of binary decision diagrams. *IEEE Transactions on Computer-Aided Design of Integrated Circuits and Systems,* 15(10):1226–1235, 1996.

[12] Cheushev V, Yanushkevich S, Shmerko V, Moraga C, and Kolodziejczyk J. Information theory method for flexible network synthesis. In *Proceedings of the 31st IEEE International Symposium on Multiple-Valued Logic,* pp. 201–206, 2001.

[13] Cortadella J. Bi-decomposition and tree-height reduction for timing optimization. In *Proceedings of the 11th IEEE/ACM International Workshop on Logic and Synthesis,* New Orleans, Louisiana, pp. 233–238, 2002.

[14] Curtis H. *A new approach to the design of switching circuits.* Princeton, Van Nostrand, 1962.

[15] Drechsler R, Sarabi A, Teobald M, Becker B, and Perkowski M. Efficient representation and manipulation of switching functions based on ordered Kronecker functional decision diagrams. In *Proceedings of the European Design Automation Conference,* pp. 191–197, 1994.

[16] Dresig F. Gruppierung - Theorie und Anwendung in der Logiksynthese. *VDI-Verlag,* Düsseldorf, Germany, 1992.

[17] Fujita M, Matsunaga Y, and Ciesielski M. Multi-level logic optimization. In Hassoun S and Sasao T, Eds., *Logic Synthesis and Verification,* Kluwer, Dordrecht, 2002.

[18] Hachtel GD and Somenzi F. *Logic Synthesis and Verification Algorithms.* Kluwer, Dordrecht, 1996.

[19] Hartmanis J and Stearns RE. *Algebraic Structure Theory of Sequential Machines.* Prentice-Hall, Englewood Cliffs, New York 1966.

[20] Jóźwiak L. Information relationships and measures in application to logic design. In *Proceedings of the 29th IEEE International Symposium on Multiple-Valued Logic*, pp. 228–235, 1999.

[21] Karplus K. Using if-then-else DAGs for multi-level logic minimization. *Technical Report UCSC-CRL-99-29*, University of California Santa Cruz, pp. 49–54, 1982.

[22] Lang Ch and Steinbach B. Bi-decompositon of function sets in multiple-valued logic for Circuit Design and Data Mining. In Yanushkevich S, Ed., *Artificial Intelligence in Logic Desing, Kluwer, Dordrecht*, pp. 73–107, 2004.

[23] Lai Y-T, Pan K-R, and Pedram M. OBDD-based functional decomposition: algorithms and implementation. *IEEE Transactions on Computer-Aided Design of Integrated Circuits and Systems,* 15(8):977–990, 1996.

[24] Lai Y-T, Pedram M, and Vrudhula S. BDD based decomposition of logic for functions with applications to FPGA synthesis. In *Proceedings of the Design Automation Conference*, pp. 642–647, 1993.

[25] Łuba T. Decomposition of multi-valued functions. In *Proceedings of the 25th IEEE International Symposium on Multiple-Valued Logic,* pp. 256–261, 1995.

[26] Matsunaga Y and Fujita M. Multi-level logic optimization using binary decision diagrams. In *Proceedings on ICCAD87*, 1987.

[27] Miller DM and Muzio JC. Two-place decomposition and the synthesis of many-valued switching circuits. In *Proceedings of the 6th International Symposium on Multiple-Valued Logic,* pp. 164–168, 1976.

[28] Minato S and DeMicheli G. Find all simple disjunctive decompositions using irredundant sum-of-products forms. In *Proceedings of the International Conference on Computer-Aided Design*, pp. 111–117, 1998.

[29] Mishchenko A, Files C, Perkowski M, Steinbach B, and Dorotska Ch. Implicit Algorithms for Multi-Valued Input Support Minimization. In Steinbach B, Ed., *Proceedings of the 4th International Workshops on Boolean Problems*, pp. 9-20, 2000.

[30] Mishchenko A, Steinbach B, and Perkowski M. An algorithm for bi-decomposition of logic functions. In *Proceedings Design Automation Conference* , pp. 103–108, 2001.

[31] Muzio JC and Miller DM. Decomposition of ternary switching functions. *Proceedings of the 3rd International Symposium on Multiple-Valued Logic,* pp. 156–165, 1973.

[32] Narayan A, Jain J, Fujita M, and Sangiovanni-Vincentelli AL. Partition ROBDDs: a compact canonical and efficiently manipulable representation for Boolean functions. In *Proceedings of the International Conference on Computer-Aided Design*, Santa-Clara, CA, pp. 547–554, 1996.

[33] Perkowski M and Grygiel S. A survey of literature on functional decomposition. In *Technical report, Department of Electrical Engineering, Portland State University*, 1995.

[34] Perkowski M, Marek-Sadowska M, Jozwiak L, Łuba T, Grygiel S, Nowicka M, Malvi R, Wang Z, and Zhang J. Decomposition of multiple-valued relations. In *Proceedings of the IEEE 27th International Symposium on Multiple-Valued Logic*, pp. 13-18, 1997.

[35] Posthoff Ch and Steinbach B. *Logic Function and Equation – Binary Models for Computer Science.* Springer, Heidelberg, 2004.

[36] Ravi K, NcMillan KL, Shiple TR, and Somenzi F. Approximation and decomposition of binary decision diagrams. In *Proceedings of the Design Automation Conference*, pp. 445–450, 1998.

[37] Rawski M, Jóźwiak L, and Łuba T. Functional decomposition with an efficient input support selection for sub-functions based on information relationship measures. *Journal of Systems Architecture*, 47:137–155, 2001.

[38] Reischer C and Simovici DA Decomposition of multiple-valued switching functions and rational databases. In *Proceedings of the 15th International Symposium on Multiple-Valued Logic,* pp. 109–114, 1985.

[39] Roth JP and Karp RM. Minimization over Boolean graphs. *IBM Journal*, pp. 227–238, 1962.

[40] Rudell R. Logic Synthesis for VLSI Design. In *PhD thesis, U.C. Berkeley*, 1989.

[41] Sasao T. Multiple-valued decomposition of generalized Boolean functions and the complexity of programmable logic arrays. *IEEE Transactions on Computers,* 30(9):635–643, 1981.

[42] Sasao T. FPGA design by generalized functional decomposition. In Sasao T, Ed., *Logic Synthesis and Optimization*, pp. 233–258, Kluwer, Dordrecht, 1993.

[43] Sasao T and Butler JT. On bi-decompositions of logic functions. In *Proceedings of the ACM/IEEE International Workshop on Logic Synthesis*, Tahoe City, CA, pp. 18–21, 2001.

[44] Selvaraj H, Łuba T, Nowicka M, and Bignall B. Multiple-valued decomposition and its applications in data compression and technology mapping of switching functions and rational databases. In *Proceedings of the International Conference on Computational Intelligence and Multimedia Applications,* Cold Coast, Australia, pp. 42–48, 1997.

[45] Sentovich E and Brand D. Flexibility in logic. In Hassoun S and Sasao T, Eds., Brayton RK, consulting Ed. *Logic Synthesis and Verification*, pp. 65–88, Kluwer, Dordrecht, 2002.

[46] Sentovich EM, Singh KJ, Lavagno L, Moon C, Murgai R, Saldanha A, Savoj H, Stephan PR, Brayton RK, and Sangiovanni-Vincentelli AL. *SIS: A System for Sequential Circuit Synthesis.* University of California, Berkeley, CA, Technical Report UCB/ERI, M92/41, ERL, Dept. of EECS, 1992.

[47] Sinha S, Mishchenko A, and Brayton RK. Topologically constrained logic synthesis. In *Proceedings of the International Conference on Computer-Aided Design*, pp. 679–686, 2002.

[48] Somenzi F. Binary decition diagram (BDD) package: CUDD v. 2.3.1. URL http://vlsi.colorado.edu /~fabio/CUDD/cuddIntro.html. University of Colorado at Boulder. 2001.

[49] Stankovic RS. Matrix-valued EXOR-TDDs in decompositon of switching functions. In *Proceedings of the 29th IEEE International Symposium on Multiple-Valued Logic*, pp. 154–159, 1999.

[50] Stanion T and Sechen C. Boolean division and factorization using binary decision diagrams. In *IEEE Transactions on Computer-Aided Design of Integrated Circuits and Systems*, 13(9):1179–1184, 1994.

[51] Steinbach B and Hesse K. Design of large digital circuits utilizing functional and structural properties. In *Proceedings of the 2nd Workshop on Boolean Problems*, Freiberg, Germany, pp. 23–30, 1996

[52] Steinbach B and Lang Ch. Exploiting functional Pproperties of Boolean functions for optimal multi-level design by bi-decomposition. In Yanushkevich S, Ed. *Artificial Intelligence in Logic Desing*, pp. 159–200, Kluwer, Dordrecht, 2004.

[53] Steinbach B and Lang Ch. Complete bi-decomposition of multiple-valued functions using MIN and MAX gates. In *Proceedings of the 35th International Symposium on Multiple-Valued Logic*, pp. 69–74, 2005.

[54] Steinbach B and Le TQ. *Entwurf Testbarer Schaltnetzwerke.* Wissenschaftliche Schriftenreihe 12/1990, Technical University Chemnitz, Germany, 1990.

[55] Steinbach B, Perkowski M and Lang Ch. Bi-decomposition of multi-valued functions for circuit design and data mining applications. In *Proceedings of the 29th IEEE International Symposium on Multiple-Valued Logic*, pp. 50–58, 1999.

[56] Steinbach B, Schuhmann F, and Stöckert M. Functional decomposition of speed optimized circuits. In Auvergne D and Hartenstein R, Eds., *Power and Timing Modelling for Performance of Integrated Circuits.* Bruchsal, IT Press, Germany, pp. 65–77, 1993.

[57] Steinbach B and Stöckert M. Design of fully testable circuits by functional decomposition and implicit test pattern generation. In *Proceedings of the 12th IEEE VLSI Test Symposium*, pp. 22–27, 1994.

[58] Steinbach B and Wereszczynski A. Synthesis of multi-level circuits using EXOR-gates. In *Proceedings of the IFIP WG 10.5 Workshop on Applications of the Reed–Muller Expansion*, Chiba, Japan, pp. 161–168, 1995.

[59] Steinbach B and Zhang Z. Synthesis for full testability of partitioned combinational circuits using Boolean differential calculus. In *Proceedings of the 6th International Workshop on Logic and Synthesis*, Granlibakken, USA, pp. 1–4, 1997.

[60] Tokmen VH. Disjoint decomposability of many-valued functions. In *Proceedings of the 10th International Symposium on Multiple-Valued Logic,* pp. 88–93, 1980.

[61] Vasudevamurthy J and Rajski J. A method for concurrent decomposition and factorization of Boolean expression. In *Proceedings of the International Conference on Computer-Aided Design,* pp. 510–513, 1990.

[62] Yamashita S, Sawada H, and Nagoya A. An efficient method for finding an optimal bi-decomposition. *IEICE Trans. Fundamentals*, E81-A(12):2529–2537, 1998.

[63] Yamashita S, Sawada H, and Nagoya A. SPFD: a method to express functional flexibility. *IEEE Transactions on Computer-Aided Design of Integrated Circuits and Systems*, 19(8):840–849, 2000.

[64] Yang C and Ciesielski M. BDS: a BDD-based logic optimization system. *IEEE Transactions on Computer-Aided Design of Integrated Circuits and Systems,* 21(7):866–876, 2002.

26

Complexity of Decision Diagrams

In this chapter, we consider the complexity of decision diagrams. We begin with complexity measured as the number of nodes. Specific types of functions are considered first to illustrate the nature of the problem. Bounds are then presented for general and totally symmetric functions. The work is presented in terms of p-valued multivalued decision diagrams (MDDs) of which BDD are a special case ($p = 2$).

Path length is a second major factor in assessing decision diagram complexity. An approach employing symmetric variable nodes to reduce both the node count and the path length is presented. Finally, a method employing multivalued logic to reduce path length is reviewed.

26.1 Introduction

A significant advantage of decision diagrams is that they often yield a representation that in both complexity and practical terms is simpler than traditional representations for binary and multivalued functions. That said, the question arises as to what the appropriate complexity measures are for decision diagrams and how they relate to practical considerations.

The number of nodes is an obvious and simple measure that is in fact quite useful. It measures the storage requirement and, for procedures that must examine the entire diagram, it also estimates the time complexity.

The maximum possible path length in a decision diagram is clearly equal to the number of variables. The actual longest and average path lengths, however, can vary in practice and can be a deciding factor in the complexity of many decision diagram based procedures.

The complexity of a decision diagram both in terms of the number of nodes and path length depends on the variable ordering underlying the diagram. In some instances, a good or even optimal variable ordering is clear from the nature of the functions, but in many cases the choice is not obvious. It is thus useful to consider the worst case against which particular instances can be judged.

26.2 Specific functions

As background for the general case and due to their practical interest, it is instructive to first consider specific functions. We begin with the basic logic functions and then consider adders and multipliers. We also examine the function class known as hidden weighted bit functions which are inherently simple but have complex BDD.

26.2.1 Basic logic functions

Analysis for the basic logic functions is straightforward but also important due to their practicality and as a starting point for considering the more complex cases.

Example 26.1 *Consider the logical AND of n variables. We observe the following for its BDD representation:*

- *The number of nonterminal nodes is n since when all variables are 1 there is a single path to the terminal node 1, and the 0 branch from every nonterminal node leads directly to the terminal 0 node.*
- *Each variable controls one node in the diagram and the growth in the number of nodes is linear as the number of variables increases.*
- *The number of paths is $n+1$, the longest path length is n, and the average path length is*

$$\frac{n + \sum_{i=1}^{n} i}{n+1} = \frac{n(3+n)}{2(n+1)}$$

- *It is readily shown that the increase in the average path length goes to $\frac{1}{2}$ as n goes to ∞. Indeed it does so quite quickly as the increase is 0.501 when $n = 30$.*

The analysis in Example 26.1 is true for any function where one input pattern leads to 1(0) and all others lead to 0(1), the so-called vertex functions including logical AND as shown, and also logical OR, NOR, NAND. The analysis is the same if complemented edges are used in the decision diagram.

Conventional representations, e.g. sum-of-products and cube notation are also linear in the number of variables for vertex functions.

> *Conventional representations grow exponentially for the exclusive-OR of n variables. A sum-of-products representation has 2^{n-1} terms. The BDD representation exhibits linear growth.*

Example 26.2 *The BDD without complemented edges for the exclusive-OR of three variables is shown in Figure 26.1(a). The number of nonterminal nodes is $2n - 1$. The number of paths is 2^n and the length of every path is n. If complemented edges are used, see Figure 26.1(b), the number of nonterminal nodes is n. The number of paths remains 2^n and again they all have length n.*

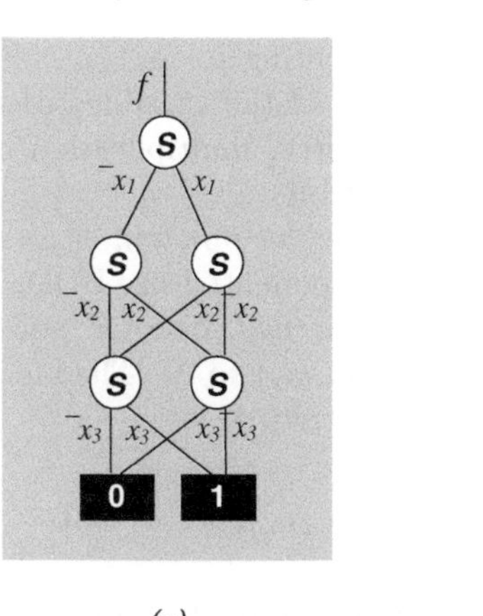

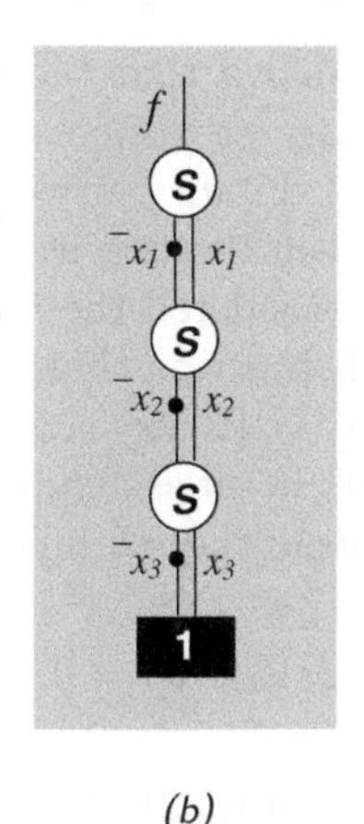

FIGURE 26.1
Decision diagrams for the exclusive-OR of 3 variables.

The exclusive-OR example shows the potential compactness of decision diagrams in an important way since exclusive-OR constructs are found in arithmetic and coding structures.

26.2.2 Adders

Addition of two n bit binary numbers $[a_{n-1}, a_{n-2}, ..., a_0]$ and $[b_{n-1}, b_{n-2}, ..., b_0]$ results in an $n+1$ bit sum $[s_n, s_{n-21}, ..., s_0]$. There are two obvious variable orderings to try for this problem:

- *Interleaved* $[a_{n-1}, b_{n-1}, a_{n-2}, b_{n-2}, ..., a_0, b_0]$ and
- *Non-interleaved* $[a_{n-1}, a_{n-2}, ..., a_0, b_{n-1}, b_{n-2}, ..., b_0]$.

It is readily verified that the BDD grows linearly as n increases for interleaved order, whereas it grows exponentially for non-interleaved order. This is a result of the fact the functional specification of an adder is symmetric in each variable pair $a_i, b_i, 0 \leq i \leq n-1$.

26.2.3 Multipliers and hidden weighted bit functions

Consider multiplication of two n bit integers yielding a $2n$ bit result:

$$\langle p_{2n-1}, ..., p_0 \rangle = \langle a_{n-1}, ..., a_0 \rangle \times \langle b_{n-1}, ..., b_0 \rangle$$

Bryant [2] has shown that the BDD representing p_n and p_{n-1} exhibit exponential complexity as n increases regardless of the variable order. Note that while addition exhibits pairwise bit symmetry, multiplication only exhibits symmetry between the two least significant bits.

Bryant also considered the *hidden weighted bit function* as an example of a relatively simple function that requires an exponential size BDD. The function has n inputs $X = \{x_1, x_2, ..., x_n\}$. Define *weight*, denoted $wt(a)$, to be the number of 1 bits in an assignment $\{a_1, a_2, ..., a_n\}$ to X. The hidden weighted bit function selects the ith input, where $i = wt(a)$:

$$HWB(a) = 0,\ wt(a) = 0,\ \ x_{wt(a)},\ wt(a) > 0.$$

Bryant showed that the BDD representation of HWB requires $\Omega(1.14^n)$ nodes:

$$BDD_{HWB} = \Omega(1.14^n).$$

Bryant shows that HWB has a VLSI implementation with low area-time complexity, which shows that BDD complexity is not necessarily directly related to implementation complexity.

26.3 Bounds

For generality, we will consider $f(x_0, x_1, ..., x_{n-1})$, a totally-specified p-valued function where the x_i are also p-valued. The variables are indexed from 0 for notational convenience below. Switching functions are the case where $p = 2$.

A function f can be represented by a reduced ordered multivalued decision diagram (MDD):

- with up to p terminal nodes, each labelled by a distinct value $0, 1, ..., p-1$;
- and where every non-terminal node is controlled by an input variable and has p outgoing edges; one corresponding to each logic value.

The MDD can be considered to have $n+1$ levels with the variables $x_0, x_1, ..., x_{n-1}$ associated with levels 0 through $n-1$ and the terminal nodes associated with level n.

Let:

- R_s denote the maximum number of nodes at level s and
- $R(n, p)$ denote the maximum number of nonterminal nodes in a p-valued, n-variable MDD.

For the special case of a totally symmetric function, we use

- Q_s to denote the maximum number of nodes at level s and
- $Q(n, p)$ to denote the maximum number of nonterminal nodes in a p-valued, n-variable MDD.

26.3.1 The general case

First, consider the general case of all p-valued functions. The decision diagram for a function has a single node at the top (root) and expands as you go down the first number of levels. It then starts to decrease in breadth down to level n which will, in general, have p terminal nodes. This structure is illustrated in Figure 26.2.

The following Lemma, which is a generalization of Heap and Mercer's Lemma 2 [7] for BDD, codifies the structure in Figure 26.2.

Lemma 26.1 *For* $0 \leq s < n$, $R_s = \min(p^s,\ p^{p^{n-s}} - p^{p^{n-s-1}})$

The proof of Lemma 26.1 and all other results can be found in the Appendix at the end of this chapter.

One can observe that:

- p^s is strictly increasing and
- $p^{p^{n-s}} - p^{p^{n-s-1}}$ is strictly decreasing with respect to s.

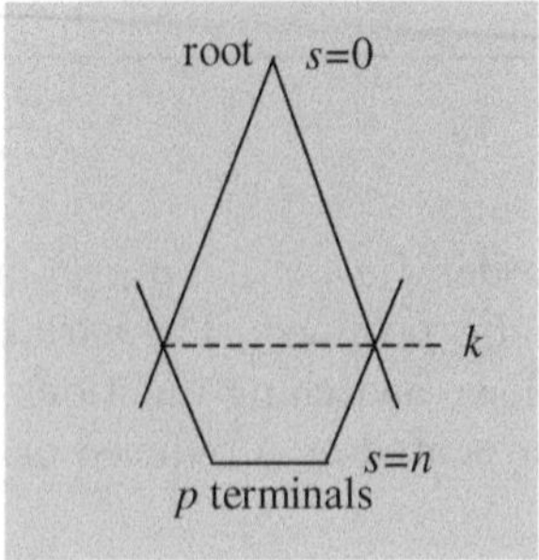

FIGURE 26.2
General structure of an MDD.

We are interested in the cross-over point k, which is defined by

$$k = \max(s|p^{p^s} - p^{p^{s-1}} \leq p^{n-s},\ 0 \leq s < n). \tag{26.1}$$

The cross-over is important since it defines the shape and the size of the worst case size MDD.

Note that:

- If $p^{p^s} - p^{p^{s-1}} \leq p^{n-s}$ holds, it holds for all smaller values of $s \geq 0$, and
- If it does not hold, it holds for no larger values of s.

Lemma 26.2 *k as defined in equation 26.1 is given by*

$$k = \lfloor \log_p(n - \lfloor log_p n \rfloor) \rfloor.$$

The principal result for general functions is

Theorem 26.1

$$R(p,n) = \frac{p^{n-k} - 1}{p - 1} + p^{p^k} - p,$$

where

$$k = \lfloor \log_p(n - \lfloor log_p n \rfloor) \rfloor.$$

Example 26.3 *Let $p = 2$, then*

$$R(2,n) = 2^{n-k} + 2^{2^k} - 3, \quad k = \lfloor \log_2(n - \lfloor log_2 n \rfloor) \rfloor.$$

26.3.2 Totally-symmetric functions

Having established the general case, it is of interest to look at what one would suspect to be a better case – totally symmetric functions.

A function is *totally symmetric* if it is unaffected by a permutation of the input variables. The number of p-valued, n-variable, totally symmetric functions is

$$p^{\binom{n+p-1}{n}}.$$

This follows from the property that when a function is totally symmetric, it is the number of occurrences of each logic value in the input pattern and not their order that determines the function value.

Note that if repetition is allowed, the number of selections of r objects that can be made from a set of m distinct objects is

$$\binom{r+m-1}{r}. \tag{26.2}$$

The number of n length selections taken from the p logic values is given by substitution into Equation 26.2. Since there are p possibilities for the function value for each distinct pattern the expression given follows. Note that p of these functions are constant while the rest depend on all n variables.

Lemma 26.3 *For* $0 \leq s < n$,

$$Q_s = \min\left(\binom{s+p-1}{s}, p^{\binom{n-s+p-1}{n-s}} - p\right).$$

Following the approach above, we now wish to determine

$$k = \max\left(s|p^{\binom{s+p-1}{s}} - p \leq \binom{n-s+p-1}{n-s}, 0 \leq s < n\right) \tag{26.3}$$

For $p = 2$, the answer is given by the following Lemma.

Lemma 26.4 *For* $p = 2$, *k as defined in Equation 26.3 is given by the largest of*

$$\begin{aligned} &q, \textit{ if } r - q + 3 \geq 2^q, \\ &q - 1, \textit{ if } r + 4 \geq q, \\ &q - 2, \textit{ otherwise.} \end{aligned}$$

This result is equivalent to Lemma 4 in [6] but is stated in a different form and in particular is applicable for all n and does not treat $n \leq 3$ as special cases.

Equation 26.3 does not have a simple solution for $p > 2$. However, we can by inspection identify the appropriate values of k for practical values of n.

Theorem 26.2

$$Q(p,n) = \sum_{s=0}^{n-k-1} \binom{s+p-1}{s} + \sum_{s=1}^{k} \left[p^{\binom{s+p-1}{s}} \right] - kp \qquad (26.4)$$

where k is as given above.

The proof is similar to the proof of Theorem 26.1 given in the Appendix.

The first summation in Equation 26.5 can be reduced to

$$\binom{n-k+p-1}{n-k-1}$$

but the second summation can not be simplified. For the case of $p = 2$, Equation 26.5 reduces to

$$Q(2,n) = \sum_{s=0}^{n-k-1} \binom{s+1}{s} + \sum_{s=1}^{k} \left[2^{\binom{s+1}{s}} \right] - kp \qquad (26.5)$$

with k given by Lemma 26.4.

26.3.3 Existence

Having established bounds in the last section, it is necessary to show that functions exist that exhibit the worst-case size MDD for both the general and the totally symmetric cases. The proof of the following theorem appears in the Appendix.

Theorem 26.3 *For all $p \geq 2$ and all $n \geq 1$, there exists a function whose MDD for at least one variable ordering has nonterminal node count given by the bound in Theorem 26.1.*

Theorem 26.4 *For all $p \geq 2$ and all $n \geq 1$, there exists a totally-symmetric function for which the MDD has the number of non-terminal nodes given by the bound in Theorem 26.2.*

The proof is analogous to the proof of Theorem 26.3 given in the Appendix.

26.4 Complemented edges

It is common to use complemented edges to reduce the size of a decision diagram. For p-valued functions one possibility is cyclic negation defined as

$$C_k(x) = (x+k)mod_p \text{ for } 0 < k < p.$$

This reduces to the normal BDD edge complement when $p = 2$.

Consider the following:

- An edge in an MDD points to a subgraph representing a subfunction g.
- An edge with a cyclic negation C_k means the destination is $C_k(g)$.
- We assume the MDD is normalized so that no cyclic negation appears on a 0-edge from any node and the MDD has a single terminal node with value 0. Note that this differs from the normalization often used in BDD packages such as CUDD where it is the 1-edge that cannot have a complement and the terminal values is 1. That corresponds to using $p - 1$.
- The value 0 is appropriate for the MDD case since that is consistent for heterogeneous MDD where the variables have differing value ranges. Using the value 0 or $p - 1$ in this context has no effect on the results presented here.
- In terms of worst case size, the top portion of the diagram, levels 0 to $n - k - 1$ remain a tree.
- The bottom portion, levels $n - k$ to n, has $1/p$ the number of nodes since only one of each of the functions equivalent under cyclic negation need be represented. It is clear using the same reasoning as above that there are enough choices to properly connect the edges from level $n - k - 1$ to nodes lower in the MDD.

For general functions represented as MDD with cyclic edge negations, we have

$$k = \max(s | p^{p^s - 1} - p^{p^{s-1} - 1} \leq p^{n-s},\ 0 \leq s < n). \tag{26.6}$$

Lemma 26.5 *k as defined in equation 26.6 is given by*

$$k = \lfloor \log_p(n - \lfloor log_p n \rfloor + 1) \rfloor.$$

Following the process in Theorem 26.1 it is readily shown that

$$R(p, n) = \frac{p^{n-k} - 1}{p - 1} + p^{p^k - 1} - 1.$$

For MDD with cyclic edge negations for totally symmetric functions,

$$Q(p, n) = \sum_{s=0}^{n-k-1} \binom{s + p - 1}{s} + \sum_{s=1}^{k} \left[p^{\binom{s+p-1}{s} - 1} - 1 \right], \tag{26.7}$$

and the existence is demonstrated in a manner analogous to the totally symmetric case without edge cycles. k is given by

$$k = \max(s | p^{\binom{s+p-1}{s} - 1} - 1 \leq \binom{n - s + p - 1}{n - s},\ 0 \leq s < n). \tag{26.8}$$

For $p = 2$, the answer is given by the following Lemma. As above, the value of k for $p > 2$ must be found by enumeration.

Lemma 26.6 *For $p = 2$, k as defined in Equation 26.8 is given by the largest of*

$$\begin{array}{l} q+1, \quad \text{if} \ \ r-q+1 \geq 2^q, \\ q, \quad \text{if} \ \ r+2 \geq q, \\ q-1, \quad \text{otherwise.} \end{array}$$

The optimal improvement for cyclic edge negations is $(p-1)/p \times 100\%$. The improvement in the worst case size is at best on the order of 1/2 the optimal improvement given for p.

26.5 Path length

Thus far, complexity of a decision diagram has been considered in terms of node count. In Chapter 12, various variable reordering techniques were reviewed. These methods, sifting and its enhancements in particular, provide practical approaches to reducing decision diagram complexity. A modification to sifting to reduce *average path length* (APL) was also discussed in Chapter 12. That work, and related work, is heuristic and no formal analysis of the complexity of decision diagrams with respect to path length appears to have been undertaken.

Here, we review two related approaches to reducing the complexity of a decision diagram. The first approach exploits the symmetry in a function to reduce the node count and the path lengths. The second strives to reduce the longest path length using heterogeneous MDD.

26.5.1 Symmetric variable nodes

Symmetry can play an important role in reducing the node count. In particular, experience shows that an effective variable ordering typically has symmetric variables adjacent. Symmetry can also be exploited by introducing a new type of node, termed a *symmetric variable node*, which captures the symmetry property over a group of variables.

Definition 26.1 *A symmetric variable node (SVN) is labelled by a set of t variables $S = [x_{i_1}, x_{i_2}, ..., x_{i_t}]$ and there is an outgoing edge for each unique assignment of values to the variables up to permutation. All variables in S must be p-valued.*

The requirement to use a SVN is that the subgraph with the SVN as its top node must be symmetric in the variables in S.

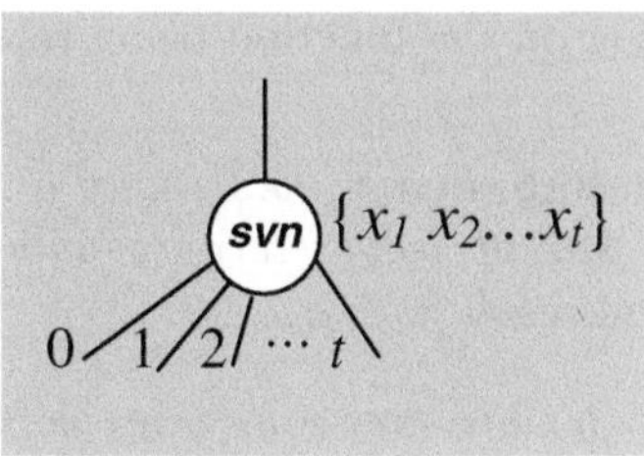

FIGURE 26.3
A binary, t variable SVN.

Lemma 26.7 *The number of outgoing edges from a SVN labelled by t p-valued variables is*

$$\frac{1}{(p-1)!}\prod_{j=1}^{p-1}(t+j)$$

> **Example 26.4** *Let $p = 2$, the number of outgoing edges is $t+1$. This case is depicted in Figure 26.3 where the edge labels denote the number of variables that take the value* 1. *t minus that number of course take the value* 0.

The notion of *ordering* is extended to diagrams with SVN such that the subsets that label nodes adhere to a fixed order. The subsets need not be disjoint, but a variable can only appear once on any path.

The normal reduction rules apply so there are no redundant nodes and there are no equivalent subgraphs. In addition, no node can be combined with its children to form a single SVN.

SVN can be identified using a top-down greedy recursive algorithm, or alternatively can be built bottom-up. Both are deterministic so they produce a unique diagram, but the two approaches do not produce the same result.

The top-down greedy algorithm begins at the root node. For a shared diagram, each root is considered in turn. For each node visited, the algorithm determines if that node and its children can be combined into a SVN. Once a SVN is formed the procedure attempts to combine it with its children to form a larger SVN. A simple marking technique can be used to ensure each node is processed only once in the recursive traversal.

The bottom-up approach operates in a manner similar to the top-down method, except that as the name implies SVN are formed at the bottom of the diagram first and then the process works towards the top forming SVN as large as possible.

Introducing SVN can reduce the node count, sometimes significantly, but can never increase it. Introducing SVN can either increase or decrease the number of edges depending on the function being represented.

Example 26.5 *Figure 26.4 shows a MDD with SVN for the sum of a pair of 2-digit ternary numbers* $[a_1, a_0]$ *and* $[b_1, b_0]$ *yielding the result* $[c_1, s_1, s_0]$.

- *SVN nodes are drawn as ovals with the variable set inside the oval.*
- *For the edge labels, the pattern before the slash shows the normalized input assignment while the value after the slash identifies the applicable cyclic edge negation.*
- *The normalization rule is that there is never a cyclic negation on an edge for the value pattern of all zeroes, and the terminal node is a 0.*
- *Note that the terminal node* 0 *is shown three times in Figure 26.4 for clarity. In practice there is of course only one terminal node.*

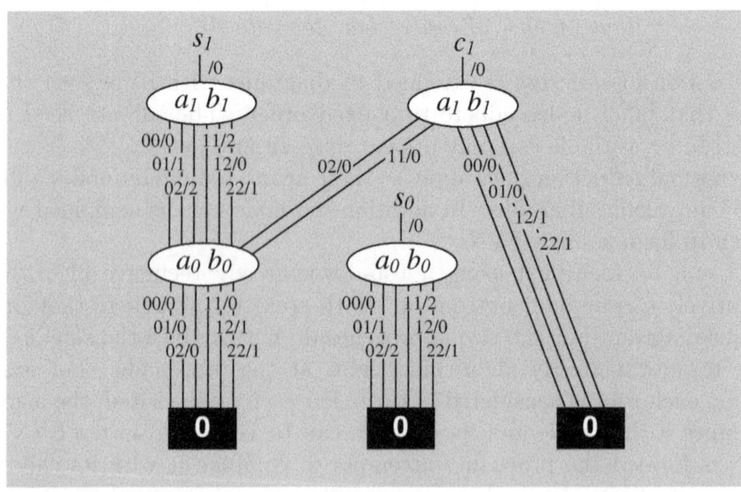

FIGURE 26.4
MDD with SVN for add function in Example 26.5.

The variable number of edges from SVN makes the implementation somewhat more difficult. Dynamic memory allocation can be used with the vector of edges at the end of the structure. The size of the structure created depends on the number of edges required for the node at hand. Reuse of freed nodes takes some care as the sizes of the nodes vary.

26.5.2 Experimental results

Experimental results for decision diagrams with SVN are presented in [11] where the top-down greedy construction approach is used. Results are presented for 8 fairly small binary benchmarks (the most complex has 25 inputs and 14 outputs). The size of the BDD and BDD with SVN are measured in terms of the actual storage area required since SVN can have varying size and generally require more space than a regular BDD node. On average the BDD with SVN are 86% the size of the BDD. In one case, the benchmark VG2, there is a 36% space reduction using SVN.

Adders are also considered in [11] for varying number of inputs and $p = 2, 3, 4$. Interleaved variable ordering is used for the decision diagrams. The size of the decision diagrams with SVN derived from the interleaved decision diagrams shows significant improvement and is likely asymptotic to a value around 52 to 53%.

A significant advantage of introducing SVN is that the longest path can be significantly shorter than the longest path in the original decision diagram. For the binary benchmarks in [11], the longest path is reduced to as little as 33% of the original. For the adders, because of their inherent pairwise bit symmetry, the longest path using SVN is exactly half of that without SVN.

Multipliers. Not surprisingly, no significant improvement is found in the length of the longest path since the only symmetry in the problem is that between the two least significant bits of the values being multiplied [11].

26.5.3 Longest path length oriented techniques

Minimizing the average path length is very effective in certain applications such as logic simulation since it minimizes the average function evaluation time. Furthermore, for other applications such as the synthesis of embedded systems and pass transistor networks, the longest path length (LPL) is the critical factor. This is readily evident in the latter case since when converting a decision diagram to a pass transistor network, the longest path determines the worst case delay through the network.

Example 26.6 *Variable reordering aimed at minimizing the number of nodes (see Chapter 12) can on occasion reduce the LPL [14] as shown in Figure 26.5:*

- *The BDD in Figure 26.5(a) minimizes the number of nodes. The LPL is 5, which is the number of variables.*
- *The BDD in Figure 26.5(b) reduces the LPL to 4, but the number of nodes is increased by 1.*

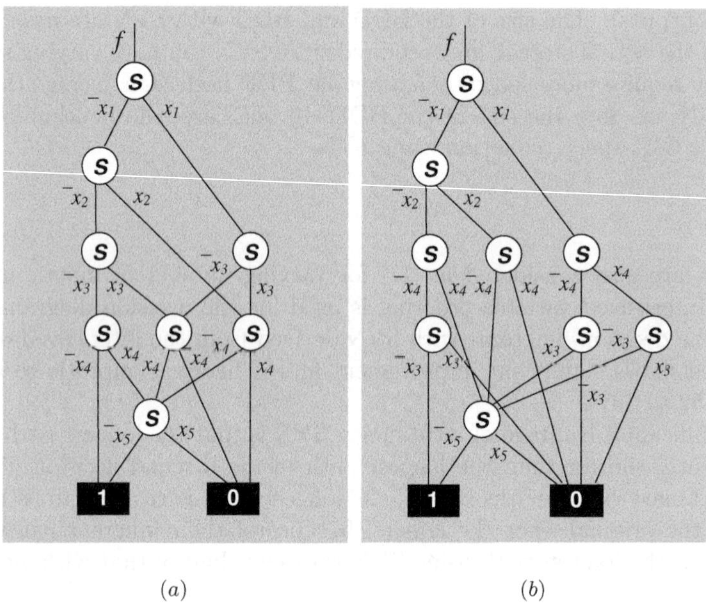

FIGURE 26.5
Example of LPL targeted sifting (Example 26.6).

An approach based on sifting. Another approach is to apply sifting but to use the LPL as the minimization criterion rather than the node count. Nagayama and Sasao [14] reported on this approach for a set of 21 benchmark problems including the ISCAS85 benchmarks. They found the average improvement to be quite small, on the order of 1% improvement in the LPL, with a corresponding average increase in the node count of 25%. The best increase is for a problem known as *i8* where the LPL was reduced from 1,044 to 853, but at a cost of increasing the node count from 1,275 to 2,195. The problem *i10* saw a reduction of the LPL from 4,643 to 4,483 with a node count increase from 20,659 to 61,815. These results show LPL targeted sifting can be effective but at a cost in storage area. This is not surprising since to reduce the LPL most often requires restructuring of the whole decision diagram and not just restructuring of a localized area.

An approach based on heterogeneous MDD. An alternative approach is to represent a switching function by a *heterogeneous MDD*. In this case, the binary variables are partitioned into disjoint sets. Each nonterminal node is labelled by one of those sets and has outgoing edges for each distinct function outcome which are identified as integers.

> **Example 26.7** *Figure 26.6 shows (a) a BDD, (b) an equivalent diagram with*
>
> $$X_1 = (x_1, x_2, x_3) \text{ and } X_2 = (x_4),$$
>
> *and (c) another equivalent diagram with*
>
> $$X_1 = (x_1) \text{ and } X_2 = (x_2, x_3, x_4).$$
>
> *Note that the diagrams in (b) and (c) both reduce the LPL from 4 to 2, but the diagram in (c) requires one more node and one more edge.*

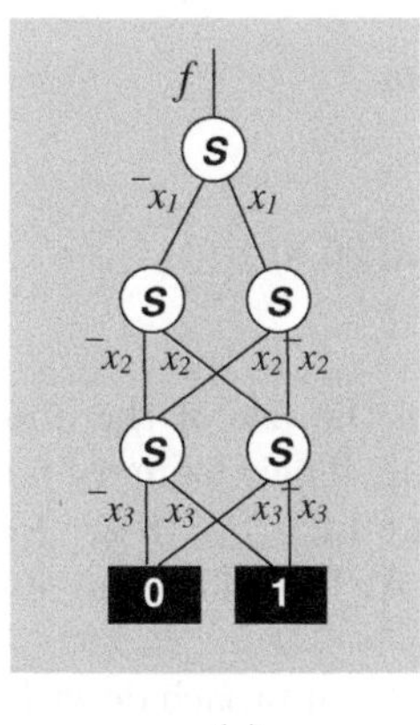

(a)

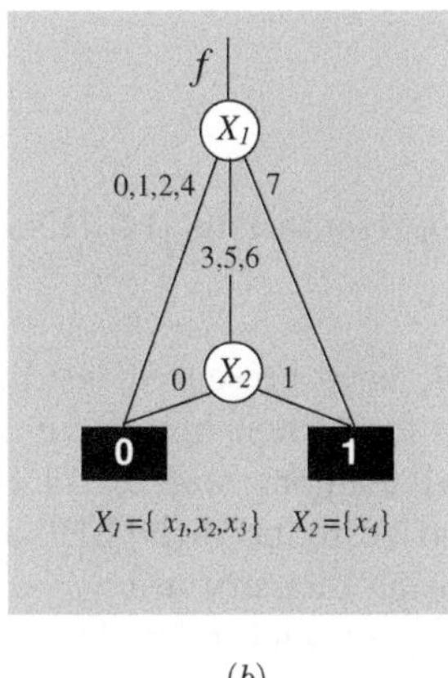

(b)

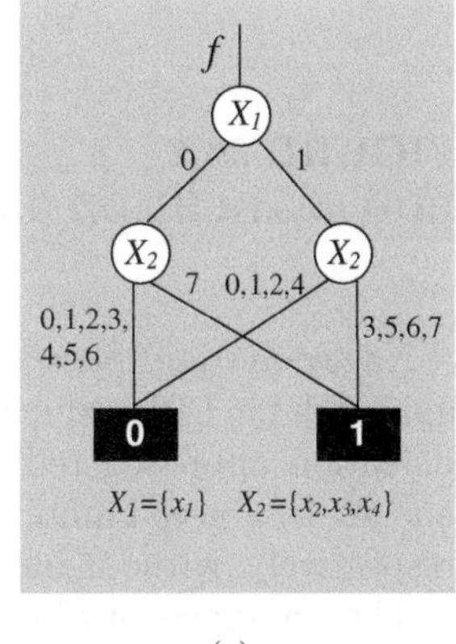

(c)

FIGURE 26.6
BDD (a) and two alternative representations: modified diagram with minimum memory size (b) and with maximum memory size (c) (Example 26.7).

Example 26.8 *Figure 26.7 shows that the heterogeneous MDD approach can reduce the node count, the LPL and the number of edges significantly in some cases. In this case,*

$$X_1 = (x_2, x_3, x_4),$$
$$X_2 = (x_1, x_5).$$

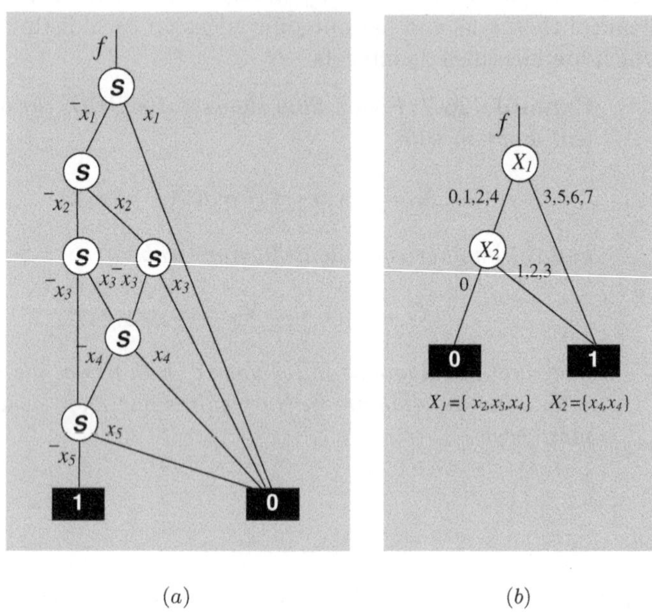

(a) (b)

FIGURE 26.7
BDD (a) and HMDD (b) representations [14] (Example 26.8).

Nagayama and Sasao [14] have presented results on the use of the above approach for LPL reduction for 21 benchmark functions. Their memory minimization approach reduced memory size by 14% and LPL by 18%. LPL and APL targeted minimization reduced the LPL and APL by 43% and 46%, respectively, without increasing memory use.

The heterogeneous MDD approach is related to the SVN approach described in the previous section in that variables are grouped and label more complex nodes, but there are differences. SVN is based on symmetry whereas heterogeneous MDD groupings do not require such a specific functional property. On the other hand, the partitioning is fixed across a heterogeneous MDD whereas SVN allows for different groupings at different points in the diagram thereby exploiting partial symmetry. A direct comparison of the two approaches has never been undertaken.

26.6 Further reading

We here considered important aspects of the complexity of BDD and MDD as these are the most fundamental and widely used decision diagram structures. Many alternative decision diagram structures for binary functions have been proposed in the literature and many of these are readily extended to the multivalued case. Reducing the complexity of the decision diagram representation for particular function types and problems has been the motivating factor for most of these alternate proposals.

It is of interest that Burch [4] has shown that exponential growth can be avoided in verifying a multiplier, i.e., verifying that a particular circuit implements multiplication. A constraint is placed on the multiplier and the number of variables is increased from $2n$ to $2n^2$. Hence, while the size of the BDD is reduced, there is a cost associated with mapping the original variables to the intermediate ones. Alternative decision diagram structures have been proposed to better handle multiplication type functions.

Symmetric variable nodes were introduced by Miller and Muranaka [11]. Details of the top-down recursive algorithm for identifying SVN can be found there together with detailed experimental results. The paper also discusses direct circuit implementation of SVN using neuron-MOSFET transistors. Symmetry is a functional property. Hence, while the work in [11] is applied to BDD and MDD, the approach can be extended to other types of decision diagrams.

Use of decision diagrams in the synthesis of embedded systems software is considered in [1, 8, 12, 15, 17] and pass transistor logic synthesis is discussed in [3, 9, 16]. Readers interested in techniques directed to minimizing longest path length are directed to [5, 10, 13, 14].

References

[1] Balarin F, Chiodo M, Giusto P, Hseih H, Jurecska A, Lavagno L, Sangiovanni-Vincentelli A, Sentovich EM, and Suzuki K. Synthesis of software programs for embedded control applications. *IEEE Transactions on Computer-Aded Design of Integrated Circuits and Systems*, 18(6):834–849, 1999.

[2] Bryant RE. On the complexity of VLSI implementations and graph representations of Boolean functions with application to integer multiplication. *IEEE Transactions on Computers*, 40(2):205–213, 1991.

[3] Buch P, Narayan A, Newton AR, and Sangiovanni-Vincentelli A. Logic synthesis for large pass transistor circuits. In *Proceedings of the International*

Conference on Computer Design, pp. 663–670 1997.

[4] Burch JR. Using BDDs to verify multipliers. In *Proceedings of the ACM/IEEE Design Automation Conference*, pp. 408–412, 1991.

[5] Ebendt R, Hoehme S, Guenther W, and Drechsler R. Minimization of the expected path length in BDDs based on local changes. In *Proceedings of the Asia and South Pacific Design Automation Conference*, pp. 866–871, 2004.

[6] Heap MA. On the exact ordered binary decision diagram size of totally symmetric functions. *JETTA*, 4:191–195, 1993.

[7] Heap MA and Mercer MR. Least upper bound on OBDD sizes. *IEEE Transactions on Computers*, 43(6):764–767, 1994.

[8] Lindgren M, Hanseon H, and Thame H. Using measurements to derive the worst-case execution time. In *Proceedings of the 7th International Conference on Real-Time Systems and Applications*, pp. 12–14, 2000.

[9] Liu TH, Ganai MK, Azie A, and Burns JL. Performance driven synthesis for pass-transistor logic. In *Proceedings of the International Workshop on Logic Synthesis*, pp, 255–259, 1998.

[10] Liu YY, Wang KH, Hwang TT, and Liu CL. Binary decision diagrams with minimum expected path length. In *Proceedings of the Conference Design and Test in Europe*, pp. 708–712, 2001.

[11] Miller DM and Muranaka N. Multiple-valued decision diagrams with symmetric variable nodes. In *Proceedings of the IEEE 26th International Symposium on Multiple-Valued Logic*, pp. 242–247, 1996.

[12] Nagayama S and Sasao T. Code generation for embedded systems using heterogeneous MDDs. In *Proceedings of the 12th Workshop on Synthesis and System Integration of Mixed Information Technologies*, pp. 258–264, 2003.

[13] Nagayama S, Mishchenko A, Sasao T, and Butler JT. Minimization of average path length in BDDs by variable reordering. In *Proceedings of the International Workshop on Logic Synthesis*, pp. 28–30, 2003.

[14] Nagayama S and Sasao T. On the minimization of longest path length for decision diagrams. In *Proceedings of the International Workshop on Logic and Synthesis*, pp. 28–35, 2004.

[15] Nolte T, Hanson H, and Norstrom C. Probabilistic worst-case response-time analysis for the controller area network. In *Proceedings of the 9th IEEE Real-Time and Embedded Technology and Applications Symposium*, pp. 200–207, 2003.

[16] Scholl C and Becker B. On the generation of multiplexor circuits for pass transistor logic. In *Proceedings of the Conference Design and Test in Europe*, pp. 372–378, 2000.

[17] Zu M and Cheng AMK. Real-time scheduling of hierarchical reward-based tasks. In *Proceedings of the 9th IEEE Real-Time and Embedded Technology and Applications Symposium*, pp. 2–9, 2003.

Appendix

This Appendix presents proofs for the principal formal results on complexity given in the Chapter. Recall that an integer n can be expressed as $n = p^q + r$ where $q = \lfloor \log_p n \rfloor$. Clearly, $r < (p-1)p^q$.

Proof of Lemma 26.1: We first show by induction that for $0 \le s < n$, $R_s \le p^s$. $R_0 = 1 \le p^0$. Assume, $R_i \le p^i$. Each node at level i has at most p children. It follows that $R_{i+1} \le pR_i \le p^{i+1}$.

Next we show $R_s \le p^{p^{n-s}} - p^{p^{n-s-1}}$. This follows since the number of nodes from level s to n is bounded above by the number of distinct p-valued functions of $n-s$ variables including functions independent of some of those variables and the constant functions. There are $p^{p^{n-s}}$ such functions. However, $p^{p^{n-s-1}}$ of these do not depend on x_s so the number of nodes at level s is bounded above by $p^{p^{n-s}} - p^{p^{n-s-1}}$.

Proof of Lemma 26.2: Observe that $n - (q-1) = p^q + r - q + 1 > p^{q-1}$ since $q < (p-1)p^{q-1} + r + 1$. Hence $p^{n-(q-1)} > p^{p^{q-1}}$ so $k \ge q - 1$. Note that $p^{p^k} - p^{p^{k-1}}$ is bounded below by $p^{p^k - 1}$. $p^q + r - q \ge p^q - 1$ only if $r \ge q$ so k can only be greater than or equal to q when $r \ge q$. In fact, $k \ge q$ when $r \ge q$ since $p^q + r - q \ge p^q$ so $p^{n-q} \ge p^{p^q}$. Since $p^q + r - q - 1 < p^{q+1} - 1$, $p^{n-(q+1)} < p^{p^{q+1}-1}$ so $k < q + 1$.

It follows that $k = q - 1$ when $r < q$ and $k = q$ when $r \ge q$. This is captured by the single expression $k = \lfloor \log_p (p^q + r - q) \rfloor$ which is $k = \lfloor \log_p (n - \lfloor \log_p n \rfloor) \rfloor$.

Proof of Theorem 26.1: Using Lemma 26.1 we have

$$R(p,n) = \sum_{s=0}^{n-1} R_s = \sum_{s=0}^{n-1} \min(p^s, p^{p^{n-s}} - p^{p^{n-s-1}})$$

Applying k as found in Lemma 26.2 gives

$$\begin{aligned} R(p,n) &= \sum_{s=0}^{n-k-1} p^s + \sum_{s=1}^{k} p^{p^s} - p^{p^{s-1}} \\ &= \frac{p^{n-k} - 1}{p-1} + p^{p^k} - p \end{aligned}$$

Proof of Lemma 26.3: Since the function is totally symmetric, the number of nodes at level s is bounded above by the number of selections of s edge values from the p logic values allowing repetition but ignoring the order since two paths with the same values must reach the same node regardless of the order of the values. It follows that,

$$Q_s \le \binom{s+p-1}{s}$$

the number of nodes from a level s less the p constant functions. Hence

$$Q_s \leq p^{\binom{n-s+p-1}{n-s}} - p$$

Proof of Lemma 26.4: For $p = 2$, equation 26.3 becomes

$$k = \max(s|2^{s+1} - 2 \leq n - s + 1, 0 \leq s < n)$$

Since $2^{q-1} \leq 2^q + r - q + 5$, $k \geq q - 2$.
Suppose:

- $k = q - 1$. $2^q \leq 2^q + r - q + 4$ if, and only if, $r \geq q - 4$, so $k \geq q - 1$ if, and only if, $r \geq q - 4$.
- $k = q$. $2^{q+1} \leq 2^q + r - q + 3$ if, and only if, $r - q + 3 \geq 2^q$.

Further, $2^{q+2} > 2^q + r - q + 2$ so $k < q + 1$. Combining the above conditions results in the value of k as stated.

Proof of Theorem 26.3: The top part of the MDD, levels 0 to $n - k - 1$ where k is given by equation 26.1, is a uniquely defined tree involving $x_0 ... x_{n-k-1}$. The bottom part of the MDD, levels $n - k$ to n has a node for every possible function involving the variables $x_{n-k} ... x_{n-1}$ including those that depend only on a subset of those variables and the p constant functions which are the terminal nodes at level n. This part of the MDD is also uniquely defined.

The only constructions to define are the edges from level $n - k - 1$ to nodes lower in the MDD. We must show that it is possible to assign these edges so that

(*a*) No two nodes at level $n - k - 1$ have the same descendants i.e., they represent distinct functions, and
(*b*) There is at least one edge leading to each node at level $n - k$.

Note we do not have to worry about edges to nodes lower than level $n-k$ since they already have edges leading to them due to the completeness of the bottom part of the MDD. (a) follows from the fact the number of possible nodes at level $n - k - 1$ is $p^{p^{n-k-1}} - p^{p^{n-k}}$ whereas the number required is p^{n-k-1}. By the definition of k the latter is less than the former since if it is not k would be at a higher level in the MDD. (b) is clearly possible since if in satisfying (a) we do not direct an edge to a node at level $n - k - 1$, we can move an edge from a lower node to the node in question. By the definition of k, there are more edges from level $n - k - 1$ than nodes at level $n - k$ so it is possible to direct at least one edge to every node at level $n - k$.

Proof of Lemma 26.5: Observe that $n - (q - 1) = p^q + r - q + 1 > p^{q-1-1}$ since $q < (p - 1)p^{q-1} + r + 2$. Hence $k \geq q - 1$.

Note that $p^{p^k - 1} - p^{p^{k-1} - 1}$ is bounded below by $p^{p^k - 2}$. $p^q + r - q \geq p^q - 2$ only if $r \geq q - 1$ so k can only be greater than or equal to q when $r \geq q - 1$. In fact, $k \geq q$ when $r \geq q - 1$ since $p^q + r - q \geq p^q - 1$ so $p^{n-q} \geq p^{p^q} - 1$.

Since $p^q + r - q - 1 < p^{q+1} - 2$, $p^{n-(q+1)} < p^{p^{q+1} - 2}$ so $k < q + 1$.

It follows that $k = q - 1$ when $r < q - 1$ and $k = q$ when $r \geq q - 1$. This is captured by the single expression $k = \lfloor \log_p(p^q + r - (q-1)) \rfloor$ which is $k = \lfloor \log_p(n - \lfloor \log_p n \rfloor + 1) \rfloor$.

Proof of Lemma 26.6: For $p = 2$, equation 26.8 becomes

$$k = \max(s | 2^s - 1 \leq n - s + 1, 0 \leq s < n)$$

Since $2^{q-1} \leq 2^q + r - q + 3$, $k \geq q - 1$.
Suppose:

- $k = q$. $2^q \leq 2^q + r - q + 2$ if, and only if, $r + 2 \geq q$, so $k \geq q$ if, and only if, $r + 2 \geq q$.
- $k = q + 1$. $2^{q+1} \leq 2^q + r - q + 1$ if, and only if, $r - q + 1 \geq 2^q$, so $k \geq q + 1$ if, and only if, $r - q + 1 \geq 2^q$.

Further, $2^{q+2} > 2^q + r - q$ so $k < q + 2$. Combining the above conditions results in the value of k as stated.

Proof of lemma 26.7: A SVN has an outgoing edge for each distinct assignment of values to the variables labeling the node up to permutation. For a given p and a SVN labeled by t variables, each unique assignment is defined by a $p - 1$ tuple $(k_1, k_2, ..., k_{p-1})$ where k_i variables take the value i. Clearly, $0 \leq k_i \leq t - \sum_{j=1}^{i-1} k_j$, for $1 \leq i \leq p - 1$. Note that the value 0 is not explicitly listed since it is defined implicitly.

The total number of assignments is given by

$$e = \sum_{k_1=0}^{t} \sum_{k_2=0}^{t-k_1} ... \sum_{k_{p-1}}^{t-q} 1, q = \sum_{j=1}^{p-2}$$

which can be shown to be

$$\frac{1}{(p-1)!} \prod_{j=1}^{p-1} (t + j)$$

27

Programming of Decision Diagrams

In previous chapters, decision diagrams and their features have been discussed from different aspects. However, to apply them in practice, decision diagrams have to be programmed. A programming implementation has to provide all the useful features offered by decision diagrams. In this chapter, we briefly discuss basic issues in programming decision diagrams.

27.1 Introduction

The tasks of programming decision diagrams consist in the following:

- To specify the data structure to represent nodes in a diagram. A decision diagram consists of non-terminal nodes and constant nodes that are connected by edges. There are also labels at the edges, and decomposition rules used to assign a given function to a decision diagram. A method has to be provided to save and subsequently use this information in manipulation and calculation with decision diagrams.
- Reduction rules must be supported by a programming package for decision diagrams. In order to control redundancy methods to periodically free the memory, usually called *garbage collection*, have to be provided. This is a suitable way to control the complexity of dealing with decision diagrams within reasonable limits.

Theoretically, decision diagrams are derived by reducing decision trees. In practice, we do not build a tree, but perform reduction during the construction of the diagram. Since there are no repeated subtrees, the redundant parts should be removed.

27.2 Representation of nodes

A node is described by:

- The decision variable assigned to the node,
- The input edges, and
- The outgoing edges.

This information is specified in different ways in different programming implementations. The structure used to represent a node differs depending on the implementation of the package, the programming environment used, and the intended applications.

Figure 27.1 shows parameters used in the description of a node in a decision diagram. Notice that the level in the diagram to which a variable is assigned is usually stored in a separate vector, the entries of which are variables assigned to the levels in the diagram. The contents of this vector changes if reordering of variables is performed. In this figure, the structure `DdChildren` consists of two fields for pointers to two successive nodes.

```
struct DdChildren
{ struct DdNode *T   Pointer to the right (Then) successor
struct DdNode *E     Pointer to the left (Else) successor
}
```

FIGURE 27.1
Parameters used in the description of a node in a decision diagram.

Description of a node in the decision diagram programming package CUDD is shown in Table 27.1.

27.3 Representation of decision diagrams

The description of nodes discussed above permits storing a decision diagram as a two-dimensional array of nodes and interconnections among them. This array, or table, is implemented as a hash table, performing the basic functions of such a table, consisting in associating a *value* to a *key*. The table is usually called the *Unique Table*, since nodes representing identical subfunctions are not generated; instead, pointers to the unique subfunctions are set [3].

TABLE 27.1
Description of node.

`struct DdNode`	
`{`	
`DdHalfWord index`	*Index of the variable assigned to the node*
`DdHalfWord` *ref*	*Counter of references, i.e., number of nodes pointing to this node*
`DdNode * next`	*Pointer to the following node in the Unique Table*
`union {`	
`CUDD_VALUE_TYPE` *value*	*For constant nodes*
`DdChildren kids`	*For internal, i.e., non-terminal nodes*
`} type`	
`}`	

The Unique Table is organized as a list of chain lists, each of them specifying nodes to which the same variable is assigned, i.e., nodes at the same level in the decision diagram. Entries of the Unique Table are pointers to the nodes. A part of memory is allocated for each node, and the corresponding address is stored in the Unique Table.

In the case of Kronecker diagrams, where different decomposition rules are performed at different levels, a *Decision Type List* saves decomposition rules selected for each level in the diagram. In the case of Pseudo-Kronecker decision diagrams, a matrix of assignment of decomposition rules is provided. It is called the *Extended Decomposition Type List.*

27.4 Construction of decision diagrams

Decision diagrams are constructed in a recursive way by the application of a set of operations suitably determined depending on:

- ▶ The domain and range of functions represented,
- ▶ The function specification given, and
- ▶ The operations used in definition of the decision diagram to be constructed.

We first construct the basic decision diagrams for variables and then combine them recursively into subdiagrams, until the final decision for the given function is constructed. The method can be illustrated by the following example.

Example 27.1 *Consider the function f of $n = 4$ variables x_1, x_2, x_3, x_4 given by the sum-of-product expression*

$$f = x_1x_2 \vee x_3 \vee x_4.$$

Figure 27.2 illustrates construction of the BDD for f:

- ▶ *First, diagrams for variables x_1 and x_2 are constructed and combined in the diagram for x_1x_2 by using the definition of the logic AND.*
- ▶ *Then, diagrams for x_3 and x_4 are constructed and combined into a diagram for $x_3 \vee x_4$ by logic OR.*
- ▶ *These two diagrams, for x_1x_2 and $x_3 \vee x_4$ are combined into the final diagram of f by logic OR.*

In this example, operations used to construct the BDD are logic operations AND and OR used in the initial specification of f, and the decomposition rule is the Shannon rule as specified in the definition of BDDs.

The construction of decision diagrams is formalized by the so-called `APPLY` procedure.

If two switching functions f and g are represented by BDDs, then the function $f < \texttt{OP} > g$, where $< \texttt{OP} >$ is a binary operation, is represented as

$$f < \texttt{OP} > g = v(f_v < \texttt{OP} > g_v) + \overline{v}(f_{\overline{v}} < \texttt{OP} > g_{\overline{v}}),$$

where v is the top variable in f and g. The cofactors of f and g with respect to different values of v are denoted by f_v, $f_{\overline{v}}$, and g_v, $g_{\overline{v}}$, respectively.

Realization of logic operations over BDDs by `APPLY` can be improved using `If-Then-Else` (ITE) operator defined as

$$\texttt{ITE}(f, g, h) = \texttt{if } f \texttt{ then } g \texttt{ else } h,$$

and expressed in terms of logic operations as [3]

$$\texttt{ITE}(f, g, h) = f \cdot g \vee \overline{f} \cdot h,$$

where f, g, and h are arbitrary switching functions and $\cdot$ and $\vee$ denote logic AND and OR, respectively.

In a way similar to that presented in the above example, decision diagrams starting from different initial descriptions of functions are constructed by using different decomposition rules.

Generalization to construction of basic multiple-place decision diagrams (MDDs) for multiple-valued logic functions are done through the `CASE` operator defined as

$$\texttt{CASE}(f, g^0, g^1, \ldots, g^{q-1}) = g^f.$$

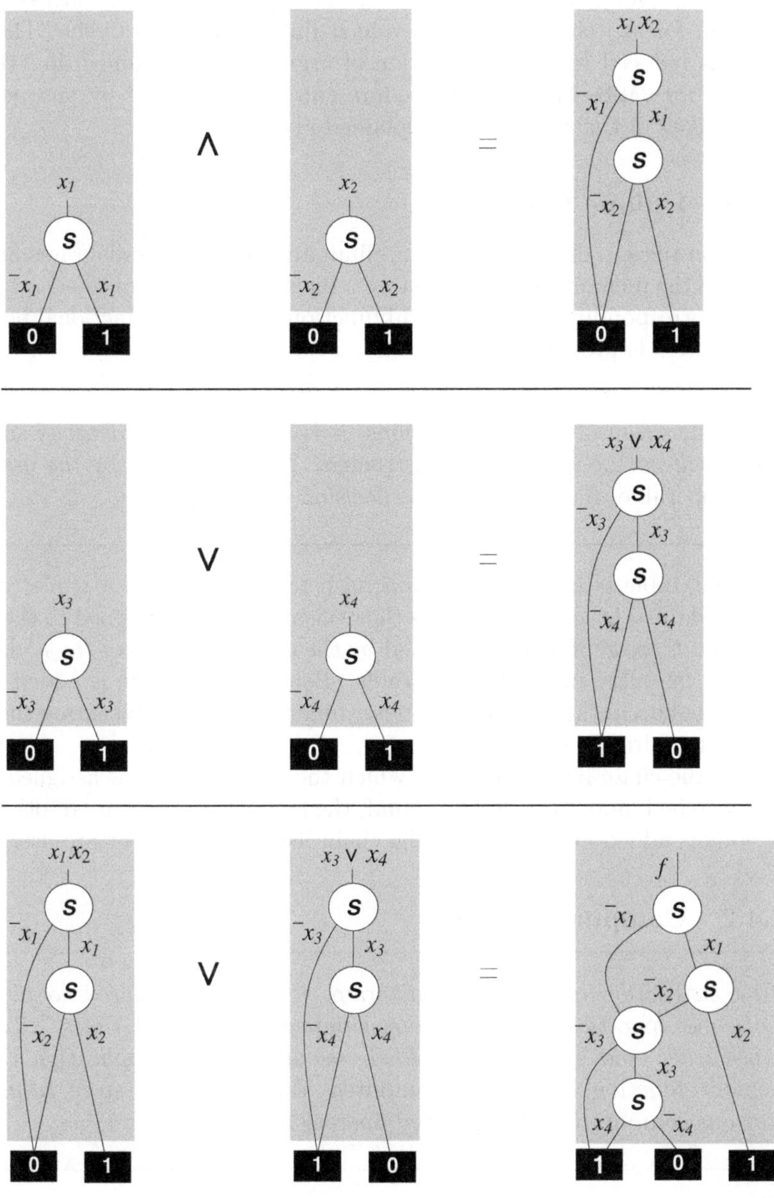

FIGURE 27.2
Construction of BDD for the function $f = x_1x_2 \vee x_3 \vee x_4$ (Example 27.1).

The procedures ITE and $CASE$ can be used to generate BDDs, respectively MDDs, for a function $f < \mathtt{OP} > g$, where $< \mathtt{OP} >$ is an arbitrary binary operation for switching and multi-valued functions, respectively. The realization is reduced to the specification of arguments f, g, and h in `ITE`, and similarly for `CASE`. The same procedure can be formulated by an operation table instead of the recursion as explained above.

27.4.1 Unique table

Reduction rules in decision diagrams eliminate identical subdiagrams in order to reduce the memory required to store a diagram. Instead of repeated subdiagrams corresponding to identical subfunctions, control of functional behavior with pointers is provided.

In construction of decision diagrams, it is assumed that identical subdiagrams will not be repeatedly constructed. This is ensured by the usage of a Unique Table *to store nodes in a decision diagram.*

Unique Table is implemented as a hash table that maps a triple (x, g, h) into a node $f = (x, g, h)$, where x is the decision variable assigned to the node, and g and h are subfunctions pointed by the outgoing edges of the node. The node has an entry in the Unique Table. Before a new node is added to the BDD, a lookup in the unique table determines if a node representing the same subfunction already exists.

Thus, the chain list for nodes to which the same variable is assigned is linearly searched, and if the node is found, the instance found is used, otherwise, the node is added at the front of the chain list for possible future use.

27.4.2 Computed table

In construction of a decision diagram, it may be required to perform the same operation over the same subdiagrams several times. This is unnecessary and should be avoided, usually by the application of the memory function called the Computed Table, *used to store results of previous computations by specified operations.*

For each operation edge pointers identifying diagrams that represent operands and a pointer to the decision diagram for the result are recorded.

In this way, before performing a computation, the search over the Computed Table is performed to check if the pointers for the operands given will correspond to some pointer to the result for the same operation specified. If there is no such pointer, computation is performed and the result stored in the corresponding position in the Computed Table; otherwise the pointer to the resulting decision diagram is returned.

27.5 Construction of decision diagrams based on truth tables and variable ordering

Another technique for constructing decision diagrams is based on sequential decomposition of the truth table using the rules of Shannon expansion*. The resulting structure is a decision tree that can be further reduced by applying reduction rules. The optimization problem arises when there is a need to select the next decomposition variable. The intuitive approach revolves around choosing the "best" variable at a given time. It means that in any step of decision tree design, we have an opportunity to choose a variable making this solution pseudo-optimal. Many heuristics have been suggested to select the "best" variable. Thus, the approach based on information measures, mainly entropies, derived from the information theory shows promising results in terms of the size of the tree and computational complexity of construction.

The result of the decision tree design is a so-called *free decision tree.* The term 'free' means that at every level different variables can occur. For some functions free decision trees and decision diagrams allow an exponential reduction with respect to the number of nodes compared to ordered ones [20]. So, the problem of variable ordering can be solved using the entropy-based criterion.

The problem can be formulated as follows. Given a logic function f of n variables in the form of a truth table (an example of a truth table is given in Table 27.2), find a quasi-optimal decision tree for the given function. In decision tree design strategy we use some basic concepts of information theory, namely, *entropy, conditional entropy,* and *mutual information.*

In order to quantify the content of information revealed by the outcome for finite field of events

$$A = \{a_1, a_2, \cdots, a_n\}$$

with the probabilities distribution

$$\{p(a_1), p(a_2), \cdots, p(a_n)\},$$

Shannon introduced the concept of entropy. *Entropy* of finite field A is given by

$$H(A) = -\sum_{i=1}^{n} p(a_i) \cdot \log p(a_i), \tag{27.1}$$

*This section was co-authored with Dr D. V. Popel.

TABLE 27.2
An example of a truth table

x_1	x_2	x_3	x_4	f	x_1	x_2	x_3	x_4	f
0	0	0	0	0	1	0	0	0	0
0	0	0	1	0	1	0	0	1	0
0	0	1	0	0	1	0	1	0	0
0	0	1	1	0	1	0	1	1	1
0	1	0	0	0	1	1	0	0	0
0	1	0	1	1	1	1	0	1	1
0	1	1	0	0	1	1	1	0	0
0	1	1	1	1	1	1	1	1	0

where logarithm is in base 2. Note that entropy $H(A)$ never is negative and is equal to zero if and only if A contains one event only (see details in Chapter 8).

Let A and B be finite fields of events with probabilities distribution $\{p(a_i)\}$, $i = 1, 2, \cdots, n$, and $\{p(b_j)\}, j = 1, 2, \cdots, m$, respectively.

Conditional entropy of A with respect to B is defined by

$$H(A|B) = -\sum_{i=1}^{n}\sum_{j=1}^{m} p(a_i, b_j) \cdot \log \frac{p(a_i, b_j)}{p(b_j)} \ . \tag{27.2}$$

Mutual information between two finite fields A and B is

$$I(A; B) = H(A) - H(A|B). \tag{27.3}$$

We deal with two finite fields:

- The set of values of switching function f for different combinations of variables values (we name such combinations as *patterns*) and
- The set of values of arbitrary variable x.

Equations 27.1, 27.2, and 27.3 are used to calculate information estimations with respect to function and its variables.

The probability, for example, for $p(f = 0)$, is calculated as follows:

$$p(f = 0) = \frac{k_{f=0}}{k},$$

where $k_{f=0}$ is the number of patterns, for which the switching function takes the value 0 and k is the total number of patterns (for completely specified switching function $k = 2^n$). Other probabilities are calculated in the same way.

Example 27.2 *Consider the function from Table 27.2. The entropy of the function is*

$$H(f) = -{}^4/_{16} \cdot \log {}^4/_{16} - {}^{12}/_{16} \cdot \log {}^{12}/_{16} = 0.81 \quad \textit{bit/pattern.}$$

The conditional entropy of function with respect to variable x_2 is

$$\begin{aligned} H(f|x_2) &= -{}^7/_{16} \cdot \log {}^7/_8 - {}^1/_{16} \cdot \log {}^1/_8 \\ &\quad -{}^5/_{16} \cdot \log {}^5/_8 - {}^3/_{16} \cdot \log {}^3/_8 \\ &= 0.75 \ \textit{bit/pattern.} \end{aligned}$$

Thus, mutual information between f and x_2 (in other words – information, which is carried out by x_2 about f) is

$$0.81 - 0.75 = 0.06 \ \textit{bit/pattern.}$$

Construction of decision trees is a *recursive decomposition* of the given function. A step of this recursive decomposition corresponds to the expansion of switching function f with respect to variable x. The Shannon expansion can be outlined as $f = \overline{x} \cdot f_{x=0} \vee x \cdot f_{x=1}$, where $f_{x=a}$ is *cofactor* of f, i.e., x is replaced with $a \in \{0, 1\}$. The following criterion is used for selecting the "best" variable. The criterion is that the conditional entropy of the function with respect to the variable x has to be minimal:

$$x = argmin\{H(f|x)\}.$$

27.6 Calculations with decision diagrams

When two functions f and g are represented by decision diagrams, the result of an operation over them $f * g$ can be determined in the form of another decision diagram constructed directly by performing the corresponding operations over nodes of decision diagrams for switching functions f and g.

That is implemented by `ITE` and `CASE` operators for binary and multiple-valued functions, respectively, with properly specified parameters.

Example 27.3 *Consider calculation of a spectral transform specified by the transform matrix that can be represented by the Kronecker product of the basic (2×2) transform matrices $\mathbf{K}_i(1)$. It follows that we perform a series of `ITE` operators whose parameters are elements of the corresponding rows of $\mathbf{K}_i(1)$.*

Extension of this method to multivalued functions by using the *CASE* is straightforward.

27.7 Further reading

The BDD packages. There are few programming packages to deal with decision diagrams, for instance, PUMA [3], BEMITA [12], and BXD [2], however, the most widely accepted is *C*olorado University Decision Diagram (CUDD) package, which nowadays almost the standard for programming decision diagrams [15].

The package CUDD by Somenzi [15] works with BDDs and algebraic decision diagrams, for example, [2], which are multiterminal BDDs (MTBDDs).

The package PUMA and the package described in [10] works with multiple-place decision diagrams (MDDs) [11] for representation of multivalued logic functions, i.e., functions whose variables can take up to $p > 2$ values, where p is specified as an input parameter. In the original implementation the value p is set to 5. PUMA is written in C++, and thus, nodes are described by classes. Since dealing with multivalued functions is assumed, labels at the edges are stored in an array *edge* instead of just as a fixed number of two edges as in CUDD.

Construction of decision diagram packages for switching and multivalued functions is considered in many publications; see, for instance, [4], [9], [10] and references therein. In [4] is given a unified way for construction and calculation with arbitrary spectral transform decision diagrams (STDDS) by using the spectral interpretation of transform decision diagrams. In [10], it is considered construction of BDDs by a recursive implementation of logic NAND instead of `ITE`, or realization of NAND through `ITE`. In [10], the extension to multivalued logic functions is provided by recursive implementation of multivalued operations MIN and MAX to construct MDDs, instead of through `CASE`.

A package for construction of linear word-level decision diagrams was developed by Yanushkevich et al. [22].

Memory function and its usage to avoid repeated computations was suggested by Bryant [4], and then used in many other decision diagram packages. In some implementations, as for example, by Brace et al. [3] for BDDs, and by Miller and Drechsler [10] for MDDs, the Computed Table approach is inserted into the *ITE* and *CASE* or *MIN* procedures. The implementation of non-terminal nodes proposed in [3] requires 22 bytes per node in a 32-bit processor machine.

References

[1] Atallah MJ. *Algorithms and Theory of Computation.* Handbook, CRC Press, 1999.

[2] Bahar RI, Frohm EA, Gaona CM, Hachtel GD, Macii E, Pardo A, and Somenzi F. Algebraic decision diagrams and their applications. In *Proceedings of the International Conference on CAD*, pp. 188–191, 1993.

[3] Brace KS, Rudell RL, and Bryant RE. Efficient implementation of a BDD package. In *Proceedings of the Design Automation Conference*, pp. 40–45, 1990.

[4] Bryant RE. Graph-based algorithms for Boolean functions manipulation. *IEEE Transactions on Computers*, 35(8):667–691, 1986.

[5] Chen YA and Bryant RE. An efficient graph representation for arithmetic circuit verification. *IEEE Transactions on CAD*, 20(12):1443–1445, 2001.

[6] Cormen TH, Leiserson CE, Rivest RL, and Stein C. *Introduction to Algorithms.* MIT Press, 2001.

[7] Drechsler R and Becker B. OKFDDs-algorithms, applications and extensions. In Sasao T and Fujita M, Eds., *Representations of Discrete Functions*, Kluwer, Dordrecht, pp. 163–190, 1996.

[8] Drechsler R, Janković D, and Stanković RS. Generic implementation of DD packages in MVL. In *Proceedings of the 25th EUROMICRO Conference*, Vol. 1, pp. 352–359, 1999.

[9] Drechsler R and Sieling D. Binary decsion diagrams in theory and practice. *International Journal on Software Tools for Technology Transfer*, 3:112–136, 2001.

[10] Drechsler R and Thorton MA. Fast and efficient equivalence checking based on NAND-BDDs. In *Proceedings of the IFIP International Conference on VLSI*, Montpelier, France, pp. 401–405, 2001.

[11] Miller DM and Drechsler R. On the construction of multiple-valued decision diagrams. In *Proceedings of the 32nd International Symposium on Multiple-Valued Logic*, pp. 245–253, 2002.

[12] Minato S. *Binary Decision Diagrams and Applictions for VLSI Synthesis.* Kluwer, Dordrecht, 1996.

[13] Perkowski MA, Chrzanowska-Jeske M, and Xu Y. Lattice diagrams using Reed–Muller logic. In *Proceedings of the IFIP WG 10.5 International Workshop on Applications of the Reed–Muller Expansion in Circuit Design.* Japan, pp. 85–102, 1997.

[14] Shen X, Hu Q, and Liang W. Embedding k-ary complete trees into hypercubes. *Journal of Parallel and Distributed Computing.* 24:100–106, 1995.

[15] Somenzi F. CUDD: CU Decision Diagram Package. University of Colorado at Boulder, http://vlsi.colorado.edu/~fabio/CUDD/, 2005.

[16] Somenzi F. Efficient manipulation of decision diagrams. *International Journal on Software Tools for Technology Transfer*, 3:171–181, 2001.

[17] Srinivasan A, Kam T, Malik Sh, and Brayant RK. Algorithms for discrete function manipulation. In *Proceedings of the International Conference on CAD*, pp. 92–95, 1990.

[18] Stanković RS. Unified view of decision diagrams for representation of discrete functions. *International Journal on Multiple- Valued Logic*, 8(2):237–283, 2002.

[19] Stanković RS and Astola JT. *Spectral Interpretation of Decision Diagrams.* Springer, Heidelberg, 2003.

[20] Stanković RS, Sasao T, and Moraga C. Spectral transforms decision diagrams. In Sasao T. and Fujita M, Eds., *Representations of Discrete Functions*, pp. 55–92, Kluwer, Dordrecht, 1996.

[21] Yanushkevich SN, Shmerko VP, and Lyshevski SE. *Logic Design of NanoICs*, CRC Press, Boca Raton, FL, 2005.

[22] Yanushkevich SN, Shmerko VP, and Dziurzanski P. LDD package. http://www.enel.ucalgary.ca/People/yanush/research.html. Release 01, 2002.

Part III

DECISION DIAGRAM TECHNIQUES FOR MULTIVALUED FUNCTIONS

- Multivalued Functions
- Libraries of Multivalued Gates
- Spectral Transforms
- Computing via Taylor Expansion
- Classification of Multivalued Decision Diagrams
- Spectral Decision Diagrams
- Event-Driven Analysis in Multivalued Systems
- Word-Level Decision Diagrams
- Information-Theoretical Measures in Multivalued Systems

28

Introduction

This part includes seven topics on decision diagram technique for multivalued functions. These topics cover the following areas:

- The basics of multivalued logic,
- Spectral techniques for multivalued functions,
- The concept of change and computing changes in multivalued systems, and
- Various types of multivalued decision diagrams.

28.1 Multivalued logic

Boolean algebra is the mathematical foundation of binary systems. Boolean algebra is defined on a set of two elements, $M = \{0, 1\}$. The operations of Boolean algebra must adhere to certain properties, called laws, or axioms used to prove more general laws about Boolean expressions to simplify expressions, or factorize them, for example.

Multivalued algebra is a generalization of Boolean algebra, based upon a set of m elements $M = \{0, 1, 2, \dots, m\}$, corresponding to multivalued signals, and the corresponding operations. This means that multivalued logic circuits operate with multivalued logic signals (Figure 28.1).

The primary advantage of multivalued signals is the ability to encode more information per variable than a binary system is capable of doing. Hence, less area is used for interconnections since each interconnection carries more information. Furthermore, reliability of systems is also relevant to the number of connections because there are are sources of wiring error and weak connections.

Thus, the adoption of a m-valued signals enables n pins to pass q^n combinations of values rather than just the 2^n limited by the binary representation.

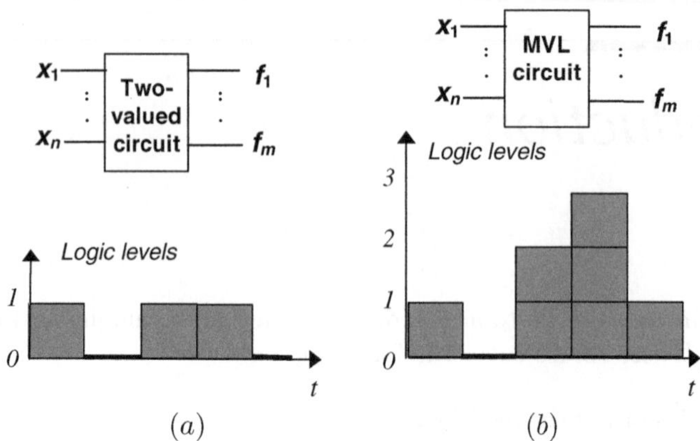

FIGURE 28.1
A switching circuit operates with two-level logic signals (a), and a quaternary circuit operates with four-valued logic signals (b).

In particular, a 32×32 bit multiplier based on quaternary signed-digit number system, consists of only three-stage signed-digit full adders using a binary-tree addition scheme, as shown by Kawahito, et. al. [27].

Another application of multivalued logic is residue arithmetic. Each residue digit can be represented by a multivalued code. For example, by this coding, *mod m* multiplication can be performed by a shift operator, and *mod m* addition can be executed using radix-arithmetic operators. Kameyama et al. [25] proposed implementation of this approach on multivalued bidirectional current-node MOS technology.

The operations of Boolean algebra have their analogs in multivalued algebra. The multivalued counterparts of binary AND, OR and NOT are the multivalued *conjunction, disjunction* and *cycling* operators introduced by Post in 1920. The other basis, *modulo* 2 was extended by Bernstein in 1928 by introducing *modulo m* addition and multiplication of the integers *modulo m*.

To synthesize logic circuits, a functionally complete, or universal, set of basic operations is needed.

Example 28.1 *There are four 1-place operations in binary logic circuits design: constants 0 and 1, identity, and complement, and 16 2-place operations. There are two operations, NOR and NAND, each one of which constitutes a functionally complete set.*
In quaternary logic, there are 256 1-place functions and $4^{4^2} = 4^{16} \approx 4.3 \times 10^9$ *2-place functions.*

There are various universal (functionally complete) sets of operations for multivalued algebra:

- *Post algebra*
- *Webb algebra*
- *Bernstein algebra*, and
- *Allen and Givone algebra*

There are a lot of other algebras oriented mostly toward circuit implementations. For example, there are algebras based on MIN, MAX, and CYCLE operations.

28.2 Representation of multivalued functions

We will use the following techniques and associated data structures for representing logic functions:

- Symbolic (algebraic) notations,
- Vector notations, i.e., truth column vector and coefficients column vectors,
- Matrix (two dimensional) notations for word-level representation,
- Graph-based representations such as direct acyclic graphs (DAG) and decision diagram techniques,
- Embedded graph-based 3D data structures.

Algebraic data structures include

- Multivalued sum-of-products
- Multivalued Reed–Muller expressions
- Multivalued arithmetic representations
- Multivalued word-level representations
- Various multivalued representations derived from spectral techniques.

Truth table and truth column vector. The simplest way to represent a multivalued logic function is the truth table. The truth table of a logic function is the representation that tabulates all possible input combinations with their associated output values.

A truth column vector of a multivalued logic function f of n m-valued variables $x_1, x_2, \ldots, x_n$ is defined as

$$\mathbf{F} = [f(0), f(1), \ldots, f(m^n - 1)]^T.$$

The index i of the element $x^{(i)}$ corresponds to the assignments $i_1 i_2 \ldots i_n$ of variables $x_1, x_2, ..., x_n$ ($i_1, i_2, ..., i_n$ is binary representation of i, $i = 0, \ldots, m^n - 1$).

Example 28.2 *The truth column vector* **F** *of a ternary MIN function of two variables is* **F**$= [000011012]^T$.

Symbolic, or algebraic notations include sum-of-products and polynomial forms. The sum-of-products form is represented as follows:

$$f = s_0\varphi_0(x_1, \ldots, x_n) + \ldots + s_{m^n-1}\varphi_{m^n-1}(x_1, \ldots, x_n),$$

where $s_i\varphi_i(x_1, \ldots, x_n)$ is a literal function such that

$$\varphi_i(x_1, \ldots, x_n) = \begin{cases} 1, \text{ if } x_1, \ldots, x_n = 1; \\ 0, \text{ otherwise.} \end{cases}$$

Appendix D contains truth tables of ternary and quaternary functions of one and two variables.

Graphical data structures, adopted for multivalued functions, include:

- ► Multivalued networks.
- ► Cube-based representation
- ► Decision trees and diagrams, and
- ► Spatial topological structures resulting from embedding decision trees and diagrams into spatial configurations.

Graph-based representations of networks such as DAGs are used at gate level design phases. Figure 28.2 shows the gate libraries for constructing both switching (a) and multivalued (b) circuits.

Graph-based representations of networks such as DAGs are used at gate level design phases.

Event-driven analysis of multivalued circuits. This part of multivalued logic design is called *logic differential calculus* and is considered in detail in Chapter 32.

28.3 Spectral theory of multivalued functions

Spectral transforms are used to make conversions of one representation forms (signal domain) to another (spectral domain). Like switching functions, spectral transforms implemented for multivalued functions can be divided into two classes:

- ► *Logic transforms,* used to generate m^n Reed–Muller expressions over $GF(m)$ for a given m-valued function.

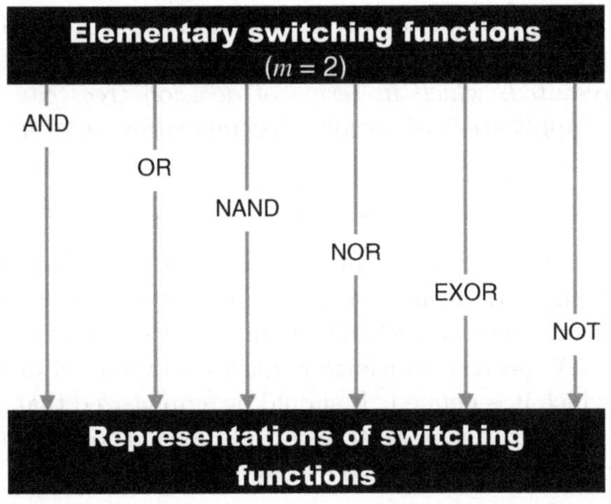

(*a*)

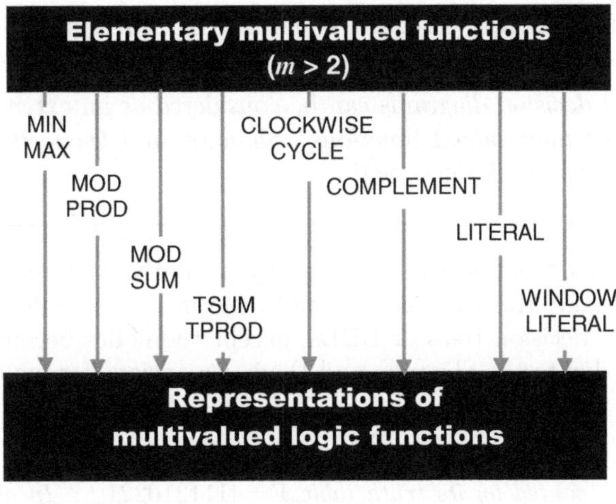

(*b*)

FIGURE 28.2
Representation of switching (a) and multivalued logic (b) functions.

- *Arithmetic transforms,* known also as *Vilenkin–Chrestenson* transforms to generate (m^n arithmetic expressions for a given m-valued function.

In spectral interpretation of decision diagrams, a given function f is assigned to a decision tree by performing a spectral transform on the function represented, which in terms of decision trees can be expressed as a recursive application of various decomposition rules at the nodes of decision trees.

Due to this interpretation, various decision diagrams defined with respect to different decomposition rules are uniformly viewed as examples of spectral transform decision diagrams (STDDs) defined with respect to different spectral transforms. A spectral transform is specified by the set of basic functions Q in terms of which it is defined. It should be emphasized that the basic functions in terms of which a spectral transform is defined correspond in some way to the recursive structure of decision trees.

28.4 Multivalued decision trees and diagrams

Multivalued decision diagrams can be considered as an extension of BDD techniques for multivalued functions. Shannon and Davio decomposition exist for multivalued logic as well.

Any m-valued function can be given by a multivalued decision tree and decision diagram. Multivalued trees and diagrams are specified in the same way as binary decision trees or BDDs, except the nodes become more complex. This is due to the Shannon and Davio expansions for multivalued logic functions.

Example 28.3 *A function of two ternary variables is represented by its truth table* $\mathbf{F}=[111210221]^T$. *In algebraic form the function is expressed by*

$$f = 0 \cdot x_1^1 x_2^2 + 1 \cdot (x_1^0 x_2^0 + x_1^0 x_2^1 + x_1^0 x_2^2 + x_1^1 x_2^1 + x_1^2 x_2^2) + 2 \cdot (x_1^1 x_2^0 + x_1^2 x_2^0 + x_1^2 x_2^1).$$

The map and the decision diagram of the function are given in Figure 28.3 (the nodes of the diagram implement the Shannon expansion for the ternary logic function).

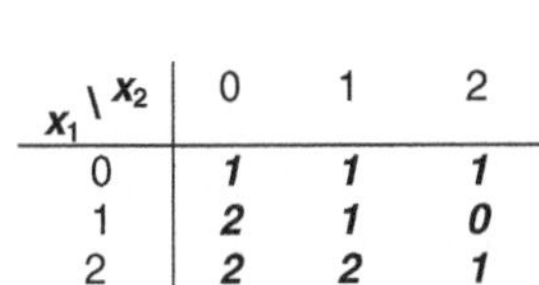

$x_1 \backslash x_2$	0	1	2
0	*1*	*1*	*1*
1	*2*	*1*	*0*
2	*2*	*2*	*1*

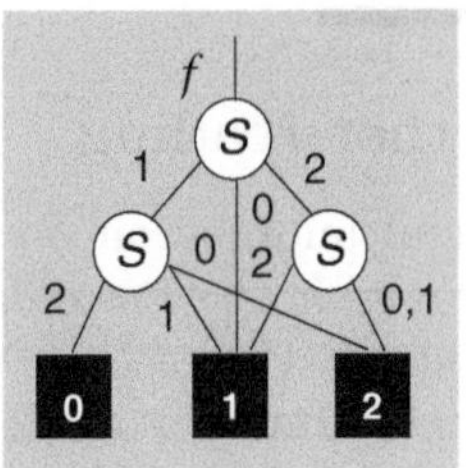

FIGURE 28.3
The map of the ternary function and its graphical representation (Example 28.3).

Compared to switching logic, multivalued logic has some advantages when designing high-speed arithmetic components in avoiding the time-consuming carry propagation inherent in switching gates. However, there are several disadvantages, since multivalued logic circuits often have to be embedded in a conventional digital system and therefore additional circuitry is needed to transform multivalued logic into digital signals and vice versa.

Multivalued logic is considered as an algebraic basis for threshold gates design. Threshold gates with various interpretations of weights and threshold functions are the components of artificial neural networks. The prototypes of these neural networks include circuits on CMOS neurons, and also recent nanoelectronic and molecular implementations.

One of the motivations to develop design techniques such as decision diagrams for multivalued functions is that nanotechnologies provide an opportunity to utilize the concept of multivalued signals in nanodevices.

For instance, multivalued logic has been used for storing synaptic weights in neural circuits (see Further Reading section). This preserves robust information processing and reduces the number of circuit components per artificial synaptic circuit. In particular, multistate Resonant Tunnelling Device (RTD) memory cells are a promising way to implement area efficient multivalued logic circuits as components for monolithically integrated neural systems. The hope is that the fault tolerance of neural circuits will compensate the errors caused by smaller noise margins.

28.5 Further reading

Plenty of useful information on the theory and application of multivalued logic can be found in the *Proceedings of the Annual IEEE International Symposium on Multiple-Valued Logic* that has been held since 1970. In addition, a good source of information on the above mentioned topics is the *International Journal on Multiple-Valued Logic and Soft Computing.* The fundamentals of multivalued logic can be found in the books by Davio [6], Epstein [9], Hurst [8], Muzio [37], and Rine [40] on the applied problems of multivalued logic. Overview of multivalued logic and its application can be found in papers by Smith [47], Hurst [20], Butler [4], Hanyu et al. [15, 16], Brayton [3], Karpovsky at al. [26] and Dubrova [8].

Spectral techniques for multivalued logic functions. In papers by Moraga [32, 33, 35, 34], various aspects of spectral techniques were developed. His fundamental study includes the complex representations of multivalued logic functions [35], spectral representation of particular classes of multivalued functions [29], Fourier spectra of multivalued functions on an Abelian group [13], computing multivalued functions using systolic arrays [36] (these ideas were later used in developing linear systolic arrays [55, 56]). Hurst [19] studied the application of Haar transform for multivalued functions analysis and representation. Stanković et al. [52] developed decision diagram technique using Haar wavelet transforms.

The book by Hurst et al. [21] is a good contribution to the fundamentals of spectral technique in multivalued logic. Sadykhov et al. [43] studied the spectral representation of signals in various Fourier-like bases, including relationships between spectral representations for switching and multivalued functions. The spectral technique has also been introduced in [31, 43, 51].

Davio [6], Green and Taylor [11, 12], studied Reed–Muller expressions of multivalued logic functions. Further development of the theory has been done by Perkowski et al. [39]. Zilic and Vranesic [61] studied Reed–Muller representation of incompletely specified multivalued functions.

The basics of word-level decision diagrams using arithmetic representation were developed by Hammer and Rudeanu [14], Rosenberg [41], and Thayse [54]. Strazdins [53] used the Pascal triangle to derive arithmetic representation of the elementary multivalued functions.

Multivalued decision diagrams are the focus of Miller's and Stankovic's papers [30, 49, 50, 51]. A survey of ternary decision diagram technique can be found in [45]. The paper by Kam et al. [23] summarizes the results for multivalued logic decision diagram techniques. Linear word-level decision diagrams (LDDs) were developed by Yanushkevich et al. [58, 59] (see Chapter 37 for details).

Nagayama et al. [38] improved quasi-reduced multivalued decision diagrams (QRMDDs) and proposed the measure of efficiency called *area-time complexity*. Iguchi et al. [22] studied paged reduced ordered multivalued decision diagrams (PMDDs) and paged quasi-reduced multivalued decision diagrams (PQMDDs) for improving several characteristics of multivalued diagrams.

Logic design. Decomposition of multivalued functions was studied by Sasao [44], Łuba [28], Brzozowski and Łuba [2], and Steinbach (see Chapter 25 for references). Various aspects of logic design of multivalued circuits and systems have been developed, in particular, linearly independent logic by Perkowski et al. [39], and logic differential calculus by Yanushkevich [28] (see Chapters 9 and 32). Computer aided design (CAD) tools of multivalued systems have been discussed by Miller [31].

Sets of pairs of functions to be distinguished (SPFDs) is a method of flexible logic design. SPFDs represent sets of allowable functions for a target network using compatible sets (see details in Chapter 8). Sinha et al. [46] introduced a generalization of SPFDs for multivalued logic functions.

Implementation. Kameyama and Higuchi [17, 24] studied the synthesis of multivalued systems, including ones based on new technologies and biochips. The usefulness of multivalued logic models in the design of *molecular devices* has been shown, in particular, in [1]. Deng et al. [20] studied a quantum device structure for the super pass transistor model using multivalued logic. Fleisher [10] studied the implementation of multivalued logic in bit-partitioned PLA structures. Implementation of multivalued functions in content-addressable memory was studied by Butler [5]. Moraga [33] developed the concept of language inference using multivalued logic. Quaternary circuits were implemented on NMOS by Kameyama et al. [25].

References

[1] Aoki T, Kameyama M, and Higuchi T. Design of interconnection-free biomolecular computing systems. In *Proceedings of the 21st IEEE International Symposium on Multiple-Valued Logic*, pp. 173–180, 1991.

[2] Brzozowski JA and Łuba T. Decomposition of Boolean functions specified by cubes. *Journal of Multiple-Valued Logic and Soft Computing,* 9(4):377–417, 2003.

[3] Brayton RK and Kharti SR. Multi-valued logic synthesis. In *Proceedings of the 12th International Conference on VLSI Design,* pp. 196–206, 1999.

[4] Butler JT. Multiple-valued logic. *IEEE Potentials,* 14(2):11–14, 1995.

[5] Butler JT. On the number of locations required in the content-addressable memory implementation of multiple-valued functions. In *Proceedings of the 13th International Symposium on Multiple-Valued Logic,* pp. 94–102, 1983.

[6] Davio M, Deschamps J-P, and Thayse A. *Discrete and Switching Functions.* McGraw–Hill, Maidenhead, 1978.

[7] Deng X, Hanyu T, and Kameyama M. Quantum-device-oriented multiple-valued logic system based on a super pass gate. *IEICE Transactions on Information and Systems,* E78-D(8):951–958, 1995.

[8] Dubrova E. Multiple-valued logic synthesis and optimization. In Hassoun S and Sasao T, Eds, Brayton RK, consulting Ed., *Logic Synthesis and Verification,* pp. 89–114, Kluwer, Dordrech, 2002.

[9] Epstein G. *Multi-Valued Logic Design.* Institute of Physics Publishing, London, UK, 1993.

[10] Fleisher H. The implementation and use of of multivalued logic in a VLSI environment. In *Proceedings of the 13th International Symposium on Multiple-Valued Logic,* pp. 138–143, 1983.

[11] Green DH and Taylor IS. Multiple-valued switching circuits by means of generalized Reed–Muller expansion. *Digital Processes,* 3:63–81, 1976.

[12] Green DH. Ternary Reed–Muller switching functions with fixed and mixed polarity. *International Journal of Electronics,* 67:761–775, 1989.

[13] Guozhen X and Moraga C. On the characterization of the Fourier spectra of functions defined on an Abelian group. In *Proceedings of the 19th International Symposium on Multiple-Valued Logic,* pp. 400–405, 1989.

[14] Hammer P and Rudeanu S. *Boolean methods in operations research and related areas.* Springer, Heildelberg, 1968.

[15] Hanyu T, Kameyma M, and Higuchi T. Prospects of multiple-valued VLSI processors. *IEIE Transactions on Electronics,* E76-C(3):383–392, 1993.

[16] Hanyu T. Challenge of a multiple-valued technology in recent deep-submicron VLSI. In *Proceedings of the 31st IEEE International Symposium on Multiple-Valued Logic,* pp. 241–247, 2001.

[17] Higuchi T and Kameyama M. Synthesis of multiple-valued logic networks based on tree-type universal logic modules. In *Proceedings of the 5th International Symposium on Multiple-Valued Logic,* pp. 121–130, 1975.

[18] Hurst S. *The Logical Processing of Digital Signals.* Arnold, London, 1978.

[19] Hurst SL. The Haar transform in digital network synthesis. In *Proceedings of the 11th International Symposium on Multiple-Valued Logic,* pp. 10–18, 1981.

[20] Hurst SL. Multiple-valued logic – its status and its future. *IEEE Transactions on Computers,* 33(12):1160–1179, 1984.

[21] Hurst S, Miller D, and Muzio J. *Spectral Technique in Digital Logic.* Academic Press, New York, 1985.

[22] Iguchi Y, Sasao T, and Matsuura M. Implementation of multiple-output functions using PQMDDs. In *Proceedings of the 32nd IEEE International Symposium on Multiple-Valued Logic,* pp. 199–205, 2000.

[23] Kam T, Villa T, Brayton RK, and Sagiovanni-Vincentelli AL. Multi-valued decision diagrams: theory and applications. *International Journal on Multiple-Valued Logic*, 4(1-2):9–62, 1998.

[24] Kameyama M and Higuchi T. Practical state assignment for multiple-valued synchronous sequential circuits. In *Proceedings of the 7th International Symposium on Multiple-Valued Logic,* pp. 70–76, 1977.

[25] Kameyama M, Hanyu T, and Higuchi T. Design and implementation of quaternary NMOS integrated circuits for pipelined image processing. In *IEEE Journal of Solid-State Circuits,* 22(1):20–27, 1987.

[26] Karpovsky MG, Stanković RS, and Moraga C. Spectral techniques in binary and multi-valued switching theory. A review of results in the decade 1991–2000. In *Proceedings of the 31st IEEE International Symposium on Multiple-Valued Logic,* pp. 41–46, 2001.

[27] Kawahito S, Kameyama M, Higuchi T, and Yamada H. A 32×32-bit multiplier using multiple-valued MOS current-mode circuits. In *IEEE Journal of Solid-State Circuits,* 23(1):124–132, 1988.

[28] Luba T. Decomposition of multiple-valued functions. In *Proceedings of the 25th IEEE International Symposium on Multiple-Valued Logic*, pp. 256–263, 1995.

[29] Luis M and Moraga C. On functions with flat Chrestenson spectra. In *Proceedings of the 19th International Symposium on Multiple-Valued Logic,* pp. 406–413, 1989.

[30] Miller DM. Spectral transformation of multiple-valued decision diagrams. In *Proceedings of the 24th IEEE International Symposium on Multiple-Valued Logic,* pp. 89–96, 1994.

[31] Miller DM. Multiple-valued logic design tools. In *Proceedings of the 23rd IEEE International Symposium on Multiple-Valued Logic*, pp. 2–11, 1993.

[32] Moraga CR and Salinas LC. On the Hilbert transform on a finite Abelian group. In *Proceedings of the 16th International Symposium on Multiple-Valued Logic,* pp. 198–203, 1986.

[33] Moraga C. Language inference using multiple-valued logic. In *Proceedings of the 10th International Symposium on Multiple-Valued Logic,* pp. 108–114, 1980.

[34] Moraga C. Extensions on multiple-valued threshold logic. In *Proceedings of the 9th International Symposium on Multiple-Valued Logic,* pp. 232–240, 1979.

[35] Moraga C. Complex spectral logic. In *Proceedings of the 8th IEEE International Symposium on Multiple-Valued Logic*, pp. 149–156, 1978.

[36] Moraga C. Systolic systems and multiple-valued logic. In *Proceedings of the 14th IEEE International Symposium on Multiple-Valued Logic*, pp. 98–108, 1984.

[37] Muzio JC and Wesselkamper TS. *Multiple-Valued Switching Theory.* Adam Higler Ltd., Bristol and Boston, 1986.

[38] Nagayama S, Sasao T, Iguchi Y, and Matsuura M. Representations of logic functions using QRMDDs. In *Proceedings of the 32nd IEEE International Symposium on Multiple-Valued Logic,* pp. 261–267, 2002.

[39] Perkowski M, Sarabi A, and Beyl F. Fundamental theorems and families of forms for binary and multiple-valued linearly independent logic. In *Proceedings of the IFIP WG 10.5 International Workshop on Applications of the Reed–Muller Expansions in Circuit Design*, pp. 288–299, Japan, 1995.

[40] Rine DC, Ed., *Computer Science and Multiple-Valued Logic. Theory and Applications.* 2nd ed., Amsterdam, North-Holland, 1984.

[41] Rosenberg I. Minimization of pseudo-Boolean functions by binary development. *Discrete Mathematics,* 7:151–165, 1974.

[42] Rudell R and Sangiovanni-Vincentelli A. ESPRESSO-MV: algorithm for multiple-valued logic minimization. In *Proceedings of the IEEE Custom Integrated Circuits Conference*, pp. 230–234, 1985.

[43] Sadykhov R, Chegolin P, and Shmerko V. *Signal Processing in Discrete Bases.* Publishing House "Science and Technics," Minsk, Belarus, 1986.

[44] Sasao T. Multiple-valued decomposition of generalized Boolean functions and the complexity of programmable logic arrays. *IEEE Transactions on Computers,* 30(9):635–643, 1981.

[45] Sasao T. Ternary decision diagram – survey. In *Proceedings of the 27th IEEE International Symposium on Multiple-Valued Logic*, pp. 241–250, 1997.

[46] Sinha S, Khatri S, Brayton RK, and Sangiovanni-Vincentelli AL. Binary and multi-valued SPFD-based wire removal in PLA network. In *Proceedings of the International Conference on Computer-Aided Design*, pp. 494–503, 2000.

[47] Smith KC. The prospects for multivalued logic: a technology and applications view. *IEEE Transactions on Computers*, 30(9):619–634, 1981.

[48] Stanković RS. Some remarks on Fourier transforms and differential operators for digital functions. In *Proceedings of the 22nd IEEE International Symposium on Multiple-Valued Logic*, pp. 365–370, 1992.

[49] Stanković R, Stankovic M, Moraga C, and Sasao T. Calculation of Reed–Muller–Fourier coefficients of multiple-valued functions through multiple-place decision diagrams. In *Proceedings of the 24th IEEE International Symposium on Multiple-Valued Logic*, pp. 82–87, 1994.

[50] Stanković RS. Functional decision diagrams for multiple-valued functions. In *Proceedings of the 25th IEEE International Symposium on Multiple-Valued Logic*, pp. 284–289, 1995.

[51] Stanković RS, Sasao T, and Moraga C. Spectral transform decision diagrams. In Sasao T, Ed., *Representations of Discrete Functions*, pp. 55–92. Kluwer, Dordrecht, 1996.

[52] Stanković RS, Stanković M, and Moraga C. Design of Haar wavelet transforms and Haar spectral transform decision diagrams for multiple-valued functions. In *Proceedings of the 31st IEEE International Symposium on Multiple-Valued Logic*, pp. 41–46, 2001.

[53] Strazdins I. The polynomial algebra of multivalued logic. *Algebra, Combinatorics and Logic in Computer Science*, 42:777–785, 1983.

[54] Thayse A. Integer expansions of discrete functions and their use in optimization problems. In *Proceedings of the 9th International Symposium on Multiple-Valued Logic*, pp. 82–87, 1979.

[55] Yanushkevich SN. Systolic arrays for multivalued logic data processing. In Kukharev GA, Shmerko VP, and Zaitseva EN. *Algorithms and Systolic Processors for Multi-Valued Data Processing*, pp. 157–252. Publishing House "Science and Technics," Minsk, Belarus, 1990.

[56] Yanushkevich SN. Systolic algorithms to synthesize arithmetical polynomial forms for k-valued logic functions. *Automation and Remote Control*, Kluwer/Plenum Publishers, 55(12):812–1823, 1994.

[57] Yanushkevich SN. *Logic Differential Calculus in Multi-Valued Logic Design.* Technical University of Szczecin Academic Publishers, Szczecin, Poland, 1998.

[58] Yanushkevich SN, Shmerko VP, and Dziurzanski P. Linearity of word-level models: new understanding. In *Proceedings of the IEEE/ACM 11th International Workshop on Logic and Synthesis*, pp. 67–72, New Orleans, LA, 2002.

[59] Yanushkevich SN, Dziurzanski P, and Shmerko VP. Word-level models for efficient computation of multiple-valued functions. Part 1: LAR based models. In *Proceedings of the IEEE 32nd International Symposium on Multiple-Valued Logic*, pp. 202–208, 2002.

[60] Yanushkevich SN, Shmerko VP, and Lyshevski SE. *Logic Design of NanoICs.* CRC Press, Boca Raton, FL, 2005.

[61] Zilic Z and Vranesic Z. Multiple valued Reed–Muller transform for incomplete functions. *IEEE Transactions on Computers*, 44(8):1012–1020, 1995.

29

Multivalued Functions

This chapter concerns multivalued logic functions that include the algebra of switching functions as a particular case of the m-valued logic when $m = 2$. The primary advantage of a multivalued system is the ability to encode more information per variable than a binary system is capable of doing. Hence, less area is used for interconnections since each interconnection has a multi-level signal carrying more information. This chapter generalizes the design paradigms of switching theory toward multivalued logic systems. The Reed–Muller expansions of switching functions can be regarded as the polynomial expansions of these functions over a Galois field $GF(2)$. The corresponding counterparts for m-valued functions are the field representations over $GF(m)$, where m is a prime number.

29.1 Introduction

Generalizations of switching theory toward multivalued logic are made in the following directions:

- Sum-of-products, Reed–Muller, and arithmetic representations of multivalued functions,
- event-driven analysis based on logic differential operations,
- Word-level representation of multivalued logic functions,
- Decision diagrams for the representation and manipulation of multivalued logic functions,
- Various techniques such as the spectral approach, minimization, the evolutionary approach in synthesis, neural and threshold nets synthesis, fuzzy-computing, learning, arithmetics, the information-theoretical approach, testing, image and language processing, verification, power estimation,
- Reversible logic and nanocomputing,
- Theorem-proving, and
- Various other applications.

The type of data structure chosen for representing multivalued systems is as critical in computing multivalued logic functions as switching ones. The choice of a convenient representation could greatly simplify the analysis and synthesis procedures. The complexity of realizing a function depends considerably upon the complexity of its algebraic or graphical representation. Spectral techniques are very flexible. Spectral coefficients carry information about the multivalued function in terms of change (logic difference, by analogy with Boolean differences in a binary system). The spectral representation is closely related to the Taylor expansion. It can be applied to generate both logic and arithmetic forms of multivalued functions.

29.2 Multivalued functions

An n-variable m-valued function f is defined as a mapping of a finite set $\{0,\ldots,m-1\}^n$ into a finite set $\{0,\ldots,m-1\}$, i.e.,

$$f:\ \{0,\ldots,m-1\}^n \rightarrow \{0,\ldots,m-1\}.$$

Example 29.1 *Function*

- $f:\{0,1,2\}^n \rightarrow \{0,1,2\}$ *is called a ternary logic function,*
- $f:\{0,1,2,3\}^n \rightarrow \{0,1,2,3\}$ *is called a quaternary logic function,*
- $f:\{0,1,2\}^n \rightarrow \{0,1\}$ *is called multivalued input binary-valued output logic functions.*

Specifically, m is considered to be a primary number. There are m^{m^n} different possible functions.

Example 29.2 *If $m = 3$ and the number of variables is equal to $n = 2$, then there are $3^{3^3} = 19.683$ possible ternary functions of two variables.*

29.3 Multivalued logic

Boolean algebra is the mathematical foundation of binary systems. Boolean algebra is defined on a set of two elements, $M = \{0,1\}$. The operations of Boolean algebra must adhere to certain properties, called laws or axioms, used to prove more general laws about Boolean expressions to simplify expressions, or factorize them, for example.

> *Multivalued algebra is a generalization of Boolean algebra towards a set of m elements $M = \{0, 1, 2, \ldots, m\}$ and corresponding operations.*

The focus of this section is

- ► Operators on m-valued logic; the set of elementary functions is larger compared to Boolean algebra,
- ► Algebras that are specified on a universal set of operations, and finally,
- ► Data structures for the representation and manipulation of multivalued logic functions.

29.3.1 Operations of multivalued logic

> *A multivalued logic function f of n variables $x_1, x_2, \ldots, x_n$ is a logic function defined on the set $M = \{0, 1, \ldots, m-1\}$. A multivalued logic circuit operates with multivalued logic signals. Each of the logic operations has a corresponding logic gate.*

Multivalued logic gates are closely linked to hardware in their implementations, given in Tables 29.1 and Table 29.2.

TABLE 29.1
Library of ternary ($m = 3$) two-variable elementary functions.

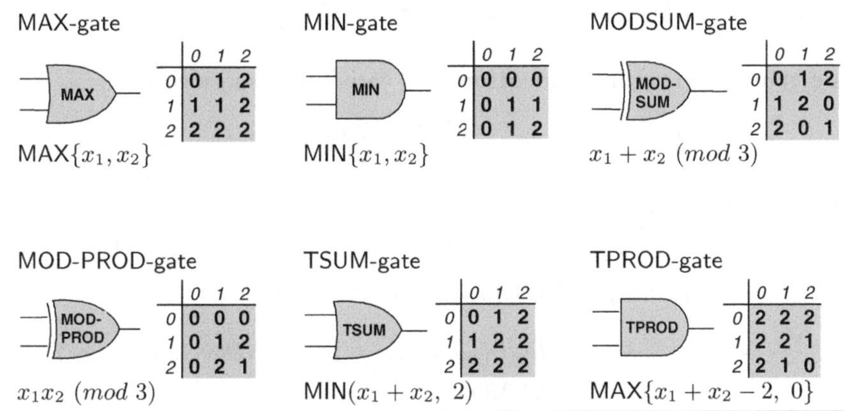

Below we list some of the implementation-oriented m-valued logic operations.

The MAX operation is defined as

$$\mathrm{MAX}(x_1, x_2) = \begin{cases} x_1 & \text{if } x_1 \geq x_2 \\ x_2 & \text{otherwise.} \end{cases}$$

When $m = 2$, this operation turns into an OR operation. The properties of MAX operations in ternary logic resemble those of Boolean algebra, i.e., $\mathrm{MAX}(x, x) = x$, that is, $x \vee x = x$ in a binary circuit, $x \vee 0 = x$, and $x \vee 2 = 2$, that is, $x \vee 0 = x$ and $x \vee 1 = 1$ in a binary case. A MAX function of n variables is written

$$\mathrm{MAX}(x_1, x_2, \ldots, x_n) = x_1 \vee x_2 \vee \ldots \vee x_n.$$

The MIN operation of x_1 and x_2 is defined as

$$\mathrm{MIN}(x_1, x_2) = \begin{cases} x_2 & \text{if } x_1 \geq x_2 \\ x_1 & \text{otherwise.} \end{cases}$$

and for n variables is written, $\mathrm{MIN}(x_1, x_2, \ldots, x_n) = x_1 \wedge x_2 \wedge \ldots \wedge x_n$.

The modulo m product operation is defined by

$$\mathrm{MOD\text{-}PROD}(x_1, x_2, \ldots, x_n) = x_1 x_2 \ldots x_n \quad mod\ (m).$$

The modulo m sum operation is defined below as

$$\mathrm{MODSUM}(x_1, x_2, \ldots, x_n) = x_1 + x_2 + \ldots + x_n \quad mod\ (m).$$

It can be shown that the modulo m sum operation MODSUM, modulo m product operation MOD-PROD, and the constant 1 constitute a universal set of operations, defined as the Galois algebra $GF(m)$.

The truncated sum operation of n variables is specified by

$$\mathrm{TSUM}(x_1, x_2, \ldots, x_n) = \mathrm{MIN}(x_1 \vee x_2 \vee \ldots \vee x_n,\ m - 1).$$

The truncated product operation is defined by

$$\mathrm{TPROD}(x_1, x_2, \ldots, x_n) = \mathrm{MIN}(x_1 \wedge x_2 \wedge \ldots \wedge x_n,\ (m - 1)).$$

Example 29.3 *(a) Let $m = 2$, then*

$$\begin{aligned} MOD\text{-}PROD(x_1, x_2, \ldots, x_n) &= AND(x_1, x_2, \ldots, x_n), \\ MODSUM(x_1, x_2, \ldots, x_n) &= x_1 \oplus x_2 \oplus \ldots \oplus x_n, \end{aligned}$$

(b) Let $m = 2,\ n = 2$, then

$$\begin{aligned} TSUM(x_1, x_2) &= MIN(x_1 \vee x_2, 1) = x_1 \vee x_2 \\ TPROD(x_1, x_2) &= MIN(x_1 x_2,\ 1) = x_1 x_2. \end{aligned}$$

The Webb function is defined below as

$$x_1 \uparrow x_2 = \mathrm{MAX}(x_1, x_2) + 1 \ \ (mod\ m).$$

The Pierce operation is a binary analog of the Webb operation.

The complement operation is specified by

$$\overline{x} = (m-1) - x,$$

where $x \in M$ is a unary operation. For example, in ternary logic, $\overline{x} = 2 - x$. Notice that the property $\overline{\overline{x}} = x$ can be used in multivalued logic. This is because $(m-1) - \overline{x} = (m-1) - ((m-1) - x) = x$.

TABLE 29.2
Library of ternary ($m = 3$) logic functions.

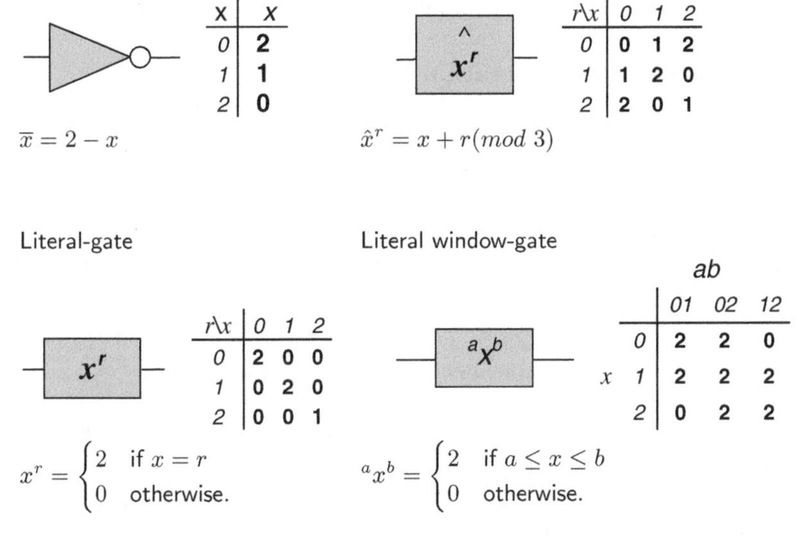

Complement-gate

x	$\overline{x}$
0	2
1	1
2	0

$\overline{x} = 2 - x$

Cyclic complement-gate

r\x	0	1	2
0	0	1	2
1	1	2	0
2	2	0	1

$\hat{x}^r = x + r(mod\ 3)$

Literal-gate

r\x	0	1	2
0	2	0	0
1	0	2	0
2	0	0	1

$$x^r = \begin{cases} 2 & \text{if } x = r \\ 0 & \text{otherwise.} \end{cases}$$

Literal window-gate

x \ ab	01	02	12
0	2	2	0
1	2	2	2
2	0	2	2

$$^a x^b = \begin{cases} 2 & \text{if } a \le x \le b \\ 0 & \text{otherwise.} \end{cases}$$

The clockwise cycle operation, or r-order cyclic complement

$$\hat{x^r} = x + r \ \ (mod\ m).$$

This implies that

$$\hat{x}^0 = x \pmod m,$$
$$\hat{x}^m = x + m = x \pmod m.$$

The operation MIN and the clockwise cycle operation form a complete system as well. Given m-valued variable, $m!$ different complements of the variable can be distinguished.

Example 29.4 *(a) Let $m = 2$, then $\overline{x} = (2-1) - x = 1 - x$. (b) Let $m = 2$, then the system is {AND, NOT}, i.e., NAND that is known to be complete.*

The literal operation is specified below

$$x^y = y^x = \begin{cases} m-1 & \text{if } x = y, \\ 0 & \text{otherwise.} \end{cases}$$

A particular case of a literal multivalued-input binary-output function

$$x^y = \begin{cases} 1 & \text{if } x = y, \\ 0 & \text{otherwise.} \end{cases}$$

The window literal operation is defined as

$${}^a x^b = \begin{cases} m-1 & \text{if } a \le x \le b, \\ 0 & \text{otherwise.} \end{cases}$$

Any m-valued single-output system can be described by a functionally complete set of primitive operations. Various algebras exist to provide functional completeness for $m > 2$.

Example 29.5 *The basic operations for $m = 3$ (ternary) and $m - 4$ (quaternary) are given in Appendix-D.*

Truth table and truth column vector. The simplest way to represent a multivalued logic function is the truth table. The truth table of a logic function is the representation that tabulates all possible input combinations with their associated output values.

A truth column vector of a multivalued logic function f of n m-valued variables $x_1, x_2, \ldots, x_n$ is defined as

$$\mathbf{F} = [f(0), f(1), \ldots, f(m^n - 1)]^T.$$

The index i of the element $x^{(i)}$ corresponds to the assignments $i_1 i_2 \ldots i_n$ of variables $x_1, x_2, ..., x_n$ ($i_1, i_2, ..., i_n$ is the binary representation of i, $i = 0, \ldots, m^n - 1$).

Example 29.6 *The truth column vector* **F** *of a ternary MIN function of two variables is* **F**= [000011012]^T *Table 29.3 contains this and other truth tables for some ternary functions.*

TABLE 29.3
Truth tables for elementary ternary logic functions.

x_1	x_2	MAX	MIN	MODSUM	MODPROD	TSUM	TPROD
0	0	0	0	0	0	0	0
0	1	1	0	1	0	1	0
0	2	2	0	2	0	2	0
1	0	1	0	1	0	1	0
1	1	1	1	2	1	2	1
1	2	2	1	0	2	2	2
2	0	2	0	2	0	2	0
2	1	2	1	0	2	2	2
2	2	2	2	1	1	2	2

Symbolic, or algebraic notations include sum-of-products and polynomial forms. The sum-of-products form is represented as follows:

$$f = s_0\varphi_0(x_1, \ldots, x_n) + \ldots + s_{m^n-1}\varphi_{m^n-1}(x_1, \ldots, x_n),$$

where $s_i\varphi_i(x_1, \ldots, x_n)$ is a literal function such that

$$\varphi_i(x_1, \ldots, x_n) = \begin{cases} 1, \text{ if } x_1, \ldots, x_n = 1; \\ 0, \text{ otherwise.} \end{cases}$$

29.3.2 Multivalued algebras

There are various universal (functionally complete) sets of operations for multivalued algebra. Some are given below.

Post algebra is based on two operations: 1-cycle inversion $(x + 1)_{mod\ m}$, and MAX operation $x \vee y = \text{MAX}(x, y)$. Using these operations, one can describe any m-valued logic function. The analogs of Post operations in Boolean algebra are NOT and OR operations, which also constitute a universal system.

Webb algebra is based on one operation, the Sheffer-Stroke operation that is specified as $x|y = \text{MAX}(x, y) + 1(mod\ m\ )$. In Webb algebra, a functionally complete set consists of one function.

Bernstein algebra or modulo-sum and modulo-product algebra includes the modulo m sum: $x_1 + x_2\ (mod\ m)$, and the modulo m product: $xy\ (mod\ m)$.

Allen and Givone algebra's universal set consists of MIN(x_1, x_2), MAX(x_1, x_2), and the *window literal operation*

$$^a x^b = \begin{cases} m-1 & \text{if } a \leq x \leq b \\ 0 & \text{otherwise.} \end{cases}$$

Allen and Givone algebra [1] is often used in multivalued logic circuit design in many variations, for example, MIN, MAX, truncated sum, and appropriate subset of 1-place functions.

There are a lot of other algebras oriented mostly toward circuit implementations. For example, there are algebras based on MIN, MAX, and CYCLE operations. Other examples are MIN, TSUM, and WINDOW LITERAL operations.

29.4 Galois fields

A Boolean algebraic system $\langle \mathbf{B}, \cdot, 0, 1\rangle$ is called a *Boolean ring* if it satisfies the conditions that there exist elements x, $(x \neq 0)$ and y such that $xy = yx = 1$. This is the condition of the field called *Galois field*, denoted by $GF(2)$. The space of the functions

$$f: \; C_2^n \rightarrow GF(2),$$

where $C_2 = (\{0, 1\}, \oplus)$ includes polynomial expressions such as sum-of-products, Reed–Muller, Kronecker expressions, etc. These expressions can be determined from the corresponding decision trees and diagrams.

29.4.1 Galois fields $GF(m)$

The generalization of the Boolean Galois field to multivalued functions

$$f: \; \{0, \ldots, m-1\}^n \rightarrow \{0, \ldots, m-1\},$$

where p is a prime, leads to $GF(m)$. Every finite set

$$Z_m = \{0, 1, 2, \ldots, m-1\}$$

with modulo m addition and multiplication is a field if and only if m is a *prime* number. Such field is called a *Galois field modulo m* denoted by $GF(m)$.

Example 29.7 *The set $\{0, 1, 2\}$ with addition and multiplication modulo 3 is a field. There is an identity 0 with respect to modulo 3 addition, and identity 1 with respect to modulo 3 multiplication. Every element has a unique additive inverse, and every element other than 0 has a multiplicative inverse.*

Shannon, Reed–Muller, and Kronecker forms for m-valued functions can be generated using various transforms over $GF(m)$.

The equivalent multivalued Reed–Muller expansion for a one-variable function over $GF(m)$ is expressed as

$$f = r_0 + r_1x + r_2x^2 + ... + r_{m-1}x^{m-1} \ over \ GF(m)$$

Given $GF(3)$ this resolves to

$$f = r_0 + r_1x + r_2x^2 \ over \ GF(3)$$

For two variables, the ternary Reed–Muller expansion is

$$\begin{aligned} f &= r_0 + r_1x_2 + r_2x_2^2 + r_3x_1 + r_4x_1x_2 \\ &+ r_5x_1x_2^2 + r_6x_1^2 + r_7x_1^2x_2 + r_8x_1^2x_2^2 \ over \ GF(3) \end{aligned}$$

Galois field $GF(3)$**.** A ternary logic function of n variables can be represented as a Galois field polynomial.

> **Example 29.8** *A ternary function $MAX(x_1, x_2)$, truth table given in Table 29.3, can be represented in algebraic (sum-of-products) form as follows:*
>
> $$\begin{aligned} MAX(x_1, x_2) &= 0 \cdot x_1^0x_2^0 + 1 \cdot x_1^0x_2^1 + 2 \cdot x_1^0x_2^2 + 1 \cdot x_1^1x_2^0 \\ &+ 1 \cdot x_1^1x_2^1 + 2 \cdot x_1^1x_2^2 + 2 \cdot x_1^2x_2^0 + 2 \cdot x_1^2x_2^1 \\ &+ 2 \cdot x_1^2x_2^2 \end{aligned}$$

Galois field $GF(4)$**.** It should be noted that *modulo* 4 addition and multiplication do not form a field, because the element 2 has no multiplicative inverse (there is no element a such that $2 \cdot a = 1$ so that $a = 2^{-1}$). However, a field $GF(2^k)$ can be introduced, where the element of the filed are k-bit binary numbers. For example, an algebra for the quaternary field $GF(4)$ with elements {0,1,A,B} can be constructed (see the Further Reading section for references), and $GF(4)$ addition and multiplication tables can be specified for that algebra.

A 4-valued logic function of n variables can be represented as a Galois field polynomial

$$f = a_0 + \sum_{i=1}^{4^n-1} a_i g(i), \tag{29.1}$$

where coefficients $a_i \in \{0, 1, 2, 3\}$, and $g(i)$ are the product terms defined as elements of the vector

$$\mathbf{X}(n) = \bigotimes_{i=1}^{n} [1 \ x_i \ x_i^2 \ x_i^3],$$

where $\bigotimes$ denotes the Kronecker product, and addition and multiplication are carried out in $GF(4)$.

Given a truth vector $\mathbf{F}[f(0)\dots f(4^n-1)]^T$ of a 4-valued logic function of n variables, the vector of coefficients $\mathbf{A}=[a_0\dots a_{4^n-1}]^T$ can be computed by the matrix equation

$$\mathbf{A}=\mathbf{GF},$$

where the $(4^n-1)\times(4^n-1)$ matrix $\mathbf{G}$ is constructed as

$$\mathbf{G}(n)=\bigotimes_{i=1}^{n}\mathbf{G}(1,\ \mathrm{G}(1=)\ \begin{bmatrix}1&0&0&0\\0&1&3&2\\0&1&2&3\\1&1&1&1\end{bmatrix}.$$

Equation 29.1 can be written in the form

$$f=\mathbf{XA}. \tag{29.2}$$

There are four permutations of the values of four-valued variables. These permutations are described by x_i+k for $k=0,1,2,3$, where the addition is in $GF(4)$, and can be identified as the four different complements of a four-valued variable.* By using these different complements of variables, k^n different Reed–Muller expressions for a given 4-valued logic function can be derived. The Reed–Muller expression with the smallest number of non-zero coefficients is called the *optimal* Reed–Muller expression for a given logic function.

Example 29.9 *Given a truth vector*

$$\mathbf{F}=[0000013203210213]^T$$

of a 4-valued logic function of two variables $n=2$. Figure 29.1 illustrates the Reed–Muller expression construction.

29.4.2 Algebraic structure for Galois field representations

For a set whose elements can be identified with first four non-negative integers $\{0,1,2,3\}$, the Galois field $GF(4)$ structure is provided by addition and multiplication defined in Table 29.4.

29.4.3 Galois field expansions

The set of elementary functions $1,x,x^2,x^3$ is a basis in the space of one-variable functions over $GF(4)$. Therefore, each quaternary function f given

*The other possible permutations up to the total number of 4! do not affect the number of coefficients.

$$\mathbf{G}(2) = \mathbf{G}(1) \otimes \mathbf{G}(1) = \begin{bmatrix} 1&0&0&0\\0&1&3&2\\0&1&2&3\\1&1&1&1 \end{bmatrix} \otimes \begin{bmatrix} 1&0&0&0\\0&1&3&2\\0&1&2&3\\1&1&1&1 \end{bmatrix}$$

$$\mathbf{A} = \mathbf{GF} = \underbrace{\left[\begin{array}{cccc|cccc|cccc|cccc}
1&0&0&0&0&0&0&0&0&0&0&0&0&0&0&0\\
0&1&3&2&0&0&0&0&0&0&0&0&0&0&0&0\\
0&1&2&3&0&0&0&0&0&0&0&0&0&0&0&0\\
1&1&1&1&0&0&0&0&0&0&0&0&0&0&0&0\\ \hline
0&0&0&0&1&0&0&0&3&0&0&0&2&0&0&0\\
0&0&0&0&0&1&3&2&0&3&1&2&0&2&2&0\\
0&0&0&0&0&1&2&3&0&3&2&1&0&2&0&2\\
0&0&0&0&1&1&1&1&3&3&3&3&2&2&2&2\\ \hline
0&0&0&0&1&0&0&0&2&0&0&0&3&0&0&0\\
0&0&0&0&0&1&3&2&0&2&2&0&0&3&1&2\\
0&0&0&0&0&1&2&3&0&2&0&2&3&3&2&1\\
0&0&0&0&1&1&1&1&2&2&2&2&3&3&3&3\\ \hline
1&0&0&0&1&0&0&0&1&0&0&0&1&0&0&0\\
0&1&3&2&0&1&3&2&0&1&3&2&0&1&3&2\\
0&1&2&3&0&1&2&3&0&1&2&3&0&1&2&3\\
1&1&1&1&1&1&1&1&1&1&1&1&1&1&1&1
\end{array}\right]}_{\mathbf{G}(2) = \mathbf{G}(1) \otimes \mathbf{G}(1)}
\begin{bmatrix}0\\0\\0\\0\\0\\1\\3\\2\\0\\3\\2\\1\\0\\2\\1\\3\end{bmatrix} = \begin{bmatrix}0\\0\\0\\0\\0\\0\\0\\0\\0\\0\\1\\0\\0\\0\\0\\0\end{bmatrix}$$

$$\begin{aligned} f &= \mathbf{XA} = [1\ x_i\ x_i^2\ x_i^3]\mathbf{A} \\ &= x_1^2 x_2^2 \end{aligned}$$

FIGURE 29.1
Deriving the Reed–Muller expression for a 4-valued logic function of two variables (Example 29.9).

TABLE 29.4
Addition and multiplication in $GF(4)$.

+	0	1	2	3
0	0	1	2	3
1	1	0	3	2
2	2	3	0	1
3	3	2	1	0

·	0	1	2	3
0	0	0	0	0
1	0	1	2	3
2	0	2	3	1
3	0	3	1	2

by the truth-vector $\mathbf{F} = [f(0), \ldots, f(3)]^T$ can be represented by Fourier-like Galois field expansion given in the matrix form by

$$f = \left[1\ x\ x^2\ x^3\right] \mathbf{G}_4(1)\mathbf{F},$$

where $\mathbf{G}_4(1) = \begin{bmatrix} 1&0&0&0\\0&1&3&2\\0&1&2&3\\1&1&1&1 \end{bmatrix}$, and the all calculations are carried out in $GF(4)$. In this notation, the basis functions $1, x, x^2, x^3$ can be considered as columns of the matrix $\mathbf{G}_4^{-1}(1)$, inverse to $\mathbf{G}_4(1)$, $\mathbf{G}_4^{-1}(1) = \begin{bmatrix} 1&0&0&0\\1&1&1&1\\1&2&3&1\\1&3&2&1 \end{bmatrix}$.

Extension of Galois field representations to n-variable functions is straightforward with the Kronecker product. GF-representation of an n-variable quaternary function f given by its truth-vector $\mathbf{F} = [f(0), \ldots, f(4^n - 1)]^T$ is given by

$$f = \left(\bigotimes_{i=1}^{n} \left[1\ x_i\ x_i^2\ x_i^3\right]\right)\left(\bigotimes_{i=1}^{n} \mathbf{G}_{4i}\right)\mathbf{F},$$

where $\otimes$ denotes the Kronecker product and $\mathbf{G}_{4i} = \mathbf{G}_4(1)$. The calculations are carried out in $GF(4)$.

29.4.4 Partial Galois field transforms

The coefficients in GF-representations of an n-variable function can be considered as the spectral coefficients of a Galois field (GF) transform defined by the transform matrix

$$\mathbf{G}_4(n) = \bigotimes_{i=1}^{n} \mathbf{G}_{4i}(1).$$

By analogy to Good–Thomas factorization in FFT-theory, $\mathbf{G}_4(n)$ can be factorized as

$$\mathbf{GF}_4(n) = \prod_{i=1}^{n} \mathbf{C}_i, \quad \mathbf{C}_i = \bigotimes_{j=1}^{n} \mathbf{C}_i^j(1),$$

$$\mathbf{C}_i^j(1) = \begin{cases} \mathbf{G}_{4i}(1), & i = j, \\ \mathbf{I}_4(1), & i \neq j, \end{cases}$$

with $\mathbf{I}_4(1)$ the identity matrix of order 4. Each matrix $\mathbf{C}_i$ defines a step in the FFT-like algorithm for the calculation of Galois field coefficients. By analogy with FFT-theory, each matrix $\mathbf{C}_i$ can be considered as a partial Galois field transform with respect to the i-th variable.

For a function f given by the truth-vector $\mathbf{F} = [f(0), \ldots, f(4^n - 1)]^T$ the vector $\mathbf{GF}_{f_i} = \mathbf{C}_i\mathbf{F}$ is the spectrum of the partial Galois field transform of f with respect to the i-th variable.

29.5 Fault models based on the concept of change

Fault models for MVL circuits depend on the style and the technology with which the circuit is implemented. There are several approaches to designing logic models for physical defects or failures. In this section, a fault is described by a set of changes of a multivalued signal. For a formal description of these changes, logic differences can be used. Note that a set of changes can be minimized. This approach can be especially useful for local testing of MVL devices.

In this chapter, direct changes to a signal are considered for the representation of various faults.

Example 29.10 *In a ternary system, signal changes from logic level 0 to logic level 1 or 2 are denoted by* $D_{0\to 1}$, $D_{0\to 2}$, *and* $D_{0\to 2}$ *for direct changes. Grouping these changes, it is possible to describe various logic models of faults in multivalued circuits.*

Below several common types of faults in multivalued system are listed:

- A *stuck-at-k* fault occurs in a line x if x generates the output signal k for all input signals.
- A $\beta_1 - \beta_2$ *window* fault occurs in a line x if x operates correctly for input signals $\beta_1 \leq t \leq \beta_2$, and x is *stuck-at-*β_1 for input signal $t < b1$ and *stuck-at-*β_2 $t > \beta_2$, where $\beta_1, \beta_2 \in \{0, 1, ..., m-1\}$.
- An *r-order input signal variation* fault occurs in a line x, $r \in \{0, 1, ..., m-1\}$, if x generates the output signal $t' = t + r$ for input signal t.
- A *shift* fault

$$p^{+},\ p \in \{0, 1, ..., m-1\} \text{ or}$$
$$p^{-},\ p \in \{1, ..., m-1\}$$

occurs in a line x if x generates the output signal $t' = (p+1)$ or $t' = (p-1)$ if $t = p$, and x operates correctly for input signals $t \neq p$.

An elementary event in a MVL gate is a direct change of the logical value on line x from α to β $\alpha, \beta \in \{0, ..., m-1\}$, and $\alpha \neq \beta$.

Example 29.11 *A fault* shift r^{+} *in a* m*-valued gate can be described by the direct change* $D_{r\to r+1}$.

Figure 29.2 illustrates the above faults by a sample of 4-valued signals.

Logic faults in a binary network are often described as *stuck-at-0* and *stuck-at-1* faults. These faults can be considered as elementary changes and described, for example, by Boolean differences. The analog of these faults in a k-valued network is *stuck-at-k* faults. They can also be described in terms of change. To describe *stuck-at-k* faults, it is necessary to analyze and take into account all possible types of changes.

Example 29.12 *In the case of a 4-valued gate the minimal number of elementary changes is 12. Hence, a set of* stuck-at-k *faults in a 4-valued combinational network is described by 12 types of various changes* $D_{r\to r+1}$.

A relationship between different types of faults can be described in terms of change. Figure 29.3 illustrates:

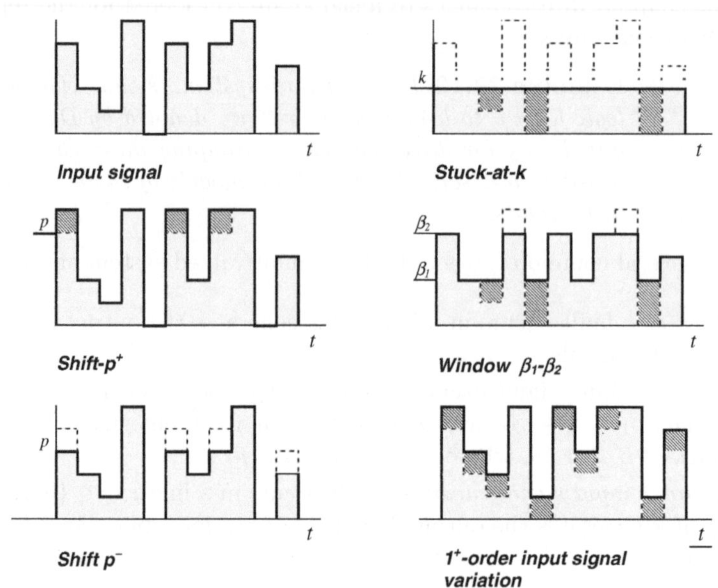

FIGURE 29.2
Changing of the 4-valued signal by various types of faults.

- The relationships between *stuck-at-k*, r-order signal variation, shift r^+ and window $\beta_1 - \beta_2$ faults for a 4-valued circuit, and
- Rules for detection of these faults.

Example 29.13 *Particular cases of fault detection in a 4-valued circuit are given below. These use the rules given in Figure 29.3:*

(a) *A test to detect the fault window 1-2 in a quaternary circuit can be derived from the following changes:* $D_{0\to1}$ *and* $D_{3\to1}$*. The same events are used to detect a* stuck-at-1 *fault.*

(b) *Tests to detect* 1^+ *input signal variation are derived from the following changes:* $D_{0\to1}$, $D_{1\to2}$, $D_{2\to3}$, *and* $D_{3\to0}$.

(c) *The changes* $D_{0\to2}$, $D_{1\to2}$, *and* $D_{3\to2}$ *are used in the logic equation for detection of stuck-at-2 faults.*

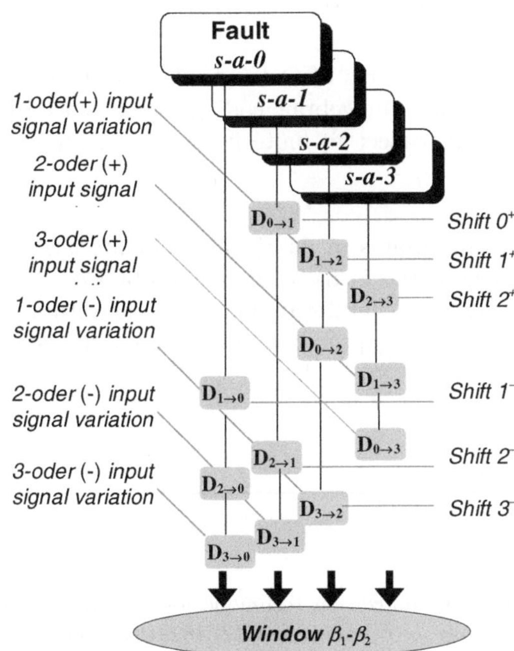

Rule 1. *Detection of a window*

$$\beta_1 - \beta_2$$

fault is equivalent to detection of the set of s-a-k *faults.*

Rule 2. *Detection of all*

$$r_1^+ - order,$$
$$r_1 = 1, 2, ..., m-1,$$

and

$$r_2^- - order,$$
$$r_2 = (m-1) - r_1,$$

input signal variation faults is equivalent to detection of all stuck-at faults.

Rule 3. *Detection of all*

$$1^+ - order,$$
$$1^- - order$$

input signal variation faults is equivalent to detection of all shift faults

$$r^+,\ r = 0, 1, ..., m-2,$$
$$r^-,\ r = 1, 2, ..., m-1.$$

Rule 4. *Detection of the shift faults*

$$r^+,\ r = 0, 1, ..., m-2$$
$$r^-,\ r = 1, 2, ..., m-1$$

is adequate for detecting the faults

$$s - a - (r+1),$$
$$s - a - (r-1).$$

FIGURE 29.3
The relationship between *s-a-k*, *r*-order input signal variation, and *window* $\beta_1 - \beta_2$ faults for a 4- valued circuit and the rules for detecting several faults.

29.6 Further reading

Spectral techniques in multivalued logic are based on spectral transformations of discrete signals [10, 7] (see Chapter 30). In [10], the Pascal triangle has been used to obtain arithmetic representations of elementary multivalued functions. Techniques for representing arithmetic forms and their manipulation based on spectral transforms have been developed by many researchers. In [13], this technique and arithmetic Taylor expansion have been used to generate arithmetic forms of multivalued functions.

Galois fields. The seminal book in the area of Galois fields is authored by Dickson [4]. Finite fields are considered in algebraic coding theory, since this theory builds on vector spaces with respect to finite fields, for example, in [2, 5].

Concept of change. The class of failures in multivalued circuits is much larger than in binary circuits. This is the focus of the studies by Coy et al. [3], Miller [10], and Wang et al. [12]. A generalization of the D-algorithm for multivalued combinational circuits has been proposed in [9], and its extensions can be found in papers [8, 11, 14].

References

[1] Allen CM and Givone DD. A minimization technique for multiple-valued logic systems. *IEEE Transactions on Computers,* 17(2):182–184, 1968.

[2] Berlekamp C. *Algebraic Coding Theory.* McGraw-Hill, New York, 1968.

[3] Coy W and Moraga C. Description and detection of faults in multiple-valued logic circuits. In *Proceedings of the 19th IEEE International Symposium on Muliple-Valued Logic*, pp. 74–81, 1979.

[4] Dickson C. *Linear Groups (with an Exposition of the Galois Field Theory)* Dover, 1958.

[5] Lidl R and Niederreiter. *Finite Fields.* Addison-Wesley, 1983.

[6] Miller DM. Spectral signature testing for multiple-valued combinational networks. In *Proceedings of the 12th IEEE International Symposium on Multiple-Valued Logic*, pp. 152–158, 1982.

[7] Miller DM. Multiple-valued logic design tools. In *Proceedings of the 23rd IEEE International Symposium on Multiple-Valued Logic*, pp. 2–11, 1993.

[8] Shmerko VP, Yanushkevich SN, and Levashenko VG. Test pattern generation for combinational MVL networks based on generalized D-algorithm. In *Proceedings of the 22nd IEEE International Symposium on Multiple-Valued Logic*, pp. 139–144, 1997.

[9] Spillman RJ and Su SYH. Detection of single, stuck-type failures in multivalued combinational networks. *IEEE Transactions on Computers*, 26(12):1242–1251, 1977.

[10] Strazdins I. The polynomial algebra of multivalued logic. *Algebra, Combinatorics and Logic in Computer Science*, 42:777–785, 1983.

[11] Tabakow IG. Using D-algebra to generate tests for m-logic combinational circuits. *International Journal of Electronics*, 75(5):897–906, 1993.

[12] Wang HM, Lee CL, and Chen JE. Complete test set for multiple-valued logic networks. In *Proceedings of the 24th IEEE International Symposium on Multiple-Valued Logic*, pp. 289–296, 1994.

[13] Yanushkevich SN. Systolic algorithms to synthesize arithmetical polynomial forms for k-valued logic functions. *Automation and Remote Control*, Kluwer/Plenum Publishers, 55(12):812–1823, 1994.

[14] Yanushkevich SN, Levashenko VG, and Moraga C. Fault models for multiple-valued combinational circuits. In *Proceedings of the International Conference on Applications of Computer System*, pp. 309–314, Szczecin, Poland, 1997.

30

Spectral Transforms of Multivalued Functions

Spectral techniques for working with multivalued logic functions are generalizations of those used for switching functions. Spectral decision diagrams of multivalued functions provide a convenient way for analyzing and synthesizing multivalued circuits. This chapter focuses on the algebraic representation of multivalued functions and their mapping onto graphical data structures.

30.1 Introduction

Because of the interdisciplinary approach to the study of logic functions, different interpretations of the same concepts are often used, and new terminology bears clarification. Below, parallels between the terminology of spectral techniques and that of classical logic design are drawn.

The Reed–Muller spectrum *is the set of coefficients of a Reed–Muller expression of a switching function as the result of transforms in the Galois field.*

An arithmetic spectrum *is the set of coefficients of an arithmetic expression of a switching function, or the coefficients of a word-level representation of several switching functions in arithmetic form. The additional term* generalized *is used to emphasize that it concerns the representation of multivalued logic functions.*

A Walsh spectrum *is the set of coefficients of a Walsh expression of a switching function, or a word-level representation of several switching functions in Walsh form.*

Spectral transforms of logic functions usually utilize a matrix technique. The spectra are also calculated using the following matrix transforms:

► A *Reed–Muller spectral transform* is the transformation of the truth table of a multivalued function to a vector of coefficients of a Reed–Muller expression.

- An *arithmetic spectral transform* is the transformation of the truth table of a multivalued function to a vector of coefficients of an arithmetic expression.
- A *Walsh transform* is the transformation of the truth table of a multivalued function to a vector of coefficients of a Walsh expression.

30.2 Reed–Muller spectral transform

Generalization of a Reed–Muller spectral transform of switching functions (see Chapter 4) for multivalued functions is quite straightforward. An m-valued logic function f of n variables is described by a sum-of-products (SOP)

$$f = \bigvee_{j=0}^{m^n-1} \varphi_{j_1}(x_1) \times \cdots \times \varphi_{j_n}(x_n), \tag{30.1}$$

where

$$\varphi_{j_i}(x_i) = \begin{cases} 0, & \text{if } j_i \neq x_i; \\ m-1, & \text{if } j_i = x_i. \end{cases}$$

30.2.1 Direct and inverse Reed–Muller transforms

Direct and inverse Reed–Muller transforms over $GF(m)$ are defined by the matrix equations (Figure 30.1)

$$\begin{aligned} \mathbf{R} &= \mathbf{R}_{m^n}^{(c)} \mathbf{F} \quad \textit{over } GF(m) \\ \mathbf{F} &= \mathbf{R}_{m^n}^{-1(c)} \mathbf{R} \quad \textit{over } GF(m) \end{aligned} \tag{30.2}$$

where $c = c_1 c_2 \ldots c_n$ is an m-valued representation of $c = 1, 2, \ldots,\ m^n$. The pair of matrices $\mathbf{R}_{m^n}^{(c)}$ and $\mathbf{R}_{m^n}^{-1(c)}$ in Equation 30.2 are calculated as

$$\begin{aligned} \mathbf{R}_{m^n}^{(c)} &= \mathbf{R}_m^{(c_1)} \otimes \mathbf{R}_m^{(c_2)} \otimes \cdots \otimes \mathbf{R}_m^{(c_n)} \\ \mathbf{R}_{m^n}^{-1(c)} &= \mathbf{R}_m^{-1(c_1)} \otimes \mathbf{R}_m^{-1(c_2)} \otimes \cdots \otimes \mathbf{R}_m^{-1(c_n)} \end{aligned} \tag{30.3}$$

The components of matrix $\mathbf{R}_m^{(c_j)}$ and $\mathbf{R}_m^{-1(c_j)}$, $j = 1, 2, \ldots, n$, are obtained as the solution to the logic equation

$$\mathbf{R}_m^{-1(c_j)} \mathbf{R}_m^{(c_j)} = \mathbf{I}_m \quad \textit{over } GF(m) \tag{30.4}$$

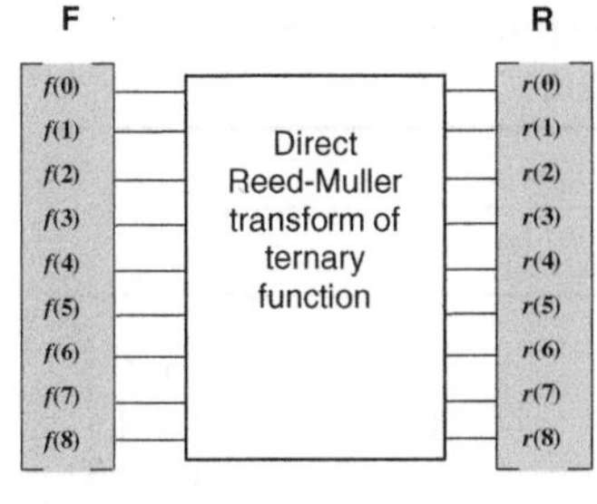

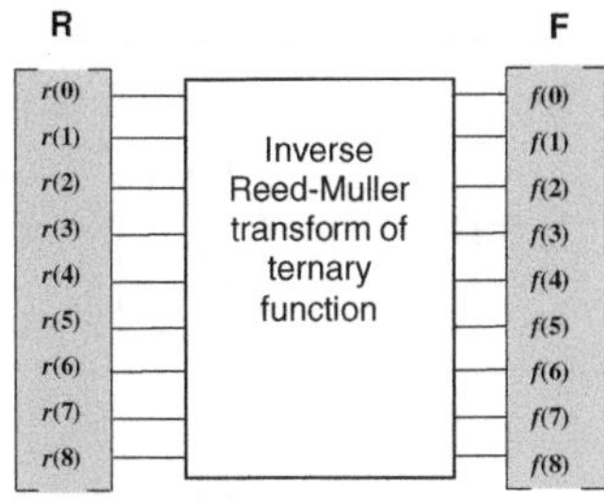

The pair of direct and inverse Reed–Muller transforms

$$\mathbf{R} = \mathbf{R}_{3^2}^{(c)}\mathbf{F} \quad over\ GF(3)$$
$$\mathbf{F} = \mathbf{R}_{3^2}^{-1(c)}\mathbf{R} \quad over\ GF(3)$$

where

$$\mathbf{R}_{3^2}^{(c)} = \mathbf{R}_3^{(c_1)} \otimes \mathbf{R}_3^{(c_2)}$$
$$\mathbf{R}_{3^2}^{-1(c)} = \mathbf{R}_3^{-1(c_1)} \otimes \mathbf{R}_3^{-1(c_2)}$$

Basic transform matrices $\mathbf{R}_{3^2}^{(c)}$ and $\mathbf{R}_{3^2}^{-1(c)}$ for polarity $c = 0, 1, 2$ are given in Table 30.1

FIGURE 30.1
Direct and inverse Reed–Muller transforms for a ternary function ($m = 3$) of two variables ($n = 2$).

where $\mathbf{I}_m$ is $m \times m$ identity matrix.

In Table 30.1, the basic arithmetic transform matrices for ternary ($m = 3$) logic functions are given.

Example 30.1 *Given $m = 3$ and $c = 2$, matrices $\mathbf{R}_3^{-1(2)}$ and $\mathbf{R}_3^{(2)}$ meets the condition from Equation 30.4*

$$\mathbf{R}_3^{-1(2)}\mathbf{R}_3^{(2)} = \begin{bmatrix} 1 & 2 & 1 \\ 1 & 0 & 0 \\ 1 & 1 & 1 \end{bmatrix} \begin{bmatrix} 0 & 1 & 0 \\ 1 & 0 & 2 \\ 2 & 2 & 2 \end{bmatrix} = \mathbf{I}_3 \quad over\ GF(3).$$

30.2.2 Polarity

Polarity plays an important role in decision diagram techniques:

- ► Equations 30.3 are a formal justification of the statement that an arbitrary logic function can be represented by m^n different Reed–Muller expressions, or *polarities.*
- ► Equations 30.3 are a formal notation of the problem of optimal representation of multivalued functions by Reed–Muller expressions (see Example 30.3). This is because it is possible to find an optimal (in terms

TABLE 30.1
Basic transform matrices of Reed–Muller for polarity $c = 0, 1, 2$, of a ternary logic function.

Polarity c	Direct	Inverse
0	$\mathbf{R}_3^{(0)} = \begin{bmatrix} 1 & 0 & 0 \\ 0 & 2 & 1 \\ 2 & 2 & 2 \end{bmatrix}$	$\mathbf{R}_3^{-1(0)} = \begin{bmatrix} 1 & 0 & 0 \\ 1 & 1 & 1 \\ 1 & 2 & 1 \end{bmatrix}$
1	$\mathbf{R}_3^{(1)} = \begin{bmatrix} 0 & 0 & 1 \\ 2 & 1 & 0 \\ 2 & 2 & 2 \end{bmatrix}$	$\mathbf{R}_3^{-1(1)} = \begin{bmatrix} 1 & 1 & 1 \\ 1 & 2 & 1 \\ 1 & 0 & 0 \end{bmatrix}$
2	$\mathbf{R}_3^{(2)} = \begin{bmatrix} 0 & 1 & 0 \\ 1 & 0 & 2 \\ 2 & 2 & 2 \end{bmatrix}$	$\mathbf{R}_3^{-1(2)} = \begin{bmatrix} 1 & 2 & 1 \\ 1 & 0 & 0 \\ 1 & 1 & 1 \end{bmatrix}$

of minimal number of literals) representation among the m^n different Reed–Muller expressions.

► Equations 30.3 are a formal description of the behavior of a multivalued function in terms of change.

The Reed–Muller expression of polarity c is described as follows:

$$R^{(c)} = \sum_{j=0}^{m^n-1} r_j (x_1 + c_1)^{j_1} (x_2 + c_2)^{j_2} \cdots (x_n + c_n)^{j_n} \quad over\ GF(m) \tag{30.5}$$

where

$$(x_i + c_i)^{j_i} = \begin{cases} x_i + c_i = \hat{x}_i^{c_i}, & j_i \neq 0; \\ 1, & j_i = 0, \end{cases} \tag{30.6}$$

where $\hat{x}_i^{c_i}$ is a c_i-order cyclic complement of the variable x_i, $c_i \in (0, 1, \ldots, m-1)$. The coefficient r_j is the j-th component of the vector $\mathbf{R}$ calculated by a matrix-vector transform (Equation 30.2). That is,

$$r_j = \sum_{k=0}^{m^n-1} f_k r_{k,j} \quad over\ GF(m), \tag{30.7}$$

where $f(k)$ is the k-th component of the truth vector $\mathbf{F}$, and $r_{j,k}$ is the $(k_{j,k})$-th element in the matrix $\mathbf{R}_{m^n}^{(c)}$.

Example 30.2 *The truth vector* $\mathbf{F} = [201000102]^T$ *represents a ternary logic function of two variables. There are nine Reed–Muller expressions generated for this function, corresponding to polarities*

$$c_1c_2 = \{00, 01, 02, 10, 11, 12, 20, 21, 22\}.$$

Figure 30.2 shows the computing of Reed–Muller expressions given $c_1c_2 = 01$*. The matrix transform implies that the coefficients* r_j*,* $j = 0, 1, 2, \ldots,$ *8, are calculated by Equation 30.7. Circuits to implement this function in SOP are*

$$\begin{aligned} f &= 2\varphi_0(x_1)\varphi_0(x_2) \vee 1\varphi_0(x_1)\varphi_2(x_2) \\ &\vee 1\varphi_2(x_1)\varphi_0(x_2) \vee 2\varphi_2(x_1)\varphi_2(x_2) \end{aligned}$$

and Reed–Muller forms are shown in Figure 30.3.

Example 30.3 *Multivalued expressions can be manipulated in different polarities; for example, the optimal form can be found using optimization techniques. Table 30.2 contains all 9 forms for the function given by truth vector* $[020120000]^T$*. The optimal (in terms of number of literals) polarity is* $c = 8$.

Example 30.4 *Table 30.3 represents a typical library of elementary multivalued gates in Reed–Muller form.*

In signal processing, the matrix representation in the form of Equation 30.3 is known as a *factorized* representation of a transform matrix. These equations play the central role in synthesis of so-called *fast algorithms*.

A ternary variable x can be represented in Reed–Muller form in one of the 28 bases which correspond to 28^n mixed polarity forms of a ternary function of n variables [5, 6] (Table 30.4, Figure 30.4).

Example 30.5 *Let*

$$\begin{aligned} \mathbf{A}_1 &= [1\ x_1\ x_1^2] \\ \mathbf{A}_2 &= [1\ \hat{x}_2\ \hat{x}_2^2]. \end{aligned}$$

Thus, the terms of a ternary function of two variables are defined as follows:

$$\begin{aligned} \mathbf{A}_1 \otimes \mathbf{A}_2 &= [1\ x_1\ x_1^2] \otimes [1\ \hat{x}_2\ \hat{x}_1^2] \\ &= [1\ \hat{x}_2\ \hat{x}_2^2\ x_1\ x_1\hat{x}_2\ x_1\hat{x}_2^2\ x_1^2\ x_1^2\hat{x}_2^2\ x_1^2\hat{x}_2^2\ ] \end{aligned}$$

Given:
A ternary function (m=3) of two variables (n=2):
$\mathbf{F} = [201000102]^T$
Polarity of a Reed–Muller expression:
$c = 1,\ c = c_1c_2 = 01,$
$c_1 = 0,\ c_2 = 1$

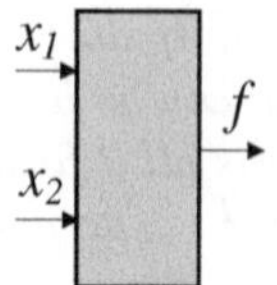

x_1	x_2	**F**
0	0	2
0	1	0
0	2	1
1	0	0
1	1	0
1	2	0
2	0	1
2	1	0
2	2	2

Elementary transform matrices for $c_1 = 0$ and $c_2 = 1$ are given in Table 30.1. Using Equation 30.3, form the transform matrix

$$\mathbf{R}^{(2)} = \mathbf{R}_{3^2}^{(2)}\ \mathbf{F} = \left(\mathbf{R}_3^{(0)} \otimes \mathbf{R}_3^{(1)}\right)\mathbf{F}$$

$$= \left(\begin{bmatrix} 1&0&0\\ 0&2&1\\ 2&2&2 \end{bmatrix} \otimes \begin{bmatrix} 0&0&1\\ 2&1&0\\ 2&2&2 \end{bmatrix}\right)\mathbf{F}$$

$$= \left[\begin{array}{ccc|ccc|ccc} 0&0&1&0&0&0&0&0&0\\ 2&1&0&0&0&0&0&0&0\\ 2&2&2&0&0&0&0&0&0\\ \hline 0&0&0&0&0&2&0&0&1\\ 0&0&0&1&2&0&2&1&0\\ 0&0&0&1&1&1&2&2&2\\ \hline 0&0&2&0&0&2&0&0&2\\ 1&2&0&1&2&0&1&2&0\\ 1&1&1&1&1&1&1&1&1 \end{array}\right] \begin{bmatrix} 2\\0\\1\\0\\0\\0\\1\\0\\2 \end{bmatrix} = \begin{bmatrix} 1\\1\\0\\2\\2\\0\\0\\0\\0 \end{bmatrix} \quad over\ GF(3)$$

Note that by Equation 30.7:

$$r_j = \sum_{k=0}^{3^2-1} f_k r_{j,k} \quad over\ GF(3),$$

$$\begin{aligned}
r_0 &= f(0)r_{0,0} + \ldots + f(8)r_{0,8} = 2\times 0 + \cdots + 2\times 0 = 1\\
r_1 &= f(0)r_{1,0} + \ldots + f(8)r_{1,8} = 2\times 2 + \cdots + 2\times 0 = 1\\
r_2 &= f(0)r_{2,0} + \ldots + f(8)r_{2,8} = 2\times 2 + \cdots + 2\times 0 = 0\\
r_3 &= f(0)r_{3,0} + \ldots + f(8)r_{3,8} = 2\times 0 + \cdots + 2\times 1 = 2\\
r_4 &= f(0)r_{4,0} + \ldots + f(8)r_{4,8} = 2\times 0 + \cdots + 2\times 0 = 2\\
r_5 &= f(0)r_{5,0} + \ldots + f(8)r_{5,8} = 2\times 0 + \cdots + 2\times 2 = 0\\
r_6 &= f(0)r_{6,0} + \ldots + f(8)r_{6,8} = 2\times 0 + \cdots + 2\times 2 = 0\\
r_7 &= f(0)r_{7,0} + \ldots + f(8)r_{7,8} = 2\times 1 + \cdots + 2\times 0 = 0\\
r_8 &= f(0)r_{8,0} + \ldots + f(8)r_{8,8} = 2\times 1 + \cdots + 2\times 1 = 0
\end{aligned}$$

Equation 30.7 for $m = 3,\ n = 2,\ c = 1$:

$$\begin{aligned}
R^{(1)} &= \sum_{j=0}^{3^2-1} r_j (x_1+0)^{j_1}(x_2+1)^{j_2} = \sum_{j=0}^{3^2-1} r_j x_1^{j_1}\hat{x}_2^{j_2}\\
&= r_0 x_1^0\hat{x}_2^0 + r_1 x_1^0\hat{x}_2^1 + r_3 x_1^1\hat{x}_2^0 + r_4 x_1^1\hat{x}_2^1\\
&= 1 + \hat{x}_2 + 2x_1 + 2x_1\hat{x}_2 \quad over\ GF(3)
\end{aligned}$$

FIGURE 30.2
Computing Reed–Muller coefficients (Example 30.2).

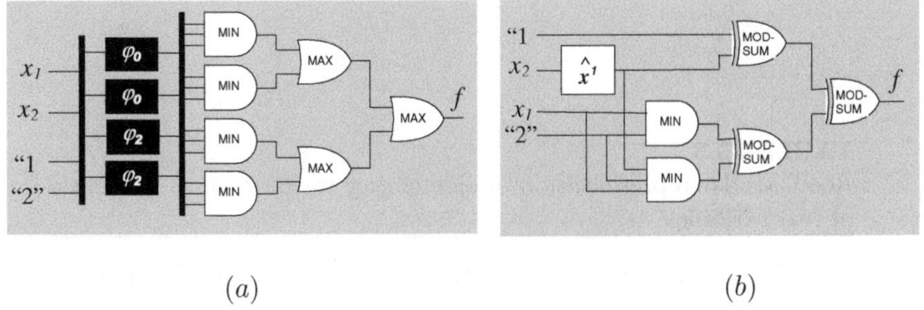

(*a*) (*b*)

FIGURE 30.3
A ternary circuit that implement an SOP expression (a) and Reed–Muller expression (b) (Example 30.2).

TABLE 30.2
Reed–Muller expression in polarities $c = 0, 1, \ldots 8$ of ternary logic functions of two variables.

Polarity	Reed–Muller expression over $GF(3)$
$c = 0\ (c_1c_2 = 00)$	$R^{(0)} = 2x_1 + x_1^2 + x_1x_2 + 2x_1^2x_2$
$c = 1\ (c_1c_2 = 01)$	$R^{(1)} = 2 + 2x_1 + 2x_1^2 + 2\hat{x}_2 + 2x_1\hat{x}_2 + 2x_1^2\hat{x}_2$
$c = 2\ (c_1c_2 = 02)$	$R^{(2)} = 2x_1\hat{\hat{x}}_2 + 2x_1^2\hat{\hat{x}}_2 + 2\hat{\hat{x}}_2$
$c = 3\ (c_1c_2 = 10)$	$R^{(3)} = 1 + 2\hat{x}_1 + \hat{x}_1^2 + 2x_2 + \hat{x}_1x_2 + 2\hat{x}_1^2x_2$
$c = 4\ (c_1c_2 = 11)$	$R^{(4)} = 2 + \hat{x}_1 + 2\hat{x}_1^2 + 2\hat{x}_2 + \hat{x}_1\hat{x}_2 + 2\hat{x}_1^2\hat{x}_2$
$c = 5\ (c_1c_2 = 12)$	$R^{(5)} = 2\hat{x}_1\hat{\hat{x}}_2 + \hat{x}_1^2\hat{\hat{x}}_2 + 2\hat{\hat{x}}_2$
$c = 6\ (c_1c_2 = 20)$	$R^{(6)} = 2\hat{\hat{x}}_1^2x_2 + \hat{\hat{x}}_1$
$c = 7\ (c_1c_2 = 21)$	$R^{(7)} = 2\hat{\hat{x}}_1^2\hat{x}_2 + 2\hat{\hat{x}}_1$
$c = 8\ (c_1c_2 = 22)$	$R^{(8)} = 2\hat{\hat{x}}_1^2\hat{\hat{x}}_1$

Example 30.6 *Given the vector of Reed–Muller polarity zero coefficients* $\mathbf{R} = [211010010]^T$ *of a ternary function of two* $(n = 2)$ *variables. We use the result of Example 30.5 to derive the Reed–Muller expression*

$$\mathbf{R} = 2 + \hat{x}_2 + \hat{x}_2^2 + x_1\hat{x}_2 + x_1^2\hat{x}_2 \quad GF(3)$$

TABLE 30.3
Reed–Muller representation of elementary ternary ($m = 3$) functions of two variables

Ternary gate	Truth table	Reed–Muller representation
NOT	x: 0, 1, 2; $\overline{x}$: 2, 1, 0	$2 + 2x + x^2$ (*mod* 3)
$\hat{x}^r$	r/x = 0 1 2; 0: 0 1 2; 1: 1 2 0; 2: 2 0 1	$\begin{cases} 1+x \ (mod\ 3), & r=1; \\ 2+x \ (mod\ 3), & r=2. \end{cases}$
MAX	0 1 2; 0: 0 1 2; 1: 1 1 2; 2: 2 2 2	$x_1 + x_2 + 2x_1x_2 + x_1x_2^2 + x_1^2x_2 + x_1^2x_2^2$ (*mod* 3)
MIN	0 1 2; 0: 0 0 0; 1: 0 1 1; 2: 0 1 2	$x_1x_2 + 2x_1x_2^2 + 2x_1^2x_2 + 2x_1^2x_2^2$ (*mod* 3)
MOD-SUM	0 1 2; 0: 0 1 2; 1: 1 2 0; 2: 2 0 1	$x_1 + x_2$ (*mod* 3)
MOD-PROD	0 1 2; 0: 0 0 0; 1: 0 1 2; 2: 0 2 1	x_1x_2 (*mod* 3)
TSUM	0 1 2; 0: 0 1 2; 1: 1 2 2; 2: 2 2 2	$x_1 + x_2 + 2x_1x_2^2 + 2x_1^2x_2 + 2x_1^2x_2^2$ (*mod* 3)
TPROD	0 1 2; 0: 0 0 0; 1: 0 1 2; 2: 0 2 2	$2x_1x_2 + x_1x_2^2 + 2x_1^2x_2 + 2x_1^2x_2^2$ (*mod* 3)

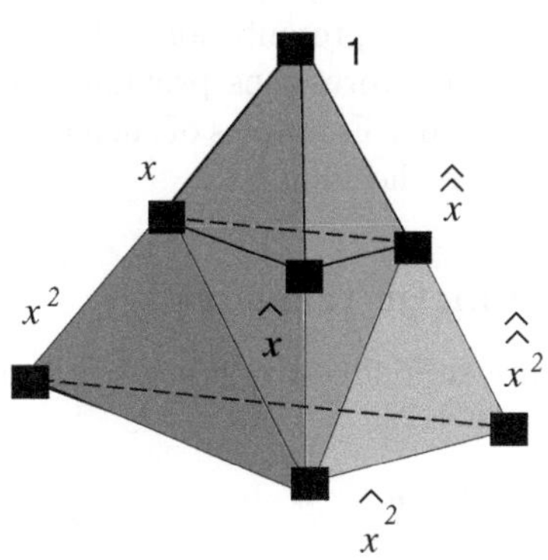

There are 28 paths between nodes pictured on pyramid, which correspond to 28 polarity vectors for ternary function of single variable:

$$\overbrace{[1\ x\ x^2], [1\ \hat{x}\ \hat{x}^2], \ldots, [1\ \hat{\hat{x}}\ \hat{x}^2],}^{12\ paths}$$

$$\overbrace{[x\ \hat{x}\ x^2], [x\ \hat{x}\ \hat{\hat{x}}^2], \ldots, [x\ x^2\ \hat{\hat{x}}^2],}^{5\ paths}$$

$$\overbrace{[\hat{x}\ \hat{\hat{x}}\ \hat{x}^2], [\hat{x}\ \hat{\hat{x}}\ x^2], \ldots, [\hat{x}\ \hat{x}^2\ x^2],}^{5\ paths}$$

$$\overbrace{[\hat{\hat{x}}\ x\ \hat{\hat{x}}^2], [\hat{\hat{x}}\ x\ x^2], \ldots, [\hat{\hat{x}}\ \hat{\hat{x}}^2\ \hat{x}^2]}^{5\ paths}$$

$$\overbrace{[x^2\ \hat{x}^2\ \hat{\hat{x}}^2]}^{1\ path}$$

FIGURE 30.4
Graphical representation of polarities of a ternary variable x.

TABLE 30.4
Possible vectors of polarities of a ternary variable x.

1	x	x^2	x	$\hat{x}$	x^2									
1	$\hat{x}$	$\hat{x}^2$	$\hat{x}$	$\hat{\hat{x}}$	$\hat{x}^2$									
1	$\hat{\hat{x}}$	$\hat{\hat{x}}^2$	$\hat{\hat{x}}$	x	$\hat{\hat{x}}^2$									
1	x	$\hat{x}^2$	x	$\hat{x}$	$\hat{x}^2$	1	x^2	$\hat{x}^2$	x	x^2	$\hat{x}^2$			
1	$\hat{x}$	$\hat{\hat{x}}^2$	$\hat{x}$	$\hat{\hat{x}}$	$\hat{\hat{x}}^2$	1	$\hat{x}^2$	$\hat{\hat{x}}^2$	$\hat{x}$	$\hat{x}^2$	$\hat{\hat{x}}^2$	x^2	$\hat{x}^2$	$\hat{\hat{x}}^2$
1	$\hat{\hat{x}}$	x^2	$\hat{\hat{x}}$	x	x^2	1	$\hat{\hat{x}}^2$	x^2	$\hat{\hat{x}}$	$\hat{\hat{x}}^2$	x^2			
1	x	$\hat{\hat{x}}^2$	x	$\hat{x}$	$\hat{\hat{x}}^2$	x	x^2	$\hat{\hat{x}}^2$						
1	$\hat{x}$	x^2	$\hat{x}$	$\hat{\hat{x}}$	x^2	$\hat{x}$	$\hat{x}^2$	x^2						
1	$\hat{\hat{x}}$	$\hat{x}^2$	$\hat{\hat{x}}$	x	$\hat{x}^2$	$\hat{\hat{x}}$	$\hat{\hat{x}}^2$	$\hat{x}^2$						

30.3 Arithmetic transform

Arithmetic expressions of multivalued functions are an alternative approach to the description of logic circuits. They share properties of Reed–Muller expressions and, at the same time, simplify representations of multioutput functions. In many applications arithmetic expressions provide a better insight into the analysis of multivalued functions. Examples of such applications are satisfiability, tautology, and equivalence checking.

30.3.1 Direct and inverse arithmetic transforms

Direct and inverse arithmetic transforms are defined as follows (Figure 30.5):

$$\begin{aligned} \mathbf{P} &= \tfrac{1}{(m-1)^n}\, \mathbf{P}^{(c)}_{m^n}\mathbf{F}, \\ \mathbf{F} &= \mathbf{P}^{-1(c)}_{m^n}\mathbf{P}. \end{aligned} \tag{30.8}$$

The pair of matrices $\mathbf{P}^{(c)}_{m^n}$ and $\mathbf{P}^{-1(c)}_{m^n}$ of arithmetic transforms in Equation 30.8 are calculated by the Kronecker product

$$\begin{aligned} \mathbf{P}^{(c)}_{m^n} &= \mathbf{P}^{(c_1)}_{m} \otimes \mathbf{P}^{(c_2)}_{m} \otimes \cdots \otimes \mathbf{P}^{(c_n)}_{m}, \\ \mathbf{P}^{-1(c)}_{m^n} &= \mathbf{R}^{-1(c_1)}_{m} \otimes \mathbf{P}^{-1(c_2)}_{m} \otimes \cdots \otimes \mathbf{P}^{-1(c_n)}_{m}. \end{aligned} \tag{30.9}$$

Elements of the matrices $\mathbf{P}^{(c_j)}_{m}$ and $\mathbf{P}^{-1(c_j)}_{m}$, $j = 1, 2, \ldots n$, are obtained as solutions of the equation

$$\mathbf{P}^{-1(c_j)}_{m}\mathbf{P}^{(c_j)}_{m} = \mathbf{I}_m. \tag{30.10}$$

In Table 30.5, the basic arithmetic transform matrices for ternary ($m = 3$) logic functions are given.

Example 30.7 *Given $m = 3$ and $c = 2$, the basic matrices of the direct and inverse transform satisfy Equation 30.10:*

$$\mathbf{P}^{-1(2)}_{3}\mathbf{P}^{(2)}_{3} = \begin{bmatrix} 1 & 2 & 1 \\ 1 & 0 & 0 \\ 1 & 1 & 1 \end{bmatrix} \begin{bmatrix} 0 & 2 & 0 \\ -1 & -3 & 4 \\ 1 & 1 & -2 \end{bmatrix} = \begin{bmatrix} 1 & 0 & 0 \\ 0 & 1 & 0 \\ 0 & 0 & 1 \end{bmatrix} = \mathbf{I}_3.$$

30.3.2 Polarity

Equation 30.9 is a formal description of forming different polarities for the manipulation of logic functions in arithmetic form:

TABLE 30.5
Basic arithmetic transform matrices for a ternary logic function of polarity $c = 0, 1, 2$.

Polarity c	Direct	Inverse
0	$\mathbf{P}_3^{(0)} = \begin{bmatrix} 2 & 0 & 0 \\ -3 & 4 & -1 \\ 1 & -2 & 1 \end{bmatrix}$	$\mathbf{P}_3^{-1(0)} = \begin{bmatrix} 1 & 0 & 0 \\ 1 & 1 & 1 \\ 1 & 2 & 1 \end{bmatrix}$
1	$\mathbf{P}_3^{(1)} = \begin{bmatrix} 0 & 0 & 2 \\ 4 & -1 & -3 \\ -2 & 1 & 1 \end{bmatrix}$	$\mathbf{P}_3^{-1(1)} = \begin{bmatrix} 1 & 1 & 1 \\ 1 & 2 & 1 \\ 1 & 0 & 0 \end{bmatrix}$
2	$\mathbf{P}_3^{(2)} = \begin{bmatrix} 0 & 2 & 0 \\ -1 & -3 & 4 \\ 1 & 1 & -2 \end{bmatrix}$	$\mathbf{P}_3^{-1(2)} = \begin{bmatrix} 1 & 2 & 1 \\ 1 & 0 & 0 \\ 1 & 1 & 1 \end{bmatrix}$

- An arbitrary m-valued logic function can be represented by m^n different generalized arithmetic expressions, or *polarities*.
- A formal description of the behavior of a multivalued function in terms of change can be derived from this equation, since rows of the transform matrices describe arithmetic analogs of logic differences, as will be shown in the section below.

$$P^{(c)} = \frac{1}{(m-1)^n} \sum_{j=0}^{m^n-1} p_j (x_1 + c_1)^{i_1} (x_2 + c_2)^{i_2} \cdots (x_n + c_n)^{i_n} \quad (30.11)$$

where

$$(x_i + c_i)^{j_i} = \begin{cases} x_i + c_i, & j_i \neq 0 \ (mod\ m); \\ 1, & j_i = 0. \end{cases} \quad (30.12)$$

Coefficient p_j is the j-th component of the vector $\mathbf{P}$ calculated by a matrix-vector transform (Equation 30.8). That is,

$$p_j = \sum_{k=0}^{m^n-1} f_k p_{k,j}, \quad (30.13)$$

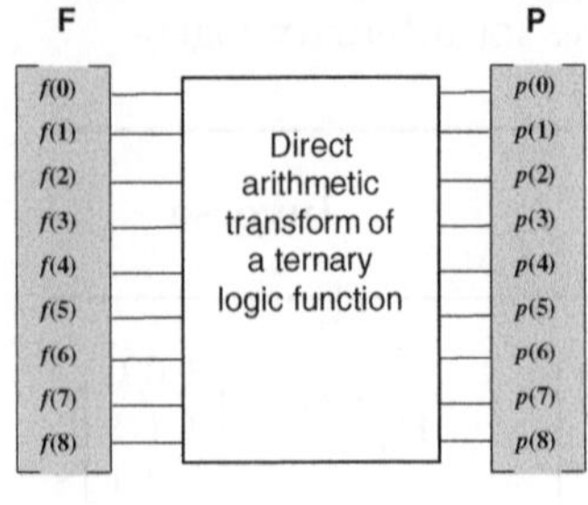

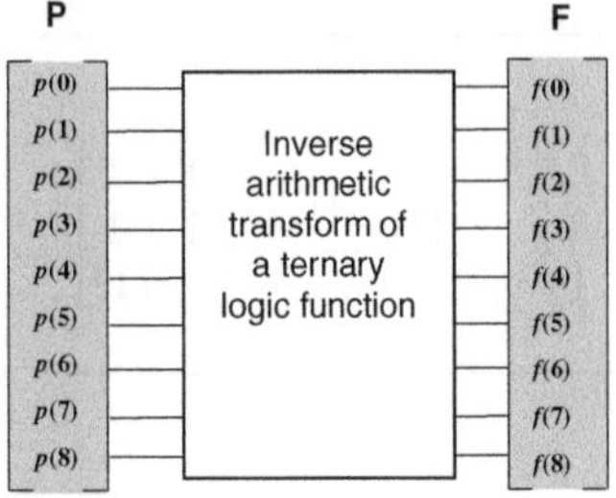

The pair of direct and inverse arithmetic transforms

$$\mathbf{P} = \frac{1}{(3-1)^2}\,\mathbf{P}_{3^2}^{(c)}\mathbf{F},$$

$$\mathbf{F} = \mathbf{P}_{3^2}^{-1(c)}\mathbf{P},$$

where

$$\mathbf{P}_{3^2}^{(c)} = \mathbf{P}_{3}^{(c_1)} \otimes \mathbf{P}_{3}^{(c_2)},$$
$$\mathbf{P}_{3^2}^{-1(c)} = \mathbf{R}_{3}^{-1(c_1)} \otimes \mathbf{P}_{3}^{-1(c_2)}.$$

Basic transform matrices $\mathbf{P}_{3^2}^{(c)}$ *and* $\mathbf{P}_{3^2}^{-1(c)}$ *for polarity* $c = 0, 1, 2$ *are given in Table 30.5*

FIGURE 30.5
Direct and inverse arithmetic transforms for a ternary function of two variables.

where $f(k)$ is the k-th component of the truth vector $\mathbf{P}$, and $p_{j,k}$ is the $(k_{j,k})$-th element in the matrix $\mathbf{P}_{m^n}^{(c)}$.

Note that the coefficients p_j are also cofactors in the Taylor expansion.

Example 30.8 *In Figure 30.6, the calculation of an arithmetic expression of polarity* $c_1c_2 = 02$ *for a ternary function of two variables given truth vector* $\mathbf{F} = [010211202]^T$ *is given. There are nine arithmetic expressions to represent this function that correspond to the polarities*

$$c_1c_2 = \{00, 01, 02, 10, 11, 12, 20, 21, 22\}.$$

Example 30.9 *Table 30.6 represents a typical library of elementary multivalued gates in arithmetic form.*

30.3.3 Word-level representation

Similar to the word-level representation of switching functions, the properties of linearity and superposition are utilized in computing word-level arithmetic expressions of multioutput logic functions.

Given:

A ternary function (m=3) of two variables (n=2):

$\mathbf{F} = [010211202]^T$

Polarity of an arithmetic expression:

$c = 2$

$c = c_1c_2 = 02,$

$c_1 = 0,\ c_2 = 2$

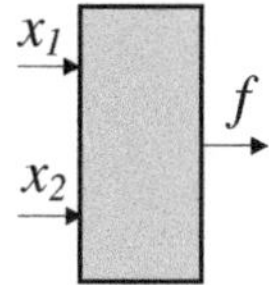

x_1	x_2	**F**
0	0	0
0	1	1
0	2	0
1	0	2
1	1	1
1	2	1
2	0	2
2	1	0
2	2	2

Elementary transform matrices for $c_1 = 0$ and $c_2 = 2$ are given in Table 30.1. Using Equation 30.3, form the transform matrix..

$$\mathbf{P}^{(2)} = {}^1/_4 \times\ \mathbf{P}_{3^2}^{(2)}\ \mathbf{F} = {}^1/_4 \times \left(\mathbf{P}_3^{(0)} \otimes \mathbf{P}_3^{(2)}\right) \mathbf{F}$$

$$= {}^1/_4 \times \left(\begin{bmatrix} 2 & 0 & 0 \\ -3 & 4 & -1 \\ 1 & -2 & 1 \end{bmatrix} \otimes \begin{bmatrix} 0 & 2 & 0 \\ -1 & -3 & 4 \\ 1 & 1 & -2 \end{bmatrix} \right) \mathbf{F}$$

$$= {}^1/_4 \times \left[\begin{array}{rrr|rrr|rrr} 0 & 4 & 0 & & & & & & \\ -2 & -6 & 8 & & & & & & \\ 2 & 2 & -4 & & & & & & \\ \hline 0 & -6 & 0 & 0 & 8 & 0 & 0 & -2 & 0 \\ 3 & 9 & -12 & -4 & -12 & 16 & 1 & 3 & -4 \\ -3 & -3 & 6 & 4 & 4 & -8 & -1 & -1 & 2 \\ \hline 0 & 2 & 0 & 0 & -4 & 0 & 0 & 2 & 0 \\ -1 & -3 & 4 & 2 & 6 & -8 & -1 & -3 & 4 \\ 1 & 1 & -2 & -2 & -2 & 4 & 1 & 1 & -2 \end{array}\right] \begin{bmatrix} 0 \\ 1 \\ 0 \\ 2 \\ 1 \\ 1 \\ 2 \\ 0 \\ 2 \end{bmatrix} = \begin{bmatrix} 4 \\ -6 \\ 2 \\ 2 \\ -1 \\ 3 \\ -2 \\ 5 \\ -3 \end{bmatrix}$$

Note that by Equation 30.13:

$$p_j = \sum_{k=0}^{3^2-1} f_k p_{j,k},$$

$$p_0 = f(0)p_{0,0} + \ldots + f(8)p_{0,8} = 0 \times 0 + \cdots + 2 \times 0 = 4$$
$$p_1 = f(0)p_{1,0} + \ldots + f(8)p_{1,8} = 1 \times (-2) + \cdots + 2 \times 0 = -6$$
$$p_2 = f(0)p_{2,0} + \ldots + f(8)p_{2,8} = 0 \times 2 + \cdots + 2 \times 0 = 2$$
$$p_3 = f(0)p_{3,0} + \ldots + f(8)p_{3,8} = 0 \times 0 + \cdots + 2 \times 0 = 2$$
$$p_4 = f(0)p_{4,0} + \ldots + f(8)p_{4,8} = 0 \times 3 + \cdots + 2 \times -4 = -1$$
$$p_5 = f(0)p_{5,0} + \ldots + f(8)p_{5,8} = 0 \times (-3) + \cdots + 2 \times 2 = 3$$
$$p_6 = f(0)p_{6,0} + \ldots + f(8)p_{6,8} = 0 \times 0 + \cdots + 2 \times 0 = -2$$
$$p_7 = f(0)p_{7,0} + \ldots + f(8)p_{7,8} = 0 \times (-1) + \cdots + 2 \times 4 = 5$$
$$p_8 = f(0)p_{8,0} + \ldots + f(8)p_{8,8} = 0 \times 1 + \cdots + 2 \times (-2) = -3$$

Equation 30.7 for $m = 3,\ n = 2,\ c = 2$:

$$P^{(2)} = {}^1/_4 \times \sum_{j=0}^{8} p_j (x_1 + 0)^{j_1} (x_2 + 2)^{j_2}$$
$$= {}^1/_4 \times (4 - 6\hat{\hat{x}}_2 + 2\hat{\hat{x}}_2^2 + 2x_1 - x_1\hat{\hat{x}}_2 + 3x_1\hat{\hat{x}}_2^2$$
$$- 2\hat{\hat{x}}_1^2 + 5x_1^2\hat{\hat{x}}_2 - 3x_1^2\hat{\hat{x}}_2^2)$$

A SOP expression can be derived from the truth table:

$$f = 1\varphi_0(x_1)\varphi_1(x_2) + 2\varphi_1(x_1)\varphi_0(x_2) + 1\varphi_1(x_1)\varphi_1(x_2)$$
$$+ 1\varphi_1(x_1)\varphi_2(x_2) + 2\varphi_2(x_1)\varphi_0(x_2) + 2\varphi_2(x_1)\varphi_2(x_2)$$

FIGURE 30.6

Computing arithmetic coefficients (Example 30.8).

TABLE 30.6
Arithmetic representation of elementary ternary ($m = 3$) two-variable functions.

Ternary gate	Truth table	Arithmetic representation
NOT	$\begin{array}{c\|c} x & \overline{x} \\ \hline \mathbf{0} & 2 \\ \mathbf{1} & 1 \\ \mathbf{2} & 0 \end{array}$	$2 - x$
$\hat{x}^r$	$\begin{array}{c\|ccc} r/x & \mathbf{0} & \mathbf{1} & \mathbf{2} \\ \hline \mathbf{0} & 0 & 1 & 2 \\ \mathbf{1} & 1 & 2 & 0 \\ \mathbf{2} & 2 & 0 & 1 \end{array}$	$\begin{cases} x, & \text{r=0;} \\ 1 + \frac{5}{2}x - \frac{3}{2}x^2, & \text{r=1;} \\ 2 - \frac{7}{2}x + \frac{3}{2}x^2, & \text{r=2.} \end{cases}$
MAX	$\begin{array}{c\|ccc} & \mathbf{0} & \mathbf{1} & \mathbf{2} \\ \hline \mathbf{0} & 0 & 1 & 2 \\ \mathbf{1} & 1 & 1 & 2 \\ \mathbf{2} & 2 & 2 & 2 \end{array}$	$x_1 + x_2 - \frac{10}{4}x_1x_2 + x_1x_2^2 + x_1^2x_2 - \frac{1}{2}x_1^2x_2^2$
MIN	$\begin{array}{c\|ccc} & \mathbf{0} & \mathbf{1} & \mathbf{2} \\ \hline \mathbf{0} & 0 & 0 & 0 \\ \mathbf{1} & 0 & 1 & 1 \\ \mathbf{2} & 0 & 1 & 2 \end{array}$	$\frac{10}{4}x_1x_2 - x_1x_2^2 - x_1^2x_2 + \frac{1}{2}x_1^2x_2^2$
MOD-SUM	$\begin{array}{c\|ccc} & \mathbf{0} & \mathbf{1} & \mathbf{2} \\ \hline \mathbf{0} & 0 & 1 & 2 \\ \mathbf{1} & 1 & 2 & 0 \\ \mathbf{2} & 2 & 0 & 1 \end{array}$	$x_1 + x_2 + \frac{21}{4}x_1x_2 - \frac{15}{4}x_1x_2^2 - \frac{15}{4}x_1^2x_2 + \frac{9}{4}x_1^2x_2^2$
MOD-PROD	$\begin{array}{c\|ccc} & \mathbf{0} & \mathbf{1} & \mathbf{2} \\ \hline \mathbf{0} & 0 & 0 & 0 \\ \mathbf{1} & 0 & 1 & 2 \\ \mathbf{2} & 0 & 2 & 1 \end{array}$	$\frac{1}{4}x_1x_2 + \frac{3}{4}x_1x_2^2 + \frac{3}{4}x_1^2x_2 - \frac{3}{4}x_1^2x_2^2$
TSUM	$\begin{array}{c\|ccc} & \mathbf{0} & \mathbf{1} & \mathbf{2} \\ \hline \mathbf{0} & 0 & 1 & 2 \\ \mathbf{1} & 1 & 2 & 2 \\ \mathbf{2} & 2 & 2 & 2 \end{array}$	$x_1 + x_2 + \frac{6}{4}x_1x_2 - x_1x_2^2 - x_1^2x_2 + \frac{1}{2}x_1^2x_2^2$
TPROD	$\begin{array}{c\|ccc} & \mathbf{0} & \mathbf{1} & \mathbf{2} \\ \hline \mathbf{0} & 0 & 0 & 0 \\ \mathbf{1} & 0 & 1 & 2 \\ \mathbf{2} & 0 & 2 & 2 \end{array}$	$\frac{6}{4}x_1x_2 - \frac{2}{4}x_1^2x_2$

Example 30.10 *A three-output ternary logic function of two variables given by truth vectors can be represented by a word-level arithmetic expression (Figure 30.7). The first method is based on the direct arithmetic transform (Equation 30.8) of truth-vectors* $\mathbf{F}_0, \mathbf{F}_1$ *and* $\mathbf{F}_2$*. The resulting vectors of coefficients* $\mathbf{P}_0, \mathbf{P}_1$ *and* $\mathbf{P}_2$ *form the vector* $\mathbf{D}$ *calculated as a weighted sum (the first method). Alternatively, the direct arithmetic transform (Equation 30.8) is applied to the vector* $\mathbf{F}$ *calculated as a weighted sum of* $\mathbf{F}_0$, $\mathbf{F}_1$, *and* $\mathbf{F}_2$ *(the second method).*

A non-linear word-level expression can be represented by a linear word-level expression (details are given in Chapter 37).

30.4 Partial Reed–Muller–Fourier transforms

By analogy to the partial Reed–Muller transforms for switching functions, partial Reed–Muller–Fourier transforms for multivalued functions are defined through the Good–Thomas factorization of the RMF-matrix $\mathbf{R}(n)$.

The partial Reed–Muller–Fourier spectrum of a function f given by the truth-vector $\mathbf{F} = [f(0), \ldots, f(4^n - 1)]^T$ is defined by

$$\mathbf{S}_{4f} = \mathbf{C}_{4i}\mathbf{F},$$

where $\mathbf{C}_{4i}$ is derived by the factorization of $\mathbf{R}_4(n)$ and is given by

$$\mathbf{C}_{4i} = \bigotimes_{j=1}^{n} \mathbf{X}_4(1), \quad \mathbf{X}_4(1) = \begin{cases} \mathbf{R}_4(1), \ i = j, \\ \mathbf{I}_4(1), \ \ i \neq j, \end{cases}$$

where $\mathbf{I}_4(1)$ is the identity matrix of order 4.

30.5 Relation of spectral representations

The goal of using spectral transforms is to "extract" information about the function, interpret this information, and utilize it for practical applications. Among these applications are:

- Technology-dependent gate-level implementations,
- Optimization of representations in the chosen domain, i.e., between a variety of bases and transforms,
- Determination of functional properties, and
- Event-driven analysis using Taylor expansion.

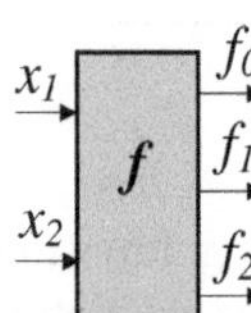

x_1	x_2	$\mathbf{F}_2$	$\mathbf{F}_1$	$\mathbf{F}_0$
0	0	0	1	2
0	1	1	1	1
0	2	0	2	1
1	0	0	2	1
1	1	2	2	2
1	2	0	0	0
2	0	2	2	1
2	1	2	1	0
2	2	1	1	1

Method 1

$$\mathbf{P}_0 = {}^1/_4 \times \mathbf{P}^{(0)}_{3^2}\mathbf{F}_0 = {}^1/_4 \times [\,8 - 6\; 2 - 6\; 3\; 3 - 17\; 2 - 17\; 9\,]^T,$$
$$\mathbf{P}_1 = {}^1/_4 \times \mathbf{P}^{(0)}_{3^2}\mathbf{F}_1 = {}^1/_4 \times [\,4\; 0\; 0\; 6\; 11 - 9\; - 2 - 7\; 5\,]^T,$$
$$\mathbf{P}_2 = {}^1/_4 \times \mathbf{P}^{(0)}_{3^2}\mathbf{F}_2 = {}^1/_4 \times [\,0\; 8 - 4 - 4\; 19 - 9\; 4 - 11\; 5\,]^T.$$
$$\mathbf{D} = 3^0\mathbf{P}_0 + 3^1\mathbf{P}_1 + 3^2\mathbf{P}_2$$
$$= {}^1/_4 \times [20\; 66 - 34 - 24\; 237 - 125\; 32 - 137\; 69]^T$$

Method 2

$$\mathbf{F}_D = [\mathbf{F}_2|\mathbf{F}_1|\mathbf{F}_0] = 3^2\mathbf{F}_2 + 3^1\mathbf{F}_1 + 3^0\mathbf{F}_0$$

$$= 3^2 \begin{bmatrix} 0\\1\\0\\0\\2\\0\\2\\2\\1 \end{bmatrix} + 3^1 \begin{bmatrix} 1\\1\\2\\2\\2\\0\\2\\1\\1 \end{bmatrix} + 3^0 \begin{bmatrix} 2\\1\\1\\1\\2\\0\\1\\0\\1 \end{bmatrix} = \begin{bmatrix} 5\\13\\4\\7\\26\\0\\25\\21\\13 \end{bmatrix}$$

$$\mathbf{D} = {}^1/_4 \times\; \mathbf{P}^{(0)}_{3^2}\mathbf{F}_D$$
$$= [20\; 66\; - 34\; - 24\; 237\; - 125\; 32\; - 137\; 69]^T$$

$$\mathbf{P}^{(0)}_{3^2} = \frac{1}{4}\left(\begin{bmatrix} 2 & 0 & 0\\ -3 & 4 & -1\\ 1 & -2 & 1 \end{bmatrix} \otimes \begin{bmatrix} 2 & 0 & 0\\ -3 & 4 & -1\\ 1 & -2 & 1 \end{bmatrix} \right)$$

$$= \frac{1}{4}\left[\begin{array}{ccc|ccc|ccc} 4 & 0 & 0 & & & & & & \\ -6 & 8 & -2 & & & & & & \\ 2 & -4 & 2 & & & & & & \\ \hline -6 & 0 & 0 & 8 & 0 & 0 & -2 & 0 & 0\\ 9 & -12 & 3 & -12 & 16 & -4 & 3 & -4 & 1\\ -3 & 6 & -3 & 4 & -8 & 4 & -1 & 2 & -1\\ \hline 2 & 0 & 0 & -4 & 0 & 0 & 2 & 0 & 0\\ -3 & 4 & -1 & 6 & -8 & 2 & -3 & 4 & -1\\ 1 & -2 & 1 & -2 & 4 & -2 & 1 & -2 & 1 \end{array}\right]$$

$$D = {}^1/_4 \times (20 + 66x_2 - 34x_2^2 - 24x_1 + 237x_1x_2 - 125x_1x_2^2 + 32x_1^2 - 137x_1^2x_2 + 69x_1^2x_2^2)$$

A SOP expression can be derived from the truth table:

$$\begin{aligned} f_0 &= 2\varphi_0(x_1)\varphi_0(x_2) + 1\varphi_0(x_1)\varphi_1(x_2) + 1\varphi_0(x_1)\varphi_2(x_2)\\ &\quad + 1\varphi_1(x_1)\varphi_0(x_2) + 2\varphi_1(x_1)\varphi_1(x_2) + 1\varphi_2(x_1)\varphi_0(x_2)\\ &\quad + 1\varphi_2(x_1)\varphi_2(x_2)\\ f_1 &= 1\varphi_0(x_1)\varphi_0(x_2) + 1\varphi_0(x_1)\varphi_1(x_2) + 2\varphi_0(x_1)\varphi_2(x_2)\\ &\quad + 2\varphi_1(x_1)\varphi_0(x_2) + 2\varphi_1(x_1)\varphi_1(x_2) + 2\varphi_2(x_1)\varphi_0(x_2)\\ &\quad + 1\varphi_2(x_1)\varphi_1(x_2) + 1\varphi_2(x_1)\varphi_2(x_2)\\ f_3 &= 1\varphi_0(x_1)\varphi_1(x_2) + 2\varphi_1(x_1)\varphi_1(x_2) + 2\varphi_2(x_1)\varphi_0(x_2)\\ &\quad + 2\varphi_2(x_1)\varphi_1(x_2) + 1\varphi_2(x_1)\varphi_2(x_2) \end{aligned}$$

FIGURE 30.7
Representation of a three-output ternary function of two variables by a word-level arithmetic expression of polarity $c = 0$ (Example 30.10).

30.5.1 Families of spectral transforms

Families of spectral transforms that are used in logic design can be divided into two classes: *logic* and *arithmetic* transforms.

Class of logic transforms. Utilizing various basic matrices allows us to generate m^n Reed–Muller expressions over $GF(m)$ for a given m-valued function.

Class of arithmetic transforms. The first family (m^n arithmetic expressions for a given m-valued function) is generated by changing the basic matrix. The next family is known as Vilenkin–Chrestenson transforms. Various forms can be derived from these transforms, including a complex representation. Here the coefficients are represented by complex numbers. This is reasonable for a large m because it allows better utilization of parallel computation. The usefulness of many other transforms has not been proven; however, it is clear that they can be used in certain areas of logic design. For example, Haar and Haar-like transforms are suitable for "catching" group behavior of logic functions.

30.5.2 Information about the behavior of logic functions in terms of change

Matrices of factorized transforms carry information about the behavior of a logic function in terms of change.

An *elementary change* of a logic function is formally described by logical or arithmetic difference. Each iteration of computing a spectrum carries information about the behavior of a logic function with respect to one variable. Since the spectral coefficients are the values of logic differences, they can be utilized in the calculation of observability and sensitivity functions.

Taylor expansion. The factorized matrix transform can be viewed as a Taylor expansion if the result of each iteration is interpreted in terms of logic differences.

Testability. The properties of a logic function can be analyzed using spectral coefficients.

30.6 Further reading

Fundamentals of spectral transforms of multivalued functions. Moraga [12] and Stanković [17] used signal processing methods for representing multivalued functions, including complex number descriptions. The book by Hurst et al. [4] presents the fundamentals of spectral techniques in multivalued logic.

Spectral techniques have also been introduced in collections of papers edited by Stanković [21] and by Rine [14].

The Reed–Muller representation of multivalued functions was studied by Green [5, 6]. In a paper by Moraga [12], formal aspects of complex representation of multivalued logic functions are developed. Stanković et al. [18, 20] used spectral methods and decision diagram techniques for representing Reed–Muller expressions. Yanushkevich [27] used several spectral techniques for representing and manipulating logic functions. In [28], a method for deriving an optimal polarity of Reed–Muller expressions for multivalued functions is developed. Zilic and Vranesic developed an interpolation technique for multivalued Reed–Muller expressions [31].

The arithmetic representation of multivalued functions is useful for word-level descriptions similar to the arithmetic word-level representations of switching functions introduced in Chapter 5. Yanushkevich [25, 27] presented various techniques for computing polynomial forms of logic functions. Details of linearization techniques are discussed in Chapter 37.

Word-level representations serves for representation of multivalued functions. It was studied by Bryant and Chen [1] for multivalued circuit verification.

Linear word-level representation is a subclass of word-level expressions. Functions in the word are located so that the linear form is obtained by arithmetic transform of the word-level truth vector. The criteria of linearity and a method to find the additional functions in the word-level expression to obtain the linear word-level form were studied in [3]. Details of linearization techniques are considered in Chapter 37.

The minimization of multiple-valued functions is developed by Sasao [15]. Various techniques for minimization of polynomial expressions of logic functions are studied by Yanushkevich [27].

Differences and logic differential equations of multivalued functions. Yanushkevich [26] developed a method to solve logical differential equations, which is useful in the analysis of the behavior of logic functions, for example, in power consumption analysis (see Chapter 21). Differences of multivalued functions are considered in Chapter 32.

Implementations of multivalued functions. A variety of implementations have been proposed during the last decades. Multivalued logic has been successfully used for many practical problems, in particular:

- 4-valued FLASH memory has been implemented by Intel.
- Multi-input binary-output PLAs have been developed by Sasao [16].
- A CAD system for multivalued logic circuit design, MV-SIS has been developed by Brayton et al. [2].
- A multivalued single electron device (SET) was developed by Inokawa and Takahashi [8].

Miller [10] proposed a spectral signature technique for testing multivalued networks.

References

[1] Bryant R and Chen Y. Verification of arithmetic functions using binary moment diagrams. In *Proceedings of the Design Automation Conference*, pp. 535–541, 1995.

[2] Brayton RK, et al. MVSIS. http:// www-cad.eecs.berkeley.edu/ Respep/ Research/ mvsis, 2004.

[3] Dziurzanski P, Malyugin VD, Shmerko VP, and Yanushkevich SN. Linear models of circuits based on the multivalued components. *Automation and Remote Control*, Plenum/Kluwer Publishers, 63(6):960–980, 2002.

[4] Epstein G. *Multi-Valued Logic Design.* Institute of Physics Publishing, London, UK, 1993.

[5] Green DH. Ternary Reed–Muller switching functions with fixed and mixed polarity. *International Journal of Electronics*, 67:761–775, 1989.

[6] Green DH. Families of Reed–Muller canonical forms. *International Journal of Electronics*, 70(2):259–280, 1991.

[7] Hurst S, Miller D, and Muzio J. *Spectral Technique in Digital Logic.* Academic Press, New York, 1985.

[8] Inokawa H and Takahashi Y. Experimental and simulation studies of single-electron-transistor-based multiple-valued logic. In *Proceedings of the 33rd IEEE International Symposium on Multiple-Valued Logic*, pp. 259–266, 2003.

[9] Kukharev GA, Shmerko VP, and Zaitseva EN. *Algorithms and Systolic Arrays for Multivalued Data Processing,* Science and Technics Publishers, Minsk, Belarus, 1990 (In Russian).

[10] Miller DM. Spectral signature testing for multiple-valued combinational networks. In *Proceedings of the 12th IEEE International Symposium on Multiple-Valued Logic*, pp. 152–158, 1982.

[11] Miller DM. Spectral transformation of multiple-valued decision diagrams. In *Proceedings of the 24th IEEE International Symposium on Multiple-Valued Logic*, pp. 89–96, 1994.

[12] Moraga C. Complex spectral logic. In *Proceedings of the 8th IEEE International Symposium on Multiple-Valued Logic*, pp. 149–156, 1978.

[13] Perkowski M, Sarabi A, and Beyl F. Fundamental theorems and families of forms for binary and multiple-valued linearly independent logic. In *Proceedings of the IFIP WG 10.5 International Workshop on Applications of the Reed–Muller Expansions in Circuit Design*, pp. 288–299, Japan, 1995.

[14] Rine DC, Ed., *Computer Science and Multiple-Valued Logic. Theory and Applications.* 2nd ed., North-Holland, Amsterdam, 1984.

[15] Sasao T. Optimization of multiple-valued AND-EXOR expressions using multiple-place decision diagrams. In *Proceedings of the 22nd IEEE International Symposium on Multiple-Valued Logic,* pp. 451–458, 1992.

[16] Sasao T. Application of multiple-valued logic to a serial decomposition of PLAs. In *Proceedings of the 19th IEEE International Symposium on Multiple-Valued Logic,* pp. 264–271, 1989.

[17] Stanković RS. Some remarks on Fourier transforms and differential operators for digital functions. In *Proceedings of the 22nd IEEE International Symposium on Multiple-Valued Logic*, pp. 365–370, 1992.

[18] Stanković R, Stanković M, Moraga C, and Sasao T. Calculation of Reed–Muller–Fourier coefficients of multiple-valued functions through multiple-place decision diagrams. In *Proceedings of the 24th IEEE International Symposium on Multiple-Valued Logic*, pp. 82–87, 1994.

[19] Stanković RS. Functional decision diagrams for multiple-valued functions. In *Proceedings of the 25th IEEE International Symposium on Multiple-Valued Logic*, pp. 284–289, 1995.

[20] Stanković RS, Sasao T, and Moraga C. Spectral transform decision diagrams. In Sasao T, Ed., *Representations of Discrete Functions*, pp. 55–92, Kluwer, Dordrecht, 1996.

[21] Stanković RS, Ed., *Recent Developments in Abstract Harmonic Analysis with Applications in Signal Processing.* Nauka, Belgrade, Yugoslavia, 1996.

[22] Stanković RS, Janković D, and Moraga C. Reed–Muller–Fourier representations versus Galois field representations of four-valued functions. In *Proceedins of the Workshop on Applications of the Reed–Muller Expansion in Circuit Design*, Oxford, UK, pp. 19–20, 1997.

[23] Stanković RS and Moraga C. Reed–Muller–Fourier representations of multiple-valued functions over Galois fields of prime cardinality. In *Proceedings of the IFIP WG 10.5 Workshop on Applications of the Reed–Muller Expansion in Circuit Design*, Kebschull U, Schubert E, and Rosenstiel W, Eds., pp. 115–124, Hamburg, Germany, 1993.

[24] Stanković RS, Moraga C, and Astola JT. Derivatives for multiple-valued functions induced by Galois field and Reed–Muller–Fourier expressions. In *Proceedings 34th IEEE International Symposium on Multiple-Valued Logic*, pp. 184–189, 2004.

[25] Yanushkevich SN. Arithmetical canonical expansions of Boolean and MVL functions as generalized Reed–Muller series. In *Proceedings of the IFIP WG 10.5 Workshop on Applications of the Reed–Muller Expansions in Circuit Design*, pp. 300–307, Japan, 1995.

[26] Yanushkevich SN. Matrix method to solve logic differential equations. *IEE Proceedings*, Pt.E, Computers and Digital Technique, UK, 144(5):267–272, 1997.

[27] Yanushkevich SN. *Logic Differential Calculus in Multi-Valued Logic Design*. Technical University of Szczecin Academic Publishers, Szczecin, Poland, 1998.

[28] Yanushkevich SN, Butler JT, Dueck GW, and Shmerko VP. Experiments on FPRM expressions for partially symmetric logic functions. In *Proceedings of the 30th International Symposium on Multiple-Valued Logic*, pp. 141–146, 2000.

[29] Yanushkevich SN, Falkowski BJ, and Shmerko VP. Spectral linear arithmetic approach for multiple-valued functions, overview of applications in CAD of circuit design. *Proceedings of the International Conference on Information, Communications and Signal Processing*, Singapore, 2001, CD publication.

[30] Yanushkevich SN. Spectral and differential methods to synthesize polynomial forms of MVL-functions on systolic arrays. In Moraga C, Gongli Z, Zhongkan L, Xinping L, and Xuefang R, Eds., *Proceedings of the 5th International Workshop on Spectral Techniques*, pp. 78–83, Beijing, China, 1994,

[31] Zilic Z and Vranesic ZG. Polynomial interpolation for Reed–Muller formas for incomplete specified functions. *International Journal on Multiple-Valued Logic*, 2:217–243, 1995.

31

Classification of Multivalued Decision Diagrams

In this chapter, we provide a classification of multivalued decision diagrams and a brief review of decision diagrams with attributed edges. Construction of both edge-valued diagrams with additive and multiplicative attributes at the edges is considered. The application of algorithms by constructing examples of decision diagrams for quaternary functions is discussed.

31.1 Introduction

Decision diagrams for representation of multivalued functions are multivalued counterparts of binary decision diagrams (BDDs), called multivalued diagrams (MDDs), and various spectral decision diagrams. Some types of diagrams serve for representation of integer-valued functions of both binary and multivalued arguments. The example is binary moment diagrams (BMDs) and their edge-valued versions $*$BMDs [1].

Instead of multiplicative factors in arithmetic transform spectra as in BMDs and $*$BMDs [1], the additive factors can also be considered, as is done in edge-valued binary decision diagrams (EVBDDs) [7] and factored edge-valued binary decision diagrams (FEVBDDs) [27]. In K$*$BMDs [3], both additive and multiplicative factors are used.

Edge-valued decision diagrams (EVDDs) are counterparts of various Kronecker transforms related decision diagrams. In this chapter, algorithms for the construction of EVDDs for multivalued functions are considered. These algorithms can be used to construct EVDDs for multivalued logic functions representing extensions of EVBDDs [7] and $*$BMDs [1] to multivalued logic functions.

Recently introduced multiplicative power hybrid decision diagrams (MPHDDs) [2] are edge-valued versions of integer Kronecker decision diagrams (KDDs) [3], with multiplicative weights restricted to powers of 2.

Determination of a compact decision diagram for a given function f consists of searching over a variety of different decision diagrams defined in the literature and subsequent optimization within the selected class of diagrams, for example, reordering of variables or linear transform over variables. From the considerations in [2], it follows that:

> *In many cases instead of searching over existing decision diagrams and their subsequent optimization, we rather construct a class of decision diagrams adapted to the class of functions to be represented, taking advantage of their possible features.*

The term "class of functions" is used to denote functions with similar behavior when represented by a particular class of decision diagrams.

In particular, edge-valued decision diagrams with multiplicative weights take into account common factors in values shown at constant nodes. In the case of arithmetic circuits, many of these factors are representable as powers of 2, due to which MPHDDs appear efficient in representation of functions describing these circuits. By using positive and negative Davio nodes, the values shown at constant nodes are converted into arithmetic transform spectra for the different polarities [17] having a larger number of common factors than the original functions. Combination of these nodes with Shannon nodes converts the values of constant nodes into various Kronecker transform spectra, increasing further the number of common factors.

31.2 Background theory

We denote by $P(G)$ the space of functions

$$f : G \rightarrow P,$$

where G is a finite group and P is a field that may be the complex field C or a finite (Galois) field $GF(p)$. The spaces of switching functions $GF_2(C_2^n)$, where n is the number of variables, and multivalued functions $GF_p(C_p^n)$ are particular examples when

$$G = C_2 \text{ and } G = C_p$$

are the support groups for $GF(2)$ and $GF(p)$.

In spectral interpretation of decision diagrams, a given function f is assigned to a decision tree by performing a spectral transform on the function represented, which in terms of decision trees can be expressed as a recursive application of various decomposition rules at the nodes of decision trees [14].

Due to this interpretation, various decision diagrams defined with respect to different decomposition rules are uniformly viewed as examples of spectral transform decision diagrams (STDDs) defined with respect to different spectral transforms. A spectral transform is specified by the set of basis functions Q in terms of which it is defined. It should be noted that the basis functions in terms of which a spectral transform is defined correspond in some way to the recursive structure of decision trees [20].

Edge-valued decision diagrams (EVDDs) are a modification of ordinary decision diagrams as follows:

- ► Attributes are assigned to the edges due to the change of values of constant nodes. The attributes are determined by referring to the partial spectral transforms defined in terms of steps in FFT-like algorithms for the calculation of related spectral transforms.
- ► For additive weights, the attributes of the edges are determined as values of some subsets of the partial transform coefficients of the functions represented.
- ► For multiplicative weights, the attributes of the edges are determined as common factors in intermediate values of spectral coefficients after performing steps of FFT-like algorithms.

Steps in FFT-like algorithms are related to the levels in decision diagrams. Therefore, the difference between EVDDs with additive and multiplicative weights is in the way of performing steps of FFT-like algorithms:

- ► *For additive weights, all steps are performed over f.*
- ► *For multiplicative weights, steps are performed recursively, just the first step over f while inputs of other steps are outputs of the previous steps.*

In steps of FFT, application of Shannon nodes in decision diagrams:

- ► Corresponds to no processing, and due to that
- ► Permits extraction of additive weights together with multiplicative weights for edges in EVDDs derived from the corresponding Kronecker decision diagrams.

Another difference is that in EVDDs with additive weights, the constant nodes are set to zero, while in those with multiplicative weights, the constant nodes show the values of the last factors in decomposition of the related spectral coefficients.

Arithmetic transform decision diagrams for multivalued functions are called arithmetic transform decision diagrams (ACDDs) [18], [20], [17]. K$*$BMDs [3] are a generalization of $*$BMDs by also using the integer counterparts of Shannon nodes, thus relating to the Kronecker transforms, and therefore, having both additive and multiplicative weights.

Figure 31.1 explains the assignment of a given function f to edge-valued decision diagrams with additive and multiplicative weights at the edges. For simplicity, we show decision diagrams with nodes having two outgoing edges. However, the same directly applies to decision diagrams for multivalued logic functions if:

- The nodes with p outgoing edges are used and
- Spectral transforms for multivalued logic functions are used.

Thus, transforms described by $(p^n \times p^n)$ transform matrices. In decision diagrams, the Kronecker transforms are usually selected. The Kronecker transform matrix is defined as

$$\mathbf{Q} = \bigotimes_{i=1}^{n} \mathbf{Q}_i,$$

where $\otimes$ denotes the Kronecker product, and $\mathbf{Q}_i$ is the basic transform matrix.

Table 31.1 systematizes EVDDs for switching functions with respect to weights at the edges. Table 31.2 shows a classification of decision diagrams for multivalued logic functions with respect to a spectral transform, in terms of which they are defined.

TABLE 31.1
Edge-valued decision diagrams for switching functions.

Weights	Decision diagram
Additive	EVBDD, FEVBDD
Multiplicative	*BMDs, *PHDDs
Additive and multiplicative	FEVBDDs, K*BMDs

31.3 Construction of EVDDs

From spectral interpretation of decision diagrams, edge-valued decision diagrams are a different notation of ordinary decision diagrams derived by manipulating common factors in the values of constant nodes. Therefore, for each decision diagram, an edge-valued version can be assigned by using the algorithms considered in this section.

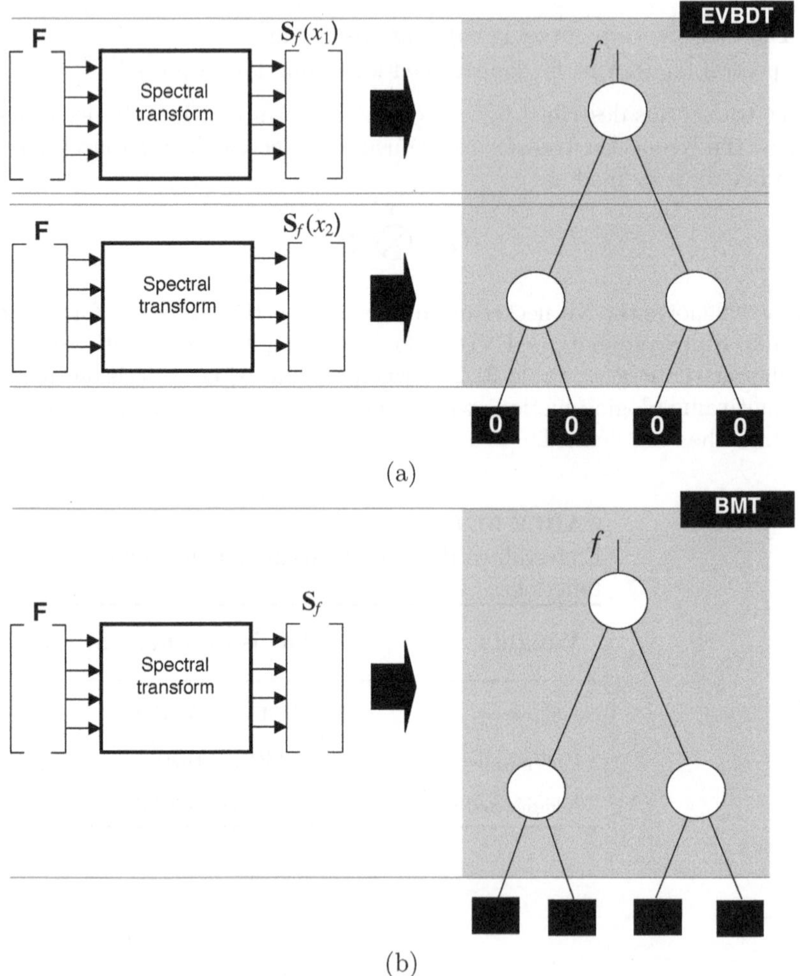

FIGURE 31.1
Assignment of f to (a) STDT with additive coefficients at edges, (b) STDT with multiplicative coefficients at edges.

TABLE 31.2
Discrete transforms and decision diagrams for multivalued logic functions.

Transform	Decision diagram
Identical	MDD [26], MD-SV [11], MTMDD [9], Quaternary [13]
GF-transform	GFDD [17], Pseudo-quaternary [13]
Reed–Muller–Fourier transform	Reed–Muller–Fourier decision diagrams [17]
Partial GF	EVGFDD [23]
Partial Reed–Muller–Fourier	EVRMFDD [19]
Arithmetic transform	BMD [1], ACDD [17], LDD [28]

31.3.1 Motivation

The motivation and applications of the algorithms proposed in this section can be found in the following considerations that can be derived from [2]:

- If we want a compact decision diagram representation for a class of functions, we first search over a variety of possible decision diagrams defined by using different decomposition rules for the most suitable class of diagrams for the given class of functions.
- We eventually perform optimization of these diagrams, for example, by manipulation with variables.

The main drawback to this is a wide search space and complete lack of guidelines in either selecting and assigning decomposition rules to nodes or manipulation of variables. Instead of that, as may be deduced from [2]:

- We can directly construct some decision diagrams by exploiting features of the class of functions under consideration.
- The guidelines for selecting a class of decision diagrams can be derived from the properties of the spectra or partial spectra of the considered functions with respect to different transforms. In particular, in the case of EVDDs, we are looking for common factors, additive or multiplicative, in the spectra.
- For some classes of functions, such spectral characterizations can be derived in a closed, analytical form, as in examples presented in [6]. In this case, we need algorithms to construct the corresponding EVDDs.

The following section provides such algorithms for EVDDs of multivalued logic functions and arbitrary spectral transforms on Abelian groups.

31.3.2 Algorithms

Algorithms for the construction of EVDDs for multivalued logic functions can be used to construct EVDDs in Tables 31.1 and 31.2, and also some new EVDDs for multivalued logic functions. In this table, MD-SV are a special class of EVDDs defined with respect to the identity transform and by exploiting symmetry properties of multivalued logic functions [11].

In EVDDs constructed by the algorithms given in Figure 31.2, labels at the edges are determined as functions $\phi(i)$ describing columns of matrices inverse to the basic transform matrices used in definition of the i-th step of FFT-like algorithms. From an , we read the function f that is represented by following labels at the edges and taking into account the assigned weights in the same way as in the case of ordinary decision diagrams or EVDDs of binary functions.

31.3.3 Complexity

EVDDs with additive weights. The complexity of the algorithm for construction of EVDDs with additive weights is mainly determined by the calculations in step 3. Since this is the it-th step of an FFT-like algorithm, it follows that the complexity of determination of weights at the i-th level is $log_p p^n$, where n is the number of variables [6]. The algorithms have to be performed over all the levels in the EVDD. Thus, the total complexity is

$$< \texttt{Complexity} > \;\; = \;\; O(n \log_p^n).$$

When f is given by a MDD, performing the i-th step of FFT-like algorithms consists of processing nodes at the i-th level of MDD. At each node we perform calculations determined by the basic transform matrix $\mathbf{Q}_i$. Therefore the complexity of step 3 in the algorithm is proportional to the width of the i-th level (number of nodes) times the number of operations in the basic transform matrix. These calculations should be repeated over all the levels in the EVDD. Thus, the complexity is the same as above.

EVDDs with multiplicative weights. In the algorithm for construction of EVDDs with multiplicative weights, the complexity is mainly due to the calculations in step 3, and we actually perform the same calculations as in the algorithm for construction of EVDDs with additive weights. The difference is that in this case, we are performing the steps of the FFT-like algorithm recursively, by taking as the input the output from the preceding step. Therefore, the complexity of the algorithm is the same as above, i.e.,

$$< \texttt{Complexity} > \;\; = \;\; O(n \log_p p^n).$$

Algorithm (EVDDs with additive weights)

1. Given a multivalued logic function f of n variables.
2. Select a Kronecker spectral transform with respect to a basis Q, spectral transform Q, in terms of which a decision diagram is required.
3. For $i = 1$ to n, apply the i-th step of an FFT-like algorithm for spectral transform Q to f. The result is denoted as $S_{i,f}$.
4. Determine common factors q_i in $S_{i,f}$.
5. Assign factors q_i as weight coefficients to the left outgoing edges of a decision tree.
6. Set the constant nodes of the decision tree to zero and perform the reduction by deleting isomorphic subtrees.

Algorithm (EVDDs with multiplicative weights)

1. Given a multivalued logic function f of n variables.
2. Select a Kronecker spectral transform with respect to a basis Q, spectral transform Q, in terms of which a decision diagram is required.
3. Perform an FFT-like algorithm of n steps to f and calculate the spectrum S_f with respect to a spectral transform Q.
4. Factorize the spectral coefficients in S_f into a product of n factors, attempting to have equal factors at the same positions.
5. If factors at the i-th position where $i = 1, \ldots n-1$ are equal, assign them as weighted coefficients at the edges at the i-th level in a decision tree.
6. The n-th factors show as the values of constant nodes in the decision tree.
7. Perform reduction of the decision tree by deleting isomorphic subtrees.

FIGURE 31.2
Algorithms for EVDDs with additive weights and EVDDs with a multiplicative weights design.

Thus in both cases, an EVDD is constructed by performing a FFT-like algorithm, and since FFT-like algorithms calculate a spectrum with the minimum number of operations, we can hardly expect lower complexity of construction of EVDDs using the spectral approach.

31.3.4 Efficiency of EVDDs

Reduction of a decision tree into a decision diagram is possible due to the existence of constant subvectors or mutually equal subvectors in the vector $\mathbf{V}$ of values of constant nodes in the decision tree. In decision diagrams, a constant subvector is represented by a single constant node, while equal subvectors result in isomorphic subtrees that can be represented by a single subtree. In reduction of EVDDs, besides values of constant and isomorphic subvectors in $\mathbf{V}$, common factors, additive or multiplicative, in elements of $\mathbf{V}$ are taken into account.

It follows from this that:

- Functions whose spectral coefficients with respect to a basis Q have many common multiplicative factors can be efficiently represented by EVDD(Q)s with multiplicative weights.
- Functions whose partial spectral coefficients with respect to a basis Q have many equal additive factors can be efficiently represented by EVDD(Q)s with additive weights.

Besides compactness, complexity of manipulation with decision diagrams is another important issue in their applications. In that respect, it should be noted that complexity of manipulation with decision diagrams constructed by the algorithms proposed is the same, or a least not greater than that of manipulation with other decision diagrams for multivalued logic functions provided that implementation is based on case structures [4] and exploiting basic principles in programming of decision diagrams for multivalued logic functions [10].

31.4 Illustrative examples

In this section, we introduce the application of the proposed algorithms for constructing two classes of edge-valued decision diagrams with respect to three different spectral transforms. These decision diagrams are multivalued logic analogs of EVBDDs [7] and *BMDs [1]. Since different edge-valued decision diagrams can be defined for different spectral transforms Q, we denote them as EVMDD(Q)s and *MMDs, with M standing for multivalued functions as in multiple-place decision diagrams (MDDs) [11], and Q specifying the transform used in definition of the particular decision diagram.

The first algorithm to construct EVMDDs. Consider the application of the first algorithm to construct EVMDDs with respect to the partial Reed–Muller–Fourier transforms and Galois field transforms. We denote these decision diagrams as edge-valued Reed–Muller–Fourier transform decision diagrams (EVRMFDDs) [19, 23]. For details about calculations for Galois field expressions and Reed–Muller–Fourier expressions to determine the corresponding decision diagrams, we refer to [9] and [22].

> **Example 31.1** *Figure 31.3 shows the edge-valued Reed–Muller–Fourier decision tree in $GF(4)$ for $p = 2$ and relationships between the weights at the edges and partial spectral Reed–Muller–Fourier coefficients. This tree represents the function in the form of a Reed–Muller–Fourier polynomial and the complexity of EVRMFDD in the number of nodes depends on the structure of the partial Reed–Muller–Fourier transforms spectra, since the values assigned to the edges are determined as the particular coefficients of the partial Reed–Muller–Fourier transforms.*
> *Under the structure of a vector, we assume*
>
> - ▶ *The number of constant and*
> - ▶ *Distribution of constant,*
>
> *and equal subvectors of order four. In EVRMFDDs, labels at the edges are*
>
> $$1,\ x_i,\ x_i^{*2},\ x_i^{*3},$$
>
> *where the exponentiation is defined as in Reed–Muller–Fourier expressions.*

> **Example 31.2** *Figure 31.4 shows an EVGFDT for $p = 4$ and $n = 2$ defined with respect to the partial Galois field transforms over $GF(4)$. It also shows relationships between the weights at the edges and partial Galois field transform coefficients. In EVGFDDs, labels at the edges are*
>
> $$1,\ x_i,\ x_i^2,\ x_i^3$$
>
> *with exponentiation in $GF(4)$.*

The following example shows an EVRMFDD for a concrete quaternary function for $n = 2$ variables.

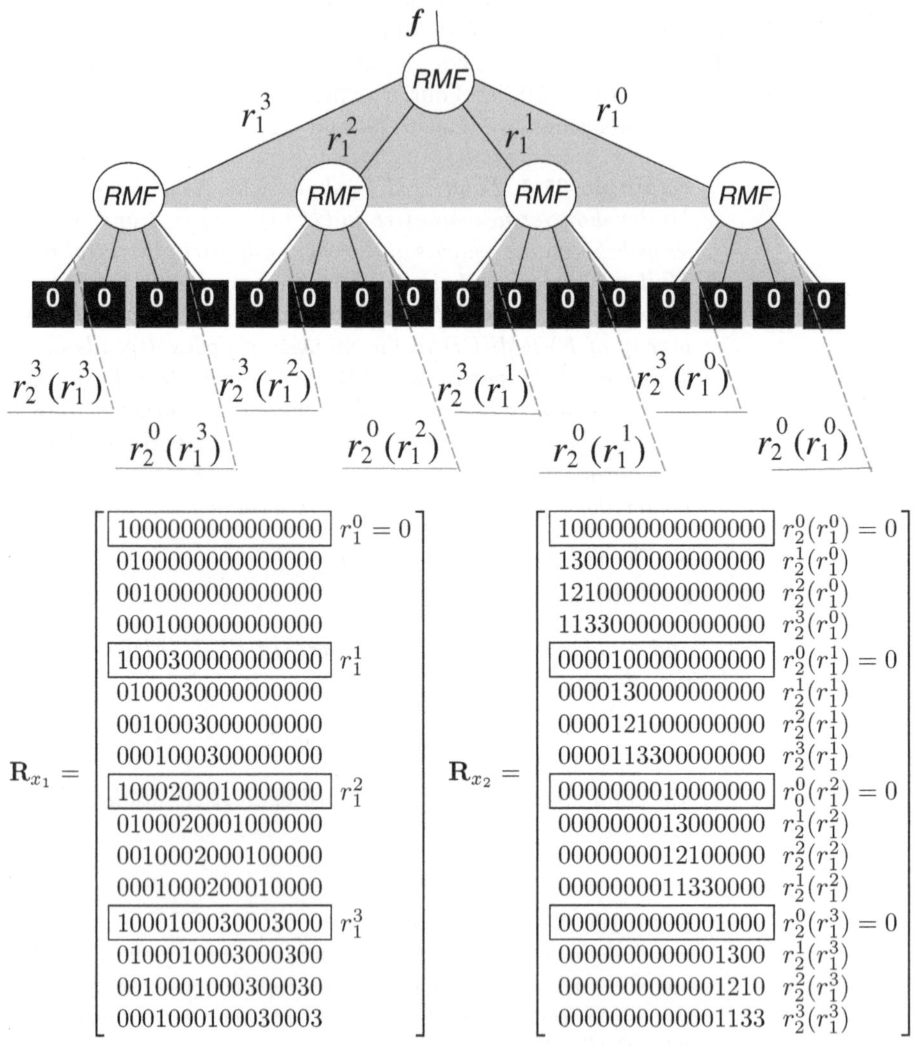

FIGURE 31.3
Edge-valued Reed–Muller–Fourier tree for $n = 2$.

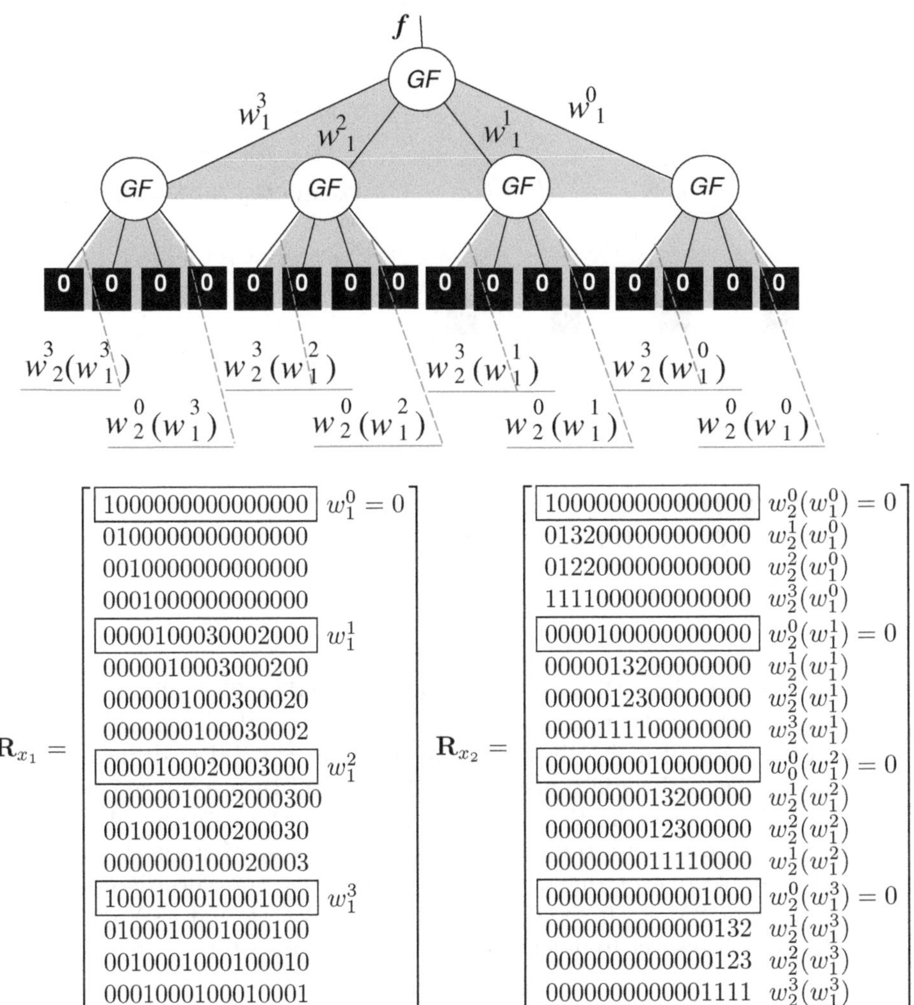

$$
\mathbf{R}_{x_1} = \begin{bmatrix}
\boxed{1000000000000000} & w_1^0 = 0 \\
0100000000000000 & \\
0010000000000000 & \\
0001000000000000 & \\
\boxed{0000100030002000} & w_1^1 \\
0000010003000200 & \\
0000001000300020 & \\
0000000100030002 & \\
\boxed{0000100020003000} & w_1^2 \\
00000010002000300 & \\
0010001000200030 & \\
0000000100020003 & \\
\boxed{1000100010001000} & w_1^3 \\
0100010001000100 & \\
0010001000100010 & \\
0001000100010001 &
\end{bmatrix}
\qquad
\mathbf{R}_{x_2} = \begin{bmatrix}
\boxed{1000000000000000} & w_2^0(w_1^0) = 0 \\
0132000000000000 & w_2^1(w_1^0) \\
0122000000000000 & w_2^2(w_1^0) \\
1111000000000000 & w_2^3(w_1^0) \\
\boxed{0000100000000000} & w_2^0(w_1^1) = 0 \\
0000013200000000 & w_2^1(w_1^1) \\
0000012300000000 & w_2^2(w_1^1) \\
0000111100000000 & w_2^3(w_1^1) \\
\boxed{0000000010000000} & w_0^0(w_1^2) = 0 \\
0000000013200000 & w_2^1(w_1^2) \\
0000000012300000 & w_2^2(w_1^2) \\
0000000011110000 & w_2^1(w_1^2) \\
\boxed{0000000000001000} & w_2^0(w_1^3) = 0 \\
0000000000000132 & w_2^1(w_1^3) \\
0000000000000123 & w_2^2(w_1^3) \\
0000000000001111 & w_2^3(w_1^3)
\end{bmatrix}
$$

FIGURE 31.4
Edge-valued GF-tree for $n = 2$.

Example 31.3 *Figure 31.5 shows the reduced EVRMFDD of a two-variable function f in $GF(4)$ given by the truth-vector* $\mathbf{F} = [0000222213022031]^T$.

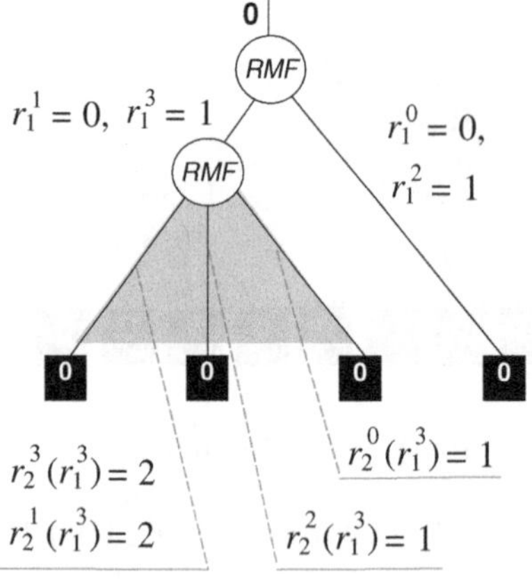

Truth-vector

$$\mathbf{F} = [0000222213022031]^T$$

The EVRMFDD represents the function f

$$f = x_1^{*2} \oplus x_1^{*3} \oplus 2x_1x_2 \oplus x_1x_2^{*2} \oplus 2x_1x_2^{*3} \oplus 2x_1^{*2}x_2^{*2}.$$

FIGURE 31.5
Reduced EVRMFDD for f (Example 31.3).

The second algorithm to construct EVMDDs. The following example illustrates the application of the second algorithm to the construction of edge valued version of multivalued moment diagrams (MVMDs) defined with respect to the arithmetic transforms. MVMDs and their edge-valued versions $*$MVMDs are generalizations of BMDs and $*$BMDs for multivalued logic functions. The arithmetic transform is a particular case of spectral transforms, and it is selected for this example since the same transform is used in the definition of BMDs. It should be noted that MVMDs and $*$MVMDs can be defined with respect to different spectral transforms with

- Real-valued coefficients or
- Complex-valued coefficients.

Thus, these decision diagrams constitute a broad family of decision diagrams that can be defined for different choices of spectral transforms. Since they are defined with respect to real-valued or complex-valued spectral transforms, MVMDs and $*$MVDs can also be used to represent integer or complex-valued functions.

Example 31.4 *Assume that Q is the arithmetic transform applied to quaternary functions. The basic transform matrix is shown in Figure 31.6. A three variable five-output quaternary function f that can be represented by the integer equivalent function derived by summation of outputs with weights 4^i, $i = 0, 1, 2, 3$, given by the vector of function values is considered. The representation of this function by a MDD [11] requires 21 non-terminal nodes and 42 constant nodes. The arithmetic spectrum given by the vector of spectral coefficients is shown in Figure 31.6.*
This vector can be used to represent the given function f by a multivalued moment diagram (MVMD), which is a multivalued logic analog of BMDs [1] having the root node at the level for x_1, three non-terminal nodes at the level for x_2, and 11 non-terminal nodes at the level for x_3. There are 16 constant nodes. Notice that five constant nodes can be saved if nodes with negated edges are allowed.
In Figure 31.6, the factorized vector is shown.

We can represent f by an $*$MVMD, which is a multivalued analog to $*$BMDs [1] with:

- ► The root node at the level for x_1,
- ► Three non-terminal nodes at the level for x_2, and
- ► 8 non-terminal nodes at the level for x_3.

First and second factors are weights at the edges of the levels for x_1 and x_2, respectively. There are 6 constant nodes showing values of the third factors, thus, -2,-2,0,1,2, and 3. Therefore, the reduction achieved by transferring from MVMD to $*$MVMD consists of three non-terminal nodes at the level for x_3 and 11 constant nodes. Figure 31.6 shows this $*$MVMD for the considered function f. In this $*$MVMD, labels at the edges are 1, w_2, w_1, and $w_1 w_2$, where w_1 and w_2 are binary valued variables. This follows from columns of the matrix $\mathbf{Q}^{-1}(1)$, as noted above.

Example 31.5 *An integer function of an even number of binary-valued variables can be converted into a function of four-valued variables by encoding pairs of binary variables by a single quaternary variable. Due to that, it can be represented by a MTBDD, or MVMD, and also $*$BMD or $*$MVMD.*

Summarizing,

- ► EVDDs are a different notation for decision diagrams without attributed edges, the ordinary decision diagrams.
- ► EVDDs can be viewed as a way to optimize ordinary decision diagrams derived by manipulation with function values and labels at the edges,

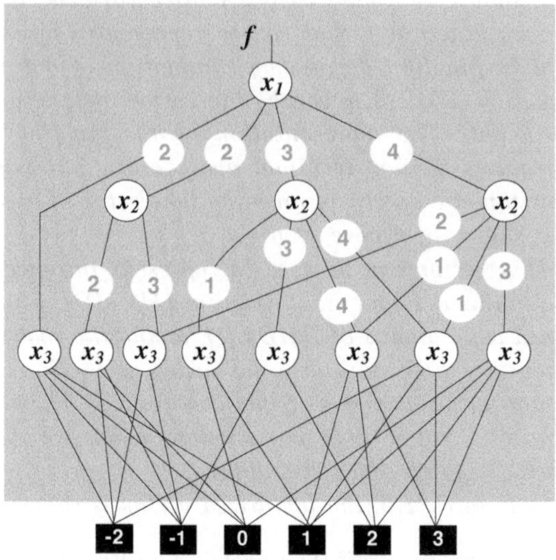

A 3-variable 5-output quaternary function f given by the vector of function values:

$$\mathbf{F} = \begin{bmatrix} 88, & 116, & 266, & 181, & 66, & 73, & 180, & 112, \\ 116, & 104, & 204, & 140, & 52, & 60, & 118, & 82, \\ 12, & 44, & 100, & 72, & 58, & 52, & 78, & 50, \\ 48, & 52, & 88, & 60, & 60, & 48, & 66, & 42, \\ 176, & 140, & 254, & 161, & 118, & 85, & 164, & 102, \\ 156, & 112, & 184, & 120, & 72, & 60, & 108, & 72, \\ 156, & 112, & 184, & 120, & 72, & 60, & 108, & 72, \\ 84, & 68, & 104, & 68, & 100, & 64, & 80, & 48, \\ 76, & 52, & 76, & 48, & 72, & 48, & 60, & 36]^T \end{bmatrix}$$

The basic transform matrix

$$\mathbf{Q}(1) = \begin{bmatrix} 1 & 0 & 0 & 0 \\ -1 & 1 & 0 & 0 \\ -1 & 0 & 1 & 0 \\ 1 & -1 & -1 & 1 \end{bmatrix}$$

The vector of arithmetic spectral coefficients:

$$\mathbf{S}_f = \begin{bmatrix} [-2, & -1, & 0, & 1, & -2, & -1, \\ 0, & 1, & -2, & -1, & 0, & 1, \\ -2, & -1, & 1, & -8, & -8, & -4, \\ -4, & -8, & -8, & -4, & -4, & -12, \\ -6, & 0, & 6, & -12, & -6, & 0, \\ 6, & 0, & 0, & 3, & 3, & -9, \\ -9, & 18, & 18, & 12, & 12, & 24, \\ 36, & -24, & -24, & 12, & 36, & -16, \\ -8, & 0, & 8, & 4, & 4, & 8, \\ 12, & -8, & -8, & 4, & 12, & 0, \\ 12, & 24, & 36 & & &]^T \end{bmatrix}$$

The factorized vector:

$$\mathbf{S}_f = \begin{bmatrix} [2\cdot 2\cdot(-2), & 2\cdot 2\cdot(-1), & 2\cdot 2\cdot 0, & 2\cdot 2\cdot 1, \\ 2\cdot 2\cdot(-2), & 2\cdot 2\cdot(-1), & 2\cdot 2\cdot 0, & 2\cdot 2\cdot 1, \\ 2\cdot 2\cdot(-2), & 2\cdot 2\cdot(-1), & 2\cdot 2\cdot 0, & 2\cdot 2\cdot 1, \\ 2\cdot 2\cdot(-2), & 2\cdot 2\cdot(-1), & 2\cdot 2\cdot 0, & 2\cdot 2\cdot 1, \\ 2\cdot 2\cdot(-2), & 2\cdot 2\cdot(-2), & 2\cdot 2\cdot(-1), & 2\cdot 2\cdot(-1), \\ 2\cdot 2\cdot(-2), & 2\cdot 2\cdot(-2), & 2\cdot 2\cdot(-1), & 2\cdot 2\cdot(-1), \\ 2\cdot 3\cdot(-2), & 2\cdot 3\cdot(-1), & 2\cdot 3\cdot 0, & 2\cdot 3\cdot 1, \\ 2\cdot 3\cdot(-2), & 2\cdot 3\cdot(-1), & 2\cdot 3\cdot 0, & 2\cdot 3\cdot 1, \\ 3\cdot 1\cdot 0, & 3\cdot 1\cdot 0, & 3\cdot 1\cdot 1, & 3\cdot 1\cdot 1, \\ 3\cdot 3\cdot(-1), & 3\cdot 3\cdot(-1), & 3\cdot 3\cdot 2, & 3\cdot 3\cdot 2, \\ 3\cdot 4\cdot 1, & 3\cdot 4\cdot 1, & 3\cdot 3\cdot 2, & 3\cdot 4\cdot 3, \\ 3\cdot 4\cdot(-2), & 3\cdot 4\cdot(-2), & 3\cdot 4\cdot 1, & 3\cdot 4\cdot 3, \\ 4\cdot 2\cdot(-2), & 4\cdot 2\cdot(-1), & 4\cdot 2\cdot 0, & 4\cdot 2\cdot 1, \\ 4\cdot 1\cdot 1, & 4\cdot 1\cdot 1, & 4\cdot 1\cdot 2, & 4\cdot 1\cdot 3, \\ 4\cdot 1\cdot(-2), & 4\cdot 1\cdot(-2), & 4\cdot 1\cdot 1, & 4\cdot 1\cdot 3, \\ 4\cdot 3\cdot 0, & 4\cdot 3\cdot 1, & 4\cdot 3\cdot 2, & 4\cdot 3\cdot 3]^T \end{bmatrix}$$

This vector can be used to represent the given function f

FIGURE 31.6
∗MVMD for f (Example 31.4).

where some weights derived from function values are assigned to the edges. Depending on the way the weights are determined, the values of constant nodes can be set to zero or may represent some factors of function values.

- ▶ EVDDs can be defined with respect to different spectral transforms and weights assigned to the edges can be determined from steps of related FFT-like algorithms.
- ▶ The efficiency of a particular class of EVDDs in the representation of a given class of functions can be estimated by the inspection of values of spectral and partial spectral coefficients with respect to a spectral transform in terms of which EVDDs are defined.
- ▶ EVDDs defined with respect to spectral transforms in terms of which spectral coefficients for a given class of functions have a large number of common additive or multiplicative factors provide for compact representations of this class of functions. This consideration determines an approach to the optimization of decision diagram representations for multivalued logic functions. For a given class of functions:

  (*a*) We first select a transform with respect to which the spectral or partial spectral coefficients have a large number of common factors, and then

  (*b*) We design the corresponding class of EVDDs.

- ▶ The algorithms for construction of EVDDs can be applied for different spectral transforms for multivalued functions and can be used to construct EVDDs with either additive or multiplicative weights. We point out that EVDDs with both additive and multiplicative weights are related to Kronecker product representable spectral transforms.

31.5 Further reading

In [24], there were enumerated 43 essentially different decision diagrams, with five more published in the meantime. Among them are recently introduced multiplicative power hybrid decision diagrams (MPHDDs) [2]. In the classification used in [24], MPHDDs are an edge-valued version of integer Kronecker decision diagrams (KDDs) [3], with multiplicative weights restricted to powers of two. Bryant introduced binary moment diagrams (BMDs) and their edge-valued versions $*$BMDs [1] to be able to represent arithmetic circuits. Arithmetic transform spectra for the different polarities were considered in [6], [8], [17].

EVBDDs [7] and $*$BMDDs [1] can be derived from arithmetic transforms. The diagrams derived directly from arithmetic transform of multivalued functions were called by Stanković et al. ACDDs [18], [20], [17]. They are generalizations of algebraic decision diagrams (ADDs) introduced for representing arithmetic expressions of switching functions (see Chapter 13. K∗BMDs [3] are a generalization of ∗BMDs by also using the integer counterparts of Shannon nodes, thus relating to the Kronecker transforms, and therefore, having both additive and multiplicative weights.

References

[1] Bryant RE and Chen YA. Verification of arithmetic functions with binary moment diagrams. In *Proceedings of the CMU-CS-94-160 Conference*, 1994.

[2] Chen YA and Bryant RE. An efficient graph representation for arithmetic circuit verification. *IEEE Transactions on Computer-Aided Design of Integrated Circuits and Systems*, 20(12):1443–1445, 2001.

[3] Drechsler R and Becker B. OKFDDs-algorithms, applications and extensions. In [14], pp. 163–190.

[4] Drechsler R, Janković D, and Stanković RS. Generic implementation of DD packages in MVL. In *Proceedings of the 25th EUROMICRO Conference*, vol. 1, pp. 352–359, 1999.

[5] Karpovsky MG. *Finite Orthogonal Series in the Design of Digital Devices*. Wiley and JUP, New York and Jerusalem, 1976.

[6] Komamiya Y. Theory of relay networks for the transformation between the decimal and binary system. *Bull. of E.T.L.*, 15(8):188–197, 1951.

[7] Lai YF, Pedram M, and Vrudhula SBK. EVBDD-based algorithms for integer linear programming, spectral transformation, and functional decomposition. *IEEE Transactions on Computer-Aided Design of Integrated Circuits and Systems*, 13(8)959–975, 1994.

[8] Malyugin VD. *Paralleled Calculations by Means of Arithmetic Polynomials*. Physical and Mathematical Publishing Company, Russian Academy of Sciences, Moscow, 1997 (in Russian).

[9] Miller DM. Spectral transformation of multivalued decision diagrams. In *Proceedings of the 24th International Symposium on Multiple-Valued Logic*, pp. 89–96, 1994.

[10] Miller DM and Drechsler R. On the construction of multiple-valued decision diagrams. In *Proceedings of the 32nd International Symposium on Multiple-Valued Logic*, pp. 245–253, 2002.

[11] Miller DM and Muranaka N. Multiple-valued decision diagrams with symmetric variable nodes. In *Proceedings of the 26th International Symposium on Multiple-Valued Logic*, pp. 242–247, 1996.

[12] Muzio JC and Wesselkamper TC. *Multiple-Valued Switching Theory.* Adam Hilger, Bristol, 1986.

[13] Sasao T and Butler JT. A design method for look-up table type FPGA by pseudo-Kronecker expansions. In *Proceedings of the 24th International Symposium on Multiple-Valued Logic*, pp. 97–104, 1994.

[14] Sasao T and Fujita M, Eds., *Representations of Discrete Functions.* Kluwer, Dordrecht, 1996.

[15] Srinivasan A, Kam T, Malik Sh, and Brayant RK. Algorithms for discrete function manipulation. In *Proceedings of the International Conference on CAD*, pp. 92–95, 1990.

[16] Stanković RS. Some remarks on Fourier transforms and differential operators for digital functions. In *Proceedings of the 22nd International Symposium on Multiple-Valued Logic*, pp. 365–370, 1992.

[17] Stanković RS. Functional decision diagrams for multiple-valued functions. In *Proceedings of the 25th International Symposium on Multiple-Valued Logic*, pp. 284–289, 1995.

[18] Stanković RS. Some remarks about spectral transform interpretation of MTBDDs and EVBDDs. In *Proceedings of the ASP-DAC Coference*, pp. 385–390, 1995.

[19] Stanković RS. Edge-valued decision diagrams based on partial Reed–Muller transforms. In *Proceedings of the Reed–Muller Colloquium*, Bristol, England, UK, pp. 9/1–9/13, 1995.

[20] Stanković RS. *Spectral Transform Decision Diagrams in Simple Questions and Simple Answers.* Nauka, Belgrade, 1998.

[21] Stanković RS and Astola JT. *Spectral Interpretation of Decision Diagrams*, Springer, Heidelberg, 2003.

[22] Stanković RS and Moraga C. Reed–Muller–Fourier representations of multiple-valued functions over Galois fields of prime cardinality. In *Proceedings of the IFIP WG 10.5 Workshop on Applications of the Reed–Muller Expansion in Circuit Design*, Hamburg, Germany, pp. 115–124, 1993.

[23] Stankovć RS and Moraga C. Edge-valued decision diagrams for multiple-valued functions. In *Proceedings of the 3rd International Conference on Application of Computer Systems*, Szczecin, Poland, 1996.

[24] Stanković RS and Sasao T. Decision diagrams for representation of discrete functions: uniform interpretation and classification. In *Proceedings of the ASP-DAC Conference*, Yokohama, Japan, pp. 13–17, 1998.

[25] Stanković RS, Sasao T, and Moraga C. Spectral transform decision diagrams. In [14], pp. 55–92.

[26] Thayse A, Davio M, and Deschamps JP. Optimization of multiple-valued decision diagrams. In *Proceedings of the 8th International Symposium on Multiple-Valued Logic*, pp. 171–177, 1978.

[27] Vrudhula SBK, Pedram M, and Lai YT. Edge valued binary decision diagrams. In [14], pp. 109–132.

[28] Yanushkevich SN, Dziurzanski P, and Shmerko VP. Word-level models for efficient computation of multiple-valued functions. Part 1: LAR based models. In *Proceedings of the IEEE 32nd International Symposium on Multiple-Valued Logic*, pp. 202–208, 2002.

32

Event - Driven Analysis in Multivalued Systems

This chapter introduces the basics of event-driven technique, the development of the binary case represented in Chapter 9. The focus of this section is:

- A formal definition of change for multivalued functions,
- The computing of change, and
- Generalization of logic Taylor expansion for multivalued functions.

32.1 Introduction

Consider a three-valued signal with three logic values 0, 2, and 3 (Figure 32.1). There are four possible situations (for simplification, the direction of change is not considered):

- Change $0 \leftrightarrow 1$,
- Change $0 \leftrightarrow 2$,
- Change $1 \leftrightarrow 2$,
- No change $(0 \leftrightarrow 0,\ 1 \leftrightarrow 1,\ 2 \leftrightarrow 2)$.

The problem is formulated as detection of changes in a ternary function f if the ternary variable x_i is changed.

In contrast to formal notation of Boolean difference, where the complement of binary variable x_i is defined as $\overline{x}_i$, in multivalued logic *the cyclic complement* of a multivalued variable x_i is used.

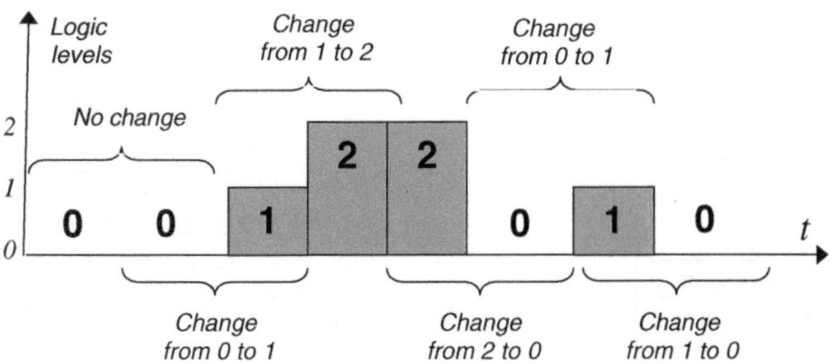

FIGURE 32.1
Change of three-valued signals.

32.2 Multivalued difference

Let f be an m-valued (m is prime) logic function of n variables. The t_i-th order cyclic complement to a variable x_i, $i = 1, 2 \ldots, n$, is

$$\overset{t_i}{\hat{x}_i} = x_i + t_i \mod (m),$$

where $t_i \in \{0, 1, 2, \ldots, \ m-1\}$. The logic difference of a function f with respect to the t_i-order cyclic complement of the variable x_i is defined as:

$$\partial f / \partial \overset{t_i}{\hat{x}_i} = \sum_{p=0}^{m-1} r_{m-t_i,p} \ f(x_1, ..., \overset{p}{\hat{x}_i}, ..., \ x_n) \quad \text{over GF}(m), \tag{32.1}$$

where $r_{m-t_i,p}$ is the $(m - t_i, p)$-th element of the multivalued Reed–Muller transform matrix $R_m^{(0)}$ (Chapter 30, Equation 30.3). It follows from Equation 32.1 that:

- ► *Logic difference reflects the change of the value of the multivalued function f with respect to t_i-th cyclic complement of the multivalued variable x_i*
- ► *There exist $m-1$ different logic differences with respect to a given variable x_i for an m-valued logic function because there exist $m-1$ complements to x_i.*
- ► *In contrast to Boolean difference, the Equation 32.1 involves m cofactors in the sum over $GF(m)$.*

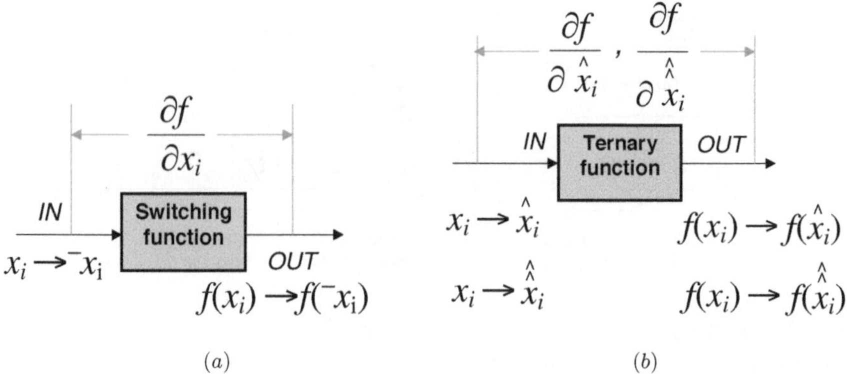

(a) (b)

FIGURE 32.2
Logic differences of switching (a) and ternary (b) functions (Example 32.1).

Given a switching function ($m = 2$), Equation 32.1 turns into a Boolean difference

$$\partial f / \partial \overset{t_i}{\hat{x}_i} = 1 \cdot f(x_1, ..., \overset{0}{\hat{x}_i}, ..., x_n) + 1 \cdot f(x_1, ..., \overset{1}{\hat{x}_i}, ..., x_n)$$

$$= \underbrace{f(x_1, ..., x_i, ..., x_n)}_{Initial\ function} \oplus \underbrace{f(x_1, ..., \overline{x}_i, ..., x_n)}_{x_i\ replaced\ by\ \overline{x}_i} = \frac{\partial f}{\partial x_i},$$

since $\overset{1}{\hat{x}_i} = \overline{x}_i = x_i \oplus 1$, and the coefficients $r_{2-1,0} = r_{2-1,1} = 1$ are taken from the matrix

$$R_2^{(0)} = \begin{bmatrix} r_{00}\ r_{01} \\ r_{10}\ r_{11} \end{bmatrix} = \begin{bmatrix} 1\ 0 \\ 1\ 1 \end{bmatrix}.$$

Example 32.1 *Figure 32.2 illustrates changes in switching and ternary functions described by Equation 32.1. The logic differences*

$$\partial f / \partial \hat{x}_i, \quad \partial f / \partial \hat{\hat{x}}_i, \quad \partial f / \partial \hat{\hat{\hat{x}}}_i$$

correspond to the behavior of a quaternary function

$$f(\hat{x}_i), \quad f(\hat{\hat{x}}_i), \quad f(\hat{\hat{\hat{x}}}_i)$$

for $x_i \rightarrow \{\hat{x}_i,\ \hat{\hat{x}}_i,\ \hat{\hat{\hat{x}}}_i\}$.

Change of a switching function f (a change in the value of f) caused by a change of the variable x_i to $\overline{x}_i$ is detected by the Boolean difference. In the ternary logic function f, a combination difference $\partial f/\partial \hat{x}_i = 2$ and $\partial f/\partial \hat{\hat{x}}_i$ recognizes the type of change.

Example 32.2 *Two logic differences with respect to a variable x_i for a ternary system are calculated by the Equation 32.1:*

$$\partial f/\partial \hat{x}_i = \sum_{p=0}^{2-1} r_{3-1,p}\ f(x_1, ..., \overset{p}{\hat{x}_i}, ..., x_n)\ \textit{over}\ GF(3),$$

$$\partial f/\partial \hat{\hat{x}}_i = \sum_{p=0}^{2-1} r_{3-2,p}\ f(x_1, ..., \overset{p}{\hat{\hat{x}}_i}, ..., x_n)\ \textit{over}\ GF(3).$$

Since

$$R_3^{(0)} = \begin{bmatrix} r_{00}\ r_{01}\ r_{02} \\ r_{10}\ r_{10}\ r_{10} \\ r_{20}\ r_{21}\ r_{22} \end{bmatrix} = \begin{bmatrix} 1\,0\,0 \\ 0\,2\,1 \\ 2\,2\,2 \end{bmatrix},$$

the coefficients $r_{m-t_i,p}$ are derived as follows:

- *1-order cyclic complement of a variable x_i: $t_i = 1$, and we take coefficients from the last row of $R_3^{(0)}$ $r_{3-1,0} = r_{3-1,1} = r_{3-1,2} = 2$;*
- *2-order cyclic complement of a variable x_i: $t_i = 2$, and we take coefficients from the middle row of $R_3^{(0)}$ $r_{3-2,0} = 0$, $r_{3-2,1} = 2$, $r_{3-2,2} = 1$.*

Figure 32.3 illustrates the changes of x_i and f that are involved in calculation of the logic differences.
Note that

$$\hat{x}_i = x_i + 1 \quad (mod\ 3)$$
$$\hat{\hat{x}}_i = x_i + 2 \quad (mod\ 3)$$

Computing logic differences. The matrix interpretation of the logic difference of an m-valued function f of n-variables with respect to a variable x_i with the t_i-order cyclic complement, $i = 1, 2, \ldots, n$, is given below:

$$\frac{\partial \mathbf{F}}{\partial \overset{t_i}{\hat{x}_i}} = \overset{t_i}{\hat{D}}{}_{m^n}^{(i)}\ \mathbf{F}, \tag{32.2}$$

where the matrix $\overset{t_i}{\hat{D}}{}_{m^n}^{(i)}$ is formed by the Kronecker product

$$\overset{t_i}{\hat{D}}{}_{m^n}^{(i)} = (m-1)I_{m^{i-1}} \otimes \Big(\sum_{p=0}^{m-1} r_{m-t_i,p}\ I_m^{(p\rightarrow)}\Big) \otimes I_{m^{n-i}}, \tag{32.3}$$

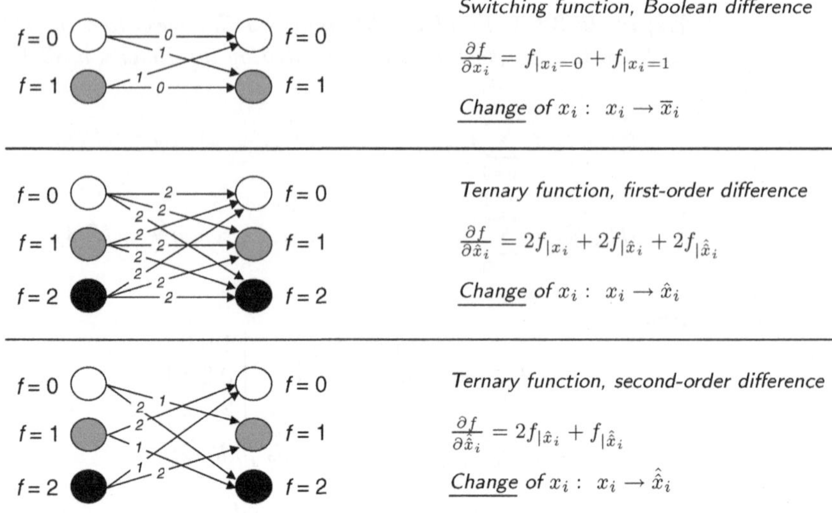

f = 0
f = 1
f = 2
Switching function, Boolean difference
$\frac{\partial f}{\partial x_i} = f_{|x_i=0} + f_{|x_i=1}$
Change of x_i : $x_i \rightarrow \overline{x}_i$
Ternary function, first-order difference
$\frac{\partial f}{\partial \hat{x}_i} = 2f_{|x_i} + 2f_{|\hat{x}_i} + 2f_{|\hat{\hat{x}}_i}$
Change of x_i : $x_i \rightarrow \hat{x}_i$
Ternary function, second-order difference
$\frac{\partial f}{\partial \hat{\hat{x}}_i} = 2f_{|\hat{x}_i} + f_{|\hat{\hat{x}}_i}$
Change of x_i : $x_i \rightarrow \hat{\hat{x}}_i$

FIGURE 32.3
Change of switching and ternary functions with respect to a variable (Example 32.2) and logic differences.

and $I_m^{(p\rightarrow)}$ is the $m \times m$ matrix generated by p-th right cyclic shift of elements of the identity matrix I_m.

Note that the denotation of matrix $\overset{t_i}{\hat{D}}{}_{m^n}^{(i)}$ carries information about

- ▶ The size of the matrix (m^n),
- ▶ The number of variables (n),
- ▶ The order of the cyclic complement (t_i), and
- ▶ The variable with respect to which the difference is calculated (x_i).

Example 32.3 *Let* $m = 3,\ t_i = 2,$ *then*

$$\begin{aligned}\sum_{p=0}^{m-1} r_{m-t_i,p}\ I_m^{(p\rightarrow)} &= \sum_{p=0}^{2} r_{1,p}\ I_3^{(p\rightarrow)} \\ &= 0 \cdot I_3^{(0\rightarrow)} + 2 \cdot I_3^{(1\rightarrow)} + 1 \cdot I_3^{(2\rightarrow)} \\ &= 2 \cdot \begin{bmatrix} 0\,1\,0 \\ 0\,0\,1 \\ 1\,0\,0 \end{bmatrix} + 1 \cdot \begin{bmatrix} 0\,0\,1 \\ 1\,0\,0 \\ 0\,1\,0 \end{bmatrix} = \begin{bmatrix} 0\,2\,1 \\ 1\,0\,2 \\ 2\,1\,0 \end{bmatrix}.\end{aligned}$$

Given a switching function ($m = 2$), Equation 32.2 is the Boolean differences in matrix form

$$\frac{\partial \mathbf{F}}{\partial x_i} = D_{2^n}^{(i)} \mathbf{F}, \tag{32.4}$$

where matrix $D_{2^n}^{(i)}$ is formed by Equation 32.3

$$D_{2^n}^{(i)} = I_{2^{i-1}} \otimes D_2 \otimes I_{2^{n-i}}, \qquad D_2 = \begin{bmatrix} 1 & 1 \\ 1 & 1 \end{bmatrix}.$$

Example 32.4 *The structure of matrix* $\overset{t_i}{\hat{D}}{}_{m^n}^{(i)}$ *for the parameters below is illustrated in Figure 32.4:*

- *Switching function of two variables,* $m = 2$, $n = 2$, *logic difference with respect to the variable* x_2, $i = 2$.
- *Ternary function of two variables,* $m = 3$, $n = 2$, *one- and two-complement logic differences with respect to the variable* x_2, $i = 2$.
- *Quaternary function of two variables,* $m = 4$, $n = 2$, *one-, two-, and three-complement logic differences with respect to the variable* x_2, $i = 2$.

Example 32.5 *Given the truth-vector* $\mathbf{F} = [0123112322233333]^T$ *of a quaternary* ($m = 4$) *logic function of two variables* ($n = 2$), *the logic difference* $\partial \mathbf{F}/\partial \hat{\hat{x}}_1$ *is calculated by Equation 32.2 and Equation 32.3 as follows:*

$$\frac{\partial \mathbf{F}}{\partial \hat{\hat{x}}_1} = \hat{\hat{D}}{}_{4^2}^{(1)} \mathbf{F}$$

$$= \left(\begin{bmatrix} 0 & 1 & 2 & 3 \\ 1 & 0 & 3 & 2 \\ 2 & 3 & 0 & 1 \\ 3 & 2 & 1 & 0 \end{bmatrix} \otimes \begin{bmatrix} 1 & & & \\ & 1 & & \\ & & 1 & \\ & & & 1 \end{bmatrix} \right) \mathbf{F}$$

$$= \begin{bmatrix}
 & & & & 1 & & & & 2 & & & & 3 & & & \\
 & & & & & 1 & & & & 2 & & & & 3 & & \\
 & & & & & & 1 & & & & 2 & & & & 3 & \\
 & & & & & & & 1 & & & & 2 & & & & 3 \\
1 & & & & & & & & 3 & & & & 2 & & & \\
 & 1 & & & & & & & & 3 & & & & 2 & & \\
 & & 1 & & & & & & & & 3 & & & & 2 & \\
 & & & 1 & & & & & & & & 3 & & & & 2 \\
2 & & & & 3 & & & & & & & & 1 & & & \\
 & 2 & & & & 3 & & & & & & & & 1 & & \\
 & & 2 & & & & 3 & & & & & & & & 1 & \\
 & & & 2 & & & & 3 & & & & & & & & 1 \\
3 & & & & 2 & & & & 1 & & & & & & & \\
 & 3 & & & & 2 & & & & 1 & & & & & & \\
 & & 3 & & & & 2 & & & & 1 & & & & & \\
 & & & 3 & & & & 2 & & & & 1 & & & &
\end{bmatrix}
\begin{bmatrix} 0 \\ 1 \\ 2 \\ 3 \\ 1 \\ 1 \\ 2 \\ 3 \\ 2 \\ 2 \\ 2 \\ 3 \\ 3 \\ 3 \\ 3 \\ 3 \end{bmatrix}
= \begin{bmatrix} 0 \\ 0 \\ 3 \\ 0 \\ 0 \\ 1 \\ 2 \\ 0 \\ 0 \\ 2 \\ 1 \\ 0 \\ 0 \\ 3 \\ 0 \\ 0 \end{bmatrix}$$

BINARY MATRIX

$$D_{2^2}^{(1)} = \begin{bmatrix} 1 & 1 & & \\ 1 & 1 & & \\ & & 1 & 1 \\ & & 1 & 1 \end{bmatrix}$$

TERNARY MATRICES

$$\hat{D}_{3^2}^{(1)} = \begin{bmatrix} 1&1&1 \\ 1&1&1 \\ 1&1&1 \\ &&&1&1&1 \\ &&&1&1&1 \\ &&&1&1&1 \\ &&&&&&1&1&1 \\ &&&&&&1&1&1 \\ &&&&&&1&1&1 \end{bmatrix} \qquad \hat{\hat{D}}_{3^2}^{(1)} = \begin{bmatrix} 1&1&1 \\ 1&1&1 \\ 1&1&1 \\ &&&1&1&1 \\ &&&1&1&1 \\ &&&1&1&1 \\ &&&&&&1&1&1 \\ &&&&&&1&1&1 \\ &&&&&&1&1&1 \end{bmatrix}$$

QUATERNARY MATRICES

$$\hat{D}_{4^2}^{(2)} = \begin{bmatrix} 1&1&1&1 \\ 1&1&1&1 \\ 1&1&1&1 \\ 1&1&1&1 \\ &&&&1&1&1&1 \\ &&&&1&1&1&1 \\ &&&&1&1&1&1 \\ &&&&1&1&1&1 \\ &&&&&&&&1&1&1&1 \\ &&&&&&&&1&1&1&1 \\ &&&&&&&&1&1&1&1 \\ &&&&&&&&1&1&1&1 \\ &&&&&&&&&&&&1&1&1&1 \\ &&&&&&&&&&&&1&1&1&1 \\ &&&&&&&&&&&&1&1&1&1 \\ &&&&&&&&&&&&1&1&1&1 \end{bmatrix}$$

$$\hat{\hat{D}}_{4^2}^{(2)} = \begin{bmatrix} 0&1&2&3 \\ 1&0&3&2 \\ 2&3&0&1 \\ 3&2&1&0 \\ &&&&0&1&2&3 \\ &&&&1&0&3&2 \\ &&&&2&3&0&1 \\ &&&&3&2&1&0 \\ &&&&&&&&0&1&2&3 \\ &&&&&&&&1&0&3&2 \\ &&&&&&&&2&3&0&1 \\ &&&&&&&&3&2&1&0 \\ &&&&&&&&&&&&0&1&2&3 \\ &&&&&&&&&&&&1&0&3&2 \\ &&&&&&&&&&&&2&3&0&1 \\ &&&&&&&&&&&&3&2&1&0 \end{bmatrix}$$

$$\hat{\hat{\hat{D}}}_{4^2}^{(2)} = \begin{bmatrix} 0&1&3&2 \\ 1&0&2&3 \\ 3&2&0&1 \\ 2&3&1&0 \\ &&&&0&1&3&2 \\ &&&&1&0&2&3 \\ &&&&3&2&0&1 \\ &&&&2&3&1&0 \\ &&&&&&&&0&1&3&2 \\ &&&&&&&&1&0&2&3 \\ &&&&&&&&3&2&0&1 \\ &&&&&&&&2&3&1&0 \\ &&&&&&&&&&&&0&1&3&2 \\ &&&&&&&&&&&&1&0&2&3 \\ &&&&&&&&&&&&3&2&0&1 \\ &&&&&&&&&&&&2&3&1&0 \end{bmatrix}$$

FIGURE 32.4
Logic difference matrices with respect to variable x_2 for switching, ternary and quaternary functions of two variables (Example 32.4).

32.3 Generation of Reed–Muller expressions

Reed–Muller representations of switching functions possess the following virtues:

- Reed–Muller expressions are associated with the analysis of switching functions in terms of change through logic Taylor expansion,
- The corresponding Reed-Muller decision tree and diagram provide a useful opportunity for detailed analysis of switching functions, including switching activity,
- The decision tree embedded into a hypercube-like structure allows word-wise computation and manipulation of Reed–Muller expressions of various polarities,
- The cost of implementation using Reed–Muller expression is often less then that of sum-of-products expressions, and
- Reed–Muller expressions can be efficiently computed using matrix transforms and, thus, the calculations are mapped onto massive parallel tools.

These attractive features of Reed–Muller expression apply to multivalued functions. The relationship between Reed–Muller expressions and generalized logic Taylor expansion, which is important for analysis of multivalued functions, is presented below.

32.3.1 Logic Taylor expansion of a multivalued function

The logic analog of the Taylor series for an m-valued function f of n variables at the point $c \in 0, 1, \ldots, m^n - 1$, is defined as

$$f = \sum_{i=0}^{m^n-1} r_i^{(c)} \underbrace{(x_1 \oplus c_1)^{i_1} \ldots (x_n \oplus c_n)^{i_n}}_{i-th\ term} \mod (m). \tag{32.5}$$

In this expression:

- m is a prime number;
- $c_1 c_2 \ldots c_n$ (polarity) and $i_1 i_2 \ldots i_n$ is the m-valued representation of c and i correspondingly;
- $r_i^{(c)}$ is the i-th coefficient, the value of the multiple (n-ordered) logic difference at the point $d = m - c$

$$r_i^{(c)} = \left. \frac{\partial^n f(d)}{\partial \overset{m-i_1}{\hat{x}_1}\, \partial \overset{m-i_2}{\hat{x}_2} \ldots \partial \overset{m-i_n}{\hat{x}_n}} \right|_{d=m-c} \tag{32.6}$$

▶ $\overset{m-i_j}{\partial}\hat{x}_i$ indicates with respect to which variables the multiple logic difference is calculated, and is defined by

$$\overset{m-i_j}{\partial}\hat{x}_i = \begin{cases} 1, & m = i_j, \\ \overset{m-i_j}{\partial}\hat{x}_j, & m \neq i_j \end{cases} \tag{32.7}$$

32.3.2 Computing Reed–Muller expressions

It follows from Equation 32.5 that:

- ▶ Logic Taylor expansion produces m^n Reed–Muller expressions that correspond to m^n polarities. In spectral interpretation this means that each of these expressions is a spectrum of the m-valued function at one of m^n polarities.
- ▶ A variable x_j is 0-polarized if it enters into the expansion uncomplemented, and c_j-polarized otherwise.
- ▶ The coefficients in the logic Taylor series are logic differences.

While the i-th coefficient r_i is described by a logical expression, it can be calculated in different ways, for example, by matrix transformations, cube-based technique, decision diagram technique, and probabilistic methods. It is possible to calculate separate coefficients or their arbitrary sets using logic differences.

Example 32.6 *By Equation 32.5, the Reed–Muller expression of an arbitrary ternary ($m = 3$) function of two variables ($n = 2$) and the 7-th polarity $c = 7$, $c_1, c_2 = 2, 1$, is defined as a logic Taylor expansion of this function (Figure 39.1).*

Example 32.7 *(Continuation of Example 32.6). Consider the function $f = MAX(x_1, x_2)$. The values of the first three coefficients at the point $c = 7$ are given in Figure 32.6. The other differences can be calculated in a similar way. Finally, the vector of coefficient is $\mathbf{R} = [200012201]^T$, which yields*

$$f = 2 + \hat{\hat{x}}_1\hat{x}_2 + 2\hat{\hat{x}}_1\hat{x}_2{}^2 + 2\hat{\hat{x}}_2{}^2 + \hat{\hat{x}}_2{}^2\hat{x}_2{}^2.$$

32.3.3 Computing Reed–Muller expressions in matrix form

The logic Taylor expansion consists of n logic differences with respect to each variable and $m^n - n - 1$ multiple logic differences.

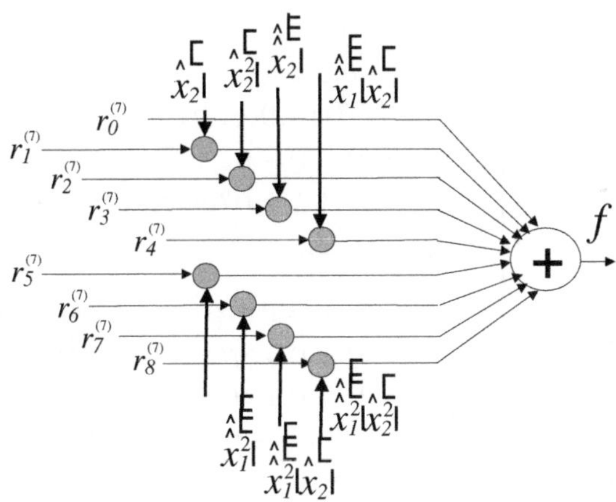

Step 1. *Apply Equation 32.5 for* $m = 2,\ n = 2:$

$$f = \sum_{i=0}^{3^2-1} r_1^{(7)}\ (x_1 \oplus 2)^{i_1}\ (x_2 \oplus 1)^{i_2} = \sum_{i=0}^{8} r_i^{(7)}\ \hat{x}_1^{i_1}\ \hat{x}_1^{i_2}$$

Step 2. *Reed–Muller expression:*

$$f = r_0^{(7)} + r_1^{(7)}\ \hat{x}_2 + r_2^{(7)}\ \hat{x}_2^2 + r3^{(7)}\ \hat{\hat{x}}_2 + r_4^{(7)}\ \hat{x}_1\hat{x}_2 + r_5^{(7)}\ \hat{x}_1\hat{x}_2^2 + r_6^{(7)}\ \hat{x}_1{}^2 + r_7^{(7)}\ \hat{x}_1{}^2\hat{x}_2 + r_8^{(7)}\ \hat{x}_1{}^2\hat{x}_2^2$$

Step 3. *Logic derivatives*

$$r_i^{(c)} = \frac{\partial^2 f(7)}{\partial \overset{3-i_1}{\hat{x}_1}\ \partial \overset{3-i_2}{\hat{x}_2}}$$

$$\partial \overset{3-i_j}{\hat{x}_i} = \begin{cases} 1, & 3 = i_j \\ \partial \overset{3-i_j}{\hat{x}_j}, & 3 \neq i_j \end{cases}$$

$r_1 = \partial f(7)/\partial \hat{\hat{x}}_2$	$r_5 = \partial^2 f(7)/\partial \hat{\hat{x}}_1 \partial \hat{x}_2$
$r_2 = \partial f(7)/\partial \hat{x}_2$	$r_6 = \partial f(7)/\partial \hat{x}_2$
$r_3 = \partial^2 f(7)/\partial \hat{\hat{x}}_1 \partial \hat{x}_2$	$r_7 = \partial^2 r(7)/\partial \hat{x}_1 \partial \hat{\hat{x}}_2$
$r_4 = \partial^2 f(7)/\partial \hat{\hat{x}}_1 \partial \hat{\hat{x}}_2$	$r_8 = \partial^2 r(7)/\partial \hat{x}_1 \partial \hat{x}_2$

FIGURE 32.5
Constructing the logic difference for a logic Taylor expansion of an arbitrary ternary ($m = 3$) function of two ($n = 2$) variables for polarity $c = 7$ (Example 32.6).

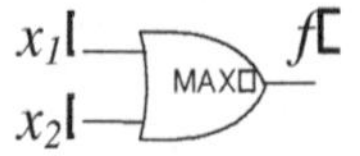

x_1 x_2 MAX f

	0	1	2
0	0	0	0
1	0	1	2
2	0	2	1

$f = \text{MAX}(x_1, x_2)$

Reed–Muller coefficients (logic differences)

$$
\begin{aligned}
r_0 &= f(7) = f(2,1) = 2,\\
r_1 &= \frac{\partial f(7)}{\partial \hat{x}_2}\\
&= 2f(x_1, \hat{x}_2) + f(x_1, \hat{\hat{x}}_2)\\
&= 2f(2, \hat{1}) + f(2, \hat{\hat{1}})\\
&= 2f(2,2) + f(2,0) = 2 \cdot 2 + 2 = 0 \pmod 3,\\
r_2 &= \frac{\partial f(7)}{\partial \hat{x}_2}\\
&= 2f(x_1, x_2) + 2f(x_1, \hat{x}_2) + 2f(x_1, \hat{\hat{x}}_2)\\
&= 2f(2,1) + 2f(2, \hat{1}) + 2f(2, \hat{\hat{1}})\\
&= 2f(2,1) + 2f(2,2) + 2f(2,0)\\
&= 2 \cdot 2 + 2 \cdot 2 + 2 \cdot 2 = 0 \ (mod\ 3)
\end{aligned}
$$

FIGURE 32.6
Taylor expansion of the ternary ($m = 3$) function MAX of two ($n = 2$) variables for polarity $c = 7$ (Example 32.7).

32.3.4 $\mathcal{N}$-hypercube representation

Let us utilize Davio tree and hypercube-like structure, which implements positive Davio expansion in the nodes (Table 32.1), to compute Boolean differences. The positive Davio expansion is given in the form

$$
\begin{aligned}
f = f_{x=0}&\\
&+ x \cdot (f_{x=1} + 3f_{x=2} + 2f_{|x=3})\\
&+ x^2 \cdot (f_{x=1} + 2f_{x=2} + 3f_{x=3})\\
&+ x^3 \cdot (f_{x=0} + f_{x=1} + f_{x=2} + f_{x=3}).
\end{aligned}
$$

Figure 32.7 illustrates the computing of logic differences by different data structures: decision tree and hypercube-like structure. It follows from this form that:

- ▶ Branches of the Davio tree carry information about logic differences;
- ▶ Terminal nodes are the values of logic differences for corresponding variable assignments;
- ▶ Computing of Reed–Muller coefficients can be implemented on the Davio tree as a data structure;
- ▶ The Davio tree includes values of all single and multiple logic differences given a variable assignment $x_1x_2...x_n = 00\ldots0$. This assignment corresponds to calculation of Reed-Muller expansion of polarity 0, so in the Davio tree, positive Davio expansion is implemented at each node;

► Representation of a logic function in terms of change is a unique representation; it means that the corresponding decision diagram is canonical;

► The values of terminal nodes correspond to coefficients of logic Taylor expansion.

The Davio tree can be embedded into a hypercube-like structure, and the above mentioned properties are valid for that data structure as well.

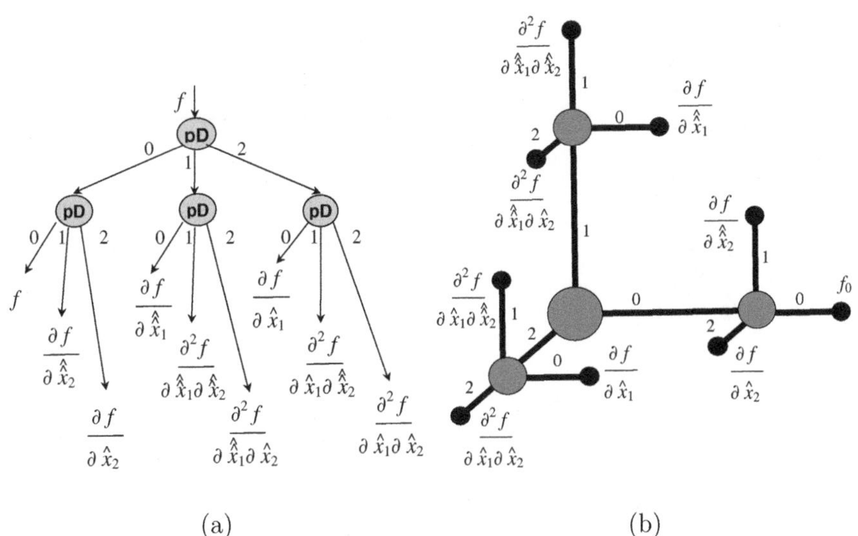

FIGURE 32.7
Computing logic differences by Davio tree (a) and hypercube-like structure for ternary logic function of two variables.

32.4 Further reading

An interpretation of multivalued functions in terms of Taylor series was studied by Thayse [10] and later by Hurst et al. [4] and Epstein [1]. This approach was developed by Yanushkevich [12].

An alternative approach to logic differential and integral calculus of multivalued logic was developed by Guima and Katbab [2, 3], Katbab and Guima

TABLE 32.1
Analogues of Shannon and Davio expansions in $GF(4)$.

Type	Rule of expansion
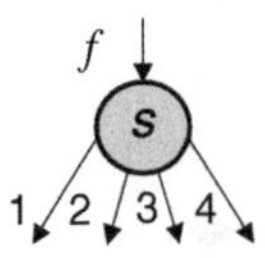	$f = \overbrace{J_0(x) \cdot f_{x=0}}^{Leaf\ 1} + \overbrace{J_1(x) \cdot f_{x=1}}^{Leaf\ 2} + \underbrace{J_2(x) \cdot f_{x=2}}_{Leaf\ 3} + \underbrace{J_3(x) \cdot f_{x=3}}_{Leaf\ 4}$
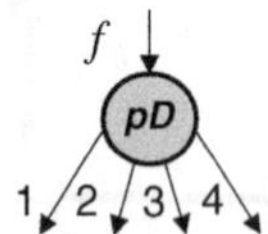	$f = f_{x=0} + x \cdot (f_{x=1} + 3f_{x=2} + 2f_{x=3}) + x^2 \cdot (f_{x=1} + 2f_{x=2} + 3f_{x=3}) + x^3 \cdot (f_{x=0} + f_{x=1} + f_{x=2} + f_{x=3})$
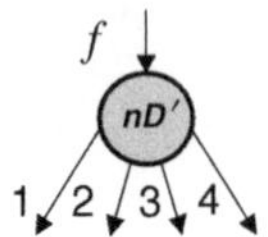	$f = f_{x=1} + {}^{1-}x \cdot (f_{x=0} + 2f_{x=2} + 3f_{x=3}) + {}^{1-}x^2 \cdot (f_{x=0} + 3f_{x=2} + 2f_{x=3}) + {}^{1-}x^3 \cdot (f_{x=0} + f_{x=1} + f_{x=2} + f_{x=3})$
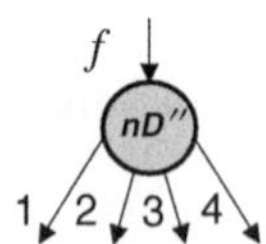	$f = f_{x=2} + {}^{2-}x \cdot (3f_{x=0} + 2f_{x=1} + f_{x=3}) + {}^{2-}x^2 \cdot (2f_{x=0} + 3f_{x=1} + f_{x=3}) + {}^{2-}x^3 \cdot (f_{x=0} + f_{x=1} + f_{x=2} + f_{x=3})$
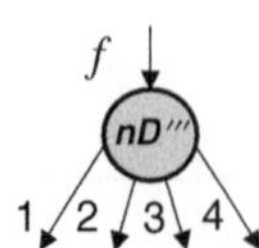	$f = f_{x=3} + {}^{3-}x \cdot (2f_{x=0} + 3f_{x=1} + f_{x=2}) + {}^{3-}x^2 \cdot (3f_{x=0} + 2f_{x=1} + f_{x=2}) + {}^{3-}x^3 \cdot (f_{x=0} + f_{x=1} + f_{x=2} + f_{x=3})$

[5], and Tapia et al. [9].

Shmerko et al. [20, 7] and Tabakow [8] studied applications of logic differences to testing of multivalued circuits.

Generalization toward arithmetic representations of multivalued functions was proposed by Yanushkevich [11].

References

[1] Epstein G. *Multi-Valued Logic Design.* Institute of Physics Publishing, London, UK, 1993.

[2] Guima TA and Katbab A. Multivalued logic integral calculus. *International Journal of Electronics*, 65:1051–1066, 1988.

[3] Guima TA and Tapia MA. Differential calculus for fault detection in multivalued logic networks. In *Proceedings of the 17th IEEE International Symposium on Multiple-Valued Logic*, pp. 99–108, 1987.

[4] Hurst S, Miller D, and Muzio J. *Spectral Technique in Digital Logic.* Academic Press, New York, 1985.

[5] Katbab A and Guima T. On multi-valued logic design using exact integral calculus. *International Journal of Electronics*, 66(1):1–18, 1989.

[6] Shmerko VP, Yanushkevich SN, and Levashenko VG. Techniques of computing logic derivatives for MVL functions. In *Proceedings of the IEEE 26th International Symposium on Multiple-Valued Logic*, pp. 267–272, 1996.

[7] Shmerko VP, Yanushkevich SN, Levashenko VG, and Bondar I. Test pattern generation for combinational MVL networks based on generalized D-algorithm. In *Proceedings of the IEEE 27th International Symposium on Multiple-Valued Logic,* pp. 139–144, 1997.

[8] Tabakow IG. Using D-algebra to generate tests for m-logic combinational circuits. *International Journal of Electronics*, 75(5):897–906, 1993.

[9] Tapia MA, Guima TA, and Katbab A. Calculus for a multivalued logic algebraic system. *Applied Mathematics and Computation*, pp. 225–285, 1991.

[10] Thayse A. Differential calculus for functions from $GF(p)$. *Philips Research Reports*, 29:560–586, 1974.

[11] Yanushkevich SN. Arithmetical canonical expansions of Boolean and MVL functions as generalized Reed–Muller series. In *Proceedings of the IFIP WG 10.5 Workshop on Applications of the Reed–Muller Expansions in Circuit Design,* Japan, pp. 300–307, 1995.

[12] Yanushkevich SN. *Logic Differential Calculus in Multi-Valued Logic Design.* Technical University of Szczecin Academic Publishers, Szczecin, Poland, 1998.

Part IV

SELECTED TOPICS OF DECISION DIAGRAM TECHNIQUES

- **Decision Diagrams in Nanoelectronics**
- **Decision Diagrams in Reversible Logic**
- **3-Dimensional Techniques**
- **Quaternion Decision Diagrams**
- **Linear Word-Level Decision Diagrams**
- **Fibonacci Decision Diagrams**
- **Technique of Computing via Taylor-like Expansion**
- **Developing New Decision Diagrams**
- **Historical Perspectives and Open Problems**

33

Introduction

This part includes eight topics on advanced decision diagram techniques and applications for micro- and nanostructure design. These topics cover the following areas of decision diagram technique:

- The extension of decision diagrams to design techniques such as reversible logic oriented toward new technologies.
- Embedding decision trees into spatial dimensions, and mapping particular decision diagrams (such as linear word-level ones) into linear parallel-pipelining topologies and spatial dimensions.
- Special types of diagrams such as diagrams on quaternion groups and Fibonacci decision diagrams,
- Hybrid computing techniques that combine various methods and strategies (Taylor-like expansions (matrix, cube-based computing, etc.)
- Techniques for discovery of new decision diagrams.

The final chapter, "Historical Perspectives and Open Problems," introduces the reader to the unsolved problems and challenges of decision diagram techniques for new technologies.

33.1 Relationship between decision diagrams and spatial structures

An efficient representation of logic functions is of fundamental importance for CAD of microelectronic devices. It is critical for the design of coming nanoscale devices. Logic design of nanodevices is being developed in two directions:

- Using advanced logic design techniques and methods from other disciplines, in particular, fault-tolerant computing.
- Development of new theory and technique for logic design in nanodimensions.

The first direction adapts architectural approaches such as, in particular, array-based logic*, parallel and distributed architectures, methods from fault-tolerant computing, and adaptive structures such as neural networks. The second direction can be justified, in particular, by nanotechnologies that implement devices on the reversibility principle. This is very relevant to the design of adiabatic circuits, which is associated not only with nanotechnology. An example is the Q3M (Quantum 3D Mesh) model of a parallel quantum computer in which each cell has a finite number of quantum bits (qubits) and interacts with neighbors, exchanging particles by means of Hamiltonian derivable from Schrödinger equation.†

In Figure 33.1, three phases of design in 3D, that is, spatial design, are shown. In the first phase, logic functions are described by algebraic equations. In the second phase, these equations are mapped into decision trees and diagrams. Finally, decision trees and diagrams are embedded into appropriate spatial graphical structures.

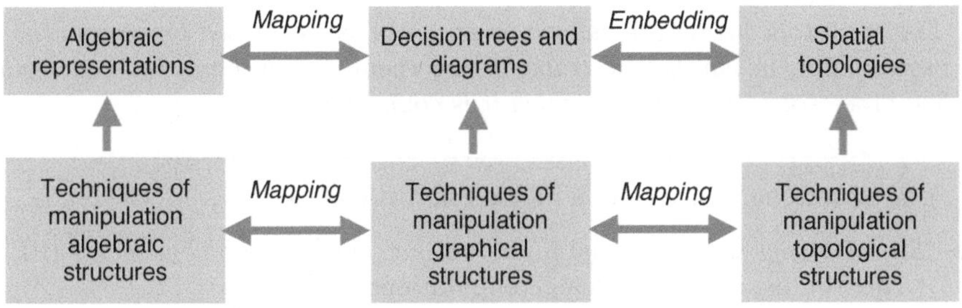

FIGURE 33.1
The relationship between algebraic representation of logic functions and decision diagrams.

33.1.1 Embedding decision trees and diagrams into different topologies

Figure 33.2 illustrates various spatial topologies developed within the theory of distributed and parallel computing, and applicable to the design of novel 3D structures in nanodimensions. Embedding decision diagrams into these topologies is considered in Chapter 34.

*DeHon A, Array-based architecture for FET-based, nanoscale electronics, *IEEE Transactions on Nanotechnology*, 2(1):23–32, 2003.
†Frank MP, *Reversibility for Efficient Computing*, MIT, 1999.

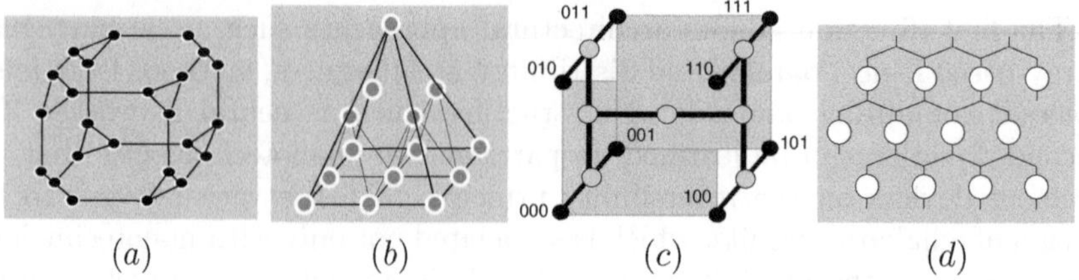

FIGURE 33.2
Spatial configurations: *CCC*-hypercube (a), pyramid (b), hypercube (c), and lattice (d).

33.1.2 Linear and multidimensional systolic arrays

System design based on the systolic principle of computing is an example of mapping high-level data structures into spatial dimensions.

Linear systolic arrays are defined as arrays of synchronized processors that process data in parallel by transmitting them, in rhythmic fashion, from one processor to the ones to which it is connected.

Linear systolic arrays can be divided into two classes:

- *Linear arrays* with bit-serial input/output, and processing elements (PEs) performing simple Boolean computations, and
- *Linear arrays* with word-level input/output and more sophisticated PE structure.

Linear arrays have a number of useful properties:

- Simple input/output organization; and
- Easy embedding into hypercube-like topology.

Design of systolic arrays includes: formal description of an algorithm (traditionally matrix-vector multiplication, convolution, discrete Fourier transform, and finite impulse response filter), the mapping of this algorithm onto 1D, 2D, or 3D arrays, and array processor design. Mapping techniques based on data-dependent graphs, signal flow graphs, and particular techniques for manipulation of these graphs are called *systolization*. In general, a systolic network with a higher dimension produces higher computation speed.[‡]

[‡]Rosenberg AI, Three-dimensional VLSI: a case study, *Journal of ACM*, 30:397–416, July, 1983;
Nudd GR, Etchells RD, and Grinberg J, Three-dimensional VLSI architecture for image understanding, *Journal of Parallel and Distributed Computing*, 2:1–29, 1985;
Inoue Y, Sugahara K, Kusunoki S, et al, A three-dimensional static RAM, *IEEE Electron Device Letters*, EDL-7:327–329, May, 1986;

In Figure 33.3, the design of spatial systolic arrays is shown. The algebraic equations of an algorithm can be mapped into 1D (linear), 2D, 3D spatial structure. For example, 3D network computations in planes and rows are carried out in parallel. The 3D array consists of $p \times q \times s = k$ cells with an area complexity of $O(k)$. This parallelism improves the computation speed. A linear systolic array (1D) can be embedded into an $\mathcal{N}$-hypercube (3D) (Figure 33.4).

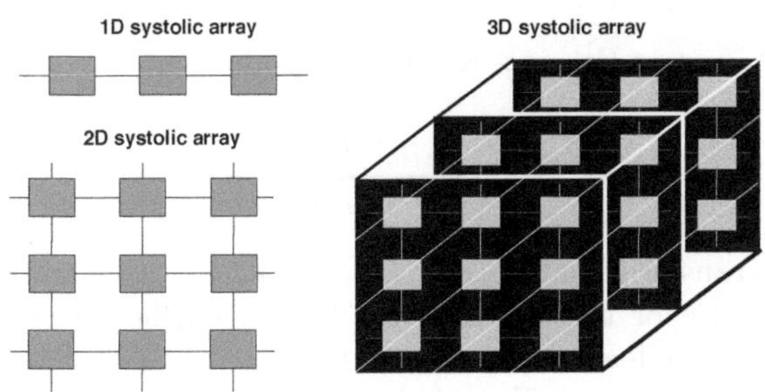

FIGURE 33.3
Spatial systolic array design.

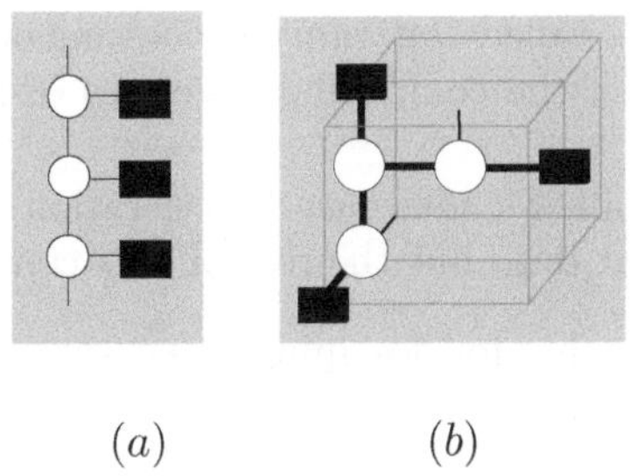

FIGURE 33.4
Embedding of a linear systolic array (a) into an $\mathcal{N}$-hypercube (b).

The systolization of an algorithm is accomplished by factorization of the transform matrix, that is, representation of the algorithm by the fast Fourier transform (FFT). For example, the implementation of the FFT means one iteration ("butterfly" operation) per processor. In order to accomplish n-iteration, or recursive, computing, the linear systolic array consists of n PEs with memory organized as a first-in-first-out (FIFO) register. The formal model of computation in the array of n PEs corresponds to the product of n matrices

$$\underbrace{\mathbf{Y} = U_{2^n}^{(1)} \times U_{2^n}^{(2)} \times \cdots \times U_{2^n}^{(n)} \times \mathbf{F}}_{n\ processing\ elements}$$

The formal model of computation in the i-th PE of the linear array is the matrix eqaution

$$\mathbf{Y}_i = U_{2^n}^{(i)} \mathbf{F}_i \tag{33.1}$$

where $2^n \times 2^n$ matrix $U_{2^n}^{(i)}$ is a Kronecker product

$$U_{2^n}^{(j)} = \underbrace{I_{2^{n-j}} \otimes U_2 \otimes I_{2^{j-1}}}_{i-th\ PEi},$$

U_2 is a 2×2 transform matrix, and $I_{2^{n-j}}$ and $I_{2^{n-j}}$ are identity matrices.

The input data for the device is the $2^n \times 1$ coefficient vector $\mathbf{F}$, its l-th components $f(l)$ are loaded at l-th time $(l = 1, ..., 2^n - 1)$. Then, PEs provide the transform of the vector $\mathbf{F}$, and the resulting coefficient vector $\mathbf{Y}$ appears as an output of n-th PE.

A linear systolic array processes data similarly to an assembly line:

- *The first phase* is called *speeding-up* (acceleration) a process whereby input data are processed by the first PE and the result is passed to the second PE, etc. In this phase, the computing resources of systolic processor are partially used;
- *The second phase* is called *stationary* processing of data. At this phase all PEs are used, i.e., the computing resources of a systolic processor are used totally;
- *The third phase* is called a *slowdown* (deceleration) process. At this phase, there are no input data but the rest of PEs continue the processing.

There are two approaches for designing linear systolic arrays for computing Reed–Muller expressions:

- Design based on factorization of the transform matrix. This technique utilizes the multiplicative property of matrices of direct and inverse transforms of logic functions.
- Design based on Taylor expansion utilizes the properties of logic Taylor expansion to represent an arbitrary switching function in terms of Boolean differences.

Figure 33.5 illustrates designing processing elements (PEs) of a linear systolic array for computing Reed-Muller expressions of three variables. The following matrix model corresponds to this iterative process:

$$\mathbf{R} = \mathbf{R}_{2^3}^{(1)} \mathbf{R}_{2^3}^{(2)} \mathbf{R}_{2^3}^{(3)} \mathbf{F} \quad (mod\ 2)$$

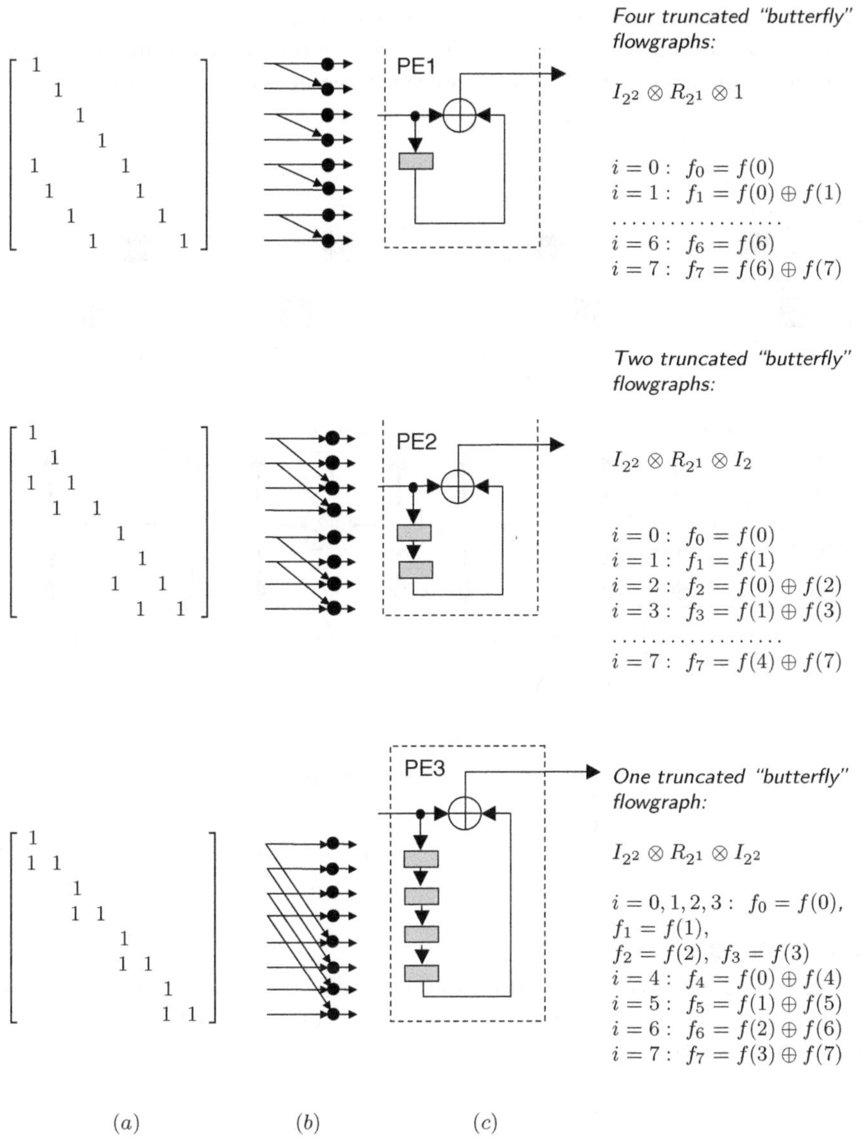

$$\begin{bmatrix} 1 &&&&&&& \\ & 1 &&&&&& \\ && 1 &&&&& \\ &&& 1 &&&& \\ 1 &&&& 1 &&& \\ & 1 &&&& 1 && \\ && 1 &&&& 1 & \\ &&& 1 &&&& 1 \end{bmatrix}$$

PE1

Four truncated "butterfly" flowgraphs:

$I_{2^2} \otimes R_{2^1} \otimes 1$

$i = 0: \ f_0 = f(0)$
$i = 1: \ f_1 = f(0) \oplus f(1)$
....
$i = 6: \ f_6 = f(6)$
$i = 7: \ f_7 = f(6) \oplus f(7)$

$$\begin{bmatrix} 1 &&&&&&& \\ & 1 &&&&&& \\ 1 && 1 &&&&& \\ & 1 && 1 &&&& \\ &&&& 1 &&& \\ &&&&& 1 && \\ &&&& 1 && 1 & \\ &&&&& 1 && 1 \end{bmatrix}$$

PE2

Two truncated "butterfly" flowgraphs:

$I_{2^2} \otimes R_{2^1} \otimes I_2$

$i = 0: \ f_0 = f(0)$
$i = 1: \ f_1 = f(1)$
$i = 2: \ f_2 = f(0) \oplus f(2)$
$i = 3: \ f_3 = f(1) \oplus f(3)$
....
$i = 7: \ f_7 = f(4) \oplus f(7)$

$$\begin{bmatrix} 1 &&&&&&& \\ 1 & 1 &&&&&& \\ && 1 &&&&& \\ && 1 & 1 &&&& \\ &&&& 1 &&& \\ &&&& 1 & 1 && \\ &&&&&& 1 & \\ &&&&&& 1 & 1 \end{bmatrix}$$

PE3

One truncated "butterfly" flowgraph:

$I_{2^2} \otimes R_{2^1} \otimes I_{2^2}$

$i = 0, 1, 2, 3: \ f_0 = f(0),$
$f_1 = f(1),$
$f_2 = f(2), \ f_3 = f(3)$
$i = 4: \ f_4 = f(0) \oplus f(4)$
$i = 5: \ f_5 = f(1) \oplus f(5)$
$i = 6: \ f_6 = f(2) \oplus f(6)$
$i = 7: \ f_7 = f(3) \oplus f(7)$

(*a*) (*b*) (*c*)

FIGURE 33.5
Design of PEs for computing Reed–Muller expressions given $n = 3$: transform matrix (a), 4-, 2-, and 1-truncated data flowgraphs (b), and PE's structure and model (c).

Figure 33.6 illustrates the second approach: embedding of a tree-like processor (a full binary tree) given $n = 2$ into a 3D $\mathcal{N}$. Four computational cycles are shown, each represents an iteration of computing (a coefficient of a fixed polarity Reed–Muller expression). The serial input data a function values. The same processor computes all Boolean differences of the function.

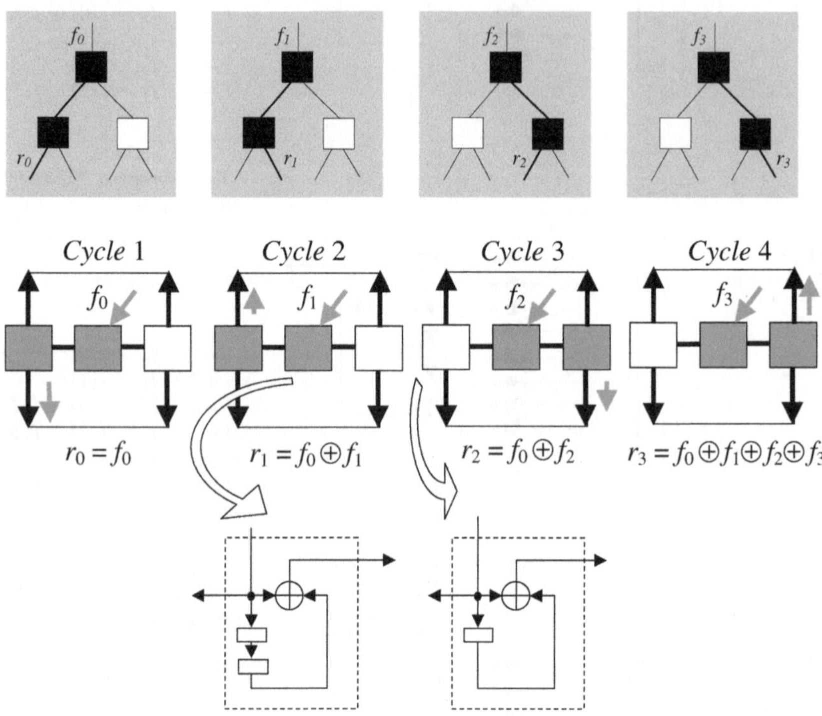

FIGURE 33.6
Design of a 2D $\mathcal{N}$-hypercube array: four computing cycles result in computing four Reed–Muller coefficients; the array consists of one 2-bit PE and two 1-bit PEs (shown at the bottom).

Both techniques can be applied to design of linear systolic arrays for computing arithmetic and Walsh expressions and corresponding logic differences. This is also suitable for processing multiple-valued functions unless the model of computation is matrix based.

33.2 Decision diagrams for reversible computation

Reversible computation studies the implementation of multivalued functions on future quantum circuits. An m-input, m-output totally-specified multi-valued logic function is reversible if it maps each input assignment to a unique output assignment. Decision diagrams can be used for the design of multi-valued reversible devices. Details are given in Chapter 35.

33.3 Special types of decision diagrams and hybrid techniques

Search for optimal topologies or adjustment of the properties of decision diagrams to the particular application has led to the invention of special types of decision diagrams.

> *Changing the domain group structure, further generalized by using finite non-Abelian groups, produces optimal decision diagrams such as quaternary decision diagrams (QDDs) and Fourier decision diagrams on finite non-Abelian groups.*

Example 33.1 *Figure 36.3 shows a transformation from a BDD into a QDD and a Fourier decision diagram on* Q_2*. Thus, the topology is changed once the domain of the group is changed. This also has implications on the 3D embedding properties.*

The search for further simplification of word-level spectra resulted in the development of linear decision diagrams (LDDs), considered in Chapter 37.

Another special class of decision diagrams are the *Fibonacci* decision diagrams and trees considered in Chapter 38. Those are defined by the analogy to BDDs, with the Shannon expansion replaced by the corresponding Fibonacci expansion. Fibonacci decision diagrams are of interest for spatial design, because they support the replacement of Boolean n-cubes by generalized Fibonacci cubes as interconnection networks that possess useful properties in network and decision diagrams design and can be embedded into other topologies.

Finally, spectral transforms such as Walsh, arithmetic, and Reed–Muller are considered within the Taylor-like expansion domain. They form a class of transforms that can be implemented through calculation of Boolean or

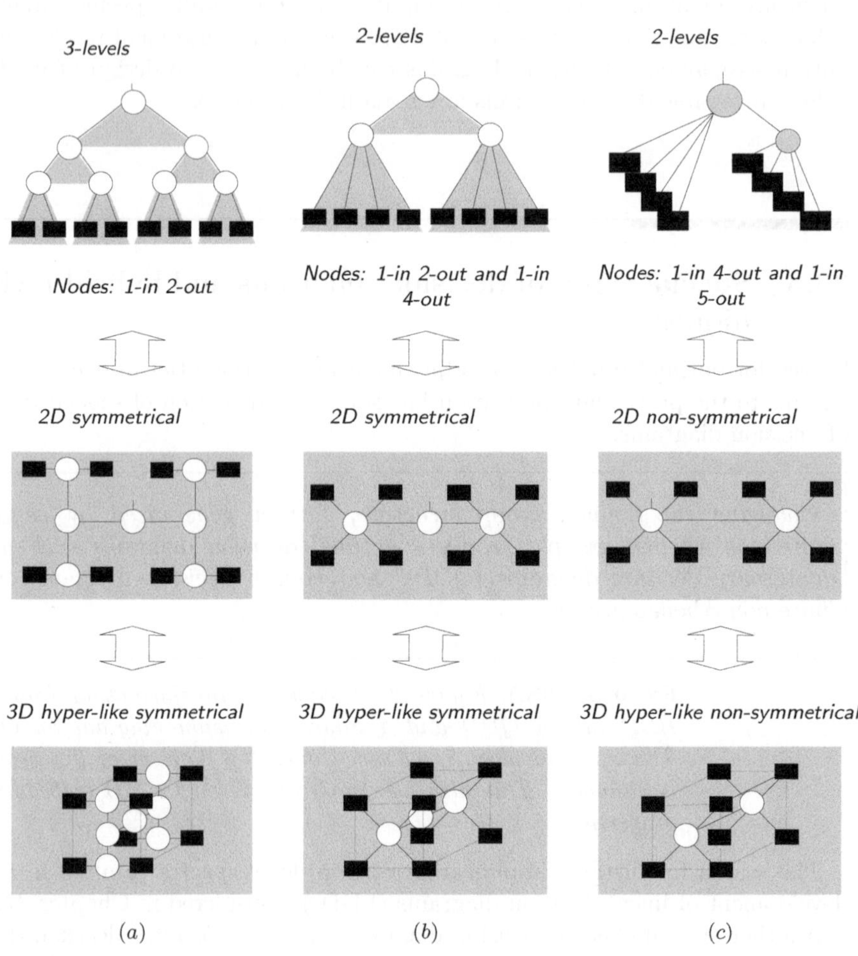

FIGURE 33.7
Topological transformations of decision diagrams: BDD (a), QDD (b), and FNADD (c) (Example 40.1).

dyadic difference and, thus, the decision diagram technique for calculation of these differences can be utilized for spectral calculations, and vice versa. This generalization is the subject of Chapter 39.

33.4 Developing of new decision diagrams

In Chapter 40 the following approaches to synthesis of various types of decision diagrams are proposed:

- An approach based on the generalization of spectral methods and corresponding decision diagrams.
- The group theoretic approach.
- The relevant methods of optimization and manipulation of these data structures.

It should be mentioned that the type of the diagrams must be chosen based on the application-specific criterion, which in general case is the Karuhnen–Loeve criterion. Karuhnen–Loeve transform specifies a set of basic functions designed specifically for each given signal to be processed. The basic functions are selected by analyzing the statistical properties of the given signal, which requires a priori knowledge of the signal that will be processed.

33.5 Further reading

Embedding decision diagrams into nanostructures. Embedding of decision diagrams into various topological structures to meet the requirements of spatial nanometric configurations were studied by Yanushkevich et al. [30]. Overview and applications of this approach to contemporary logic design is presented in [31, 32].

Advanced spectral analysis. Spectral techniques for logic synthesis has been first explored by Karpovsky [10] and Lechner [15], and later developed by Hurst et al. [8, 9]. These techniques are build upon so-called abstract harmonic analysis, and many algorithms are adopted from traditional harmonic analysis and its applications to digital signal processing, including implementation on specialized signal processors. The latter reason, and also because spectral techniques provide simple and elegant analytic solutions to many problems of logic design, they are of increasing interest, as can be seen from the review by Karpovsky et al. [11]. The comprehensive references to the state-of-the-art spectral techniques for VLSI design can be found in the book by Thornton et al. [25].

Spectral interpretation of decision diagram. The fundamentals and the latest advances in spectral interpretation of decision diagram are presented, in the book by Stanković and Astola [23]. A brief overview of these advances can be also found in the paper by Falkowski and Stanković [7].

Advanced decision diagram techniques. The latest results in application of the group-theoretic approach (i.e., non-Abelian groups, quaternations groups) to synthesis of decision diagrams are reported in papers by Stanković et al. [20, 21, 22, 24]. Linear word-level arithmetic representations of multi-valued functions have been introduced by Dziurzanski et al. [5], Yanushkevich et al. [29] and logic ones have been studied by Tomaszewska et al. [26]. Interpretation and manipulation of Taylor-like representations of logic functions using decision diagram techniques was studied by Yanushkevich [28].

Parallel architectures. Leighton's book [16] is a comprehensive general reference to parallel algorithms and architectures.

Systolic arrays. Systolic 1D and 2D arrays have been proposed by Kung [14]. In a certain sense, they resemble cellular automata. This idea, combined with RISC architectures based on pipeline paradigm, gave birth to parallel-pipelined computations. The processors that implement such computing have been called *systolic* by Kung in 1980. Later, however, the modest term *parallel-pipelined* has been rehabilitated. A 2D architecture has been called *matrix systolic arrays*, or simply systolic arrays. They can be two-dimensional (rectangular, triangular, tree-like) and data may flow between the cells (which can be programmed) in different directions and at different speeds. The algorithms that specify operations of each cell in systolic arrays are called *systolic algorithms.* These algorithms have been suggested for solving many problems such as binary and polynomial arithmetic, solution of linear systems, geometric problems, and matrix operations. The latter, especially matrix multiplication, filtering and convolution, have become the basis for systolic signal processors. These processors are based on the mapping of fast digital signal processing algorithms (FFT) to parallel-pipelined structures of the single-input and single-output systolic arrays.

Linear systolic arrays. A pipeline is an example of a *linear systolic array* in which data flows only in one direction. A design of linear systolic arrays have been considered in [12]. Also a multidimensional analysis of signal based on multiple application of one-dimensional analysis has been introduced in by Kukharev et al. [13]. Linear systolic arrays have also been developed for transforms of Boolean functions [13, 18], including solutions of Boolean equations [17] and arithmetic and other transforms [27].

References

[1] Ancona MG. Systolic processor design using single-electron digital circuits. *Superlattices and Microstructures*, 20(4):461–472, 1996.

[2] Chaudhuri PP, Chowdhury DP, Nandi S, and Chattopadhyay S. *Additive Cellular Automata: Theory and Applications.* IEEE Computer Society – Wiley, 1997.

[3] Delorme M, and Mazoyer J. *Cellular Automata: A Parallel Model.* Kluwer Academic Publishers, Dordrecht, 1998.

[4] Danielsson PE. Serial parallel convolvers. In Swartzlander EE Jr., Ed., *Systolic Signal Processing Systems.* Dekker, NY, 1987.

[5] Dziurzanski P, Malyugin VD, Shmerko VP, and Yanushkevich SN. Linear models of circuits based on the multivalued components. *Automation and Remote Control*, Kluwer/Plenum Publishers, 63(6):960–980, 2002.

[6] Jones SR, Sammut KM, and Hunter J. Learning in linear systolic neural network engines: analysis and implementation. *IEEE Transactions on Neural Networks*, 5(4):584-593, 1994.

[7] Falkowski BJ and Stanković RS. Spectral interpretation and applications of decision diagrams. *VLSI Design International Journal of Custom Chip Design, Simulation and Testing*, 11(2):85–105, 2000.

[8] Hurst SL. Logical Processing of Digital Signals, Crane Russak and Edward Arnold, London and Basel, 1978.

[9] Hurst SL, Miller DM, and Muzio JC. *Spectral Techniques in Digital Logic*, Academic Press, Bristol, 1985

[10] Karpovsky MG. *Finite Orthogonal Series in the Design of Digital Devices.* Wiley and JUP, New York and Jerusalem, 1976.

[11] Karpovsky MG, Stanković RS, and Astola JT. Spectral techniques for design and testing of computer hardware. In *Proceedings of the Workshop on Spectral Transforms and Logic Design for Future Digital Systems*, Tampere, Finland, pp. 1–34, 2000.

[12] Kumar VKP, and Tsai Y-C. Designing linear systolic arrays. *Journal of Parallel and Distributed Computing*, 7:441–463, 1989.

[13] Kukharev GA, Tropchenko AY, and Shmerko VP. *Systolic Signal Processors.* Publishing House "Belarus", Minsk, Belarus, 1988 (In Russian).

[14] Kung SY. *VLSI Array Processors.* Prentice Hall, New York, 1988.

[15] Lechner R. A transform theory for functions of binary variables in theory of switching. *Harvard Computation Lab., Cambridge, Mass.*, Progress Rept. BL-30, Sec-X, pp. 1–37, 1961.

[16] Leighton FT. *Introduction to Parallel Algorithms and Architectures: Arrays, Trees and Hypercubes.* Kaufmann, San Mateo, CA, 1992.

[17] Levashenko VG, Shmerko VP, and Yanushkevich SN. Solution of Boolean Differential Equations on Systolic Arrays. *Cybernetics and Systems Analysis*, Kluwer/Plenum Publishers, 32(1):26–40, 1996.

[18] Shmerko VP, and Yanushkevich SN. Algorithms of Boolean differential calculus for systolic arrays. *Cybernetics and Systems Analysis*, Kluwer/Plenum Publishers, 3:38–47, 1990.

[19] Sinha BP, and Srimani PK. Fast parallel algorithms for binary multiplication and their implementation on systolic architectures. *IEEE Transactions on Computers*, 38(3):424–431, 1989.

[20] Stanković RS, Milenović D, and Janković D. Quaternion groups versus dyadic groups in representations and processing of switching functions. In *Proceedings of the IEEE 29th International Symposium on Multiple-Valued Logic*, pp. 18–23, 1999.

[21] Stanković RS, Moraga C, and Astola JT. From Fourier expansions to arithmetic-Haar expressions on quaternion groups. *Applicable Algebra in Engineering, Communications and Computing*, AAECC12:227–253, 2001.

[22] Stanković RS. Non-Abelian groups in optimization of decision diagrams representations of discrete functions. *Formal Methods in System Design*, 18:209–231, 2001.

[23] Stanković RS and Astola JT. *Spectral Interpretation of Decision Diagrams.* Springer, Heidelberg, 2003.

[24] Stanković RS and Astola JT. Design of decision diagrams with increased functionality of nodes through group theory. *IEICE Transactions Fundamentals*, E86-A(3):693–703, 2003.

[25] Thornton MA, Drechsler R, and Miller DM. *Spectral Techniques in VLSI CAD, Kluwer Academic Publishers*, Kluwer, Dordrecht, 2001.

[26] Tomaszewska A, Yanushkevich SN, and Shmerko VP. Word-level models for efficient computation of multiple-valued functions. Part 2: LWL based models. In *Proceedings of the IEEE 32nd International Symposium on Multiple-Valued Logic*, pp. 209–214, 2002.

[27] Yanushkevich SN. Systolic algorithms to synthesize arithmetical polynomial forms for k-valued logic functions. *Automation and Remote Control*, Kluwer/Plenum Publishers, 55(12):1812–1823, 1994.

[28] Yanushkevich SN. *Logic Differential Calculus in Multi-Valued Logic Design.* Technical University of Szczecin Academic Publishers, Szczecin, Poland, 1998.

[29] Yanushkevich SN, Dziurzanski P, and Shmerko VP. Word-level models for efficient computation of multiple-valued functions. Part 1: LAR based models. In *Proceedings of the IEEE 32nd International Symposium on Multiple-Valued Logic*, pp. 202–208, 2002.

[30] Yanushkevich SN, Shmerko VP, and Lyshevski SE. *Logic Design of NanoICs.* CRC Press, Boca Raton, FL, 2005.

[31] Yanushkevich SN. Logic design of nanodevices, In Rieth M and Schommers W, Eds., *Handbook of Theoretical and Computational Nanotechnology*, Scientific American Publishers, 2005.

[32] Yanushkevich SN and Shmerko VP. Decision diagram technique. In *Handbook for Electrical Engineers*, Second Editions, CRC Press, Boca Raton, FL, 2005.

34

Three - Dimensional Techniques

Logic design of nanodevices in spatial dimensions is based on selected methods of advanced logic design, and appropriate spatial topologies. This chapter focuses on the relationship of decision diagrams and 3D graph models such as hypercube data structures, and the hypercube-like topology, $\mathcal{N}$-hypercube.

34.1 Introduction

Three-dimensional design has been explored at the macro-level for a long time, for example, in the design of distributed systems. It was inspired by nature; for example, the brain, with its "distributed computing and memory", was the prototype in the case of the "connection machine"*. The 3D computing architecture concept has been employed by the creators of the supercomputers Cray T3D, NEC's Ncube, and cosmic cube. The components of these supercomputers, as systems, are designed based on classical paradigms of 2D architecture that becomes 3D because of the *3D topology* (of interconnects), or *3D data structures* and corresponding algorithms, or *3D communication flow.*

Example 34.1 *The topology of today's silicon integrated circuits at a system level varies:*

- 1D arrays, *e.g., pipelines, linear systolic processors,*
- 2D arrays, *e.g., matrix processors, systolic arrays, and*
- 3D arrays, *e.g., of hypercube architecture.*

At the physical level, very large scale integration circuits, for instance, are 3D devices, because they have a layered structure, i.e., interconnection between layers while each layer has a 2D layout. On the way to the top VLSI hierarchy (the most complex VLSI systems), linear and 2D arrays eventually evolved to multi-unit architectures such as 3D arrays. Their properties can be summarized as follows:

*Hillis WD. The connection machine, MIT Press, 1985

► As stated in the theory of parallel and distributed computing, processing units are packed together and communicate best only with their nearest neighbors.

► In the optimal organization, each processing unit will have a diameter comparable to the maximum distance a signal can travel within the time required for some reasonable minimal amount of processing of a signal, for example to determine how it should be routed or whether it should be processed further.

► 3D architectures need to contain a fair number of cells before the advantages of the multi-cell organization become significant compared to competing topologies.

In current multiprocessor designs, 3D structures are not favored, since they suffer from gate and wire delay. Internally to each processing unit, 2D architecture is preferable, since at that smaller scale, communication times will be short compared with the cost of computation.

Nanostructures are associated with a molecular/atomic physical platform. This has a truly 3D structure instead of the 3D layout of silicon integrated circuits composed of 2D layers with interconnections forming the third dimension. At the nanoscale level, the advantages of the 3D architecture will begin to become apparent for the following reasons (see, for example, study by Frank [28, 10]):

► *The distance light can travel in 1* ns *in a vacuum is only around 30* cm *and two times less in a solid (for example, 1* ns *is the cycle of a computer with clock speed of 1GHz), which means that components of such a computer must be packed in a single chip of several* cm *size; thus, a reasonable number of 3D array elements of nanometer size must be integrated on a single tiny chip.*

► *They are desirable for their ideal nature for large scale computing.*

► *There are many 3D algorithms and designs for existing microscale components that are arranged in 3D space, which computer designers already have experience with.*

► *There are limits to information density as well, that imply a direct limit on the size of, in particular, a memory cell.*

Thus, the speed of light limit (that is information transfer speed limit) and information density pose the following implications for computer architecture:

► Traditional architecture will be highly inefficient, since most of the processor and memory will not be accessed at any given time due to the size limit,

► Interconnection topology must be scalable; most of the existing multiprocessor topologies (hypercubes, binary trees) don't scale, since communication times start to dominate as the machine sizes are scaled up.

The solutions to this are:

- Use of a parallel, multiprocessing architecture, where each processor is associated with some local memory, in contrast to the traditional von Neumann architecture.
- The number of processing nodes reachable in n hops from any node cannot grow faster than order n^3 and still embed the network in 3D space with a constant time per hop[†].

This leads us to the conclusion that there is only one class of network topologies that is asymptotically optimal as the machine size is scaled up: namely, a 3D structure, where each processing element is connected to a constant number of physical neighbors. In fact the processing elements of this 3D mesh can simulate both the processors and wires of the alternative architecture, such as, in particular, a randomly connected network of logic gates. The processing elements must be spread through the structure at a constant density.

Therefore, as processor speeds increase, the speed-of-light limit will cause communication distances to shrink, and the idea of mesh-connected processing elements and memory is, perhaps, the most reasonable and feasible solution.

34.2 Spatial structures

Several network topologies have been developed to fit different styles of computation, including a massive parallel computation in spatial dimensions.

A spatial network topology intended for implementing switching functions should possess several characteristics, in particular, minimal degree, ability to extend the size of structure with minimal changes to the existing configuration, ability to increase reliability and fault tolerance with minimal changes to the existing configuration, good embedding capabilities, flexibility of design methods, and flexibility of technology.

Based on the above criteria, a number of topologies can be considered as potentially useful for solving the problems of spatial logic design, namely: hypercube topology, cycle-connected cube known as CCC-topology (hypercube), X-hypercube topology, and specific topologies (hyper-Peterson, hyperstar, Fibonacci cube, etc.). Some are shown in Figure 34.1.

[†]Vitanyi P.M.B. Multiprocessor architectures and physical law, Proc. 2nd IEEE Workshop on Physics and Computation, PhysComp'94, Dallas (Texas), November 17-20, 1994, pp. 24-29

Hypercube topology (Figure 34.1a) has received considerable attention in classical logic design due mainly to its ability to interpret logic formulas (small diameter, high regularity, high connectivity, and good symmetries). Hypercube-based structures are at the forefront of massive parallel computation because of the unique characteristics of hypercubes (fault tolerance, ability to efficiently permit the embedding of various topologies, such as lattices and trees).

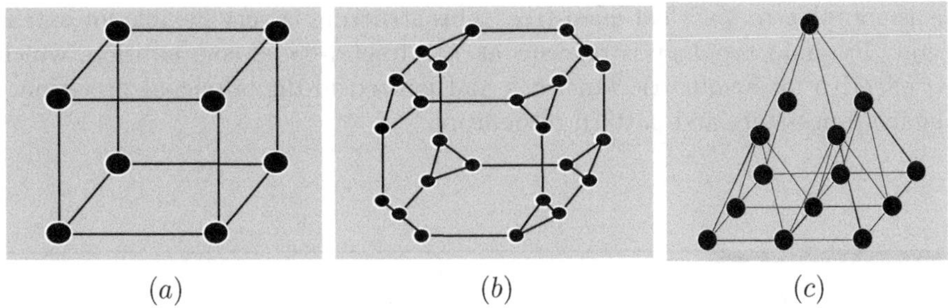

(a) (b) (c)

FIGURE 34.1
Spatial configurations: hypercube (a), CCC-hypercube (b), and pyramid (c).

The binary n-hypercube is a special case of the family of k-ary n-hypercubes, which are hypercubes with n dimensions and k nodes in each dimension. The total number of nodes in such a hypercube is $N = k^n$. Parallel computers with direct-connect topologies are often based on k-ary n-cubes or isomorphic structures such as rings, meshes, tori, and direct binary n-cubes. A node in a k-ary n-hypercube can be represented by an n-digit address $d_{n-1} \dots d_0$ in radix k. The i-th digit, d_i, of the address represents the node's position in the i-th dimension, where $0 \le d_i \le k-1$. Two nodes with addresses $(d_{n-1}, d_{n-2}, \dots, d_0)$ and $(d'_{n-1}, d'_{n-2}, \dots, d'_0)$ are neighbors in the ith dimension if, and only if, either

$$d_i = (d'_i + 1)\ mod\ k, \qquad \text{or}$$
$$d_j = d'_j, \quad \forall j \neq i.$$

The CCC-hypercube is created from a hypercube by replacing each node with a cycle of s nodes (Figure 34.1b). Hence it increases the total number of nodes from 2^n to $s \cdot 2^n$ and preserves all features of the hypercube. The CCC-hypercube is closely related to the butterfly network. As has been shown in the previous sections, "butterfly" data flowgraphs are the "nature" of most transforms of switching functions in matrix form.

Pyramid topology (Figure 34.1c) is suitable for computations that are based on the principle of hierarchical control, for example, decision trees and decision diagrams. An arbitrarily large pyramid can be embedded into the hypercube with a minimal load factor. The dilation is two and the congestion is $\Theta(\sqrt{l})$ where l is the load factor. Pyramid P_n has nodes on levels $0, 1, \ldots, n$. The number of nodes on level i is 4^i. So, the number of nodes of P_n is $(4^{n+1}/3)$. The unique node on level 0 is called the *root* of the pyramid. The subgraph of P_n induced by the nodes on level i is isomorphic to a mesh $2^i \times 2^i$. The subgraph of P_n induced by the edges connecting different levels is isomorphic to a 4^n-leaf quad-tree. This structure is very flexible for extension. Pyramid topology is relevant also to fractal-based computation, which is effective for symmetric functions and is used in digital signal processing, image processing, and pattern recognition.

34.3 Hypercube data structure

Traditionally, hypercube topology is used in:

- ▶ Logic design for switching function manipulation (minimization, representation),
- ▶ Communication problems for traffic representation and optimization, and
- ▶ Multiprocessor system design for optimization of parallel computing.

In the first approach, each variable of a switching function is associated with one dimension in hyperspace. The manipulation of the function is based on a special encoding of the vertices and edges in the hypercube. In the second approach used in communication problems and multiprocessor systems design, the hypercube is the underlying computational model. To design this model, a decision tree or a decision diagram must be constructed and embedded into a hypercube. In this approach the hypercube is utilized as a topological structure for computing in 3D space.

The problem of assembling a hypercube from a number of topological components is introduced. An alternative approach based on embedding graphs in hypercubes is also discussed.

Hypercube definition and characteristics. Hypercubes of different dimensions are shown in Figure 34.2). Hypercube properties are given in Table 34.1.

Example 34.2 *The string* $A = a_1a_2a_3a_4 = \{01 * *\}$ *represents the 2-subcube of* Q_4 *with the node set* $\{0100, 0101, 0110, 0111\}$.

TABLE 34.1
Hypercube properties.

Property	Definition
Extension	A hypercube is an extension of a graph. The dimensions are specified by the set $\{0, 1, \ldots, n-1\}$. An n-dimensional binary hypercube is a network with $N = 2^n$ nodes and diameter n. There are $d \times 2^{d-1}$ edges in a hypercube of d dimensions.
Specification	Each node of an n-dimensional hypercube can be specified by the binary address $(g_{n-1}, g_{n-2}, \ldots, g_0)$ of length n, where the bit g_i corresponds to the i-th dimension in a Boolean space. Two nodes with addresses $(g_{n-1}, g_{n-2}, \ldots, g_0)$ and $(g'_{n-1}, g'_{n-2}, \ldots, g'_0)$ are connected by an edge (or link) if and only if their addresses differ by exactly one bit.
Hamming distance	There are $\binom{n}{x}$ nodes at Hamming distance of x from a given node, and n node-disjoint paths between any pair of nodes of the hypercube.
Recursion	Hypercube Q_n can be defined recursively.
Fan-out	The *fan-out* (i.e., degree) of every node is n, and the total number of communication links is $\frac{1}{2}N \log N$.
Decomposition	A k-dimensional subcube (k-subcube) of hypercube Q_n, $k \leq n$, is a subgraph of Q_n, that is, a k-dimensional hypercube. A k-subcube of Q_n is represented by a ternary vector $A = a_1 a_2 \ldots, a_n$, where $a_i \in \{0, 1, *\}$, and $*$denotes an element that can be either 0 or 1.
Intercube distance	Given two subcubes $A = a_1 a_2 \ldots, a_n$ and $B = b_1 b_2 \ldots, b_n$, the *intercube distance* $D_i(A, B)$ between A and B along the i-th dimension is 1 if $\{a_i, b_i = \{0, 1\}\}$; otherwise, it is 0. The distance between two subcubes A, B is given by $D(A, B) = \sum_{i=1}^{n} D_i(A, B)$.
Path	A *path* P of length l is an ordered sequence of nodes $x_{i_0}, x_{i_1}, x_{i_2}, \ldots, x_{i_l}$, where the nodes are labeled with x_{i_j}, $0 \leq j \leq l$, and $x_{i_k} \neq x_{i_{k+1}}$, for $0 \leq k \leq l-1$.

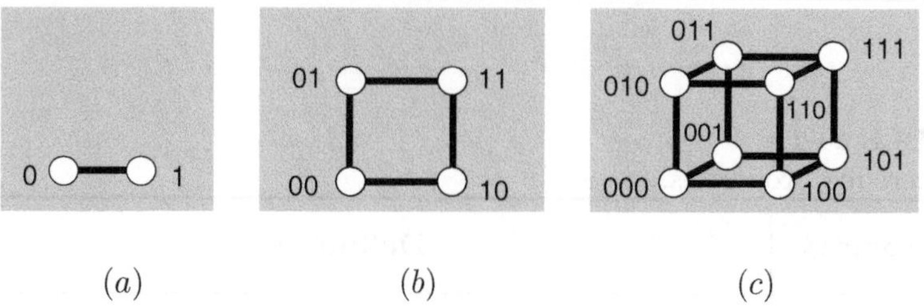

(*a*) (*b*) (*c*)

FIGURE 34.2
1D hypercube (a), 2D hypercube (b), and 3D hypercube (c).

Gray code is used for encoding the indexes of the nodes. There are several reasons to encode the indexes this way. The most important of them is to simplify analysis, synthesis and embedding of topological structures. Gray code is referred to as a *unit-distance* code. Let $b_n...b_1b_0$ be a binary representation of an integer positive number B and $g_n...g_1g_0$ be its Gray code.

$$\texttt{Binary representation}\ b_n...b_1b_0 \iff \texttt{Gray code}\ g_n...g_1g_0$$

Suppose that $B = b_n...b_1b_0$ is given, then the corresponding binary Gray code representation is

$$g_i = b_i \oplus b_{i+1} \tag{34.1}$$

where $b_{n+1} = 0$. Consider the inverse problem: given the Gray code $G = g_n...g_1g_0$, the corresponding binary representation is derived by

$$b_i = g_0 \oplus g_1 \oplus \ldots g_{n-i} = \bigoplus_{i=0}^{n-i} g_i. \tag{34.2}$$

Table 34.2 illustrates the above transformation for $n = 3$.

Example 34.3 *Binary to Gray and Gray to binary transformation is illustrated in Figure 34.3 for* $n = 3$.

Useful rule. To build a Gray code for d dimensions, one takes the Gray code for $d - 1$ dimensions, reflects it top to bottom across a horizontal line just below the last element, and adds a leading one to each new element below the line of reflection.

The Hamming distance is a useful measure in hypercube topology. The Hamming sum is defined as the bitwise operation

$$(g_{d-1} \cdots g_0) \oplus (g'_{d-1} \cdots g'_0) = (g_{d-1} \oplus g'_{d-1}), \ldots, (g_1 \oplus g'_1), (g_0 \oplus g'_0) \tag{34.3}$$

TABLE 34.2
Relationships for binary and Gray code for $n = 3$.

Binary code	Gray code
000	000
001	001
010	011
011	010
100	110
101	111
110	101
111	100

Binary code	Gray code
000	000
001	001
011	010
010	011
110	100
111	101
101	110
100	111

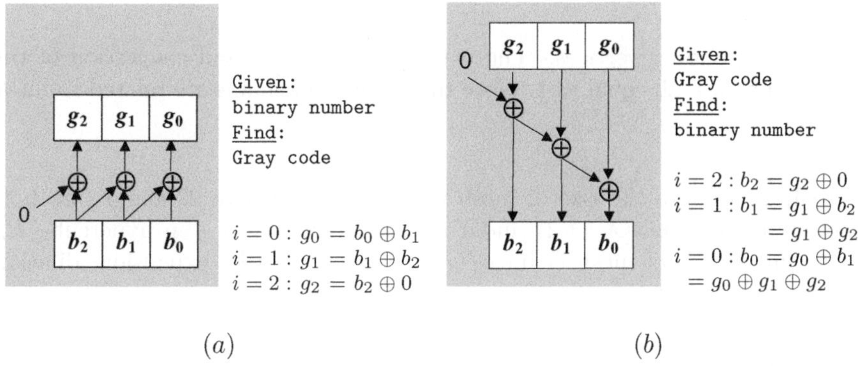

FIGURE 34.3
Data flowgraph and formal equation for binary to Gray code (a) and inverse transformation (b) (Example 34.3).

where $\oplus$ is an exclusive or operation.

In the hypercube, two nodes are connected by a link (edge) if and only if they have labels that differ by exactly one bit. The number of bits by which labels g_i and g_j differ is denoted by $h(g_i, g_j)$; this is the Hamming distance between the nodes.

Example 34.4 *Hamming sum operation on two hypercubes for 3-variable switching functions is illustrated in Figure 34.4.*

Rings and chains. Let G be a ring (chain) with 2^d vertices. The ring (chain) vertices are numbered 0 through $2^d - 1$ in Gray code. Map vertices of G to vertices of H, and map i, j-th edge of G to the unique edge in H_d that connects

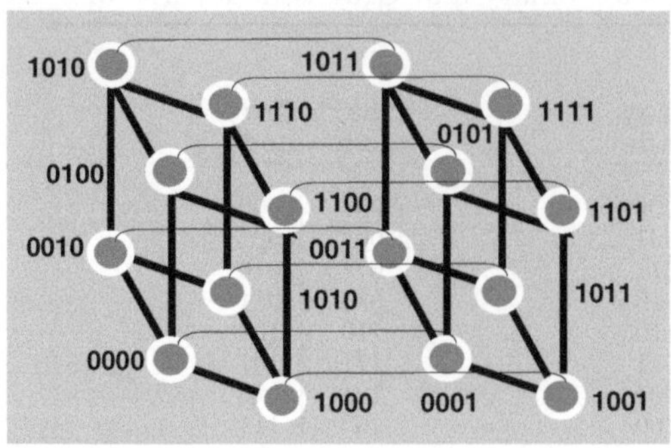

$0000 \oplus 0001 = 0001$
$0010 \oplus 0011 = 0001$
$0100 \oplus 0101 = 0001$
$0110 \oplus 0111 = 0001$
$1000 \oplus 1001 = 0001$
$1010 \oplus 1011 = 0001$
$1100 \oplus 1101 = 0001$
$1110 \oplus 1111 = 0001$

FIGURE 34.4
Hamming sum operation (Example 34.4).

the corresponding vertices. The expansion, dilation, and congestion of this embedding are all equal to 1. Note that rings and chains are related to linear arrays.

Meshes. Denote the size of mesh G by $X \times Y$, where X and Y are both powers of two. Let $X = 2^x$ and $Y = 2^y$. Let $d = x + y$ in hypercube H_d. The embedding of mesh G into H_d is characterized by expansion, dilation, and congestion of 1.

Complete binary trees. Let T_i be the complete binary tree of height i. This tree consists of $2^i - 1$ vertices. The following statements are useful for embedding:

- Let $i > 0$, then there is a dilation of 1 when embedding a complete binary tree T_i into the hypercube H_{i+1}.
- Let $i > 0$, then there is a dilation of 2 when embedding a complete binary tree T_i into the hypercube H_i.
- Let $i > 2$, then there is no dilation when embedding a complete binary tree T_i into the hypercube H_i. One can justify this statement by construction: T_1 is trivially embedded into H_1, T_2 is embedded into H_2 with a dilation and congestion of one, and for $i > 2$ there is no dilation. Note that the condition of embedding with dilation 1 includes two requirements:

  (a) vertices of T_i that are on an odd level of tree hierarchy ($V_{odd\ level} = 2^0 + 2^2 + 2^4 + \cdots + 2^{i-1} = \frac{2^{i+1}-1}{3}$) are mapped onto hypercube H_i vertices that have an even number of ones, and

(b) vertices of T_i that are on an even level of tree ($V_{even\ level} = 2^0 + 2^3 + 2^5 + \cdots + 2^{i-1} = \frac{2(2^i-1)}{3}$) are mapped to hypercube H_i vertices that have an odd number of ones.

Because $V_{odd\ level} > 2^{i-1}$ for $i > 1$ and $V_{even\ level} > 2^{i-1}$ for $i > 2$, the hypercube h_i does not have enough vertices with odd and even numbers of ones to host the vertices of tree T_i on odd and even levels of tree hierarchy.

Rings, meshes, pyramids, shuffle-exchange networks, and complete binary trees can be embedded into hypercubes.

34.4 Assembling of hypercubes

Assembling is the basic topological operation that we apply to synthesize hypercube and hypercube-like data structures. Assembling is the first phase of the development of self-assembly, that is, the process of constructing a unity from components acting under forces/motives internal or local to the components themselves.

To apply an assembly procedure, the following items must be defined:

- The structural topological components,
- Formal interpretation of the structural topological components in terms of the problem, and
- The rules of assembly.

Assembling is a key philosophy of building complex systems. In this section, the assembling of classical hypercubes is considered. Assembling a hypercube of switching functions is accomplished by:

- Generating the products as enumerated points (nodes) in the plane,
- Encoding the nodes by Gray code,
- Generating links using Hamming distance,
- Assembling the nodes and links, and
- Joining a topology of hypercube in n dimensions.

Let $x_j^{i_j}$ be a literal of a Boolean variable x_j such that $x_j^0 = \overline{x}_j$, and $x_j^1 = x_j$, and $x_1^{i_1} x_2^{i_2} \ldots x_n^{i_n}$ is a product of literals. Topologically, it is a set of points on the plane numerated by $i = 0, 1, \ldots, n$. To map this set into the hypercube, the numbers must be encoded by Gray code and represented by the corresponding graphs based on Hamming distance. The example below demonstrates the assembly procedure.

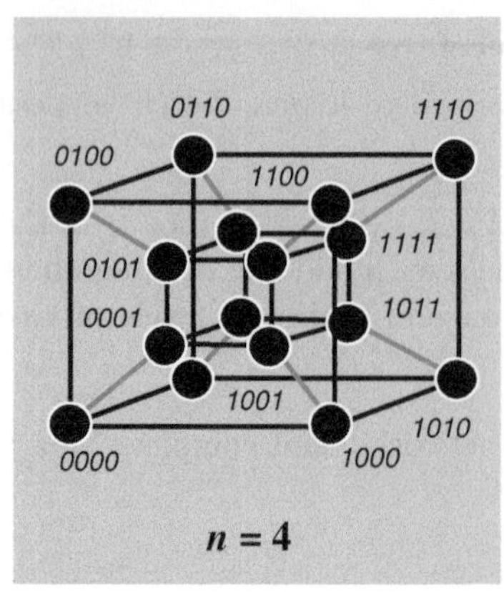

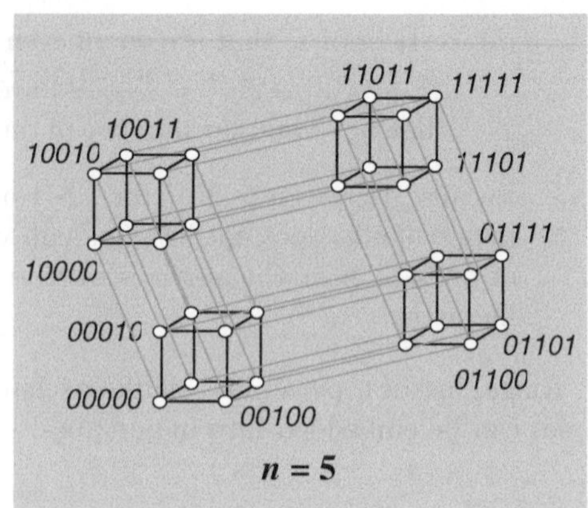

FIGURE 34.5
Assembling 3D hypercubes to represent the product term of four $n = 4$ and five $n = 5$ variables (Example 34.5).

Example 34.5 *The assembly of hypercubes for products of one, two and three variables is as follows:*

Product $x_1^{i_1}$ $\Longleftrightarrow$ 2 points $\Longleftrightarrow$ 1D hypercube ($n = 1$)
Product $x_1^{i_1}x_2^{i_2}$ $\Longleftrightarrow$ 4 points $\Longleftrightarrow$ 2D hypercube ($n = 2$)
Product $x_1^{i_1}x_2^{i_2}x_3^{i_3}$ $\Longleftrightarrow$ 8 points $\Longleftrightarrow$ 3D hypercube ($n = 3$)

Products with more than three variables are represented by assembling 3D hypercubes (Figure 34.5b):

Product $x_1^{i_1}x_2^{i_2}x_3^{i_3}x_4^{i_4}$ $\Longleftrightarrow$ 16 points $\Longleftrightarrow$ 4D hypercube ($n = 4$)
Product $x_1^{i_1}x_2^{i_2}x_3^{i_3}x_4^{i_4}x_5^{i_5}$ $\Longleftrightarrow$ 32 points $\Longleftrightarrow$ 5D hypercube ($n = 5$)

Notice that the 0-dimensional hypercube ($n = 0$) represents the constant 0. The line segment connects vertices 0 and 1, and these vertices are called the *face* of 1D hypercube and denoted by **x**. A 2D hypercube has four faces, 0**x**, 1**x**, **x**0, and **x**1. The total 2D hypercube can be denoted by **xx**.

Example 34.6 *Six faces of the 3D hypercube,* **xx**0*,* **xx**1*,* 0**xx***,* 1**xx***,* **x**1**x***, and* **x**0**x** *(Figure 34.6) represent 1-term products for a switching function of three variables.*

Assembling hypercubes for switching functions. The most useful property of a sum-of-products representation of switching functions for the hypercube assembly is that it can be derived directly from the switching function.

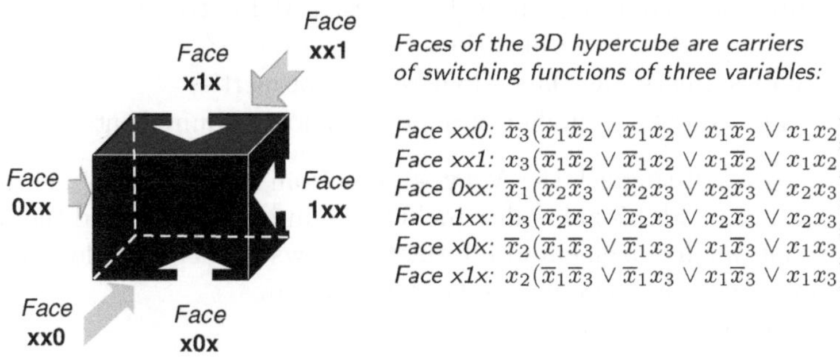
Face xx1
Face x1x
Face 0xx
Face 1xx
Face xx0
Face x0x
Faces of the 3D hypercube are carriers of switching functions of three variables:
Face xx0: $\overline{x}_3(\overline{x}_1\overline{x}_2 \vee \overline{x}_1 x_2 \vee x_1\overline{x}_2 \vee x_1 x_2)$
Face xx1: $x_3(\overline{x}_1\overline{x}_2 \vee \overline{x}_1 x_2 \vee x_1\overline{x}_2 \vee x_1 x_2)$
Face 0xx: $\overline{x}_1(\overline{x}_2\overline{x}_3 \vee \overline{x}_2 x_3 \vee x_2\overline{x}_3 \vee x_2 x_3)$
Face 1xx: $x_3(\overline{x}_2\overline{x}_3 \vee \overline{x}_2 x_3 \vee x_2\overline{x}_3 \vee x_2 x_3)$
Face x0x: $\overline{x}_2(\overline{x}_1\overline{x}_3 \vee \overline{x}_1 x_3 \vee x_1\overline{x}_3 \vee x_1 x_3)$
Face x1x: $x_2(\overline{x}_1\overline{x}_3 \vee \overline{x}_1 x_3 \vee x_1\overline{x}_3 \vee x_1 x_3)$

FIGURE 34.6
Faces of the hypercube interpretation in the sum-of-products of a switching function of three variables (Example 34.6).

If the variable x_j is not present in the hypercube, then $c_j = \mathbf{x}$ (don't care), i.e., $x_j^{\mathbf{x}} = 1$. In hypercube notation, a term is described by a hypercube that is a ternary vector $[i_1 i_2 \ldots i_n]$ of components $i_j \in \{0,\ 1,\ \mathbf{x}\}$. A set of cubes corresponding to the true values of a switching function f represent the sum-of-products for this function. The hypercube $[i_1 i_2 \ldots i_n]$ is called an n-hypercube or n-cube specifying the size of the hypercube. Cube algebra includes a set of elements as $C = \{1, 0, *\}$, and basic operations to find new cubes. Directly adjacent elements of the on-set are called the *adjacency plane.* Each adjacency plane corresponds to a product term.

Details of representation of switching functions by hypercubes are given in Chapters 2 and 4.

34.5 $\mathcal{N}$-hypercube definition

In this section the extension of the traditional hypercube is considered. This extension is called the $\mathcal{N}$-hypercube. The classic hypercube structure is the basis for $\mathcal{N}$-hypercube design and inherits most of the properties of classic hypercubes.

34.5.1 Extension of a hypercube

Based on the above, the new, hypercube-like topology called the $\mathcal{N}$-hypercube is introduced. The following reasons advocate developing $\mathcal{N}$-hypercube-based topologies:

▶ $\mathcal{N}$-hypercubes ideally reflect all properties of decision trees and decision diagrams, popular in advanced logic design data structure, enhancing them to more than two dimensions.

- $\mathcal{N}$-hypercubes inherit the classic hypercube's properties.
- $\mathcal{N}$-hypercubes satisfy a number of nanotechnology requirements.

Several features distinguish the $\mathcal{N}$-hypercube from a hypercube, in particular, the existence of additional nodes, including a unique node called the root. Thanks to that, an arrangement of information flows that is suitable from the point of view of certain technologies can be achieved.

The extension of a hypercube is made by:

- Embedding additional nodes,
- Distinguishing the types of nodes,
- Special space coordinate distribution of the additional nodes,
- New link assignments.

Additional embedded node and link assignments correspond to embedding decision trees in a hypercube and thus, convert a hypercube from the passive representation of a function to a connection-based structure, i.e., a structure in which calculations can be accomplished. In other words, information connectivity is introduced into the hypercube. Distinguishing the type of nodes satisfies the requirements of graphical data structures of switching functions.

An $\mathcal{N}$-hypercube's components are shown in Figure 34.7.

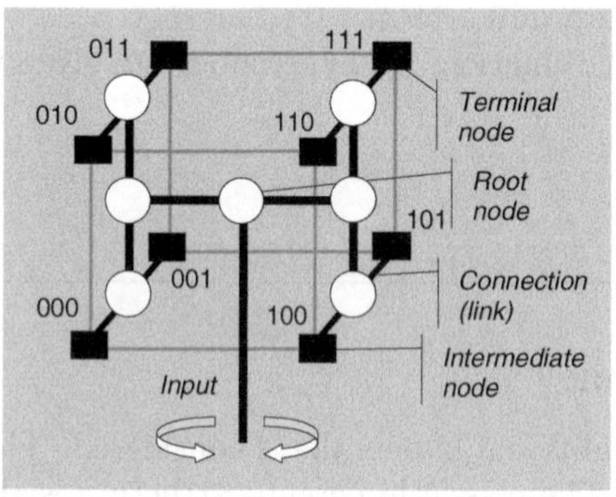

- *The node is a demultiplexor element, i.e., the node performs Shannon expansion*
- *Each intermediate node has a degree of freedom*
- *Terminal nodes carry the results of computing*
- *The root node resembles the root node of a decision tree*
- *The nodes (their functions and coordinates), and links carry information about the function implemented by the $\mathcal{N}$-hypercube.*

FIGURE 34.7
An $\mathcal{N}$-hypercube.

34.5.2 Degree of freedom and rotation

The degree of freedom of each intermediate node can be used for variable order manipulation, as the order of variables is a parameter to adjust in decision trees and diagrams. Additional intermediate nodes, the root node and corresponding links in the $\mathcal{N}$-hypercube are associated with

- ► The polarity of variables in a switching function representation,
- ► The structure of the decision tree and decision diagram, and the variable order, and
- ► The degree of freedom and rotation of the $\mathcal{N}$-hypercube based topology.

Consider the switching function of a single variable x. Corresponding 1D $\mathcal{N}$-hypercubes are shown in Figure 39.2.

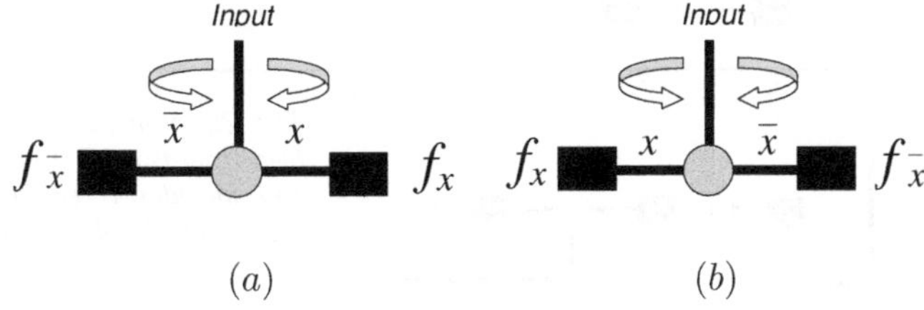

FIGURE 34.8
The switching function of an uncomplemented and complemented variable x, and corresponding 1D $\mathcal{N}$-hypercubes (a) and (b).

Only the intermediate and root nodes in an $\mathcal{N}$-hypercube have freedom. An intermediate node in the 1D $\mathcal{N}$-hypercube has two degrees of freedom (Figure 34.9). The 2D $\mathcal{N}$-hypercube is assembled from two 1D $\mathcal{N}$-hypercubes, and includes three intermediate nodes. The $\mathcal{N}$-hypercube in 2D has $2 \times 2 \times 2 = 8$ degrees of freedom. There are four decision trees with different orders of variables.

Consider an $\mathcal{N}$-hypercube in 3D. This $\mathcal{N}$-hypercube is assembled from two two-dimensional $\mathcal{N}$-hypercubes and includes seven intermediate nodes having $8 \times 8 \times 2 = 128$ degrees of freedom. The degree of freedom of an intermediate node in i-th dimension, $i = 2, 3, \ldots, n$, is equal to $DF_i = 2^{n-i} + 1$. In general, the degree of freedom of the n-dimensional $\mathcal{N}$-hypercube is defined as

$$\texttt{Degree of freedom} = \sum_i DF_i = \sum_i (2^{n-i} + 1). \qquad (34.4)$$

34.5.3 Coordinate description

There are two possible configurations of the intermediate nodes. The first configuration (planes) is defined as (Figure 34.10a)

$$\texttt{x00} \Longleftrightarrow \texttt{x01} \Longleftrightarrow \boxed{\text{xx1}} \Longleftrightarrow \texttt{x11} \Longleftrightarrow \texttt{x10} \Longleftrightarrow \boxed{\text{xx0}} \Longleftrightarrow \texttt{x00}.$$

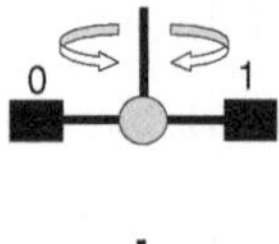

An intermediate node in the 1D $\mathcal{N}$-hypercube has two degrees of freedom.

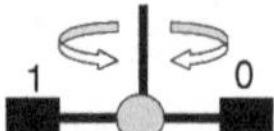

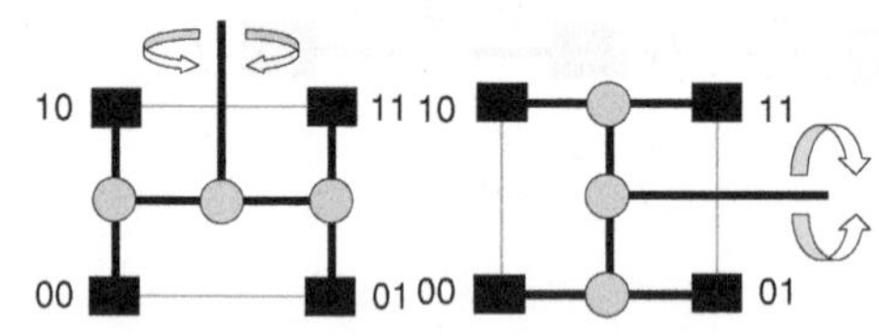

The 2D $\mathcal{N}$-hypercube is assembled from two 1D $\mathcal{N}$-hypercubes, and includes three intermediate nodes The $\mathcal{N}$-hypercube in 2D has

$2 \times 2 \times 2 = 8$ *degrees of freedom.*

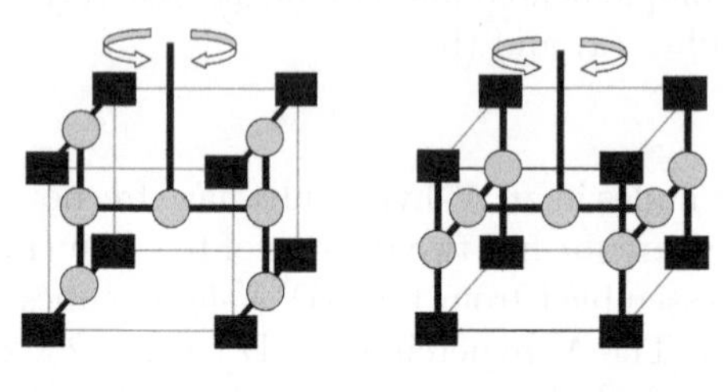

This $\mathcal{N}$-hypercube is assembled from two two-dimensional $\mathcal{N}$-hypercubes, includes 7 intermediate nodes, and has

$8 \times 8 \times 2 = 128$ *degrees of freedom.*

The degree of freedom of an intermediate node at i-th dimension, $i = 2, 3, \ldots n$, is equal to

$$DF_i = 2^{n-i} + 1.$$

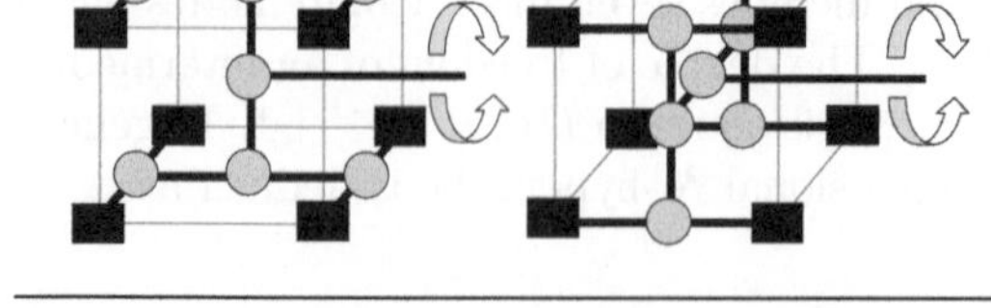

In general, the degree of freedom DF of the n-dimensional $\mathcal{N}$-hypercube is defined as

$$DF = \sum_i DF_i = \sum_i (2^{n-i} + 1).$$

FIGURE 34.9
Degree of freedom and rotation of the $\mathcal{N}$-hypercube in 1D and 2D.

The second possible configuration (planes) is related to the symmetric faces (Figure 34.10b)

$$\texttt{00x} \Longleftrightarrow \boxed{\textsf{0xx}} \Longleftrightarrow \texttt{01x} \Longleftrightarrow \texttt{11x} \Longleftrightarrow \boxed{\textsf{1xx}} \Longleftrightarrow \texttt{10x} \Longleftrightarrow \texttt{00x}.$$

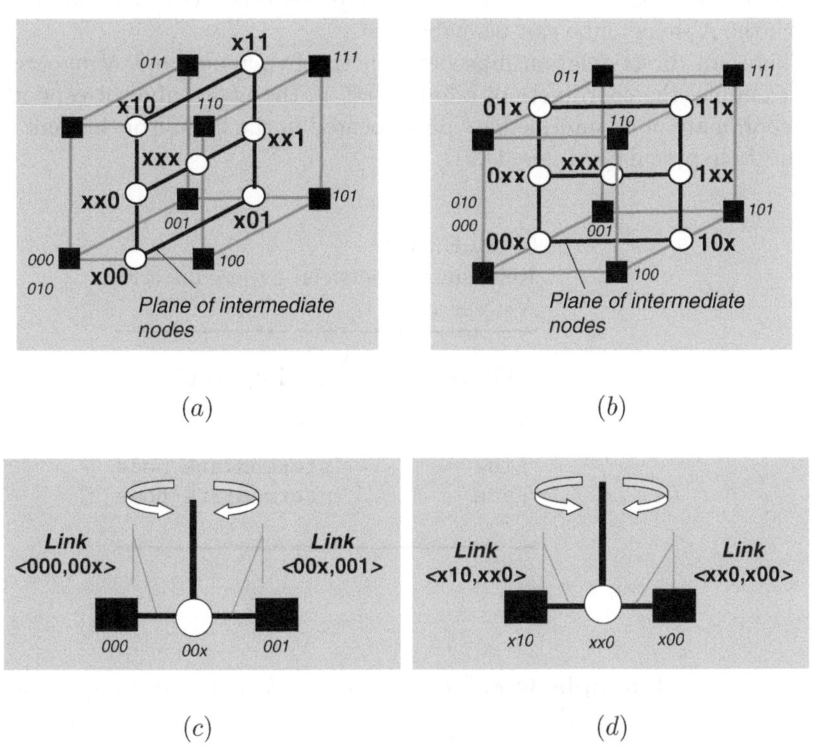

(*a*) (*b*)

(*c*) (*d*)

FIGURE 34.10
Coordinate description of the $\mathcal{N}$-hypercube: (a) the first plane, (b) the second plane, (c) the links of an intermediate node, and (d) links of the root.

An $\mathcal{N}$-hypercube includes two types of links with respect to the root node:

$$\text{Link 1: } \boxed{\textsf{xx0}} \Longleftrightarrow \texttt{xxx} \Longleftrightarrow \boxed{\textsf{xx1}},$$
$$\text{Link 2: } \boxed{\textsf{0xx}} \Longleftrightarrow \texttt{xxx} \Longleftrightarrow \boxed{\textsf{1xx}}.$$

The root node coordinate is xxx. There are two types of link in an $\mathcal{N}$-hypercube: links between terminal nodes and intermediate nodes, and links between intermediate nodes, including the root node.

Example 34.7 *In Figure 34.10, link <000,00x> indicates the connection of the terminal node 000 and intermediate node 00x. By analogy, if two intermediate nodes x10 and xx0 are connected, we indicate this fact by <x10,xx0> (Figure 34.10d).*

The number of terminal nodes in the $\mathcal{N}$-hypercube is always equal to the number of nodes in the hypercube. Therefore, the classic hypercube can be considered the basic data structure for representing switching functions in which the $\mathcal{N}$-hypercube can be embedded.

There are direct relationships between the hypercube and $\mathcal{N}$-hypercube. For example, the coordinate of a link (face) in the hypercube corresponds to the coordinate of an intermediate node located in the middle of this link (face in the $\mathcal{N}$-hypercube) (Table 34.3).

TABLE 34.3
Relationship between hypercube and $\mathcal{N}$-hypercube.

Hypercube	**$\mathcal{N}$-hypercube**
Link	Intermediate node
Face	Intermediate node

Example 34.8 *The relationship between coordinate description of the hypercube and the $\mathcal{N}$-hypercube is given in Table 34.4.*

TABLE 34.4
Relationship of the components of the hypercube and $\mathcal{N}$-hypercube (Example 34.8).

Hypercube	**$\mathcal{N}$-hypercube**
Links x00, 0x0, x10, 10x	Intermediate nodes x00, 0x0, x10,10x
Faces xx0, xx1, 0xx, 1xx, x1x	Intermediate nodes 0xx, 1xx, x1x,x0x

34.5.4 $\mathcal{N}$-hypercube design for $n > 3$ dimensions

Consider the two 3D $\mathcal{N}$-hypercubes shown in Figure 34.10b. To design a 4D $\mathcal{N}$-hypercube, two $\mathcal{N}$-hypercubes must be joined by links. There are seven possibilities for connecting two $\mathcal{N}$-hypercubes since links are allowed between intermediate nodes, intermediate nodes and root nodes, and between the root nodes. The new root node is embedded in the link <xxx0, xxx1>.

Therefore, the number of bits in the coordinate description of both $\mathcal{N}$-hypercubes must be increased by one bit. Suppose that $\mathcal{N}$-hypercubes are connected via link <xxx0, xxx1> between the root nodes xxx0 and xxx1. The resulting topological structure is called a 4D $\mathcal{N}$-hypercube.

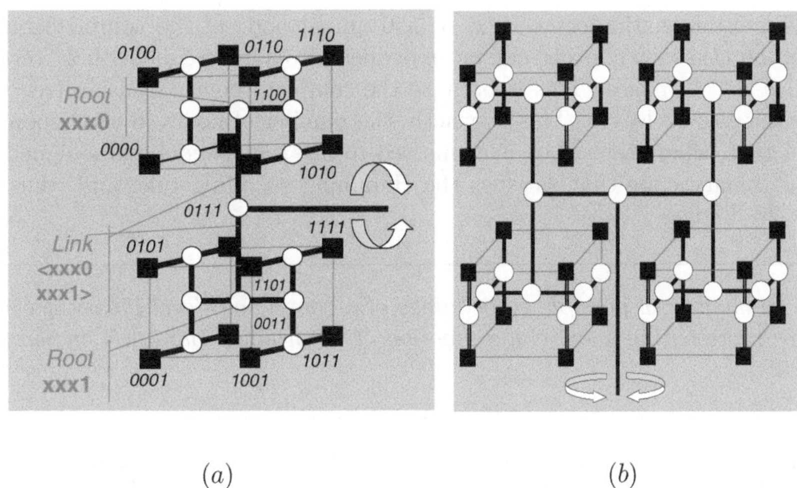

(a) (b)

FIGURE 34.11
Connections between $\mathcal{N}$-hypercubes in n-dimensional space (Example 34.9).

> **Example 34.9** *Figure 34.11 shows the possibility of connecting a given $\mathcal{N}$-hypercube to another $\mathcal{N}$-hypercube. This connection property follows from the properties of intermediate nodes.*

34.6 Embedding a binary decision tree into an $\mathcal{N}$-hypercube

A binary tree that represents a switching function can be embedded into a hypercube that also represents this function. The $\mathcal{N}$-hypercube can be specified as a hypercube with the following properties:

- An n-dimensional $\mathcal{N}$-hypercube is derived by embedding an n-level 2^n-leaf complete binary tree into an n-dimensional hypercube.
- An $\mathcal{N}$-hypercube includes $k = 2^n$ terminal nodes labelled from 0 to $2^n - 1$ so that there is an edge and an intermediate node between any two terminal nodes if the binary representations of their labels differ by only one bit.
- Each intermediate node is assigned a label that corresponds to a binary representation of the adjacent nodes with the don't care value for the only different bit.

For example, the leaf vertex, or a terminal node of the complete binary decision tree with n levels, can be embedded into a hypercube with 2^n vertices and $n \times 2^{n-1}$ edges. This is because the complete binary decision tree with n levels has 2^n leaves. This is exactly the number of nodes in the hypercube structure, where each node is connected to $n - 1$ neighbors and assigned the n-bit binary code that satisfies the Hamming encoding rule, and, thus has $n \times 2^{n-1}$ edges.

The number of possible embeddings of a complete n-level binary decision tree representing a switching function of n variables into an $\mathcal{N}$-hypercube is $n!$

The number of possible variable orders given n variables is equal to $n!$, so the possible number of embeddings is $n!$

Recurrence is a general strategy to generate spatial and homogeneous structures. This strategy can be used for embedding a binary decision tree into a $\mathcal{N}$-hypercube:

Step 1. Embed 2^n leaves of the binary tree (nodes of the level $n + 1$) into the 2^n-node of the n-dimensional hypercube; assign a code to the node so that each node is connected to q Hamming-compatible nodes.

Step 2. Embed 2^{n-1} nodes of the binary tree (nodes of the level n) into edges connecting the existing nodes of the hypercube, taking into account the polarity of the variable.

Step 3. Repeat recursively till we embed the root of the tree into the center of the hypercube. The structure obtained is a $\mathcal{N}$-hypercube.

Example 34.10 *Let $n = 1$, then a binary decision tree represents a switching function of one variable (Figure 34.12). The function takes a value of 1 while $x = 0$ and value 0 while $x = 1$. These values are assigned to two leaves of the binary decision diagram.*

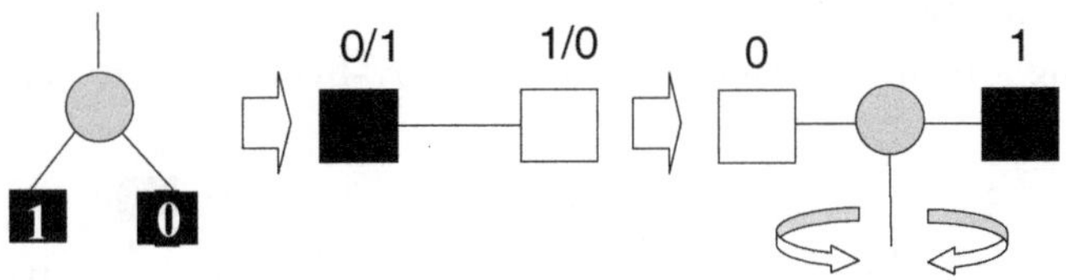

FIGURE 34.12
Embedding a binary decision tree of one variable into the $\mathcal{N}$-hypercube (Example 34.10).

In more detail for Example 34.10, the embedding is as follows (Figure 34.13):

Step 1. Embed 4 leaves of the binary tree into a 4-node $\mathcal{N}$-hypercube; assign the codes 00,01,10,11 so that each node is Hamming-compatible with neighbor nodes. The Hamming distance between the neighbor nodes is equal to one.

Step 2. Embed 2 inner nodes of the binary tree into edges connecting the existing nodes of the $\mathcal{N}$-hypercube; note that two of the edge-embedded nodes must be considered at a time (Figure 34.13, the left figure corresponds to the order x_2, x_1, and the right figure describes the order x_1, x_2; the axes are associated with the polarity of variables (complemented, uncomplemented) and explain the meaning of the bold edges).

Step 3. Embed the root of the tree into the center of the facet of the $\mathcal{N}$-hypercube and connect it to the edge-embedded nodes.

Example 34.11 *Let $n = 2$, the four leaf nodes of the complete binary decision tree of a two-variable function can be embedded into a $\mathcal{N}$-hypercube with one root and four intermediate nodes and four leaves (Figure 34.13, second hypercube). To implement a binary decision tree, the root and two intermediate nodes are used, so that there are $n! = 2! = 2$ possible embeddings (last two hypercubes in Figure 34.13).*

Example 34.12 *Let $n = 3$, then 8 leaves of the complete binary decision tree of the three-variable function are embedded into the 3D $\mathcal{N}$-hypercube (Figure 34.14).*

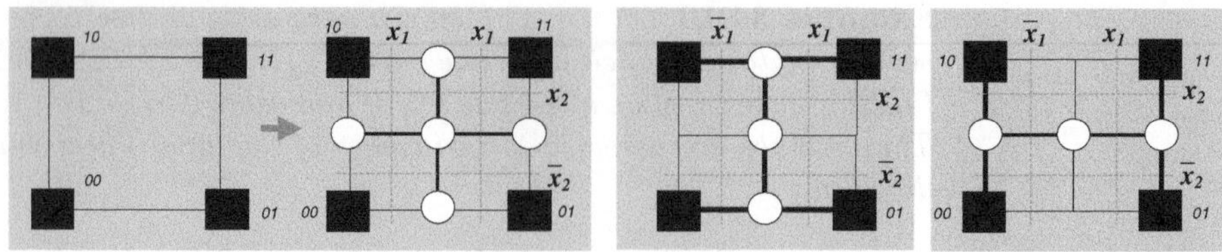

FIGURE 34.13
Embedding a binary decision tree into a 2D $\mathcal{N}$-hypercube (Example 34.11).

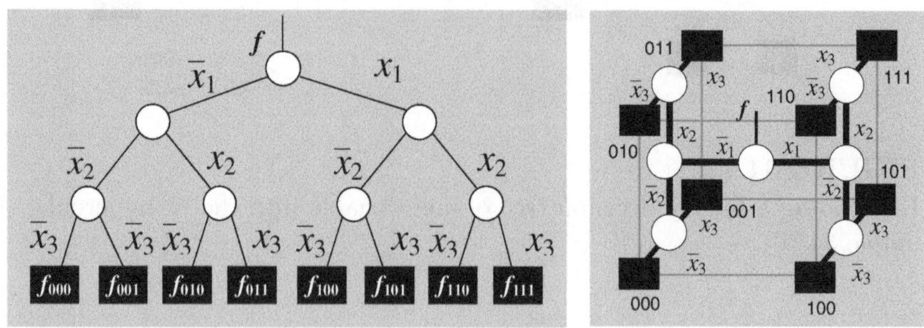

FIGURE 34.14
Embedding the complete binary tree into the 3D $\mathcal{N}$-hypercube (Example 34.12).

> *The total number of nodes and the total number of edges (connections) between nodes in the $\mathcal{N}$-hypercube are*
>
> $$N_d = \sum_{i=0}^{n} 2^{n-i} C_i^n \qquad (34.5)$$
>
> $$N_c = \sum_{i=0}^{n} 2i \cdot 2^{n-i} C_i^n. \qquad (34.6)$$

Assembling $\mathcal{N}$-hypercubes Assembling is one possible approach to $\mathcal{N}$-hypercube design. There are two assembly procedures for the $\mathcal{N}$-hypercube design:

- Assembling an $\mathcal{N}$-hypercube from $\mathcal{N}$-hypercubes of smaller dimensions; this is a recursive procedure based on several restrictions (rules), and
- Assembling a shared set of $\mathcal{N}$-hypercube based structures; in this approach, some extensions of the above-mentioned rules are used.

The following rules are the basis of the assembly procedure.

Rule 1 (Connections of leaves). A terminal node is connected to one intermediate node only. In Figure 34.15, 32 paired terminal nodes are connected to 16 intermediate nodes.

Rule 2 (Connections). Each intermediate node is connected to a node in the upper and to the lower dimension node, and the root node is connected to two intermediate nodes located symmetrically in opposite faces. Figure 34.15 explains this for a 5-dimensional structure.

Rule 3 (Symmetry). Configurations of the terminal and intermediate nodes on the opposite faces are symmetric. In the assembly, Figure 34.15, two pairs of 3D $\mathcal{N}$-hypercubes are connected via their root nodes, forming two new root nodes, and then two pairs are symmetrically connected via a new root nodes, etc.

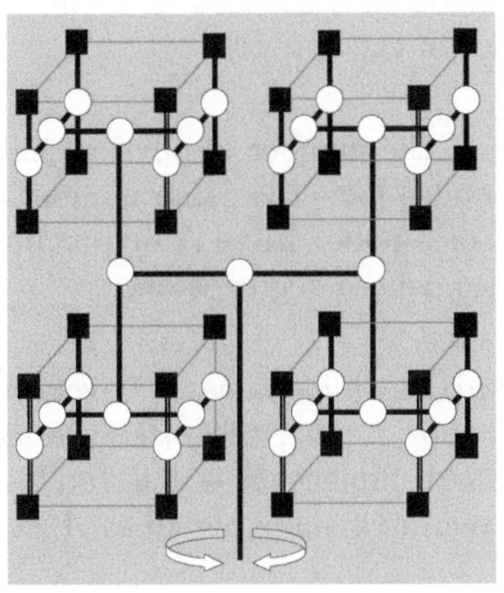

<u>Rule 1</u> *A terminal node is connected to one intermediate node only.*

<u>Rule 2</u> *The root node is connected to two intermediate nodes located symmetrically in opposite faces.*

<u>Rule 3</u> *Configurations of the terminal and intermediate nodes on the opposite face are symmetric. Two symmetrical planes include terminal nodes only.*

FIGURE 34.15
Assembly rules for $\mathcal{N}$-hypercube design.

If the values of some terminal nodes in two $\mathcal{N}$-hypercubes assigned with the same codes are equal, then these $\mathcal{N}$-hypercubes can share some nodes. The $\mathcal{N}$-hypercubes are called *shared* $\mathcal{N}$-hypercubes

Fibonacci cubes. Reasons to study Fibonacci cubes can be found in the following features:

1. The Boolean n-cube is involved into the set of generalized Fibonacci cubes.

2. The order of a generalized Fibonacci cube that can be embedded into a Boolean n-cube with $k = 1, 2$ faulty nodes is greater than 2^{n-1}.

3. A k dimensional Fibonacci cube of the order $n + k$ is equivalent to a Boolean n-cube of order $0 \leq n < k$.

4. Algorithms developed for a generalized Fibonacci cube are executable on the Boolean cube of the corresponding order.

34.7 Spatial topological measurements

There exist a number of basic measures for describing the $\mathcal{N}$-hypercube based structures (Table 34.5).

Diameter and link complexity. The diameter of a network is defined as the maximum distance between any two nodes in the network.

Link complexity or node degree is defined as the number of physical links per node. For a regular network, where all nodes have the same number of links, the node degree of the network is that of a node. In an $\mathcal{N}$-hypercube, the node degree is 3, except for terminal nodes, whose degree is one.

Distance between sub-hypercubes (or sub$\mathcal{N}$ − *hypercubes*). Given two sub$\mathcal{N}$-hypercubes $A = a_1a_2\ldots, a_n$ and $B = b_1b_2\ldots, b_n$, the *intercube distance* $D_i(A, B)$ between A and B along the i-th dimension is 1 if $\{a_i, b_i = \{0, 1\}\}$; otherwise, it is 0. The distance between two subhypercubes A, B is given by $D(A, B) = \sum_{i=1}^{n} D_i(A, B)$.

Embedding. Given graphs G and H, an embedding of a graph G into a graph H is a one-to-one mapping $\alpha : V(G) \rightarrow V(H)$, along with a mapping of β an edge $(u, v) \in E(G)$ to a path between $\alpha(u)$ and $\alpha(v)$ in H.

Example 34.13 *Embedding multiterminal decision diagrams of different shape into $\mathcal{N}$-hypercubes is shown in Figure 34.16:*

(a) 2-level $S_2 - 2S_4$ tree into 2D $\mathcal{N}$-hypercube,
(b) 3-level $S_2 - 2S_2 - 4S_4$ tree into 3D $\mathcal{N}$-hypercube,
(c) 2-level $S_2 - (S_2 + S_4)$ tree into 2D $\mathcal{N}$-hypercube, and
(d) 2-level $S_4 - 4S_2$ tree into 2D $\mathcal{N}$-hypercube.

Dilation. The *dilation* of an edge $dil \in E(G)$ is the length of the path $\beta(dil)$ in H. The dilation of an embedding is the maximum dilation with respect to all edges in G.

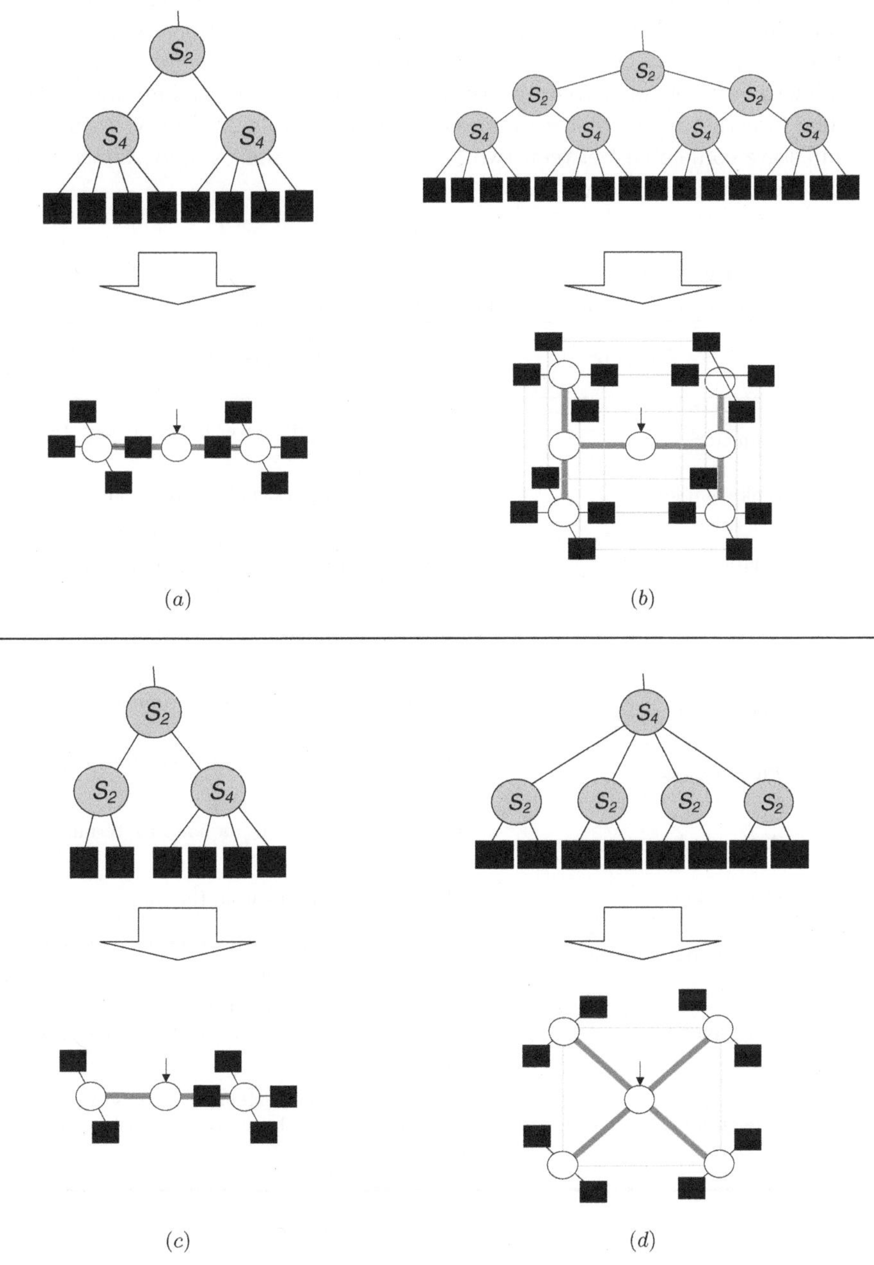

FIGURE 34.16
Embedding multiterminal decision diagrams into $\mathcal{N}$-hypercubes (Example 34.13).

Bisection width. The bisection width of a network is defined as the minimum number of links that have to be removed to partition the network into two equal halves. The bisection width indicates the volume of communication allowed between any two halves of the network with an equal number of nodes. The bisection width of a n-dimensional hypercube is $2^{n-1} = N/2$ since many links are connected between two $(n-1)$-dimensional $\mathcal{N}$-hypercubes to form an n-dimensional $\mathcal{N}$-hypercube.

TABLE 34.5
Metrics on $\mathcal{N}$-hypercube based structures.

Metric	Characteristic
Diameter	The maximum distance between any two nodes in the network
Link complexity	The number of physical links per node
Dilation of an edge	The length of the path $\alpha(e)$ in H. The dilation of an embedding is the maximum dilation with respect to all edges in G
Average message distance	The average number of links that a carrier should travel between any two nodes
Total number of primitives	The number of $\mathcal{N}$-hypercube primitives in the network
Effectiveness	The average number of variables that represent an $\mathcal{N}$-hypercube
Active nodes	The nodes connected to nonzero terminals through a path
Connectivity	The number of paths from the root
Average path length	The number of links connecting the root node to a nonzero terminal
Bisection width	The minimum number of links to be removed in order to partition the network into two equal halves
The granularity of size scaling	The ability of the system to increase in size with minor or no change to the existing configuration

Granularity of size scaling. The granularity of size scaling is the ability of the system to increase in size with minor or no change to the existing configuration, and with an expected increase in performance proportional to the extent of the increase in size. The size of a hypercube can only be increased by doubling the number of nodes; that is, the granularity of size scaling in an n-dimensional hypercube is 2^n.

Average message distance. The average distance in a network is defined as the average number of links that a carrier (e.g., an electron) should travel between any two nodes. Let N_i represent the number of nodes at a distance i, then the average distance $\overline{L}$ is defined as

$$\overline{L} = \frac{1}{N-1}\sum_{i=1}^{n} iN_i$$

where N is the total number of nodes and n is the degree.

34.8 Embedding decision diagrams into incomplete $\mathcal{N}$-hypercubes

The problem with the $\mathcal{N}$-hypercube topology is that the number of terminal nodes in a system must be a power of 2. Incomplete $\mathcal{N}$-hypercubes can be constructed with any number of terminal nodes.

34.8.1 Incomplete $\mathcal{N}$-hypercubes

Embedding technique uses the properties of *host* and *guest* graphs. The host graph describes the interaction between the computing nodes; this is a decision tree or diagram. The topology of the computer system is captured by the host graph. The embedding function maps each node and edge in the guest graph (decision tree or diagram) into a unique node and edge, respectively, in the host graph (computer system configuration).

Example 34.14 *Incomplete $\mathcal{N}$-hypercubes are constructed by deleting given nodes (Figure 34.17).*

34.8.2 Embedding technique

Embedding decision diagrams into incomplete $\mathcal{N}$-hypercubes consists of the following steps:

- Representation of original decision diagram in the form of multiterminal decision diagrams,
- Embedding multiterminal decision diagrams into $\mathcal{N}$-hypercubes, and

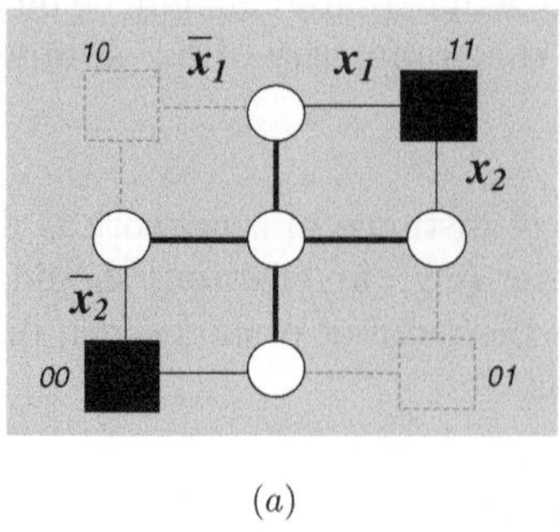

(a)

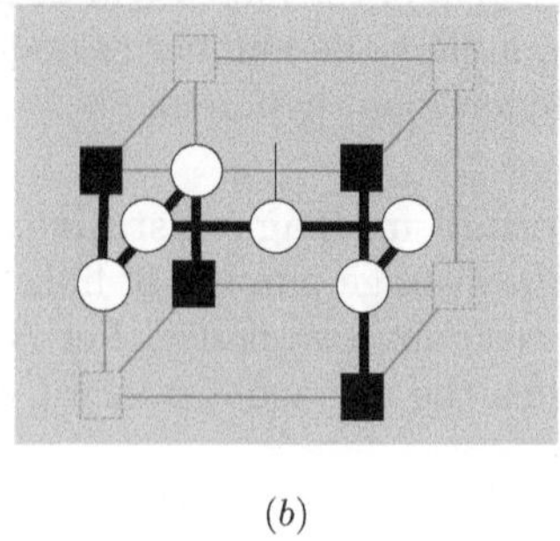

(b)

FIGURE 34.17
Incomplete 2D (a) and 3D (b) $\mathcal{N}$-hypercubes (Example 34.14).

▶ Deleting nonused intermediate and terminal nodes in an $\mathcal{N}$-hypercube.

Example 34.15 *Figure 34.18 shows the technology of embedding of decision diagrams with different topologies into incomplete 3D $\mathcal{N}$-hypercubes.*

(a) *In Figure 34.18a, a decision tree is transformed into a decision diagram, and then to a linear decision diagram that is embedded into an incomplete 3D $\mathcal{N}$-hypercube (see detail in Chapter 37).*

(b) *In Figure 34.18b,c a decision tree is transformed into a decision diagram, then it is restructured to a multiterminal diagram, and embedded into an incomplete 3D $\mathcal{N}$-hypercube.*

34.9 Further reading

3D macro-scale topology. 3D topology has been explored for the design of macro-scale distributed systems [3], [15], [16], [27].

The 3D computing architecture concept has been employed by the creators of the supercomputers Cray T3D, NEC's Ncube, and cosmic cube [34]. The components of these supercomputers, as systems, are designed based on classical paradigms of 2D architecture that becomes 3D because of the *3D topology* (of interconnects), or *3D data structures* and corresponding algorithms, or *3D communication flow.* An example is Q3M (Quantum 3D Mesh) model of a parallel quantum computer in which each cell has a finite number of quantum

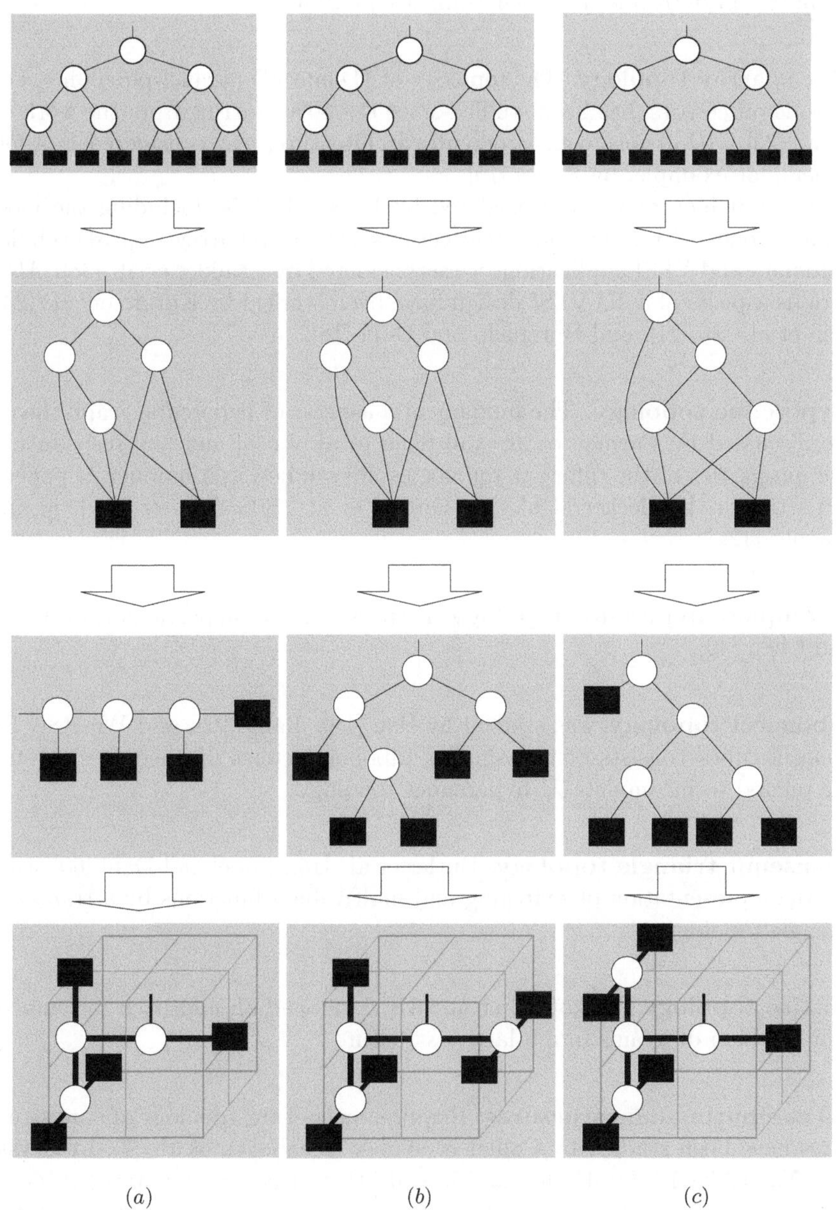

FIGURE 34.18
Some examples of embedding binary reduced decision trees into incomplete $\mathcal{N}$-hypercubes (Example 34.15).

bits (qubits) and interacts with neighbors, exchanging particles by means of Hamiltonian derivable from Schrödinger equation [28].

Linear array topology. The topology of 1D and 2D parallel-pipelining, i.e., systolic processors, has been studied since the 80's, starting from the work by Kung [23]. *3D arrays,* e.g., hypercube architecture, was considered in a few papers, for example, by Hayes [15].

3D VLSI hierarchy was considered by Lai et.al. [25], including the most complex ones with multi-unit architectures such as 3D arrays [8]. 3D cellular automata and VLSI applications were considered by Tsalides et al. [38]. Also, various aspects of a 3D VLSI design have been studied by Kurino et. al. [24], Jain et al. [20, 21], and Horiguchi and Ooki [18].

Hypercube topology. The fundamental aspects of hypercube graph theory are discussed by Cormen et al. and Saad et al. in [5] and [33] respectively. The properties of the different topologies are studied in a number of papers, in particular, by Becker et al. [1], Ohring et al. [30], Shen et al. [35], and Wagner [42].

Incomplete hypercube topology has been studied, in particular, by Tzeng and Chen [39].

Fibonacci topology was studied by Hsu [19], Jiang [22], and Wu [18]. Fibonacci cubes consists of two smaller Fibonacci cubes of unequal sizes and are related to incomplete $\mathcal{N}$-hypercubes topology.

Transeunt triangle topology. Butler et al. [15], Popel and Dani [60] studied the representation of switching and multivalued functions by a transeunt triangle topology.

Lattice topology. Perkowski et al. [31] developed an approach for embedding decision diagrams into a lattice structure.

3D computing nanostuctures. Requirements of the topology of nanostructures have been studied by Collier et al. [4], Ellenbogen et al. [7], Lyshevski [28], Yanushkevich et al. [46], and in collection of papers [6]. Margolis et al. [29], Smith [37] and Frank [10, 11] advocated for 3D memory array as the way to accommodate the requirements for speed of a nanometer size computing structure to be integrated on a single tiny chip. In addition, there are many 3D algorithms and designs for existing microscale components that are arranged in 3D space, which computer designers already have experience with (see, for example, works by Greenberg et al.[14] and Leiserson et al.[26].

A hyperpyramid is a level structure of k hypercubes. Ho and Johnsson [17] developed an algorithm for embedding hyperpyramids into hypercubes.

$\mathcal{N}$-hypercube topology. Yanushkevich et al. developed the approach for embedding decision diagrams into $\mathcal{N}$-hypercubes [45, 69].

Fault tolerance topology. In the probability fault model, the reliability of each node at time t is a random variable. The probability that a subhypercube is operational is represented by the reliability of data processing in the subhypercubes. The hypercube reliability can be formulated as the union of probabilistic events that all the possible hypercubes are operational.

References

[1] Becker B and Simon HU. How Robust is the n-cube? *Information and Computation,* 77:162–178, 1988.

[2] Butler JT, Dueck GW, Yansushkevich SN, and Shmerko VP. On the number of generators for transeunt triangles. *Discrete Applied Mathematics*, 108:309–316, 2001.

[3] Campbell ML, Toporg ST, and Taylor SL. 3D wafer stack neurocomputing. In *Proceedings of the IEEE International Conference on Wafer Scale Integration,* pp. 67–74, 1993.

[4] Collier CP, Wong EW, Belohradsk M, Raymo FM, Stoddart JF, Kuekes PJ, Williams RS, and Heath JR. Electronically configurable molecular-based logic gates. *Science*, 285:391–394, July 1999.

[5] Cormen TH, Leiserson CE, Riverst RL, and Stein C. *Introduction to Algorithms.* MIT Press, 2001.

[6] Crawley D, Nikolić K, and Forshaw M., Eds., *3D Nanoelectronic Computer Architecture and Implementation.* Institute of Physics Publishing, UK, 2005.

[7] Ellenbogen JC and Love JC. Architectures for molecular electronic computers: logic structures and an adder designed from molecular electronic diodes. In *Proceedings of the IEEE*, 88:386–426, 2000.

[8] Endoh T, Sakuraba H, Shinmei K, and Masuoka F. New three dimensional (3D) memory array architecture for future ultra high density DRAM. In *Proceedings of the 22nd International Conference on Microelectronics*, 442:447–450, 2000.

[9] Frank MP and Knight TF Jr. Ultimate theoretical models of nanocomputers. *Nanotechnology*, 9:162–176, 1998.

[10] Frank MP. Physical Limits of Computing. *Computing in Science and Engineering*, 4(3):16–25, 2002.

[11] Frank MP. Approaching the physical limits of computing. In *Proceedings of the IEEE 35th International Symposium on Multiple-Valued Logic*, pp. 168–185, 2005.

[12] Gerousis C, Goodnick SM, and Porod W. Toward nanoelectronic cellular neural networks. *International Journal of Circuits Theory and Applications* 28(6):523-535, 2000.

[13] Goser K, Pacha C, Kanstein A, and Rossmann ML. Aspects of systems and circuits for nanoelectronics. *Proceedings of the IEEE*, 85:558–573, April, 1997.

[14] Greenberg RI. The fat-pyramid and universal parallel computation independent of wire delay. *IEEE Transactions on Computers*, 43(12):1358–1364, 1994.

[15] Hayes JP, Mudge T, Stout QF, Colley S, and Palmer J. A microprocessor-based hypercube supercomputer. *IEEE Micro*, 6(5):6–17, 1986.

[16] Hillis WD. *The connection machine.* MIT Press, 1985.

[17] Ho C-T and Johnsson SL. Embedding hyperpyramids into hypercubes. *IRM Journal Res. Develop.*, 38(1):31–45 1994.

[18] Horiguchi S and Ooki T. Hierarchical 3D-torus interconnection network. In *Proceedings of the International Simposium on Parallel Architectures, Algorithms, and Networks*, Dallas, TX, pp. 50–56, 2000.

[19] Hsu WJ. Fibonacci cubes – a new interconnection topology. *IEEE Transactions on Parallel and Distributed Systems*, 4(1):3–12, 1993.

[20] Jain VK, Ghirmai T, and Horiguchi S. TESH: a new hierarchical interconnection network for massively parallel computing. *IEICE Transactions on Information and Systems*, E80D(9):837–846,1997.

[21] Jain VK and Horiguchi S. VLSI considerations for TESH: a new hierarchical interconnection network for 3D integration. *IEICE Transactions on VLSI Systems*, 6(3):346–353,1998.

[22] Jiang FS. Embedding of generalized Fibonacci cubes in hybercubes with faulty nodes. *IEEE Transactions on Parallel and Distributed Systems*, 8(7):727–737, 1997.

[23] Kung HT and Leiserson CE. Systolic arrays for VLSI. In Mead C and Conway L, Eds., *Introduction to VLSI Systems*, Addison-Wesley, pp. 260–292, 1980.

[24] Kurino H, Matsumoto T, Yu KH, Miyakawa N, Itani H, Tsukamoto H, and Koyanagi M. Three-dimensional integration technology for real time micro-vision systems. In *Proceedings of the IEEE International Conference on Innovative Systems in Silicon*, pp. 203–212, 1997.

[25] Lai YT and Wang PT. Hierarchical interconnection structures for field programmable gate arrays. *IEEE Transactions on VLSI Systems*, 5(2):186–196, 1997.

[26] Leiserson CH. Fat-trees: universal networks for hardware-efficient supercomputing. *IEEE Transactions on Computers*, 34(10):892–901, 1985.

[27] Little MJ, Grinberg J, Laub SP, Nash JG, and Jung MW. The 3D computer. In *Proceedings of the IEEE International Conference on Wafer Scale Integration*, pp. 55–64, 1989.

[28] Lyshevski SE. 3D multi-valued design in nanoscale integrated circuits. In *Proceedings of the 35th IEEE International Symposium on Multiple-Valued Logic*, pp. 82–87, 2005.

[29] Margolus N and Levitin L. The maximum speed of dynamical evolution. *Physica D*, 120:188–195, 1998.

[30] Ohring S and Das SK. Incomplete hypercubes: embeddings of tree-related networks. *Journal of Parallel and Distributed Computing*, 26:36–47, 1995.

[31] Perkowski MA, Chrzanowska-Jeske M, and Xu Y. Lattice diagrams using Reed–Muller logic. In *Proceedings of the IFIP WG 10.5 International Workshop on Applications of the Reed–Muller Expansion in Circuit Design.* Japan, pp. 85–102, 1997.

[32] Popel DV and Dani A. Sierpinski gaskets for logic function representation. In *Proceedings of the IEEE 32nd International Symposium on Multiple-Valued Logic*, pp. 39–45, 2002.

[33] Saad Y and Schultz MH. Topological Properties of Hypercubes. *IEEE Transactions on Computers*, 37(7):867–872, 1988.

[34] Seitz CL. The cosmic cube. *Communications of the ACM*, 28(1):22–33, 1985.

[35] Shen X, Hu Q, and Liang W. Embedding k-ary complete trees into hypercubes. *Journal of Parallel and Distributed Computing*. 24:100–106, 1995.

[36] Shmerko VP and Yanushkevich SN. Three-dimensional feedforward neural networks and their realization by nano-devices. *Artificial Intelligence Review International Journal*, Special Issue on Artificial Intelligence in Logic Design, pp. 473–494, 1994.

[37] Smith W. Fundamental physical limits on computation. *Technical report, NECI*, May 1995. http://www.neci.nj.nec.com/- homepages/wds/fundphys.ps., 1995

[38] Tsalides Ph, Hicks PJ, and York TA. Three-dimensional cellular automata and VLSI applications. *IEE Proceedings*, Pt.E, 136(6):490–495, 1989.

[39] Tzeng N-F and Chen H-L. Structural and tree embedding aspects of incomplete hypercubes. *IEEE Transactions on Computers*, 43(312):1434–1439, 1994.

[40] Vichniac G. Simulating physics with cellular automata. *Physica D*, 10:96–115, 1984.

[41] Vitanyi PMB. Multiprocessor architectures and physical law. *Proceedings of the 2nd IEEE Workshop on Physics and Computation*, Dallas, TX, pp. 24–29, 1994.

[42] Wagner AS. Embedding the complete tree in hypercube. *Jouranal of Parallel and Distributed Computing*, 26:241–247, 1994.

[43] Wu AY. Embedding tree networks into hypercubes. *Journal of Parallel and Distributed Computing*, 2:238–249, 1985.

[44] Wu J. Extended Fibonacci cubes. *IEEE Transactions on Parallel and Distributed Systems*, 8(12):1203–1210, 1997.

[45] Yanushkevich SN, Shmerko VP, and Lyshevski SE. *Logic Design of NanoICs*, CRC Press, Boca Raton, FL, 2004.

[46] Yanushkevich SN, Shmerko VP, Guy L, and Lu DC. Three dimensional multiple-valued circuit design based on single-electron logic. In *Proceedings of the 34th IEEE International Symposium on Multiple-Valued Logic*, pp. 275–280, 2004.

35

Decision Diagrams in Reversible Logic

In this chapter, we consider decision diagram techniques for reversible logic and for quantum circuits that are implementations of reversible logic functions. Representation, synthesis, and simulation aspects are considered.

35.1 Introduction

An m-input, m-output switching function is *reversible* if it maps each input pattern to a unique output pattern. A reversible function can be realized by a cascade of reversible gates in which there is no fan-out or feedback.

The notion of reversibility arises from the fact that, since each input pattern is associated with a unique output pattern, a reversible function always has an inverse. Frequently, the function is its own inverse, particularly for primitive reversible gates. If one replaces each gate in the cascade with the gate realizing the inverse of the function realized by the original gate (it may be itself), then the cascade applied in reverse order realizes the inverse of the function realized by the original cascade.

The interest in reversible logic started with Landauer's seminal paper [14] in which he proved that using traditional irreversible logic gates (e.g. AND, OR, NAND, NOR) necessarily leads to power dissipation regardless of the underlying technology. Bennett [4] subsequently showed that for power not to be dissipated in an arbitrary circuit, it must be built from reversible gates, i.e., gates that implement reversible functions. Hence there are compelling reasons to consider circuits composed of reversible gates. Reversible circuits are of particular interest in low-power CMOS design, optical computing, quantum computing, and nanotechnology.

The interest here is in how decision diagram techniques can be used for the representation, synthesis and simulation of reversible functions and circuits. Particular attention is paid to the case of quantum circuits. This is of considerable interest since, while conventional reversible gates and circuits are two-valued, quantum circuits operate in a significantly different manner that must be modelled using complex numbers. Applying decision diagram techniques to the quantum case thus requires different techniques from those used in conventional logic design.

35.2 Reversible and quantum circuits

In this section, we provide a brief overview of the concepts of reversible and quantum circuits. Readers seeking more background are directed to [19].

35.2.1 Reversible circuits

The formal definition of a reversible switching function is as follows:

Definition 35.1 *An m-input, m-output, totally-specified switching function* $f(x_1, x_1, ..., x_m)$ *is reversible if it maps each input assignment to a unique output assignment. Note that for* $B = \{0, 1\}$*:*

$$f : B^m \rightarrow B^m,$$
$$f^{-1} : B^m \rightarrow B^m.$$

A reversible function:

- ▶ Can be written as a standard truth table, as shown in Example 35.1, and
- ▶ Can also be viewed as a bijective mapping of the set of integers $\{0, 1, ..., 2^m - 1\}$ onto itself.

A reversible function is thus a permutation of its domain, which can be expressed as an ordered list of the integers 0 to $2^m - 1$.

Example 35.1 *A 3-input, 3-output reversible function that specifies cyclic increment (Table 35.1). The inverse function is cyclic decrement as can be seen by reading the truth table from right to left, rather than the normal left to right. This function realizes the permutation* $\{1, 2, 3, 4, 5, 6, 7, 0\}$.

Definition 35.2 *An n-input, n-output gate is reversible if it realizes a reversible function.*

The most commonly used set of reversible gates is the Toffoli family defined as follows.

TABLE 35.1
A 3-input, 3-output reversible function (Example 35.1).

Inputs			Outputs		
x_1	x_2	x_3	x'_1	x'_2	x'_3
0	0	0	0	0	1
0	0	1	0	1	0
0	1	0	0	1	1
0	1	1	1	0	0
1	0	0	1	0	1
1	0	1	1	1	0
1	1	0	1	1	1
1	1	1	0	0	0

Definition 35.3 *An $n \times n$ Toffoli gate:*

- *Passes the first $n-1$ lines (controls) through unchanged, and*
- *Inverts the nth line (target) if the control lines are all 1, otherwise the target line is passed through unchanged.*

Toffoli gates are clearly self-inverse.

An $n \times n$ Toffoli gate will be denoted $TOFn(x_1, x_2, ..., x_n)$ where x_n is the target line. Using a prime symbol to denote the value of a line after passing through the gate we have

$$x'_i = x_i, \quad i < n, \tag{35.1}$$

$$x'_n = x_1 x_2 ... x_{n-1} \oplus x_n \tag{35.2}$$

$TOF1(x_1)$ is the special case where there are no control inputs, so x_1 is always inverted, i.e., it is a NOT gate. $TOF2(x_1, x_2)$ has been termed a Feynman or controlled-NOT (CNOT) gate. $TOF3(x_1, x_2, x_3)$ is often referred to simply as a Toffoli gate, and sometimes as a controlled-controlled-NOT (CCNOT) gate. Toffoli gates are often drawn as shown in Figure 35.1.

A second family of interest is the set of *Fredkin gates.* Each such gate interchanges two target lines if all the control lines have the value 1. Such a gate with no control lines is a swap gate. Fredkin gates are also self-inverse. We do not elaborate further on Fredkin gates, or other possible reversible gates, since the decision diagram techniques illustrated for Toffoli gates extend to other reversible gates in a straightforward manner.

Example 35.2 *A reversible function can be realized by a circuit that is a cascade of reversible gates. The circuit in Figure 35.2 realizes the function in Example 35.1.*

While certain functions are inherently reversible, code converters and counters for example, the majority of practical functions are not. An irreversible

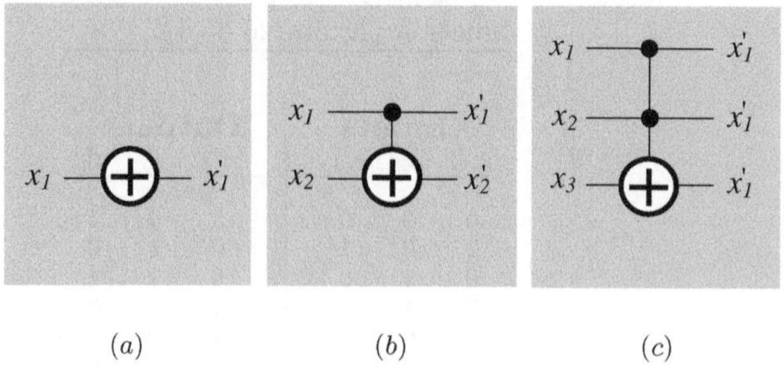

FIGURE 35.1
Toffoli gates: $TOF1(x_1)$ (a), $TOF2(x_1, x_2)$ (b), and $TOF3(x_1, x_2, x_3)$ (c).

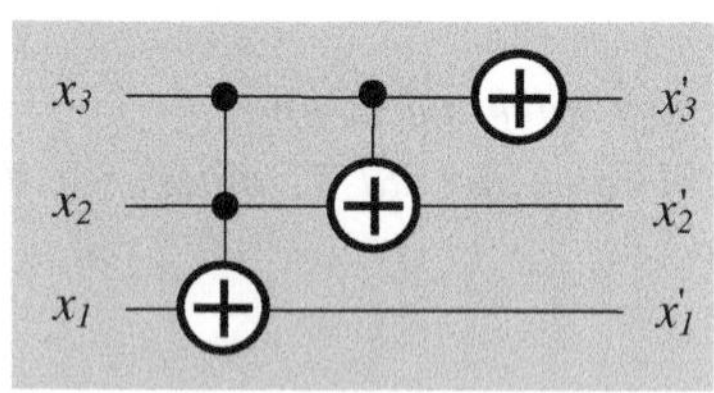

- ▶ *From left to right, this circuit realizes the cyclic increment function.*
- ▶ *Since each Toffoli gate is self-inverse, when applied from right to left the circuit realizes cyclic decrement.*

FIGURE 35.2
Toffoli gate circuit for cyclic increment reversible function (Example 35.1).

function can be embedded in a larger reversible specification. Finding an optimal embedding is a difficult and as yet unsolved problem.

> **Example 35.3** *The specification of a 3-bit full adder given in Table 35.2 and its embedding into a reversible specification given in Table 35.3 are shown below. Input d is a constant input that is set to 0 to realize the adder function. Output p is a garbage output, which in this case is the adder propagate signal $a \oplus b$. g is a garbage output which in this case is the input a. The circuit in Figure 35.3 was found from specification given in Table 35.3 using the method described in [16].*

As seen in Example 35.3, garbage outputs and constant inputs must generally be added in transforming an irreversible specification to a reversible one. The minimum number of garbage outputs required is $\lceil \log_2 q \rceil$, where q is the maximum output pattern multiplicity of the original irreversible function, i.e., the maximum number of times a single output pattern is repeated in the specification. Constant inputs must be added so that the reversible specification has the same number of inputs and outputs.

TABLE 35.2
Irreversible specification.

c	b	a	$carry$	sum
0	0	0	0	0
0	0	1	0	1
0	1	0	0	1
0	1	1	1	0
1	0	0	0	1
1	0	1	1	0
1	1	0	1	0
1	1	1	1	1

TABLE 35.3
Reversible specification.

d	c	b	a	$carry$	sum	p	g
0	0	0	0	0	0	0	0
0	0	0	1	0	1	1	1
0	0	1	0	0	1	1	0
0	0	1	1	1	0	0	1
0	1	0	0	0	1	0	0
0	1	0	1	1	0	1	1
0	1	1	0	1	0	1	0
0	1	1	1	1	1	0	1
1	0	0	0	1	0	0	0
1	0	0	1	1	1	1	1
1	0	1	0	1	1	1	0
1	0	1	1	0	0	0	1
1	1	0	0	1	1	0	0
1	1	0	1	0	0	1	1
1	1	1	0	0	0	1	0
1	1	1	1	0	1	0	1

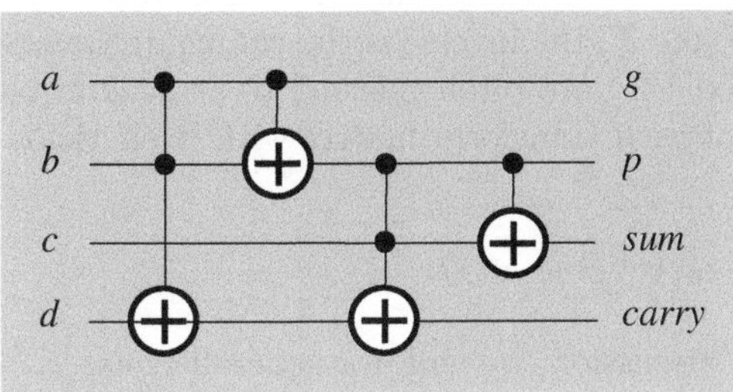

- $carry = ab + ac + bc$
- $sum = a \oplus b \oplus c$
- Constant input d is set to 0 to realize the adder.
- g and p are garbage outputs that in this case realize useful functions. That is not generally the case.

FIGURE 35.3
Reversible circuit for a 3-bit full adder using specification (b) in Example 35.3.

35.2.2 Quantum circuits

The reversible circuits discussed thus far operate on values 0 and 1 throughout the cascade. Quantum technology gates exhibit quite different behaviour that must be modelled in the complex domain. We consider quantum circuits realizing reversible functions for which the primary inputs and outputs are 0 or 1. Such a circuit is again a simple cascade of gates with no feedback or fan-out.

For quantum circuits, the notion of a bit value in a switching circuit is replaced by a quantum bit (qubit). Since the circuit is a cascade, there are consistently m lines throughout the circuit. The state of the m qubits at any point in the circuit is described as an element of the tensor (Kronecker) product of the single state spaces and is expressed as a normalized state vector of length 2^m. As will be shown below, the operation of each gate can be expressed as a $2^m \times 2^m$ complex matrix, and the transformation performed by a sequence of gates can be expressed as the product of the corresponding matrices.

A single qubit has two observable values depicted as $|0\rangle$ and $|1\rangle$, respectively. Due to properties of the physical phenomena representing qubits, the observation of $|0\rangle$ or $|1\rangle$ upon measurement of a qubit is probabilistic. In particular, the state of a single qubit is a linear combination

$$\alpha|0\rangle + \beta|1\rangle,$$

where α and β are complex numbers called the amplitudes such that

$$|\alpha|^2 + |\beta|^2 = 1.$$

The real numbers $|\alpha|^2$ and $|\beta|^2$ represent the probabilities of reading the states $|0\rangle$ and $|1\rangle$ upon measurement.

Table 35.4 shows the matrix transform definitions for a number of basic single qubit gates and their inverses. For those that are self-inverse, the matrices are Hermitian. For, V, S and T, the inverse is the conjugate transpose. Gates V and V^{-1} (also denoted V^+) are often referred to as *square root of NOT* gates since the product of each transform matrix with itself yields the matrix defining NOT, i.e.,

$$V^2 = (V^+)^2 = NOT$$

Also, $S = T^2$. In the following discussion, we will, for generality, use

$$U = \begin{bmatrix} \alpha & \beta \\ \gamma & \delta \end{bmatrix}$$

to denote an arbitrary single qubit transform.

Example 35.4 *Assume a 3-qubit circuit. Applying transform U to the least significant qubit requires it be expressed as an 8×8 matrix given by*

$$I3 \otimes I3 \otimes U,$$

where $I3$ is the identity matrix of dimension 2^3 and $\otimes$ denotes the Kronecker product. The resulting matrix is shown in Figure 35.4(a). If U is applied to the most significant bit, the matrix is

$$U \otimes I3 \otimes I3$$

The result is shown in Figure 35.4(b).

TABLE 35.4
Single qubit quantum gate transforms.

Name	Symbol	Transform	Inverse
NOT	N	$\begin{bmatrix} 0 & 1 \\ 1 & 0 \end{bmatrix}$	self-inverse
V	V	$\frac{1}{2}\begin{bmatrix} 1+i & 1-i \\ 1-i & 1+i \end{bmatrix}$	$\frac{1}{2}\begin{bmatrix} 1-i & 1+i \\ 1+i & 1-i \end{bmatrix}$
Hadamard	H	$\frac{1}{\sqrt{2}}\begin{bmatrix} 1 & 1 \\ 1 & -1 \end{bmatrix}$	self-inverse
Pauli-X	X	$\begin{bmatrix} 0 & 1 \\ 1 & 0 \end{bmatrix}$	self-inverse
Pauli-Y	Y	$\begin{bmatrix} 0 & -i \\ i & 0 \end{bmatrix}$	self-inverse
Pauli-Z	Z	$\begin{bmatrix} 1 & 0 \\ 0 & -1 \end{bmatrix}$	self-inverse
Phase	S	$\begin{bmatrix} 1 & 0 \\ 0 & i \end{bmatrix}$	$\begin{bmatrix} 1 & 0 \\ 0 & -i \end{bmatrix}$
$\pi/8$	T	$\begin{bmatrix} 1 & 0 \\ 0 & \frac{1+i}{\sqrt{2}} \end{bmatrix}$	$\begin{bmatrix} 1 & 0 \\ 0 & \frac{1-i}{\sqrt{2}} \end{bmatrix}$
X rotation	RX	$\begin{bmatrix} \cos\frac{\Theta}{2} & -i\sin\frac{\Theta}{2} \\ -i\sin\frac{\Theta}{2} & \cos\frac{\Theta}{2} \end{bmatrix}$	$\begin{bmatrix} \cos\frac{-\Theta}{2} & -i\sin\frac{-\Theta}{2} \\ -i\sin\frac{-\Theta}{2} & \cos\frac{-\Theta}{2} \end{bmatrix}$
Y rotation	RY	$\begin{bmatrix} \cos\frac{\Theta}{2} & -\sin\frac{\Theta}{2} \\ \sin\frac{\Theta}{2} & \cos\frac{\Theta}{2} \end{bmatrix}$	$\begin{bmatrix} \cos\frac{-\Theta}{2} & -\sin\frac{-\Theta}{2} \\ \sin\frac{-\Theta}{2} & \cos\frac{-\Theta}{2} \end{bmatrix}$
Z rotation	RZ	$\begin{bmatrix} e^{-i\theta/2} & 0 \\ 0 & e^{i\theta/2} \end{bmatrix}$	$\begin{bmatrix} e^{i\theta/2} & 0 \\ 0 & e^{-i\theta/2} \end{bmatrix}$

$$\begin{bmatrix} \alpha & \beta & 0 & 0 & 0 & 0 & 0 & 0 \\ \gamma & \delta & 0 & 0 & 0 & 0 & 0 & 0 \\ 0 & 0 & \alpha & \beta & 0 & 0 & 0 & 0 \\ 0 & 0 & \gamma & \delta & 0 & 0 & 0 & 0 \\ 0 & 0 & 0 & 0 & \alpha & \beta & 0 & 0 \\ 0 & 0 & 0 & 0 & \gamma & \delta & 0 & 0 \\ 0 & 0 & 0 & 0 & 0 & 0 & \alpha & \beta \\ 0 & 0 & 0 & 0 & 0 & 0 & \gamma & \delta \end{bmatrix} \qquad \begin{bmatrix} \alpha & 0 & 0 & 0 & \beta & 0 & 0 & 0 \\ 0 & \alpha & 0 & 0 & 0 & \beta & 0 & 0 \\ 0 & 0 & \alpha & 0 & 0 & 0 & \beta & 0 \\ 0 & 0 & 0 & \alpha & 0 & 0 & 0 & \beta \\ \gamma & 0 & 0 & 0 & \delta & 0 & 0 & 0 \\ 0 & \gamma & 0 & 0 & 0 & \delta & 0 & 0 \\ 0 & 0 & \gamma & 0 & 0 & 0 & \delta & 0 \\ 0 & 0 & 0 & \gamma & 0 & 0 & 0 & \delta \end{bmatrix}$$

(a) (b)

FIGURE 35.4
Two sample U transforms for an $m = 3$ circuit (Example 35.4).

The quantum gates discussed thus far can be extended to have a control input, which determines whether the transform is applied to the target input. We adopt the normal assumption that the control input is restricted to 0 and 1.

Example 35.5 *In a 2 qubit circuit if the transform U is to be applied to the least significant qubit using the other as control, the resulting transform matrix is as shown in Figure 35.5(a). If the roles are reversed the matrix is given by Figure 35.5(b). This extends to higher values of m. For example, for $m = 3$, applying U to the middle qubit using the least significant qubit as control yields the transform matrix in Figure 35.5(c), which is readily verified to affect the middle qubit only when the least significant qubit is 1.*

Examples 35.4 and 35.5 illustrate the construction of matrices for individual transforms. The transformation carried out by a reversible circuit is computed as the product of the transform matrices for the gates in the cascade. This is illustrated in Example 35.6.

$$\begin{bmatrix} 1 & 0 & 0 & 0 \\ 0 & 1 & 0 & 0 \\ 0 & 0 & \alpha & \beta \\ 0 & 0 & \gamma & \delta \end{bmatrix} \qquad \begin{bmatrix} 1 & 0 & 0 & 0 \\ 0 & \alpha & 0 & \beta \\ 0 & 0 & 1 & 0 \\ 0 & \gamma & 0 & \delta \end{bmatrix} \qquad \begin{bmatrix} 1 & 0 & 0 & 0 & 0 & 0 & 0 & 0 \\ 0 & \alpha & 0 & \beta & 0 & 0 & 0 & 0 \\ 0 & 0 & 1 & 0 & 0 & 0 & 0 & 0 \\ 0 & \gamma & 0 & \delta & 0 & 0 & 0 & 0 \\ 0 & 0 & 0 & 0 & 1 & 0 & 0 & 0 \\ 0 & 0 & 0 & 0 & 0 & \alpha & 0 & \beta \\ 0 & 0 & 0 & 0 & 0 & 0 & 1 & 0 \\ 0 & 0 & 0 & 0 & 0 & \gamma & 0 & \delta \end{bmatrix}$$

(*a*) (*b*) (*c*)

FIGURE 35.5
Sample controlled-U transforms for a 3 qubit cicruit (Example 35.5).

Example 35.6 *Consider the cyclic increment function given in Example 35.1. This function can be realized by the circuit in Figure 35.6(a). The individual gate transforms are shown in Table 35.5. The gates are denoted* $name(target)$ *and* $name(control, target)$ *as appropriate. The lines are numbered 1 to 3 with line 1 the least significant. Taking the product of these matrices in order yields the transform matrix in Figure 35.6(b), which is clearly the appropriate permutation matrix for the cyclic increment function, since*

$$P \times (0, 1, 2, 3, 4, 5, 6, 7)^T = (1, 2, 3, 4, 5, 6, 7, 0)^T.$$

The examples above show important properties of the matrices encountered in quantum circuit design and analysis. They are potentially quite large, having dimension $2^n \times 2^n$. However, they have quite regular structures since they are formed from 2×2 base matrices in a very regular way using the Kronecker product. In the next section, it is shown how decision diagrams can be used to represent and manipulate the necessary matrices taking their structure into account in a very effective manner.

The matrix approach to describing quantum gate behaviour is clearly also applicable to reversible switching circuits, the difference being that in that case all matrices are 0-1 matrices and are in fact permutation matrices corresponding to the permutation operation performed by the Boolean reversible gate.

TABLE 35.5
Gate transform matrices for quantum cyclic increment circuit.

Gate	Gate transform matrix matrix	Gate	Gate transform matrix matrix
$V(2,3)$	$\begin{bmatrix} 1 & 0 & 0 & 0 & 0 & 0 & 0 & 0 \\ 0 & 1 & 0 & 0 & 0 & 0 & 0 & 0 \\ 0 & 0 & \frac{1+i}{2} & 0 & 0 & 0 & \frac{1-i}{2} & 0 \\ 0 & 0 & 0 & \frac{1+i}{2} & 0 & 0 & 0 & \frac{1-i}{2} \\ 0 & 0 & 0 & 0 & 1 & 0 & 0 & 0 \\ 0 & 0 & 0 & 0 & 0 & 1 & 0 & 0 \\ 0 & 0 & \frac{1-i}{2} & 0 & 0 & 0 & \frac{1+i}{2} & 0 \\ 0 & 0 & 0 & \frac{1-i}{2} & 0 & 0 & 0 & \frac{1+i}{2} \end{bmatrix}$	$NOT(1)$	$\begin{bmatrix} 0 & 1 & 0 & 0 & 0 & 0 & 0 & 0 \\ 1 & 0 & 0 & 0 & 0 & 0 & 0 & 0 \\ 0 & 0 & 0 & 1 & 0 & 0 & 0 & 0 \\ 0 & 0 & 1 & 0 & 0 & 0 & 0 & 0 \\ 0 & 0 & 0 & 0 & 0 & 1 & 0 & 0 \\ 0 & 0 & 0 & 0 & 1 & 0 & 0 & 0 \\ 0 & 0 & 0 & 0 & 0 & 0 & 0 & 1 \\ 0 & 0 & 0 & 0 & 0 & 0 & 1 & 0 \end{bmatrix}$
$NOT(1,2)$	$\begin{bmatrix} 1 & 0 & 0 & 0 & 0 & 0 & 0 & 0 \\ 0 & 0 & 0 & 1 & 0 & 0 & 0 & 0 \\ 0 & 0 & 1 & 0 & 0 & 0 & 0 & 0 \\ 0 & 1 & 0 & 0 & 0 & 0 & 0 & 0 \\ 0 & 0 & 0 & 0 & 1 & 0 & 0 & 0 \\ 0 & 0 & 0 & 0 & 0 & 0 & 0 & 1 \\ 0 & 0 & 0 & 0 & 0 & 0 & 1 & 0 \\ 0 & 0 & 0 & 0 & 0 & 1 & 0 & 0 \end{bmatrix}$	$V^{+}(2,3)$	$\begin{bmatrix} 1 & 0 & 0 & 0 & 0 & 0 & 0 & 0 \\ 0 & 1 & 0 & 0 & 0 & 0 & 0 & 0 \\ 0 & 0 & \frac{1-i}{2} & 0 & 0 & 0 & \frac{1+i}{2} & 0 \\ 0 & 0 & 0 & \frac{1-i}{2} & 0 & 0 & 0 & \frac{1+i}{2} \\ 0 & 0 & 0 & 0 & 1 & 0 & 0 & 0 \\ 0 & 0 & 0 & 0 & 0 & 1 & 0 & 0 \\ 0 & 0 & \frac{1+i}{2} & 0 & 0 & 0 & \frac{1-i}{2} & 0 \\ 0 & 0 & 0 & \frac{1+i}{2} & 0 & 0 & 0 & \frac{1-i}{2} \end{bmatrix}$
$V(1,3)$	$\begin{bmatrix} 1 & 0 & 0 & 0 & 0 & 0 & 0 & 0 \\ 0 & \frac{1+i}{2} & 0 & 0 & 0 & \frac{1-i}{2} & 0 & 0 \\ 0 & 0 & 1 & 0 & 0 & 0 & 0 & 0 \\ 0 & 0 & 0 & \frac{1+i}{2} & 0 & 0 & 0 & \frac{1-i}{2} \\ 0 & 0 & 0 & 0 & 1 & 0 & 0 & 0 \\ 0 & \frac{1-i}{2} & 0 & 0 & 0 & \frac{1+i}{2} & 0 & 0 \\ 0 & 0 & 0 & 0 & 0 & 0 & 1 & 0 \\ 0 & 0 & 0 & \frac{1-i}{2} & 0 & 0 & 0 & \frac{1+i}{2} \end{bmatrix}$		

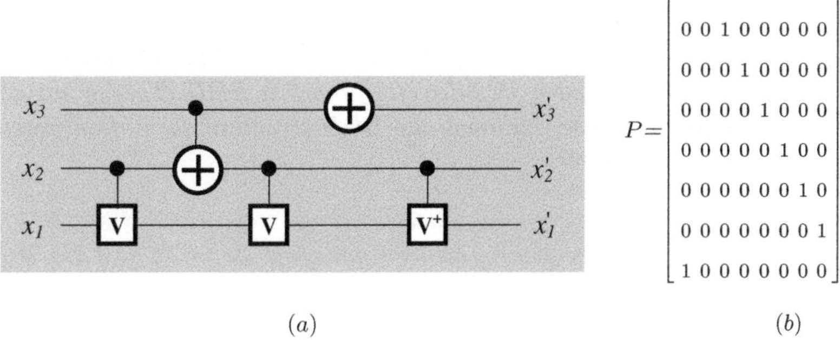

FIGURE 35.6
Quantum circuit for cyclic increment function (Example 35.6).

Example 35.7 *The transformation performed by a Toffoli gate in a 3 line reversible circuit with the least significant bit as target and the other lines as controls is given by the following matrix*

$$\begin{bmatrix} 1&0&0&0&0&0&0&0\\ 0&1&0&0&0&0&0&0\\ 0&0&1&0&0&0&0&0\\ 0&0&0&1&0&0&0&0\\ 0&0&0&0&1&0&0&0\\ 0&0&0&0&0&1&0&0\\ 0&0&0&0&0&0&0&1\\ 0&0&0&0&0&0&1&0 \end{bmatrix}$$

which defines the permutation $\{0, 1, 2, 3, 4, 5, 7, 6\}$.

35.3 Decision diagram techniques

A reversible function is a special form of switching function that can clearly be represented using a shared BDD as demonstrated in Example 35.8.

Example 35.8 *Consider the 3-input, 3-output cyclic increment function given in Example 35.1. The shared BDD for this function is shown in Figure 35.7. The use of complemented edges makes the exclusive-OR structures much more evident. Indeed it is quite straightforward to extract the structure of the circuit in Figure 35.2 directly from this BDD. This is not generally the case, although the BDD structure does often provide some guidance.*

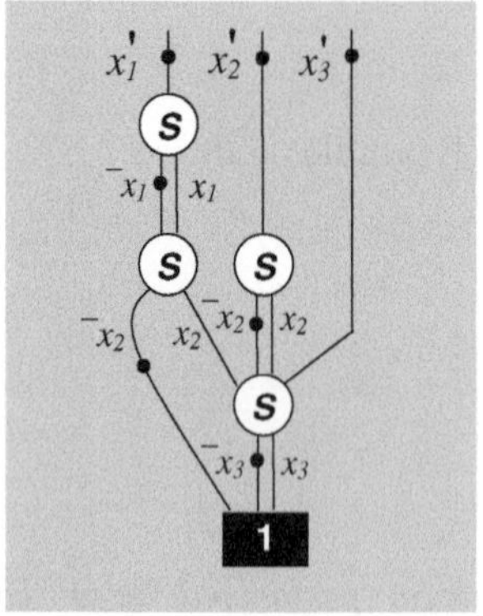

- *A small black circle indicates a complemented edge which points to the complement of the function represented by the sub-BDD to which the edge points.*
- $x_3' = x_3$
- $x_2' = x_2 \oplus x_3$
- $x_1' = x_1 x_2 \oplus x_3$

FIGURE 35.7
Shared BDD for cyclic increment function (Example 35.8).

> *Reversible logic synthesis procedures, which are typically described in terms of truth tables, are readily adapted to use BDD.*

It is also of interest to use properties of the BDD directly. Kerntopf [10] has presented a reversible logic synthesis method that uses the size of the shared BDD (number of nodes) as a criterion in searching for a circuit. The method is a greedy algorithm that exhaustively explores ties at each stage. The goal is to transform the initial shared BDD to the shared BDD representing the identity transform. The latter has a shared BDD with m non-terminal nodes. No reversible function can have a BDD representation with less than m nonterminal nodes.

In other approaches, the reversible function is transformed to a positive polarity Reed–Muller representation and the complexity of that representation is used to guide the search for a circuit implementation. A switching function, including a reversible function, represented as a BDD is readily transformed to its positive polarity Reed–Muller representation directly in the decision diagram domain with the Reed–Muller spectra being represented as a decision diagram. The transformation between the Boolean and Reed–Muller domains using decision diagram techniques is discussed elsewhere in this book. Details of reversible logic synthesis is beyond the discussion here. The interested reader should consult the references noted for further reading at the end of this chapter.

35.3.1 Representing matrices as decision diagrams

A particularly interesting application of decision diagram techniques is their use in representing and manipulating the matrices encountered in the synthesis and analysis of reversible and quantum circuits. As noted above, these matrices are quite large even for modest values of m, but they are highly structured and only basic matrix operations are generally required.

The idea of representing a matrix as a decision diagram is straightforward. It is convenient that in this application each matrix has dimension $2^m \times 2^m$. In many other applications it is necessary to pad the matrix with dummy entries to make its dimensions powers of 2.

Example 35.9 *Consider the matrix shown in Figure 35.8(a), which defines the transform for a two input Toffoli gate with the least significant input as the target. The columns and rows can be labelled as shown in Figure 35.8(b) with column and row selection variables added as indicated.*
Having done that, the matrix can be represented as a BDD as shown in Figure 35.8. The symmetry of the matrix is reflected in the symmetrical structure of the diagram. Note the interleaved order of the column and row selection variables.

The decision diagram in Figure 35.8 treats the given matrix as if its elements define a switching function, which is of course appropriate for many applications. However, the matrices encountered for reversible and particularly quantum circuits must be treated as arithmetic.

A BDD represents a mapping $f{:}\{0,1\}^n \rightarrow \{0,1\}$. An algebraic decision diagram extends this idea to allow arithmetic values at the terminal nodes.

Definition 35.4 *An algebraic decision diagram (ADD), also called arithmetic transform decision diagram (ACDD) is a decision diagram where each nonterminal node is controlled by a binary variable and for which there can be a number of terminal nodes, each assigned a distinct element from a range of arithmetic values. ADD belong to a class of multiterminal BDD (MTBDD). An ADD represents a mapping $f{:}\{0,1\}^n \rightarrow R$ where R is the range of terminal arithmetic values.*

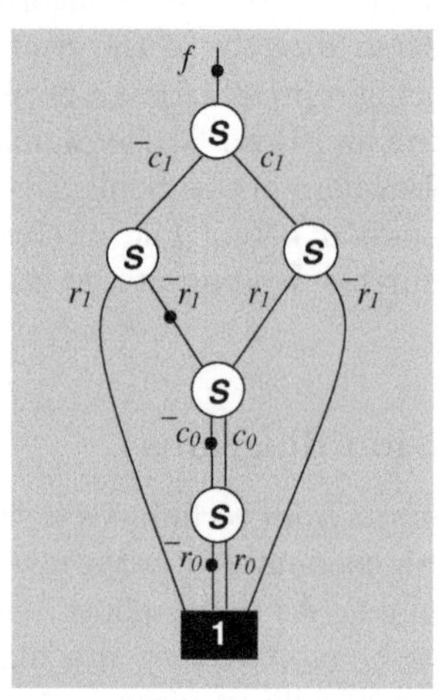

$$\begin{bmatrix} 1\,0\,0\,0 \\ 0\,1\,0\,0 \\ 0\,0\,0\,1 \\ 0\,0\,1\,0 \end{bmatrix}$$

(a)

r_1r_0 \ c_1c_0	00	01	10	11
00	1	0	0	0
01	0	1	0	0
10	0	0	0	1
11	0	0	1	0

(b)

FIGURE 35.8
BDD representation of the transform matrix for a controlled-NOT gate (Example 35.9).

> **Example 35.10** *The ADD for the matrix in Figure 35.8(b) is shown in Figure 35.9. The Diagram is somewhat larger than the BDD in Figure 35.8 as complemented edges are not applicable for ADD.*

Complemented edges as used in BDD are not applicable to ADD. However, edge attribute values can be used leading to *edge-valued decisions diagrams* (EVDD) and *factored* (FEVDD).

Applying ADD to the matrices encountered for quantum circuits requires the nonterminal value range be a subset of the complex numbers. Such an ADD will here be referred to as a *quantum decision diagram* (QDD) due to the application under consideration.

> **Example 35.11** *Consider the matrix shown in Figure 35.10, which is the transform over 2 qubits performed by a V gate where the most significant qubit is the target and the least significant qubit is the control. The QDD representation for this matrix using the same row and column selection variable scheme as in Example 35.9 is also shown in Figure 35.10.*

The block structure induced by the Kronecker product is illustrated in Figure 35.10 where two blocks involving c_0 and r_0 are implemented and then each used twice as selected by c_1 and r_1.

Most ADD manipulation programs target the integers. A very effective approach used in the QuIDDPro package [29] uses integer terminal values but

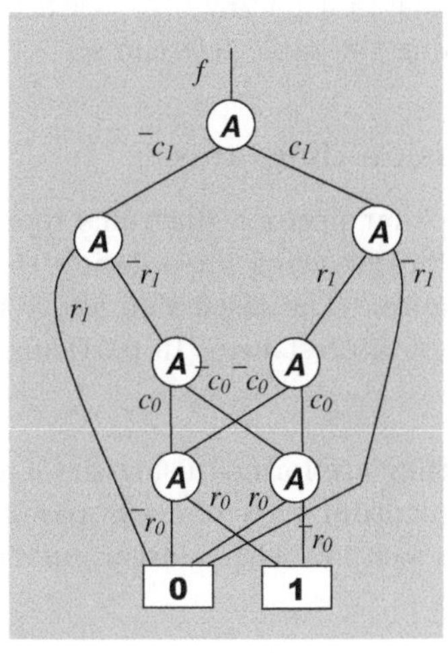

- *The nonterminal nodes are labelled A to denote selection of one of two arithmetic values rather than a Shannon expansion.*
- *The terminal nodes are shown in white to emphasize they are arithmetic rather than logical values.*
- *Note that the diagram is not quite symmetric.*
- *As in the BDD case, interleaving the column and row selection variables yields the most compact diagram.*

FIGURE 35.9
ADD representation of the transform matrix for a controlled-NOT gate (Example 35.10).

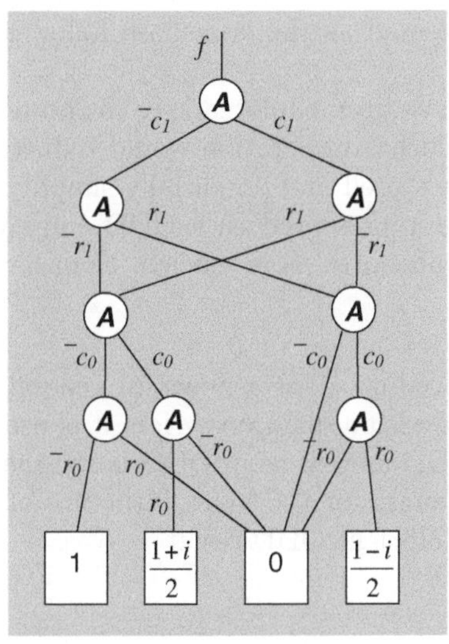

Matrix transform for a $V(1,2)$ gate.

$$\begin{bmatrix} 1 & 0 & 0 & 0 \\ 0 & \frac{1+i}{2} & 0 & \frac{1-i}{2} \\ 0 & 0 & 1 & 0 \\ 0 & \frac{1-i}{2} & 0 & \frac{1+i}{2} \end{bmatrix}$$

FIGURE 35.10
Matrix transform and QDD for a V gate applied as described in Example 35.11.

treats those as indices into a table of complex values. This allows quite direct use of the ADD code in the CUDD decision diagram package [23] with the trivial additional computation cost of performing the table references.

35.3.2 Matrix operations using decision diagrams

Variable ordering. It has been observed [6] that decision diagrams representing certain types of matrices are most efficient using an ordering that interleaves the column and row selection variables. The discussion above illustrates that this is the case for the matrices of interest here. In particular:

- The matrices are square and non-singular,
- The matrices are also highly regular since they are formed using the Kronecker product either to allow multiple quantum gates to act in parallel or to extend quantum gates to operate in a circuit with a larger number of qubits.

The interleaved variable ordering is inherent in the problem at hand and the potentially expensive operation of dynamic variable ordering can thus be avoided.

Matrix transpose and complex conjugate. Matrix transpose is readily implemented by swapping the row and column selection variable for the QDD. The conjugate transpose can be implemented as the transpose followed by conjugating the terminal values.

An alternative that seems not yet to have been explored is to introduce an edge operation for complex conjugate. Such an operation would reduce the number of distinct values that need to be stored and potentially simplify the structure of QDD. Conjugating a matrix represented as a QDD employing conjugate edge operations should be significantly more efficient than having to conjugate the individual terminals.

Apply procedure. Bryant's *Apply* procedure [5] is a powerful general approach to performing operations on two decision diagrams. Here it is used in forming the Kronecker product of two QDD as well as in QDD-based matrix multiplication. Proper use of *Apply* accounts for the block structure of the matrices of interest and thus leads to an efficient QDD result.

Apply accepts three arguments:

- Two decision diagrams, and
- An operation (*op*) to be performed on them.

Apply recursively traverses the decision diagram and performs *op* on each terminal node pair encountered, one from each of the decision diagram operands. Depending on the nature of the decision diagram, *op* can be a Boolean, arithmetic, or some hybrid operation.

The following notation is useful in explaining *Apply*:

- *Apply*(f, g, op) denotes an application of *Apply*.
- *top*(f) denotes the variable associated with the top node in the decision diagram f.
- $T(f)$ denotes the sub-decision diagram found by following the 1-edge from the top node in f.
- $F(f)$ denotes the sub-decision diagram found by following the 0 -edge from the top node in f.
- $DD(x, f_1, f_0)$ is the decision diagram whose top node has control variable x and whose 1 and 0 edges point to the sub-decision diagram f_1 and f_0, respectively.
- $x \prec y$ means x precedes y in the variable ordering, i.e., x will be closer to the top if both appear in the same decision diagram.

Apply involves the recursive application of the four rules in Table 35.6. The arguments to the initial call to *Apply* are the two decision diagrams to be operated on and the *op* to be applied to the terminal nodes.

TABLE 35.6
Rules defining *Apply*(f, g, op) operation.

Condition	Return result
f and g are both terminals	A terminal node $op(f, g)$
g is a terminal or $top(f) \prec top(g)$	$DD(top(f), Apply(T(f), g, op), Apply(F(f), g, op))$
f is a terminal or $top(g) \prec top(f)$	$DD(top(g), Apply(f, T(g), op), Apply(f, F(g), op))$
$top(f) = top(g)$	$DD(top(f), Apply(T(f), T(g), op),$ $Apply(F(f), F(g), op))$

Apply can be made much more efficient by recognizing applicable *early* evaluation cases. For example, if *op* is OR, the result is immediate if one of the decision diagrams is constant 1 or 0. The result is 1 in the first case, and the other decision diagram in the second. Likewise, if *op* is multiplication, the result is known if one of the decision diagrams represents arithmetic 0 or 1. Here the result is 0 in the first case and the other decision diagram in the second.

The Kronecker product of two matrices is denoted $A \otimes B$ and is defined as

$$A \otimes B = \begin{pmatrix} a_{11}B & a_{12}B & \cdots & a_{1n}B \\ a_{21}B & a_{22}B & & a_{2n}B \\ \vdots & & & \vdots \\ a_{m1}B & a_{m2}B & \cdots & a_{mn}B \end{pmatrix}$$

The key to using *Apply* to implement the Kronecker product is to arrange the variable ordering so that all the column and row selection variables for A appear earlier in the order than the column and row selection variables for B. This ensures *Apply* will first recurse to a terminal in the decision diagram representing A, and then apply that terminal across the complete decision diagram for B. The *op* is multiplication. The effect is to perform a multiplication of each element of A and the complete matrix B, which is what is required for the Kronecker product.

It happens that if the decision diagram for A and B have appropriate interleaved column and row selection variable orderings, the decision diagram constructed as described also has an appropriate interleaved variable ordering that takes proper advantage of the block structure of the Kronecker product.

Matrix multiplication. We consider the multiplication of two square matrices of dimension 2^m, which includes the case of interest for quantum circuit matrices. The concept is readily extended to non-square matrices. Dimensions that are not a power of 2 can be accommodated by extending the matrix with zero rows and columns as needed.

A square matrix A of dimension 2^m can be partitioned into four submatrices, each square and of dimension 2^{m-1} as follows:

$$A = \begin{pmatrix} A_{11} & A_{12} \\ A_{21} & A_{22} \end{pmatrix}$$

Using that partitioning, matrix multiplication of two such matrices can be expressed as

$$A \times B = \begin{pmatrix} A_{11} & A_{12} \\ A_{21} & A_{22} \end{pmatrix} \times \begin{pmatrix} B_{11} & B_{12} \\ B_{21} & B_{22} \end{pmatrix} = \begin{pmatrix} A_{11}B_{11} + A_{12}B_{21} & A_{11}B_{12} + A_{12}B_{22} \\ A_{21}B_{11} + A_{22}B_{21} & A_{21}B_{12} + A_{22}B_{22} \end{pmatrix}$$

This partitioning can be applied recursively until individual matrix elements are reached. Applications of *Apply* with *op* set to multiply are required for each multiplication of pairs of matrix elements. Each addition involves a straightforward application of *Apply* with *op* set to add. This includes addition of single values and addition of matrices. It is critical to note that this partitioning and subsequent partitionings of submatrices adhere to the interleaved column and row selection variable ordering discussed above.

Given the block structure of the matrices involved, and the frequency of repeated blocks, it is not uncommon to encounter the same submatrix multiplications during the course of a matrix multiplication. This duplication of effort is easily avoided using a computed table technique where the results of recent computations are recorded and subsequently retrieved when the same computation is required at a later time.

An empirical study. Techniques along the lines of those described above are incorporated in the QuIDDPro package [29], which is built on top of the CUDD decision diagram package [23]. To assess the effectiveness of the approach, Viamontes et al. [27] used QuIDDPro to simulate instances of Grover's algorithm [8], which is one of the foundational algorithms that has motivated the interest in quantum computation and quantum circuits.

The first important observation in this study is that memory usage grows linearly with the number of qubits. This is in contrast to the exponential growth of the matrices themselves.

Viamontes *et al.* compared the operation of QuIDDPro to three other approaches: use of the well known software package MATLAB, use of Octave (a package similar to MATLAB), and use of Blitz++ (a highly optimized numerical linear algebra package usable with C++). QuIDDPro is also implemented in C++. Octave and MATLAB could not handle problems beyond 15 qubits. Indeed, it is rather impressive they can work to that level, since they operate on matrix representations, albeit using very sophisticated packed representations to do so. Using Blitz++, simulation up to 21 qubits was possible. Results were successfully found using QuIDDPro for up to 40 qubits. The results presented show the QuIDDPro approach significantly outperforms the others in terms of both execution and memory usage. The latter is particularly dramatic. Readers interested in full detail of this study should consult [27].

35.4 Further reading

Reversibility Implications of the reversible computing paradigm for the future of computing have recently been discussed by Frank [7]. Reversible functions and circuits are discussed in detail in [19]. Numerous synthesis methods have been proposed for reversible circuits, particularly circuits composed of Toffoli gates [10, 16, 21]. These techniques are generally described as tabular but are readily transformed to use BDD for the basic representation of the reversible functions. The notion of reversibility extends readily to the multivalued case [1, 17]. MDD can be employed in this instance.

Quantum circuits, including related synthesis and simulation techniques, are discussed in detail in [19], which includes an extensive bibliography on those and related subjects. An evolutionary approach is introduced in [15].

Multivalued quantum circuits have been considered in [17, 18]. So-called ternary field sum-of-products expressions and decision diagrams are introduced for the design of ternary reversible devices by Khan et al. [12]. Some properties of multivalued reversible gates related to completeness were studied by Kerntopf et al. [11] (a set of m-valued logic functions (or elementary functions) is called universal (or completeness) if an arbitrary m-valued logic function can be represented by this finite set of elementary functions).

Algebraic decision diagrams were introduced in [2]. ADD are fully implemented in the CUDD decision diagram package [23]. Edge-valued decision diagrams are discussed in [13] and factored EVDD are considered in [24].

Matrix multiplication using ADD is discussed in [2, 3] where techniques are give for performing many other matrix operations using ADD.

QDD and their use in representing and simulating quantum circuits are discussed in [9, 20, 22, 25, 26, 27]. Details of the QuIDDPro package can be found at [29]. Use of decision diagram techniques for the simulation of quantum computation in the density matrix representation is discussed in [28].

Developments and studies to date, including those cited here, clearly show ADD and QDD to be very promising approaches to representing, manipulating and simulating reversible and quantum circuits. This will be of increasing importance as the size of such circuits grows.

References

[1] Al-Rabadi A. New multiple-valued Galois field Sum-of-product cascades and lattices for multiple-valued quantum logic synthesis. In *Proceedings of the 6th International Symposium on Representations and Methodology of Future Computing Technologies*, pp. 171–182, 2003.

[2] Bahar RI, Frohm EA, Gaona CM, Hachtel GD, Macii E, Pardo A, and Somenzi F. Algebraic decision diagrams and their applications. In *Proceedings of the International Conference on Computer-Aided Design,*, pp. 188–191, Santa Clara, CA, 1993.

[3] Bahar RI, Frohm EA, Gaona CM, Hachtel GD, Macii E, Pardo A, and Somenzi F. Algebraic decision diagrams and their applications. *Journal of Formal Methods in Systems Design*, 10(2/3):171–206, 1997.

[4] Bennett C. Logical reversibility of computation. *I.B.M. Journal of Research and Development*, 17:525–532, 1973.

[5] Bryant RE. Graph-based algorithms for boolean function manipulation. *IEEE Transactions on Computers*, 35(8):677–691, 1986.

[6] Clarke E, Fujita M, McGeer PC, McMillan K and Yang J. Multi-terminal decision diagrams: an efficient data structure for matrix representation. In *Proceedings of the International Workshop on Logic Synthesis*, pp. 6a:1–15, 1993.

[7] Frank MP. Approaching the physical limits of computing. In *Proceedings of the 35th IEEE Symposium on Multiple-Valued Logic*, invited address, pp. 168–185, 2005.

[8] Grover L. Quantum mechanics helps in searching for a needle in a haystack. *Physics Review Letters*, 79:325–328, 1997.

[9] Homeister M. Quantum ordered decision diagrams with repeated tests. *Technical Report IFI-TB-2004-03*, Georg-August-Universität, Göttingen, 2004.

[10] Kerntopf P. A new heuristic algorithm for reversible logic synthesis. In *Proceedings of the IEEE/ACM Design Automation Conference*, pp. 834–837, 2004.

[11] Kerntopf P, Perkowski M, and Khan MHA, On universality of general reversible multi-valued logic gates. In *Proceedings of the 34th IEEE International Symposium on Multiple-Valued Logic*, pp. 68–73, 2004.

[12] Khan MHA, Perkowski M, and Khan MR. Ternary Galois field expansions for reversible logic and Kronecker decision diagrams for ternary GFSOP minimization. In *Proceedings of the 34th IEEE International Symposium on Multiple-Valued Logic*, pp. 58–67, 2004.

[13] Lai Y-T and Sastry S. Edge-valued binary decision diagrams for multi-level hierarchical verification. In *Proceedings of the IEEE/ACM Design Automation Conference*, pp. 608–613, 1992.

[14] Landauer R. Irreversibility and heat generation in the computing process. *I.B.M. Journal of Research and Development*, 5:183-191, 1961.

[15] Lukac M, Perkowski M, Goi H, Pivtoraiko M, Yu CH, Chung K, Jee H, Kim B-G, and Kim Y-D. Evolutionary approach to quantum and reversible circuits synthesis. In Yanushkevich SN, Ed., *Artificial Intelligence in Logic Design*, pp. 201–257, Kluwer, Dordrecht, 2004.

[16] Miller DM, Maslov D, and Dueck GW. A transformation based algorithm for reversible logic synthesis. In *Proceedings of the IEEE/ACM Design Automation Conference*, pp. 318–323, 2003.

[17] Miller DM, Dueck GW and Maslov D. A synthesis method for MVL reversible logic. In *Proceedings of the 34th IEEE Symposium on Multiple-Valued Logic*, pp. 74–80, 2004.

[18] Miller DM, Maslov D, and Dueck GW. Synthesis of quantum multiple-valued circuits. *Journal of Multiple-Valued Logic and Soft Computing,* Special Issue on Nano MVL Structures, in press.

[19] Nielsen M and Chuang I. *Quantum Computation and Quantum Information*, Cambridge University Press, 2000.

[20] Sauerhoff M and Sieling D. Quantum branching programs and space-bounded nonuniform quantum computation. *Quanth-ph/0403164*, http://de.arxiv.org/abs/quant-ph/0403164, submitted to Theoretical Computer Science, 2004.

[21] Shende VV, Prasad AK, Markov IL, and Hayes JP. Synthesis of Reversible Logic Circuits. *IEEE Transactions on CAD*, 22(6):710–722, 2003.

[22] Shende VV, Bullock SS, and Markov IL. A practical top-down approach to quantum circuit synthesis. In *Proceedings of the Asia and South Pacific Design Automation Conference*, pp. 272–275, 2005.

[23] Somenzi F. CUDD: CU Decision Diagram package. University of Colorado at Boulder, http://vlsi.colorado.edu/~fabio/CUDD/, 2005.

[24] Tafertshofer P and Pedram M. Factored edge-valued decision diagrams. *Formal Methods in System Design*, 10(2/3)243–270, 1997.

[25] Viamontes GF, Rajagopalan M, Markov IL, and Hayes JP. High-performance simulation of quantum computation using QuIDD. In *Proceedings of the 6th Intlernational Conferemce on Quantum Communication, Measurement and Computing (QCMC)*, 2002.

[26] Viamontes GF, Rajagopalan M, Markov IL, and Hayes JP. Gate-level simulation of quantum circuits. In *Proceedings of the Asia South Pacific Design Automation Conference*, pp. 295–301, 2003.

[27] Viamontes GF, Markov IL, and Hayes JP. Improving gate-level simulation of quantum circuits. *Quantum Information Processing*, 2(5):347–380, 2003.

[28] Viamontes GF, Markov IL, and Hayes JP. Graph-based simulation of quantum computation in the density matrix representation. *Quantum Information and Computation*, 5(2):113–130, 2005.

[29] Viamontes GF, Markov IL, and Hayes JP. QuIDDPro: High-performance quantum circuit simulation. University of Michigan, http://vlsicad.eecs.umich.edu/Quantum/qp/, 2005.

36

Decision Diagrams on Quaternion Groups

The optimization of decision diagrams by changing the domain group structure is further generalized by using finite non-Abelian groups. In this respect, the quaternion group Q_2 appears as very convenient, since it has the order 8, which corresponds to the domain for switching functions of three variables.

For the domain group of a switching function defined in 2^n points, can the following be used:

- The direct product of cyclic groups C_2 and C_4, of orders 2 and 4 respectively, and
- The quaternion group Q_2,

each of them possibly raised to some integer power, depending on the number of variables in a given switching function f.

A similar approach, although interpreted as an encoding of pairs of variables, has been applied by Sasao and Butler [66] where the pairs of subgroups C_2 are replaced by a single group C_4. In this way, a BDD with n levels is transferred into a Quaternary decision diagram with $n/2$ levels. In a Quaternary decision diagram,

- Each node represents a variable taking four different values. This replacement reduces the number of variables that should be controlled, which further reduces the depth and in many cases the size of the decision diagram, as well as complexity of interconnections.
- From QDDs, networks having a simplified and possibly regular structure can be produced.

This method can be extended by using non-Abelian groups as, for instance, the quaternion group Q_2, as subgroups in the factorization of the domain group [14].

The following decision trees and diagrams are considered in this chapter:

- *Quaternary decision*,
- *Fourier decision diagrams on finite non-Abelian groups* (FNA-decision diagrams), and

- *Fourier decision diagrams on finite non-Abelian groups with preprocessing* (FNAP-decision diagrams).

36.1 Terminology

The following terminology is used in discussion of decision diagram techniques on the quaternion group.

Non-Abelian groups are groups where the commutativity law for the group operation is not satisfied, i.e., $x \circ y \neq y \circ x$ for every $x, y \in G$, where $\circ$ is the group operation of the group G.

Quaternion group is a particular example of finite non-Abelian groups.

Multiple-place decision diagrams are decision diagrams to represent p-valued functions of p-valued variables, i.e., they are diagrams defined on groups C_p^n where n is the number of variables.

Quaternary decision diagrams (QDDs) are a particular case of MDDs introduced to represent functions of four-valued variables. Thus, they are decision diagrams for $p = 4$, i.e., diagrams defined on the group C_4^n.

Kronecker QDDs and *pseudo-Kronecker QDDs.* These QDDs are introduced by allowing to freely choose for each level in Kronecker and for each node in pseudo-Kronecker QDDs, any of different (4×4) non-singular binary-valued matrices.

Multiterminal decision diagrams (MTDDs) are a generalization of binary decision diagrams (BDD), multiterminal binary decision diagrams (MTBDDs), and multiple-place decision diagrams (MDDs) to represent functions defined on finite groups, which are representable as the direct product of subgroups of various orders, i.e., not restricted to groups C_2^n and C_p^n. Thus, BDDs and MTBDDs are particular cases of MTDDs on groups C_2^n, where C_2 is the basic cyclic group of order 2. Similarly, MDDs are also a particular case of MTDDs.

Fourier decision diagrams on finite non-Abelian groups (FNA-decision diagrams) are decision diagrams defined on finite non-Abelian groups by using a decomposition rule derived from the Fourier transform on finite groups.

Fourier decision diagrams on finite non-Abelian groups with preprocessing (FNAP-decision diagrams) are decision diagrams to represent matrix-valued functions defined on finite, not necessarily Abelian groups.

36.2 Introduction

In a decision diagram, the number of levels is denoted as the *depth* of the decision diagram. The maximal number of nodes at a level is denoted as the *width* of the decision diagram. Both parameters characterizing a decision diagram are important, for instance, in decision diagram based circuit design methods, since they directly relate or even determine the propagation delay and the area occupied by the circuit designed from a decision diagram.

> Fourier decision diagrams on finite non-Abelian groups *(FNA-decision diagrams) are decision diagrams that permit reduction of the depth without necessarily increasing the number of non-terminal nodes per levels.*

However, they are defined for a general class of discrete functions, and values of constant nodes can be taken over a field P that may be a finite (Galois) field or the complex field C. In application to switching functions, in FNA-decision diagrams:

- ▶ The number of constant nodes is increased compared to that in bit-level decision diagrams. However, this does not necessarily imply increase of the total number of nodes.
- ▶ In many cases FNA-decision diagrams permit a simultaneous reduction of the depth, the width, and the size of the decision diagram.

> *As in QDDs, in FNA-decision diagrams:*
>
> (*a*) *The structure of the basic decision tree is changed by changing decomposition of the domain group G of f.*
>
> (*b*) *The non-Abelian groups are permitted as the constituent subgroups in decomposition of G in definition of* the Fourier decision trees on finite non-Abelian groups *(FNA-decision trees).*
>
> *Unlike BDDs and QDDs, in FNA-decision diagrams:*
>
> (*a*) The Fourier decomposition rule, *derived from the Fourier transform on groups, is performed at the nodes of the FNA-decision trees instead the Shannon decomposition rule in BDDs and its generalization used in QDDs.*
>
> (*b*) *Since FNA-decision diagrams are defined on non-Abelian groups in terms of the Fourier basis, the advantages are taken from the matrix notation of group representations of non-Abelian groups.*

FNA-decision diagrams correspond to the Walsh decision diagrams, which are Fourier decision diagrams on a particular Abelian group, the finite dyadic group C_2^n.

Fourier decision diagrams with preprocessing (FNAPDDs) are a generalization of FNA-decision diagrams defined to further exploit properties of matrix representations of finite non-Abelian groups. They permit the representation of matrix-valued functions, but can also be used to represent number-valued functions.

In application of FNAP-decision diagrams to number-valued functions, with switching functions as a particular case, these functions should be first transferred into their matrix-valued equivalents. This can be performed in different ways, and probably the simplest is to split the vector of function values into subvectors that are written as columns of a matrix. The order of these subvectors should be equal to the dimension of the group representation.

A rationale for introduction of FNAPDDs and possible advantages of their application can be found in the following considerations:

- ▶ *A given function is first transferred into its matrix-valued equivalent and represented by the matrix-valued FNAP-decision diagram on a non-Abelian group of small order.*
- ▶ *Matrices representing values of constant nodes are transformed back into the number-valued functions and represented by decision diagrams of small order.*

In that way, by using advantages of the structure of Fourier decision diagrams on non-Abelian groups that will be discussed below, depth reduction is provided without increase to the width and size of the produced decision diagrams. Moreover, in many cases reduction of all three parameters, the depth, width, and size, is achieved.

As in other world-level decision diagrams, multiple-output switching functions are represented by their integer-valued equivalents derived by adding the outputs multiplied with the weighting coefficients 2^i.

36.3 Group-theoretic approach to decision diagrams

Group-theoretic approach to switching functions and related decision diagrams consist of assuming a group structure for the domain of switching functions. Thus, switching functions are considered as functions on finite dyadic groups C_2^n. However, some other groups can be also used, which may result in certain advantages of interest in particular applications.

Example 36.1 *Two dyadic groups C_2 as the domain groups for a pair of binary variables can be replaced by the group C_4 which is the domain group for quaternary variables.*

36.3.1 Decision diagrams for LUT FPGAs design

In circuit synthesis with programmable logic devices, as LUT FPGAs, realization of any function up to the number of variables equal to the number of inputs in LUTs, has the same cost. The delay time for interconnections is often larger than that for LUTs. Thus, it is more important to reduce the propagation delay than to reduce the number of LUTs. To reduce the propagation delay, we have to reduce the number of levels in the network, which directly corresponds to the number of levels in the decision diagram. The propagation delay is smaller in networks with regular interconnections and layout. With this motivation, the Quaternary decision diagrams have been proposed by Sasao and Butler [66].

Structural properties of QDDs. QDDs consist of nodes with four outgoing edges corresponding to four-valued decision variables. Switching functions are represented by QDDs through:

- ▶ Pairing binary variables and
- ▶ Encoding each pair with a quaternary variable.

In QDDs, a pair of binary variables

$$(x_i,\ x_k),\ x_i, x_k \in \{0, 1\}$$

is encoded by a four-valued variable

$$X \in \{0, 1, 2, 3\}.$$

In this way, the number of levels in related decision diagrams is reduced from n to $n/2$ by increasing the number of outgoing edges of nodes from two to four. The resulting diagram has smaller depth and usually simplified interconnections.

Example 36.2 *Multiple-place decision diagrams for functions of four-valued variables can be viewed as a particular example of QDD, where the mapping performed at each node to assign a given f with four-valued variables to the multiple-place decision diagram is the identity transform described by the (4×4) identity matrix. Therefore, MDDs are a straightforward generalization of BDDs. QDDs are a large class of decision diagrams where any of (4×4) possible non-singular binary-valued matrices can be used to assign a given function to a decision diagram.*

Extension of QDDs to represent complex-valued functions on C_2^n is directly possible. In this case, they are a generalization of MTBDDs and we denoted them as *multiterminal decision diagrams* (MTDDs). In the replacement of C_2^2 by C_4, a subtree consisting of three non-terminal nodes with two outgoing edges is replaced by a node with four outgoing edges. Further generalization, where C_2^r for an arbitrary r is replaced by a group of order 2^r, is straightforward.

Example 36.3 *Figure 40.3 shows this transformation of decision diagrams.*

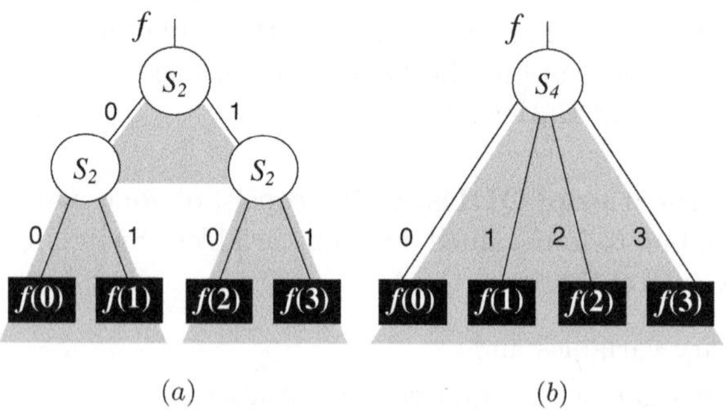

FIGURE 36.1
BDD and quaternary decision diagram (Example 40.12).

36.3.2 Spectral interpretation

In spectral interpretation, MDDs are the simplest example of QDDs, since they are defined with respect to the identity transform. A further step towards optimization of QDDs in the number of nodes count is through defining Kronecker and pseudo-Kronecker QDDs. These QDDs are introduced by allowing a free choice

- For each level in Kronecker and
- For each node in pseudo-Kronecker QDDs,

any of different (4×4) non-singular binary-valued matrices. Thus, matrices (with identical entries) that differ in the order of rows and columns are not distinguished.

Example 36.4 *Figure 36.2 shows examples of matrices describing nodes in different QDDs used as motivating examples in [66].*

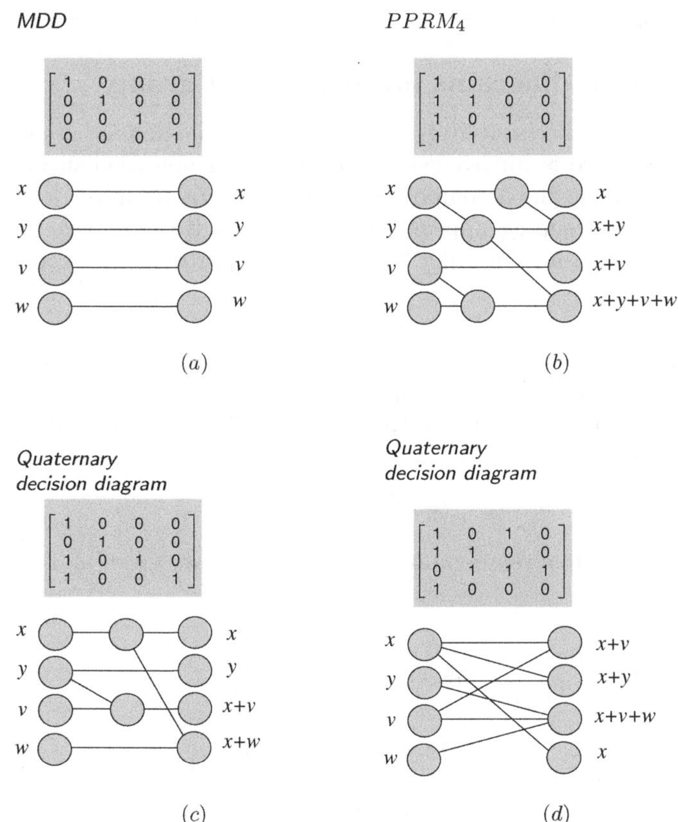

FIGURE 36.2
Nodes in decision diagrams on C_4^n (Example 36.4).

Classification of QDDs. QDDs belong to the broad class of Kronecker decision diagrams, which can be simply derived from MDDs by performing at each node of the MDD operations described by a matrix determining the nodes at the corresponding level in the Kronecker QDDs. This variety of possible assignments of matrices to the nodes in a decision diagram provides for a large choice of different decision diagrams for a given function f. However, the same as in Kronecker and pseudo-Kronecker decision diagrams on C_2^n , the key problem in practical applications is the lack of a criterion or an efficient algorithm to choose a suitable combination of matrices for nodes in a QDD for f.

36.4 Decision diagrams on quaternion group

The optimization of decision diagrams by changing the domain group structure discussed above is further generalized by using non-Abelian groups as possible subgroups in the decomposition of the domain group G.

The quaternion group Q_2 appears convenient since it can replace the product of three cyclic groups of order 2, i.e., in this approach C_2^3 is replaced by Q_2.

Quaternion group. The quaternion group Q_2 is generated by two elements a and b, for example, and the group identity is denoted by e. If the group operation is written as abstract multiplication, the following relations hold for the group generators:

$$b^2 = a^2, \; bab^{-1} = a^{-1}, \; a^4 = e.$$

The unitary irreducible representation of Q_8 over C are given in Table 36.1, where

$$\mathbf{I} = \begin{bmatrix} 1 & 0 \\ 0 & 1 \end{bmatrix}, \quad \mathbf{A} = \begin{bmatrix} 1 & 0 \\ 0 & -1 \end{bmatrix}, \quad \mathbf{B} = \begin{bmatrix} -1 & 0 \\ 0 & 1 \end{bmatrix},$$

$$\mathbf{C} = \begin{bmatrix} 0 & -1 \\ 1 & 0 \end{bmatrix}, \quad \mathbf{D} = \begin{bmatrix} 0 & 1 \\ 1 & 0 \end{bmatrix}, \quad \mathbf{E} = \begin{bmatrix} 0 & 1 \\ -1 & 0 \end{bmatrix}.$$

When written as columns of a matrix $\mathbf{Q}$ with matrix-valued entries, the columns in this table define the matrix of basis functions used in definition of the Fourier transform on Q_2.

36.4.1 Fourier transform on Q_2

In matrix notation, the Fourier transform on Q_2 is defined by the matrix inverse to the matrix $\mathbf{Q}$ of basis functions, i.e., the Fourier transform matrix on Q_2 is

$$\mathbf{Q}^{-1} = \frac{1}{8} \begin{bmatrix} 1 & 1 & 1 & 1 & 1 & 1 & 1 & 1 \\ 1 & -1 & 1 & -1 & 1 & -1 & 1 & -1 \\ 1 & 1 & 1 & 1 & -1 & -1 & -1 & -1 \\ 1 & -1 & 1 & -1 & -1 & 1 & -1 & 1 \\ 2\mathbf{I} & 2i\mathbf{B} & -2\mathbf{I} & 2i\mathbf{A} & 2\mathbf{E} & 2i\mathbf{D} & 2\mathbf{C} & -2i\mathbf{D} \end{bmatrix}$$

If for a function defined in eight points and, thus, given by the vector of function values $\mathbf{F} = [f(0), f(1), \ldots, f(7)]^T$, the group Q_2 is assumed as the domain group, the Fourier spectrum will be a vector

$$\mathbf{S}_f = [S_f(0), S_f(1), S_f(2), S_f(3), \mathbf{S}_f(4)]^T,$$

TABLE 36.1
Irreducible unitary representation of Q_2 over C.

x	Reed–Muller coefficients $\mathbf{R}_0$	$\mathbf{R}_1$	$\mathbf{R}_2$	$\mathbf{R}_3$	$\mathbf{R}_4$
0	1	1	1	1	$\mathbf{I}$
1	1	1	-1	-1	$i\mathbf{A}$
2	1	1	1	1	$-\mathbf{I}$
3	1	1	-1	-1	$i\mathbf{B}$
4	1	-1	1	-1	$\mathbf{C}$
5	1	-1	-1	1	$-i\mathbf{D}$
6	1	-1	1	-1	$\mathbf{E}$
7	1	-1	-1	1	$i\mathbf{D}$
	$r_0 = 1$	$r_1 = 1$	$r_2 = 1$	$r_3 = 1$	$r_4 = 2$

where the fifth entry is the matrix-valued coefficient $\mathbf{S}_f(4)$ as is clear from the definition of the matrix $\mathbf{Q}^{-1}$.

36.4.2 Decision diagrams on finite non-Abelian groups

Fourier decision diagrams on finite non-Abelian groups (FNA-decision diagrams) on Q_2 are defined as decision diagrams having:

- This Fourier spectrum as the values of constant nodes and
- Labels at the edges are determined in the same way as in other decision diagrams.

Thus, by following labels at the edges, the inverse transform is performed to read the function f from the Fourier spectrum shown in constant nodes. Therefore, the labels at the edges are determined from the definition of the inverse Fourier transform on Q_2 as

$$\mathbf{f}(x) = \sum_{w=0}^{K-1} Tr(\mathbf{S}_f(w)\mathbf{R}_w(x)), \tag{36.1}$$

where $Tr(\mathbf{X})$ is the trace of the matrix $\mathbf{X}$.

The matrix-valued node for $\mathbf{S}_f(4)$ can be represented by a subtree, when matrix entries are written as a vector, for instance, by concatenating rows or columns of this matrix.

Example 36.5 *Figure 36.3 shows the Fourier decision tree on Q_2 derived in this way and by using a generalized Shannon node for QDDs to represent the matrix-valued Fourier coefficient $\mathbf{S}_f(4)$. This tree can be reduced into the corresponding Fourier decision diagram, if there are constant nodes showing equal values.*

Fourier decision trees on Q_2 can be used as subtrees in other decision diagrams.

Example 36.6 *Fourier decision trees can be used to replace subtrees corresponding to a triple of binary valued variables in BDDs, consisting of seven nodes with two outgoing edges.*

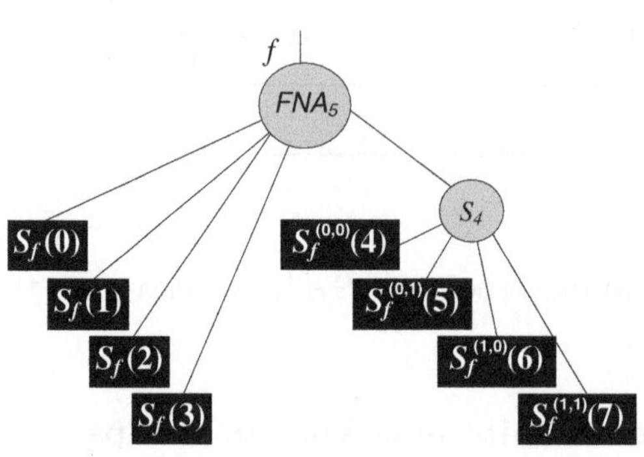

- *FNADT on Q_2 consists of a Fourier node with five edges and Shannon node with four edges.*
- *A BDT for functions defined in eight points has 7 Shannon nodes distributed over three levels.*
- *A corresponding MTDD with two levels has a Shannon node with two outgoing edges and two Shannon nodes with four outgoing edges.*

FIGURE 36.3
Fourier decision tree on Q_2 (Example 40.1).

36.4.3 Decision diagrams on non-Abelian groups with preprocessing

The Fourier transform matrix on Q_2 can be applied to calculate the Fourier spectrum of functions with matrix-valued elements, thus, given by a vector $[\mathbf{F}] = [\mathbf{f}(0), \dots \mathbf{f}(7)]^T$, where the entries $\mathbf{f}(i)$, $i = 1, \dots, 7$, are (2×2) matrices. In this case:

- All spectral coefficients will be matrix-valued,
- The first four coefficients are (2×2) matrices,
- The fifth coefficient is a (2×2) matrix whose entries are also (2×2) matrices.

In this way, the *matrix-valued Fourier decision trees on* Q_2 are derived. They have the same form as the Fourier diagram in Figure 36.3 but in this case, all the coefficients are matrix-valued.

If the matrix-vector $[\mathbf{F}]$ is derived from a number-valued vector $\mathbf{F}$ defining a function specified in 32 points, by writing pairs of function values as columns or rows in the matrix-valued entries of $[\mathbf{F}]$, the corresponding matrix-valued

Fourier decision tree is called the *Fourier decision tree with preprocessing.* The term preprocessing shows that the matrix-valued vector [**F**] is derived from a vector of function values **F**.

Similar as above, matrix-valued Fourier decision diagrams can be used as subtrees in other decision diagrams to replace some subdiagrams, when this reduces the overall complexity of the entire diagram for a given function that should be compactly represented.

36.4.4 Advantages of non-Abelian groups in optimization of decision diagrams

Decision diagrams on Abelian groups. With respect to the decomposition rules, decision diagrams can be classified into basic diagrams as BDD, MTBDD, MDD, or MTDD, then Fourier, and various Fourier-like diagrams.

In all these decision diagrams on Abelian groups:

> *The number of outgoing edges of nodes at a level i is always equal to the cardinality of the subgroup G_i where the variables assigned to this level take their values.*

This feature is due to the property of the Fourier analysis on finite groups that the dual object of the domain group G is an Abelian group isomorphic to it.

Decision diagrams on non-Abelian groups. In the case of finite non-Abelian groups, single dimensional group representations are not sufficient to define the Fourier transform, and higher-dimensional group representations are used. Due to this,

- The dual object for the domain group G is not a group, but a set of matrices, whose cardinality is smaller than the order of G.
- In Fourier decision diagrams on finite non-Abelian groups, the number of outgoing edges of Fourier nodes is smaller than the number of values a variable can take. Since in a decision tree, the number of outgoing edges of nodes at a level determines the number of nodes at the next level, this property permits reducing the width of the decision diagrams for some functions.

36.5 Further reading

Fundamentals of group-theoretic approach. The reader can find the basics of group theory, including group representations and Fourier transforms in the book by Cantrell [1]. Computing aspects of a group theory are given in the book by Cormen et al. [2]. Some engineering applications of Fourier analysis on groups can be found in [18].

Fourier decision diagrams. on Abelian groups to represent multi-valued logic functions have been introduced in [9]. These diagrams can be also called the *Vilenkin-Chrestenson decision diagrams*, since the corresponding Fourier transform is the Vilenkin-Chrestenson transform. In the same setting, the *Walsh decision diagrams* [17] are a particular case of Fourier decision diagrams, for functions of binary-valued variables.

Fourier decision diagrams on finite non-Abelian groups and their applications to represent some arithmetic functions are considered in [12], [13], and [14].

The depth reduction problem. A solution of the depth reduction problem is proposed by Minato et al. [8]. The method is, in essence, a direct realization of polynomial representation of switching functions, with the Boolean polynomial derived from the BDD of f through *the reachability matrices* [8]. Advantages are taken from the associativity properties of Boolean matrix multiplication. However, a disadvantage is the increased complexity of the hardware needed for such realizations. Also, in many practical examples, the width of BDDs for n-variable switching functions approximates $O(2^n/n)$.

The method [8] was extended to Kronecker decision diagrams (KDDs) by Drechsler and Becker [4]. A generalization of the method to multivalued functions is proposed by Stanković and Drechsler [15]. However, the same comment about the complexity of the required hardware may be given for KDDs in both switching and multivalued logic functions.

Another solution of the depth reduction problem is proposed by Sasao and Butler [66] through the quaternary decision diagrams. For a given switching function f, a Quaternary decision diagram is derived from BDD by recoding pairs of binary-valued variables by four-valued variables. In a group theoretic interpretation, the method is based upon the change of the underlined algebraic structure for the decision trees used though the change of the decomposition of the domain group of f. BDDs that are an example of decision diagrams on groups of order 2^n are replaced by quaternary decision diagrams that are decision diagrams on groups of order $C_4^{n/2}$. The decomposition rule performed at the nodes in quaternary decision diagrams is a direct extension of the Shannon decomposition rule used at the nodes in BDDs. Therefore, in spectral interpretation, both BDDs and quaternary decision diagrams may be uniformly considered as decision diagrams on Abelian groups defined in terms of the so-called *trivial basis*. This result was obtained by Stanković et al. [17]. A disadvantage of the method is that use of nodes with four outgoing edges increase the number of nodes per levels, thus, the width of the decision diagram in many examples.

A solution of the depth reduction problem has been proposed through the introduction of FNA-decision diagrams by Stanković [12].

Multiterminal and multiple-place decision diagrams. In the case of the Shannon decomposition and its generalizations to multivalued and other discrete functions, the values of constant nodes in the corresponding decision trees are the function values. Examples of decision trees derived from such trees are

- BDDs and quaternary decision diagrams for switching functions, multiterminal binary decision diagrams proposed by Clarke et al. [3],
- Multiple-place decision diagrams developed by Srinivasan et al. [11],
- Various forms of decision diagrams for binary-valued functions proposed by Sasao [7],
- Complex-valued functions on groups C_p^n Miller [9], Stanković et al. [16].

Group-theoretic approach. Some heuristic to reduce the search space for Kronecker and pseudo-Kronecker quaternary decision diagrams with the minimum number of nodes is proposed by Sasao and Butler [66]. In the group-theoretic approach, QDDs and methods for pairing of variables producing MDDs as a particular example of QDDs can be uniformly considered as optimization of decision diagrams by changing the domain group for the represented functions from C_2^n into C_4^n. If n is an odd number, then the root node is chosen as a node on C_2.

References

[1] Cantrell CD. *Modern Mathematical Methods for Physicists and Engineers.* Cambridge University Press, Cambridge, England, 2000.

[2] Cormen TH, Leiserson CE, Rivest RL, and Stein C. *Introduction to Algorithms.* MIT Press, Boston, 2001.

[3] Clarke EM, McMillan KL, Zhao X, and Fujita M. Spectral transforms for extremely large Boolean functions. In Kebschull U, Schubert E, and Rosenstiel W, Eds., *Proceedings of the IFIP WG 10.5 Workshop on Applications of the*

Reed–Muller Expansion in Circuit Design, pp. 86–90, Hamburg, Germany, 1993.

[4] Drechsler R and Becker B. OKFDDs-algorithms, applications and extensions. In Sasao T and Fujita M, Eds., *Representations of Discrete Functions*, Kluwer, Dordrecht, pp. 163–190, 1996.

[5] Drechsler R and Becker B. Overview of decision diagrams, *IEE Proceedings Comput. Digit. Tech.*, 144(3):187–193, 1997.

[6] Drechsler R and Becker B. *Binary Decision Diagrams, Theory and Implementation.* Kluwer, Dordrecht, 1998.

[7] Sasao T. AND-EXOR expressions and their optimization. In Sasao T, Ed., *Logic Synthesis and Optimization*, Kluwer, Dordrecht, 1993.

[8] Minato S, Ishiura N, and Yajima S. Shared binary decision diagrams with attributed edges for efficient Boolean functions manipulation. In *Proceedings of the 27th IEEE/ACM DAC*, pp. 52–57, 1990.

[9] Miller DM. Spectral transformation of multiple-valued decision diagrams. In *Proceedings of the IEEE 24th International Symposium on Multiple-Valued Logic*, pp. 89–96, 1994.

[10] Sasao T and Butler JT. A design method for look-up table type FPGA by pseudo-Kronecker expansions. In *Proceedings of the IEEE 24th International Symposium on Multiple-Valued Logic*, pp. 97–104, 1994.

[11] Srinivasan A, Kam T, Malik Sh, and Brayant RK. Algorithms for discrete function manipulation. In *Proceedings of the International Conference on CAD*, pp. 92–95, 1990.

[12] Stanković RS. Fourier decision diagrams on finite non-Abelian groups with preprocessing. In *Proceedings of the IEEE 27th International Symposium on Multiple-Valued Logic*, pp. 281–286, 1997.

[13] Stanković RS and Astola JT. Design of decision diagrams with increased functionality of nodes through group theory. *IEICE Transactions Fundamentals*, E86-A(3):693–703, 2003.

[14] Stanković RS and Astola JT. *Spectral Interpretation of Decision Diagrams.* Springer, Heidelberg, 2003.

[15] Stanković RS and Drechsler R. Circuit design from Kronecker Galois field decision diagrams for multiple-valued functions. In *Proceedings of the 27th International Symposium on Mutiple-Valued Logic*, 275–280, 1997.

[16] Stanković RS, Stanković M, Moraga C, and Sasao T. Calculation of Vilenkin–Chrestenson transform coefficients of multiple-valued functions through multiple-place decision diagrams. In *Proceedings of the 5th International Workshop on Spectral Techniques*, pp. 107–116, Beijing, China, 1994.

[17] Stanković RS, Sasao T, and Moraga C. Spectral transform decision diagrams. In *Proceedings of the IFIP WG 10.5 Workshop on Application of the Reed–Muller Expansion in Circuit Design*, pp. 46–53, Chiba, Japan, 1995.

[18] Stanković RS, Stojić MR and Stanković MS, Eds., *Recent Developments in Abstract Harmonic Analysis with Applications in Signal Processing.* Nauka, Belgrade and Elektronski fakultet, Niš, Serbia, 1996.

37

Linear Word - Level Decision Diagrams

This chapter extends the word-level technique presented in Chapter 17. A subclass of linear word-level arithmetic, sum-of-products, and Reed–Muller expressions are considered.

37.1 Introduction

The purpose of linearization is to find the sum of unary literals, for example, $x_1 + x_2 + 1$. This sum is an arithmetic form of a multi-output, or word-level function. Not every arrangement of bits (functions) at the word-level results in a linear arithmetic form. Thus, some manipulation is required. Word-level manipulation of switching functions $f_1, f_2, \ldots, f_r$ are based on Shannon, Davio and the arithmetic analogs of Davio expansion.

37.2 Linearization

Criteria for linearization. There is a large group of word-level expressions that cannot be linearized with traditional approaches. Criteria for linearization will identify the switching functions that cannot be represented by linear expressions. These conditions give an understanding of the limitations of the word-level format.

Synthesis methods are understood here as approaches to constructing word-level models of switching functions and circuits under conditions of linearity. In this chapter, we focus on representing arbitrary multilevel circuits by linear expressions and linear decision diagrams level by level. The final result is a set of linear expressions and linear decision diagrams.

Grouping and masking methods. The order of switching functions in a word is dependent on certain criteria. In this chapter, linearization is achieved through grouping and masking over a standard library of gates. This approach is useful in hypercube space as well.

> *A linear decision diagram is the result of the direct mapping of a linear expressions into a word-level decision diagram. A linear decision diagram can be considered as a boundary case of word-level diagrams and has a number of useful features.*

The most important and promising property for space representation is the simple embedding procedure of linear decision diagrams into 3D structures (hypercubes, hypercube-like topology, pyramids, etc.).

Methods of computing coefficients. For arithmetic expressions, the main goal is to minimize the effects of large value coefficients in linear arithmetic expressions. The crucial idea is to replace the computation by the manipulation of codes of coefficients. This is possible in some cases because of the regular structure of coefficients. The problem is simplified significantly for word-level sum-of-products and Reed–Muller expressions.

37.3 Linear arithmetic expressions

An arbitrary switching function can be represented by a unique arithmetic expression. For example, $x_1 \vee x_2$ corresponds to $x_1+x_2-x_1x_2$. The remarkable property of arithmetic expressions is that they can be applied to an r-output (word-level) function f with outputs $f_1, f_2, ..., f_r$. In this section we focus on the effects of grouping the functions with the goal of representing a word-level expression in linear form.

37.3.1 Grouping

Consider the problem of grouping several switching functions in a word-level format. Let an r-output function f with outputs $f_1, f_2, ..., f_r$ be given. This function is described by the word-level arithmetic expression

$$f = 2^{r-1} f_r + \ldots + 2^1 f_2 + 2^0 f_1, \tag{37.1}$$

and the outputs can be restored in a unique way. Therefore, the outputs of a circuit can be grouped together by using a weighted sum of the outputs.

Given the simplest commutator function, the direct transmission of input data to outputs is described by

$$f = 2^{r-1}x_1 + 2^{r-2}x_2 + \ldots + x_r.$$

Example 37.1 *Assume* $n = 2$, *then* $f = 2x_1 + x_2$ *(Figure 37.1). This expression does not include product terms of variables, therefore, it is* linear.

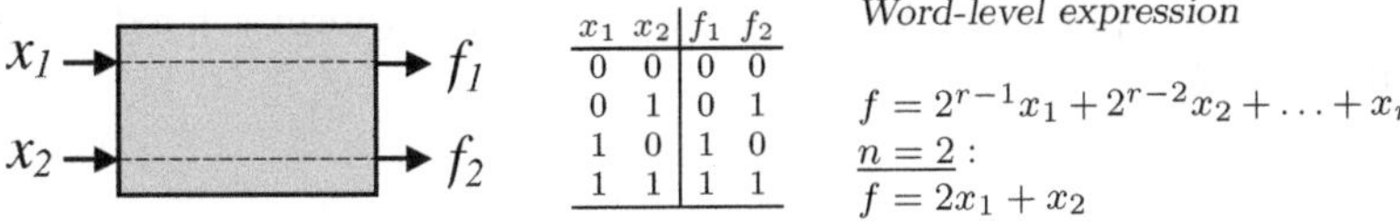

x_1	x_2	f_1	f_2
0	0	0	0
0	1	0	1
1	0	1	0
1	1	1	1

FIGURE 37.1
The direct transmission of input data to outputs, truth table, and word-level representation (Example 37.1).

The linear arithmetic expression of a switching function f of n variables $x_1, \ldots, x_n$ is an expression with $(n+1)$ integer coefficients $d_0^*, d_1^*, \ldots, d_n^*$

$$f = d_0^* + \sum_{i=1}^{n} d_i^* x_i = d_0^* + d_1^* x_1 + \ldots + d_n^* x_n. \tag{37.2}$$

Note that the word-level arithmetic expression

$$f = \sum_{i=0}^{2^n-1} d_i \cdot (x_1^{i_1} \cdots x_n^{i_n})$$

can be linear (Equation 37.2) in two cases: either the arithmetic expression of each f_j is linear, or no f_j generates linear expressions separately, but their combination produces a linear arithmetic expression.

Linearization generally means the transformation of a nonlinear expression to a linear arithmetic expression (Equation 37.2), with no more than $(n+1)$ nonzero coefficients. Briefly, the idea of linearization can be explained by a simple example. The function

$$f = x_1 \vee x_2 = x_1 + x_2 - x_1 x_2$$

is extended to the 2-output switching function

$$f_1 = 1 \oplus x_1 \oplus x_2,$$
$$f_2 = x_1 \vee x_2,$$

that derives from the linear word-level representation $f = 2^1 f_2 + 2^0 f_1 = x_1 + x_2 + 1$. The position of f_2 (the most significant bit) in this linear expression is indicated by the masking operator

$$\Xi^2\{f\} = \Xi^2\{x_1 + x_2 + 1\}.$$

In other words, to obtain a linear arithmetic expression given the switching function $f_2 = x_1 \vee x_2$, a garbage function $f_1 = 1 \oplus x_1 \oplus x_2$ has to be added. Then, f_2 can be extracted using the masking operator. The problem is how to find this additional function. In the absence of such a technique, a small amount of multioutput functions can generate linear arithmetic expressions using the naive approach. We use such functions to form a fixed library of primitive cells.

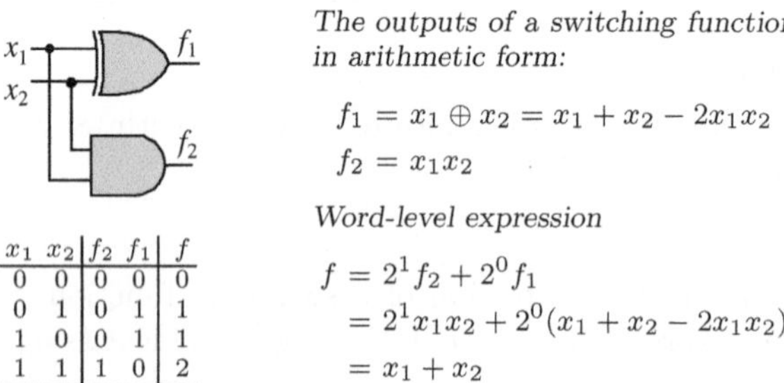
x_1
x_2
f_1
f_2
The outputs of a switching function in arithmetic form:
$f_1 = x_1 \oplus x_2 = x_1 + x_2 - 2x_1x_2$
$f_2 = x_1x_2$
Word-level expression
$f = 2^1 f_2 + 2^0 f_1$
$= 2^1 x_1x_2 + 2^0(x_1 + x_2 - 2x_1x_2)$
$= x_1 + x_2$

x_1	x_2	f_2	f_1	f
0	0	0	0	0
0	1	0	1	1
1	0	0	1	1
1	1	1	0	2

FIGURE 37.2
Half-adder circuit, its truth table, and the word-level representation (Examples 37.2 and 37.5).

Example 37.2 *The half-adder (Figure 37.2) can be represented by the linear arithmetic expression $f = x_1 + x_2$. Permutation of the outputs f_1 and f_2 generates the nonlinear expression: $f = 2x_1 + 2x_2 - 3x_1x_2$.*

The above example demonstrates the high sensitivity of a linear arithmetic expression (Equation 37.2) to any permutation of outputs in the word-level description. On the other hand, it is a unique representation given the order of the switching function f_1 or f_2.

37.3.2 Computing the coefficients in the linear expression

As an example, Equation 37.2 describes a set of switching functions. If $n = 1$, then $f = d_0^* + d_1^* x_1$, and the function f is single-output. The coefficients d_0^* and d_1^* can be calculated by the equation

$$\mathbf{D} = \mathbf{P}_{2^1} \cdot \mathbf{F} = \begin{bmatrix} 1 & 0 \\ -1 & 1 \end{bmatrix} \begin{bmatrix} 0 \\ 1 \end{bmatrix} = \begin{bmatrix} 0 \\ 1 \end{bmatrix},$$

i.e., $d_0^* = 0,\ d_1 = 1$. In general, f is the r-output switching function $f_1, \ldots, f_r$. Let f be a 3-output switching function: $f_1 = x_1,\ f_2 = \overline{x}_1,\ f_3 = x_1$, with the truth vector $\mathbf{F} = [2\ 5]^T$. Calculation of the coefficients implies:

$$\mathbf{D} = \mathbf{P}_{2^1} \cdot \mathbf{F} = \begin{bmatrix} 1 & 0 \\ -1 & 1 \end{bmatrix} \begin{bmatrix} 2 \\ 5 \end{bmatrix} = \begin{bmatrix} 2 \\ 3 \end{bmatrix},$$

and $f = 2 + 3x_1$. Assuming $n = 2$, Equation 37.2 yields $f = d_0^* + d_1^* x_1 + d_2^* x_2$.

> **Example 37.3** *The linear word-level arithmetic expression for a half adder function is defined as shown in Figure 37.3.*

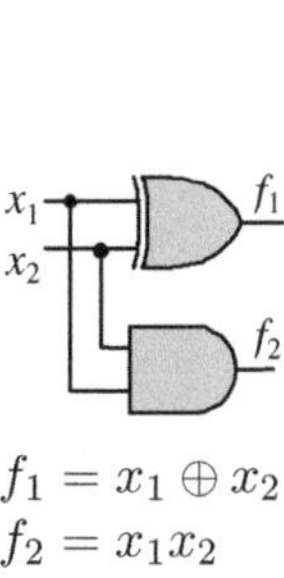

Truth vector

$$\mathbf{F} = [\ \mathbf{F_2} | \mathbf{F_1}] = \begin{bmatrix} 0 & 0 \\ 0 & 1 \\ 0 & 1 \\ 1 & 0 \end{bmatrix} = \begin{bmatrix} 0 \\ 1 \\ 1 \\ 2 \end{bmatrix}$$

Vector of coefficients

$$\mathbf{D} = \mathbf{P}_{2^2} \cdot \mathbf{F} = \left[\begin{array}{rr|rr} 1 & 0 & 0 & 0 \\ -1 & 1 & 0 & 0 \\ \hline -1 & 0 & 1 & 0 \\ 1 & -1 & -1 & 1 \end{array}\right] \begin{bmatrix} 0 \\ 1 \\ 1 \\ 2 \end{bmatrix} = \begin{bmatrix} 0 \\ 1 \\ 1 \\ 0 \end{bmatrix}$$

Word-level linear arithmetic expression $f = x_1 + x_2$

FIGURE 37.3
Constructing the linear word-level expression for a half adder by the matrix method (Example 37.3).

37.3.3 Weight assignment

There are two kinds of weight assignments in a linear arithmetic expression in the following formulation:

- The weight assignment to each function in a set of switching functions, and
- The weight assignment to each linear arithmetic expression in a set of expressions.

The weight assignment to the set of switching functions is defined by Equation 37.1. The weight assignment to the set of linear arithmetic expressions is defined as follows. Let f_i be the i-th, $i = 1, 2, \ldots, r$, linear word-level arithmetic expression of n_i variables. A linear expression of an elementary switching function is represented by

$$t_i = \lceil \log_2 n_i \rceil + 1 \ \ bits, \tag{37.3}$$

where $\lceil x \rceil$ is a ceiling function (the least integer greater than or equal to x). Suppose that f_i is the description of some primitive. In this formulation, the problem is to find a representation of an arbitrary level of a combinational circuit from a linear arithmetic expression as an n-input r-output switching function. To construct a word-level expression of r linear arithmetic expressions, the weight assignment must be made appropriately. This means that applying any pattern to the inputs of f, an output of each expression f_i cannot affect the outputs of others.

Figure 37.4a illustrates the problem. Formally, the weight assignments such that functions do not overlap are determined by the equation

$$f = \sum_{i=0}^{r-1} 2^{T_i} f_{i+1}, \tag{37.4}$$

where

$$T_i = \begin{cases} 0 & \text{for } i = 0 \\ T_{i-1} + t_i & \text{for } i > 0 \end{cases}$$

and t_i is calculated by Equation 37.3.

Example 37.4 *Let the word-level arithmetic expression f consist of three linear arithmetic expressions f_1, f_2 and f_3 (Figure 37.4b). The expressions are constructed as follows*

$$\begin{aligned} f_1 &= 2^0 f_{11} + 2^1 f_{12} + 2^2 f_{13}, \\ f_2 &= 2^3 f_{21} + 2^4 f_{22}, \\ f_3 &= 2^5 f_{31} + 2^6 f_{32} + 2^7 f_{33} + 2^8 f_{34}. \end{aligned}$$

by Equation 37.4, the weight assignment of linear expressions f_1, f_2 and f_3 results in $f = 2^0 f_1 + 2^3 f_2 + 2^5 f_3$.

37.3.4 Masking

The masking operator $\Xi^t\{f\}$ indicates the position $t \in (1, \ldots, r)$ of a given function f in the word-level expression that represents a set of r switching functions.

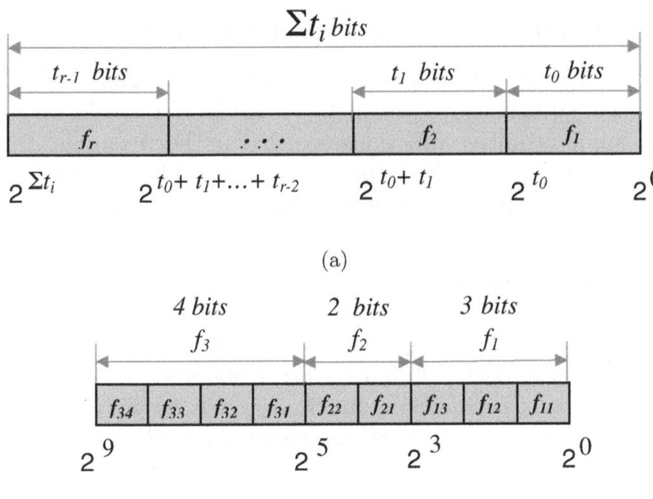

(b)

FIGURE 37.4
The word-level format for the set of r linear arithmetic expressions f_i (a) and an example for f_1, f_2, and f_3 that are respectively 3rd, 2nd and 4th subfunctions (b).

Example 37.5 *The arithmetical expression of a half adder $f = x_1 + x_2$ is a two-output switching function: $f_1 = x_1 \oplus x_2$, $f_2 = x_1 x_2$ (Figure 37.2). The most significant bit (f_2) can be extracted by the masking operator*

$$\Xi^2\{x_1 + x_2\},$$

whereas the function f_1 is encoded by the least significant bit that can be recovered by the masking operator

$$\Xi^1\{x_1 + x_2\}.$$

37.4 Linear arithmetic expressions of elementary functions

As was shown earlier, the majority of functions cannot be converted to a linear arithmetic expression, since their arithmetic equivalent includes non-linear products. In this section, we focus on linearizing elementary switching functions.

37.4.1 Functions of two and three variables

An arbitrary switching function of two variables can be described, in a unique way, by a linear arithmetic expression. In Table 37.1 the linear expressions for the two-input primitive functions are shown. For example, the function AND is the second function described by $x_1 + x_2$. The linear expressions for the three-input primitives are given in Table 37.1.

Example 37.6 *A switching function $x_1 \oplus x_2$ is represented by the nonlinear arithmetic expression $x_1 + x_2 - 2x_1x_2$. The nonlinear part is the product term $2x_1x_2$. To linearize, this function is expanded to the 2-input function $f_1 = x_1 \oplus x_2$ and $f_2 = x_1x_2$. The switching function f_1 is extracted from the linear expression by the masking operator $f = \Xi^1\{x_1 + x_2\}$.*

TABLE 37.1
Linear arithmetic expressions for two-input and three-input gates.

Function	2-input	3-input
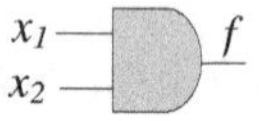 $f = x_1x_2$	$\Xi^2\{x_1 + x_2\}$	$\Xi^3\{x_1 + x_2 + x_3\}$
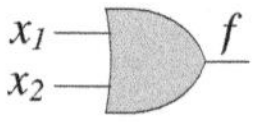 $f = x_1 \vee x_2$	$\Xi^2\{1 + x_1 + x_2\}$	$\Xi^3\{3 + x_1 + x_2 + x_3\}$
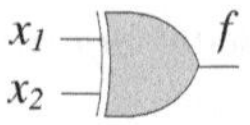 $f = x_1 \oplus x_2$	$\Xi^1\{x_1 + x_2\}$	$\Xi^1\{x_1 + x_2 + x_3\}$
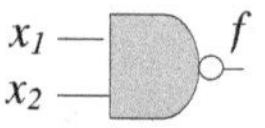 $f = \overline{x_1x_2}$	$\Xi^2\{3 - x_1 - x_2\}$	$\Xi^3\{6 - x_1 - x_2 - x_3\}$
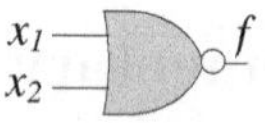 $f = \overline{x_1 \vee x_2}$	$\Xi^2\{2 - x_1 - x_2\}$	$\Xi^3\{4 - x_1 - x_2 - x_3\}$

37.4.2 AND, OR, and EXOR functions of n variables

The linear combination of linear expressions for some elementary switching functions produces linear expressions. An approach to designing linear expressions for two-input and three-input elementary switching functions is discussed in the previous section. An elegant method for designing linear arithmetic expressions for many-input elementary switching functions has been developed by *Malyugin* [2]. We introduce *Malyugin's theorem's* without proof.

Let the input variable of a primitive gate be x_j or $\overline{x}_j$. Denote the j-th input, $j = 1, 2, \ldots, n$, as

$$x_j^{i_j} = \begin{cases} x_j \text{ if } j_i = 0, \\ \overline{x}_j \text{ if } i_j = 1. \end{cases}$$

Theorem 37.1 *The n-variable AND function $x_1^{i_1} \ldots\ x_n^{i_n}$ can be represented by the linear arithmetic expression*

$$f = 2^{t-1} - n + \sum_{j=1}^{n} (i_j + (-1)^{i_j} x_j), \tag{37.5}$$

generated by an r-output function, in which the function AND is the most significant bit, as indicated by the masking operator $\Xi^r\{f\}$.

Theorem 37.2 *The n-variable OR function $x_1^{i_1} \vee \ldots \vee\ x_n^{i_n}$ can be represented by the linear arithmetic expression*

$$f = 2^{t-1} - 1 + \sum_{j=1}^{n} (i_j + (-1)^{i_j} x_j) \tag{37.6}$$

of an r-output function, so that the function OR is the most significant bit $f = \Xi^r\{f\}$.

Theorem 37.3 *The n-variable EXOR function $x_1^{i_1} \oplus \ldots \oplus\ x_n^{i_n}$ can be represented by the linear arithmetic expression*

$$f = \sum_{j=1}^{n} (i_j + (-1)^{i_j} x_j) \tag{37.7}$$

of an r-output function, in which the function EXOR is in the least significant bit, $\Xi^1\{f\}$.

In the above statements, the parameter t (the number of bits in a linear word-level representation of a given switching function) is defined by Equation 37.3: $t = \lceil \log_2 n \rceil + 1$. Note that the expression $1 \oplus x_j^{i_j}$ must be avoided in Equation 37.7. Before applying Equation 37.7, we have to replace $\overline{x}_j$ with $x_j \oplus 1$, or replace $x_j \oplus \overline{x}_j$ with 1 in order to cancel 1's.

TABLE 37.2
Linear arithmetic expressions for the n-input AND, OR, and EXOR functions.

Function	Linear arithmetic expression
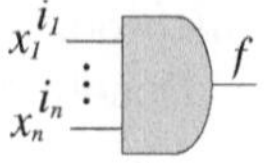 $f = x_1^{i_1} \ldots x_n^{i_n}$	$\Xi^r\{2^{t-1} - n + \sum_{j=1}^{n}(i_j + (-1)^{i_j} x_j)\}$
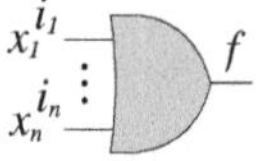 $f = x_1^{i_1} \vee \ldots \vee x_n^{i_n}$	$\Xi^r\{2^{t-1} - 1 + \sum_{j=1}^{n}(i_j + (-1)^{i_j} x_j)\}$
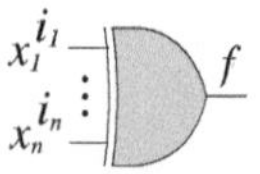 $f = x_1^{i_1} \oplus \ldots \oplus x_n^{i_n}$	$\Xi^1\{\sum_{j=1}^{n}(i_j + (-1)^{i_j} x_j)\}$

Table 37.2 contains three n-input primitives and corresponding linear expressions. For a NOT function, the corresponding linear arithmetic expression equals $\Xi^1\{1 - x\}$. Based on the expressions from Table 37.2, it is possible to describe modified gates, including, for example,

$$\begin{aligned}\overline{x}_1 \vee x_2 &= \Xi^2\{1 + (1 - x_1) + x_2\}\\ &= \Xi^2\{2 - x_1 + x_2\},\\ \overline{x}_1 \oplus x_2 &= \Xi^1\{(1 - x_1) + x_2\}\\ &= \Xi^1\{1 - x_1 + x_2\}.\end{aligned}$$

37.4.3 "Garbage" functions

Linear arithmetic expressions are word-level arithmetic expressions that possess specific properties. First, the linear expression involves extra functions called *garbage functions*. The number of garbage functions G increases with the number of variables in a function that have been linearized:

$$G = t - 1 = \lceil \log_2 n \rceil. \qquad (37.8)$$

Example 37.7 *The linear expression for a two-input AND function (Table 37.1) is $f = x_1 + x_2$. To derive this linear form, the garbage function f_1 has been added, so that the given function AND is the most significant bit in the word-level description, f_2 (Figure 37.5a). To derive a linear representation of a three-input AND function, two garbage functions have been added through the two least significant bits of the 3-bit word (Figure 37.5b).*

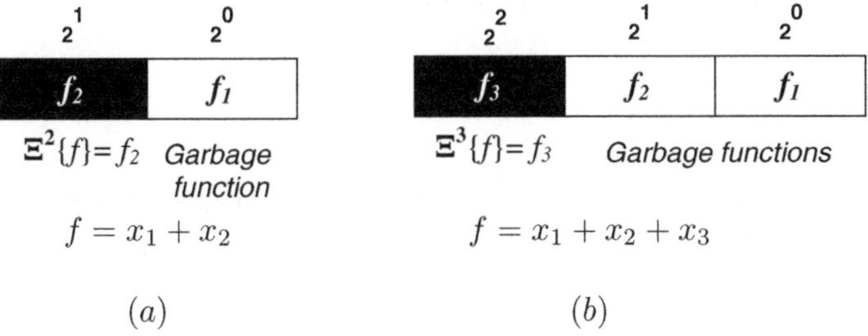

FIGURE 37.5
Garbage functions in linear arithmetic expression of a two-input (a) and three-input AND function (Example 37.7).

37.5 Linear decision diagrams

A *linear* decision diagram is used to represent a multioutput (word-level) switching function.

> *A set of linear diagrams is a formal model used to represent a multilevel circuit.*

In linear word-level diagrams, a node realizes the *arithmetic analog* of a positive Davio expansion

$$pD_A : f = \underbrace{f_{x_i=0}}_{left\ term} + \underbrace{x_i(-f_{x_i=0} + f_{x_i=1})}_{right\ term}$$

and terminal nodes correspond to integer-valued coefficients of the switching function f. thus, linear decision diagrams consists of pD_A nodes and binary moment decision diagrams (BMDs) (see Chapter 17).

A linear decision diagram is used to represent an arbitrary network described by a linear word-level arithmetic expression; the nodes correspond to a pD_A expansion and the terminal nodes are assigned to the coefficients of the linear expression. A linear decision diagram for n-input linear arithmetic expression includes n nonterminal and $n+1$ terminal nodes.

Example 37.8 *Figure 37.6a,b shows a linear decision diagram for an AND function. This diagram includes three nodes. Lexigraphic order of variables is used:* x_1, x_2.

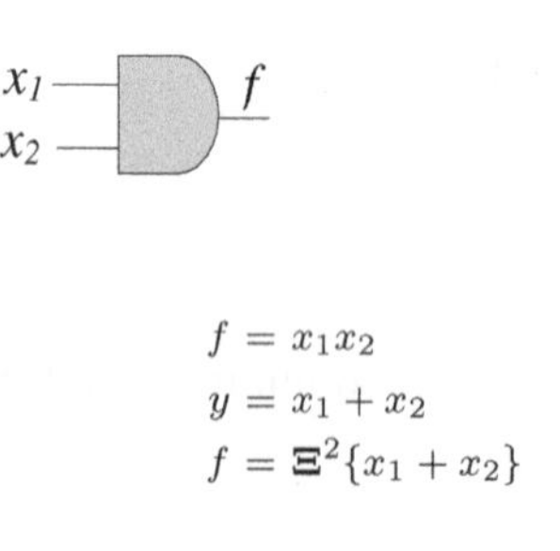

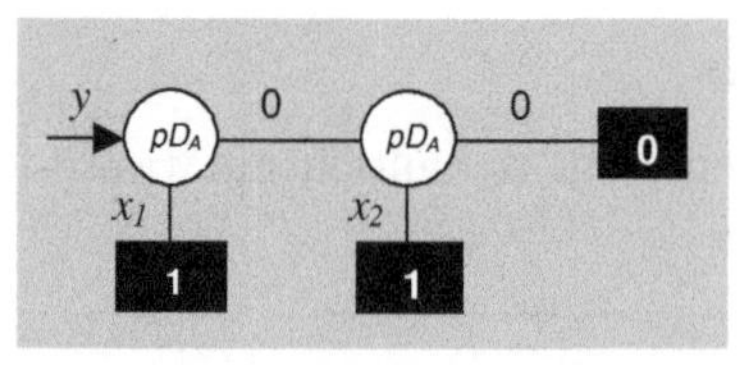

Step 1 : Compute $f_{x_1=0}$ and $f_{x_1=1}$:

$$f_{x_1=0} = 0 \cdot x_2 = 0,$$
$$f_{x_1=1} = 1 \cdot x_2 = x_2.$$

Step 2 : A terminal node with the value 1 is generated since the left term of the pD_A expansion is the constant $f_{x_1=0} = 0$. The right term is equal to

$$f_{x_1=1} - f_{x_1=0} = x_2 - 0 = x_2$$

and requires further decomposition.

Step 3 : Compute

$$f\Big|_{\substack{x_1=0\\x_2=0}} = 0, \quad f\Big|_{\substack{x_1=0\\x_2=1}} = 0.$$

The terminal node is equal to

$$x_2(-f_{x_2=0} + f_{x_2=1})$$
$$= -f\Big|_{\substack{x_1=0\\x_2=0}} + f\Big|_{\substack{x_1=0\\x_2=1}}$$
$$= -0 + 0 = 0.$$

FIGURE 37.6
Design of a linear decision diagram for the AND function (Example 37.8).

The linear decision diagram embedded in a 2D $\mathcal{N}$-hypercube is given in Figure 37.6c. In Table 37.3, the linear decision diagrams for two-input gates from the typical gate library are given.

Summarizing,

- Elementary switching functions can be represented by linear arithmetic expressions as shown in Table 37.1 and Table 37.2,
- Linear compositions of linear expressions (gates in the level of a circuit) produce linear arithmetic expressions,
- Linear expressions directly map into linear decision diagrams (Table 37.3).

37.6 Representation of a circuit level by linear word-level expression

Suppose a multilevel circuit with respect to a typical library of gates is given. The problem is formulated as follows: represent an arbitrary level of this circuit by

- Linear word-level equation and
- Linear word-level decision diagram.

Let a circuit level is described by n inputs $x_1, ..., x_n$ and r gates. The solution is based on the following theorem.

Theorem 37.4 *A circuit level with n inputs $x_1, ..., x_n$ and r gates (r outputs) is modeled by a linear diagram with n nodes assigned input variables, and $n+1$ terminal nodes, assigned coefficients of the linear expression.*

The proof follows immediately from the fact that an arbitrary n-input r-output function can be represented by a weighted arithmetic expression.

Example 37.9 *A level of a circuit is shown in Figure 37.7. The linear arithmetic expressions describing the first, the second, and the third gate are given in Table 37.1. Combining these expressions, we compile f where parameters T_0, T_1, T_2 are calculated by (37.4). The final result is*

$$\begin{aligned} f &= 2^0(x_1 + x_2) + 2^2(\overline{x}_1 + x_2) + 2^4(\overline{x}_2 + x_3) \\ &= 2^0(x_1 + x_2) + 2^2(1 - x_1 + x_2) + 2^4(1 - x_2 + x_3) \\ &= -3x_1 - 12x_2 + 17x_3 + 20. \end{aligned}$$

Let us design a set of linear decision diagrams for the circuit from Example 37.9, i.e., $f_1 = x_1 + x_3$, $f_2 = 1 - x_1 + x_2$, and $f_3 = 1 - x_2 + x_3$. Note that the order of variables in the diagram can be arbitrary. Let us choose lexigraphical order: x_1, x_2.

TABLE 37.3
Linear decision diagrams derived from linear word-level arithmetic expressions for two-variable functions.

Function	Linear decision diagram	Mask
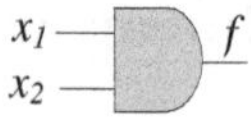 x_1, x_2, f $f = x_1 x_2$ $y = x_1 + x_2$	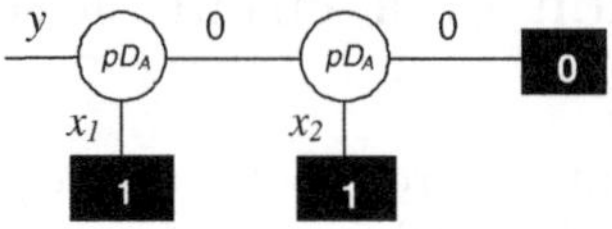 y; pD_A (x_1: 1), 0; pD_A (x_2: 1), 0; 0	$f = \Xi^2\{x_1 + x_2\}$
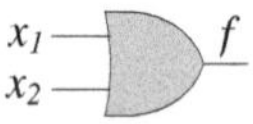 x_1, x_2, f $f = x_1 \vee x_2$ $y = 1 + x_1 + x_2$	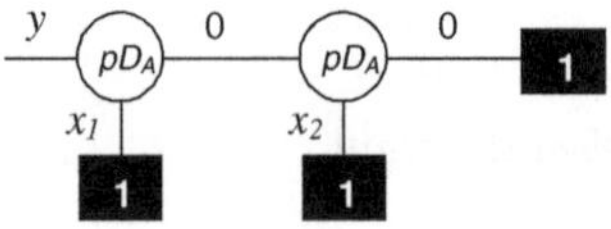 y; pD_A (x_1: 1), 0; pD_A (x_2: 1), 0; 1	$f = \Xi^2\{1 + x_1 + x_2\}$
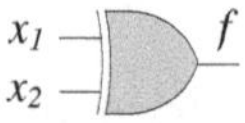 x_1, x_2, f $f = x_1 \oplus x_2$ $y = x_1 + x_2$	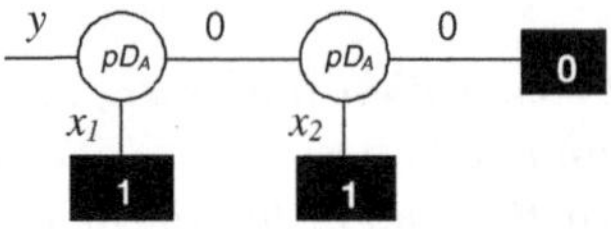 y; pD_A (x_1: 1), 0; pD_A (x_2: 1), 0; 0	$f = \Xi^1\{x_1 + x_2\}$
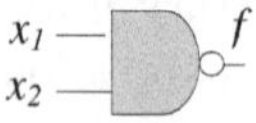 x_1, x_2, f $f = \overline{x_1 x_2}$ $y = 3 - x_1 - x_2$	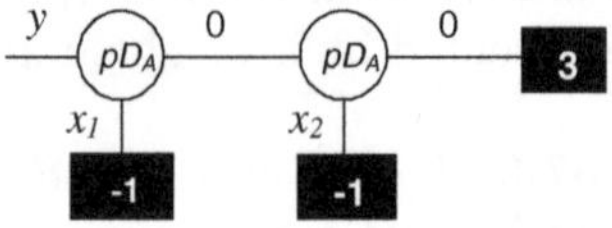 y; pD_A (x_1: -1), 0; pD_A (x_2: -1), 0; 3	$f = \Xi^2\{3 - x_1 - x_2\}$
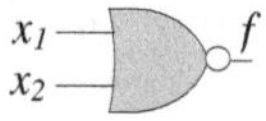 x_1, x_2, f $f = \overline{x_1 \vee x_2}$ $y = 2 - x_1 - x_2$	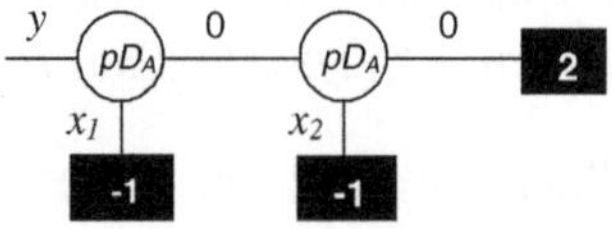 y; pD_A (x_1: -1), 0; pD_A (x_2: -1), 0; 2	$f = \Xi^2\{2 - x_1 - x_2\}$

$$f_1 = x_1 + x_3$$
$$f_2 = 1 - x_1 + x_2$$
$$f_3 = 1 - x_2 + x_3$$

Outputs

$$y_1 = \Xi^2\{x_1 + x_2\}$$
$$y_2 = \Xi^2\{1 - x_1 + x_2\}$$
$$y_3 = \Xi^2\{1 - x_2 + x_3\}$$

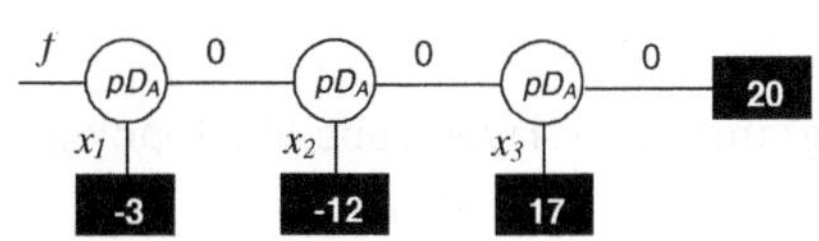

Level description

$$f = 2^{T_2} f_3 + 2^{T_1} f_2 + 2^{T_0} f_1$$

Masking parameters

$$t_0 = \lceil log_2 2\rceil + 1 = 2 \text{ bits}$$
$$t_1 = \lceil log_2 2\rceil + 1 = 2 \text{ bits}$$
$$t_2 = \lceil log_2 2\rceil + 1 = 2 \text{ bits}$$
$$T_0 = 0$$
$$T_1 = 2$$
$$T_2 = 4$$

Linear word-level arithmetic expression

$$f = -3x_1 - 12x_2 + 17x_3 + 20$$

Arithmetic positive Davio expansion

$$pD_A : f = f_{x_i=0} + x_i(-f_{x_i=0} + f_{x_i=1})$$

FIGURE 37.7
Technique for representing a level of a circuit by linear word-level arithmetic expression and decision diagrams (Examples 37.9 and 37.10).

Example 37.10 *The linear decision diagram for the expression* $-3x_1 - 12x_2 + 17x_3 + 20$ *consists of three nodes. There are three steps to designing the diagram:*

Step 1 : *Compute*

$$f_{x_1=0} = -3 \cdot 0 - 12x_2 + 17x_3 + 20,$$
$$f_{x_1=1} = -3 \cdot 1 - 12x_2 + 17x_3 + 20.$$

The terminal node is equal to -3 *because the right product is a constant* $f_{x_1=1} - f_{x_1=0} = -3$*. The left product needs further decomposition.*

Step 2 : *Compute*

$$f\Big|_{\substack{x_1=0\\x_2=0}} = 17x_3 + 20, \qquad f\Big|_{\substack{x_1=0\\x_2=1}} = -12 + 17x_3 + 20.$$

The terminal node equals

$$f\Big|_{\substack{x_1=0\\x_2=1}} - f\Big|_{\substack{x_1=1\\x_2=0}} = -12.$$

Step 3 : *By analogy, the terminal node for variable* x_3 *is* 17*, and the free terminal node is* 20*.*

From Examples 37.9 and 37.10, one can observe that the coefficients in linear expressions are quite large even for the small circuits. Therefore, a special technique is needed to alleviate this effect.

37.7 Linear decision diagrams for circuit representation

In this section, an arbitrary r-level combinational circuit is represented by r linear decision diagrams, i.e., for each level of a circuit, a linear diagram is designed. The complexity of this representation is $O(G)$, where G is the number of gates in the circuit. The outputs of this model are calculated by transmitting data through this set of diagrams. This approach is the basis for representing circuits in spatial dimensions:

< 2D circuit> ⇒ < A set of linear diagrams>⇒ <Hypercube-like topology>.

> *An arbitrary m-level switching network can be uniquely described by a set of m linear decision diagrams, and, vice versa, this set of linear decision diagrams corresponds to a unique network.*

To prove this statement, let one of m levels with r n-input gates from a fixed library be described by one linear arithmetic expression. Fixing the order of the gates in the level, i.e., keeping unambiguity in the structure, we can derive the unique linear decision diagram for this level, as well as for other $m-1$ levels of the network.

From this statement it follows that:

- The order of gates in a level of circuit must be fixed.
- The complexity of linear decision diagram does not depend on the order of variables.
- Data transmission through linear decision diagrams must be provided.

Example 37.11 *Figure 37.8 shows the* 3*-level switching network and its representation by a set of* 3 *linear decision diagrams.*

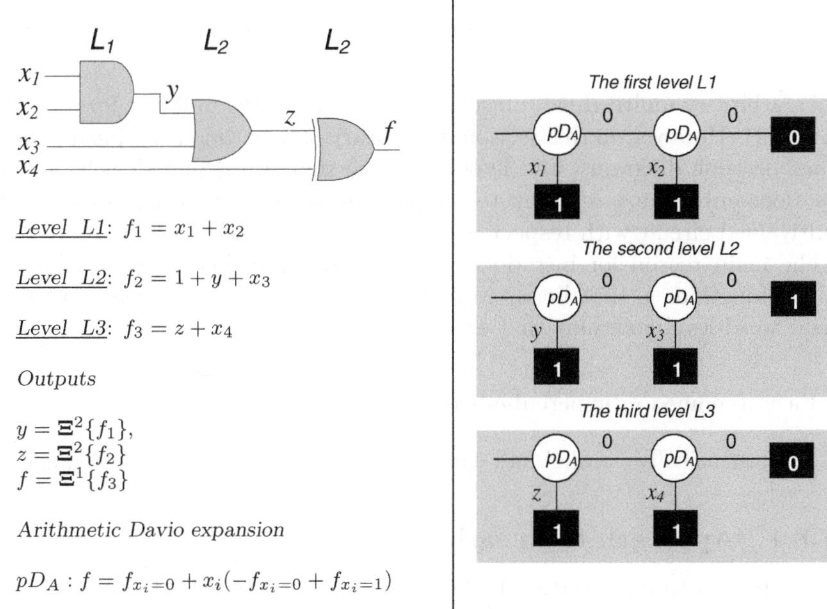

FIGURE 37.8
Representation of a three-level circuit by a set of linear decision diagrams (Example 37.11).

37.8 Linear word-level expressions of multivalued functions

In this section, the generalization of linear word-level data structures toward multivalued functions is considered. In a similar manner to binary functions, linear word-level expressions and decision diagrams are distinguished by their type of decomposition (expansion). There exist three linear word-level forms for multivalued valued functions:

- Linear word-level arithmetic expressions;
- Linear word-level Reed–Muller (modulo m) expressions; and
- Linear word-level sum-of-products expressions.

The last two forms are considered in the next sections. The focus of this section is an approach to the representation of m-valued functions of n variables by the linear word-level expression

$$f = d_0 + d_1 x_1^{\circ} + d_2 x_2^{\circ} + \cdots + d_n x_n^{\circ}. \tag{37.9}$$

An arbitrary multivalued function can be represented in linear form (Equation 37.9). However, in this section, the library of linear models (linear expressions, decision diagrams, and hypercube-like structures) includes elementary functions only. Thus, different techniques can be used to design an arbitrary multivalued circuit with respect to this library of gates.

The main reason for introducing this approach to linearization of multivalued functions is that linear word-level expressions can be represented by linear word-level diagrams that are:

- Easy to embed in hypercube-like structures, and
- Are intrinsically parallel and can be calculated by massive parallel arrays.

37.8.1 Approach to linearization

The approach to linearization includes the following phases (Figure 37.9):

Step 1. Partitioning of the truth vector **F** of the m-valued function f of n m-valued variables to a set of subvectors $\mathbf{F}_j^{\circ}$,
Step 2. Encoding the multivalued variables x_i. The new, binary variables x_i° are called *pseudo-variables*, and
Step 3. Representation of the multivalued function f by a linear word-level arithmetic expression that depends on binary pseudo-variables x_i°.

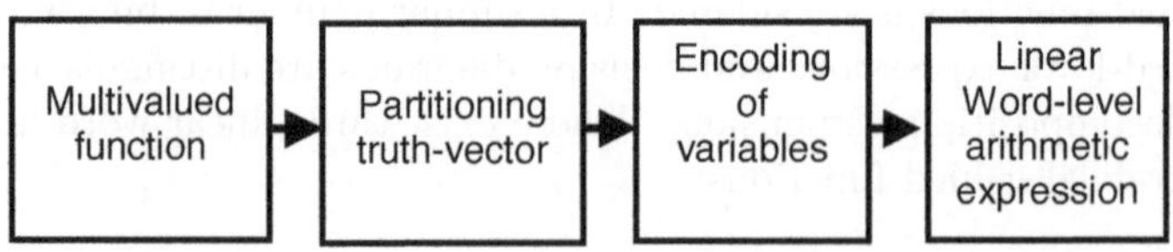

FIGURE 37.9
An algorithm for representing a multivalued function by a linear word-level expression.

37.8.2 Algorithm for linearization of multivalued functions

Step 1: Partition. Given the truth vector $\mathbf{F}$ of an m-valued n-variable logic function f. Let us partition this vector into τ subvectors,

$$\tau = \left\lceil \frac{m^n}{n+1} \right\rceil, \tag{37.10}$$

where $\lceil a \rceil$ denotes the least integer greater than or equal to a. The order of the partition is fixed (with respect to assignments of variables). The index μ of subvector $\mathbf{F}_\mu$ that contains the i-th element of the initial truth table is equal to

$$\mu = \left\lfloor \frac{i}{n+1} \right\rfloor, \tag{37.11}$$

where $\lfloor a \rfloor$ is the greatest integer less than or equal to a.

Example 37.12 *Partitioning the truth vector* $\mathbf{F}$ *of lengh* $3^3 = 27$ *of a ternary* ($m = 3$) *function of three variables* ($n = 3$) *is illustrated in Figure 37.10. The location of the 20-th element of the truth vector* $\mathbf{F}$ *is determined by the index* $\mu = \left\lfloor \frac{20}{3+1} \right\rfloor = 5$ *of subvector* $\mathbf{F}_\mu$*. This element belongs to subvector* $\mathbf{F}_5$.

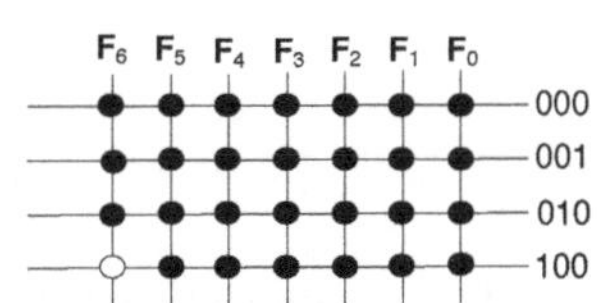

The vector $\mathbf{F}$ *is partitioned to*

$$\tau = \left\lceil \frac{3^3}{3+1} \right\rceil = \left\lceil \frac{27}{4} \right\rceil = \lceil 6.7 \rceil = 7$$

subvectors $\mathbf{F}_0, \mathbf{F}_1, \ldots, \mathbf{F}_6$
The 21-th element is located in the truth-vector $\mathbf{F}_\mu$,

$$\mu = \left\lfloor \frac{21}{3+1} \right\rfloor = 5$$

FIGURE 37.10
Representation of truth-vector of a multivalued function by 2D data structure (Example 37.12).

Step 2: Encoding. Consider the μ-th subvector $\mathbf{F}_\mu$, $\mu = 0, 1, \ldots, \tau - 1$. The length of the subvector $\mathbf{F}_\mu$ is $n+1$. Hence, the i-th element is allocated in the subvector $\mathbf{F}_\mu$. Its position inside μ is specified by the index $j = Res\left(\frac{i}{n+1}\right) = 5$. Assignments of n variables $x_1^\circ, x_2^\circ, \ldots, x_n^\circ$ in $\mathbf{F}_\mu$ are called *pseudo-variables.* The pseudo-variables are the *binary* variables valid for assignments with at most one 1.

Example 37.13 *Assignments of pseudo-variables of a three-valued function of two, three, and four variables are given below:*

(a) $x_1^\circ x_2^\circ = \{00, 01, 10\}$*; given* $i = 1$, $\mu = \left\lfloor \frac{1}{2+1} \right\rfloor = 0$*;*

(b) $x_1^\circ x_2^\circ x_3^\circ = \{000, 001, 010, 100\}$*; given* $i = 20$, $\mu = \left\lfloor \frac{20}{3+1} \right\rfloor = 5$*;*

(c) $x_1^\circ x_2^\circ x_3^\circ x_4^\circ = \{0000, 0001, 0010, 0100, 1000\}$*; given* $i = 10$, $\mu = \left\lfloor \frac{10}{4+1} \right\rfloor = 2$.

Step 3: Representation of a function by linear word-level arithmetic expression. This phase consists of:

(a) Forming a word-level vector $\mathbf{F}^\circ$ from subvectors $\mathbf{F}_1^\circ$, $\mathbf{F}_2^\circ$, $\dots$ $\mathbf{F}_\tau^\circ$, and

(b) A truncated arithmetic transform of vector $\mathbf{F}^\circ$.

Let $\mathbf{W} = \left[m^{\tau-1}\ m^{\tau-2}\ \cdots\ m^1\ m^0\right]^T$ be the weight vector. A truth vector $\mathbf{F}^\circ$ of a function f of n pseudo-variables $x_1^\circ, \dots, x_n^\circ$ includes $n+1$ elements and is calculated by

$$\mathbf{F}^\circ = [\mathbf{F}_{\tau-1}|\dots|\mathbf{F}_1|\mathbf{F}_0]\mathbf{W}, \qquad (37.12)$$

The truncated transform of $\mathbf{F}^\circ$ yields the vector of arithmetic coefficients $\mathbf{D}$. The relationship between the $\mathbf{F}^\circ$ and vector of coefficients $\mathbf{D} = [d_0 d_1 \dots d_n]$ is defined by the pair of transforms

$$\mathbf{D} = \mathbf{T}_{n+1} \cdot \mathbf{F}^\circ, \qquad (37.13)$$

$$\mathbf{F}^\circ = \mathbf{T}_{n+1}^{-1} \cdot \mathbf{D}, \qquad (37.14)$$

where $(n+1) \times (n+1)$ direct $\mathbf{T}_{n+1}$ and inverse $\mathbf{T}_{n+1}^{-1}$ truncated arithmetic transform matrices are formed by truncation of $2^n \times 2^n$ arithmetic transform matrices P_{2^n} and $P_{2^n}^{-1}$ respectively. The truncated rule is as follows:

(a) Remove all rows that contain more than one 1;

(b) Remove the remaining columns that consist of all 0s.

The vector of coefficients $\mathbf{D}$ yields the linear word-level arithmetic expression

$$D = d_0 + d_1 x_1^\circ + d_2 x_2^\circ + \dots + d_n x_n^\circ$$

Example 37.14 *Given a two-variable three-valued function, the* 3×3 *direct and inverse arithmetic truncated matrices are equal to*

$$\mathbf{T}_3 = \begin{bmatrix} 1 & 0 & 0 \\ -1 & 1 & 0 \\ -1 & 0 & 1 \end{bmatrix}, \qquad \mathbf{T}_3^{-1} = \begin{bmatrix} 1 & 0 & 0 \\ 1 & 1 & 0 \\ 1 & 0 & 1 \end{bmatrix}.$$

Example 37.15 *(Continuation of Example 37.14). Arithmetic expressions for subvectors* $\mathbf{D_0}$, $\mathbf{D_1}$ *and* $\mathbf{D_2}$ *in Table 37.4 are calculated by the direct truncated transform (Equation 37.13). For example,* $\mathbf{D_1}$ *is calculated as follows:*

$$\mathbf{D_1} = \mathbf{T_3 F_1} = \begin{bmatrix} 1 & 0 & 0 \\ -1 & 1 & 0 \\ -1 & 1 & 1 \end{bmatrix} \begin{bmatrix} 1 \\ 1 \\ 2 \end{bmatrix} = \begin{bmatrix} 1 \\ 0 \\ 1 \end{bmatrix},$$

which yields the algebraic form $d_1 = 1 + x_1^{\circ}$.

Example 37.16 *(Continuation of Example 37.15) The truth vector* $\mathbf{F}^{\circ}$ *of the two-input* $MAX(x_1, x_2)$ *function is calculated as shown in Figure 37.11. The direct truncated transform (Equation 37.13) is used for transformation. The final result is the linear expression*

$$\begin{aligned} D &= 3^2 D_2 + 3^1 D_1 + 3^1 D_0 \\ &= 21 + 5x_1^{\circ} + x_2^{\circ} \end{aligned}$$

TABLE 37.4
Partitioning of the truth vector **F** and deriving the linear word-level arithmetic expression for a ternary $MAX(x_1, x_2)$ function.

Function MAX			Linear model		
x_1x_2	$\mathbf{F}$	$\mathbf{F}_\mu$	$x_1^{\circ}x_2^{\circ}$	$\mathbf{D}_\mu$	D_μ
00 01 02	0 1 2	$\mathbf{F_0} = \begin{bmatrix} 0 \\ 1 \\ 2 \end{bmatrix}$	00 01 10	$\mathbf{D_0} = \begin{bmatrix} 0 \\ 1 \\ 2 \end{bmatrix}$;	$D_0 = x_2^{\circ} + 2x_1^{\circ}$
10 11 12	1 1 2	$\mathbf{F_1} = \begin{bmatrix} 1 \\ 1 \\ 2 \end{bmatrix}$	00 01 10	$\mathbf{D_1} = \begin{bmatrix} 1 \\ 0 \\ 1 \end{bmatrix}$;	$D_1 = 1 + x_1^{\circ}$
20 21 22	2 2 2	$\mathbf{F_2} = \begin{bmatrix} 2 \\ 2 \\ 2 \end{bmatrix}$	00 01 10	$\mathbf{D_2} = \begin{bmatrix} 2 \\ 0 \\ 0 \end{bmatrix}$;	$D_2 = 2$

37.8.3 Manipulation of the linear model

A linear word-level expression of elementary multivalued functions is a form of representation and computation due to the following properties:

- It is convertible to the initial function by way of an operator (a control parameter of the linear model);

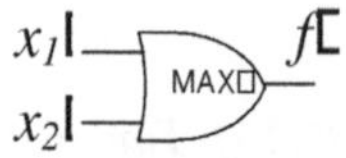

x_1, x_2 → MAX → f

	0	1	2
0	0	0	0
1	0	1	2
2	0	2	1

$f = MAX(x_1, x_2)$

The truth table is partitioned to

$\tau = \lceil 3^2/(2+1) \rceil = 3$ *vectors*

Truth vector $\mathbf{F}^\circ$:

$$\mathbf{F}^\circ = [\mathbf{F}_2|\mathbf{F}_1|\mathbf{F}_0]\mathbf{W} = \begin{bmatrix} 2\ 1\ 0 \\ 2\ 1\ 1 \\ 2\ 2\ 2 \end{bmatrix} \begin{bmatrix} 3^2 \\ 3^1 \\ 3^0 \end{bmatrix} = \begin{bmatrix} 21 \\ 22 \\ 26 \end{bmatrix}$$

Vector of coefficients:

$$\mathbf{D} = \mathbf{T}_3 \cdot \mathbf{F}^\circ = \begin{bmatrix} 1\ 0\ 0 \\ -1\ 1\ 0 \\ -1\ 0\ 1 \end{bmatrix} \begin{bmatrix} 21 \\ 22 \\ 26 \end{bmatrix} = \begin{bmatrix} 21 \\ 1 \\ 5 \end{bmatrix}$$

Linear expression:

$D = 21 + 5x_1^\circ + x_2^\circ,\ \ x_1^\circ, x_2^\circ \in \{0, 1\}$

Calculation of $f = MAX(2, 1)$, $\mu = 2$:

$x_1 = 2 \rightarrow x_1^\circ = 0$
$x_2 = 1 \rightarrow x_2^\circ = 1$

$MAX(2, 1) = \Xi^2\{21 + 5x_1^\circ + x_2^\circ\}$
$= \Xi^2\{22\}$
$= \Xi^2\{211_3\} = 2$

FIGURE 37.11
Representation of the 3-valued 2-variable logic function $f = MAX(x_1, x_2)$ by a linear word-level arithmetic expression (Examples 37.16 and 37.17).

- It is an intrinsically parallel model because it is at word-level; and
- It is extendable to arbitrary logic functions.

The example below illustrates some of these properties by calculation of the function using the linear model given the input assignments. Let a three-valued ($m = 3$) two-input ($n = 2$) elementary logic function be given by the linear expression D. The masking operation

$$f = \Xi^\mu\{D\} \tag{37.15}$$

is used to recover the value of the logic function.

Example 37.17 *(Continuation of Example 37.16.) Calculation of values of $MAX(x_1, x_2)$ given the linear model and $x_1 = 2$, $x_2 = 1$ involves several steps (Figure 37.11):*

(a) *Find the index μ of subvector $\mathbf{F}_\mu$ in a word-level representation. Here, the parameter μ is determined as follows: assignment $x_1 x_2 = 21$ corresponds to the 7-th element of the truth-vector $\mathbf{F}$; hence $\mu = \lfloor 7/3 \rfloor = 2$.*

(b) *Use the encoding rule given in Table 37.4: $x_1 x_2 = 21 \rightarrow x_1^\circ, x_2^\circ = 01$.*

(c) *Calculate the value of $MAX(2,1)$ for the assignment of pseudo-variables $x_1^\circ = 0, x_2^\circ = 1$: $MAX(2,1) = D_2(0,1) = 2$.*

37.8.4 Library of linear models of multivalued gates

Table 37.5 contains the linear arithmetic expressions of various ternary gates from a library of gates. The linear models from Table 37.5 can be extended to an arbitrary logic function.

TABLE 37.5
Library of linear word-level arithmetic models of three-valued gates.

Function			Vector of coefficients
$\overline{x}$	$=$	$2 - x$	$\mathbf{D} = [2\ -1]^T$
$x_1 \cdot x_2 \pmod 3$	$=$	$15x_1^\circ + 21x_2^\circ$	$\mathbf{D} = [0\ 21\ 15]^T$
$MIN(x_1, x_2)$	$=$	$21x_1^\circ + 12x_2^\circ$	$\mathbf{D} = [0\ 12\ 21]^T$
$TSUM(x_1, x_2)$	$=$	$21 + 5x_1^\circ + 4x_2^\circ$	$\mathbf{D} = [21\ 4\ 5]^T$
$MAX(x_1, x_2)$	$=$	$21 + 5x_1^\circ + x_2^\circ$	$\mathbf{D} = [21\ 1\ 5]^T$
$TPROD(x_1, x_2)$	$=$	$21x_1^\circ + 9x_2^\circ$	$\mathbf{D} = [0\ 9\ 21]^T$
$(x_1 + x_2) \pmod 3$	$=$	$21 - 10x_1^\circ - 14x_2^\circ$	$\mathbf{D} = [21\ -14\ -10]^T$
$x_1 \vert x_2$	$=$	$1 - x_1^\circ + 5x_2^\circ$	$\mathbf{D} = [1\ 5\ -1]^T$

Example 37.18 *The ternary function $\overline{x_1 + x_2}$ can be represented by a linear expression as follows*

$$\begin{aligned}\overline{x_1 + x_2} \pmod 3 &= \Xi^\mu\{3^0(2 - x_2^\circ - 2x_1^\circ) \\ &\quad + 3^1(1 - x_2^\circ + x_1^\circ) \\ &\quad + 3^2(2x_2^\circ + x_1^\circ)\} \\ &= \Xi^\mu\{5 + 14x_2^\circ + 10x_1^\circ\}.\end{aligned}$$

37.8.5 Representation of a multilevel, multivalued circuit

Let D be a level of a multivalued, multilevel circuit and consist of r two-input multivalued gates. The level implements an n-input r-output logic function, or subcircuit over the library of gates. Since each gate is described by a linear arithmetic expression, this subcircuit can be described by a linear expression too. The strategy for representation of a multivalued logic circuit by a set of linear expressions is as follows:

$$\texttt{Gate model } D_j \Longleftrightarrow f = \Xi^{\mu}\{D\}$$

$$\texttt{Circuit level model D} \Longleftrightarrow f_j = \Xi^{3(j-1)+\mu}\{L\}$$

$$\texttt{Circuit model (set of D)} \Longleftrightarrow \texttt{ Set of level outputs}$$

To simplify the formal notation, let us consider the library of ternary gates given in Table 37.5.

Let D_j, $j = 1, 2, \ldots, r$, be a linear arithmetic representation of the j-th gate and its output correspond to the j-output of a subcircuit. Assume that the order of gates in the subcircuit is fixed. A linear word-level arithmetic of an n-input r-output of a ternary subcircuit (level) is defined as

$$D = \sum_{j=1}^{r} 3^{3(j-1)} D_j. \tag{37.16}$$

Example 37.19 *Let a level of a ternary circuit be given as shown in Figure 37.12. This figure explains the calculation of the linear expression using Equation 37.16.*

To calculate the value of the j-th output f_j, $j \in \{1, \ldots, r\}$, a masking operator is utilized:

$$f_j = \Xi^{\xi}\{D\}, \tag{37.17}$$

where $\xi = 3(j-1) + \mu$. This recovers the ξ-th digit in a word-level value D.

Example 37.20 *(Continuation of Example 37.19.) Given the assignment*

$$x_1x_2x_3x_4x_5x_6 = 201112 \Longrightarrow x_1^{\circ}x_2^{\circ}x_3^{\circ}x_4^{\circ}x_5^{\circ}x_6^{\circ} = 000110,$$

the outputs f_j, $j \in \{1, 2, 3\}$, are calculated as follows: $f_1 = 2$, $f_2 = 1$, and $f_3 = 2$.

37.8.6 Linear decision diagrams

There are three hierarchical levels in the representation of multivalued functions. The first level corresponds to the description of a gate:

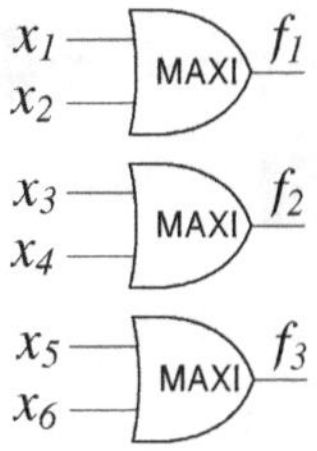

$x_1, x_2 \to$ MAXI $\to f_1$
$x_3, x_4 \to$ MAXI $\to f_2$
$x_5, x_6 \to$ MAXI $\to f_3$

$f_1 \to D_1$
$f_2 \to D_2$
$f_3 \to D_3$
$D_1 = 21 + 5x_1^\circ + x_2^\circ$
$D_2 = 21 + 5x_3^\circ + x_4^\circ$
$D_3 = 21 + 5x_5^\circ + x_6^\circ$

$$
\begin{aligned}
D &= \sum_{j=1}^{3} 3^{3(j-1)} D_j = 3^0 D_1 + 3^1 D_2 + 3^2 D_3 \\
&= 3^0(21 + 5x_1^\circ + x_2^\circ) \\
&+ 3^3(21 + 5x_3^\circ + x_4^\circ) \\
&+ 3^6(21 + 5x_5^\circ + x_6^\circ) \\
&= 15897 + 5x_1^\circ + x_2^\circ + 135x_3^\circ + 27x_4^\circ + 3645x_5^\circ + 729x_6^\circ.
\end{aligned}
$$

The relationship of the assignments of variables and pseudo-variables:

$$x_1x_2x_3x_4x_5x_6 = 201112 \Longrightarrow x_1^\circ x_2^\circ x_3^\circ x_4^\circ x_5^\circ x_6^\circ = 000110$$

Given the assignments $\mu_1 = 2$, $\mu_2 = 1$, $\mu_3 = 1$,
$D = 15897{+}5{\cdot}0{+}1{\cdot}0{+}135{\cdot}0{+}27{\cdot}1{+}3645{\cdot}1{+}729{\cdot}0 = 19569$

The outputs f_j, $j \in \{1, 2, 3\}$, *are recovered by*

$$f_1 = \Xi^{3\cdot 0+2}\{19569\} = \left\lfloor \frac{19569}{3^2} \right\rfloor \ (mod3) = 2$$

$$f_2 = \Xi^{3\cdot 1+1}\{19569\} = \left\lfloor \frac{19569}{3^4} \right\rfloor \ (mod3) = 1$$

$$f_3 = \Xi^{3\cdot 2+1}\{19569\} = \left\lfloor \frac{19569}{3^7} \right\rfloor \ (mod3) = 2$$

FIGURE 37.12
Recover of the MAX function from a word-level linear arithmetic expression (Examples 37.19 and 37.20).

```
Gate ⟺
     Linear expression ⟺
                                       Linear decision diagram
```

The second level corresponds to the description of a level in a multilevel circuit:

```
Circuit level ⟺
              Linear expression ⟺
                                      Linear decision diagram
```

The third level corresponds to the description of the circuit:

```
Circuit ⟺
        Set of linear expressions ⟺
                                        Set of linear decision diagrams
```

Based on the above statements, an arbitrary multivalued network can be modeled by a set of linear word-level decision diagrams.

Example 37.21 *The linear decision diagram and its embedding in a $\mathcal{N}$-hypercube for the ternary MAX gate are represented in Figure 37.13.*

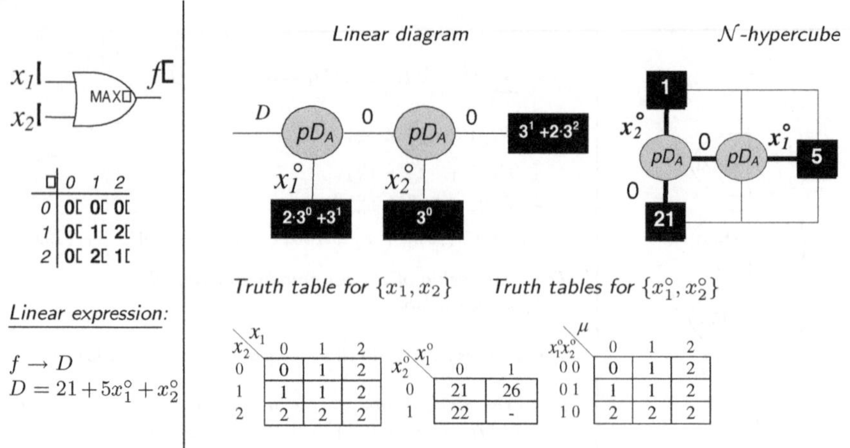

FIGURE 37.13
Representation of the ternary function MAX by a linear decision diagram and $\mathcal{N}$-hypercube (Example 37.21).

37.8.7 Remarks on computing details

One of the problems of word-level representation, including linear forms, is the exponential values of terminal nodes. To calculate these, so-called ZBDD-like trees can be used [3]. A special encoding scheme must be applied in order to achieve reasonable memory usage [10] (see additional information in the "Further reading" Section).

37.9 Linear nonarithmetic word-level representation of multivalued functions

It has been shown in Chapter 7 that an arbitrary switching function can be represented by a linear nonarithmetic word-level expression. In this section, an extension of this technique to multivalued functions is presented.

37.9.1 Linear word-level for MAX expressions

Let us denote:

Variable x_i by $x_{i,0}$ $(q = 0)$,
The complement of variable $\overline{x}_i = (m-1) - x_i$ by $x_{i,1} (q = 1)$,
The cyclic complement of variable $\widehat{x}_i = x_i + 1$ by $x_{i,2}$ $(q = 2)$,

The integer positive values that correspond to the i-th variable x_i by $w_{i,q}$, *MAX* function by $\vee$.

A linear word-level expression for the MAX operation of a n-variable multivalued function is defined by

$$f = \widehat{\bigvee_{i=1}^{n}} w_{i,q} x_{i,q}, \qquad (37.18)$$

Example 37.22 *Examples of word-level representation are given below. For a single-output ternary MAX function of two variables* $f = x_1 \,\widehat{\vee}\, x_2 = x_1 \vee x_2$. *It is a linear expression because it does not contain any product of variables. A two-output ternary function* $f_1 = \overline{x}_1 \vee \overline{x}_2$, $f_2 = x_1 \vee x_2$ *of two variables can be represented by the linear expression* $f = 3\overline{x}_1 \,\widehat{\vee}\, 3\overline{x}_2 \,\widehat{\vee}\, x_1 \,\widehat{\vee}\, x_2$. *Details are given in Figure 37.14.*

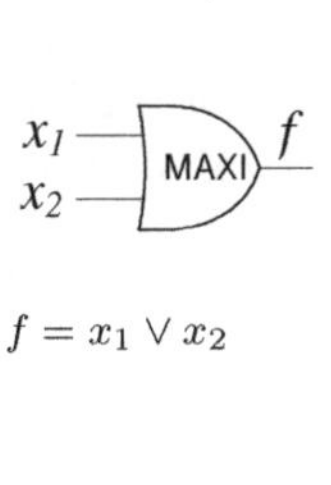

Word-level representation

$$f = w_{1,0}x_{1,0} \,\widehat{\vee}\, w_{2,0}x_{2,0} = x_1 \,\widehat{\vee}\, x_2 = x_1 \vee x_2,$$

where $q = 0,\ i = 1, 2,\ w_{1,0} = w_{2,0} = 1$

(a)

$f_1 = \overline{x}_1 \vee \overline{x}_2$
$f_2 = x_1 \vee x_2$

Word-level representation

$$f = 3(\overline{x}_1 \vee \overline{x}_2) \,\widehat{\vee}\, (x_1 \vee x_2) = w_{1,0}x_{1,0} \,\widehat{\vee}\, w_{2,0}x_{2,0}$$
$$= 3\overline{x}_1 \,\widehat{\vee}\, 3\overline{x}_2 \,\widehat{\vee}\, x_1 \,\widehat{\vee}\, x_2,$$

where $q = \{0, 1\},\ i = 1, 2,\ w_{1,0} = w_{2,0} = 1,\ w_{1,1} = w_{2,1} = 3$

(b)

FIGURE 37.14
Linear word-level nonarithmetic representation of a ternary MAX function (a) and two ternary MAX functions (Example 37.22).

To recover the initial data from the linear expression, we apply a masking operator. A value f_j of a j-th multivalued function, $j \in \{1,\dots,r\}$, can be recovered from a linear expression (Equation 37.18) by the masking operator $f_j = \Xi^j\{f\}$.

Example 37.23 *(Continuation of Example 37.22).*
(a) A single-output ternary MAX function of two variables is recovered:
$f = \Xi^1\{x_1 \,\widehat{\vee}\, x_2\} = x_1 \vee x_2$.
(b) A two-output ternary function of two variables is recovered:
$f_1 = \Xi^1\{3\overline{x}_1 \,\widehat{\vee}\, 3\overline{x}_2 \,\widehat{\vee}\, x_1 \,\widehat{\vee}\, x_2\} = \overline{x}_1 \vee \overline{x}_2$.
$f_2 = \Xi^2\{3\overline{x}_1 \,\widehat{\vee}\, 3\overline{x}_2 \,\widehat{\vee}\, x_1 \,\widehat{\vee}\, x_2\} = x_1 \vee x_2$.

37.9.2 Network representation by linear models

A multilevel multivalued logic network can be described by a linear word-level logic, once each level consists of gates of the same type.

Example 37.24 *The two-input, three-output level of a ternary circuit given in Figure 37.15 is described by the linear expression*

$$0x_1 \,\widehat{\vee}\, 4x_2 \,\widehat{\vee}\, 3\overline{x}_1 \,\widehat{\vee}\, 9\overline{x}_2.$$

The linear decision diagram that corresponds to this expression consists of four nodes implementing the multiple ternary Shannon expansion. The values of the outputs given truth vectors $\mathbf{F_1}$, $\mathbf{F_2}$ *and* $\mathbf{F_3}$ *are calculated in Figure 37.15. Given the assignment*

$$x_1x_2\overline{x}_1\overline{x}_2 = \{0022\},$$

the outputs are equal to

$$\begin{aligned} f_1(0022) &= 0\vee0\vee0\vee0 = 0(x_1 \vee x_2 = 0 \vee 0 = 0)\\ f_2(0022) &= 0\vee0\vee2\vee0 = 2(\overline{x}_1 \vee x_2 = 2 \vee 0 = 2)\\ f_3(0022) &= 0\vee0\vee0\vee2 = 2(x_1 \vee \overline{x}_2 = 0 \vee 2 = 2). \end{aligned}$$

37.10 Further reading

Linearization technique. An elegant method for the linearization of AND, OR and EXOR functions of arbitrary number of input variables was intro-

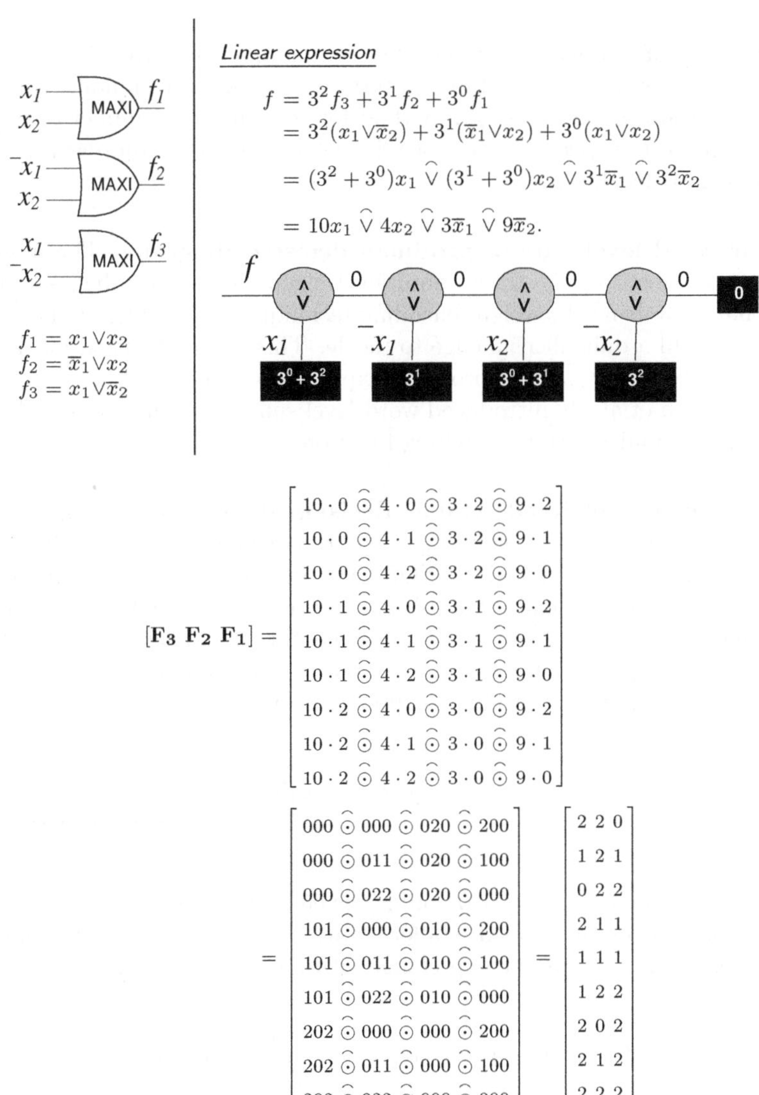

x1
x2
MAXI
f1
¯x1
x2
MAXI
f2
x1
¯x2
MAXI
f3
f1 = x1∨x2
f2 = ¯x1∨x2
f3 = x1∨¯x2
Linear expression
f = 3^2 f3 + 3^1 f2 + 3^0 f1
= 3^2(x1∨¯x2) + 3^1(¯x1∨x2) + 3^0(x1∨x2)
= (3^2 + 3^0)x1 ∨̂ (3^1 + 3^0)x2 ∨̂ 3^1¯x1 ∨̂ 3^2¯x2
= 10x1 ∨̂ 4x2 ∨̂ 3¯x1 ∨̂ 9¯x2.
f
0
0
0
0
0
x1
¯x1
x2
¯x2
3^0 + 3^2
3^1
3^0 + 3^1
3^2
[F3 F2 F1] =
10·0 ⊙̂ 4·0 ⊙̂ 3·2 ⊙̂ 9·2
10·0 ⊙̂ 4·1 ⊙̂ 3·2 ⊙̂ 9·1
10·0 ⊙̂ 4·2 ⊙̂ 3·2 ⊙̂ 9·0
10·1 ⊙̂ 4·0 ⊙̂ 3·1 ⊙̂ 9·2
10·1 ⊙̂ 4·1 ⊙̂ 3·1 ⊙̂ 9·1
10·1 ⊙̂ 4·2 ⊙̂ 3·1 ⊙̂ 9·0
10·2 ⊙̂ 4·0 ⊙̂ 3·0 ⊙̂ 9·2
10·2 ⊙̂ 4·1 ⊙̂ 3·0 ⊙̂ 9·1
10·2 ⊙̂ 4·2 ⊙̂ 3·0 ⊙̂ 9·0
=
000 ⊙̂ 000 ⊙̂ 020 ⊙̂ 200
000 ⊙̂ 011 ⊙̂ 020 ⊙̂ 100
000 ⊙̂ 022 ⊙̂ 020 ⊙̂ 000
101 ⊙̂ 000 ⊙̂ 010 ⊙̂ 200
101 ⊙̂ 011 ⊙̂ 010 ⊙̂ 100
101 ⊙̂ 022 ⊙̂ 010 ⊙̂ 000
202 ⊙̂ 000 ⊙̂ 000 ⊙̂ 200
202 ⊙̂ 011 ⊙̂ 000 ⊙̂ 100
202 ⊙̂ 022 ⊙̂ 000 ⊙̂ 000
=
2 2 0
1 2 1
0 2 2
2 1 1
1 1 1
1 2 2
2 0 2
2 1 2
2 2 2

FIGURE 37.15
A two-level ternary circuit and its linear diagram (Example 37.24).

duced by Malyugin [2]. The method is based on the so-called *algebra of corteges*. Different aspects of this linearization technique can be found in [80]. Additional references can be found in Chapter 9.

Linear transformation of variables is a method for optimizing the representation of a switching function. In terms of a spectral technique, the linear transformation of variables is a method for reducing the number of nonzero coefficients in the spectrum of a switching function. This approach has been developed in [18, 4, 7].

Linear word-level sum-of-products decision diagrams The most important drawback of the linear word-level algebraic decision diagrams is the fact that even using the weight encoding technique, the problem of large coefficients is still a difficult challenge to tackle. This is a motivation for defining the linear word-level sum-of-products expressions and study their properties. Yanushkevich et al. [9] introduced word-level sum-of-products *linear* decision diagrams for multi-output switching functions.

Linear word-level Reed–Muller decision diagrams. The linear word-level sum-of-products decision diagrams are not efficient for EXOR circuits. However, in some technologies EXOR logic was more efficient to implement than NAND and NOR logic. Yanushkevich et al [9] developed the modification of a linear sum-of-products model to avoid this drawback. The word-level Reed–Muller expression for an n-input r-output switching function f is defined as the bitwise of sum-of-products expressions of f_j, $j = 1, \ldots, r$,

$$f = 2^{r-1} f_r \mathbin{\widehat{\oplus}} \cdots \mathbin{\widehat{\oplus}} 2^1 f_2 \mathbin{\widehat{\oplus}} 2^0 f_1 \quad = \quad \widehat{\bigoplus_{i=0}^{2^n-1}} \quad w_i \cdot \underbrace{(x_1^{i_1} \cdots \; x_n^{i_n \cdot})}_{i-th \ product}$$

where $x_i^{i_j}$ is equal to 1 if $i_j = 0$, and $x_i^{i_j}$ is equal to x_i if $i_j = 1$.

The *linear* word-level Reed–Muller expression of a switching function f of n variables $x_1, \ldots, x_n$ is the expression with integer coefficients $w_1^*, \ldots, w_n^*$

$$f = w_1^* x_1^{i_1} \mathbin{\widehat{\oplus}} \ldots \mathbin{\widehat{\oplus}} w_n^* x_n^{i_n} \quad = \quad \widehat{\bigoplus_{i=0}^{n}} \; w_i^* \cdot x_i^{i_j}. \qquad (37.19)$$

where

$$w_i^* x_i^{i_j} = \begin{cases} w_i^{'} x_i, \; i_j = 1; \\ w_i^{''} \overline{x}_i, \; i_j = 0. \end{cases}$$

The above properties are similar to those of linear word-level sum-of-products expressions. Grouping, weight assignment, and masking in the format of linear word Reed–Muller expressions are similar to the linear word-level sum-of-products model. For instance, the linearity property does not depend on the order of the functions in a word-level expression. A linear decision diagram derived from the linear word-level Reed–Muller expression (Equation 37.19) is a linear tree with nodes in which the *multiple* Davio expansion is implemented:

$$f = f_j(x_i = 0) \oplus x_i f_j(x_i = 1), \tag{37.20}$$

where $j = 1, 2, \ldots, r$. The term $w_i^* x_i^{i_j}$ of a word-level sum-of-products carries information about:

- The number of required Davio expansions with respect to variables x_i,
- The functions f_j to which Davio expansion has been applied.

For example, consider the term $5x_1$ in a word-level Reed–Muller expression. We observe that the Davio expansion with respect to variable x_1 is used twice as indicated by the coefficient 5: the number of 1s in $5 = 101$ is equal to 2, $1 + 0 + 1 = 2$. This coefficient also carries information about the function of action: Davio expansion is applied to f_1 and f_3.

In general, a linear decision diagram includes a $2n$ nonterminal nodes and $2n + 1$ terminal nodes as follows from Equation 37.19.

A package for construction of linear word-level decision diagrams was developed by Yanushkevich et al. [10].

References

[1] Karpovsky MG. *Finite Orthogonal Series in the Design of Digital Devices.* John Wiley and Sons, New York, 1976.

[2] Malyugin VD. Realization of Boolean function's corteges by means of linear arithmetical polynomial. *Automation and Remote Control*, Kluwer/Plenum Publishers, 45(2):239–245, 1984.

[3] Minato S. *Binary Decision Diagrams and Applications for VLSI CAD*, Kluwer, Dordrecht, 1996.

[4] Moraga C. On some applications of the Chrestenson functions in logic design and data processing. *Mathematics and Computers in Simulation*, 27:431–439, 1985.

[5] Shmerko VP. Synthesis of arithmetic forms of Boolean functions using the Fourier transform. *Automation and Remote Control*, Plenum/Kluwer Publishers, 50(5):684–691, Pt2, 1989.

[6] Shmerko VP. Malyugin's theorems: a new concept in logical control, VLSI design, and data structures for new technologies. *Automation and Remote*

Control, Plenum/Kluwer Publishers, Special Issue on Arithmetical Logic in Control Systems, 65(6):893–912, 2004.

[7] Stanković RS and Astola JT. Some remarks on linear transform of variables in representation of adders by word-level expressions and spectral transform decision diagrams. In *Proceedings of the IEEE 32nd International Symposium on Multiple-Valued Logic*, pp. 116–122, 2002.

[8] Yanushkevich SN, Shmerko VP, and Dziurzanski P. Linearity of word-level models: new understanding. In *Proceedings of the IEEE/ACM 11th International Workshop on Logic and Synthesis*, pp. 67–72, New Orleans, LA, 2002.

[9] Yanushkevich SN, Shmerko VP, and Lyshevski SE. *Logic Design of NanoICs*, CRC Press, Boca Raton, FL, 2005.

[10] Yanushkevich SN, Shmerko VP, and Dziurzanski P. LDD package. http://www.enel.ucalgary.ca/People/yanush/research.html. Release 01, 2002.

38

Fibonacci Decision Diagrams

In this chapter, a particular class of decision diagrams, *Fibonacci* decision diagrams, are considered.

Fibonacci diagrams are useful in representation and manipulation of discrete functions for the following reasons:

- *Permit representation of functions defined in a number of points different from $N = 2^n$ by decision diagrams still consisting of nodes with two outgoing edges.*
- *Support replacement of Boolean n-cubes by generalized Fibonacci cubes as the algebraic structures suitable for 3D modelling and hosting Fibonacci diagrams in various topological structures.*

Rationales to study Fibonacci cubes can be found in the following features:

(*a*) Boolean n-cube is involved into the set of generalized Fibonacci cubes.

(*b*) The order of a generalized Fibonacci cube that can be embedded into a Boolean n-cube with $k = 1, 2$ faulty nodes is greater than 2^{n-1}.

(*c*) The k dimensional Fibonacci cube of the order $n + k$ is equivalent to a Boolean n-cube of order $0 \leq n < k$.

(*d*) Algorithms developed for a generalized Fibonacci cube are executable on the Boolean cube of the corresponding order.

In this chapter, considered are

- *Fibonacci* decision diagrams.
- *Spectral Fibonacci* decision diagrams.

38.1 Introduction

The sequence whose terms are $1, 1, 2, 3, 5, 8, 13, 21, 34\ldots$ is called the *Fibonacci sequence*. This sequence has the property that after starting with two 1's, each term is the sum of the preceding two, i.e., the Fibonacci numbers are defined by the recurrence

$$F_0 = 0,\ F_1 = 1,\ F_i = F_{i-1} + F_{i-2}$$

for $i \geq 2$. Fibonacci numbers are related to the *golden ratio* and its conjugate.

Fibonacci decision diagrams (F-decision diagrams) consist of nodes with two outgoing edges, and permit representation of functions defined on sets of points whose cardinalities are equal to the generalized Fibonacci p-numbers.

For instance, with Fibonacci decision diagrams for different values of p, we can represent functions defined, for example in 5,7,13,19,55, etc., points, which cannot be done by using other decision diagrams that all can be uniformly viewed as diagrams based upon the multiplicative decomposition for cardinality of the domain of definition for switching function f.

This property justified introduction of Fibonacci decision diagrams by referring to spectral interpretation of decision diagrams as graphical representations of some functional expressions for discrete functions. In the same way, Fibonacci decision trees have been introduced as graphic representations of contracted Fibonacci p-codes.

Fibonacci decision diagrams can be used as a data structure suitable for efficient calculation of the generalized Fibonacci transforms, and in this way permit us to calculate the generalized Fibonacci spectra of functions defined in a number of points equal to large Fibonacci numbers, where the FFT-like algorithms can hardly be used for their exponential complexity in terms of both time and space.

38.2 Terminology and abbreviations for Fibonacci decision trees and diagrams

Fibonacci decision trees are defined by the analogy to the BDDs with the Shannon expansion replaced by the corresponding Fibonacci expansion.

Consider a function defined in a set of points $0, 1, \ldots, \phi_p(i) - 1$, where $\phi_p(i)$ is a generalized Fibonacci number!Fibonacci!number. These points can be represented by n-tuples $(w_1, \ldots, w_n)$ corresponding to the Fibonacci p-codes.

For given values of p and i, the Fibonacci decision tree

$$Fib_pDT(i)$$

for $f(w_1, \ldots, w_n)$ is defined by the recursive application of the Fibonacci-Shannon decomposition rule defined by

$$f = \overline{w}_j f_0 + w_j f_1,$$

where

$$f_0 = f(w_1, \ldots, w_{i-1}, 0, w_{i+1}, \ldots, w_n),$$
$$f_1 = f(w_1, \ldots, w_{i-1}, 1, w_{i+1}, \ldots, w_n),$$

to all the variables w_j in f.

Fibonacci decision diagrams (FibDDs) are derived by the reduction of Fibonacci decision trees with the generalized BDD reduction rules.

Spectral Fibonacci decision trees and diagrams are defined as extensions of the notion of spectral decision diagrams to Fibonacci interconnection topologies.

Spectral Fibonacci decision tree (FibSTDT) is a generalization of the Fibonacci decision tree derived by allowing the Fibonacci spectral coefficients as the values of constant nodes. In a Fibonacci decision tree, products of labels at the edges represent columns of a transform $\mathbf{T}_p(i)^{-1}$ inverse to the Fibonacci transform matrix $\mathbf{T}_p(i)$ used in calculation of the values of constant nodes.

Fibonacci–Walsh decision trees and diagrams are defined as spectral Fibonacci decision trees where the underlying spectral transform is the Fibonacci–Walsh transform.

Definition 38.1 *The Fibonacci–Walsh p-transform in the Hadamard ordering is defined by the matrix*

$$\mathbf{W}^{(p,n)} = \left[\begin{array}{c|c} \bar{\mathbf{W}}^{(p,n-1)} \quad \sqrt{2}\hat{\mathbf{W}}^{(p,n-1)} & \bar{\mathbf{W}}^{(p,n-1)} \\ \hline (\sqrt{2})^p \mathbf{W}^{(p,n-p-1)} & \mathbf{0} - (\sqrt{2})^p \mathbf{W}^{(p,n-p-1)} \end{array}\right], \tag{38.1}$$

for $n > p$, and

$$\mathbf{W}^{(p,m)} = [1], \text{ for } m \leq p, \tag{38.2}$$

$$\mathbf{W}^{(p,p+1)} = \begin{bmatrix} 1 & 1 \\ 1 & -1 \end{bmatrix}, \tag{38.3}$$

where $\bar{\mathbf{W}}^{(p,n-1)}$ and $\hat{\mathbf{W}}^{(p,n-1)}$ are the rectangular matrices formed from the matrix $\mathbf{W}^{(p,n-1)}$ by taking its first

$$\phi_p(n-p-1)$$

columns, and its last

$$\phi_p(n-1) - \phi_p(n-p-1)$$

columns, respectively.

Example 38.1 *Using Definition 38.1, the matrix of Fibonacci–Walsh p-transform for $\phi_1(5)$ is constructed as follows:*

$$\mathbf{W}^{(1,5)} = \left[\begin{array}{cc|c} \bar{\mathbf{W}}^{(1,5-1)} & \sqrt{2}\hat{\mathbf{W}}^{(1,5-1)} & \bar{\mathbf{W}}^{(1,5-1)} \\ \hline (\sqrt{2})^1\mathbf{W}^{(1,5-1-1)} & & \mathbf{0} - (\sqrt{2})^1\mathbf{W}^{(1,5-1-1)} \end{array}\right],$$

$$\mathbf{W}^{(1,1)} = [1], \quad \mathbf{W}^{(1,2)} = \begin{bmatrix} 1 & 1 \\ 1 & -1 \end{bmatrix}$$

$$\mathbf{W}^{(1,3)} = \begin{bmatrix} \overline{\mathbf{W}}^{(,2)} & \sqrt{2}\widehat{\mathbf{W}}^{(1,2)} & \overline{\mathbf{W}}^{(1,2)} \\ \sqrt{2}\mathbf{W}^{(1,1)} & \mathbf{0} & -\sqrt{2}\mathbf{W}^{(1,1)} \end{bmatrix} = \begin{bmatrix} 1 & \sqrt{2} & 1 \\ 1 & -\sqrt{2} & 1 \\ \sqrt{2} & 0 & -\sqrt{2} \end{bmatrix}$$

$$\mathbf{W}^{(1,4)} = \begin{bmatrix} \overline{\mathbf{W}}^{(1,3)} & \sqrt{2}\widehat{\mathbf{W}}^{(1,3)} & \overline{\mathbf{W}}^{(1,3)} \\ \sqrt{2}\mathbf{W}^{(1,2)} & \mathbf{0} & -\sqrt{2}\mathbf{W}^{(1,2)} \end{bmatrix} = \begin{bmatrix} 1 & \sqrt{2} & \sqrt{2} & 1 & \sqrt{2} \\ 1 & -\sqrt{2} & \sqrt{2} & 1 & -\sqrt{2} \\ \sqrt{2} & 0 & -2 & \sqrt{2} & 0 \\ \sqrt{2} & \sqrt{2} & 0 & -\sqrt{2} & -\sqrt{2} \\ \sqrt{2} & -\sqrt{2} & 0 & -\sqrt{2} & \sqrt{2} \end{bmatrix}$$

$$\mathbf{W}^{(p,n-1)} = \begin{bmatrix} \overline{\mathbf{W}}^{(p,n-1)} & \sqrt{2}\widehat{\mathbf{W}}^{(p,n-1)} & \overline{\mathbf{W}}^{(p,n-1)} \\ (\sqrt{2})^p\mathbf{W}^{(p,n-p-1)} & \mathbf{0} & -(\sqrt{2})^p\mathbf{W}^{(p,n-p-1)} \end{bmatrix}$$

$$\mathbf{W}^{(1,5)} = \left[\begin{array}{cc|cc|cc|cc} 1 & \sqrt{2} & \sqrt{2} & \sqrt{2} & 2 & 1 & \sqrt{2} & \sqrt{2} \\ 1 & -\sqrt{2} & \sqrt{2} & \sqrt{2} & -2 & 1 & -\sqrt{2} & \sqrt{2} \\ \hline \sqrt{2} & 0 & -2 & 2 & 0 & \sqrt{2} & 0 & -2 \\ \sqrt{2} & \sqrt{2} & 0 & -2 & -2 & \sqrt{2} & \sqrt{2} & 0 \\ \hline \sqrt{2} & -\sqrt{2} & 0 & -2 & 2 & \sqrt{2} & -\sqrt{2} & 0 \\ \sqrt{2} & 2 & \sqrt{2} & 0 & 0 & -\sqrt{2} & -2 & -\sqrt{2} \\ \hline \sqrt{2} & -2 & \sqrt{2} & 0 & 0 & -\sqrt{2} & 2 & -\sqrt{2} \\ 2 & 0 & -2 & 0 & 0 & -2 & 0 & 2 \end{array}\right].$$

Example 38.2 *The direct and inverse Fibonacci–Walsh transform, FWHT, for $\phi_1(5)$ and switching function f of three variables given by a truth vector $\mathbf{F} = [01101001]^T$ is given in Figure 38.1.*

Direct Fibonacci–Walsh transform of the switching function given by the truth vector

$$\mathbf{F} = [\,0\,1\,1\,0\,1\,0\,0\,1\,]^T$$

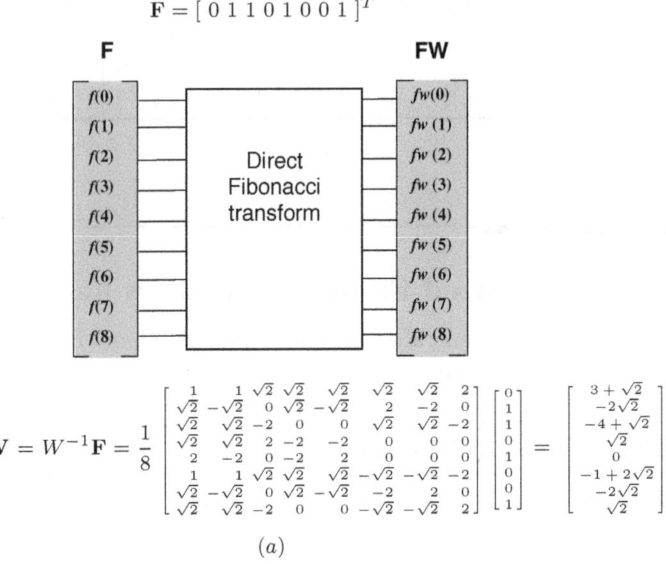

$$\mathbf{FW} = W^{-1}\mathbf{F} = \frac{1}{8}\begin{bmatrix} 1 & 1 & \sqrt{2} & \sqrt{2} & \sqrt{2} & \sqrt{2} & \sqrt{2} & 2 \\ \sqrt{2} & -\sqrt{2} & 0 & \sqrt{2} & -\sqrt{2} & 2 & -2 & 0 \\ \sqrt{2} & \sqrt{2} & -2 & 0 & 0 & \sqrt{2} & \sqrt{2} & -2 \\ \sqrt{2} & \sqrt{2} & 2 & -2 & -2 & 0 & 0 & 0 \\ 2 & -2 & 0 & -2 & 2 & 0 & 0 & 0 \\ 1 & 1 & \sqrt{2} & \sqrt{2} & \sqrt{2} & -\sqrt{2} & -\sqrt{2} & -2 \\ \sqrt{2} & -\sqrt{2} & 0 & \sqrt{2} & -\sqrt{2} & -2 & 2 & 0 \\ \sqrt{2} & \sqrt{2} & -2 & 0 & 0 & -\sqrt{2} & -\sqrt{2} & 2 \end{bmatrix} \begin{bmatrix} 0 \\ 1 \\ 1 \\ 0 \\ 1 \\ 0 \\ 0 \\ 1 \end{bmatrix} = \begin{bmatrix} 3+\sqrt{2} \\ -2\sqrt{2} \\ -4+\sqrt{2} \\ \sqrt{2} \\ 0 \\ -1+2\sqrt{2} \\ -2\sqrt{2} \\ \sqrt{2} \end{bmatrix}$$

(a)

Inverse transform Fibonacci–Walsh transform of the function given by the vector of Fibonacci coefficients

$$\mathbf{FW} = [\,3+2\sqrt{2},\ \ -2\sqrt{2},\ \ -4+2\sqrt{2},\ \sqrt{2},\ 0,\ \ -1+2\sqrt{2},\ \ -2\sqrt{2},\ \sqrt{2}\,]^T$$

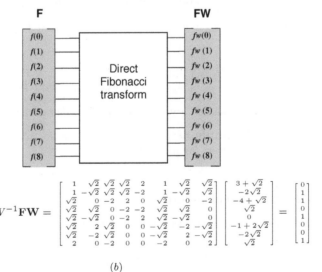

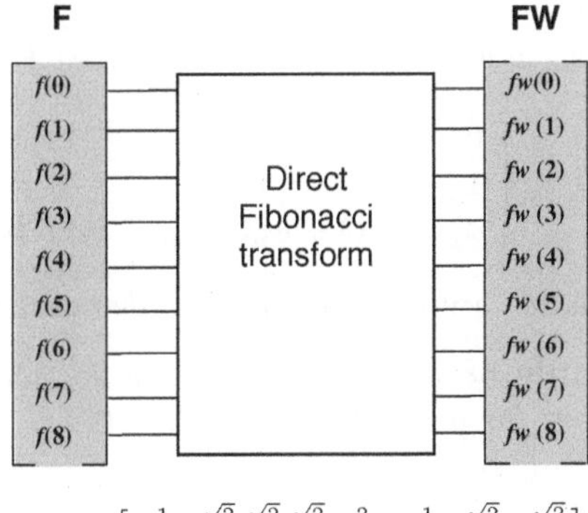

$$\mathbf{F} = W^{-1}\mathbf{FW} = \begin{bmatrix} 1 & \sqrt{2} & \sqrt{2} & \sqrt{2} & 2 & 1 & \sqrt{2} & \sqrt{2} \\ 1 & -\sqrt{2} & \sqrt{2} & \sqrt{2} & -2 & 1 & -\sqrt{2} & \sqrt{2} \\ \sqrt{2} & 0 & -2 & 2 & 0 & \sqrt{2} & 0 & -2 \\ \sqrt{2} & \sqrt{2} & 0 & -2 & -2 & \sqrt{2} & \sqrt{2} & 0 \\ \sqrt{2} & -\sqrt{2} & 0 & -2 & 2 & \sqrt{2} & -\sqrt{2} & 0 \\ \sqrt{2} & 2 & \sqrt{2} & 0 & 0 & -\sqrt{2} & -2 & -\sqrt{2} \\ \sqrt{2} & -2 & \sqrt{2} & 0 & 0 & -\sqrt{2} & 2 & -\sqrt{2} \\ 2 & 0 & -2 & 0 & 0 & -2 & 0 & 2 \end{bmatrix} \begin{bmatrix} 3+\sqrt{2} \\ -2\sqrt{2} \\ -4+\sqrt{2} \\ \sqrt{2} \\ 0 \\ -1+2\sqrt{2} \\ -2\sqrt{2} \\ \sqrt{2} \end{bmatrix} = \begin{bmatrix} 0 \\ 1 \\ 1 \\ 0 \\ 1 \\ 0 \\ 0 \\ 1 \end{bmatrix}$$

(b)

FIGURE 38.1
Direct and inverse Fibonacci transforms for switching function of three variables (Example 38.2).

TABLE 38.1
Generalized Fibonacci numbers.

$\phi_p(i)$	*i=0*	*1*	*2*	*3*	*4*	*5*	*6*	*7*	*8*	*9*
p=0	1	2	4	8	16	32	64	128	256	512
1	1	1	2	3	5	8	13	21	34	55
2	1	1	1	2	3	4	6	9	13	19
3	1	1	1	1	2	3	4	5	7	10

38.3 Generalized Fibonacci numbers and codes

In this section, the generalized Fibonacci numbers and codes are derived from the Fibonacci sequence.

38.3.1 Fibonacci p-numbers

Definition 38.2 *A sequence $\phi(n)$ is the Fibonacci sequence if for each $n \geq 1$,*

$$\phi(n) = \phi(n-1) + \phi(n-2),$$

with initial values $\phi(0) = 1, \phi(n) = 0, n < 0$. Elements of this sequence are the Fibonacci numbers.

A generalization of Fibonacci numbers is as follows.

Definition 38.3 *A sequence $\phi_p(i)$ is the generalized Fibonacci p-sequence if*

$$\phi_p(i) = \begin{cases} 0, & i < 0, \\ 1. & i = 0, \\ \phi_p(i-1) + \phi_p(i-p-1), & i > 0. \end{cases}$$

Elements of this sequence are the generalized Fibonacci p-numbers.

Example 38.3 *Table 38.1 shows the generalized Fibonacci p-numbers for $p = 0, 1, 2, 3,$ and $i = 0, 1, \ldots, 9$.*

38.3.2 Fibonacci p-codes

The Fibonacci p-representation of a natural number B is defined as

$$B = \sum_{i=p}^{n-1} w_i \phi_p(i).$$

The sequence $\mathbf{w} = (w_{n-1}, \ldots, w_p)_p$ is the Fibonacci p-code for B. Since with thus defined weighting coefficients, a given number B may be represented by few different code sequences, the normal unique Fibonacci p-code is introduced by the requirement

$$B = \phi_p(n-1) + m,$$

where $\phi(n-1)$ is the greatest Fibonacci p-number smaller or equal to B, and $0 \leq m < \phi_p(n-p-1)$.

38.4 Fibonacci decision trees

The Fibonacci decision trees can be constructed by:

- Using the decomposition of the cardinal numbers of the domains for the represented functions,
- Using the generalized Fibonacci numbers, and
- Techniques for binary decision trees and multiterminal binary decision trees design.

Notice that binary decision trees and MTBDTs are the Fibonacci decision trees for $p = 0$.

38.4.1 Binary decision trees and multiterminal binary decision trees

Binary decision trees are defined by using the Shannon expansion

$$f = \overline{x}_i f_0 \oplus x_i f_1,$$

where the cofactors f_0 and f_1 for f are defined as

$$f_0 = f(x_1, \ldots, x_{i-1}, 0, x_{i+1}, \ldots, x_n)$$
$$f_1 = f(x_1, \ldots, x_{i-1}, 1, x_{i+1}, \ldots, x_n)$$

to assign a given f to a binary decision tree. The assignment is done by the decomposition of f by the recursive application of the Shannon expansion to all the variables.

The same expansion is used to define the multiterminal binary decision trees (MTBDT), since there is no restriction to the range for f. In these decision trees, it is assumed that the cardinality of the domain for the represented functions is of the form $N = 2^n$. Thus, the Shannon expansion assumes the multiplicative decomposition of N into a product of equal factors.

In the vector representation of f, such decomposition assumes that the vector $\mathbf{F}$ of order 2^n representing values of f is recursively split into subvectors of orders 2^k, $k = 0, \ldots, n-1$.

Example 38.4 *Figure 38.2a shows a node in the decision diagrams for functions with two-valued variables, and Figure 38.2b shows a binary decision tree for functions with $n = 3$ variables. In this tree, the decomposition*

$$N = 8 = 2^3 = 2 \cdot 2 \cdot 2$$

is used. Since the basic factor is 2, this decomposition assumes that the binary tree is generated from the basic binary decision trees which consists of a node with two outgoing edges pointing to two constant nodes.

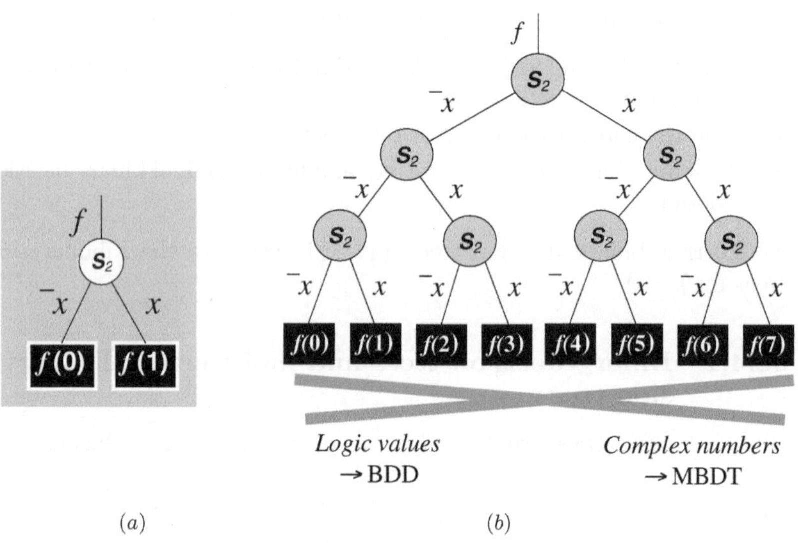

FIGURE 38.2
Basic binary decision tree for the factor 2 (Examples 38.4).

A BDT has the same form as a multiterminal binary decision tree (MTBDT) (see details in Chapter 13), except that in a BDT values of constant nodes are the logic values 0 and 1, while in the MTBDT, they are arbitrary complex numbers. In both cases, the decision tree for a given function f is built by the recursive combination of the basic decision trees for the factor 2.

The Fibonacci decision trees are defined by using the decomposition of the cardinal numbers of the domains for the represented functions in terms of the generalized Fibonacci numbers.

This notion will be introduced by the following example.

Example 38.5 *Consider the generalized Fibonacci number for $p = 1$ and $i = 5$,*

$$N = 8 = \phi_1(5)$$

Since,

$$\phi_1(5) = \phi_1(4) + \phi_1(3) = (\phi_1(3) + \phi_1(2)) + \phi_1(3)$$

we have the decomposition

$$8 = 5 + 3 = (3 + 2) + 3.$$

With this decomposition, the Fibonacci decision tree for $p = 1$ and $i = 5$ is built up from the basic Fibonacci decision trees corresponding to the additive factors 3 and 2, respectively. Figure 38.3 shows the basic Fibonacci decision trees for the factors 2 and 3, respectively.

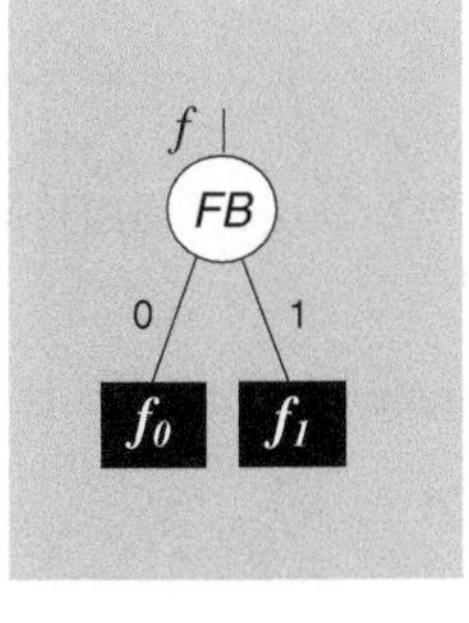

(a)

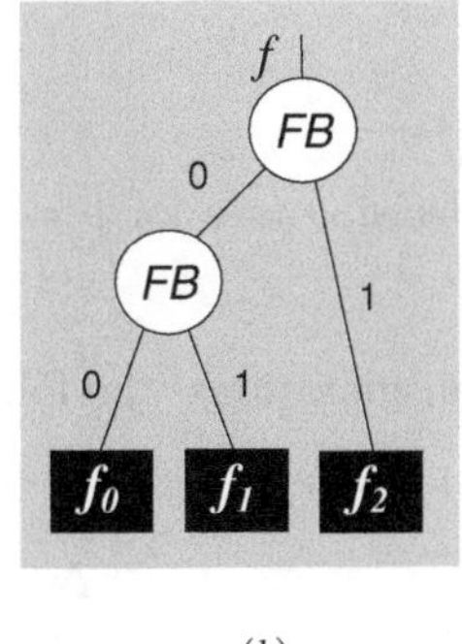

(b)

FIGURE 38.3
Basic Fibonacci decision trees (Example 38.5).

Example 38.6 *Figure 38.4 shows the Fibonacci decision tree for $p = 1$ and $i = 5$ built up as a combination of the basic Fibonacci decision trees. This combination is determined by the assumed decomposition of N in terms of the generalized Fibonacci numbers. The constant nodes represent values of f at particular points in the domain D for the function represented. Thus, $Fib_1DT(5)$ has four levels, since the decomposition of 8 in terms of the generalized Fibonacci 1-numbers is done in four steps.*

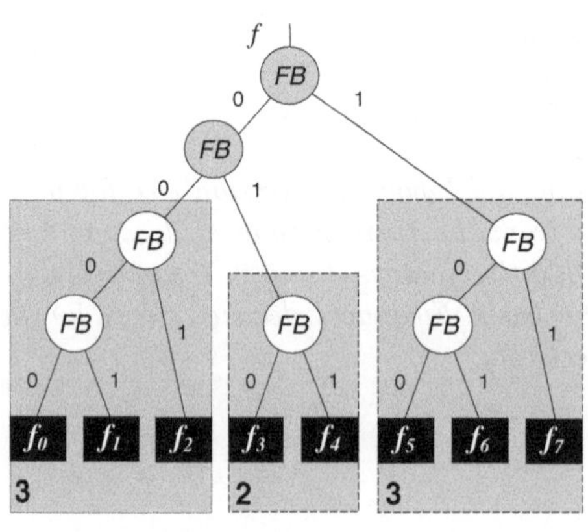

- *The Fibonacci decision trees is a combination of the basic Fibonacci trees*
- *The combination is determined as decomposition of N in terms of the generalized Fibonacci numbers*
- *The constant nodes represent values of f at particular points in D.*
- *The decomposition of 8 in terms of the generalized Fibonacci 1-numbers*

$$
\begin{aligned}
8 &= 5 + 3 \\
&= (3 + 2) + 3 \\
&= (2 + 1) + 2) + (2 + 1) \\
&= (((1 + 1) + 1) \\
&\quad + (1 + 1) + ((1 + 1) + 1))
\end{aligned}
$$

- *$Fib_1DT(5)$ has four levels*

FIGURE 38.4
The Fibonacci decision tree for $p = 1$ and $i = 5$ $\mathrm{Fib}_1\mathrm{DT}(5)$ (Example 38.6).

From these considerations, the following general definition is possible.

Definition 38.4 *For a function f defined in*

$$N = \phi_p(i)$$

points, the Fibonacci decision tree, $Fib_pDT(i)$, is a decision tree derived by using the additive decomposition of N in terms of the generalized Fibonacci p-numbers. The vector $\mathbf{F}$ of order $\phi_p(i)$ showing values for f is recursively split into subvectors of orders

$$\phi_p(i) \quad and \quad \phi_p(i - p - 1).$$

38.4.2 Properties of Fibonacci decision diagrams

In this section, basic properties of Fibonacci decision diagrams are discussed through a comparison with properties of BDDs and MTBDTs. BDTs and MTBDTs consist of basic decision trees for the smallest factors in the multiplicative decomposition for N derived by the Shannon decomposition rule. Similarly, Fibonacci decision trees consist of:

- ► *The basic Fibonacci decision trees for the smallest factors 2 and 3 used in the generalized Fibonacci numbers.*
- ► *A binary decision tree BDT and a MTBDT for* $N = 2^n$ *has* n *levels. Each level corresponds to a multiplicative factor in the decomposition of* N*.*
- ► *A* $Fib_pDT(i)$ *has* $i - p$ *levels, each level corresponding to a step in the recursive determination of* $\phi_p(i)$ *through addition of* $\phi_p(i-1)$ *and* $\phi_p(i-p-1)$*.*
- ► *Since the step in this recursion is equal to* p*, the outgoing edges of a node at the* j*-th level point to the nodes at the* $(j-1)$*-th level and the* $(j-p-1)$*-th level in the* $Fib_pDT(i)$*.*
- ► *In a* Fib_p *decision tree, the left outgoing edges are of the length 1, since they connect the succussive levels.*
- ► *The right outgoing edges are of the length* $p+1$*. This makes an important difference with respect to binary decision trees and MTB decision trees, where both outgoing edges of a node at the* j*-th level point to the nodes at the* $(i-1)$*-th level in the decision tree.*

In binary decision trees and MTBDTs, as in all other decision trees for functions with two-valued variables, based on the multiplicative decomposition for N, both outgoing edges are of the length 1. However, the numbers of the form $N = 2^n$ can be considered as the generalized Fibonacci numbers for $p = 0$ and $i = n$.

Example 38.7 *In terms of the Fibonacci numbers, $N = 8$ can be written as*

$$N = 4 + 4 = (2 + 2) + (2 + 2)$$
$$= ((1 + 1) + (1 + 1)) + ((1 + 1) + (1 + 1)).$$

There are three steps in this decomposition of 8 in terms of the generalized Fibonacci numbers. Each step corresponds to a level in the binary decision tree for $n = 3$. Both outgoing edges of a node at the j-th level point to the nodes at the $(j - 1)$-th level, since $p = 0$.

A binary decision trees and MTBDTs are the Fibonacci decision trees for $p = 0$.

As noticed above, the introduction of Fibonacci decision trees is justified by the property that they permit representing functions defined in a number of points different from $N = 2^n$ by decision trees with two outgoing edges. As is noted above, for a given N, the number of levels in the $\text{Fib}_p\text{DT}(i)$ is $(i - p)$. Different choices for p require different values for i, and thus produce Fibonacci decision trees with different numbers of levels. However, these Fibonacci decision trees have the same number of nodes.

Example 38.8 *Figure 38.5 compares Fibonacci decision trees for $N = 13$ with $p = 1$, and $p = 2$:*

- *$p = 1$ From Table 38.1, $i = 6$. The number of levels of $Fib_1DT(6)$ is equal to*

$$i - p = 6 - 1 = 5.$$

- *$p = 2$ From Table 38.1, $i = 8$. The number of levels of $Fib_2DT(8)$ is equal to*

$$i - p = 8 - 2 = 6.$$

Both Fibonacci decision trees have 12 non-terminal nodes.

For a given N, the $\text{Fib}_p\text{DT}(i)$ with the smallest value for p has the minimum number of levels. In that respect, binary decision trees and MTB decision trees are the optimal decision trees.

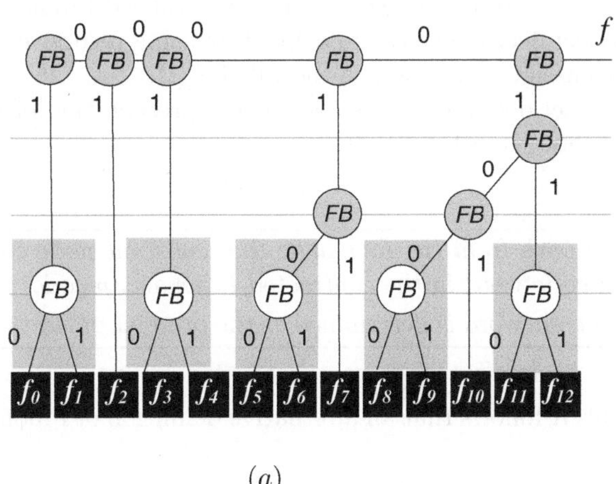

(a)

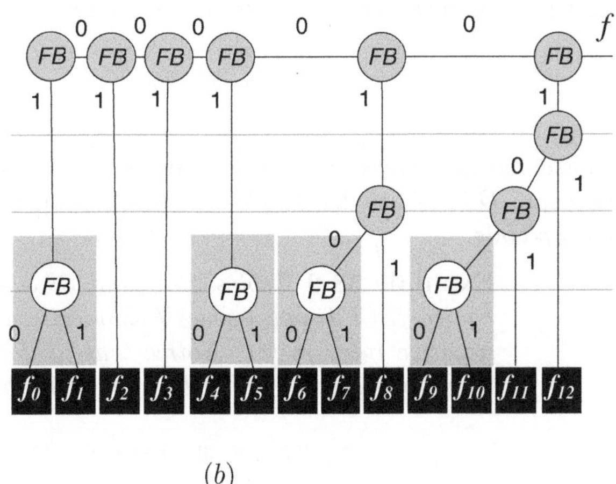

(b)

FIGURE 38.5
Fibonacci decision tree for $N = 13$ with $p = 2$ (Example 38.8).

38.5 Fibonacci decision trees and contracted Fibonacci p-codes

The contracted Fibonacci p-code is defined by the decomposition of the given number in terms of the largest possible generalized Fibonacci numbers. Since the Fibonacci decision trees are defined by using recursively the same decomposition to assign a function to the $\text{Fib}_p\text{DT}(i)$.

A Fibonacci decision tree is a graphic representation of the contracted Fibonacci p-code. In a $\text{Fib}_p\text{DT}(i)$:

- ► *Each path from the root node to a constant node corresponds to a code word in the contracted Fibonacci p-code.*
- ► *This code word is determined as the product of labels at the edges.*

From this it follows that an alternative definition of Fibonacci decision trees is possible.

Definition 38.5 *Binary decision trees are a graphic representation of switching function f when the points x_i, $i = 0, \ldots, N-1$ in the domain D for f are denoted by the binary sequences determined by using the mapping*

$$x_i = \sum_{j=1}^{n} 2^{n-j} x_{i,j}.$$

Fibonacci decision trees are graphic representations for f, when the points in the domain D for f are denoted by the binary sequences in the generalized Fibonacci p-code.

Example 38.9 *Tables 38.2 and 38.3 show the normal and contracted Fibonacci codes!Fibonacci!code for the first 12 non-negative integers [5]. Notice that in the normal Fibonacci p-code for a given number B, if $w_i = 1$, then*

$$w_{i-1} = w_{i-2} = \cdots = w_{i-p} = 0$$

Utilizing this property, the contracted Fibonacci p-code is defined by deleting p zeros after each 1 in the Fibonacci p-code for B, except for the rightmost 1, in which case we should delete max(i, p) zeros, where there are i zeros to the right of the rightmost 1.

The contracted Fibonacci 2-codes given in Table 38.3 are determined by descending the paths in the Fib_2decision tree in Figure 38.5.

TABLE 38.2
Normal Fibonacci 2-code.

B	**Code**
0	000000
1	000001
2	000010
3	000100
4	001000
5	001001
6	010000
7	010001
8	010010
9	100000
10	100001
11	100010
12	100100

TABLE 38.3
Contracted Fibonacci 2-code.

B	**Code**
0	000000
1	000001
2	00001
3	0001
4	0010
5	0011
6	0100
7	0101
8	011
9	1000
10	1001
11	101
12	11

Example 38.10 *Table 38.4 shows functions P_r, $r = 1, 2, 3, 4$, whose values at the point w are equal to the the r-th bits in the normal Fibonacci 1-code word w for $i = 5$. Figure 38.6 compares representation of these functions by BDDs and $Fib_1DD(i)$.*

TABLE 38.4
Function P_i, $i = 1, 2, 3, 4$.

	0	1	2	3	4	5	6	7
P_1	0	0	0	0	0	1	1	1
P_2	0	0	0	1	1	0	0	0
P_3	0	0	1	0	0	0	0	1
P_4	0	1	0	0	1	0	1	0

38.6 Spectral Fibonacci decision diagrams

We use the spectral interpretation of decision diagrams (see chapter 13), to define the *spectral Fibonacci* decision diagrams. In this interpretation, the

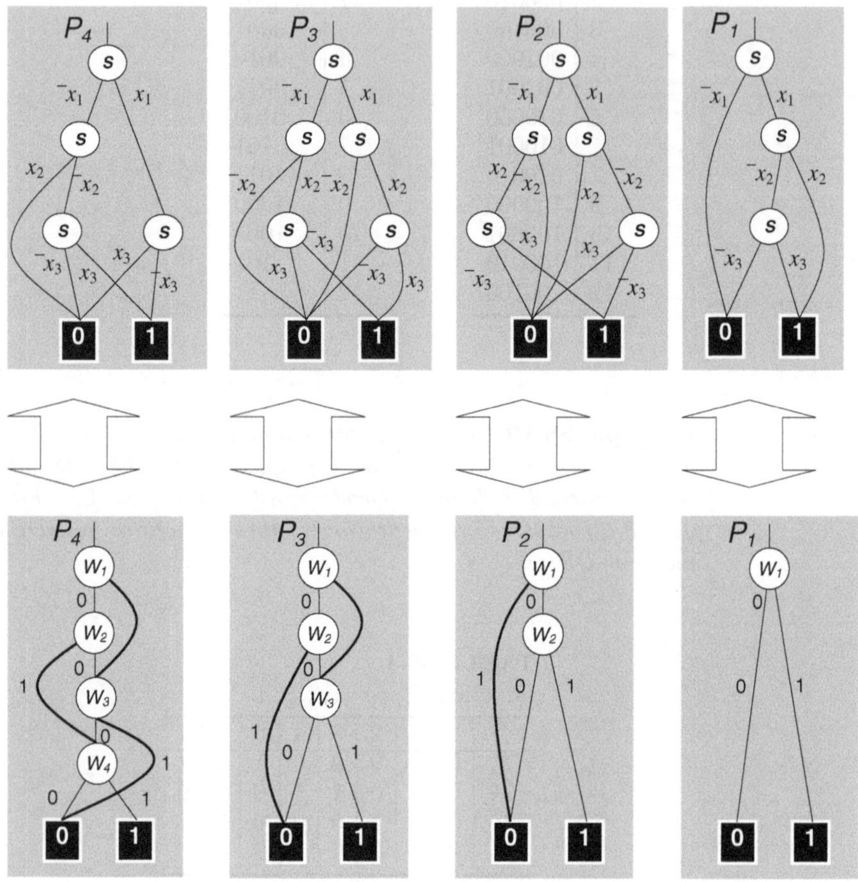

FIGURE 38.6
BDDs and $\text{Fib}_p\text{DT}(i)$s for P_r (Example 38.10)

same as binary decision trees and MTB-decision diagrams, Fibonacci decision diagrams are defined by using the identical mapping to assign a given f to a decision tree. This statement appreciates difference in the multiplicative and the additive decompositions of the cardinality N used in definition of these decision trees.

Example 38.11 *If $N = 8$, the multiplicative decomposition used in binary decision trees and MTBDTs generates the identity matrix $\mathbf{I}_8$ as*

$$\mathbf{I}_8 = \mathbf{I}_2 \times \mathbf{I}_2 \times \mathbf{I}_2.$$

In the Fibonacci decision trees, the additive decomposition of $\mathbf{I}_8$ is given as

$$\mathbf{I}_8 = diag\{\mathbf{I}_3, \mathbf{I}_2, \mathbf{I}_3\}.$$

Different decision diagrams are defined by using different spectral transforms to assign f to the decision tree. Spectral transform Fibonacci decision diagrams (FibST-decision diagrams) in terms of the generalized Fibonacci transforms are defined as Fibonacci decision trees whose constant nodes show values of the Fibonacci spectrum for the function represented. In these decision diagrams, labels at the edges are determined such that the product of labels along paths from the root node to the constant nodes describe columns in the matrix inverse to the transform matrix of the Fibonacci transform used in definition of the decision tree.

The following example taken from [73] illustrates a spectral Fibonacci decision diagram.

Example 38.12 *Figure 38.7 shows a Fibonacci–Walsh decision tree, Fib_1 WTDT(5), for a switching function f with a flat Fibonacci–Walsh spectrum*

$$\mathbf{X}_{w,f} = [0, 0, 0, 0, 1, 0, 1, 1]^T.$$

38.7 Further reading

Fibonacci heaps and cubes A Fibonacci heap is a collection of trees. Fibonacci heaps were introduced Fredman and Tarjan [10]. Details on applications of Fibonacci heaps the reader can find in book by Cormen et al. [8]. Fibonacci cubes are interconnections networks that possess useful properties in networks and decision diagrams design. The Fibonacci cube proposed by Hsu [37] is a special subcube of a hypercube based on Fibonacci numbers.

In the original domain, this function is given by

$$\mathbf{F} = [2+2\sqrt{2},\ -2-\sqrt{2},\ -2+\sqrt{2},\ -2,\ 2,\ -2,\ 2-\sqrt{2},\ 2-\sqrt{2}]^T$$

$$q = 1-2w_1$$
$$v = 1-2w_2$$
$$r = 1-2w_3$$
$$z = 1-2w_3$$

The Fib$_1$*WDD represents* f *in the form of the Fibonacci–Walsh polynomial in terms of the functions* P_i *for* $\phi_1(5)$

$$\begin{aligned} f &= (\overline{w}_2+\sqrt{2}w_2)\sqrt{2}(1-2w_2)\overline{w}_3(1-2w_4) \\ &\quad +\sqrt{2}(1-2w_1)\overline{p}_2(\overline{w}_4+\sqrt{2}w_4)(1-2w_4) \\ &\quad +2\overline{w}_2\overline{w}_4(1-2w_1)(1-2w_3) \\ &= \sqrt{2}\overline{w}_2\overline{w}_3 - 2\sqrt{2}\overline{w}_2\overline{w}_3w_4 - 2w_2\overline{w}_3 \\ &\quad +4w_2\overline{w}_3w_4 + (2+\sqrt{2})\overline{w}_2\overline{w}_4 \\ &\quad -(4+2\sqrt{2})w_1\overline{w}_2\overline{w}_4 - 2\overline{w}_2w_4 \\ &\quad +4w_1\overline{w}_2w_4 - 4\overline{w}_2w_3\overline{w}_4 + 8w_1\overline{w}_2w_3\overline{w}_4 \end{aligned}$$

Fibonacci–Walsh spectrum is

$$\mathbf{X}_{w,f} = [0,0,0,0,1,0,1,1]^T$$

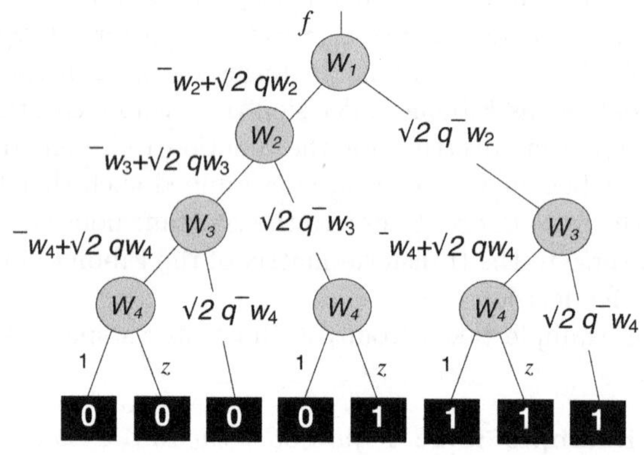

FIGURE 38.7
Fibonacci–Walsh decision tree Fib$_1$WTDT(5) (Example 38.12).

It has been shown that Fibonacci cubes can efficiently emulate various algorithms. Wu [18] extended a class of Fibonacci cubes and proposed *extended Fibonacci cubes*, which are better embedding in other topologies.

Fibonacci decision diagrams. Fibonacci decision diagrams are introduced in [5]. We refer to [1], [2], [6], [7] for applications of such functions and related spectral transforms.

Spectral transform Fibonacci decision diagrams. Different decision diagrams are defined by using different spectral transforms to assign f to the decision tree. The spectral transform Fibonacci decision diagrams (FibST-decision diagrams) in terms of the Generalized Fibonacci transforms are discussed in [6], [5], [9].

Polynomial expressions for functions in Fibonacci topologies. In [4], [15] and [16] various Boolean representations for switching functions, such as sum-of-products, Reed–Muller expressions, Kronecker, and pseudo-Kronecker AND-EXOR expressions, are extended to functions used in Fibonacci interconnection topologies. Both bit-level and world-level expressions, as well as arithmetic and Walsh expressions, to these functions are considered. In this way, a theoretical base to extend application of powerful representation methods and related design tools for switching functions to functions in Fibonacci interconnection topologies are provided.

Circuit synthesis from Fibonacci decision diagrams. Synthesis of circuits from Fibonacci decision diagrams has been considered in [17].

Calculations of Fibonacci transforms. Calculation of various generalizations of Fibonacci transforms by Fibonacci decision diagrams has been discussed in [3].

References

[1] Agaian S, Astola J, and Egiazarian K. *Binary Polynomial Transforms and Nonlinear Digital Filters.* Marcel Dekker, 1995.

[2] Agaian S, Astola J, Egiazarian K, and Kuosmanen P. Decompositional methods for stack filtering using Fibonacci p-codes. *Signal Processing*, 41(1):101–110,1955.

[3] Aizenberg NN, Aizenberg IN, Stanković M, Stanković RS, Astola JT, and Egiazarian K. Calculation of golden Fibonacci transforms. In *Proceedings of the International TICSP Workshop on Spectral Methods and Multirate Signal Processing*, Pula, Croatia, pp. 25–32, 2001.

[4] Astola JT, Egiazarian K, Stanković M, and Stanković RS. Fibonacci arithmetic expressions. *Automation and Remote Control*, Kluwer/Plenum Publishers, 65(6):842–856, 2004.

[5] Egiazarian K, Astola J, and Agaian S. Orthogonal transforms based on generalized Fibonacci recursions. In *Proceedings of the 2nd International Workshop on Spectral Techniques and Filter Banks*, Brandenburg, Germany, 1999.

[6] Egiazarian K and Astola J. Discrete orthogonal transforms based on Fibonacci-type recursion. In *Proceedings of the IEEE Digtal Signal Processing Workshop*, Norway, 1996.

[7] Egiazarian K, Gevorkian D, and Astola J. Time-varying filter banks and multiresolution transforms based on generalized Fibonacci topology. In *Proceedings of the 5th IEEE International Workshop on Intelligent Signal Processing and Communication Systems*, Kuala Lumpur, Malaysia, pp. S16.5.1–S16.5.4, 1997.

[8] Cormen TH, Leiserson CE, Rivest RL, and Stein C. *Introduction to Algorithms.* MIT Press, 2001.

[9] Egiazarian K and Astola J. On generalized Fibonacci cubes and unitary transforms *Applicable Algebra in Engineering, Communication and Computing,* AAECC8:371–377, 1997.

[10] Fredman ML and Tarjan RE. Fibonacci heaps and their uses in improved network optimization algorithms. *Journal of the ACM*, 34(3):596–615, 1987.

[11] Hsu WJ. Fibonacci cubes – a new interconnection topology. *IEEE Transactions on Parallel and Distributed Systems*, 4(1):3–12, 1993.

[12] Muroga S. *Threshold Logic and its Applications.* John Wiley and Sons, New York, 1971.

[13] Stakhov AP. *Algorithmic Measurement Theory,* Publishing Houes "Znanie", Moscow, Issue 6, 1979 (In Russian).

[14] Stanković RS, Stanković M, Astola JT, and Egiazarian,K. Fibonacci decision diagrams and spectral transform Fibonacci decision diagrams. In *Proceedings of the 30th International Symposium on Multiple-Valued Logic*, pp. 206–211, 2000.

[15] Stanković RS, Stanković M, Astola JT, and Egiazarian K. *Fibonacci Decision Diagrams.* TICSP Series # 8, ISBN 952-15-0385-8, Tampere, Finland, 2000.

[16] Stanković RS, Stanković M, Astola JT, and Egiazarian K. Bit-level and word-level polynomial expressions for functions in Fibonacci interconnection topologies. In *Proceedings of the 31st International Symposium on Multiple-Valued Logic*, pp. 305–310, 2001.

[17] Stanković RS, Astola J, Stanković M, and Egiazarian K. Circuit synthesis from Fibonacci decision diagrams. *VLSI Design, Special Issue on Spectral Techniques and Decision Diagrams*, 14(1):23–34, 2002.

[18] Wu J. Extended Fibonacci cubes. *IEEE Transactions on Parallel and Distributed Systems*, 8(12):1203–1210, 1997.

39

Techniques of Computing via Taylor - Like Expansions

The aim of this chapter is to introduce various techniques used for implementing Reed–Muller, arithmetic and Walsh representations of switching functions. These representations are based on *polynomial-like* descriptions, where information about switching functions is carried by the values of the coefficients. When extracting this information, various properties of these polynomial-like representations can be used. For example, switching funtions can be recovered using the inverse Reed–Muller, arithmetic, and Walsh transforms. In practice, a local level of analysis is needed. The *Taylor-like* expansions can be used on both levels of analysis. These expansions are based on the classical Taylor expansion. The modifications are made with respect to particular properties of logic functions and are related to the definition of differences: *Boolean differences** for Reed–Muller expansion, *arithmetic differences* for arithmetic expansion, and *Walsh differences* for Walsh expressions.

The basic idea is that:

- ▶ *Coefficients of the Reed–Muller expression are values of Boolean differences, the components of the Taylor logic expansion.*
- ▶ *Coefficients of the arithmetic expression are the values of the arithmetic analogs of the Boolean differences.*
- ▶ *Coefficients of the Walsh expression are values of "Walsh" differences, the arithmetic analogs of Boolean differences.*

*The term "logic differences" is for switching and multivalued functions. For switching functionality, the term "Boolean differences" or "Boolean derivative" is used. In this chapter, Taylor-like expansions are considered for switching functions. Logic differences for multivalued functions are introduced in Chapter 32.

39.1 Terminology

Taylor-like expansion is based on the classical Taylor expansion, which is modified using some particular properties of switching functions. In Taylor-like expansion, the coefficient corresponds to the differences.

Differences of switching functions are the coefficients in Taylor-like expansions. These differences are described in terms of switching functions.

Polynomial-like representations are Reed–Muller, arithmetic, Walsh, and word-level expansions. There are many other types of polynomial-like representations such as *Walsh-like*, and *Haar-like*, and their generalizations for multiple-valued functions.

Spectral coefficients. This term is used when stressing the spectral nature of polynomial-like expressions of logic funtions. Equivalent terms are Reed–Muller, arithmetic, and Walsh coefficients, or logic, arithmetic, and Walsh differences when Taylor expansion is used.

Distribution. Information about a switching function is contained in the values of the coefficients of a Taylor-like expansion. The distribution is defined by grouping coefficients. Each polynomial-like representation is characterized by a particular and unique grouping of coefficients and distribution.

Global and local levels of extracting information from coefficients. A global level refers to the methods for recovering a switching function or computing the behavioral characteristics of a complete switching function. In this level, all coefficients of polynomial-like expressions are processed. At the local level, information about the local properties of switching function is extracted.

39.1.1 Additive structure of spectral coefficients

The examples below demonstrate the additive structure of Reed–Muller coefficients.

Example 39.1 *There are three nonzero binary coefficients, r_1, r_2, and r_3 in Reed–Muller spectrum*

$$\begin{aligned} f &= r_0 \oplus r_1 x_2 \oplus r_2 x_1 \oplus r_3 x_1 x_2 \\ &= x_2 \oplus x_1 \oplus x_1 x_2 \end{aligned}$$

of the OR switching function. The indices and values of these coefficients carry information about the OR function. The simplest method to extract the information is to utilize the additive property of the Reed–Muller transform:

$$\begin{aligned} r_1 &= f(0) \oplus f(1), \quad r_2 = f(0) \oplus f(2), \\ r_3 &= f(0) \oplus f(1) \oplus f(2) \oplus f(3). \end{aligned}$$

The usefulness of this property is limited in practice because of its high computational complexity, unless the transform is converted to the fast Fourier-like transform.

39.1.2 Multiplicative structure of spectral coefficients

In this chapter, the Taylor-like expansion is used to extract information in the form of *multiplicative* data from spectral coefficients. The information is extracted in the form of multiplicative data that characterizes the properties of a switching function in terms of change.

Example 39.2 *Applying logic Taylor expansion to an OR function of two variables* $f = x_1 \vee x_2$ *given the assignment* $x_1x_2 = 00$ *yields:*

$$r_0 = f(0) = f(00) = 0,$$

Changing x_2 *causes the change of* f*:* $f(00) \rightarrow f(01)$, *i.e.,*

$$r_1 = \frac{\partial f(0)}{\partial x_2} = 1.$$

The same with coefficient r_2*: changing* x_1 *causes the change of* f*:* $f(00) \rightarrow f(10)$, *i.e.,* $r_2 = \frac{\partial f(0)}{\partial x_1} = 1$. *In contrast to coefficients* r_1 *and* r_2*, which are described by a single change, coefficint* r_3 *is represented by the product of two changes: separate and simultaneous changing of* x_1 *and* x_2 *causes change of* f*, i.e.,* $r_3 = \frac{\partial^2 f(0)}{\partial x_1 \partial x_2} = 1$.

39.1.3 Relationship of polynomial-like representations

Several forms of representing switching or multivalued functions utilize a polynomial structure for description. This means that the values of the coefficients in polynomial forms carry information about the logic function.

The main property of polynomial representations is that information about a logic function is *distributed* in the coefficents. The character of distribution is defined by the corresponding transforms: Reed–Muller transform for Reed–Muller or EXOR representation, arithmetic transform for arithmetic representation, and Walsh transform for Walsh representation. Polynomial-like representations of switching functions require different strategies for extracting the information.

The efficiency of polynomial-like methods for manipulating logic functions is defined by the efficiency of representation, and the efficiency of extraction of information.

Given a switching function, a polynomial-like representation with a minimal number of non-zero coefficients corresponds to an efficient representation of a switching function. Given a polynomial-like representation of a switching function, a method that is able to extract information about the properties of a switching function from a few coefficients of a large group of non-zero coefficients is referred to as an efficient method.

39.2 Computing Reed–Muller expressions

In a classical Taylor series, the coefficients are calculated as derivatives of the initial function at a certain point. By analogy, the coefficients of the logic Taylor series are Boolean differences (derivatives) with respect to a variable, and multiple Boolean differences at certain points (assignments of Boolean variables). The logic Taylor series for a switching function f of n variables at the point $c \in 0, 1, \ldots, 2^n - 1$, is defined by the equation

$$f = \bigoplus_{i=0}^{2^n-1} r_i^{(c)} \underbrace{\overbrace{(x_1 \oplus c_1)}^{c_1-polarity}{}^{i_1} \ldots \overbrace{(x_n \oplus c_n)}^{c_n-polarity}{}^{i_n}}_{i-th\ product}, \tag{39.1}$$

where $c_1, c_2, \ldots, c_n$ and $i_1, i_2, \ldots, i_n$ is the binary representation of c and i correspondingly, and the i-th coefficient $r_i^{(c)}(d)$ is calcutated as follows:

$$r_i^{(c)}(d) = \left.\frac{\partial^n f(c)}{\partial x_1^{i_1} \partial x_2^{i_2} \ldots \partial x_n^{i_n}}\right|_{d=c} \quad \text{and} \quad \partial x_i^{i_j} = \begin{cases} 1, & i_j = 0, \\ \partial x_j & i_j = 1. \end{cases}$$

The $r_i^{(c)}(d)$ is the value of the n-order Boolean difference of f where $x_1 = c_1, \ldots, x_n = c_n$.

It follows from Equation 39.1 that

- ► *A variable x_j is 0-polarized if it enters into the expansion uncomplemented, and 1-polarized otherwise.*
- ► *Coefficients of Reed–Muller expression are described in terms of change formally defined as Boolean derivatives.*
- ► *The logic Taylor expansion of an n variable switching function f produces 2^n Reed–Muller expressions of 2^n polarities.*

While the i-th spectral coefficient $r_i^{(c)}(d)$ is described by a Boolean expression, it can be calculated in different ways, for example, matrix transformations, cube-based technique, decision diagram technique, and probabilistic methods (see details in Chapters 20 and 24).

Example 39.3 *According to Equation 39.1, the Reed–Muller spectrum of an arbitrary switching function of two variables ($n = 2$) and the polarity $c = 1$ (x_1 is uncomplemented and x_2 is complemented) is a logic Taylor expansion of this function (Figure 39.1). The corresponding Davio tree can be embedded into an $\mathcal{N}$-hypercube, and computing the coefficients $r_i(1)$ can be accomplished in four systolic cycles on this $\mathcal{N}$-hypercube structure. Details on embedding techniques are given in Chapter 34 and on systolic computing – in Chapter 33.*

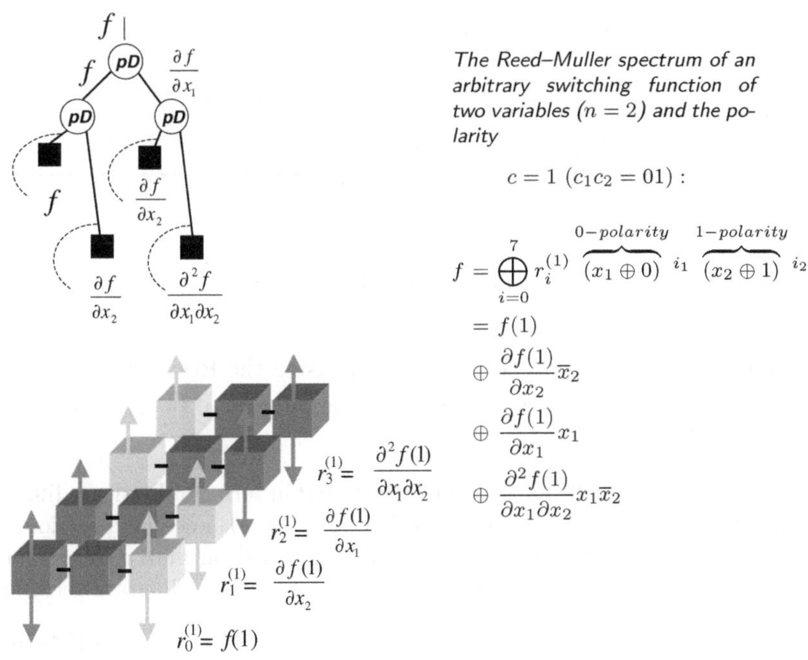
pD
The Reed–Muller spectrum of an arbitrary switching function of two variables ($n = 2$) and the polarity
$c = 1 \ (c_1c_2 = 01)$:
0−polarity
1−polarity
$f = \bigoplus_{i=0}^{7} r_i^{(1)} \overbrace{(x_1 \oplus 0)}^{0-polarity}{}^{i_1} \overbrace{(x_2 \oplus 1)}^{1-polarity}{}^{i_2}$
$= f(1) \oplus \frac{\partial f(1)}{\partial x_2}\overline{x}_2 \oplus \frac{\partial f(1)}{\partial x_1}x_1 \oplus \frac{\partial^2 f(1)}{\partial x_1 \partial x_2}x_1\overline{x}_2$
$r_3^{(1)} = \frac{\partial^2 f(1)}{\partial x_1 \partial x_2}$
$r_2^{(1)} = \frac{\partial f(1)}{\partial x_1}$
$r_1^{(1)} = \frac{\partial f(1)}{\partial x_2}$
$r_0^{(1)} = f(1)$

FIGURE 39.1
Representing a switching function of two variables by a logic Taylor expansion (Example 39.3).

Table 39.1 represents a fragment of a library of primitive gates in terms of Boolean differences. For example, the OR gate in polarity $c = 3$ ($c_1c_2 = 11$) is represented by two non-zero spectral components $f^{(0)}(3)$ and $f^{(3)}(3)$,

$f = 1 \oplus \overline{x}_1\overline{x}_2$. This is an optimal spectral representation of the OR function (optimal polarity).

TABLE 39.1
Reed–Muller expressions as logic Taylor expansions of elementary switching functions.

Function	Reed–Muller spectrum F	$\frac{\partial f}{\partial x_2}$	$\frac{\partial f}{\partial x_1}$	$\frac{\partial^2 f}{\partial x_1 \partial x_2}$	Reed–Muller expression
$f = x_1 \wedge x_2$	0	0	0	1	x_1x_2
	0	0	1	1	$x_1 \oplus x_1\overline{x}_2$
	0	1	0	1	$x_2 \oplus \overline{x}_1x_2$
	1	1	1	1	$1 \oplus \overline{x}_2 \oplus \overline{x}_1 \oplus \overline{x}_1\overline{x}_2$
$f = x_1 \vee x_2$	0	1	1	1	$x_2 \oplus x_1 \oplus x_1x_2$
	1	1	0	1	$1 \oplus \overline{x}_2 \oplus x_1\overline{x}_2$
	1	0	1	1	$1 \oplus \overline{x}_1 \oplus \overline{x}_1x_2$
	1	0	0	1	$1 \oplus \overline{x}_1\overline{x}_2$
$f = x_1 \oplus x_2$	0	1	1	0	$x_2 \oplus x_1$
	1	1	1	0	$1 \oplus \overline{x}_2 \oplus x_1$
	1	1	1	0	$1 \oplus x_2 \oplus \overline{x}_1$
	0	1	1	0	$\overline{x}_2 \oplus \overline{x}_1$

There are two ways to calculate the Reed–Muller spectrum of a switching function:[†]

- ▶ The direct and inverse transformations using the Reed–Muller basis, and
- ▶ Logic Taylor series at each point; the components of the logic Taylor series are Boolean derivatives.

The i-th spectral coefficient can be interpreted in terms of Boolean differences. While the i-th spectral coefficient is described by a Boolean expression, it can be calculated in different ways, for example, based on matrix representation of Boolean differences, cube representation, or decision diagram technique.

Example 39.4 *In Figure 39.2, a three-input logic circuit is represented by Reed–Muller expressions of different polarities.*

It follows from this example that it is possible to calculate separate spectral coefficients or their arbitrary sets to characterize the pruned property of a logic Taylor expansion (Equation 39.1). The logic Taylor expansion generates a family of 2^n Reed–Muller spectra of a switching function. In terms of signal

[†]We do not discuss approaches to spectral computing based on the probabilistic method and information-theoretic measures.

		Reed–Muller spectral coefficients $r_i^{(C)}$ Boolean differences							
		000	*001*	*010*	*011*	*100*	*101*	*110*	*111*
c	$\mathbf{R}_c$	$\mathbf{F}$	$\frac{\partial f}{\partial x_3}$	$\frac{\partial f}{\partial x_2}$	$\frac{\partial^2 f}{\partial x_2 \partial x_3}$	$\frac{\partial f}{\partial x_1}$	$\frac{\partial^2 f}{\partial x_1 \partial x_3}$	$\frac{\partial^2 f}{\partial x_1 \partial x_2}$	$\frac{\partial^3 f}{\partial x_1 \partial x_2 \partial x_3}$
0	$\mathbf{R}_0$	*1*	*1*	*0*	*0*	*0*	*1*	*0*	*1*
1	$\mathbf{R}_1$	*0*	*1*	*0*	*0*	*1*	*1*	*1*	*1*
2	$\mathbf{R}_2$	*1*	*1*	*0*	*0*	*0*	*0*	*0*	*1*
3	$\mathbf{R}_3$	*0*	*1*	*0*	*0*	*0*	*0*	*1*	*1*
4	$\mathbf{R}_4$	*1*	*0*	*0*	*1*	*0*	*1*	*0*	*1*
5	$\mathbf{R}_5$	*1*	*0*	*1*	*1*	*1*	*1*	*1*	*1*
6	$\mathbf{R}_6$	*1*	*1*	*0*	*1*	*0*	*0*	*0*	*1*
7	$\mathbf{R}_7$	*0*	*1*	*1*	*1*	*0*	*0*	*1*	*1*

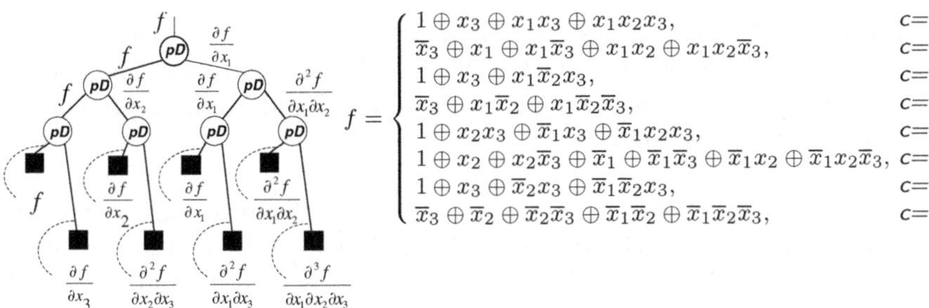

f
pD
$\frac{\partial f}{\partial x_1}$
$\frac{\partial f}{\partial x_2}$
$\frac{\partial^2 f}{\partial x_1 \partial x_2}$
$\frac{\partial f}{\partial x_3}$
$\frac{\partial^2 f}{\partial x_2 \partial x_3}$
$\frac{\partial^2 f}{\partial x_1 \partial x_3}$
$\frac{\partial^3 f}{\partial x_1 \partial x_2 \partial x_3}$
$f = \begin{cases} 1 \oplus x_3 \oplus x_1 x_3 \oplus x_1 x_2 x_3, & c=0 \\ \overline{x}_3 \oplus x_1 \oplus x_1 \overline{x}_3 \oplus x_1 x_2 \oplus x_1 x_2 \overline{x}_3, & c=1 \\ 1 \oplus x_3 \oplus x_1 \overline{x}_2 x_3, & c=2 \\ \overline{x}_3 \oplus x_1 \overline{x}_2 \oplus x_1 \overline{x}_2 \overline{x}_3, & c=3 \\ 1 \oplus x_2 x_3 \oplus \overline{x}_1 x_3 \oplus \overline{x}_1 x_2 x_3, & c=4 \\ 1 \oplus x_2 \oplus x_2 \overline{x}_3 \oplus \overline{x}_1 \oplus \overline{x}_1 \overline{x}_3 \oplus \overline{x}_1 x_2 \oplus \overline{x}_1 x_2 \overline{x}_3, & c=5 \\ 1 \oplus x_3 \oplus \overline{x}_2 x_3 \oplus \overline{x}_1 \overline{x}_2 x_3, & c=6 \\ \overline{x}_3 \oplus \overline{x}_2 \oplus \overline{x}_2 \overline{x}_3 \oplus \overline{x}_1 \overline{x}_2 \oplus \overline{x}_1 \overline{x}_2 \overline{x}_3, & c=7 \end{cases}$

FIGURE 39.2
The calculation of eight Reed–Muller spectra via logic Taylor expansion (Boolean differences) for the function $f = x_1 \overline{x}_2 \vee \overline{x}_3$ (Example 39.4).

processing theory, we can implement a transform in one of several different 2^n polarities.

39.2.1 Cube-based logic Taylor computing

Let $x_i^{c_i}$ be a literal of a Boolean variable x_i such that $x_i^{c_i} = \overline{x}_i$ if $c_i = 0$, and $x_i^{c_i} = x_i$ if $c_i = 1$. A product of literals $x_1^{c_1} x_2^{c_2} \dots x_1^{c_n}$ is called a *term.* If the variable x_i is not represented in a cube, $c_i = \mathbf{x}$ (don't care), i.e., $x_i^{\mathbf{x}} = 1$. In cube notation, a term is described by a cube that is a ternary vector $[c_1 c_2 \dots c_n]$ of components $c_i \in \{0,\ 1,\ \mathbf{x}\}$. A set of cubes corresponding to the true values of a switching function f represents the sum-of-products (SOP) for this function.

Example 39.5 *In Figure 39.3 (left), various representations of a circuit are given: algebraic (SOP and Reed–Muller), graphical and cube, and two types of decision diagram (with Shannon and positive Davio expansions in the nodes). The right part of Figure 39.3 shows the manipulation of various data structures using logic Taylor expansion (Equation 39.1). Details of computing Boolean differences are given in Figure 39.4.*

It follows from Example 39.5 that:

- Manipulation of cubes for computing of Boolean differences is restricted to sum-of-products form only.
- In terms of decision diagrams, Boolean differences can be computed on ROBDD or positive Davio decision diagrams.
- Boolean differences cannot be represented directly by the cubes: we need to transform them into SOPs. This is an incentive to apply special operations to cubes that do not correspond to a Reed-Muller representation of Boolean differences.

To utilize cube-based computing of Boolean differences, the cubes must be converted to the EXOR domain. Manipulation of the cubes involves the operations OR, AND and EXOR, applied to the appropriate literals by the rules given in Table 39.2.

TABLE 39.2
AND, OR and EXOR operation over literals of a cubes.

AND C_i/C_j	0	1	x	*OR* C_i/C_j	0	1	x	*EXOR* C_i/C_j	0	1	x
0	0	∅	0	0	0	x	x	0	0	x	1
1	∅	1	1	1	x	1	1	1	x	1	0
x	0	1	x	x	x	1	x	x	1	0	x

Suppose a sum-of-product expression for a switching function f is given by cubes. To represent this function in a Reed–Muller form, we have to generate cubes based on the equation $x \vee y = x \oplus y \oplus xy$ that can be written in cube notation as

$$[C_1] \vee [C_2] = [C_1] \oplus [C_2] \oplus [C_1][C_2]. \tag{39.2}$$

Three-input logic circuit

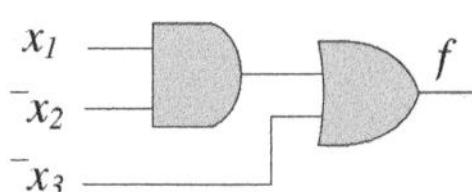

Formal (algebraic) logic description

$$f = x_1\overline{x}_2 \vee \overline{x}_3$$

Graphical cube-based representation

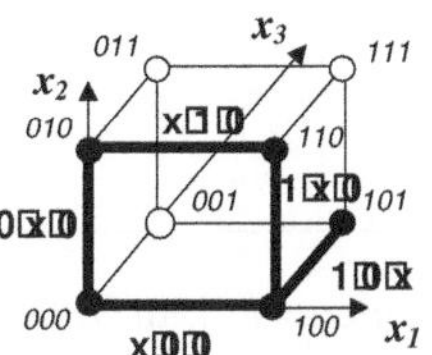

Formal cube-based description

$$f = \begin{bmatrix} \mathbf{x}\ \mathbf{x}\ 0 \\ 1\ 0\ \mathbf{x} \end{bmatrix}$$

Reed–Muller representation

$$f = x_1\overline{x}_2 \oplus \overline{x}_3 \oplus x_1\overline{x}_2\overline{x}_3$$

Representation by ROBDD (left) and Reed–Muller (right) decision diagrams

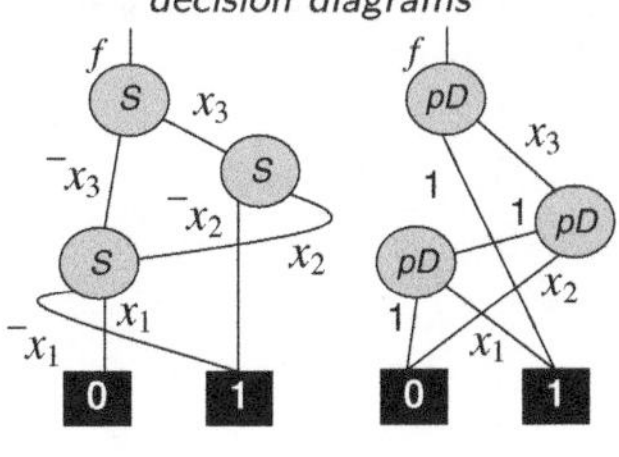

The positive Davio expansion

$$f = f_{x_i=0} \oplus x_i\ (f_{x_i=0} \oplus f_{x_i=1}) = f_{x_i=0} \oplus x_i\ \frac{\partial f}{\partial x_i}$$

$$\frac{\partial f}{\partial x_i} = f_{x_i=0} \oplus f_{x_i=1}$$

For $f = x_1\overline{x}_2 \vee \overline{x}_3$, the Boolean difference with respect to variable x_1

$$\frac{\partial f}{\partial x_1} = (f_{x_i=0} = \overline{x}_3) \oplus (f_{x_i=1} = \overline{x}_2 \vee \overline{x}_3) = \overline{x}_2 x_3$$

Manipulation of ROBDD for computing of $\frac{\partial f}{\partial x_1}$

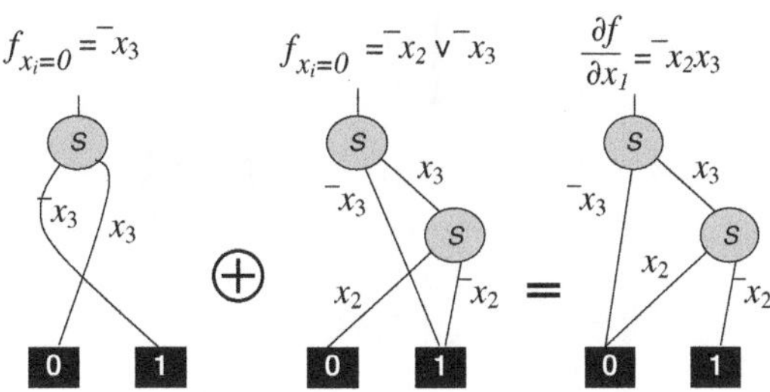

Manipulation of cubes

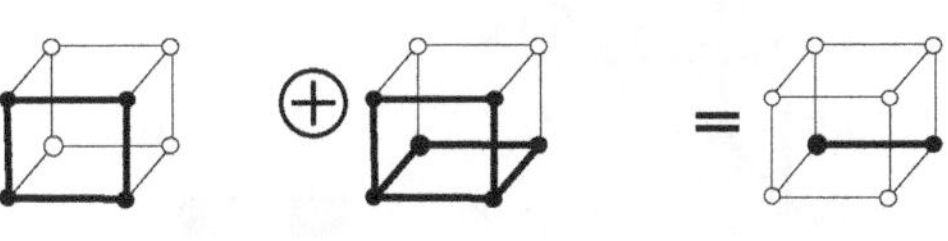

Manipulation of Reed-Muller decision diagrams

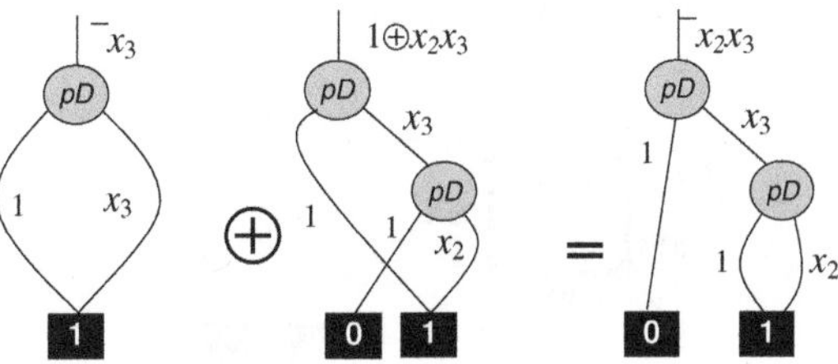

FIGURE 39.3
Relationship between various data structures in circuit description (Example 39.5).

$\frac{\partial f}{\partial x_1} = [\mathbf{x}\ 0\ 1] = \overline{x}_2 x_3$

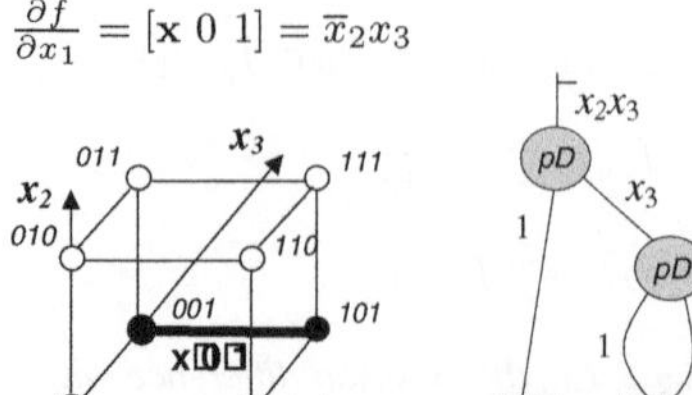

Boolean difference with respect to variable x_1*:*

$$\frac{\partial f}{\partial x_1} = (x_1\overline{x}_2 \vee \overline{x}_3) \oplus (\overline{x}_1\overline{x}_2 \vee \overline{x}_3) = \overline{x}_2 x_3$$

$\frac{\partial f}{\partial x_2} = [1\ \mathbf{x}\ 1] = x_1 x_3$

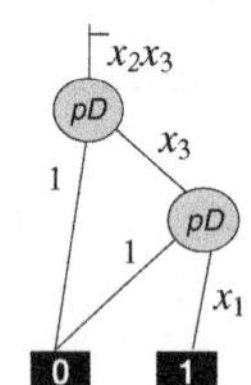

Boolean difference with respect to variable x_2*:*

$$\frac{\partial f}{\partial x_2} = (x_1\overline{x}_2 \vee \overline{x}_3) \oplus (x_1 x_2 \vee \overline{x}_3) = x_1 x_3$$

$$\frac{\partial f}{\partial x_3} = \begin{bmatrix} \mathbf{x} & 1 & \mathbf{x} \\ 0 & \mathbf{x} & \mathbf{x} \end{bmatrix} = x_2 \vee \overline{x}_1$$

Boolean difference with respect to variable x_3*:*

$$\frac{\partial f}{\partial x_3} = (x_1\overline{x}_2 \vee \overline{x}_3) \oplus (x_1\overline{x}_2 \vee x_3) = \overline{x}_1 \vee x_2$$

$\frac{\partial^2 f}{\partial x_1 \partial x_2} = [\mathbf{x}\ \mathbf{x}\ 1] = x_3$

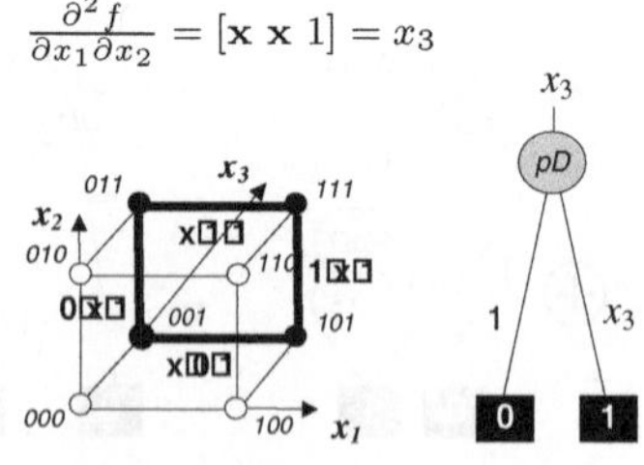

Boolean difference with respect to variable x_1 *and* x_2*:*

$$\frac{\partial^2 f}{\partial x_1 \partial x_2} = \frac{\partial(\overline{x}_2 x_3)}{\partial x_2} = x_3$$

FIGURE 39.4
The representation of Boolean differences by cubes and decision diagrams (Example 39.5).

Example 39.6 *Table 39.2 is used in the following examples: given the cubes* $[1\ 1\ \mathbf{x}]$ *and* $[1\ 0\ \mathbf{x}]$, *we derive*

$$[1\ 1\ \mathbf{x}] \wedge [1\ 0\ \mathbf{x}] = [1\ \emptyset\ \mathbf{x}] = \emptyset$$
$$[1\ 1\ \mathbf{x}] \vee [1\ 0\ \mathbf{x}] = [1\ \mathbf{x}\ \mathbf{x}] \text{ and } [1\ 1\ \mathbf{x}] \oplus [1\ 0\ \mathbf{x}] = [1\ \mathbf{x}\ \mathbf{x}]$$

Example 39.7 *A switching function is given in a sum-of-product form by four cubes,*

$$f = [\mathbf{x}\ 1\ 0\ 1] \vee [1\ 0\ 0\ \mathbf{x}] \vee [0\ \mathbf{x}\ \mathbf{x}\ 0] \vee [\mathbf{x}\ \mathbf{x}\ 1\ 0].$$

To find its ESOP expression, replace $\vee$ *by* $\oplus$ *and compute AND for each pair of cubes distinguished by only one literal:*

$$\begin{aligned} f &= [\mathbf{x}\ 1\ 0\ 1] \oplus [1\ 0\ 0\ \mathbf{x}] \oplus [0\ \mathbf{x}\ \mathbf{x}\ 0] \oplus [\mathbf{x}\ \mathbf{x}\ 1\ 0] \oplus [0\ \mathbf{x}\ 1\ 0] \\ &= x_2\overline{x}_3x_4 \oplus x_1\overline{x}_2\overline{x}_3 \oplus \overline{x}_1\overline{x}_4 \oplus x_3\overline{x}_4 \oplus \overline{x}_1x_3\overline{x}_4. \end{aligned}$$

ESOP is a mixed-polarity form where a variable may be both complemented and uncomplemented, whereas fixed polarity forms, called Reed–Muller expansions of a given polarity, comprise variables that are either complemented or uncomplemented, but not both.

39.2.2 Properties of Boolean difference in cube notation

Three properties of Boolean difference are useful in manipulation by cubes:

- ► $\frac{\partial f}{\partial x_i}$ does not depend on x_i,
- ► $\frac{\partial f}{\partial x_i} = 0$ if f does not depend on x_i
- ► $\frac{\partial (f \oplus g)}{\partial x_i} = \frac{\partial f}{\partial x_i} \oplus \frac{\partial y}{\partial x_i}$

Property 1. The Boolean difference of cube $[C_j]_{x_i=0}$ ($[C_j]_{x_i=1}$) while $[C_j]_{x_i=1}$ ($[C_j]_{x_i=0}$) is not present in the function, i.e., $f_{x_i=0} = 1$ ($f_{x_i=0}$) with respect to variable x_i is calculated by replacing the variable $x_i = 0$ ($x_i = 1$) by $x_i = \mathbf{x}$; the resulting cube:

$$\frac{\partial}{\partial x_i}[C_j]_{x_i=0} = [C_k]_{x_i=\mathbf{x}} \text{ and } \frac{\partial}{\partial x_i}[C_j]_{x_i=1} = [C_k]_{x_i=\mathbf{x}}$$

Property 2. The Boolean difference with respect to variable x_i of the cube $[C_j]_{x_i=\mathbf{x}}$ is an empty cube

$$\frac{\partial}{\partial x_i}[C_j]_{x_i=\mathbf{x}} = \emptyset.$$

Note, that the cube $[C_j]_{x_i=\mathbf{x}}$ means that f does not depend on x_i, which yields $\frac{\partial}{\partial x_i} = 0$.

Property 3. A Boolean difference of a switching function with respect to variable x_i is a modulo 2 sum of the Boolean differences derived from the cubes C_j representing ESOP of this function

$$\frac{\partial f}{\partial x_i} = \bigoplus_j ([C_j]_{x_i=0} \oplus [C_j]_{x_i=1}) = \bigoplus_j \frac{\partial}{\partial x_i}[C_j]$$

Example 39.8 *Figure 39.5 demonstrates the results of the computations for a 4-variable switching function. A cube with one* **x** *is represented by an edge between two nodes (since this cube covers two terms), and a cube with two* **x***'s is represented by a face (the cube covers four terms).*

Using the above properties, a cube-based algorithm for calculating a Boolean difference with respect to the variable x_i can be derived, given the cubes that represent a Reed–Muller expression. If one of these cubes is present in the expression while another one is not (i.e., if $f_{x_i=0} = 0$ while $f_{x_i=1} = 1$, or $f_{x_i=1} = 1$ while $f_{x_i=1} = 0$), then replace c_i with a don't care **x** while leaving the remaining literals unchanged:

$$\frac{\partial}{\partial x_i}[C_j] = f_{x_i=0} \oplus f_{x_i=1} = 1.$$

Based on the technique for calculating Boolean differences with respect to a single variable, we are able to calculate the multiple Boolean difference $\frac{\partial^k f}{\partial x_i \ldots \partial x_j}$.

Example 39.9 *Figure 39.5 shows the computing Boolean difference* $\frac{\partial^2 f}{\partial x_2 \partial x_3}$ *Boolean difference with respect to variable* x_3 *of a switching function* f *given in the Reed–Muller form.*

39.2.3 Numerical example

In Table 39.3, the first and second columns include the benchmark **#Name**, the number of inputs and outputs **#In/#Out**, and the number of cubes in sum-of-product form. The next three columns contain the number of cubes in ESOP form required to calculate first 256, 1024, and 4096 coefficients.

39.2.4 Matrix form of logic Taylor expansion

In the matrix form of logic Taylor expansion, the coefficients are computed using a matrix representation of Boolean differences (see details in Chapter 9).

Switching function

$$f = \begin{bmatrix} x & 1 & 0 & 1 \\ 1 & 0 & 0 & x \\ 1 & x & 1 & 0 \\ 0 & x & 1 & 1 \end{bmatrix}$$

Reed-Muller form

$$f = \begin{bmatrix} x & 1 & 0 & 1 \\ 1 & 0 & 0 & x \\ x & x & 1 & 0 \\ 0 & x & 1 & x \end{bmatrix}$$

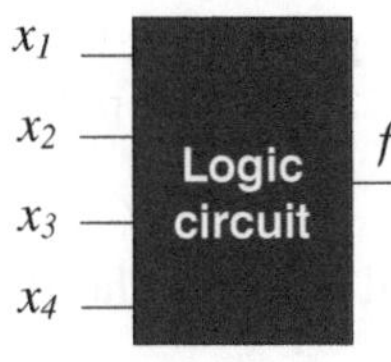

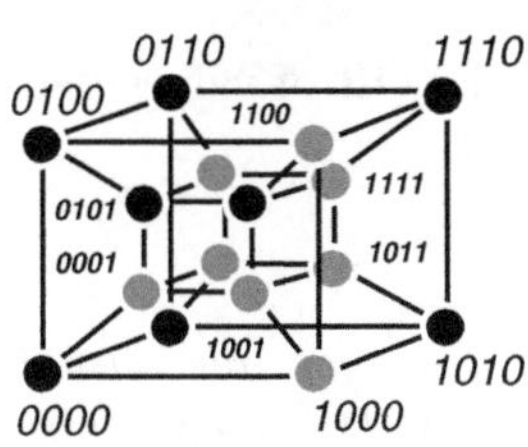

Sum-of-product of a switching function

$$f = x_2\overline{x}_3x_4 \vee x_1\overline{x}_2\overline{x}_3 \vee x_1x_3\overline{x}_4 \vee \overline{x}_1x_3x_4$$

Reed–Muller representation of a switching function

$$f = x_2\overline{x}_3x_4 \oplus x_1\overline{x}_2\overline{x}_3 \oplus x_3\overline{x}_4 \oplus \overline{x}_1x_3$$

The Boolean difference from the cubes:

$$\frac{\partial}{\partial x_3}[\mathrm{x}\ 1\ 0\ 1] = [\mathrm{x}\ 1\ \mathrm{x}\ 1], \qquad \frac{\partial}{\partial x_3}[\mathrm{x}\ \mathrm{x}\ 1\ 0] = [\mathrm{x}\ \mathrm{x}\ \mathrm{x}\ 0],$$

$$\frac{\partial}{\partial x_3}[1\ 0\ 0\ \mathrm{x}] = [1\ 0\ \mathrm{x}\ \mathrm{x}], \qquad \frac{\partial}{\partial x_3}[0\ \mathrm{x}\ 1\ \mathrm{x}] = [0\ \mathrm{x}\ \mathrm{x}\ \mathrm{x}].$$

The modulo 2 sum of cubes:

$$\frac{\partial f}{\partial x_3} = [\mathrm{x}\ 1\ \mathrm{x}\ 1] \oplus [1\ 0\ \mathrm{x}\ \mathrm{x}] \oplus [\mathrm{x}\ \mathrm{x}\ \mathrm{x}\ 0] \oplus [0\ \mathrm{x}\ \mathrm{x}\ \mathrm{x}]$$
$$= x_2x_4 \oplus x_1\overline{x}_2 \oplus \overline{x}_4 \oplus \overline{x}_1.$$

Computation in symbolic form:

$$\frac{\partial f}{\partial x_3} = (f_{x_3=0}) \oplus (f_{x_3=1}) = (x_2x_4 \vee x_1\overline{x}_2) \oplus (x_1\overline{x}_4 \vee \overline{x}_1x_4)$$
$$= \overline{\overline{x_2x_4}\,\overline{x_1\overline{x}_2}} \oplus \overline{x}_1 \oplus \overline{x}_4$$
$$= (x_2x_4 \oplus 1)(x_1\overline{x}_2 \oplus 1) \oplus 1 \oplus \overline{x}_1 \oplus \overline{x}_4$$
$$= x_2x_4 \oplus x_1\overline{x}_2 \oplus \overline{x}_1 \oplus \overline{x}_4$$

The second order Boolean difference with respect to variables x_2 and x_3:

$$\frac{\partial^2 f}{\partial x_2 \partial x_3} = \frac{\partial}{\partial x_2}\begin{bmatrix} \mathrm{x} & 1 & \mathrm{x} & 1 \\ 1 & 0 & \mathrm{x} & \mathrm{x} \\ \mathrm{x} & \mathrm{x} & \mathrm{x} & 0 \\ 0 & \mathrm{x} & \mathrm{x} & \mathrm{x} \end{bmatrix} = \begin{bmatrix} \mathrm{x} & \mathrm{x} & \mathrm{x} & 1 \\ 1 & \mathrm{x} & \mathrm{x} & \mathrm{x} \\ \mathrm{x} & \emptyset & \mathrm{x} & 0 \\ 0 & \emptyset & \mathrm{x} & \mathrm{x} \end{bmatrix} = \begin{bmatrix} \mathrm{x} & \mathrm{x} & \mathrm{x} & 1 \\ 1 & \mathrm{x} & \mathrm{x} & \mathrm{x} \end{bmatrix}$$
$$= [\mathrm{xxx}1] \oplus [1\mathrm{xxx}] = x_4 \oplus x_1$$

FIGURE 39.5
Computing the Boolean difference of switching function by cube manipulation (Example 39.8 and 39.9).

39.2.5 Computing logic Taylor expansion by decision diagram

The coefficients of logic Taylor expansion can be computed using decision diagrams. Since $\frac{\partial f}{\partial x_i} = f_{x_i=0} \oplus f_{x_i=1}$, the calculation can be implemented with respect to the lowest-level (in ROBDD) variable x_i by manipulating the terminal nodes corresponding to $f_{x_i=0}$ and $f_{x_i=0}$. The algorithm is given below.

Step 1. Find the paths to constant (terminal) nodes:

(*a*) For a node assigned with x_i, iterate through left (*low*) and right (*high*) branches.

TABLE 39.3
Cube-based Reed–Muller spectrum computation.

Name	Test In/Out	cubes	RM spectra [cubes] at p=0 256	1024	4096
alu4	14/8	1028	63	246	3023
misex3	14/14	1848	52	222	2900
add6	13/7	607	141	213	417
duke2	22/29	242	29	37	37
vg2	25/8	110	11	21	24
misex2	25/18	29	4	4	4
e64	65/65	65	65	65	65

(*b*) For other nodes, check the polarity of the variable. If the polarity is 0, the *low* branch will be searched. If the polarity is 1, the *high* branch will be searched. This results in 2^{n-1} pairs of paths from the root to a constant value.

Step 2. Iterate to calculate the intermediate and final value of the Boolean difference given p:

(*a*) For a node assigned with x_i, we calculate EXOR of values derived from both *high* and *low* outgoing branches.

(*b*) For other nodes, no calculations are needed.

Example 39.10 *Calculate* $\frac{\partial^2 f}{\partial x_1 \partial x_3}|_{p=101}$ *for the switching function* $f = x_1\overline{x}_2 \vee \overline{x}_3$*, given by a ROBDD (Figure 39.6).*

39.3 Computing arithmetic expressions via arithmetic Taylor expansion

Arithmetic spectrum coefficients carry additional information about function behavior compared to Reed-Muller coefficients.

39.3.1 Arithmetic analog of logic Taylor expansion

An analog of the logic Taylor series for a switching function, called the *arithmetical Taylor series*, is expressed by the equation

$$f = \sum_{i=0}^{2^n-1} p_i^{(c)} \underbrace{\overbrace{(x_1 \oplus c_1)}^{c_1\ polarity}{}^{i_1} \ldots \overbrace{(x_n \oplus c_n)}^{c_n\ polarity}{}^{i_n}}_{i-th\ product}, \tag{39.3}$$

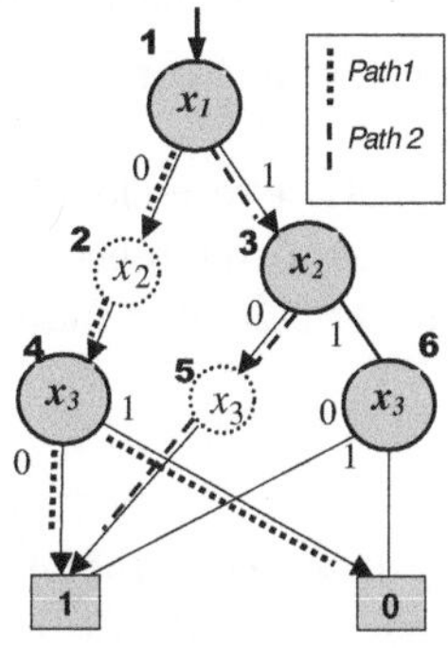

Searching results in two paths:
1-2-4-constant 0 $\Longleftrightarrow$ *1-2-4-constant 1 and*
1-3-5-constant 1 $\Longleftrightarrow$ *1-3-5-constant 1*
Iterate results in the following manner:

- *Calculate the value*

$$1 = 1 \oplus 0$$

from the constant values 1 and 0 that are the left and right successors of node 4. Similarly, calculate the value

$$0 = 1 \oplus 1$$

from the 1's that are the successors of node 5. The obtained values 1 and 0 are the substitutes for nodes 4 and 5 respectively.

- *Calculate the value*

$$1 = 1 \oplus 0$$

from the substitutes obtained above, 1 and 0, which the successors of node 1 (nodes 2 and 3 are transparent since the polarity of x_2 *is 0). The resulting value is*

$$\frac{\partial^2 f}{\partial x_1 \partial x_3}|_{p=101} = (0 \oplus 1) \oplus (1 \oplus 1) = 1$$

Processing of node #4 and # 5 processing

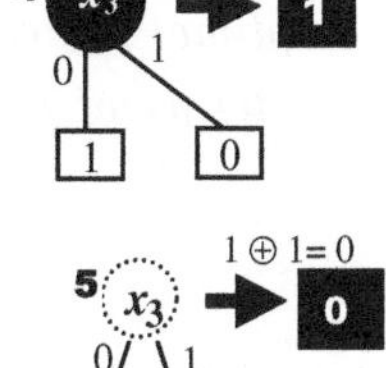

Processing of node #1

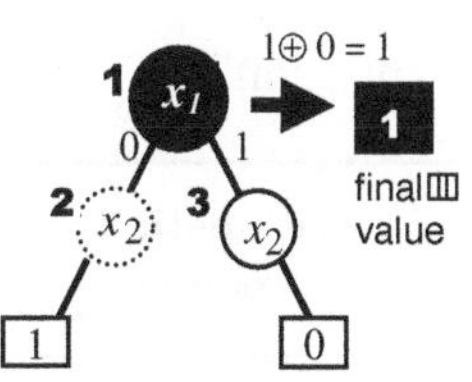

FIGURE 39.6
Calculation of the second order Boolean difference at point $p = 101$ $\frac{\partial^2 f}{\partial x_1 \partial x_3}|_{p=101}$ given the switching function $f = x_1\overline{x}_2 \vee \overline{x}_3$ and $p = 101$ (Example 39.10).

where $c_1c_2 \ldots c_n$ and $i_1, i_2, \ldots, i_n$ are the binary representations of c (polarity) and i respectively, and the i-th coefficient is defined as

$$p_i^{(c)} = \frac{\widetilde{\partial}^n f(c)}{\widetilde{\partial} x_1^{i_1} \widetilde{\partial} x_2^{i_2} \ldots \widetilde{\partial} x_n^{j_n}}\Bigg|_{d=c} \quad \text{and} \quad \widetilde{\partial} x_i^{i_j} = \begin{cases} 1, & i_j = 0, \\ \widetilde{\partial} x_i, & i_j = 1. \end{cases}$$

The $p_i^{(c)}$ is a value of the arithmetical analog of an n-ordered Boolean difference of f given c, i.e., $x_1 = c_1, ..., x_n = c_n$. The coefficients $p_i^{(c)}$ compose the *arithmetic spectrum* of a switching function f.

The arithmetic Taylor expansion produces 2^n arithmetic expressions, corresponding to 2^n polarities. Similarly to multiple Boolean differences, one can draw the data flowgraph for any subset of variables to calculate multiple arithmetic differences.

Example 39.11 *In Figure 39.7, the arithmetic Taylor series in polarity $c = 3$ of an arbitrary 3-variable ($n = 3$) switching function is given. Details are illustrated for polarity $c = 3$ ($c_1c_2c_3 = 011$). The arithmetic spectrum in 0-polarity is represented by the four non-zero spectral components $f^{(000)}$, $f^{(001)}$, $f^{(101)}$, and $f^{(111)}$. This is an optimal spectral representation of the given switching function (optimal polarity). Each spectral coefficient $p_i^{(c)}$ is calculated in terms of arithmetic analogs of Boolean differences.*

Arithmetic Taylor expansions for several elementary switching functions are given in Table 39.4.

TABLE 39.4
Arithmetic spectrum as arithmetic Taylor expansion of elementary switching functions.

Function	Arithmetic spectrum F	$\frac{\partial f}{\partial x_2}$	$\frac{\partial f}{\partial x_1}$	$\frac{\partial^2 f}{\partial x_1 \partial x_2}$	Arithmetic expression
x_1, x_2 → f; $y = x_1 \wedge x_2$	0	0	0	1	x_1x_2
	0	0	1	-1	$x_1 - x_1\overline{x}_2$
	0	1	0	-1	$x_2 - \overline{x}_1x_2$
	1	-1	-1	1	$1 - \overline{x}_2 - \overline{x}_1 + \overline{x}_1\overline{x}_2$
x_1, x_2 → f; $y = x_1 \vee x_2$	0	1	1	-1	$x_2 + x_1 - x_1x_2$
	1	-1	0	1	$1 - \overline{x}_2 + x_1\overline{x}_2$
	1	0	-1	1	$1 - \overline{x}_1 + \overline{x}_1x_2$
	1	0	0	-1	$1 - \overline{x}_1\overline{x}_2$
x_1, x_2 → f; $y = x_1 \oplus x_2$	0	1	1	-2	$x_2 + x_1 - 2x_1x_2$
	1	-1	-1	2	$1 - \overline{x}_2 - x_1 + 2x_1\overline{x}_2$
	1	-1	-1	2	$1 - x_2 - \overline{x}_1 + 2\overline{x}_1x_2$
	0	1	1	-2	$\overline{x}_2 + \overline{x}_1 - 2\overline{x}_1\overline{x}_2$

		Arithmetic spectral coefficients / Arithmetical differences							
		000	*001*	*010*	*011*	*100*	*101*	*110*	*111*
c	$\mathbf{P}_c$	**F**	$\frac{\widetilde{\partial} f}{\widetilde{\partial} x_3}$	$\frac{\widetilde{\partial} f}{\widetilde{\partial} x_2}$	$\frac{\widetilde{\partial}^2 f}{\widetilde{\partial} x_2 \widetilde{\partial} x_3}$	$\frac{\widetilde{\partial} f}{\widetilde{\partial} x_1}$	$\frac{\widetilde{\partial}^2 f}{\widetilde{\partial} x_1 \widetilde{\partial} x_3}$	$\frac{\widetilde{\partial}^2 f}{\widetilde{\partial} x_1 \widetilde{\partial} x_2}$	$\frac{\widetilde{\partial}^3 f}{\widetilde{\partial} x_1 \widetilde{\partial} x_2 \widetilde{\partial} x_3}$
0	$\mathbf{P}_0$	1	-1	0	0	0	1	0	-1
1	$\mathbf{P}_1$	0	1	0	0	1	-1	-1	1
2	$\mathbf{P}_2$	1	-1	0	0	0	0	0	1
3	$\mathbf{P}_3$	0	1	0	0	0	0	1	-1
4	$\mathbf{P}_4$	1	0	0	-1	0	-1	0	1
5	$\mathbf{P}_5$	1	0	-1	1	-1	1	1	-1
6	$\mathbf{P}_6$	1	-1	0	1	0	0	0	-1
7	$\mathbf{P}_7$	0	1	1	-1	0	0	-1	1

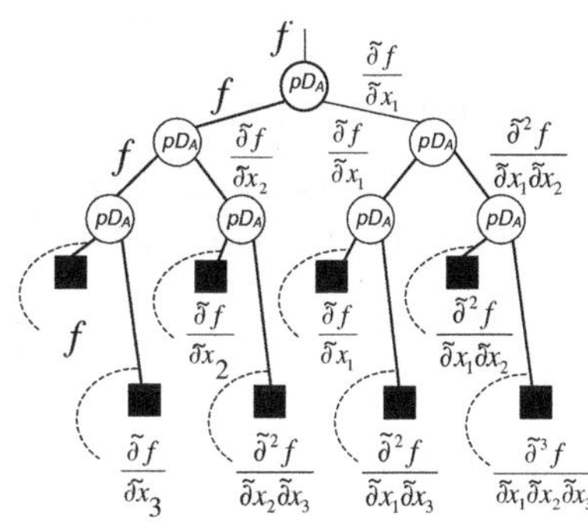

Arithmetic Taylor expansion of a 3-variable ($n = 3$) switching function in polarity $c = 3$ ($c_1c_2c_3 = 011$):

$$P_3 = \sum_{j=0}^{7} p_3^{(j)} (x_1 \oplus 0)^{j_1} (x_2 \oplus 1)^{j_2} (x_3 \oplus 1)^{j_3}$$

$$= (f(3) + \frac{\widetilde{\partial} f(3)}{\widetilde{\partial} x_3}\overline{x}_3 + \frac{\widetilde{\partial} f(3)}{\widetilde{\partial} x_2}\overline{x}_2 + \frac{\widetilde{\partial}^2 f(3)}{\widetilde{\partial} x_2 \widetilde{\partial} x_3}\overline{x}_2\overline{x}_3$$

$$+ \frac{\widetilde{\partial} f(3)}{\widetilde{\partial} x_1}x_1 + \frac{\widetilde{\partial}^2 f(3)}{\widetilde{\partial} x_1 \widetilde{\partial} x_3}x_1\overline{x}_3 + \frac{\widetilde{\partial}^2 f(3)}{\widetilde{\partial} x_1 \widetilde{\partial} x_2}x_1\overline{x}_2$$

$$+ \frac{\widetilde{\partial}^3 f(3)}{\widetilde{\partial} x_1 \widetilde{\partial} x_2 \widetilde{\partial} x_3}x_1\overline{x}_2\overline{x}_3)$$

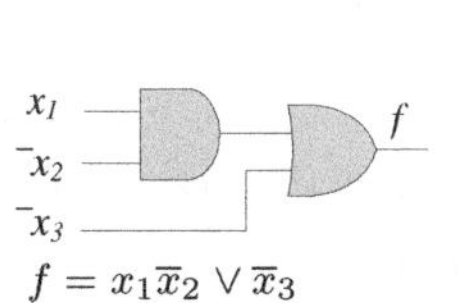

$f = x_1\overline{x}_2 \vee \overline{x}_3$

$$f = \begin{cases} 1 - x_3 + x_1x_3 - x_1x_2x_3, & c=0 \\ \overline{x}_3 + x_1 - x_1\overline{x}_3 - x_1x_2 + x_1x_2\overline{x}_3, & c=1 \\ 1 - x_3 + x_1\overline{x}_2x_3, & c=2 \\ \overline{x}_3 + x_1\overline{x}_2 - x_1\overline{x}_2\overline{x}_3, & c=3 \\ 1 - x_2x_3 - \overline{x}_1x_3 + \overline{x}_1x_2x_3, & c=4 \\ 1 - x_2 + x_2\overline{x}_3 - \overline{x}_1 + \overline{x}_1\overline{x}_3 + \overline{x}_1x_2 - \overline{x}_1x_2\overline{x}_3, & c=5 \\ 1 - x_3 + \overline{x}_2x_3 - \overline{x}_1\overline{x}_2x_3, & c=6 \\ \overline{x}_3 + \overline{x}_2 - \overline{x}_2\overline{x}_3 - \overline{x}_1\overline{x}_2 + \overline{x}_1\overline{x}_2\overline{x}_3, & c=7 \end{cases}$$

FIGURE 39.7
Calculation of eight arithmetic spectra via Taylor expansion (arithmetic differences) for the switching function $f = x_1\overline{x}_2 \vee \overline{x}_3$ (Example 39.11).

39.3.2 Cube based arithmetic spectrum computing

To represent a function given by a SOP expression of its cubes, as an arithmetic expression, an algorithm similar to the one to derive its ESOP form can be applied. However, it must be taken into account that operations over cubes in an arithmetic form are specific. The new cube generation procedure is based on the following equation

$$x \vee y = x + y - xy.$$

Given the cubes $[C_1]$ and $[C_2]$ of the SOP expression, the cubes to be included in the arithmetical expression are derived by the equation

$$[C_1] \vee [C_2] = [C_1] + [C_2] - [C_1][C_2]. \tag{39.4}$$

Example 39.12 *Given two cubes*

$$[C_1] = [\mathbf{x}\ 1\ 0\ \mathbf{x}\ 0\ 1\ 1\ \mathbf{x}\ \mathbf{x}] \quad \textit{and} \quad [C_2] = [\mathbf{x}\ 1\ 0\ 1\ \mathbf{x}\ \mathbf{x}\ 1\ \mathbf{x}\ 1],$$

three cubes are produced: $[C_1]$, $[C_2]$, *and the new cube*

$$-[C_1][C_2] = -[\mathbf{x}\ 1\ 0\ 1\ 0\ 1\ 1\ \mathbf{x}\ 1].$$

A cube that corresponds to a product in the arithmetic expression of a switching function is composed of the components: $\{0,\ 1,\ \mathbf{x}\}$, a, and b, where

$$a = -\overline{x}_i + x_i = (-1)^{\overline{x}_i}$$
$$b = -x_i + \overline{x}_i = (-1)^{x_i}.$$

Example 39.13 *Given a function represented by the SOP cubes,*

$$f = [\mathbf{x}\ 1\ 0\ 1] \vee [1\ 0\ 0\ \mathbf{x}] \vee [0\ \mathbf{x}\ \mathbf{x}\ 0] \vee [\mathbf{x}\ \mathbf{x}\ 1\ 0].$$

To derive cubes to be included in an arithmetic expression, Equation 39.4 is be applied:

$$f = [\mathbf{x}\ 1\ 0\ 1] + [1\ 0\ 0\ \mathbf{x}] + [0\ \mathbf{x}\ \mathbf{x}\ 0] + [\mathbf{x}\ \mathbf{x}\ 1\ 0] - [1\ \mathbf{x}\ 1\ 0]$$
$$= x_2\overline{x}_3x_4 + x_1\overline{x}_2\overline{x}_3 + \overline{x}_1\overline{x}_4 + x_3\overline{x}_4 - x_1x_3\overline{x}_4.$$

39.3.3 Properties of arithmetic differences in cube notation

Three properties of arithmetic difference are useful for manipulation in cube form.

Property 1. Arithmetic differences of $[C_j]_{x_i=0}$ while $f_{x_i=1} = 0$ (cube $[C_j]_{x_i=1}$ is not present) or of $[C_j]_{x_i=1}$ while $f_{x_i=0} = 0$ (cube $[C_j]_{x_i=0}$ is not present) with respect to variable x_i is calculated by replacing $x_i = 0$ ($x_i = 1$) with $a = (-1)^{\overline{x}_i}$ ($b = (-1)^{x_i}$)

$$\frac{\widetilde{\partial}}{\widetilde{\partial} x_i}[C]_{x_i=1} = [C]_{x_i=a} \text{ and } \frac{\widetilde{\partial}}{\widetilde{\partial} x_i}[C]_{x_i=0} = [C]_{x_i=b}.$$

It is follows from the below statement:

> $\frac{\widetilde{\partial} f}{\widetilde{\partial} x_i}$ `depends on the variable` x_i, `since`
>
> $$-\overline{x}_i + x_i = (-1)^{\overline{x}_i},$$
> $$-x_i + \overline{x}_i = (-1)^{\overline{x}_i}.$$

Property 2. The arithmetic difference with respect to variable x_i of the cube $[C_j]_{x_i=\mathbf{x}}$ is an empty cube,

$$\frac{\widetilde{\partial}}{\widetilde{\partial} x_i}[C]_{x_i=\mathbf{x}} = \emptyset.$$

This is due the fact that:

> $\frac{\widetilde{\partial} f}{\widetilde{\partial} x_i} = 0$ `if` f `does not depend on` x_i.

Property 3. The arithmetic difference with respect to a variable x_i of a switching function is the sum of arithmetic differences of cubes that represent this function in arithmetic form

$$\begin{aligned}\frac{\widetilde{\partial} f}{\widetilde{\partial} x_i} &= \sum_{j=1}^{t} \{(-1)^{\overline{x}_i}[C_j]_{x_i=0} + (-1)^{x_i}[C_j]_{x_i=1}\} \\ &= \sum_{j=1}^{t} \frac{\widetilde{\partial}}{\widetilde{\partial} x_i}[C_j] \qquad (39.5)\end{aligned}$$

A technique for calculating the multiple arithmetic difference $\frac{\widetilde{\partial}^k f}{\widetilde{\partial} x_i \ldots \widetilde{\partial} x_j}$ is considered in the example below.

Example 39.14 *Figure 39.8 illustrates the computing of the arithmetic difference with respect to variable x_3 of a switching function f given a sum-of-product expression.*

39.3.4 Matrix form of arithmetic Taylor expansion

Here we introduce a matrix based algorithm to calculate the coefficients of the arithmetic Taylor expansion.

Formal cube-based description

$$f = \begin{bmatrix} \mathbf{x} & 1 & 0 & 1 \\ \mathbf{x} & \mathbf{x} & 1 & 0 \\ 0 & \mathbf{x} & \mathbf{x} & 0 \\ 1 & \mathbf{x} & 1 & 0 \end{bmatrix}$$

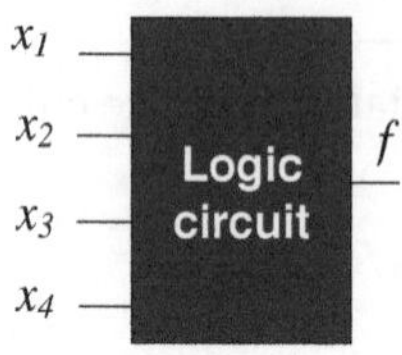

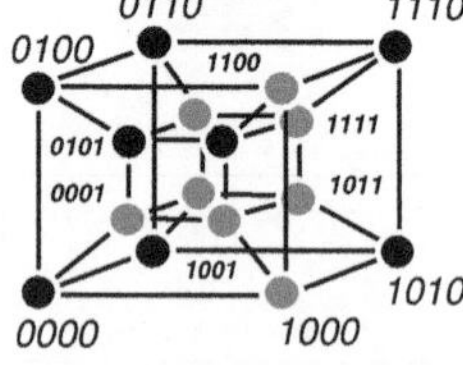

Computing arithmetic differences for the given cubes

$$\frac{\widetilde{\partial}}{\widetilde{\partial} x_3}[\mathbf{x}\ 1\ 0\ 1] = [\mathbf{x}\ 1\ b\ 1]$$

$$\frac{\widetilde{\partial}}{\widetilde{\partial} x_3}[\mathbf{x}\ \mathbf{x}\ 1\ 0] = [\mathbf{x}\ \mathbf{x}\ a\ 0]$$

$$\frac{\widetilde{\partial}}{\widetilde{\partial} x_3}[\mathbf{x}\ \mathbf{x}\ 0] = \emptyset$$

$$\frac{\widetilde{\partial}}{\widetilde{\partial} x_3}[1\ 0\ 0\ \mathbf{x}] = [1\ 0\ b\ \mathbf{x}]$$

$$\frac{\widetilde{\partial}}{\widetilde{\partial} x_3}[1\ \mathbf{x}\ 1\ 0] = [1\ \mathbf{x}\ a\ 0]$$

Arithmetic difference in Reed–Muller form:

$$\begin{aligned} \frac{\widetilde{\partial}}{\widetilde{\partial} x_3} &= [\mathbf{x}\ 1\ b\ 1] + [1\ 0\ b\ \mathbf{x}] \\ &+ [\mathbf{x}\ \mathbf{x}\ a\ 0] + [1\ \mathbf{x}\ a\ 0] \\ &= (-1)^{x_3} x_2 x_4 + (-1)^{x_3} x_1 \overline{x}_2 \\ &+ (-1)^{\overline{x}_3} \overline{x}_4 + (-1)^{\overline{x}_3} x_1 \overline{x}_4 \end{aligned}$$

Computing in symbolic form yields

$$\begin{aligned} \frac{\widetilde{\partial}}{\widetilde{\partial} x_3} &= -f_{\overline{x}_3} + f_{x_3} \\ &= x_2\overline{x}_3x_4 + x_3\overline{x}_4 + x_1\overline{x}_2\overline{x}_3 + x_1x_3\overline{x}_4 \\ &- x_2x_3x_4 - \overline{x}_3\overline{x}_4 - x_1x_2x_3 - x_1\overline{x}_3\overline{x}_4 \\ &= (\overline{x}_3 - x_3)x_2x_4 + (x_3 - \overline{x}_3)x_1\overline{x}_2 \\ &+ (x_3 - \overline{x}_3)\overline{x}_4 + (x_3 - \overline{x}_3)x_1\overline{x}_4 \\ &= (-1)^{x_3} x_2x_4 + (-1)^{x_3} x_1\overline{x}_2 \\ &+ (-1)^{\overline{x}_3}\overline{x}_4 + (-1)^{\overline{x}_3} x_1\overline{x}_4 \end{aligned}$$

Arithmetic difference $\frac{\widetilde{\partial} f^2}{\widetilde{\partial} x_2 \widetilde{\partial} x_4}$ *is calculated as follows:*

$$\begin{aligned} \frac{\widetilde{\partial}^2 f}{\widetilde{\partial} x_2 \widetilde{\partial} x_4} &= \frac{\widetilde{\partial}}{\widetilde{\partial} x_2}\left(\frac{\widetilde{\partial}}{\widetilde{\partial} x_4}\begin{bmatrix} 0 & \mathbf{x} & \mathbf{x} & 0 & 1 & \mathbf{x} \\ 0 & 1 & 0 & \mathbf{x} & \mathbf{x} & 1 \\ 0 & 0 & 1 & \mathbf{x} & \mathbf{x} & 1 \\ 0 & 0 & 1 & \mathbf{x} & 0 & 0 \\ 1 & \mathbf{x} & 0 & \mathbf{x} & 1 & 1 \end{bmatrix}\right) \\ &= \frac{\widetilde{\partial}}{\widetilde{\partial} x_2}\begin{bmatrix} 0 & \mathbf{x} & \mathbf{x} & 0 & a & \mathbf{x} \\ 0 & 0 & 1 & \mathbf{x} & b & 0 \\ 1 & \mathbf{x} & 0 & \mathbf{x} & a & 1 \end{bmatrix} = \begin{bmatrix} 0 & 0 & a & \mathbf{x} & b & 0 \\ 1 & \mathbf{x} & b & \mathbf{x} & a & 1 \end{bmatrix} \\ &= \sum_{t=1}^{5} \frac{\widetilde{\partial}^2}{\widetilde{\partial} x_2 \widetilde{\partial} x_4}[C_t] \\ &= (-1)^{\overline{x}_2 + x_4}\overline{x}_1\overline{x}_5(-1)^{x_2+\overline{x}_4}x_5 \\ &= (x_2 - \overline{x}_2)\overline{x}_1\overline{x}_5 + (\overline{x}_4 - x_4)\overline{x}_1\overline{x}_5 \\ &+ (\overline{x}_2 - x_2)x_5 - (x_4 - \overline{x}_4)x_5 \end{aligned}$$

FIGURE 39.8
Computing arithmetic differences of switching function by algebraic and cube manipulations (Example 39.14).

39.3.5 Computing arithmetic Taylor expansion by decision diagram

The algorithm for arithmetic spectrum computation is similar to the case of Reed–Muller except that we use an addition or subtraction operation instead of XOR. The following rules are applied:

(a) For a node assigned with x_i, if the polarity of this variable is 1, we subtract the value derived from the high outgoing branch and the low one; otherwise (if the polarity of this variable is 0), the subtracted values are swapped.

(b) For the nodes assigned with other variables, no operation is needed.

Example 39.15 *(Continuation of Example 39.10.) In Figure 39.9, the computing of arithmetical difference* $\frac{\widetilde{\partial}^2 f}{\widetilde{\partial} x_1 \widetilde{\partial} x_3}|_{p=101}$ *is given. The final result is*

$$\frac{\widetilde{\partial}^2 f}{\widetilde{\partial} x_1 \widetilde{\partial} x_3}|_{p=101} = (1-1) - (0-1) = 1.$$

39.4 Computing Walsh expressions via Taylor expansion

In this section we introduce the Walsh spectrum in the form of a Taylor series.

39.4.1 Matrix form of Walsh differences

While Boolean differences and its arithmetical analog are well-known in logic design, Walsh differences are not. Here we define the Walsh differential matrix and operators related to the matrix formalization of the Boolean and arithmetic differences. This approach is based on a factorized representation of the Walsh matrix $W_{2^n}^{(p_i)}$ that is used in digital signal processing for fast algorithm design.

Theorem 39.1 *The Walsh difference of switching function of n variables with respect to the variable x_i is defined as*

$$\frac{\widetilde{\widetilde{\partial}}\mathbf{F}}{\widetilde{\widetilde{\partial}} x_i} = W_{2^n}^{(p)}\mathbf{F}, \qquad (39.6)$$

where $2^{n-i} \times 2^{n-i}$ matrix $W_{2^n}^{(p)}$ is called the Walsh differential matrix *and is defined as the product of the n matrices*

$$W_{2^n}^{(p)} = \frac{1}{2^n} W_{2^n}^{(p_1)} W_{2^n}^{(p_2)} \dots W_{2^n}^{(p_n)}, \qquad (39.7)$$

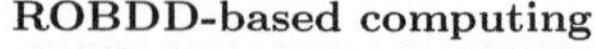

ROBDD-based computing

1°. Find the paths in ROBBD: $\overline{x}_1 - x_2 - x_3$ *and* $x_1 - x_2 - \overline{x}_3$

2°. Calculate the intermediate values -1 and 0, i.e.,

$$\frac{\widetilde{\partial} f}{\widetilde{\partial} x_3} = x_1\overline{x}_2(-1)^{\overline{x}_3} + (-1)^{x_3} = \begin{cases} -1, & \text{for } \overline{x}_1 - x_2; \\ 0, & \text{for } x_1 - x_2. \end{cases}$$

3°. Calculate

$$\frac{\widetilde{\partial}^2 f}{\widetilde{\partial} x_1 \widetilde{\partial} x_3}|_{p=101} = -(-1) + 0 = 1$$

$$\begin{aligned} f &= x_1\overline{x}_2 \vee \overline{x}_3 = x_1\overline{x}_2 + \overline{x}_3 - x_1\overline{x}_2\overline{x}_3 \\ &= x_1\overline{x}_2(1 - \overline{x}_3) + \overline{x}_3 = x_1\overline{x}_2 x_3 + \overline{x}_3 \end{aligned}$$

Computing in algebraic form

$$\begin{aligned} \frac{\widetilde{\partial}^2 f}{\widetilde{\partial} x_1 \widetilde{\partial} x_3} &= \frac{\widetilde{\partial}}{\widetilde{\partial} x_1}(-x_1\overline{x}_2\overline{x}_3 - x_3 + x_1\overline{x}_2 x_3 + \overline{x}_3) \\ &= -x_1\overline{x}_2\overline{x}_3 + x_1\overline{x}_2 x_3 + \overline{x}_1\overline{x}_2\overline{x}_3 - \overline{x}_1\overline{x}_2 x_3 \\ &= -x_1\overline{x}_2(\overline{x}_3 - x_3) + \overline{x}_1\overline{x}_2(\overline{x}_3 - x_3) \\ &= (x_1\overline{x}_2 + \overline{x}_1\overline{x}_2)(\overline{x}_3 - x_3) \\ &= \overline{x}_2(-x_1 + \overline{x}_1)(\overline{x}_3 - x_3) \\ &= (-1)^{\overline{x}_1}\overline{x}_2(-1)^{\overline{x}_3} \end{aligned}$$

Computing in cube form

Given a switching function in a sum-of-product cube form f=$\begin{bmatrix} 1\ 0\ \mathbf{x} \\ \mathbf{x}\ 0\ 0 \end{bmatrix}$. *The arithmetic cube-based representation is f=*$\begin{bmatrix} 1\ 0\ 1 \\ \mathbf{x}\ 0\ 0 \end{bmatrix}$. *Cube-based computing yields*

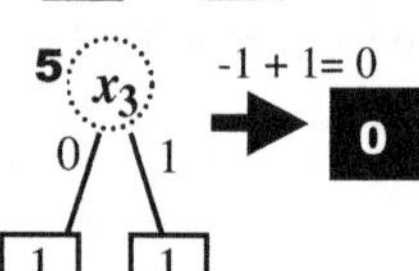

$$\frac{\widetilde{\partial}^2}{\widetilde{\partial} x_1 \widetilde{\partial} x_3}\begin{bmatrix} 1\ 0\ 1 \\ \mathbf{x}\ 0\ 0 \end{bmatrix} = \frac{\widetilde{\partial}}{\widetilde{\partial} x_1}\begin{bmatrix} 1\ 0\ a \\ \mathbf{x}\ \mathbf{x}\ b \end{bmatrix} = \begin{bmatrix} 0\ 0\ a \end{bmatrix}, i.e.,$$

$$\frac{\widetilde{\partial}^2 f}{\widetilde{\partial} x_1 \widetilde{\partial} x_3} = (-1)^{\overline{x}_1}\overline{x}_2(-1)^{\overline{x}_3} = 1$$

$$\frac{\widetilde{\partial}^2 f}{\widetilde{\partial} x_1 \widetilde{\partial} x_3}|_{p=101} = (-1)^{\overline{0}_1}\overline{x}_2(-1)^{\overline{1}_3} = 1$$

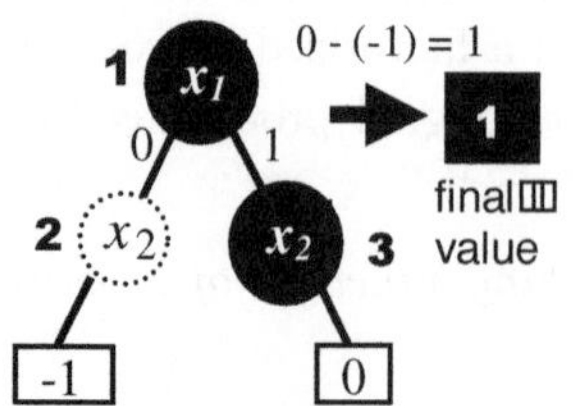

FIGURE 39.9

Calculation of arithmetic difference $\frac{\widetilde{\partial}^2 f}{\widetilde{\partial} x_1 \widetilde{\partial} x_3}|_{p=101}$ given the switching function $f = x_1\overline{x}_2 \vee \overline{x}_3$ and $p = 101$ (Example 39.15).

matrix $W_{2^n}^{(p_i)}$ *is formed by the rule*

$$W_{2^n}^{(p_i)} = \frac{1}{2}\begin{cases} I_{2^{i-1}} \otimes \left[\begin{smallmatrix} 1 & 1 \\ 1 & 1 \end{smallmatrix}\right] \otimes I_{2^{n-i}}, & p_i = 0 \\ I_{2^{i-1}} \otimes \left[\begin{smallmatrix} 1 & -1 \\ -1 & 1 \end{smallmatrix}\right] \otimes I_{2^{n-i}}, & p_i = 1 \end{cases} \tag{39.8}$$

where $p_1p_2\ldots p_n$ *is a binary code of* p.

The proof is given at the end of this chapter.

Example 39.16 *Figure 39.10 shows the flow graph of computing the Walsh difference, where matrices* $W_{2^3}^{(p_1)}, W_{2^3}^{(p_2)}$, *and* $W_{2^3}^{(p_3)}$ *are formed by Equation 39.8 for* $p_1 = 0,\ p_2 = 1,\ p_3 = 1$ *as given in Figure 39.11. The details of computation are illustrated with respect to variable* x_2 *of a 3-variable switching function* f *given by the truth vector* $\mathbf{F} = [11101100]^T$. *Since* $i = 2$, *then* $p_1p_2p_3 = 010$.

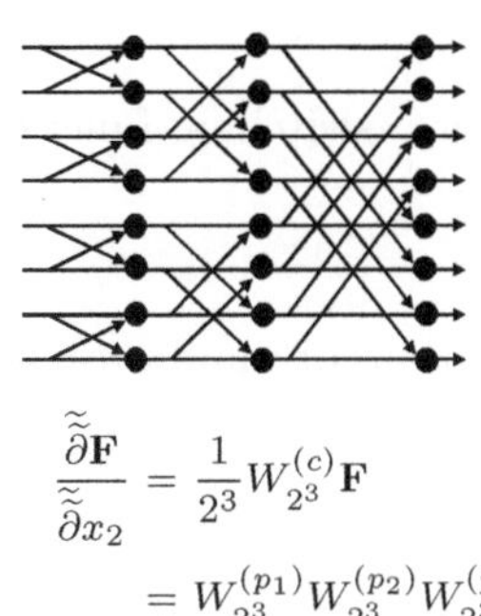

$$\frac{\tilde{\tilde{\partial}}\mathbf{F}}{\tilde{\tilde{\partial}} x_2} = \frac{1}{2^3} W_{2^3}^{(c)} \mathbf{F} = W_{2^3}^{(p_1)} W_{2^3}^{(p_2)} W_{2^3}^{(p_3)} \mathbf{F}$$

$$\frac{\tilde{\tilde{\partial}}\mathbf{F}}{\tilde{\tilde{\partial}} x_2} = \frac{1}{8}\,[3\ 3 - 3 - 3\ 3\ 3 - 3 - 3]^T$$

A 0-polarity Walsh expression of the initial function is

$$W_0 = \frac{1}{2^3}(5 + (-1)^{x_3} + 3(-1)^{x_2} - (-1)^{x_2+x_3} + (-1)^{x_1} + (-1)^{x_1+x_3}(-1)^{x_1+x_2} - (-1)^{x_1+x_2+x_3})$$

For instance $w_0^{(2)} = \frac{\tilde{\tilde{\partial}} f(0)}{\tilde{\tilde{\partial}} x_2} = 3.$

FIGURE 39.10
Computing Walsh difference for a switching function (Example 39.16).

Figure 39.11 contains the Walsh differential matrices and corresponding flow graphs of the Walsh differences for a 3-variable switching function f. To calculate the Walsh differences with respect to a variable x_i, three iterations are needed. The same Boolean and arithmetic differences are calculated via one matrix. The Walsh multiple differences of a switching function are defined by analogy with the multiple Boolean and arithmetic differences.

An analog of the logic Taylor series of a switching function f of n variables in the Walsh domain is defined as

$$f = \frac{1}{2^n}\sum_{i=0}^{2^n-1} w_i^{(c)} \underbrace{(-1)^{\overbrace{(x_1 \oplus c_1)}^{c_1\ polarity}\,^{i_1}} \ldots \overbrace{(x_n \oplus c_n)}^{c_n\ polarity}\,^{i_n}}_{i-th\ product}, \tag{39.9}$$

$$\frac{\tilde{\tilde{\partial}} f}{\tilde{\tilde{\partial}} x_1} = \overset{W_{2^3}^{(p_1)}}{\begin{bmatrix} 1 & & & & -1 & & & \\ & 1 & & & & -1 & & \\ & & 1 & & & & -1 & \\ & & & 1 & & & & -1 \\ -1 & & & & 1 & & & \\ & -1 & & & & 1 & & \\ & & -1 & & & & 1 & \\ & & & -1 & & & & 1 \end{bmatrix}} \overset{W_{2^3}^{(p_2)}}{\begin{bmatrix} 1 & & 1 & & & & & \\ & 1 & & 1 & & & & \\ 1 & & 1 & & & & & \\ & 1 & & 1 & & & & \\ & & & & 1 & & 1 & \\ & & & & & 1 & & 1 \\ & & & & 1 & & 1 & \\ & & & & & 1 & & 1 \end{bmatrix}} \overset{W_{2^3}^{(p_3)}}{\begin{bmatrix} 1 & 1 & & & & & & \\ 1 & 1 & & & & & & \\ & & 1 & 1 & & & & \\ & & 1 & 1 & & & & \\ & & & & 1 & 1 & & \\ & & & & 1 & 1 & & \\ & & & & & & 1 & 1 \\ & & & & & & 1 & 1 \end{bmatrix}} \overset{\mathbf{F}}{\begin{bmatrix} f(0) \\ f(1) \\ f(2) \\ f(3) \\ f(4) \\ f(5) \\ f(6) \\ f(7) \end{bmatrix}}$$

$$\frac{\tilde{\tilde{\partial}} f}{\tilde{\tilde{\partial}} x_2} = \begin{bmatrix} 1 & & & & 1 & & & \\ & 1 & & & & 1 & & \\ & & 1 & & & & 1 & \\ & & & 1 & & & & 1 \\ 1 & & & & 1 & & & \\ & 1 & & & & 1 & & \\ & & 1 & & & & 1 & \\ & & & 1 & & & & 1 \end{bmatrix} \begin{bmatrix} 1 & & -1 & & & & & \\ & 1 & & -1 & & & & \\ -1 & & 1 & & & & & \\ & -1 & & 1 & & & & \\ & & & & 1 & & -1 & \\ & & & & & 1 & & -1 \\ & & & & -1 & & 1 & \\ & & & & & -1 & & 1 \end{bmatrix} \begin{bmatrix} 1 & 1 & & & & & & \\ 1 & 1 & & & & & & \\ & & 1 & 1 & & & & \\ & & 1 & 1 & & & & \\ & & & & 1 & 1 & & \\ & & & & 1 & 1 & & \\ & & & & & & 1 & 1 \\ & & & & & & 1 & 1 \end{bmatrix} \begin{bmatrix} f(0) \\ f(1) \\ f(2) \\ f(3) \\ f(4) \\ f(5) \\ f(6) \\ f(7) \end{bmatrix}$$

$$\frac{\tilde{\tilde{\partial}} f}{\tilde{\tilde{\partial}} x_3} = \begin{bmatrix} 1 & & & & 1 & & & \\ & 1 & & & & 1 & & \\ & & 1 & & & & 1 & \\ & & & 1 & & & & 1 \\ 1 & & & & 1 & & & \\ & 1 & & & & 1 & & \\ & & 1 & & & & 1 & \\ & & & 1 & & & & 1 \end{bmatrix} \begin{bmatrix} 1 & & 1 & & & & & \\ & 1 & & 1 & & & & \\ 1 & & 1 & & & & & \\ & 1 & & 1 & & & & \\ & & & & 1 & & 1 & \\ & & & & & 1 & & 1 \\ & & & & 1 & & 1 & \\ & & & & & 1 & & 1 \end{bmatrix} \begin{bmatrix} 1 & -1 & & & & & & \\ -1 & 1 & & & & & & \\ & & 1 & -1 & & & & \\ & & -1 & 1 & & & & \\ & & & & 1 & -1 & & \\ & & & & -1 & 1 & & \\ & & & & & & 1 & -1 \\ & & & & & & -1 & 1 \end{bmatrix} \begin{bmatrix} f(0) \\ f(1) \\ f(2) \\ f(3) \\ f(4) \\ f(5) \\ f(6) \\ f(7) \end{bmatrix}$$

FIGURE 39.11
Matrix based computation of the Walsh differences for a switching function of three variables.

where $c_1c_2\dots c_n$ and $i_1, i_2, \dots, i_n$ are binary representations of c and i respectively, and the i-th coefficient is defined as

$$w_i^{(c)}(d) = \frac{\tilde{\tilde{\partial}}^n f(c)}{\tilde{\tilde{\partial}} x_1^{i_1} \tilde{\tilde{\partial}} x_2^{i_2} \dots \tilde{\tilde{\partial}} x_n^{i_n}} \bigg|_{d=c} \quad \text{and} \quad \partial x_i^{i_j} = \begin{cases} 1, & i_j = 0, \\ \partial x_j. & i_j = 1 \end{cases}$$

The coefficient $w_i^{(c)}(d)$ is the value of a n-ordered Walsh difference of f given c i.e., ($x_1 = c_1, \dots, x_n = c_n$). The coefficients $w_i^{(c)}(d)$ (Equation 39.4.1) are called the *Walsh spectrum* of the switching function f.

Example 39.17 *The coefficients of a Walsh Taylor series of 2^3 polarities for the switching function $f = x_1\overline{x}_2 \vee \overline{x}_3$ is shown in Figure 39.12. For example, the Walsh spectrum in 0-polarity is represented by the four non-zero spectral components $f^{(000)}, f^{(001)}, f^{(101)}$, and $f^{(111)}$. Similarly to the Reed–Muller spectrum, we observe that each spectral coefficient carries implicit information about the function, and can be explained in terms of Walsh analogs of Boolean difference, calculated using matrix-vector multiplications (Equation 39.6).*

Polarity		**Walsh spectral coefficients / Walsh differences**								
	$j_1j_2j_3$		*000*	*001*	*010*	*011*	*100*	*101*	*110*	*111*
c	$\mathbf{W}_c$	w	$\mathbf{F}$	$\frac{\tilde{\tilde{\partial}} f}{\tilde{\tilde{\partial}} x_3}$	$\frac{\tilde{\tilde{\partial}} f}{\tilde{\tilde{\partial}} x_2}$	$\frac{\tilde{\tilde{\partial}}^2 f}{\tilde{\tilde{\partial}} x_2 \tilde{\tilde{\partial}} x_3}$	$\frac{\tilde{\tilde{\partial}} f}{\tilde{\tilde{\partial}} x_1}$	$\frac{\tilde{\tilde{\partial}}^2 f}{\tilde{\tilde{\partial}} x_1 \tilde{\tilde{\partial}} x_3}$	$\frac{\tilde{\tilde{\partial}}^2 f}{\tilde{\tilde{\partial}} x_1 \tilde{\tilde{\partial}} x_2}$	$\frac{\tilde{\tilde{\partial}}^3 f}{\tilde{\tilde{\partial}} x_1 \tilde{\tilde{\partial}} x_2 \tilde{\tilde{\partial}} x_3}$
0	$\mathbf{W}_0$	5	1	3	1	-1	-1	1	-1	1
1	$\mathbf{W}_1$	5	0	-3	1	1	-1	-1	-1	-1
2	$\mathbf{W}_2$	5	1	3	-1	1	-1	1	1	-1
3	$\mathbf{W}_3$	5	0	-3	-1	-1	-1	-1	1	1
4	$\mathbf{W}_4$	5	1	3	1	-1	1	-1	1	-1
5	$\mathbf{W}_5$	5	1	-3	1	1	1	1	1	1
6	$\mathbf{W}_6$	5	1	3	-1	1	1	-1	-1	1
7	$\mathbf{W}_7$	5	0	-3	-1	-1	1	1	-1	-1

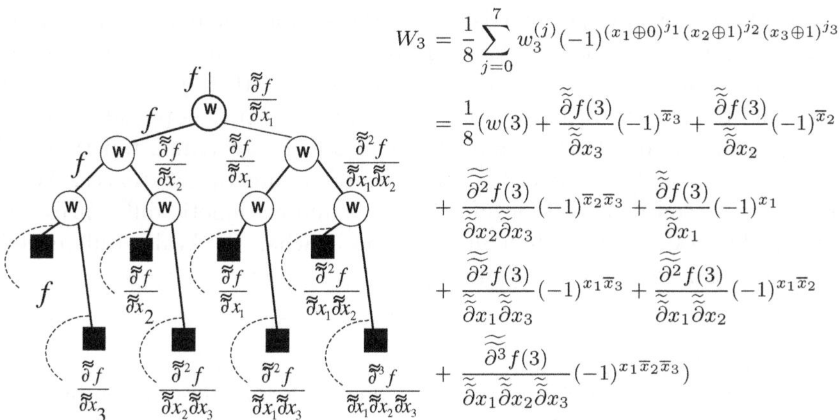

$$W_3 = \frac{1}{8}\sum_{j=0}^{7} w_3^{(j)}(-1)^{(x_1\oplus 0)^{j_1}(x_2\oplus 1)^{j_2}(x_3\oplus 1)^{j_3}}$$
$$= \frac{1}{8}(w(3) + \frac{\tilde{\tilde{\partial}} f(3)}{\tilde{\tilde{\partial}} x_3}(-1)^{\overline{x}_3} + \frac{\tilde{\tilde{\partial}} f(3)}{\tilde{\tilde{\partial}} x_2}(-1)^{\overline{x}_2}$$
$$+ \frac{\tilde{\tilde{\partial}}^2 f(3)}{\tilde{\tilde{\partial}} x_2 \tilde{\tilde{\partial}} x_3}(-1)^{\overline{x}_2\overline{x}_3} + \frac{\tilde{\tilde{\partial}} f(3)}{\tilde{\tilde{\partial}} x_1}(-1)^{x_1}$$
$$+ \frac{\tilde{\tilde{\partial}}^2 f(3)}{\tilde{\tilde{\partial}} x_1 \tilde{\tilde{\partial}} x_3}(-1)^{x_1\overline{x}_3} + \frac{\tilde{\tilde{\partial}}^2 f(3)}{\tilde{\tilde{\partial}} x_1 \tilde{\tilde{\partial}} x_2}(-1)^{x_1\overline{x}_2}$$
$$+ \frac{\tilde{\tilde{\partial}}^3 f(3)}{\tilde{\tilde{\partial}} x_1 \tilde{\tilde{\partial}} x_2 \tilde{\tilde{\partial}} x_3}(-1)^{x_1\overline{x}_2\overline{x}_3})$$

FIGURE 39.12
Computing eight Walsh spectra via Walsh–Taylor expansion (Walsh differences) for the switching function $f = x_1\overline{x}_2 \vee \overline{x}_3$ (Example 39.17).

39.4.2 Walsh differences in symbolic form

The differential operator $\frac{\mathcal{L}^p}{\mathcal{L}x_i}$ ($\mathcal{L}$–operator for short) of a switching function f with respect to the variable x_i and parameter $p = p_1p_2\ldots p_n$

$$\frac{\mathcal{L}^{p_i} f}{\mathcal{L}x_i} = f(x_1\ldots x_i\ldots x_n) + (-1)^{p_i} f(x_1\ldots \overline{x}_i \ldots x_n)$$
$$= \begin{cases} f(x_1\ldots 1\ldots x_n) + f(x_1\ldots 0\ldots x_n), & if \;\; p_i = 0, \\ f(x_1\ldots 1\ldots x_n) - f(x_1\ldots 0\ldots x_n), & if \;\; p_i = 1. \end{cases} \quad (39.10)$$

This operator detects all the possible changes of a switching function with respect to a given variable: $\frac{\mathcal{L}^1 f}{\mathcal{L}x_i} = -\frac{\tilde{\partial} f}{\tilde{\partial} x_i}$ (which is an arithmetic analog of the Boolean difference) recognizes changes of function, and difference $\frac{\mathcal{L}^0 f}{\mathcal{L}x_i}$

distinguishes two cases when the function is unchanged while variable x_i is changed.

The Walsh difference with respect to a variable x_i of an n-variable switching function f given by truth vector $\mathbf{F}$ is the multiple differential operator

$$\frac{\tilde{\tilde{\partial}} f}{\tilde{\tilde{\partial}} x_i} = \frac{\mathcal{L}^{p_1}}{\mathcal{L}x_1}\left(\cdots\left(\frac{\mathcal{L}^{p_n} f}{\mathcal{L}x_n}\right)\right), \tag{39.11}$$

where $p_i = 1$ and $p_j = 0$ for $j \neq i$.

Equation 39.11 follows immediately from the n-iteration transform (Equation 39.6).

The Walsh difference with respect to x_i corresponds to a "butterfly" configuration of the flow graph or matrix $\widetilde{D}_{2^{n-i}}$, the standard matrix representation of the i-th iteration of the fast Walsh transform. A particular case of Equation 39.11 is when $p = 1 \ldots 1$, i.e., the Walsh difference with respect to a variable x_i is equal to the n-th order arithmetic difference with the opposite sign.

Table 39.5 contains four cases of the behavior of a switching function with respect to a variable x_i, described in terms of arithmetic differences. Walsh differences, in contrast to Boolean differences and arithmetic differences, distinguish the cases of an unchanging function (0 when the Walsh difference is equal to 0, and 1 when it is equal to 2).

TABLE 39.5
Behavior of the switching function f and its formal description by Walsh difference in cube form.

Behavior of f	$\frac{\tilde{\tilde{\partial}} f(0)}{\tilde{\tilde{\partial}} x_i} = \frac{\mathcal{L}^0 f}{\mathcal{L}x_i}$	$\frac{\tilde{\tilde{\partial}} f(1)}{\tilde{\tilde{\partial}} x_i} = \frac{\mathcal{L}^1 f}{\mathcal{L}x_i}$	**Cube form**
$f_{x_i=0} = 0$ *and* $f_{x_i=1} = 0$	0	0	$\frac{\tilde{\tilde{\partial}}}{\tilde{\tilde{\partial}} x_i}[C] = \emptyset$
$f_{x_i=0} = 0$ *and* $f_{x_i=1} = 1$	1	1	$\frac{\tilde{\tilde{\partial}} f}{\tilde{\tilde{\partial}} x_i}[C] = [C]_{x_i=b}$ *or* $[C]_{x_i=a}$
$f_{x_i=0} = 1$ *and* $f_{x_i=1} = 0$	1	1	$\frac{\tilde{\tilde{\partial}} f}{\tilde{\tilde{\partial}} x_i}[C] = [C]_{x_i=b}$
$f_{x_i=0} = 1$ *and* $f_{x_i=1} = 1$	2	0	$\frac{\tilde{\tilde{\partial}} f}{\tilde{\tilde{\partial}} x_i}[C] = 2[C]$ *or* $\emptyset$

Table 39.6 contains the Walsh coefficients and symbolic forms for AND, OR and EXOR functions.

TABLE 39.6
Walsh spectrum as a Walsh–Taylor expansion of elementary switching functions.

Function	Walsh spectrum F	$\frac{\tilde{\partial} f}{\tilde{\partial} x_2}$	$\frac{\tilde{\partial} f}{\tilde{\partial} x_1}$	$\frac{\tilde{\partial}^2 f}{\tilde{\partial} x_1 \tilde{\partial} x_2}$	Walsh expression
x_1, x_2, f	1	−1	−1	1	$\frac{1}{4}[1-(-1)^{x_2}-(-1)^{x_1}+(-1)^{x_1 x_2}]$
	1	1	−1	−1	$\frac{1}{4}[1+(-1)^{\overline{x}_2}-(-1)^{x_1}-(-1)^{x_1 \overline{x}_2}]$
	1	−1	1	−1	$\frac{1}{4}[1-(-1)^{x_2}+(-1)^{\overline{x}_1}-(-1)^{\overline{x}_1 x_2}]$
$y = x_1 \wedge x_2$	1	1	1	1	$\frac{1}{4}[1+(-1)^{\overline{x}_2}+(-1)^{\overline{x}_1}+(-1)^{\overline{x}_1 \overline{x}_2}]$
x_1, x_2, f	3	−1	−1	−1	$\frac{1}{4}[3-(-1)^{x_2}-(-1)^{x_1}-(-1)^{x_1 x_2}]$
	3	1	−1	1	$\frac{1}{4}[3+(-1)^{\overline{x}_2}-(-1)^{x_1}+(-1)^{x_1 \overline{x}_2}]$
	3	−1	1	1	$\frac{1}{4}[3-(-1)^{x_2}+(-1)^{\overline{x}_1}+(-1)^{\overline{x}_1 x_2}]$
$y = x_1 \vee x_2$	3	1	1	−1	$\frac{1}{4}[3+(-1)^{\overline{x}_2}+(-1)^{\overline{x}_1}+(-1)^{\overline{x}_1 \overline{x}_2}]$
x_1, x_2, f	2	0	0	−2	$\frac{1}{4}[2-2(-1)^{x_1 x_2}]$
	2	0	0	2	$\frac{1}{4}[2+2(-1)^{x_1 \overline{x}_2}]$
	2	0	0	2	$\frac{1}{4}[2+2(-1)^{\overline{x}_1 x_2}]$
$y = x_1 \oplus x_2$	2	0	0	−2	$\frac{1}{4}[2-2(-1)^{\overline{x}_1 \overline{x}_2}]$

39.4.3 Properties of Walsh differences in cube notation

The following properties of Walsh differences of switching functions are useful in the manipulation of cubes:

- ▶ Walsh difference with respect to variable x_i depends on this variable.
- ▶ Walsh difference $\frac{\tilde{\partial} f}{\tilde{\partial} x_i}$ of switching function f that does not depend on variable x_i is equal to 0.
- ▶ Walsh difference distinguishes not only cases $\{f_{\overline{x}_i} = 0,\ f_{\overline{x}_i} = 1\}$ and $\{f_{\overline{x}_i} = 1,\ f_{x_i} = 0\}$ but also the case $f_{\overline{x}_i} = f_{x_i} = 1$.

Properties of operators

Property 1. The operator $\frac{\mathcal{L}^0}{\mathcal{L}x_i}$ over the cube $[C_j]_{x_i=0}$ ($[C_j]_{x_i=1}$) with respect to variable x_i yields

$$\frac{\mathcal{L}^0}{\mathcal{L}x_i}[C]_{x_i=1} = [C]_{x_i=a} \text{ and } \frac{\mathcal{L}^0}{\mathcal{L}x_i}[C]_{x_i=0} = [C]_{x_i=b}.$$

Property 2. The operator $\frac{\mathcal{L}^0}{\mathcal{L}x_i}$ over the cube $[C_j]_{x_i=\mathbf{x}}$ with respect to the variable x_i yields an empty cube

$$\frac{\mathcal{L}^0}{\mathcal{L}x_i}[C]_{x_i=\mathbf{x}} = \emptyset.$$

Property 3. If both cubes $[C]_{x_i=0}$ and $[C]_{x_i=1}$ are present in a function cover, the operator $\frac{\mathcal{L}^1}{\mathcal{L}x_i}$ over the cube $[C_j]$ with respect to variable x_i yields

$$\frac{\mathcal{L}^1}{\mathcal{L}x_i}[C]_{x_i=0} = \frac{\mathcal{L}^1}{\mathcal{L}x_i}[C]_{x_i=1} = [C]_{x_i=\mathbf{x}}.$$

Property 4. The operator $\frac{\mathcal{L}^1}{\mathcal{L}x_i}$ over the cube $[C_j]$ with respect to variable x_i yields

$$\frac{\mathcal{L}^1}{\mathcal{L}x_i}[C]_{x_i=1} = 2[C]_{x_i=\mathbf{x}},$$

where $[C]_{x_i=\mathbf{x}}$ means that the function gets value 1 in both cases: $f_{x_i=0}$ and $f_{x_i=1}$.

The Walsh difference with respect to the variable x_i of a switching function is a sum of the Walsh differences derived from the cubes $[C_i]$ representing the Walsh form of this function

$$\frac{\tilde{\tilde{\partial}} f}{\tilde{\tilde{\partial}} x_i} = \sum_{i=1}^{t} \frac{\tilde{\tilde{\partial}}}{\tilde{\tilde{\partial}} x_i}[C_i] \tag{39.12}$$

It follows from this property that:

- Each cube is derived from Equation 39.11 in n iterations (applying the operator $\mathcal{L}^{p_i}$ n times) and
- Each cube must be assembled by Equation 39.12 in one Walsh expression using the sum operation.

Example 39.18 *Figure 39.13 shows the computation of a Walsh spectral coefficient $w^{(9)}$ for switching function f for $p = 0$, given by cubes corresponding to an arithmetic expression by using the $\mathcal{L}^0$-operator.*

The calculation of Walsh spectral coefficients formed from multiple Walsh differences is illustrated in the next example.

Example 39.19 *Figure 39.13 shows the computation of a Walsh difference $\frac{\partial^2 f}{\partial x_1 \partial x_3}$ for the switching function given in arithmetic form. Operators $\mathcal{L}^0$ to variable x_2, and $\mathcal{L}^1$ to variables x_1, x_3 are used.*

Given:
The cubes corresponding to the arithmetic expression of a switching function f of five variables

$$\begin{bmatrix} 0 & \mathrm{x} & 0 & 1 & \mathrm{x} \\ 1 & \mathrm{x} & \mathrm{x} & 1 & 1 \\ 0 & 1 & \mathrm{x} & \mathrm{x} & 1 \\ 1 & 1 & \mathrm{x} & \mathrm{x} & 0 \\ \mathrm{x} & 0 & \mathrm{x} & 0 & 0 \end{bmatrix}$$

The Walsh spectral coefficient $w^{(9)}$

$$\frac{\mathcal{L}^0 f}{\mathcal{L}x_4} = \sum_{t=1}^{3} \frac{\mathcal{L}^0}{\mathcal{L}x_4}[C_t]$$
$$= \frac{\mathcal{L}^0}{\mathcal{L}x_4}[\mathrm{0x01x}] + \frac{\mathcal{L}^0}{\mathcal{L}x_4}[\mathrm{1xx11}] + \frac{\mathcal{L}^0}{\mathcal{L}x_4}[\mathrm{01xx1}]$$
$$+ \frac{\mathcal{L}^0}{\mathcal{L}x_4}[\mathrm{11xx0}] - \frac{\mathcal{L}^0}{\mathcal{L}x_4}[\mathrm{x0x00}]$$
$$= \overline{x}_1\overline{x}_3 + x_1x_5 + 2\overline{x}_1x_2x_5 + 2x_1x_2\overline{x}_5\overline{x}_2\overline{x}_5 - \overline{x}_2\overline{x}_5.$$

$$w^{(2)} = \frac{\mathcal{L}^0 f}{\mathcal{L}x_4}|_{p=00010} = 1 + 0 + 0 + 0 - 1 = 0$$

Given:
The cubes corresponding to the arithmetic expression of a switching function

$$\begin{bmatrix} 1 & 0 & \mathrm{x} \\ \mathrm{x} & \mathrm{x} & 0 \\ 1 & 0 & 0 \end{bmatrix}$$

The Walsh difference $\frac{\partial^2 f}{\partial x_1 \partial x_3}$ for switching function $f =$ $[10\mathrm{x}] + [\mathrm{xx}0] - [100]$:

$$\frac{\widetilde{\partial}^2 f}{\widetilde{\partial} x_1 \widetilde{\partial} x_3} = \frac{\mathcal{L}^1}{\mathcal{L}x_1}\left(\frac{\mathcal{L}^0}{\mathcal{L}x_2}\left(\frac{\mathcal{L}^1 f}{\mathcal{L}x_3}\right)\right)$$
$$= \frac{\mathcal{L}^1}{\mathcal{L}x_1}\left(\frac{\mathcal{L}^0}{\mathcal{L}x_2}([\mathrm{xx}b] - [10b])\right)$$
$$= \frac{\mathcal{L}^1}{\mathcal{L}x_1}(2[\mathrm{xx}b] - [1\mathrm{x}b])$$
$$= -[\ a\mathrm{x}b\ ]$$
$$= -(-1)^{\overline{x}_1 + x_3}$$

FIGURE 39.13
Operators $\mathcal{L}^0$ and $\mathcal{L}^1$ for computing the Walsh differences of a switching function (Examples 39.18 and 39.19).

In Table 39.7, the first and second columns include the benchmark **#Name** and the number of inputs and outputs **#In/#Out**. The next column shows the number of cubes in the initial sum-of-products representation. The next three columns contain the number of cubes obtained after calculation of first 256, 1024, and 4096 Walsh coefficients (at polarity $p = 0$).

39.4.4 Computing Taylor expansion by decision diagram

To compute a given value of a Walsh spectrum, we apply the top-down search to find the paths, and then perform bottom-up search. The only difference is that we apply addition or subtraction over all nodes, despite the variables of differentiation. Information about the sign of the intermediate components must be taken into account.

TABLE 39.7
Cube-based Walsh spectrum computation for switching functions.

Name	Test In/Out	cubes	Walsh spectra in cubes at p=0 256	1024	4096
alu4	14/8	1028	2567	5444	9270
misex3	14/14	1848	1071	3732	8200
add6	13/7	607	133	607	2537
duke2	22/29	242	148	189	218
vg2	25/8	110	104	163	261
misex2	25/18	29	26	26	26
e64	65/65	65	65	65	65

Example 39.20 *(Continuation of Example 39.15). Calculate the Walsh difference* $\frac{\tilde{\tilde{\partial}}^2 f}{\tilde{\tilde{\partial}} x_1 \tilde{\tilde{\partial}} x_3}|_{p=101}$ *given the function* $f = x_1\overline{x}_2 \vee \overline{x}_3$ *and polarity* $p = 101$.

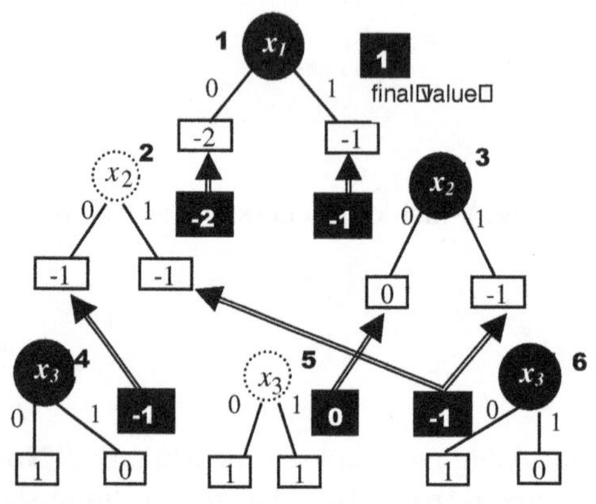

1°. *Find the paths in ROBBD given in Figure 39.6*

2°. *Calculate the intermediate values derived from the outgoing branches of nodes 4, 5 and 6.*

3°. *Calculate the intermediate values derived from the outgoing branches of nodes 2 and 3.*

4°. *Calculate the final values derived from the outgoing branches of node 1.*

FIGURE 39.14
Computing of Walsh difference $\frac{\tilde{\tilde{\partial}}^2 f}{\tilde{\tilde{\partial}} x_1 \tilde{\tilde{\partial}} x_3}|_{p=101}$ for the switching function $f = x_1\overline{x}_2 \vee \overline{x}_3$ via ROBDD (Example 39.20).

In the case of computing a Walsh coefficient, all possible paths are found in the top-down search. The bottom-up iterating results in the intermediate (value -1 derived from the outgoing values 1 and 0 of node 4). Also, values 0 and -1 are derived (node 5 and 6, Figure 39.14). At the next ROBDD level, this results in the values -2 and -1 (nodes 2 and 3 respectively, ((0-1)+(0-1)=-2 and (0-1)+(1-1)=-1). The final value

$$\frac{\tilde{\tilde{\partial}}^2 f}{\tilde{\tilde{\partial}} x_1 \tilde{\tilde{\partial}} x_3}|_{p=101} = [(0-1)+(1-1)] - [(0-1)+(0-1)] = 1$$

is obtained at node 1.

39.5 Further reading

Historical remarks. Lee [13] developed the method of representing switching functions by an n-dimensional cube.

The fundamentals of logic differential calculus are developed by Akers [1], Bochmann and Posthoff [2], Davio et al. [5], Posthoff and Steinbach [16], Thayse and Davio [24]. Yanushkevich [30] developed algorithms for the solution of Boolean differential equations. Gibbs and Millard [8] proposed another definition of differences (recently called Gibbs differences or Gibbs derivatives) that differs from Akers Boolean differences [1]. Edwards [6] and Stanković et al. [23] studied the relationship of Boolean and Gibbs differences. Karpovsky [10, 11] studied the spectral properties of switching functions.

Taylor-like expansions of switching functions were introduced by Akers [1] and Davio et al. [5]. Later, many algorithms were developed based on the unique properties of the Taylor expansion including the generalization of the notion of change for multivalued functions [29, 52]. Data structures in the form of change are related to properties of logic circuits such as reachability and observability. Yanushkevich [52] studied the multiplicative properties of differences in Taylor-like expansions.

Spectral technique is introduced in Chapters 7, 13, and 30.

Relationships between various techniques. In [22], the relationship among the Walsh, the arithmetic, and the Reed–Muller transform is established by the following statements: the arithmetic transform is equivalent to the integer-valued Reed–Muller transform and the Walsh transform is equivalent to the Reed–Muller transform in (1,-1) coding.

Computing Boolean differences. In [19], techniques for computing Boolean differences are introduced.

Differences for multivalued functions. Shmerko and Yanushkevich [18] extended Akers Boolean differences as coeficients of logic Taylor expansion for multivalued functions. Later, they developed techniques for computing logical derivatives for multivalued functions based on logic Taylor expansion [19]. Stanković et al. [23] studied Boolean and Gibbs differences in Galois fields.

Applications. Sellers [17] and Marinos [14] used Boolean differences in testing. Shmulevich et al. [21] applied probabilistic Boolean differences to the study of gene regulatory networks. Karpovsky et al. [12] used spectral methods for design and testing. Larrabee [9] employed Boolean differences in testing using Boolean satisfiability.

Concept of change in nanostructures. Yanushkevich et al. [32] introduced logic and arithmetic differences in spatial dimensions (hypercube topology of nanostructures).

A review on techniques for representing and manipulating logic functions can be found in a paper by Falkowski [7]. Results and trends in logic differential calculus are analyzed in a paper by Bochmann et al. [4].

References

[1] Akers SB. On a theory of Boolean functions. *Society for Industrial and Applied Mathematics*, 7(4):487–498, 1959.

[2] Bochmann D and Posthoff Ch. *Binäre Dynamishe Systeme*, Akademieverlag, Berlin, 1981.

[3] Bochmann D, Posthoff Ch, Shmerko VP, Stanković RS, Tosić Ž, and Yanushkevich SN. State-of-the-art of logic differential calculus. In *Proceedings of the International Workshop on Boolean Problems*, Germany, pp. 117–124, 2000.

[4] Bochmann D, Stanković RS, Tosić Ž, Shmerko VP, and Yanushkevich SN. Logic differential calculus: progress, tendencies and applications. *Automation and Remote Control*, Plenum/Kluwer Publishers, 61(6):1033–1047, 2000.

[5] Davio M, Deschamps J and Thayse A. *Discrete and Switching Functions.* McGraw-Hill, 1978.

[6] Edwards CR. The Gibbs dyadic differentiator and its relationship to the Boolean difference. *Computers and Electronic Engineering,* 5(4):335–344, 1978.

[7] Falkowski BJ. A note on the polynomial form of Boolean functions and related topics. *IEEE Transactions on Computers*, 48(8):860–863, 1999.

[8] Gibbs JE and Millard MS. Walsh functions as solutions of a logical differential equation. *DES Report No.1, National Physical Laboratory* Middlesex, England, 1969.

[9] Larrabee T. Test pattern generation using Boolean satisfiability. *IEEE Transactions on Computer-Aided Design of Integrated Circuits and Systems*, 11(1):4–15, 1992.

[10] Karpovsky MG. *Finite Orthogonal Series in the Design of Digital Devices.* Wiley and JUP, New York and Jerusalem, 1976.

[11] Karpovsky MG, Ed. *Spectral Techniques and Fault Detection.* Academic Press, New York, 1985.

[12] Karpovsky MG, Stanković RS, and Moraga C. Spectral technique for design and testing of computer hardware. In *Proceedings TICSP Workshop on Spectral Transforms and Logic Design for Future Digital Systems*, Tampere, Finland, pp. 1–34, 2000.

[13] Lee CY. Switching functions on an n-dimensional cube. *Transactions of the AIEE*, 73:289–291, Sept. 1954.

[14] Marinos P. Derivation of minimal complete sets of test-input sequences using Boolean differences. *IEEE Transactions on Computers*, 20(1):25–32, 1981.

[15] Najm FN. A Survey of power estimation techniques in VLSI circuits. *IEEE Transactions on VLSI*, 2(4):446–455, 1994.

[16] Posthoff Ch and Steinbach B. *Logic Functions and Equations.* Springer, Heidelberg, 2004.

[17] Sellers FF, Hsiao MY, and Bearson LW. Analyzing errors with the Boolean difference. *IEEE Transactions on Computers*, no. 1, pp. 676–683, 1968.

[18] Shmerko VP and Yanushkevich SN. Fault detection in multivalued logic networks by new type of derivatives of multivalued functions. In *Proceedings of the European Conference on Circuit Theory and Design*, Switzerland, pp. 643–646, 1993.

[19] Shmerko VP, Yanushkevich SN, and Levashenko VG. Techniques of computing logical derivatives for MVL functions. *Proceedings of the 26th IEEE International Symposium on Multiple-Valued Logic*, pp. 267–272, 1996.

[20] Shmerko VP, Yanushkevich SN, and Malecki K. A Class of logic design problems solved based on parallel computations of "butterfly" configurations. *Proceedings of the IEEE International Conference on Parallel and Distributed Processing Techniques and Applications*, Sunnyvale, CA, pp. 1589–1592, 1996.

[21] Shmulevich I, Dougherty ER, Kim S, and Zhang W. Probabilistic Boolean networks: a rule-based uncertainty model for gene Rregulatory networks. *Bioinformatics*, (18):274–277, 2002.

[22] Stanković RS, Shmerko VP, and Moraga C. From Fourier expansions to arithmetic expressions on finite Abelian groups. In *Proceedings of the 6th International Conference on Advanced Computer Systems*, Szczecin, Poland, pp. 483–493, 1999.

[23] Stanković RS, Moraga C, and Astola JT. Derivatives for multiple-valued functions induced by Galois field and Reed–Muller–Fourier expressions. In *Proceedings of the 34th IEEE International Symposium on Multiple-Valued Logic*, pp. 184–189, 2004.

[24] Thayse A and M. Davio M. Boolean differential calculus and its application to switching theory. *IEEE Transactions on Computers*, (22):409–420, 1973.

[25] Tosić Ž. Arithmetical representation of logic functions. In *Discrete Automatics and Networks,* USSR Academy of Sciences/Nauka, Moscow, pp. 131–136, 1970.

[26] Tucker JH, Tapia MA, and Bennet AW. Boolean integral calculus for digital systems. *IEEE Transactions on Computers,* 34:78–81, 1985.

[27] Yanushkevich SN. Systolic algorithms for arithmetic polynomial forms of k-valued functions of Boolean algebra. *Automation and Remote Control,* Kluwer/Plenum Publishers, 55(12):1812–1823, 1994.

[28] Yanushkevich SN. Matrix method to solve logic differential equations. *IEE Proceedings, Pt.E, Computers and Digital Technique*, 144(5):267–272, 1997.

[29] Yanushkevich SN. Development of the methods of Boolean differential calculus for arithmetic logic. *Automation and Remote Control (Kluwer/Plenum Publishers)*, 55(5):715–729, Pt. 2, 1994.

[30] Yanushkevich SN. Matrix and combinatorics solution of Boolean differential equations. *Discrete Applied Mathematics*, (117):279–292, 2001.

[31] Yanushkevich SN. Multiplicative properties of spectral Walsh coefficients of Boolean funtions. *Automation and Remote Control (Kluwer/Plenum Publishers)*, 64(12):1933–1947, 2003.

[32] Yanushkevich SN, Shmerko VP, and Lyshevski SE. *Logic Design of NanoICs*, CRC Press, Boca Raton, FL, 2005.

[33] Yanushkevich SN. Analogues of Boolean differences and differentials in arithmetical logic. In *Proceedings of the International Workshop on Boolean Problems*, Freiberg, Germany, pp. 115–120, 1996.

[34] Wesselkamper TC. Divided difference method for Galois switching functions. *IEEE Transactions on Computers*, 27(3):232–238, 1978.

Appendix

Prove the Theorem 39.1

The original Hadamard matrix (a basis for Walsh transform) is defined as

$$H_2 = \begin{bmatrix} 1 & 1 \\ 1 & -1 \end{bmatrix}, \quad H_{2^n} = \otimes^n \begin{bmatrix} 1 & 1 \\ 1 & -1 \end{bmatrix}.$$

However, we can determine the so-called Walsh transform of **polarity** c by analogy with Reed–Muller and arithmetic transforms. The polarity c Walsh matrix is specified as below:

$$H_{2^n}^{(c)} = \frac{1}{2^n} H_2^{(c_1)} \otimes ... \otimes H_2^{(c_n)},$$

where the elementary matrices are $H_2^{(0)} = \begin{bmatrix} 1 & 1 \\ 1 & -1 \end{bmatrix}$, $H_2^{(1)} = \begin{bmatrix} 1 & 1 \\ -1 & 1 \end{bmatrix}$. This transform can be represented in factorized form known to be used for the fast Walsh transform:

$$H_{2^n}^{(c)} = \frac{1}{2^n} H_{2^n}^{(c_1)} H_{2^n}^{(c_2)} \dots H_{2^n}^{(c_n)},$$

matrix $H_{2^n}^{(c_i)}$ is formed by the rule

$$H_{2^n}^{(c_i)} = \frac{1}{2} \begin{cases} I_{2^{i-1}} \otimes \begin{bmatrix} 1 & 1 \\ 1 & -1 \end{bmatrix} \otimes I_{2^{n-i}}, \; c_i = 0 \\ \\ I_{2^{i-1}} \otimes \begin{bmatrix} 1 & 1 \\ -1 & 1 \end{bmatrix} \otimes I_{2^{n-i}}, \; c_i = 1 \end{cases}$$

where $c_1 c_2 \dots c_n$ is a binary code of c. Obviously, there are 2^n separate polarity Walsh matrices for any given n.

The Boolean difference matrices (2^n distinguished matrices, including an identity matrix and all possible one-order, two-order, and n-order matrices) have been derived from the family of 2^n Reed–Muller transform matrices. This holds for their arithmetical analogs as well. One can generalize this approach to Walsh transforms. For instance, the elementary 0-polarity Walsh differential matrix is equal to $\begin{bmatrix} 1 & 1 \\ 1 & 1 \end{bmatrix}$, and the elementary 1-polarity Walsh differential matrix is equal to $\begin{bmatrix} 1 & -1 \\ -1 & 1 \end{bmatrix}$. Note that the rows of those matrices form the rows of 0- and 1-polarity elementary matrices $\begin{bmatrix} 1 & 1 \\ 1 & -1 \end{bmatrix}$, and $\begin{bmatrix} 1 & 1 \\ -1 & 1 \end{bmatrix}$ from Equation 39.5. If we apply the same rule to the $2^n \times 2^n$ Walsh transform matrices, we can derive the Walsh differential matrices with respect to variables $x_1^{p_1}, ..., x_n^{p_n}$, where $x_i^0 = 1$, and $x_i^1 = x_i$. For example, the Walsh differential matrix $W_{23}^{(101)}$ corresponds to the calculation of a two-order Walsh difference with respect to variables x_1, x_3. The equation to calculate the Walsh differential matrix is:

$$W_{2^n}^{p} = \frac{1}{2^n} W_2^{p_1} \otimes ... \otimes W_2^{p_n},$$

where

$$W_2^0 = \begin{bmatrix} 1 & 1 \\ 1 & 1 \end{bmatrix}, \quad W_2^1 = \begin{bmatrix} 1 & -1 \\ -1 & 1 \end{bmatrix}$$

The factorized form derived from this equation is Equation 39.6.

40

Developing New Decision Diagrams

The challenges of digital system design, and new technologies and principles of information processing (nano and biocomputing) inspire the devlopment of new types of data representation and processing, among them new decision diagrams. Decision diagrams present the information about the functions in a different way and in a different (spectral) domains, which may facilitate observation of peculiar properties of the functions or calculations with them. In this chapter, the synthesis of various types of decision diagrams is revisited, based on (*i*) generalization of spectral methods and corresponding decision diagrams (Reed–Muller, arithmetic, Walsh, Haar, and Fibonacci), (*ii*) the group theoretic approach, and on (*iii*) the relevant methods of optimization and manipulation of these data structures.

40.1 Introduction

Given a logic design problem described in terms of logic functions, there is typically a diagram providing the most suitable representation with respect to an a *priori* determined criterion. Conversely, for each given type of decision diagram, it is possible to specify for which applications this diagram is the optimal in some specified sense. Therefore, the derivation of modified or possibly new diagrams will frequently be useful in finding representations suited for particular applications.

40.2 Spectral tools for generation of decision diagrams

Spectral methods offer techniques and tools for the *spectral interpretation* of logic functions, including the generation of algebraic forms and methods for construction of spectral decision diagrams. Spectral interpretations of data structures such as decision diagrams, derived to represent a function, reveal additional information about this function. On the other hand, construction of various spectra can be accomplished by changing the transform basis or by mutual transforms between spectra in different bases, in particular, between Walsh, Haar, and arithmetic transforms. This can also be accomplished through manipulation of spectral decision diagrams.

40.2.1 Approaches to the construction of decision diagrams

Figure 40.1 illustrates two approaches to the construction of decision diagrams. Given a logic function and application requirements, its decision diagram:

- Is chosen from a fixed library of several types of decision diagrams,
- Is generated using *application-specific criteria.* This generator is called *decision diagram design tools.*

Application-specific criteria may vary for specific tasks. The more general case is the Karuhnen–Loeve transform, where a specific set of basis functions is designed for each given signal to be processed. The basis functions are selected by analyzing statistical properties of the given signal, which requires a priori knowledge of the signal that will be processed. However, a considerable disadvantage is the lack of fast calculation algorithms, sice the basis functions do not necessarily support the Kronecker product structure or some other features convenient to developing fast Fourier transform-like algorithms. Due to that the application of the Karhunen–Loeve transform in practice is restricted and the transform is often approximated by some other related transforms, as, for instance, the discrete cosine transform.

A library of decision diagrams. This approach involves the selection of a known type of decision diagram from a library. In practice, an acceptable BDD based solution may not exist. In this case, another representation of a given function, or another method for construction of a decision diagram should be chosen.

Generation of a decision diagram. The second approach is more flexible and more complex. This approach may result in the construction of decision diagrams that are not included in the existing library or discussed in the literature. The method for construction can be based on the spectral interpretation of decision diagrams. In spectral techniques, Karuhnen–Loeve criteria can be used for choosing optimal spectral representation of logic functions, under restrictions that the transform selected must express a structure corresponding to the recursive structure of decision trees. A related method would be to use the methods for constructing basis functions for spectral transforms in discrete wavelet packet analysis.

In addition, a strategy for generating decision diagrams can be implemented using artificial intelligence methods, including bio-inspired methods (see Chapter 41).

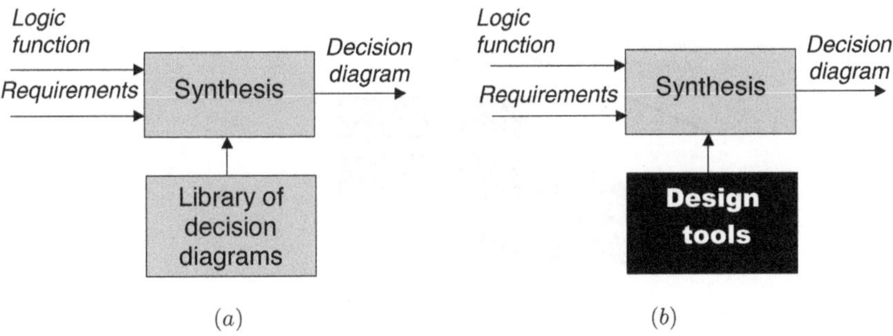

FIGURE 40.1
Two approaches to the construction of decision diagrams: by using a library of decision diagrams (a), and decision diagram design tools (b).

Relationship between various representations. Figure 40.2 shows the relationship between the logic function representations in various domains:

- *Algebraic* representations and *spectral* representations of logic functions, and
- Decision trees and diagrams, and spectral trees and diagrams.

These relationships can be used in the synthesis of new algebraic representations of logic functions, and in the design of decision diagrams with new topological characteristics, in particular,

- Classification of decision trees and diagrams,
- Studying functional and topological properties,
- Development of new methods of choosing an appropriate decision diagram for a given logic function, and
- For the development of new decision diagrams appropriate to particular tasks of logic function processing and embedding into other topological structures.

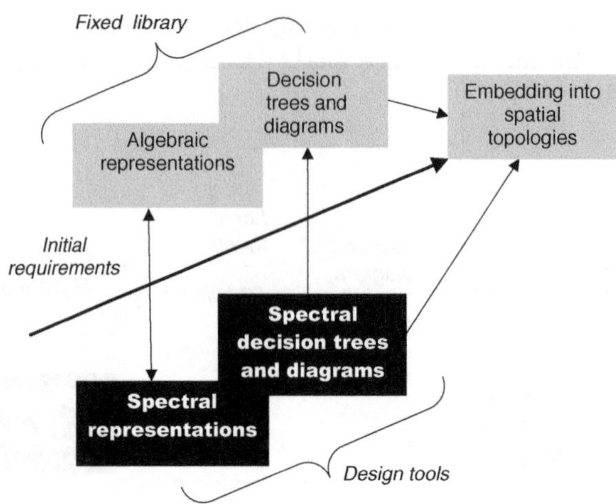

FIGURE 40.2
The relationship between the two approaches to decision diagram design.

> **Example 40.1** *Figure 40.3 shows a binary decision tree, chosen from a library of diagrams, and a spectral approach (Figure 40.3a and 40.3b respectively). These types lead to different topologies, which are used, in particular, at the layout design stage.*

40.2.2 The spectral approach

The theoretical basis for the generation of new types of decision trees and diagrams includes:

- Direct and inverse Fourier-like transforms (see details in Chapter 7),
- Algebraic description of the spectrum of a switching function derived from the result of inverse spectral transform,
- Mapping the spectra in algebraic form into a spectral decision diagram and vice versa, and
- Reduction of a spectral decision tree resulting in a spectral decision diagram.

40.2.3 Basic theorems

The following theorems constitute the background for the spectral synthesis of any decision tree and diagram [70, 47, 48].

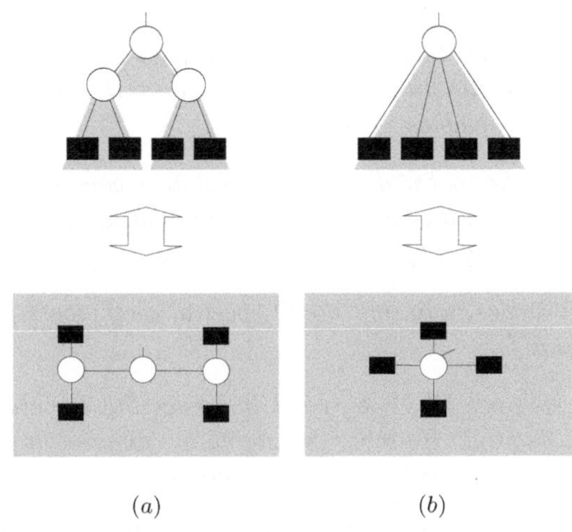

Each decision tree representing a function f can be considered to be a Shannon tree representing the Q-spectrum of f. In a decision tree,

- ▶ *The path from the root node to the terminal nodes corresponds to a basic function in Q.*
- ▶ *The terminal nodes represent the Q-spectral coefficients for the function f.*

FIGURE 40.3
Types of decision diagrams used to represent logic functions: (a) binary tree and (b) Q-diagram leading to different topologies (Examples 40.12 and 40.2).

Theorem 40.1 *Consider the space $P(G)$ of functions $f : G \in P$, where G is a finite group and P is a field that can be the complex field C or a finite field $GF(P)$. Decision trees are graphical representations of spectral transform expansions of function $f \in P(G)$ with respect to a basis Q in $P(G)$. In a decision tree,*

- ▶ *Each path from the root node to a terminal node corresponds to a basic function in Q.*
- ▶ *The terminal nodes represent the Q-spectral coefficients for the function f.*

The proof is based on the definition of a decision tree and its relationship to the sum-of-product expressions, which is then extended to other functional expressions.

It follows from Theorem 40.1 that changing the basis results in changing the labels along the path, and this property can be used in the construction of decision trees and diagrams.

Theorem 40.2 *A decision tree defined with respect to a basis Q represents both a switching function f and the Q-spectrum of f.*

The proof is based on the analysis of the edge labels and the recursive derivation of a decision tree.

Example 40.2 *In Figure 40.3, the explanations of Theorems 40.1 and 40.2 are given.*

Example 40.3 *Figure 40.4 shows that the path corresponds to the application of function decomposition over the basis Q with respect to the variables x_1, x_2, x_3:*

- *Each path from the root node to a terminal node corresponds to a component of the spectral expression that is a product of some literals and Q-spectral coefficients.*
- *The terminal nodes represent the Q-spectral coefficients for the function f.*

A change of basis over P(G) does not influence the structure of the decision diagram. However, changing the domain group G changes the structure of the decision diagram.

In Figure 40.4, the labels at the edges are defined from a description of basis functions q_i used in the definition of the decision tree considered. For instance, if a basis function q_i is written as a function ϕ_i in terms of the cofactros $q_{0,i}$ and $q_{1,i}$ with respect to a binary-valued variable x_i, then the labels at the left and the right edges are $q_{i,0}$ and $q_{i,1}$, respectively. For the nodes at the level i, they can be expressed as a function ϕ_i:

$$q_i = \phi_i(q_{i,0}, q_{i,1}),$$

where $q_{i,0}$ and $q_{i,1}$ are the co-factors of q_i with respect to the variable x_i. Conversely, q_i is as a function ϕ_i of these co-factors.

In Figure 40.4, the labels at edges per levels would be

$$q_{00},\ q_{01}$$
$$q_{1,0},\ q_{1,1},\ q_{1,0},\ q_{1,1}$$
$$q_{2,0},\ q_{2,1},\ q_{2,0},\ q_{2,1},\ q_{2,0},\ q_{2,1}.$$

The letters A in Figure 40.4 will be as follows, in their appearance from top to the bottom and from the left to the right:

$$A = q_0,\ A = q_1,\ A = q_2,\ A = q_3,\ A = q_4,\ A = q_5,\ A = q_6,\ A = q_7.$$

40.2.4 Decision diagram and the spectrum of a switching function

In the spectral representation of signals, we distinguished *direct* and *inverse* transforms as used to decompose a signal in terms of a given set of elementary signals, and to recover the signal from its spectral coefficients, respectively. (see Chapter 7). This property is used to develop tools for the synthesis of the decision diagrams.

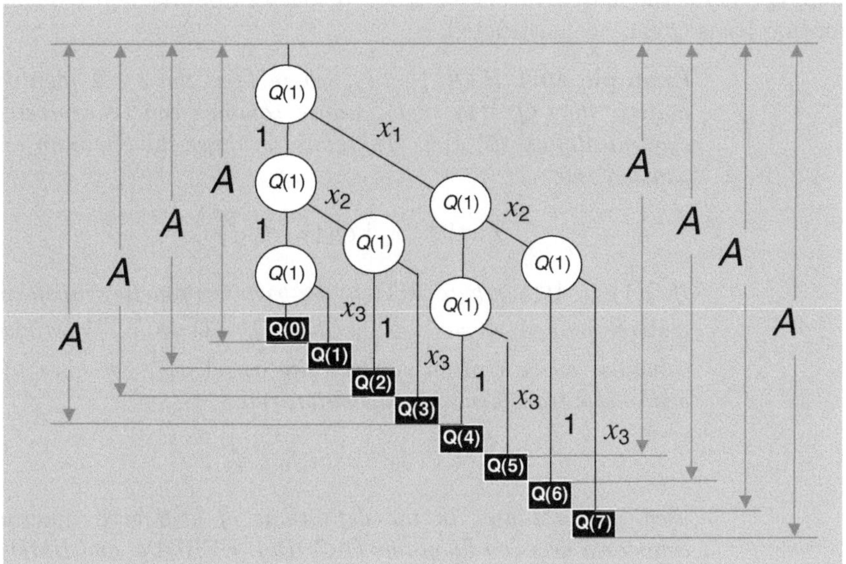

FIGURE 40.4
Each path from the root node to the terminal nodes of decision tree corresponds to a basic function in Q (Example 40.3).

> *The spectral interpretation of BDDs and functional decision diagrams (FDDs) implies the definition of* spectral transform *decision trees (STDTs). This concept captures various types of decision diagrams, and relates each to a particular specification of the basis Q.*

Let $\mathbf{Q}(n)$ be a $(2^n \times 2^n)$ non-singular matrix with elements in P. Thus, columns of $\mathbf{Q}$ form a set of linearly independent functions. Since there are 2^n such functions, it follows that $\mathbf{Q}(n)$ determines a basis in $P(C_2^n)$.

Suppose that $\mathbf{Q}(n)$ is represented as the Kronecker product of n factors $\mathbf{Q}(1)$,

$$\mathbf{Q}(n) = \bigotimes_{i=1}^{n} \mathbf{Q}(1).$$

Given a basis Q, the i-th basic matrix $\mathbf{Q}(1)$ defines an expansion of f with respect to the i-th variable

$$f = \mathbf{Q}^{-1}(1)\mathbf{Q}(1) \begin{bmatrix} f_0 \\ f_1 \end{bmatrix},$$

where $\mathbf{Q}^{-1}(1)$ is the matrix inverse of $\mathbf{Q}(1)$. It follows from the above definition that basis Q can be constructed.

Example 40.4 *If $Q(1) = I_2$, where I_2 is the 2×2 identity matrix, then $Q^{-1}(1) = I_2$, whose columns can be expressed symbolically as $[\ \overline{x}_i\ x_i\ ]$. Therefore, we have the Shannon expansion rule*

$$f = [\ \overline{x}_i\ x_i\ ]\ Q(1)\ [\ f_0\ f_1\ ].$$

If $Q(1) = A(1)$, where $A(1)$ is the basis arithmetic transform matrix, i.e., $A(1) = \begin{bmatrix} 1 & 0 \\ -1 & 1 \end{bmatrix}$, then $Q^{-1}(1) = \begin{bmatrix} 1 & 0 \\ 1 & 1 \end{bmatrix}$, whose columns can be written symbolically as $[\ 1\ x_i\ ]$, we have the arithmetic transform decomposition rule,

$$f = 1 \cdot f_0 + x_i(-f_0 + f_1),$$

used, for instance, in the definitions of arithmetic spectral transform decision diagrams (ACDDs), EVBDDs, and BMDs.

The bases of decision diagram design are *spectral transform* decision trees (STDTs) and diagrams (STDDs).

Definition 40.1 *A spectral transform decision tree, STDT, is a decision tree assigned to a switching function f by the decomposition of f with respect to the basis Q.*

Definition 40.2 *A spectral transform decision diagram, STDD, is a decision diagram derived by the reduction of the corresponding STDT.*

Reduction of a STDT is performed by deleting redundant nodes and sharing isomorphic subtrees.

The general form of decomposition is defined as

$$f = x_i^0(q_{0,0}f_0 \odot q_{0,1}f_1) \odot x_i^1(q_{1,0}f_0 \odot q_{1,1}f_1),$$

where x_i^0 and x_i^1 can be treated in various ways. For example, in the case of Shannon expansion, $x_i^0 = \overline{x}_i$ and $x_i^1 = x_i$, and in the case of Davio expansion, $x_i^0 = 1$ and $x_i^1 = x_i$. The i-th basic matrix is $Q(1) = \begin{bmatrix} q_{0,0} & q_{0,1} \\ q_{1,0} & q_{1,1} \end{bmatrix}$.

Example 40.5 *Figure 40.5 illustrates two approaches to decision diagram design for the switching function $f = \overline{x}_1\overline{x}_2\overline{x}_3 \vee x_1\overline{x}_3$:*

- *The first approach results in a ROBDD,*
- *The second approach results in a regular linear-like topology, a Fibonacci decision diagram (see details in Chapter 38).*

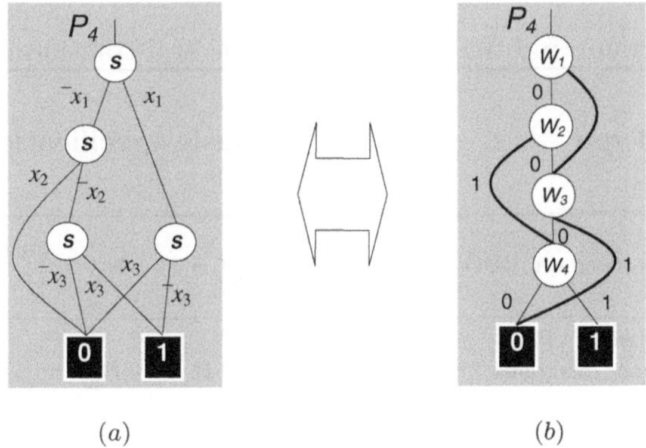

(*a*) (*b*)

FIGURE 40.5
Given the requirements of linear-like topology. Design using the fixed library of decision diagrams results in a ROBDD (a) and spectral approach results in a Fibonacci decision diagram (b) (Example 40.5).

Table 40.1 summarizes the allowed bases and compares them in basic functions in terms of which various AND-EXOR expressions and corresponding decision trees for switching functions in $GF(C_2^n)$ are defined.

Requirements for decision diagrams to be generated are formulated in terms of of implementation, computational and topological criteria. These requirements, in particular, include:

- Functions of non-terminal nodes and parameters of their implementation relevant to the technological requirements, which might be area- or speed-driven, etc.
- Topological characteristics of decision diagrams such as dimensionality (2D, 3D), type of topology, etc. (see details in Chapters 2, 3, and 34)
- Parallelism of processing (single or multiple-output, shared, binary or multivalued implementation, etc.)

The example below explains the simplest criteria of choosing the type of a decision diagram.

> **Example 40.6** (*a*) *Given a Reed–Muller expression of a switching function, the PPRM, FPRM, PSRM, or GRM can be derived. However, a BDD cannot be directly constructed from the Reed–Muller expression.*
> (*b*) *Given a sum-of-products of a switching function. a BDD can be generated directly.*

However, for some classes of functions, for example, symmetric functions, the size of ROBDD is exponential. Thus, for such functions, the SOP should

TABLE 40.1
Decision diagrams and the corresponding spectral transform bases.

Type	Basis specification
Binary decision diagram, BDD	$\mathbf{S}_{2^n} = \bigotimes \mathbf{S}_{2^1}, \quad \mathbf{S}_{2^1} = \begin{bmatrix} 1 & 0 \\ 0 & 1 \end{bmatrix}$
Positive polarity Reed–Muller decision diagrams, PPRM	$\mathbf{R}_{2^n} = \bigotimes \mathbf{R}_{2^1}^{(0)}, \quad \mathbf{R}_{2^1}^{(0)} = \begin{bmatrix} 1 & 0 \\ 1 & 1 \end{bmatrix}$
Fixed polarity Reed–Muller decision diagrams, FPRM	$\mathbf{R}_{2^n} = \bigotimes_{i=1}^{n} \mathbf{R}_{2^1}^{(j_i)}, \quad \mathbf{R}_{2^1}^{(0)} = \begin{bmatrix} 1 & 0 \\ 1 & 1 \end{bmatrix}, \quad \mathbf{R}_{2^1}^{(1)} = \begin{bmatrix} 0 & 1 \\ 1 & 1 \end{bmatrix}$
Pseudo–Reed–Muller decision diagram	$\mathbf{R}_{2^1}^{(0)} = \begin{bmatrix} 1 & 0 \\ 1 & 1 \end{bmatrix}, \quad \mathbf{R}_{2^1}^{(1)} = \begin{bmatrix} 0 & 1 \\ 1 & 1 \end{bmatrix}$
Generalized Reed–Muller decision diagram	$\mathbf{R}_{2^1}^{(0)} = \begin{bmatrix} 1 & 0 \\ 1 & 1 \end{bmatrix}, \quad \mathbf{R}_{2^1}^{(1)} = \begin{bmatrix} 0 & 1 \\ 1 & 1 \end{bmatrix}, \quad \mathbf{S}_{2^1} = \begin{bmatrix} 1 & 0 \\ 0 & 1 \end{bmatrix}$
Kronecker decision diagram	$\mathbf{Q}_{2^n} = \bigotimes_{i=1}^{n} \mathbf{Q}_{2^j}^{(i_j)}, \quad \mathbf{Q}_{2^1}^{(0)} = \mathbf{R}_{2^1}^{(0)}$ $\mathbf{Q}_{2^1}^{(1)} = \mathbf{R}_{2^1}^{(1)}, \quad \mathbf{Q}_{2^1}^{(2)} = \mathbf{S}_{2^1}$
Pseudo–Kronecker decision diagram	$\mathbf{Q}_{2^1}^{(0)}, \mathbf{Q}_{2^1}^{(1)}, \mathbf{Q}_{2^1}^{(2)}$

be transformed into a Reed–Muller expression and then a KDD, PKDD, or PSKDD can be synthesized from it.

Figure 40.6 introduces the problem in the following formulation: given the requirements of the data structure, construct an appropriate decision diagram.

Example 40.7 *The following examples illustrate the bases of some spectral transforms:*

(a) *The trivial basis is defined as the identity transform. This basis is used for deriving sum-of-products expressions of switching functions.*

(b) *The Kronecker basis is formed from the identity transform and the positive and the negative polarity basis Reed–Muller transforms under the restriction that the same basis transform is performed at all the nodes at a level in the decision tree. It can be used for partially symmetric functions.*

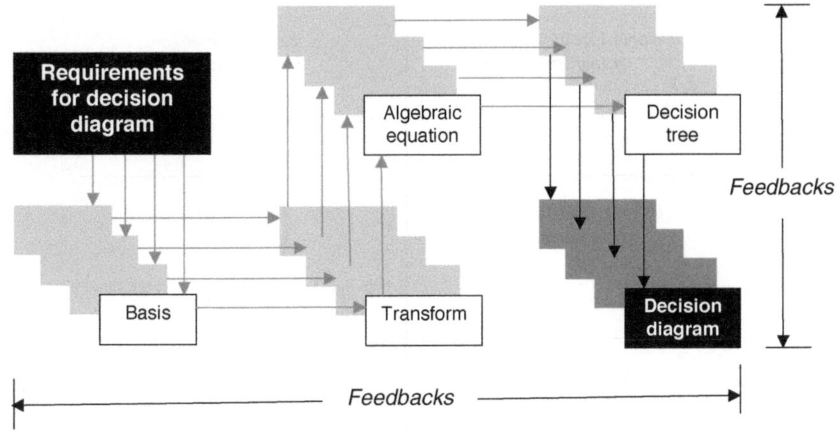

FIGURE 40.6
Design flow for the synthesis of decision diagrams.

A spectral transform is defined in terms of algebraic expressions or by a transform matrix. If a discrete function is defined by a vector of its values, the matrix formalization expresses performing a spectral transform as a matrix-vector multiplication. In addition, spectral transform matrices can be factored. This results in fast Fourier transform-like algorithms. This is convenient, since the computation of the spectrum can be performed by using computational architecture designed to implement matrix and fast algorithms.

Algebraic expressions. Description of a spectrum of a switching function in algebraic form is derived from the result of the matrix spectral transform as follows:

Step 1 The vector

$$\mathbf{X} = [1\ x_1] \otimes [1\ x_2] \otimes \cdots \otimes [1\ x_n]$$

is formed

Step 2 The vector of spectral coefficients is calculated using the matrix transform, and

Step 3 The algebraic expression is formed.

Example 40.8 *The above algorithm for deriving an algebraic equation from a matrix of an inverse spectral transform is illustrated in Figure 40.7.*

Inverse Reed–Muller transform	*Reed–Muller expression*
$[1\ x_1] \otimes [1\ x_2] = [\ 1\ x_2\ x_1\ x_1x_2\ ]$ $\mathbf{R} = [\ r_0\ r_1\ r_2\ r_3\ ]^T$	$f = r_0 \oplus r_1x_2 \oplus r_2x_1 \oplus r_3x_1x_2$
Inverse arithmetic transform	*Arithmetic expression*
$[1\ x_1] \otimes [1\ x_2] = [\ 1\ x_2\ x_1\ x_1x_2\ ]$ $\mathbf{P} = [\ p_0\ p_1\ p_2\ p_3\ ]^T$	$f = p_0 + p_1x_2 + p_2x_1 + p_3x_1x_2$
Inverse Walsh transform	*Walsh expression*
$[1\ x_1] \otimes [1\ x_2] = [\ 1\ x_2\ x_1\ x_1x_2\ ]$ $\mathbf{W} = [\ w_0\ w_1\ w_2\ w_3\ ]^T$	$f = \frac{1}{4}(w_0 + w_1(-1)^{x_2} + w_2(-1)^{x_1}$ $+ w_3(-1)^{x_1+x_2})$
Inverse Haar transform	*Haar expression*
$[1\ x_1] \otimes [1\ x_2] = [\ 1\ x_2\ x_1\ x_1x_2\ ]$ $\mathbf{H} = [\ h_0\ h_1\ h_2\ h_3\ ]^T$	$f = \frac{1}{4}(h_0 + h_1(-1)^{x_1} + \sqrt{2}h_2(-1)^{x_2}\overline{x}_1$ $+ h_3(-1)^{x_2}x_1)$

FIGURE 40.7
Deriving an algebraic expression from the matrix of a spectral transform for a switching function of two variables (Example 40.8).

A decision tree is derived from the matrix expression of a switching function. In constructing a decision tree, the following rules are applied:

- The function associated with a node is specified by the elementary transform matrix,
- recursive decomposition is associated with the iterations of the fast transform,
- The terminal nodes correspond to the resulting spectrum (vector of the spectral coefficients),

Details of the decision tree design are given in Chapter 3, and particular cases can be found in most of chapters of this book.

A decision diagram is obtained from a decision tree by applying reduction rules.

Example 40.9 *Figure 40.8 illustrates several phases of the design of the Haar decision diagram.*

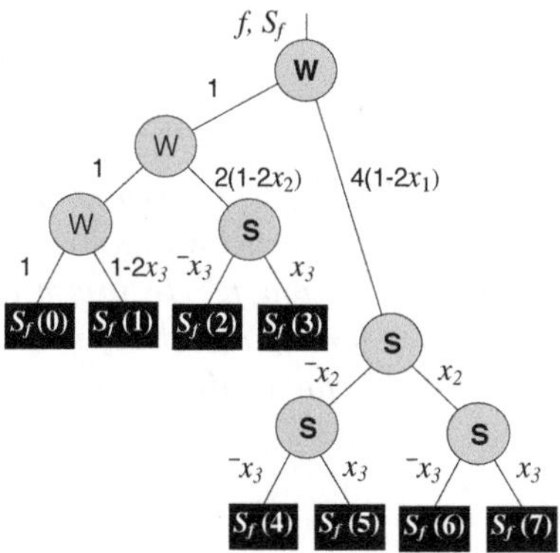

The HSTDT represents f in the form of the Haar expression for f:

$$
\begin{aligned}
f = \frac{1}{8}(&S_f(0) + S_f(1)(1 - 2x_3) \\
&+ 2S_f(2)(1 - 2x_2)\overline{x}_3 \\
&+ 2S_f(3)(1 - 2x_2)x_3 \\
&+ 4Sf(4)(1 - 2x_1)\overline{x}_2\overline{x}_3 \\
&+ 4S_f(5)(1 - 2x_1)\overline{x}_2 x_3 \\
&+ 4S_f(6)(1 - 2x_1)x_2\overline{x}_3 \\
&+ 4S_f(7)(1 - 2x_1)x_2 x_3)
\end{aligned}
$$

FIGURE 40.8
Design of Haar decision trees using spectral techniques (Example 40.9).

The development of a new type of decision diagram given algebraic and topological requirements is accomplished in the following design steps:

- ► *The constant nodes show the values of spectral coefficients for the represented function in terms of spectral transforms used in the definition of related decision trees.*
- ► *The basic functions are defined by products of the labels at the edges. Therefore, to assign a given switching function f to a decision tree, we perform the direct transform.*
- ► *By reading a switching function f from a given decision tree, we perform an inverse transform. The same applies while restoring the function from a decision diagram, since the reduction rules do not destroy the information content of the decision tree.*
- ► *A spectral decision diagram represents, at the same time, a switching function f and its spectrum.*

40.3 Group theoretic approach to designing decision diagrams

The group theoretic approach assumes a group structure for the domain of switching functions. Thus, switching functions are considered as functions on finite dyadic groups C_2^n. However, some other groups can also be used, which may result in advantages in particular applications.

Example 40.10 *By pairing input variables,*

$$(x_i, x_j) \rightarrow X_q$$

where $x_i, x_j \in \{0,1\}$, and $X_q \in \{0,1,2,3\}$, a switching function f_n of $n = 2r$ variables can be converted into an r-variable four-valued input, binary output function f_r. In this way, the dyadic group C_2 as the domain for binary variables, is replaced by the cyclic group of order four C_4 for quaternary variables.

40.3.1 Basic theorems

The main principle of the group-theoretic approach is:

> *The recursive structure of decision diagrams originates in the decomposition of the domain group G for the represented functions into the direct product of some subgroups G_i, thus,*
>
> $$G = \mathop{\times}_{i=1}^{n} G_i,$$
>
> *where n is the number of variables in f.*

For BDDs, $G_i = C_2$, and for QDDs, $G_i = C_4$ for each i. Spectral interpretation shows that for each decision diagram, a given function f is assigned to the decision diagram through a spectral transform on the not necessarily Abelian domain group G for f. The number of edges is equal to the number of different values x_i can take.

Example 40.11 *Figure 40.9 illustrates several steps of the Fibonacci decision diagram design.*

40.3.2 Group-theoretic approach and topology

Group-theoretic approach assumes a group structure for the domain of switching functions. Thus, switching functions are considered as functions on finite dyadic groups C_2^n. However, some other groups can also be used, which may result in advantages in particular applications. The dyadic group C_2 as the domain for binary variables can be replaced by the cyclic group of order four C_4 for quaternary variables (Figure 40.10).

In the replacement of C_2^2 by C_4, a subtree consisting of three non-terminal nodes with two outgoing edges is replaced by a node with four outgoing edges.

Example 40.12 *Figure 40.3 shows this transformation of decision diagrams.*

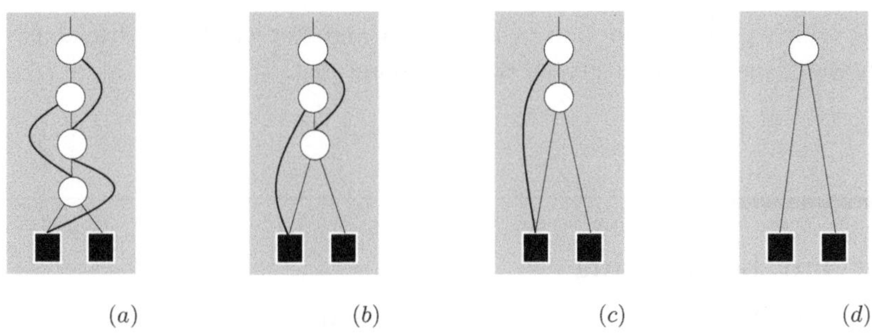

FIGURE 40.9
Example of the Fibonacci decision diagram design using spectral techniques (Example 40.11).

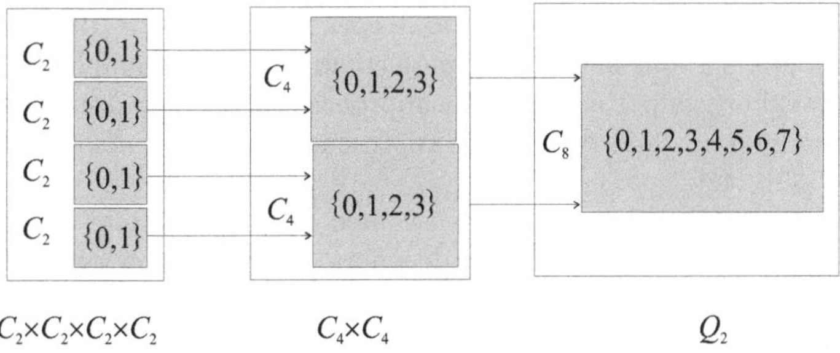

FIGURE 40.10
Domain groups.

For decision diagrams on Q_2, five outgoing edges for FNA_5 nodes are used (see details in Chapter 38). Hence, we can exploit advantages of the reduction of both depth and width offered by group representations of non-Abelian groups.

In this way, a BDD with n levels can be transferred into a Q-decision diagram with $n/2$ levels by utilizing the increased functionality of nodes. In a Q-decision diagram, each node represents a function taking four different values. This replacement reduces:

- The number of variables that must be controlled,
- The depth and the size of the decision diagram,
- The complexity of interconnections.

In these cases, given Q-decision diagrams, networks with a regular structure can be generated. The price is the increased cost of nodes expressed in terms of the number of edges, which in many cases results in the increased width of the Q-decision diagram. This is the motivation for the use of non-Abelian groups, which permit reduction of the depth and the width simultaneously, and in many cases also the size of decision diagrams.

40.4 Further reading

Spectral analysis and synthesis using various discrete bases. Fundamentals of spectral theory using various bases can be found, in particular, in books by Sadykhov et al. [25], Trachtenberg [50], Hurst [13, 14], and Hurst, Miller, and Muzio [15].

The group-theoretic approach to switching theory and logic design dates back to the beginning of these areas and work by Komamiya [19]. It was used in different applications in this area, in particular, by Sasao and Kinoshita [31], and Trachtenberg [50]. Recently, this approach has been used in optimization of decision diagrams and polynomial representations by Stanković et al. [33], [35], [36], [37], [42].

The derived multivalued input two-valued output functions are useful for simplification of PLAs with input decoders, simplification of multi-output functions, state assignment, code assignment, and design of multi-level logic networks, as shown by Sasao [28, 29, 30]. This approach has been initially developed, by, in particular, Karpovsky [18], and Lechner [21], which resulted in the spectral techniques for logic synthesis developed by Hurst [13, 14], and Hurst *et al.* [15], since switching functions are considered as elements of vector spaces on finite groups. The advances in these areas have been periodically reviewed in overviews, for example, by Moraga [23] and Karpovsky, Stanković and Astola [17]. Currently renewed interest in spectral and related group-theoretic approaches to logic design is dictated by the ever-increasing demands of technology and practice, which can hardly be met by classical approaches. In particular, spectral techniques may provide us with simple and elegant analytic solutions, while the traditional approaches reduce to brute force search methods [17, 23]. At the same time, this approach appears convenient to provide a uniform and consistent theoretic environment for recent achievements in this area, which appear divergent and disperse in many other approaches studied by Stanković et al. [41, 43, 44]. The prospects of decision diagrams for the generation of new devices have been considered by Yanushkevich and Shmerko [53, 55].

Spectral decision diagrams. Fundamentals of spectral decision diagrams and spectral interpretation of decision diagrams based on Shannon and Davio expansion are given in Chapter 13.

The cost of a decision diagram representation is defined as

$$\mathtt{cost(DD)} = \overbrace{\mathtt{cost(node)}}^{Optimization} \times \overbrace{\mathtt{size(DD)}}^{Optimization} + \overbrace{\mathtt{cost(interconnections)}}^{Optimization}$$

Each of the cost parameters can be optimized. Spectral decision diagrams are directly mapped into networks consisting of simple two-input modules realizing the operations determined by the matrices describing the nodes. In such networks, the area is proportional to the size and the width of the related decision diagrams. In decision diagrams on C_2^n , the depth is equal to the number of variables n in the represented functions. The propagation delay is proportional to the depth of the decision diagrams. In many cases, besides the size and depth, the price paid for simplicity of modules is complexity of interconnections. Therefore, in optimization of decision diagram representations, some efforts have been devoted to the reduction of the cost of interconnections, which has resulted in planar decision diagrams, for example, by Sasao and Butler [26], and decision diagrams with regular layout by Perkowski et al. [24].

Development of new decision diagrams. Various aspects of decision diagram synthesis using spectral techniques have been discussed, in particular, by Stanković et al. [70, 47, 48] and Falkowski and Stanković [3].

Embedding decision diagrams into nanostructures. Yanushkevich et al. [79, 54] studied an embedding of decision diagrams into various topological structures to meet the requirements of spatial nanometric configurations.

References

[1] Agayan SS. Hadamard Matrices and Their Applications. Springer, Berlin, 1985.

[2] Ahmed N and Rao KR. *Orthogonal Transform for Digital Signal Processing.* Springer, Berlin, 1975.

[3] Falkowski BJ and Stanković RS. Spectral interpretation and applications of decision diagrams. *VLSI Design International Journal of Custom Chip Design, Simulation and Testing*, 11(2):85–105, 2000.

[4] Falkowski BJ. Properties and ways of calculation of multi-polarity generalized Walsh transform. *IEEE Transactions on Circuits and Systems*, 41(6):380–391, 1994.

[5] Falkowski BJ. A note on the polynomial form of Boolean functions and related topics. *IEEE Transactions on Computers*, 48(8):860–864, 1999.

[6] Falkowski BJ. Relations between arithmetic and Haar wavelet transforms in the form of layered Kronecker matrices. *Electronics Letters*, 35(10):799–800, 1999.

[7] Falkowski BJ and Chang CH. Forward and inverse transformations between Haar spectra and ordered binary decision diagrams of Boolean functions. *IEEE Transactions on Computers*, 46(11):1272–1279, 1997.

[8] Falkowski BJ and Chang CH. Mutual conversions between generalized arithmetic expressions and free binary decision diagrams. *IEE Proceedings on Circuits, Devices, and Systems*, 145(4):219–228, 1998.

[9] Falkowski BJ and Chang CH. Hadamard–Walsh spectra characterization of Reed-Muller expansions. *Computers and Electrical Engineering*, 25(2):111–134, 1999.

[10] Hansen JP and Sekine M. Synthesis by spectral transformation using Boolean decision diagrams In *Proceedings of the 33th ACM/IEEE Design Automation Conference*, Las Vegas, NV, pp. 248–253, 1996.

[11] Hansen JP and Sekine M. Decision diagrams based techniques for the Haar wavelet transform. In *Proceedings of the IEEE International Conference on Information, Communications and Signal Processing*, Singapore, Vol. 1, pp. 59–63, 1997.

[12] Ho P and Perkowski MA. Free Kronecker decision diagrams and their application to Actel 6000 FPGA mapping. In *Proceedings of the European Design Automation Conference*, pp. 8–13, 1994.

[13] Hurst SL. Logical Processing of Digital Signals, Crane Russak and Edward Arnold, London and Basel, 1978.

[14] Hurst SL. The Haar transform in digital network synthesis. In *Proceedings of the 11th International Symposium on Multiple-Valued Logic*, pp. 10–18, 1981.

[15] Hurst SL, Miller DM, and Muzio JC. *Spectral Techniques in Digital Logic*, Academic Press, Bristol, 1985.

[16] Kam T, Villa T, Brayton RK, and Sagiovanni-Vincentelli AL. Multi-valued decision diagrams: theory and applications. *International Journal on Multiple-Valued Logic*, 4(1-2):9–62, 1998.

[17] Karpovsky MG, Stanković RS, and Astola JT. Spectral techniques for design and testing of computer hardware. In *Proceedings of the Workshop on Spectral Transforms and Logic Design for Future Digital Systems*, Tampere, Finland, pp. 1–34, 2000.

[18] Karpovsky MG. *Finite Orthogonal Series in the Design of Digital Devices.* Wiley and JUP, New York and Jerusalem, 1976.

[19] Komamiya Y. Theory of relay networks for the transformation between the decimal and binary system. *Bull. of E.T.L.*, 15(8):188–197, 1951.

[20] Khan MHA, Perkowski M, and Khan MR. Ternary Galois field expansions for reversible logic and Kronecker decision diagrams for ternary GFSOP min-

imization. In *Proceedings of the IEEE 34th International Symposium on Multiple-Valued Logic*, pp. 58–67, 2004.

[21] Lechner R. A transform theory for functions of binary variables in theory of switching. *Harvard Computation Lab., Cambridge, Mass.*, Progress Rept. BL-30, Sec-X, pp. 1–37, 1961.

[22] Miller DM. Spectral transformation of multiple-valued decision diagrams. In *Proceedings of the 24th IEEE International Symposium on Multiple-Valued Logic*, pp. 89–96, 1994.

[23] Moraga C. A decade of spectral techniques. In *Proceedings of the 21st International Symposium on Multiple-Valued Logic*, pp. 182–188, 1991.

[24] Perkowski MA, Pierzchala E, and Drechsler R. Ternary and quaternary lattice diagrams for linearly independent logic, multiple-valued logic and analog synthesis. In *Proceedings of the International Conference on Information, Communications and Signal Processing*, Singapore, Vol. 1, pp. 269–273, 1997.

[25] Sadykhov RCh, Chegolin PM, and Shmerko VP. *Signal Processing in Discrete Bases.* "Science and Technics" Publishers, Minsk, Belarus, 1986.

[26] Sasao T and Butler JT. A method to represent multiple-output functions by using multi-valued decision diagrams. In *Proceedings of the 26th International Symposium on Multiple-Valued Logic*, pp. 248–254, 1996.

[27] Sasao T and Kinoshita K. Conservative logic circuits and their universality. IEEE Transactions on Computers, 28(9):682–685, 1979.

[28] Sasao T. An application of multiple-valued logic to a design of programmable logic arrays. In *Proceedings of the International Symposium on Multiple-Valued Logic*, pp. 65–72, 1978.

[29] Sasao T. Application of multiple-valued logic to a serial decomposition of PLAs. In *Proceedings of the International Symposium on Multiple-Valued Logic*, Zougzhan, China, pp. 264–271, 1989.

[30] Sasao T. A transformation of multiple-valued input two-valued output functions and its application to simplification of exclusive-OR sum-of-products expressions. In *Proceedings of the International Symposium on Multiple-Valued Logic*, pp. 270–279, 1991.

[31] Sasao T and Kinoshita K. Conservative logic circuits and their universality. *IEEE Transactions on Computers*, 28(9):682–685, 1979.

[32] Shmerko VP. Synthesis of arithmetic forms of Boolean functions using the Fouier transform. *Automation and Remote Control*, Kluwer/Plenum Publishers, 50(5):684–691, Pt2, 1989.

[33] Stanković RS. Fourier decision diagrams on finite non-Abelian groups with preprocessing. In *Proceedings of the IEEE 27th International Symposium on Multiple-Valued Logic*, pp. 281–286, 1997.

[34] Stanković RS. Non-Abelian groups in optimization of decision diagrams representations of discrete functions. *Formal Methods in System Design*, 18:209–231, 2001.

[35] Stanković RS, Milenović D, and Janković D. Quaternion groups versus dyadic groups in representations and processing of switching functions. In *Proceedings of the IEEE 29th International Symposium on Multiple-Valued Logic*, pp. 18–23, 1999.

[36] Stanković RS, Shmerko VP, and Moraga C. From Fourier expansions to arithmetic expressions on finite Abelian groups. In *Proceedings of the 6th International Conference on Advanced Computer Systems,* Szczecin, Poland, pp. 483–493, 1999.

[37] Stanković RS, Moraga C, and Astola JT. From Fourier expansions to arithmetic-Haar expressions on quaternion groups. *Applicable Algebra in Engineering, Communications and Computing,* AAECC12:227–253, 2001.

[38] Stanković RS, Stojić MR, and Stanković MS, Eds., *Recent Developments in Abstract Harmonic Analysis with Applications in Signal Processing.* Nauka and Elektronski fakultet, Belgrade and Niś, 1995.

[39] Stanković RS. Fourier decision diagrams on finite non-Abelian groups with preprocessing. In *Proceedings of the 27th International Symposium on Multiple-Valued Logic*, pp. 281–286, 1997.

[40] Stanković RS. *Spectral Transform Decision Diagrams in Simple Questions and Simple Answers*, Nauka, Belgrade, 1998.

[41] Stanković RS. Some remarks on basic characteristics of decision diagrams. In *Proceedings of the 4th International Workshop on Applications of Reed–Muller Expansion in Circuit Design,* pp. 139–146, 1999.

[42] Stanković RS. Non-Abelian groups in optimization of decision diagrams representations of discrete functions. *Formal Methods in System Design,* 18:209–231, 2001.

[43] Moraga C, Sasao T, and Stanković RS. A unified approach to edge-valued and arithmetic transform decision diagrams. *Automation and Remote Control,* Kluwer/Plenum Publishers, 63(1):125–138, 2002.

[44] Stanković RS and Sasao T. Decision diagrams for discrete functions: Classification and unified interpretation. In *Proceedings of ASP-DAC'98,* pp. 349–446, 1998.

[45] Stanković RS and Astola JT. *Spectral Interpretation of Decision Diagrams.* Springer, Heidelberg, 2003.

[46] Stanković RS. Functional decision diagrams for multiple-valued functions. In *Proceedings of the IEEE 25th International Symposium on Multiple-Valued Logic*, pp. 284–289, 1995.

[47] Stanković RS, Sasao T, and Moraga C. Spectral transform decision diagrams. In Sasao T and Fujita M, Eds., *Representations of Discrete Functions,* Kluwer, Dordrecht, pp. 55–92, 1996.

[48] Stanković RS. *Spectral Transform Decision Diagrams in Simple Questions and Simple Answers.* NAUKA Publishers, Belgrade, Yugoslavia, 1998.

[49] Stanković RS. Non-Abelian groups in optimization of decision diagrams representations of discrete functions. *Formal Methods in System Design*, 18:209–231, 2001.

[50] Trachtenberg EA. Application of Fourier analysis on groups in engineering practice. In [38], pp. 331–403.

[51] Yanushkevich SN, Shmerko VP, and Lyshevski SE. *Logic Design of NanoICs*, CRC Press, Boca Raton, FL, 2005.

[52] Yanushkevich SN. Multiplicative properties of spectral Walsh coefficients of Boolean funtions. *Automation and Remote Control (Kluwer/Plenum Publishers)*, 64(12):1933–1947, 2003.

[53] Yanushkevich SN. Computer arithmetic. In Dorf R, Ed., *The Electrical Engineering Handbook*, 3rd Edition, CRC Press, Boca Raton, FL, 2005.

[54] Yanushkevich SN. Logic design of nanodevices, In Rieth M and Schommers W, Eds., *Handbook of Theoretical and Computational Nanotechnology*, Scientific American Publishers, 2005.

[55] Yanushkevich SN and Shmerko VP. Decision diagram technique. In *Dorf R, Editor. The Electrical Engineering Handbook*, Third Edition, CRC Press, Boca Raton, FL, 2005.

41

Historical Perspectives and Open Problems

In this chapter, the progress in decision diagram techniques is analyzed, and new frontiers are indicated for applications in nanotechnologies.

41.1 Trends in decision diagram techniques

Decision diagram techniques gained practical microelectronic design consideration due to the extensive work on graphical data structures for switching function representation arising from Bryant's seminal 1986 paper [12]. Since then, considerable progress has been made on the manipulation, characterization, and development of new types of decision diagrams. Currently, there are several trends in decision diagram techniques, in particular (Figure 41.1):

- New design and optimization strategies, including bio-inspired and artificial intelligence, probabilistic and information-theoretical approaches,
- Extension of the area of application to various stages of electronic design, and
- Decision diagrams as vehicles for modeling nanocomputing structures, including topology!indexTopology controlled design of decision diagrams (sizes, topological characteristics, embedding properties).

41.2 New design and optimization strategies

The design of various types of decision diagrams is accomplished by using methods of generation for the corresponding algebraic expressions, i.e., spectral techniques and other methods of manipulation of switching function representations which can be translated or have a mapping into decision diagrams.

Design strategies	▪ The fixed library of decision diagrams ▪ Design methods for generation of decision diagrams
Optimization strategy	▪ Bio-inspired techniques ▪ Based on advanced methods of description and utilization of flexibility
Adaptable structures and flexibility	▪ Evolutionary strategies ▪ Machine learning techniques
Embedding properties	▪ Optimal embedding into spatial configurations ▪ Modeling 3D computing structures (delay, clock, power analysis, etc.) ▪ Embedding into particular topologies (lattices, hexagonal, transient triangles, etc.)
Decomposition techniques	▪ Using new description of flexibility ▪ Using information-theoretical criteria ▪ Decomposition with respect to embedding characteristics
Probabilistic techniques and	▪ Modeling probabilistic behavior of circuits (switching activity, etc.)
Information-theoretical approaches	▪ Measures in terms of entropy and information ▪ Optimization based on information criteria
Stochastic computing	▪ Modeling various physical phenomena in VLSI, ULSI, and nanoelectronics ▪ Error correction codes

FIGURE 41.1
Strategies in advanced decision diagram techniques.

This strategy is not new, but leaves much space for optimization during design, or fitting and adaptation of the diagrams to the task or technology requirements.

Today's intensively investigated methods for optimization or adaptation of decision diagrams to implementation include

- Bio-inspired approaches,
- Adaptive methods, and
- Artificial intelligence techniques.

41.2.1 Bio-inspired strategies

Rather than rely on any central control, a natural immune system has a distributed task force that has the intelligence to take action from a local and also a global perspective using its network of messengers for communication. In decision diagrams design, *phylogenesis* (evolution), *ontogenesis* (development), and *epigenesis* (learning) methods can be used.

Phylogenesis methods are used for adaptive reconstruction of the system at runtime. To be capable of self-repair and growth, a model needs to be able to create new cells and configure them.

Ontogenesis methods are used in the developing self-repair, or robust, circuits when their functionality can be changed at runtime.

Learning, or epigenesis, is a widely used strategy applied to design artificial neural networks, and self-learning systems. For example, epigenesis methods using neural networks can create new neurons at runtime. Decision diagrams have been applied in machine learning, in particular, for verification and validation of knowledge-based systems, and knowledge discovery (see, for example, works by Mues et al. [53, 54, 54], Anderson et al. [2], Gupta et al. [30])).

The anticipated advantages of a bio-inspired approach in terms of decision diagrams process are as follows:

- When a new node of a diagram is created, it can start a design process without the need of recalculating all paths already created, and
- The process is distributed without any global control.

Bio-inspired technologies include *evolutionary strategies.* The principle of evolution is the primary concept of biology, linking every organism together in a historical chain of events. For example, every creature (circuit) in the chain is the product of a series of events (subcircuits) that have been sorted out thoroughly under selective pressure (correct or incorrect subcircuits) from the environment (design technique). Over many generations, random variation and natural selection Haar decision the behaviors of individuals (circuits) to fit the demands of their surroundings. This strategy has been applied as heuristics for optimization of BDDs for large circuits where exact methods are unacceptably expensive (see the Further Reading section).

41.2.2 Adaptive reconfiguration

High-performance, highly-constrained applications must be adaptable to their requirements and environment. FPGAs are the enabling technology for a computing platform that is able to adapt to changes in the processing algorithm. Extremely large design spaces can result from the freedom given in defining the algorithm structure alternatives in such systems. Symbolic manipulation employing BDDs has been used to manage these large state spaces and to provide a way to prune the design space without examining each design alternative individually.

Reconfigurable computing device technology provides a way for fast, dynamically adaptable systems. The essence of an adaptive computing approach is to create several different configurations of the hardware architecture and the software topology for a system, each tailored to a specific set of operational requirements (see the Further reading section).

Example 41.1 *Cheushev et al. [17, 18] used an adaptive reconfiguration using the concept of changing a model and evolutionary strategy with respect to factors affecting logic cells. Using an information theoretical approach, it is possible to verify not only that an affected logic cell achieves the target functionality, but also that this cell can be automatically* corrected *to achieve this.*
Let G_f be a logic cell implementing the target function f and G_g be the given cell with the set X of primary inputs. Denote by $V(G_g)$ the set of output, internal outputs, and primary inputs of the cell G_g and constants. Conditional entropy $\mathbf{H}(f|V(G_g)$ is an information measure, reflecting an ambiguity of values of the target function f given cell G_g. Given function f and logic cell G_g, there exists a pair $v_i, v_j \in V(G_g)$ such that $\mathbf{H}(f|v_i, v_j) = 0$, then there exists a logic function ϕ such that $f = \phi(v_i, v_j)$ and G_g is a 1-neighbor of G_f. During entropy computation, the necessary data to form the truth table of the function ϕ is obtained.

41.2.3 Artificial intelligence

Decision table and diagram techniques have been originally employed in decision making strategies. They remain an efficient instrument in knowledge engineering and business intelligence as well as knowledge discovery.

Yet another application of decision diagrams is in artificial intelligence planning, because of their ability to efficiently represent huge sets of states commonly encountered in state-space search (see the Further Reading section).

41.2.4 Neural networks

Implementation of artificial neural networks on available digital architecture with reconfigurable structures has been studied for many years. The most prominent approach is based on FPGA. Artificial neural networks were implemented using conventional FPGA circuits (see, for example, the paper by Gustin [29]). Durand et. al. [23] proposed a prototype of the multiplexer tree based cell, using standard FPGA and RAM circuits but with reprogramming by genome rewriting. Therefore, these biodules represented a type of evolvable structure. Multiplexer-based tree cells were used to verify the self-reproduction and self-repair processes. This and other approaches are capable of implementing a digital realization of an artificial neural network.

Neural networks have been used for BDD optimization, in particular, for ordering in converting a fault tree to a BDD (Barlett and Andrews [8]), and for BDD construction (Barlett [9]).

Shape of decision diagrams. Stanković [72] showed that the information content of a decision diagram is coded in its Haar decision. There are decision diagrams that have the same depth, size and width, and even identical distributions of nodes at each level, but with different interconnections. Thus, these BDDs are different, but not distinguished by the mentioned characteristics. Therefore, these characteristics are not sufficient to precisely characterize decision diagrams.

The shape is an invariant characteristic of decision diagrams, since by choosing different interpretations of nodes and labels at the edges, we can read different information from the decision diagram.

Symmetries are both *functional* and a *topological* property in decision diagram techniques. A function exhibits symmetry in two variables if exchanging these variables leaves the function invariant. In decision diagram design, symmetries are used to simplify synthesis and improve circuit characteristics (see, for example, Miller and Muranaka's paper [51]).

Example 41.2 *Symmetries establish relations between decomposition of a circuit and a library of primitives that make this decomposition possible.*

Linear topology is defined as a stream of connected nodes where each node has an edge pointing to a terminal node. A decision diagram with linear topology is called a *linear decision diagram.* There are several approaches to deriving linear decision diagrams. They can be used at a local level (local linearity) or a global level (linear decision diagram).

Planar topology. Networks without crossings are useful in synthesis with field programmable gate arrays (FPGAs), since crossings produce considerable delays. A planar networks are desirable in sub-micron LSIs, since delays in the interconnections and crossings are comparable to the delays for logic circuits. Decision diagrams provide a simple mapping to technology, since a network is easily derived from a decision diagram by replacing each node with the logic element realizing the decomposition rule.

A *planar decision diagram* is one where all the intersections of two edges are a vertex. Since this graph is located within a plane, its topology is two-dimensional. A planar decision diagram can be derived by the reduction of a decision tree if sharing of isomorphic subtrees is restricted to subtrees rooted at neighboring nodes at the same level in the decision tree.

Planar decision diagrams were studied by Sasao and Butler [65]. The design of planar BDDs by using Walsh coefficients has been considered by Karpovsky et al. [44].

Lattice topology is constructed by aggregation of connection arcs between nodes in 2D graphs or edge nodes into "tunnels" between those edge nodes in 3D. 2D lattice decision diagrams have been studied for cell-based structure design (see the Further Reading section) and are considered promising for hexagonal wrap-gate quantum nanowire circuits.

Example 41.3 *Binary decision diagrams for totally symmetric functions form the lattice topology. The number of nodes at each level is worst case linear.*

Example 41.4 *In bimolecular electronics, each computational cell is a collection of molecules. Cells are the basic blocks of a circuit. Each molecule is connected to its 4 neighbors and to a routing unit and can contain, for example, an m-bit look-up table.*

Hypercube-like topology. Given a decision diagram, the quality of embedding into a hypercube-like topology can be measured by the ratio of the size of this diagram to the size of the embedding cube. This ratio reflects the utilization, for example, of nodes in the embedding hypercube-like structures.

The Fibonacci cube proposed by Hsu [37] is a hyper-like topological structure that can emulate many hypercube algorithms and is characterized by fewer links than the comparable hypercube. The Fibonacci decision diagrams proposed by Stanković et al. [73] can be embedded into Fibonacci cubes (see details in Chapter 38).

41.2.5 Flexibility

Flexibility in logic design is understood as relaxing the requirements to local functionality of a circuits to achieve better area, delay, power or testability of the circuit, while preserving its global functionality (see, for example, the study by Sentovich and Brand [71]). The source of flexibility is a don't-care set of the circuit functionality.

> **Example 41.5** *An incompletely-specified switching function of a subcircuit provides flexibility in its implementation.*

Decision diagrams for incompletely specified functions (see Chapter 19) exploit the don't cares as the source of flexibility to achieve optimality of data structures given an optimization criterion.

SPFD concept. *Set of Pairs of Functions to be Distinguished,* (SPFD) is a method to represent the flexibility of a logic cell developed by Yamashita *et al.* [78]. An SPFD attached to a logic cell specifies which pairs of primary input minterms can be or have to be distinguished. This can be understood as the information content of the node, since it indicates what information the node contributes to the network. Study of the relationship of SPFDs to BDD techniques, aimed at reduction of calculation for large circuits, has not been reported as yet.

41.2.6 Probabilistic techniques and information-theoretical measures

It has obtained practical consideration due to paper by Najm [57], where the BDD were used to compute the probability of a switching functions for further calculation of transition density that is the probability of Boolean difference also calculated using BDD (see Chapter 21). The other applications to switching theory is considered in Chapter 20.

> **Example 41.6** *In modeling nanodevices, probabilistic models of failures are used because of the probabilistic nature of nanophenomena. Also, the dynamic behavior of discrete devices and systems is described by stochastic models.*

The basics of probabilistic decision diagrams are probabilistic graphical structures, in particular, probabilistic graphs and trees. For computing entropy and information, decision diagrams must be interpreted in information-theoretical notation, i.e., input, outputs, and functions of nodes must be described in terms of entropy. Nodes with Shannon and Davio expansions are described by different information-theoretical equations called information-theoretical analogs of Shannon and Davio expansion, correspondingly. This approach is discussed in Chapter 20.

It was shown in Chapter 8 that models of nodes of decision diagrams can be described in terms of Shannon information. Information-theoretical notation of various decision diagrams is used for manipulation switching and multivalued functions, in particular, for heuristic finding optimal variable ordering.

Study of information contents of the BDD based networks in terms of information theory also has a prospective application to the design of 3D nanostructure, and for stochastic modeling of nanodevices.

Probabilities of attributes have been used to design sequential testing programs, also called decision trees, in many problems of software engineering, pattern recognition, and management. Incorporated in Shannon entropy evaluation, they represent powerful tools for finding optimal decision tree representation with respect to some criteria (see Chapter 24).

Yet another aspect of probabilistic techniques considered promising for modeling nanodevices is Markov chains and random fields. Markov chains have been used as a mathematical model for describing how a probabilistic method finds a global minimum in optimization problems. Markov random fields have been recently considered as a formal probabilistic framework for a reconfigurable architecture without training [6]. In this model (not implemented in a physical device structures as yet), a logic circuit is mapped into a Markov random field, which is a graph indicating the neighborhood relations of the circuit nodes. This graph is used in the probability maximization process, aimed at characterization of circuit configurations for the best thermal noise reliability (expressed in terms of logic signal errors).

41.3 Extension of the area of application

From a computational point of view, decision diagrams are relevant, by genesis, to sequential programs and search algorithms. Hence, they are applicable to many problems of logic design which are search-with-optimization problems. As data structures, they describe functional dependencies, and are thus technology independent (this applies, at least, to microelectronics). However, new technologies, which often offer the merging of functional and technological aspects, may consider decision diagram the most appropriate data structure for some or all steps of design of computing structures.

41.3.1 Implementation aspects

Contemporary logic design of VLSI circuits uses BDDs primarily at the logic optimization step, and also at the verification step of the transistor-level implementation after technology mapping or block-level verification [32]. However, certain CMOS technologies such as pass-transistor logic can be designed using so-called one-pass synthesis, which relies on the direct mapping of the BDD topology into a circuit topology (in particular, multiplexer based, which requires two pass transistors per multiplexer, or a universal logic module based circuit). What is of importance is that the direct mapping of BDD topology, is accommodated by some nanotechnologies, in particular, BDD based wrap-gate quantum nanowire circuits, which cannot be designed using traditional gate-level techniques because of the non-unilateral structure of these quantum devices (see the Further reading section below).

41.3.2 Topology and embedding properties

The topology of a decision diagram is an important characteristic at the implementation phase. Topology of decision diagrams can be of particular 2D and 3D configurations: binary tree, , linear, symmetrical, hexagonal, etc.

Topology has an impact on implementation and embedding. In simplest terms, the problem can be formulated as follows:

- *Given a decision diagram, find an appropriate topology for spatial representation.*
- *Given a spatial topology, find an appropriate decision diagram.*

41.4 Nanocircuit design – a new frontier for decision diagrams

New technologies, approaching nanoscale, may bring radically different methods of information processing and information carrier understanding. An example is the charge state logic in single-electron parametron (by Likharev et al.[49]), as an alternative to traditional voltage-state logic. Another example is the hexagonal wrap-gate quantum network where the topology and information principle resemble those of the decision diagrams (see the paper by Yamada et al. [77]).

41.4.1 Modeling spatial dimensions

The problem is formulated as embedding decision trees and decision diagrams into nanodimensions. Decision trees and diagrams can be embedded into various spatial dimensions (see details in Chapter 34).

41.4.2 Modeling spatial computing structures

Physical perfection in nanocircuits is hard to achieve: defects and faults arise from instability and noise-proneness on nanometer scales. This leads to unreliable and undesirable computational results. In order to ensure more reliable computation, techniques are necessary to cope with such errors. This can be achieved in nanotechnology using probabilistic models. Such models assume that input signals are applied to gates with some level of probability and correct output signals are calculated with some level of probability. When noise is allowed, the switching function is replaced with a random function and the configuration is a set of random variables.

41.4.3 Modeling nanocomputing

Physical effects that are exploited in nanotechnologies (molecular, quantum, etc.) provide different constraints and possibilities for design of computing elements. The effectiveness of a model depends on how it well satisfy these requirements and utilize properties of physical phenomena. Hence, different models with respect to various phenomenological paradigms are needed.

Basic components of logic nanocircuits that are logic nanocells which are the simplest elements of logic circuit. The set of logic gates that can be used for the effective design of arbitrary circuits is called a *library* of logic gates. Libraries are created for each phenomenological paradigm, and contain a small set of logical gates (10-20) with extensions for placement, topology, *etc.* Logic gates are functionally very simple and even methods of high complexity can be used for modeling. This is because these methods can be drastically simplified for cell models.

Using decision diagrams in reversible device design is discussed in Chapter 35.

41.5 Further reading

The underlying model of decision diagrams was developed by Lee [48] (1959) and Akers [1] (1978). Bryant [12] (1986) substantially improved this model and made it extremely useful for practical problems in logic design.

Overviews. Bryant and Meinel [14] discussed state-of-the-art and some prominent applications of decision diagrams. The state-of-the-art spectral technique for design and manipulation of decision diagrams, as well as prospective approaches, are reviewed by Stanković and Astola [70].

In Brayton's overview [13] of logic synthesis and verification, BDDs are assigned a significant roles in the resolution of a number of forecasted problems:

- ▶ The state space exploration, which is a determination of all the states that can be reached from a set of initial states.

- An efficient data structure implementation of verification and symbolic trajectory evaluation, that is the computation of the reached state set in property checking.
- Semi-canonical BDDs can be considered as a possible way of inventing new types of diagrams for application where BDDs are not efficient because of their canonicity.

Evolutionary strategies. Soon after the introduction of genetic algorithms, evolutionary strategies were utilized for evolvable hardware design by Iba et al. [38]. The interpretation of evolvable circuits as artificial immune systems was considered by Dasgupta et al. [19]. Applications of evolutionary strategy to the design of switching and multivalued circuits with information-theoretical measures embedded in fitness function evaluation was considered by Cheushev et al. [17, 18].

A heuristic approach based on evolutionary algorithms for determining a good ordering for a BDD has been studied, for example by Drechsler [22]. The BDD of a switching function is optimized in such a way that:

- The corresponding disjunctive SOP of the function is minimized.
- The objective, or fitness function in these evolutionary algorithms assigns to each instance a fitness that measures the quality of the variable ordering.
- The BDD is constructed using the variable ordering given by the individual, then the number of reduced one-paths are counted.
- The selection is performed by a selection, normally roulette wheel selection, and a steady-state reproduction is employed: the best individuals of the old population are included in the new one of equal size. This strategy guarantees that the best element never gets lost and a fast convergence is obtained.

Other tasks of optimization on BDDs can be resolved using a similar strategy.

Adaptive reconfiguration. FPGAs are the enabling technology for a computing platform that is able to adapt to changes in the processing algorithm, and ordered BDDs is one of the tools applied for such symbolic manipulation, including FPGA implementation, as studied by Marchal and Stauffer [50]*.

*Adaptive computing systems adapt and evolve in response to the changing environment while operational, without compromising the consistency and real-time properties of the system, as stated by Howard and Taylor [36]. The biggest challenge is runtime reconfiguration. An example of this is given by Eldredge and Hutchings [26]. This problem has been approached at two levels: the design level and the runtime level. At the design level, high-level design tools for representation, analysis, and synthesis of dynamically reconfigurable systems are needed. At the runtime level, a uniform execution environment for execution of dynamically reconfigurable systems are required. Babty et al. [7] introduced a run-

The flexible design of FPGA using an information-theoretical approach has been studied by Cheushev at. al. [17]. This approach combined:

- ▶ The evolutionary approach to the best cell-based implementation of a given logic function, and
- ▶ An information-theoretic approach for fitness function evaluation.

Such interpretation of flexibility can also be combined with information-theoretical optimization of a BDD. Such an optimization employing decision trees was started by Hartman [31] and developed by Shmerko et al. [67] and [68] for minimization of switching and multivalued functions. Watanabe's basic theory of information networks [83] could be applied to information flow estimation of logic networks at the system level.

Artificial intelligence. Mues et al. have investigated the use of decision table and decision diagram techniques in knowledge engineering and business intelligence, most notably the verification and validation of knowledge-based systems [55], business rule modelling (cf. the 'Prologa' tool home page) [54], as well as knowledge discovery [53].

Horiyama and Ibaraki [35] considered the use of OBDDs for implementing knowledge bases. They showed that the OBDD-based representation is more efficient and suitable in some cases, compared against the traditional representations. They presented polynomial time algorithms for the two problems of testing whether a given OBDD represents a unate switching function, and of testing whether it represents a Horn function.

Hoey et al. [34] considered Markov decision processes (MDPs), which are popular as models of decision theoretic planning, using decision diagrams. As an alternative to traditional dynamic programming methods, which perform well for problems with small state spaces, a value iteration algorithm for MDPs was proposed that uses algebraic decision diagrams to represent value functions and policies. An MDP is represented using Bayesian networks and ADDs, and dynamic programming is applied directly to these ADDs. This method was demonstrated on large MDPs (up to 63 million states).

Yet another application of decision diagrams is artificial intelligence planning, because of the ability to efficiently represent huge sets of states commonly encountered in a state-space search. A model-checking integrated planning system called MIPS has been developed by Edelkamp et al. [24]. This

time reconfiguration approach incorporates a multi-stage multi-resolution design analysis in the environment that starts with a large number of design alternatives and progressively refines the design space to a small set of best choices. This component of the analysis is an analytical tool based on OBDD. This tool symbolically evaluates the design space against user-defined constraints. The constraints are selectively applied to eliminate the designs that fail to meet the system requirements, thereby pruning the design space. The next stage of the analysis process is a multi-resolution performance simulation facility using a VHDL simulator. The results of the simulation are translated back to the modeling environment for use in determining if the design satisfies performance specifications.

system uses formal verification techniques based on BDDs, as opposed to the various graph-plan or SAT-based approaches on a broad spectrum of domains. The BDDs allow compact storing and operating on sets of concise states. These states represent exhibiting knowledge that is implicit in the description of the planning domain as has been shown by Edelkamp and Helmert [25]. This representation is then used to carry out an accurate reachability analysis without necessarily encountering exponential explosion of the representation.

Fuzzy decision diagrams were proposed by Strehl and Thiele [74] and developed by Moraga and Strehl et al. [52, 75].

Probabilistic techniques. Qi et al. [61] used Markov chains in modeling multiplexed circuits[†]. A Markov random fields are considered as a formal probabilistic framework for a reconfigurable nanocircuit architecture without training [6][‡]. Fault-tolerant computing in future nanocircuits based on probabilistic technique is considered by Peper et al. [58] and Sadek et al. [63].

Reducing the size of decision diagrams. Nagayama and Sasao [56] proposed exact and heuristic algorithms for minimizing the average path length of so-called *heterogeneous* multivalued decision diagrams. Janković et al. [40] studied the reduction of the size of multivalued decision diagrams by copy properties (symmetry and shift). Copy decision diagrams exploit an extended library of relationships between sequences producing isomorphic subtrees. Karpovsky et al. [43] used a total autocorrelation function for reducing the size of decision diagrams. Sasao and Butler [65] estimated the average and worst case number of nodes in binary decision diagrams of symmetric multivalued functions. Partitioning of decision diagrams is considered as a way to resolve the size explosion problem (see, for example, the work by Cabodi et al. [16]).

Delaunay trees. Delaunay triangulations can be organized into a hierarchical data structure called *Delaunay trees* (see, in particular the paper by

---

[†]Markov chains have been used as a mathematical model for describing how a probabilistic method finds a global minimum in optimization problems. Any Markov model is defined by a set of probabilities p_{ij}, which define the probability of transition from any state i to any state j. The transition probability p_{ij} depends only on states i and j and is completely independent of all past states except the last one, state i. Starting from the chosen initial state and an initial value T, a sequence of states is generated. These states form the first Markov chain. The probabilistic method lowers the value of T and generates a new Markov chain whose initial state is the final state of the previous chain. Similarly, the value of T is successively lowered and the corresponding Markov chains are generated. This continues until T has assumed a sufficiently low value. Under certain restrictions on the modeling process, the probabilistic method asymptotically produces an optimal solution to the combinatorial optimization problem.

[‡]Let $F = \{F_1, \dots, f_m\}$ be a family of random variables defined on the set S, in which each random variable F_i takes a value f_i in $\mathcal{L}$. The family of F is called a *random field* (see, in particular, [21, 47, 76]).

Boissonnat and Teilland [11]). Delaunay trees generate Voronoi diagrams that can describe 2D and 3D topologies. The relationship between Voronoi diagrams and decision trees considered, in particular, by Aurenhammer [5], is a subject of further study promising for description of 3D topologies in nanodimensions (see Chapter 3).

Graph embedding techniques. Rings and linear arrays can be mapped into hypercubes. For example, a linear array of length k can be mapped into a hypercube of dimension $n = \log_2(l+1)$. Linear array and linear word-level decision diagrams have the same configuration. Meshes that represent cell-based computing using lattice decision diagrams have been considered by Perkowski et al. [59]. Sasao and Butler [66] developed a method for representation look-up table type FPGA by decision diagrams. For this, so-called *pseudo Kronecker* decision diagrams were used (see Chapter 3).

Triangle topology. Popel and Dani [60] interpreted Sierpinski gaskets using Shannon and Davio expansions. Butler et al. [15] considered transeunt triangles and Sierpinski triangles for calculation of optimal polarity of Reed–Muller expressions (see Chapters 3 and 4).

Decision diagrams in nanocomputing. Shmerko and Yanushkevich [69] mapped feedforward neural networks into a hypercube-like structure. For this each computing element was represented by a decision tree. Design technique for nanoscale dimensions is developed by Yanushkevich et al. [79]. Khan et al. [46] used decision diagram technique for designing computing structures based on reversible devices. The technique of linearization of decision diagrams for embedding in nanostuctures was developed by Yanushkevich et al. [80, 82]. There are several results on mapping decision diagrams into computing nanostructures. Asachi et al. [4], Kasai et al. [45], and Yamada et al. [77] used decision diagrams for design single electron devices.

Homogenous highly parallel arithmetic circuits, in particular, systolic structures and error correction coding developed in the last decade, are a major focus in nanotechnology. Scenarios where errors may occur include nanodevices, data storage, interconnections, etc. The design technique of 1D interleaving and 2D error burst correction codes have been well studied and documented. The multi-dimensional interleaving technique followed by a random error correction code has become the most common approach to correcting multi-dimensional error bursts as stated by Blaum et al. [10]. For example, size 2 means that the error burst is a 3D hypercube volume $2 \times 2 \times 2$ blocks. Random error correction codes can be used efficiently to correct bursts of errors.

Plenty of useful information on the application of multivalued logic in nanosystem design can be found in the *Proceedings of the Annual IEEE International Symposium on Multiple-Valued Logic* that has been held since 1971.

In addition, a good source of information on decision diagrams for multivalued systems design is the *International Journal on Multiple-Valued Logic and Soft Computing.*

Aoki et al. [3] developed an algebraic system called a set-valued logic, which is a special class of multivalued logic for biomolecular computing. Deng et al. [20] studied characteristics of a super-pass transistor as a quantum device candidate for VLSI systems based on multivalued logic. Higuchi and Kameyama [33, 41] proposed a T-gate, a tree type universal logic module to implement multivalued logic networks. Kameyama and Higuchi discussed a multivalued coding of organic molecules and basic switching models [42].

References

[1] Akers SB, Binary decision diagrams. *IEEE Transactions on Computers,* 27(6):509–516, 1978.

[2] Anderson G, Bjesse P, Cook B, and Hanna Z. A proof engine approach to solving combinational design automation problems. In *Proceedings of the ACM/IEEE DAC*, New Orleans, LA, pp. 725–731, 2002.

[3] Aoki T, Kameyama M, and Higuchi T. Interconnection-free biomolecular computing. *Computer*, pp. 41–50, Nov., 1992.

[4] Asahi N, Akazawa M, and Amemiya Y. Single-electron logic device based on the binary decision diagram. *IEEE Transactions on Electron Devices*, 44(7):1109–1116, 1997.

[5] Aurenhammer F. Voronoi diagrams – a survey of a fundamental geometric data structure. *ACM Computing Surveys*, 23(3):345–405, 1991.

[6] Bahar RI, Chen J, and Mundy J. A probabilistic-based design for nanoscale computation. In Shukla S and Bahar RI, Eds., *Nano, Quantum and Molecular Computing: Implications to High Level Design and Validation*, Kluwer, Dordrecht, 2004.

[7] Bapty T, Neema S, Scott J, Sztipanovits J, and Asaad S. Model-integrated tools for the design of dynamically reconfigurable systems. *Technical Report ISIS-99-01*, Institute for Software-Integrated Systems, Vanderbilt University, TN, 1999.

[8] Bartlett LM and Andrews JD. Selecting an ordering heuristic for the fault tree to binary decision diagram conversion process using neural networks. *IEEE Transactions on Reliability* , 51(3):344–349, 2002.

[9] Bartlett LM. Improving the neural network selection mechanism for BDD construction. *Quality Reliability Engineering International*, 20:217–223, April 2003.

[10] Blaum M, Bruck J, and Vardy A. Interleaving schemes for multidimensional cluster errors. *IEEE Transactions on Information Theory*, 44(2):730–743, 1998.

[11] Boissonnat JD and Teilland M. A hierarchical representation of objects: the Delaunay tree. In *Proceedings of the 2nd Annual ACM Symposium on Computational Geometry*, pp. 260–268, 1986.

[12] Bryant RE. Graph-based algorithms for Boolean function manipulation. *IEEE Transactions on Computers*, 35(6):677–691, 1986.

[13] Brayton RK The future of logic synthesis and verification. In [32], pp. 403–434.

[14] Bryant RE and Meinel C. Ordered binary decision diagrams. Foundations, applications and innovations. In [32], pp. 285–307.

[15] Butler JT, Dueck GW, Yansushkevich SN, and Shmerko VP. On the number of generators for transeunt triangles. *Discrete Applied Mathematics*, 108:309–316, 2001.

[16] Cabodi G, Camurati P, and Quer S. Improving the efficiency of BDD-based operators by means of partitioning. *IEEE Transactions on Computer Aided Design of Integrated Circuits and Systems*, 18(5):545–556, 1999.

[17] Cheushev VA, Yanushkevich SN, Moraga C, and Shmerko VP. Flexibility in logic design. An approach based on information theory methods. *Research Report 741*, Forschungsbericht, University of Dortmund, Germany, 2000.

[18] Cheushev VA, Yanushkevich SN, Shmerko VP, Moraga C, and Kolodziejczyk J. Information theory methods for flexible network synthesis. In *Proceedings of the IEEE 31st International Symposium on Multiple-Valued Logic*, pp. 201–206, 2001.

[19] Dasgupta D. *Artificial Immune Systems and Their Applications.* Springer, Heidelberg, 1999.

[20] Deng X, Hanyu T, and Kameyama M. Multiple-valued logic network using quantum-device-oriented superpass gates and its minimization. *IEE Proceedings, Circuits, Devices and Systems*, 142(5):299–306, 1995.

[21] Derin H and Kelly PA. Discrete-index Markov-type random fields. *Proceedings of the IEEE*, 77(10):1485–1510, 1989.

[22] Drechsler R. *Evolutionary Algorithms for VLSI CAD.* Kluwer, Dordrecht, 1998.

[23] Durand S, Stauffer A, and Mange D. Biodule: An Introduction to Digital Biology. *Technical Report, Logic Systems Laboratory, EPFL*, Lausanne, 1994.

[24] Edelkamp S and Helmert M. On the implementation of MIPS. In *Proceedings of the 5th International Conference on Artificial Intelligence Planning and Scheduling, Workshop on Model-Theoretic Approaches to Planning*, Breckenridge, Colorado, pp. 8–15, April 2000.

[25] Edelkamp S and Helmert M. Exhibiting knowledge in planning problems to minimize state encoding length. In Fox M and Biundo S, Eds., *Recent Advances in AI Planning. Lecture Notes in Artificial Intelligence*, Vol. 1809, Springer, Heidelberg, pp. 135–147, 1999.

[26] Eldredge JG and Hutchings BL. Density enhancement of a neural network using FPGAs and run-time reconfiguration. In *Proceedings of IEEE Workshop on FPGAs for Custom Computing Machines*, Napa, CA, pp. 180–188, April 1994.

[27] Evans WS and Schulman LJ. Signal propagation and noisy circuits. *IEEE Transactions on Information Theory*, 45(7):2367–2373, 1999.

[28] Frank MP and Knight TFJr. Ultimate theoretical models of nanocomputers. *Nanotechnology*, 9:162–176, 1998.

[29] Gustin V. Artificial neural network realization with programmable logic circuit. *Microprocessing and Microprogramming*, 35:187–192, 1992.

[30] Gupta A, Ganai M, Wang C, Yang Z, and Ashar P. Learning from BDDs in SAT-based bounded model checking. In *Proceedings of the ACM/IEEE DAC*, Anaheim, CA, June 2-6, pp. 824–830.

[31] Hartmann CRP, Varshney PK, Mehrotra KG, and Gerberich CL. Application of information theory to the construction of efficient decision trees. *IEEE Transactions on Information Theory*, 28(5):565–577, 1982.

[32] Hassoun S and Sasao T, Eds., Brayton RK, consulting Ed. *Logic Synthesis and Verification.* Kluwer, Dordrecht, 2002.

[33] Higuchi T and Kameyama M. Static-hazard-free T-gate for ternary memory element and its application to ternary counters. *IEEE Transactions on Computers*, 26(12):1212-1221, 1977.

[34] Hoey J, St-Aubin R, Hu A, and Boutilier C. SPUDD: stochastic planning using decision diagrams. In *Proceedings of the 15th Annual Conference on Uncertainty in Artificial Intelligence (UAI-99)*, Morgan Kaufmann Publishers, San Francisco, CA, pp. 279–288, 1999.

[35] Horiyama T, and Ibaraki T. Ordered binary decision diagrams as knowledge-bases. *Artificial Intelligence*, 136(2):189–213, 2002.

[36] Howard N and Taylor RW. Reconfigurable logic: technology and applications. *Computing Control Engineering Journal*, pp. 235–240, Sept. 1992.

[37] Hsu WJ. Fibonacci cubes – a new interconnection topology. *IEEE Transactions on Parallel and Distributed Systems*, 4(1):3–12, 1993.

[38] Iba H, Iwata M, and Higuchi T. Machine learning approach to gate level evolvable hardware. In *Lecture Notes in Computer Science*, 1259:327–393, Springer, Heidelberg, 1997.

[39] Jain J, Bitner J, Fussell DS, and Abraham JA. Probabilistic verification of Boolean functions. In *Formal Methods in System Design*, 1:61–115, Kluwer, Dordrecht, 1992.

[40] Janković D, Stanković RS, and Drechsler R. Reduction of sizes of multi-valued decision diagrams by copy properties. In *Proceedings of the IEEE 34th International Symposium on Multiple-Valued Logic*, pp. 223–228, 2004.

[41] Higuchi T, and Kameyama M. Synthesis of multiple-valued logic networks based on tree-type universal logic module. *IEEE Transactions on Computers*, 26(12):1297–1302, December 1977.

[42] Kameyama M and Higuchi T. Prospects of multiple-valued bio-information processing systems. In *Proceedings of the 18th International Symposium on Multiple-Valued Logic*, pp. 237–242, 1988.

[43] Karpovsky MG, Stanković RS, and Astola JT. Reduction of sizes of decision diagrams by autocorrelation functions. *IEEE Transactions on Computers*, 52(5):592–606, 2003.

[44] Karpovsky MG, Stanković RS, and Astola JT. Construction of linearly transformed planar BDD by Walsh coefficients. In *Proceedings of the International Symposium on Circuits and Systems*, Vol. 4, pp. 517–520, 2004.

[45] Kasai S and Hasegawa H. A single-electron binary-decision-diagram quantum logic circuit based on Schottky wrap gate control of a GaAs nanowire hexagon. *IEEE Electron Device Letters*, 23(8):446–448, 2002.

[46] Khan MHA, Perkowski M, and Khan MR. Ternary Galois field expansions for reversible logic and Kronecker decision diagrams for ternary GFSOP minimization. In *Proceedings of the IEEE 34th International Symposium on Multiple-Valued Logic*, pp. 58–67, 2004.

[47] Kindermann S and Snell JL. Markov random fields and their applications. *Providence, R.I., American Mathematical Society*, 1980.

[48] Lee CY. Representation of switching circuits by binary decision programs. *Bell System Technical Journal*, 38:985–999, 1959.

[49] Likharev KK. Single-electron devices and their applications. In *Proceedings of the IEEE*, 87(4):606–632, 1999.

[50] Marchal P and Stauffer A. Binary decision diagram oriented FPGAs. In *Proceedings of the ACM International Workshop on Field-Programmable Gate Arrays*, Berkeley, pp. 108–114, Feb., 1994.

[51] Miller DM and Muranaka N. Multiple-valued decision diagrams with symmetric variable nodes. In *Proceedings of the IEEE 26th International Symposium on Multiple-Valued Logic*, pp. 242–247, 1996.

[52] Moraga C, Trillas E, and Guadarrama S. Multiple-valued logic and artificial intelligence fundamentals of fuzzy control revisited. In Yanushkevich SN, Ed., *Artificial Intelligence in Logic Design.* Kluwer, Dordrecht, pp. 9–37, 2004.

[53] Mues C, Baesens B, Files CM, and Vanthienen J. Decision diagrams in machine learning. *Lecture Notes in Computer Science*, vol. 2980, pp. 395–397, 2004.

[54] Mues C, Baesens B, Files CM, and Vanthienen J. Decision diagrams in machine learning: an empirical study on real-life credit-risk data. *Expert Systems with Applications*, 27(2):257–264, 2004.

[55] Mues C and Vanthienen J. Efficient rule base verification using binary decision diagrams. *Lecture Notes in Computer Science*, vol. 3180, pp. 445–454, 2004.

[56] Nagayama S and Sasao T. On the minimization of average path length for heterogeneous MDDs. In *Proceedings of the IEEE 34th International Symposium on Multiple-Valued Logic*, pp. 216–222, 2004.

[57] Najm F. Transition density: a new measure of activity in digital circuits. *IEEE Transactions on Computer Aided Design of Integrated Circuits and Systems*, 12(2):310–323, 1993.

[58] Peper F, Lee J, Abo F, Isokawa T, Adachi S, Matsui N, and Mashiko S. Fault-tolerance in nanocomputers: a cellular array approach. *IEEE Transactions on Nanotechnology*, 3(1):187–201, 2004.

[59] Perkowski M, Chrzanowska-Jeske M, and Xu Y. Lattice diagrams using Reed-Muller Llogic. In *Proceedings of the IFIP 10.5 Reed–Muller Workshop*, pp. 85–102, Oxford University, UK, Sept. 1997.

[60] Popel DV and Dani A. Sierpinski gaskets for logic function representation. In *Proceedings of the IEEE 32th International Symposium on Multiple-Valued Logic*, pp. 39–45, 2002.

[61] Qi Y, Gao J, and Fortes AB. Markov chains and probabilistic computation – a general framwork for multiplexed nanoelectronic system. *IEEE Transactions on Nanotechnology*, 4(2):194–205, 2005.

[62] Rao TRN. *Error Coding for Arithmetic Processors*, Academic Press, New York, 1974.

[63] Sadek AS, Nikolić K, and Forshaw M. Parallel information and compuation with restriction for noise-tolerant nanoscale logic networks. *Nanotechnology*, 15:192–210, 2004.

[64] Sakanashi H, Higuchi T, Iba H, and Kakaza Y. Evolution of binary decision diagrams for digital circuits design using genetic programming. In *Evolvable Systems: From Biology to Hardware. Lecture Notes in Computer Science*, vol. 1259, pp. 470–481, Springer, Heidelberg, 1997.

[65] Sasao T and Butler JT. Planar decision diagrams for multiple-valued functions. *International Journal on Multiple-Valued Logic,* 1:39–64, 1996.

[66] Sasao T and Butler JT. A design method for look-up table type FPGA by pseudo-Kronecker expansions. In *Proceedings of the IEEE 24th International Symposium on Multiple-Valued Logic*, pp. 97–106, 1994.

[67] Shmerko VP, Popel DV, Stanković RS, Cheushev VA, and Yanushkevich SN. Information theoretical approach to minimization of AND/EXOR expressions of switching functions. In *Proceedings IEEE International Conference on Telecommunications*, pp. 444–451, Yugoslavia, 1999.

[68] Shmerko VP, Popel DV, Stanković RS, Cheushev VA, and Yanushkevich SN. Entropy based algorithm for 4-valued functions minimization. In *Proceedings of the IEEE 30th International Symposium on Multiple-Valued Logic*, pp. 265–270, 2000.

[69] Shmerko VP and Yanushkevich SN. Three-dimensional feedforward neural networks and their realization by nano-devices. *Artificial Intelligence Review, International Journal*, Special Issue on Artificial Intelligence in Logic Design, 20(3-4):473–494, 2003.

[70] Stanković RS and Astola JT. *Spectral Interpretation of Decision Diagrams.* Springer, Heidelberg, 2003.

[71] Sentovich E and Brand D. Flexibility in logic. In [32], pp. 65–88.

[72] Stanković RS. Simple theory of decision diagrams for representation of discrete functions. In *Proceedings of the 4th International Workshop on Applications of the Reed–Muller Expansion in Circuit Design.* University of Victoria, BC, Canada, pp. 161–178, 1999.

[73] Stanković RS, Stanković M, Astola JT, and Egiazarian K. Fibonacci decision diagrams and spectral transform Fibonacci decision diagrams. In *Proceedings of the 30th International Symposium on Multiple-Valued Logic*, pp. 206–211, 2000.

[74] Strehl K and Thiele L. Symbolic model checking of process networks using interval diagram techniques. newblock In *Proceedings of the IEEE/ACM International Conference on Computer Aided Design*, pp. 686–692, ACM Press, 1998.

[75] Strehl K, Moraga C, Temme KH, and Stanković RS. Fuzzy decision diagrams for the representation, analysis and optimization of rule bases. In *Proceedings of the IEEE 30th International Symposium on Multiple-Valued Logic*, pp. 127–132, 2000.

[76] Woods JW. Two-dimensional discrete Markovian fields. *IEEE Transactions on Information Theory*, 18:232–240, 1972.

[77] Yamada T, Kinoshita Y, Kasai S, Hasegawa H, and Amemiya Y. Quantum-dot logic circuits based on shared binary decision diagram. *Journal of Applied Physics*, 40(7):4485–4488, 2001.

[78] Yamashita S, Sawada H, and Nagoya A. SPFD: a method to express functional flexibility. *IEEE Transactions on Computer-Aided Design of Integrated Circuits and Systems*, 19(8):840–849, 2000.

[79] Yanushkevich SN, Shmerko VP, and Lyshevski SE. *Logic Design of NanoICs*, CRC Press, Boca Raton, FL, 2005.

[80] Yanushkevich SN, Shmerko VP, Malyugin VD, and Dziurzanski P. Linearity of word-level models: new understanding. In *Proceedings of the IEEE/ACM 11th International Workshop on Logic and Synthesis,* pp. 67–72, New Orleans, LA, 2002.

[81] Yanushkevich SN. Logic design of nanodevices. In *Rieth M and Schommers W, Eds., Handbook of Theoretical and Computational Nanotechnology*, American Scientific Publishers, 2005.

[82] Yanushkevich SN. *Logic Design of Micro and NanoICs*. Lecture Notes for a graduate course. University of Calgary, 2005.

[83] Watanabe H. A basic theory of information network. *IEICE Transactions Fundamentals*, E76-A(3):265–276, 1993.

Part V

APPENDIX

- Group Representation
- Fourier Analysis on Finite Abelian Groups
- Fourier Analysis on Non-Abelian Groups
- Fourier - like Series
- Discrete Walsh Transform
- Relationship between Walsh and switching functions
- Basic Operations for Ternary Logic
- Basic Operations for Quaternary Logic

Appendix-A: Algebraic Structures for the Fourier Transform on Finite Groups

This appendix presents basic notions and definitions about Fourier transform on finite, not necessarily Abelian, groups.

Group

A group is a basic algebraic structure used in mathematical modelling of signals. It consists of a set equipped with a binary operation over elements of it.

Definition 41.1 *An algebraic structure* $\langle G, \circ, 0\rangle$ *with the following properties is a* group*:*

- *Associative law:*

$$(x \circ y) \circ z = x \circ (y \circ z), \quad x, y, z \in G.$$

- *Identity: for all* $x \in G$*, the unique element 0 (identity) satisfies*

$$x \circ 0 = 0 \circ x = x.$$

- *Inverse element: for any* $x \in G$*, there exists an element* x^{-1} *such that*

$$x \circ x^{-1} = x^{-1} \circ x = 0.$$

The group is usually written as G instead of $\langle G, \circ, 0\rangle$. A group G is an *Abelian group* if

$$x \circ y = y \circ x$$

for each $x, y \in G$. Otherwise, G is a *non-Abelian group*.

Definition 41.2 *Let $\langle G, \circ, 0\rangle$ be a group and $0 \in H \subseteq G$. If $\langle H, \circ, 0\rangle$ is a group, it is called a subgroup of G.*

Example 41.7 *The following structures are groups:*

- $\mathbf{Z}_q = \langle\{0, 1, \ldots, q-1\}, \oplus_q\rangle$, *the group of integers modulo q,*
- $\mathbf{Z}_2 = \langle\{0, 1\}, \oplus\rangle$, *the simplest nontrivial group, and*
- $\mathbf{Z}_6 = \langle\{0, 1, 2, 3, 4, 5\}, \oplus\rangle$, *the additive group of integers modulo 6.*

Notice that addition modulo 2 is equivalent to the logic operation EXOR usually denoted in switching theory simply as $\oplus$. Likewise, multiplication modulo 2 is equivalent to logic AND.

Vector space

Definition 41.3 *Given an Abelian group $G = (\mathcal{G}, \oplus)$ and a field F, the pair (G, F) is a linear vector space over F, $\langle V, \oplus\rangle$. The pair (G, F) is a short vector space, if the multiplication of elements of G with elements of F is defined such that for each $x, y \in G$ and $\lambda, \mu \in F$:*

- $\lambda(x \oplus y) = \lambda x \oplus \lambda y$,
- $(\lambda + \mu)x = \lambda x \oplus \mu x$,
- $\lambda(\mu x) = (\lambda\mu)x$,
- $e \cdot x = x$, *where e is the identity element in P.*

Functions on finite discrete groups can be viewed as elements of some vector spaces. Let

- $F(G)$ be the set of all functions $f : G \to F$, where G is a finite group of order g,
- F be a field that may be the complex-field C,
- R be the real-field,
- Q be the field of rational numbers, or a finite (Galois) field $GF(p)$.

Definition 41.4 *$F(G)$ endorses the structure of a vector space if*

- *For $f, z \in F(G)$, addition of f and z is defined by*

$$(f + z)(x) = f(x) + z(x), \quad \forall f, \quad z \in F(G),$$

- *Multiplication of $f \in P(G)$ by an $\alpha \in F$ is defined as*

$$(\alpha f)(x) = \alpha f(x), \quad \forall f \in F(G), \quad \alpha \in F.$$

Since the elements of $F(G)$ are vectors of order g, it follows that the multiplication of f by $\alpha \in F$ is actually the componentwise multiplication with constant vectors in $F(G)$.

Example 41.8 *Consider the set $GF(C_2^n)$:*

- *This set in $GF(2)$ is the vector space over $GF(2)$ under the operations in $GF(2)$ applied componentwise to the truth-vectors of functions.*
- *This set $C(C_2^n)$ in the complex-field C is the vector space over C under the operations in C applied componentwise to the vectors of functions.*

Group representations

Various groups can be defined over different sets and by using different operations. For a uniform consideration of different groups, the theory of *group representations* has been developed to study properties of abstract groups via their representations as *linear transformations* of *vector spaces.* This theory is important, since through group representations many group-theoretic problems are reduced to problems in linear algebra that permit exploiting of many useful mathematical tools developed in this area.

Definition 41.5 *A representation of a group G on a vector space V over a field F is a group homomorphism from G to $GL(V)$, the* general linear group *on V.*

When V is of a finite n-dimension, it is convenient to choose a *basis* for V and identify $GL(V)$ with the group of $(n \times n)$ invertible matrices with the group operation as ordinary matrix multiplication. This group is usually denoted by $GL(n, F)$.

A subspace V' of V that is fixed under the group action of g is called a *subrepresentation* of G on V over F. If V has a non-zero proper subrepresentation, the representation is said to be *reducible.* Otherwise, it is *irreducible.*

A representation $x \rightarrow R_w(x)$ of G over V is equivalent to the representation $x \rightarrow R_q(x)$ over V', which we denote by $R_w(x) \equiv R_q(x)$, if there is bounded isomorphism S of V into V such that

$$SR_w(x) = R_q(x)S, \quad \text{for each } x \in G.$$

Definition 41.6 *The set of all nonequivalent irreducible representations of G will be denoted by Γ and will be called the dual object of G. Denote by K the cardinal number of Γ.*

A representation is:

- *Unitary* if each $R(x)$, $x \in G$ is a unitary matrix over V, and
- *Trivial* if $R(x) = I$ for each $x \in G$, where I is the identity matrix.

In the case of compact groups, the study of group representations can be restricted to unitary representations without loss of generality.

Each unitary irreducible representation R for a compact group G is finite dimensional. We denote by r_w the dimension of the representation R_w.

Notice that finite groups, which are considered in this book, are compact groups. Therefore, we will deal with finite-dimensional unitary irreducible representations.

Definition 41.7 *The* group character $\chi(w, x)$ *of a finite-dimensional representation* R_w *of a group* G *is defined as the trace of the operator* $R_w(x)$, *i.e.,*

$$\chi_w(x) = Tr(R_w(x)).$$

In the case of Abelian groups, all the representations are one-dimensional, thus, reduce to group characters. In this case, Γ exhibits the structure of an Abelian group isomorphic to G.

The Peter–Weyl theorem provides a basis for extension of the Fourier analysis to compact non-necessarily Abelian groups.

Definition 41.8 *In the case of finite compact groups, the Fourier transform pair is defined as*

$$f(x) = \sum_{w=0}^{K-1} Tr(\mathbf{S}_f(w)\mathbf{R}_w(x)),$$

$$\mathbf{S}_f(w) = r_w g^{-1} \sum_{u=0}^{g-1} f(u)\mathbf{R}_w(u^{-1}).$$

Definition 41.9 *For Abelian groups, group representations are group characters, and, thus, the Fourier transform pair is*

$$f(x) = \sum_{w=0}^{g-1} S_f(w)\chi_w(x),$$

$$\mathbf{S}_f(w) = g^{-1} \sum_{u=0}^{g-1} f(u)\chi_w(u^{-1}).$$

Further reading

The theory of group characters was greatly elaborated by Pontryagin [8]. Algebraic structures are discussed by Batisda [1], Bell [2], Cantrell [3], Edward

[5], and Winter [10].

Classical books in group representations and Fourier analysis are [4], [6], and [7]. A comprehensive theoretical background and some engineering applications of Fourier analysis on finite groups are presented in [9].

References

[1] Batisda JR. *Field Expansions and Galois Theory*, Cambridge University Press, New York, 1984.

[2] Bell AW. *Algebraic Structures.* John Wiley and Sons, New York, 1966.

[3] Cantrell CD. *Modern Mathematical Methods for Physicists and Engineers.* Cambridge University Press, New York, 2000.

[4] Curtis CN and Rainer I. *Representation Theory of Finite Groups and Associateive Algebras*, Halsted, New York, 1962.

[5] Edward HM. *Galois Theory.* Springer, Heidelberg, 1993.

[6] Hewitt E and Ross KA. *Abstract Harmonic Analysis.* Part I, Springer, Heilderberg, 1963.

[7] Hewitt E and Ross KA. *Abstract Harmonic Analysis.* part II, Springer, Heilderberg, 1970.

[8] Pontryagin LS. *Group Theory.* State Publisher, 1954, (in Russian).

[9] Stanković RS, Moraga C and Astola JT. *Readings in Fourier Analysis on Finite Non-Abelian Groups.* TICSP Series #5, ISBN 952-15-0284-3, Tampere, Finland, 1999.

[10] Winter DJ. *The Structure of Fields.* Springer, Heidelberg, 1974.

Appendix-B: Fourier Analysis on Groups

In the spectral techniques approach to switching theory, logic values 0 and 1 are formally interpreted as elements of a field F, and due to that, switching functions are considered as elements of a Hilbert space consisting of functions on finite dyadic groups with values in F. A finite dyadic group of order 2^n, C_2^n is defined as the direct product on n cyclic groups of order 2, i.e.,

$$C_2^n = \times_{i=1}^n C_2,$$

where $C_2 = (\{0, 1\}, \oplus)$. If logic values 0 and 1 are interpreted as integers 0 and 1, switching functions are viewed as elements of a complex-valued function space on C_2^n.

> ***Spectral techniques*** *for switching functions are defined as a particular example in* abstract harmonic analysis, *that is a mathematical discipline derived from the classical Fourier analysis by the replacement of the real group R by an arbitrary locally compact Abelian or compact non-Abelian group. Finite dyadic groups are a particular case of Abelian groups.*

Algebraic structures for Fourier analysis on finite Abelian groups

Let G be

- A finite Abelian group of order g with elements in some fixed order,
- A field P that could be the complex field C, and
- The real field R or a finite field admitting the existence of a Fourier transform.

A known theorem in algebra states that each non-cyclic group G can be represented as the direct product of some cyclic subgroups G_i of order g_i, i.e.,

$$G = \bigotimes_{i=1}^{n} G_i, \quad \text{and} \quad g = \prod_{i=1}^{n} g_i, \tag{1}$$

where the value of n is determined by the values of g and the corresponding g_i's. The group operation $\circ$ can be expressed in terms of group operations of G_i as

$$x \circ y = (x_1 \overset{\circ}{1} y_1, \ldots, x_n \overset{\circ}{n} y_n),$$

where $x, y \in G, x_i, y_i \in G_i$. Also, each $x \in G$ can be defined in terms of x_i as follows,

$$x = \sum_{i=1}^{n} a_i x_i, \;\; x_i \in G_i, \quad \text{with} \quad a_i = \begin{cases} \prod_{j=i+1}^{n} g_j, & i = 1, \ldots, n-1, \\ 1, & i = n, \end{cases} \tag{2}$$

where g_j is the order of G_j.

Example 41.9 *As noticed above, the finite dyadic group C_2^n is an example of a group representable as the direct product of some subgroups. The elements of C_2^n are binary n-tuples $(x_1, \ldots, x_n)$, $x_i \in \{0, 1\}$ and the group operation is the componentwise addition modulo 2, that is, the componentwise EXOR.*

Any function $f : G \to P$ can be considered as a function of n-variables, i.e., $f(x) = f(x_1, \ldots, x_n)$, $x \in G$, $x_i \in G_i$, $i = 1, \ldots, n$.

Example 41.10 *In the case of switching functions,*

$$G = C_2^n \quad and \quad P = GF(2).$$

However, the switching functions can be alternatively regarded as functions of a single variable x corresponding to the binary n-tuples

$$(x_1, \ldots, x_n),$$

i.e.,

$$f(x), \ \ x \in \{0, \ldots, 2^n - 1\}.$$

In that case, switching functions can be defined by enumerating their values at the points in

$$C_{2^n} = \{0, \ldots, 2^n - 1\}$$

denoted by the decimal indexes defined by Equation 2. Thanks to the simplicity of their range, the switching functions can be alternatively regarded as functions on C_{2^n} into C if the elements of $GF(2)$, logical 0 and 1, are interpreted as the integers 0 and 1.

The set $P(G)$ of functions on G into P is an Abelian group under the pointwise addition defined by

$$(f + g)(x) = f(x) + g(x), \ \ f, g \in P(G), \ \ x \in G.$$

By introducing the multiplication by a scalar defined by

$$(\alpha f)(x) = \alpha f(x), \ \ \alpha \in P,$$

$P(G)$ becomes a linear space.

If in $P(G)$ the inner product is defined as

$$\langle f, h \rangle = g^{-1} \sum_{x \in G} f(x) \stackrel{\sim}{h} (x), \ \ \forall f, h \in P(G),$$

where $\stackrel{\sim}{h} = h^*$ is the complex-conjugate of h for P to be the complex field C, and $\stackrel{\sim}{h} = h$ for P to be the real field R or a finite field, then $P(G)$ becomes a Hilbert space with the norm

$$\|f\| = (\langle f, f \rangle)^{\frac{1}{2}} = g^{-1} \sum_{x \in G} (|f(x)|^2)^{\frac{1}{2}}, \ \ \forall f \in P(G).$$

The space $P(G)$ becomes a linear commutative function algebra by adjunction the componentwise multiplication of functions through

$$(fh)(x) = f(x)g(x), \ \ \forall x \in G, \ \ \forall f, h \in P(G).$$

The convolution product on G is defined by

$$(f * h)(y) = g^{-1} \sum_{x \in G} f(x)h(x^{-1} \circ y), \ \ \forall x, y \in G, \ \ \forall f, h \in P(G).$$

Example 41.11 *If $G = C_2^n$, the convolution product is called the dyadic convolution and is defined by*

$$(f \star h)(y) = 2^{-n} \sum_{x=0}^{2^n-1} f(x)h(x \oplus y),$$

where $x, y \in G$, $f, h \in P(C_2^n)$ with calculations in P and $P(G)$ is a linear commutative function algebra.

Series expansions

A *basis* in the linear space $P(G)$ is a set $\{e_r, r = 0, \ldots, g-1\}$ of linearly independent elements of $P(G)$. Such the linear independence is expressed by asserting that the equation

$$\sum_{r=0}^{g-1} c(r)e_r(x) = 0 \in P(G),$$

where, for each $r = 0, \ldots, g-1, c(r) \in P$, has no solution other than $c = (c(0), \ldots, c(g-1)) = 0 \in P(G)$. In matrix notation, the linear independence is equivalently stated as a request that a matrix over P, whose columns are the set of functions $\{e_r, r = 0, \ldots, g-1\}$, is nonsingular.

Each function $f \in P(G)$ can be represented in terms of a given basis $\{\phi_i\}$, $i = 0, \ldots, g-1$, as a linear combination

$$f = \sum_{x \in G} a_i \phi_i(x). \tag{3}$$

The basis $\{e_r,\ r = 0, \ldots, g-1\}$ is an orthonormal system in $P(G)$ if

$$\langle e_i, e_j \rangle = \begin{cases} 1, \ i = j, \\ 0, \ i \neq j. \end{cases} \ \ i, j \in \{0, \ldots, g-1\}.$$

The elements of an orthonormal system are linearly independent and, thus, form a basis in $P(G)$. The coefficients in Equation 3 with respect to an orthonormal basis are the inner products of f with the basic functions, i.e., $a_i = \langle f, e_i \rangle$. The corresponding series are called the ***Fourier-like*** *series for the resemblance to the Fourier series on the real line R.*

Example 41.12 *The set*

$$E = \{e_{r,j}\} \ \ r, j \in \{0, \ldots, g-1\}, \tag{4}$$

where

$$e_{r,j} = \begin{cases} 1, \ r = j, \\ 0, \ r \neq j, \end{cases}$$

is a basis in $P(G)$ called the trivial basis.

> *In matrix notation, the **trivial basis** functions are represented by columns of the identity matrix. In engineering practice, this basis is identified with the block pulse functions. If $P = GF(2)$, this basis is defined by minterms, i.e., complete products of switching variables for different polarities of literals.*

The sum-of-products form of switching functions can be considered as the Fourier-series like expansion with respect to a basis (Equation 4).

Another example of a basis for switching functions is the set of functions that consists of the Reed–Muller functions.

Example 41.13 *The set of functions on C_2^n into $GF(2)$ consists of the Reed–Muller functions, i.e., columns of the Reed–Muller matrix $\mathbf{R}(n)$. If $P = C$, the corresponding basis consists of the integer Reed–Muller functions, i.e., columns of the arithmetic transform matrix $\mathbf{A}^{-1}(n)$.*

Fourier series

A complete orthonormal set in $P(G)$ can be obtained by referring to the dual object Γ of G represented by a transversal $\Gamma = \{R_0, \ldots, R_{K-1}\}$ of all equivalent classes of unitary irreducible representations of G over P. These representations may be given a fixed order.

In the case of Abelian groups, all unitary irreducible representations are one dimensional, that is, they reduce to group characters. If $P = C$, the group characters are defined as the homomorphisms of G into the unit circle [10], or equivalently to the multiplicative group T of complex numbers with modulo equal 1.

Definition 41.10 *Complex-valued function $\chi_w(x)$ on G is the group character of G if $|\chi_w(x)| = 1, \ \ \forall x \in G$ and*

$$\chi_w(x \circ y) = \chi_w(x)\chi_w(y), \forall x, y \in G.$$

The character $\chi_0(x) = 1, \ \forall x \in G$ is called the *principal character*. The set $\Gamma = \{\chi_w(x)\}$ of necessarily continuous characters expresses the structure of a multiplicative group called the *dual group* for G, with the group operation denoted by $\circ$ and defined as

$$(\chi_1 \circ \chi_2)(x) = \chi_1(x)\chi_2(x), \ \ \forall x \in G, \ \ \chi_1, \chi_2 \in \Gamma.$$

- If G is a discrete group, Γ is compact and vice versa.
- If G is finite, Γ is also finite and isomorphic to G.

It follows that for finite Abelian groups, Γ can be represented by Equation 1 and each $w \in \Gamma$ can be described by Equation 2. Thanks to that duality, the index w and argument x in $\chi_w(x)$ have the equivalent roles and it is convenient to express that property through the notation $\chi(w, x), x \in G, w \in \Gamma$.

Properties of the group characters

The properties of the group characters follow from their definition:

- Property 1: $\chi(w, x_1 \circ x_2) = \chi(w, x_1)\chi(w, x_2)$,
- Property 2: $\chi(x, w_1 \odot w_2) = \chi(x, w_1)\chi(x, w_2)$,
- Property 3: $\chi(0, x) = \chi(x, 0) = 1, \ \ \forall x \in G, w \in \Gamma$,
- Property 4: $\chi(x^{-1}, w) = \chi(x, w^{-1}) = \chi^{-1}(x, w) = \chi^*(x, w)$,

where χ^* is the complex-conjugate of χ. The group characters of a cyclic Abelian group are defined by

$$\chi(w, x) = \omega^{wx}$$

where $\omega = e^{\frac{2\pi i}{g}}$. Since each non-cyclic Abelian group G can be represented by Equation 1, the dual object Γ of a finite Abelian group is uniquely determined by G and consists of the set of group characters defined by

$$\begin{aligned}\chi(w, x) &= \chi((w_1, \ldots w_n), (x_1, \ldots, x_n)) \\ &= \prod_{i=1}^{n} \omega^{wx} = e^{(2\pi j \sum_{i=1}^{n} \frac{w_i x_i}{g_i})}. \end{aligned} \tag{5}$$

The set of characters $\{\chi(w, x)\}$ is a complete orthonormal set for $C(G)$, i.e.,

$$\langle \chi(w, \cdot), \chi(\psi, \cdot) \rangle = \begin{cases} g, \ w = k, \\ 0, \ w \neq k, \end{cases}$$

and $\langle f, \chi(w, \cdot) \rangle = 0, \ \ \forall w \in G$, implies that $f = 0$.

Therefore, $f \in C(G)$, G-finite Abelian group, is conveniently represented in the form of a series expansion in terms of the group characters and, thus, it is called the *Fourer series*

$$f = \sum_{w \in \Gamma} S_f(w)\chi(w,x), \tag{6}$$

where the coefficients S_f are given by

$$S_f(w) = g^{-1} \sum_{x \in G} f(x)\chi^*(w,x). \tag{7}$$

Fourier transform

Given discrete structures, the concept of the sum and integral coincides, the Equation 6 and Equation 7 are usually reported as the Fourier transform pair in $C(G)$ consisting of the direct (Equation 7) and inverse (Equation 6) Fourier transform for f.

Therefore, in the matrix notation, Fourier transform on finite Abelian groups is given by

$$\mathbf{S}_f = g^{-1}\left[\chi^*\right],$$
$$\mathbf{F} = [\chi]\,\mathbf{S}_f,$$

where $[\chi]$ is the matrix of group characters. The terminology is taken by the analogy to the classical Fourier analysis on the real line R, since it is shown that if $G = R$, then $\Gamma = R$ and the group characters have the form $\chi(w,x) = e^{jwx}$. Thus, the Fourier transform is defined by $S_f(w) = \int_{-\infty}^{\infty} f(x)e^{-jwx}dx$.

If G is a cyclic group of integers modulo less than a given integer g, i.e., $G = Z_g$ under the addition modulo g, then $\Gamma = Z_g$, and we have the discrete Fourier transform (DFT), given by

$$S_f = g^{-1} \sum_{n=0}^{g-1} f(n)e^{-j\frac{2\pi nk}{g}}, \quad 0 \leq k \leq g-1.$$

The functions $\exp -j\frac{2\pi nk}{g}$ are the discrete exponential functions. Therefore, the Discrete Fourier transform (DFT) is defined with respect to the discrete exponential functions.

If G is a finite Abelian group and P is a finite Galois field $GF(p^r)$, the existence conditions for the Fourier transform in $P(G)$ reduce to the requirement η divide $p^r - 1$, where η is the smallest common divisor of $g_1, \ldots, g_n$. In that case, as in Equation 5, the group characters are also squares of 1 of order g_i, but over $GF(p^r)$, i.e., they are given by

$$\chi(w,x) = \prod_{i=0}^{n-1} \xi_i^{w_i x_i}, \quad w_i, x_i \in \{0, \ldots, g-1\},$$

where $\xi_i = \sqrt[g_i]{1} \in GF(p^r)$, and all calculations are done in $GF(p^r)$. It is shown that if $P = C$, $\xi_i = e^{\frac{2\pi j}{g_i}}$ with $j = (-1)^{\frac{1}{2}}$, it results in Equation 5.

Finite dyadic group

In the case of finite dyadic group $g_1 = g_2 = \ldots = g_n = 2$, so that the character group of the dyadic group over $GF(2)$ seems to be trivially the singleton group whose only element is the principal character. An approach to the Fourier analysis for switching functions through the group characters over $GF(2)$ appears therefore to be excluded.

It follows that:

- For a proper counterpart of the Fourier analysis for switching functions, these functions have to be considered as elements of the Hilbert space of functions on C_2^n into C.
- Walsh functions are the group characters of finite dyadic groups. Therefore, the Fourier analysis on dyadic groups is performed in terms of Walsh functions. In the case of finite groups, it is done in terms of the discrete Walsh functions.

Fourier transform on non-Abelian groups

Let

- $C(G)$ be the space of complex functions on a finite not necessarily Abelian group G of order g, and
- Γ be the dual object of G. Γ consists of K unitary irreducible representations $\mathbf{R}_w(x)$ of G over C.

Note that $\mathbf{R}_w(x)$ stands for a non-singular $r_w \times r_w$ matrix over C. For each non-Abelian group, at least one of the representations is of the order $r_w > 1$.

Definition 41.11 *The direct and inverse Fourier transforms of a function $f \in C(G)$ are defined respectively by,*

$$\mathbf{S}_f(w) = r_w g^{-1} \sum_{u=0}^{g-1} f(u)\mathbf{R}_w(u^{-1}), \tag{8}$$

$$f(x) = \sum_{w=0}^{K-1} Tr(\mathbf{S}_f(w)\mathbf{R}_w(x)), \tag{9}$$

where for a matrix $\mathbf{Q}$, $Tr(\mathbf{Q})$ denotes the trace of $\mathbf{Q}$, i.e., the sum of elements on the main diagonal of $\mathbf{Q}$.

From this definition, the Fourier coefficients for a function f on a non-Abelian group are $r_w \times r_w$ matrices when $r_w \geq 1$. The same definition applies to Abelian groups. In that case, all the unitary irreducible representations are one-dimensional, i.e., they reduce to the group characters.

Further reading

Terminology. There are some differences in terminology of signal processing and spectral theory of logic functions. For example, Beauchamp [1] identified the trivial basis defined in term of minterms with block pulse functions.

Spectral theory of switching functions. Sadykhov et al. [11] and Stojić et al. [15] studied spectral transforms in various orthogonal bases. Shmerko [12] and Stanković [13] showed that the disjunctive normal form of switching functions can be considered as the Fourier-series like expansion with respect to the trivial basis. Details can also be found in [14].

Spectral theory of multivalued functions. Moraga [7, 8] studied applications of digital signal processing methods for multivalued functions: Chrestenson functions and linear p-adic systems. Muzio and Wesselkamper [9] discussed multivalued switching theory.

Historical remarks on discrete Fourier transform. Details of discrete Fourier transform (DFT) and related fast calculation algorithms (FFT) can be found in [2] [15].

Fourier transform on non-Abelian groups. We refer to classical references such as the books by Hewitt and Ross [4, 5], or the more recent publications such as [15].

References

[1] Agaev GN, Vilenkin NYa, Dzafarli GM, Rubinstein AI. *Multiplicative Systems of Functions and Harmonic Analysis on Zero-Dimensional Groups.* Elm Publisher, Baku, 1981.

[2] Beauchamp KG. *Applications of Walsh and Related Functions with an Introduction to Sequency Theory.* Academic Press, Bristol, 1984.

[3] Chrestenson HE. A class of generalized Walsh functions. *Pacific J. Math.*, 5:17–31, 1955.

[4] Hewitt E and Ross KA. *Abstract Harmonic Analysis.* Springer, Heidelberg, vol. 1, 1963.

[5] Hewitt E and Ross KA. *Abstract Harmonic Analysis.* Springer, Heidelberg, vol. II, 1970.

[6] Karpovsky MG. *Finite Orthogonal Series in the Design of Digital Devices.* Wiley and JUP, New York and Jerusalem, 1976.

[7] Moraga C. On some applications of the Chrestenson functions in logic design and data processing. *Mathematics and Computers in Simulation*, (27):431–439, 1985.

[8] Moraga C. Introduction to linear *p*-adic systems. In Trappl R, Ed., *Cybernetics and Systems Research, 2,* North-Holland, 1984.

[9] Muzio JC and Wesselkamper TC. *Multiple-Valued Switching Theory.* Adam Hilger, 1986.

[10] Rudin W. *Fourier Analysis on Groups.* Interscience Publisher, New York, 1960.

[11] Sadykhov R, Chegolin P, and Shmerko V. *Signal Processing in Discrete Bases.* Publishing House "Science and Technics," Minsk, Belarus, 1986 (In Russian).

[12] Shmerko VP. Synthesis of arithmetic forms of Boolean functions using the Fourier transform. *Automation and Remote Control*, 50(5):684–691, Pt2, 1989.

[13] Stanković RS. Some remarks about spectral transform interpretation of MTBDDs and EVBDDs. In *Proceedings of the ASP-DAC'95*, Makuhari Messe, Chiba, Japan, pp. 385–390, 1995.

[14] Stanković RS, Stanković M, and Janković D. *Spectral Transforms in Switching Theory, Definitions and Calculations.* Nauka, Belgrade, 1998.

[15] Stojić MR, Stanković MS, and Stanković RS. *Discrete Transforms in Applications.* Nauka, Beograd, 1993.

Appendix-C: Discrete Walsh Functions

Discrete Walsh functions are a discrete version of the functions introduced by J. L. Walsh [6] in solving some problems in uniform approximation of square-integrable functions on the interval $[0, 1)$. In the original definition by Walsh, these functions are defined by the following recurrence relations:

$$\phi_0 = 1, \quad \phi_1(x) = \begin{cases} 1, & 0 \leq x < \frac{1}{2}, \\ -1, & \frac{1}{2} < x \leq 1. \end{cases}$$

For $n \geq 1, n \in N$ represented as $n = 2^{m-1} + l - 1, \ \ l = 1, 2, \ldots, 2^{m-1}; m = 1, 2 \ldots$

$$\phi_n^{(l)} = \phi_m^{(l)}(x),$$

where it is assumed $\phi_1 = \phi_1^{(1)}$. For $m \geq 2$,

$$\phi_{m+1}^{(2l-1)}(x) = \begin{cases} \phi_m^{(l)}(2x), & 0 \neq x < \frac{1}{2}, \\ (-1)^{l+1}\phi_m^{(l)}(2x-1), & \frac{1}{2} < x \neq 1, \end{cases}$$

$$\phi_{m+1}^{(2l)}(x) = \begin{cases} \phi_m^{(l)}(2x), & 0 \neq x < \frac{1}{2}, \\ (-1)^{l}\phi_m^{(l)}(2x-1), & \frac{1}{2} < x \neq 1, \end{cases}$$

with

$$\phi_2^{(l)}(x) = \begin{cases} \phi_m^{(l)}(2x), & 0 \neq x < \frac{1}{2}, \\ -\phi_m^{(l)}(2x-1), & \frac{1}{2} < x \neq 1, \end{cases}$$

$$\phi_2^{(2l)}(x) = \begin{cases} \phi_1^{(l)}(2x), & 0 \neq x < \frac{1}{2}, \\ \phi_1^{(l)}(2x-1), & \frac{1}{2} < x \neq 1. \end{cases}$$

It is assumed that

$$\phi_n(x) = \frac{1}{2}[\phi_n(x+1) + \phi_n(x-0)]$$

at the dyadic rational points. The set of the defined functions is now called the Walsh functions and usually denoted by

$$\text{wal}(w, x), w \in N.$$

Kronecker product

The discrete Walsh functions can be viewed as columns (or rows) of the Walsh matrix defined as

$$\mathbf{W}(n) = \bigotimes_{i=1}^{n} \mathbf{W}(1), \tag{10}$$

where $\mathbf{W}(1) = \begin{bmatrix} 1 & 1 \\ 1 & -1 \end{bmatrix}$.

Thanks to this Kronecker product representation, $\mathbf{W}(n)$ can also be represented as

$$\mathbf{W}(n) = \begin{bmatrix} \mathbf{W}(n-1) & \mathbf{W}(n-1) \\ \mathbf{W}(n-1) & -\mathbf{W}(n-1) \end{bmatrix}, \tag{11}$$

which is a relation characterizing a subset of the *Hadamard* matrices introduced by M. J. Hadamard by a generalization of the results by J. J. Sylvester. Thus, the Walsh matrices are the Hadamard matrices of order 2^n. Therefore, the Walsh matrix whose columns are the discrete Walsh functions ordered in this way, is a Hadamard matrix, and, thus, ordered Walsh functions are reported as the Walsh–Hadamard functions.

The Kronecker product structure of the Walsh matrix, expressed by (10) and (11), and used to define the discrete Walsh functions, follows from the property that the Walsh functions are characters of the dyadic group, which is representable as the direct product of cyclic groups of order 2.

Ordering

Walsh functions are a special case of the Vilenkin–Chrestenson functions derived from the group characters of Abelian groups. A general theory of harmonic analysis on topological Abelian groups including the Walsh–Fourier analysis on the dyadic groups was also given by Paley and Wiener.

Symmetry property

The Walsh functions, besides the properties for group characters, possess the symmetry property. Due to that, the Walsh matrix is orthogonal and symmetric and, therefore, self-inverse over C up to the normalization constant 2^{-n}.

Sequency and frequency

In the original definition by Walsh [6], the Walsh functions were ordered with respect to the increasing number of zero-crossings called the *sequency*, a concept analogous in some aspects to the frequency in the classical Fourier analysis. This ordering of Walsh functions is called the *Walsh ordering.* However, in some other aspects, sequency can hardly be accepted as a generalization of the frequency.

Rademacher functions

The term Walsh–Kaczmarz functions is also used, since Kaczmarz studied thus ordered Walsh functions and proved that they can be considered as the extension of the Rademacher functions into a complete system. It is convenient to define the Rademacher functions by

$$rad(j,x) = (-1)^{x_i} = cos(\pi x_i), \quad x \in \{0,\ldots,2^n-1\}, \quad j = 0,1,\ldots,n,$$

where x_i is the i-th coordinate in the binary representation of x.

In the Hadamard ordered Walsh functions,

$$wal(2^j,x) = rad(j,x).$$

The Rademacher functions $rad(j,x)$, $j = 0,1,\ldots,n$, $x = 0,1,\ldots,2^n-1$, are related to the switching variables

$$rad(j,x) = 1 - 2x_j(x),$$

where $x_j(x)$ is the switching function $f(x_1,\ldots,x_n) = x_j$. It is assumed that logic values 0 and 1 for switching variables and functions are interpreted as integers 0 and 1. Therefore, Rademacher functions are equal to switching variables in

$$(0,1)_{GF(2)} \to (1,-1)_Z$$

encoding, where GF(2) is the Galois field of order 2, and Z is the set of integers.

Walsh–Paley functions

The Walsh functions of the index $w = 2^j$ are called the *basic* Walsh functions, since Paley proved that the Walsh functions can be defined as products of Rademacher functions. The Walsh functions ordered with respect to the

Paley definition are called the Walsh–Paley or *dyadically ordered* Walsh functions. For a Walsh function in ordering $wal(p, x)$, the constituent Rademacher functions are defined by the position of the non-zero coordinates in the dyadic expansion of the order index p.

Example 41.14 *If* $p = 13 = (1101)_2$,

$$wal(13, x) = rad(4, x) \times rad(3, x) \times rad(1, x)$$

since 1 appears in the dyadic expansion of 13 at the coordinates 1, 3 and 4. Walsh–Paley functions are the most important special case of the Takayasu Ito functions.

The relationship between the discrete Walsh functions and the switching functions

Thanks to the Paley definition, the relationship between the discrete Walsh functions and the switching functions is simple and can be established by the following consideration. For a given switching function f, the inner product with basic Walsh functions

$$s_i = \langle f, wal(2^{n-i}, x)\rangle$$

determines the correlation of f with the switching variables. This property has been proven useful in some applications in logic design. In particular, the coefficients s_i, ordered in the descending order of the absolute values, can be identified to the Chow parameters used in characterization of threshold functions. For that reason, the use of Walsh functions ordered with respect to the increasing number of component Rademacher functions and their indexes is suggested. By referring to the intended applications, that ordering is reported as the logical ordering of Walsh functions, or Rademacher–Walsh ordering.

Summarizing,

- The Rademacher functions $rad(j, x)$, $j = 0, 1, \ldots, n$, $x = 0, 1, \ldots, 2^n - 1$, are related to switching variables $rad(j, x) = 1-2x_j(x)$. The Rademacher functions can be considered as switching variables.
- The Walsh functions are group characters of C_2^n and can be expressed as products of the Rademacher functions

$$wal(r, x) = \prod_{j=1}^{n}(1 - 2x_j)^{r_j}$$

  where r_j are coordinates in the binary representation of the index $r = (r_1, \ldots, r_n)$.
- The set of 2^n Walsh functions endorse the structure of a multiplicative group Γ with the group operations defined as multiplication in Z. This multiplicative group is isomorphic to the finite dyadic group C_2^n.

- There is a relation ... Walsh functions and line... func-tions (EXOR sum ... variables).

> *For a switching function* ... $, x_n)$ *the Walsh spectr...* $S_f(w)$, $w = 0, 1, \ldots, 2^n - 1$, *is d...*
>
> $$S_f(w) = \sum_{x=0}^{2^n-1} f(x)w...$$
>
> *where* $x_1, \ldots, x_n$ *is binary representation of* ...

Matrix form

The matrix obtained by replacing +1 values in the Rademacher–Walsh matrix by a logical 0 and -1 by a logical 1 is the binary Walsh matrix.

> *The rows 1 to n in this matrix are the Rademacher functions in 0,1 coding, i.e., the switching variables, and thus can be used to form all subsequent rows of this matrix by the componentwise EXOR. Therefore, the rows of this matrix are linear switching functions.*

The use of binary Walsh matrices was suggested by Beauchamp and Hurst. In the Rademacher–Walsh matrix, if +1 is replaced by a logical 0 and -1 by a logical 1, the Besslich–Rademacher–Walsh matrix is obtained. The rows 1 to n of this matrix generate all subsequent rows through the componentwise AND. Such matrices were used in the final determination of an irredundant prime implicant cover of a switching function f.

Walsh–Fourier transform

Since the Walsh transform is the Fourier transform on a particular group, it possesses all the properties of the Fourier transform on groups, and the name Walsh–Fourier transform is used somewhere in the literature. In this notation, the coefficients of fixed polarity Walsh polynomials correspond to some different orderings of the Walsh transform matrix. Referring to the discussed orderings of the Walsh matrix, the Walsh–Kaczmarz, Walsh–Paley and Walsh–Hadamard transforms are distinguished. Moreover, somewhere the Hadamard ordered Walsh transform is simply called the Hadamard transform.

...the term Hadamard transform ...d instead of the Hadamard Walsh transform, it should ... at a subset of Hadamard trans- assumed. In general, the ... matrices are a subset of Hadamard ...es of order $2n^k$.

Example 41. ... is a Hadamard matrix of order 12 and 956.

Further reading

Walsh [6] defined Walsh functions by solving problems of uniform approximation of functions on the interval $(0, 1)$. Fine [3] proved the relationships between the Walsh functions and group characters of infinite dyadic groups. The Fourier analysis on dyadic groups is performed in terms of Walsh functions. In the case of finite groups, it is done in terms of the discrete Walsh functions (see, for example, study by Beauchamp [1], Hurst et al. [2], Karpovsky and Moskalev [4]. Vilenkin [5] considered Fourier transforms on locally compact Abelian groups.

References

[1] Beauchamp KG. *Applications of Walsh and Related Functions with an Introduction to Sequency Theory*. Academic Press, Bristol, 1984.

[2] Hurst SL, Miller DM, Muzio JC. *Spectral Techniques in Digital Logic*. Academic Press, Bristol, 1985.

[3] Fine NJ. On the Walsh functions. *Trans. Amer. Math. Soc.*, (3):372–414, 1949.

[4] Karpovsky MG and Moskalev ES. *Specral Methods in Analysis and Synthesis of Discrete Devices*. "Energy" Publisher, Leningrad, 1973 (in Russian).

[5] Vilenkin NYa. Concerning a class of complete orthogonal system. *Proceedings of the USSR Academy of Sciences, Mathematics*, No. 11, 1947 (in Russian).

[6] Walsh JL. A closed set of orthogonal functions. *American Journal of Mathematics*, 55:5–24, 1923.

Appendix-D: The Basic Operations for Ternary and Quaternary Logic

TABLE 41.1
Basic operations over three-valued arguments.

COMPLEMENT

	x = 0	1	2
$\overline{x}$	2	1	0

r-CYCLIC COMPLEMENT

r	x = 0	1	2
0	0	1	2
1	1	2	0
2	2	0	1

LITERAL

	x = 0	1	2
x^0	2	0	0
x^1	0	2	0
x^2	0	0	2

WINDOW LITERAL

	x = 0	1	2
${}^0x^0$	2	0	0
${}^0x^1$	2	2	0
${}^0x^2$	2	2	2
${}^1x^1$	0	2	0
${}^1x^2$	0	2	2
${}^2x^2$	0	0	2

MAX

	0	1	2
0	0	1	2
1	1	1	2
2	2	2	2

$\overline{MAX}$

	0	1	2
0	2	1	0
1	1	1	0
2	0	0	0

TSUM

	0	1	2
0	0	1	2
1	1	2	2
2	2	2	2

$\overline{TSUM}$

	0	1	2
0	2	1	0
1	1	0	0
2	0	0	0

TPROD

	0	1	2
0	0	0	0
1	0	0	1
2	0	1	2

$\overline{TPROD}$

	0	1	2
0	2	2	2
1	2	2	1
2	2	1	0

MIN

	0	1	2
0	0	0	0
1	0	1	1
2	0	1	2

$\overline{MIN}$

	0	1	2
0	2	2	2
1	2	1	1
2	2	1	0

MODSUM

	0	1	2
0	0	1	2
1	1	2	0
2	2	0	1

$\overline{MODSUM}$

	0	1	2
0	2	1	0
1	1	0	2
2	0	2	1

MODPROD

	0	1	2
0	0	0	0
1	0	1	2
2	0	2	1

$\overline{MODPROD}$

	0	1	2
0	2	2	2
1	2	1	0
2	2	0	1

TABLE 41.2
Basic operations over four-valued arguments.

COMPLEMENT

x	0	1	2	3
$\overline{x}$	3	2	1	0

r-CYCLIC COMPLEMENT

r \ x	0	1	2	3
0	0	1	2	3
1	1	2	3	0
2	2	3	0	1
3	3	0	1	2

LITERAL

x	0	1	2	3
x^0	3	0	0	0
x^1	0	3	0	0
x^2	0	0	3	0
x^3	0	0	0	3

WINDOW LITERAL

x	0	1	2	3
${}^0x^0$	3	0	0	0
${}^0x^1$	3	3	0	0
${}^0x^2$	3	3	3	0
${}^0x^3$	3	3	3	3
${}^1x^1$	0	3	0	0
${}^1x^2$	0	3	3	0
${}^1x^3$	0	3	3	3
${}^2x^2$	0	0	3	0
${}^2x^3$	0	0	3	3
${}^2x^3$	0	0	3	3
${}^3x^3$	0	0	0	3

MAX

	0	1	2	3
0	0	1	2	3
1	1	1	2	3
2	2	2	2	3
3	3	3	3	3

$\overline{MAX}$

	0	1	2	3
0	3	2	1	0
1	2	2	1	0
2	1	1	1	0
3	0	0	0	0

TSUM

	0	1	2	3
0	0	1	2	3
1	1	2	3	3
2	2	3	3	3
3	3	3	3	3

$\overline{TSUM}$

	0	1	2	3
0	3	2	1	0
1	2	1	0	0
2	1	0	0	0
3	0	0	0	0

TPROD

	0	1	2	3
0	0	0	0	0
1	0	0	0	1
2	0	0	1	2
3	0	1	2	3

$\overline{TPROD}$

	0	1	2	3
0	3	3	3	3
1	3	3	3	2
2	3	3	2	1
3	3	2	1	0

MIN

	0	1	2	3
0	0	0	0	0
1	0	1	1	1
2	0	1	2	2
3	0	1	2	3

$\overline{MIN}$

	0	1	2	3
0	3	3	3	3
1	3	2	2	2
2	3	2	1	1
3	3	2	1	0

MODSUM

	0	1	2	3
0	0	1	2	3
1	1	2	3	0
2	2	3	0	1
3	3	0	1	2

$\overline{MODSUM}$

	0	1	2	3
0	3	2	1	0
1	2	1	0	3
2	1	0	3	2
3	0	3	2	1

MODPROD

	0	1	2	3
0	0	0	0	0
1	0	1	2	3
2	0	2	0	2
3	0	3	2	1

$\overline{MODPROD}$

	0	1	2	3
0	3	3	3	3
1	3	2	1	0
2	3	1	3	1
3	3	0	1	2

Index

B

C

D

E

H

I

K

L

T - #0207 - 101024 - C0 - 234/156/51 [53] - CB - 9780849334245 - Gloss Lamination